Matrix Structural Analysis

Second Edition

William McGuire
Professor of Civil Engineering, Emeritus
Cornell University

Richard H. Gallagher
Late Professor and President
Clarkson University

Ronald D. Ziemian
Associate Professor of Civil Engineering
Bucknell University

John Wiley & Sons, Inc.

ACQUISITIONS EDITOR	Wayne Anderson
MARKETING MANAGER	Katherine Hepburn
PRODUCTION EDITOR	Ken Santor
COVER DESIGNER	Lynn Rogan
ILLUSTRATION COORDINATOR	Sigmund Malinowski/Eugene Aiello
ELECTRONIC ILLUSTRATIONS	Precision Graphics, Inc.

This book was set in 10/12 Times Ten by UG / GGS Information Services, Inc., and printed and bound by Hamilton Printing Company, Inc. The cover was printed by Phoenix Color Corporation.

This book is printed on acid-free paper. ∞

The paper in this book was manufactured by a mill whose forest management programs include sustained yield harvesting of its timberlands. Sustained yield harvesting principles ensure that the numbers of trees cut each year does not exceed the amount of new growth.

Library of Congress Cataloging-in-Publication Data

McGuire, William, 1920–
Matrix structural analysis / by William McGuire, Richard H. Gallagher, Ronald D. Ziemian. — 2nd ed.
p. cm.
Includes index.
ISBN 978-0-471-12918-9 (cloth : alk. paper)
1. Structural analysis (Engineering)—Matrix methods.
I. Gallagher, Richard H. II. Ziemian, Ronald D. III. Title.
TA642.M25 2000
624.1′7—dc21 98-53457
CIP

Printed in the United States of America

20 19 18 17 16 15 14 13

To

Barbara, Terry, and Constance

Preface to the Second Edition

The first edition of this book was written 20 years ago, but our aims remain the same: to place proper emphasis on the methods of matrix structural analysis used in current practice and to provide the groundwork for forthcoming practice and allied, more advanced subject matter. The present edition accounts for changes in practice that have taken place in the intervening years, and it incorporates advances in the art of analysis we regard as suitable for application now and of increasing importance to practitioners in the years just ahead.

Among the major changes from the first edition is the addition of chapters on the nonlinear analysis of framed structures, treatment of the solution of nonlinear problems, and packaging of a compact disk containing the computer program, MASTAN2. The major reductions include the deletion of Chapter 7 on the flexibility method, Section 10.7 on the transfer matrix method, and Chapter 12 on the finite element method. The entire text and illustrative examples have also been edited extensively, and several chapters have been repositioned. These changes deserve some explanation.

The reduction in coverage of the flexibility method was presaged by a comment in the preface to the first edition: "It is hardly used in practice." Although the method is advantageous in certain circumstances and we believed it might receive greater attention if cast in a more efficient form, such has not been the case. Therefore, the development of flexibility method equations of global analysis has been deleted. But extensive coverage of the flexibility approach to the determination of element force-displacement relationships has been retained because of the value that it has in many instances. Elimination of the global flexibility method formulation also led to reconsideration of the chapter on the equations of statics and kinematics (Chapter 6 of the first edition) which was in large measure preparation for the identification of redundant forces in the flexibility method. That material has been deleted also. The other deletions—material on the transfer matrix method and the general finite element concept—were made in the interest in retaining a reasonably wieldy volume.

The additions represent some departures from the spirit of the first edition. In that edition's preface it was noted that, in concentrating on the direct stiffness method, we were dealing with an accredited, widely practiced procedure, whereas an earlier generation of texts frequently dealt with subject matter that was not fully crystallized. Nonlinear structural analysis has a long history and it is the subject of a number of books and some highly sophisticated computer programs, but it can hardly be called "crystallized." Many of its features are ad hoc—in the good sense of the term—and are in need of reduction to more routine procedures through further research and generalization. Others may always require the expertise of specialists in analysis, as opposed to the design engineer who has many additional concerns. Nevertheless, it is believed to be timely, and indeed necessary, to include an introduction to nonlinear analysis in a text on the analysis of framed structures, particularly those of the civil engineering variety. Techniques for its practical application are at hand, and there is a growing awareness of their place in design.

Similarly in the first edition we decided to exclude computer programs, citing in its preface their rapid obsolescence and the widespread availability of commercial programs such as STRUDL. Advances in all areas of computer technology over the past twenty years have changed this. Particularly relevant to our subject is the capacity for packaging structural analysis programs of significant power in an interactive graphic-supported medium suitable for running on readily available personal computers. The computer is not a substitute for mastery of the subject matter, but in the ability for the user to control the terms of the analysis and to picture the results that it now provides, it can be an invaluable aid to understanding the theory and appreciating the physical significance of the outcome. We believe these pedagogical advantages outweigh the inevitability of obsolescence and are therefore packaging with every copy of the text MASTAN2[1], an interactive graphics program with provisions for geometric and material nonlinear analysis, as well as conventional linear elastic anlysis. MASTAN2 is based on MATLAB[2], a premier software package for numeric computing and data analysis.

Also, whereas we used SI units exclusively in the first edition, in this one many of the examples and problems—particulary those of a nonlinear nature that have strong implications for design—are in Imperial units. We do this because the SI system has not been fully assimilated in practice, and we believe that appreciation of the physical significance of the results is essential to their understanding.

Our aims for the present edition are explained further in the following discussion of the motivation and salient features of the respective chapters:

Three purposes are served by Chapter 1, the introduction. First, we present a concise sketch of the development of the subject. One intent of this history is to emphasize that computerized structural analysis methods are merely one part of a continuing process that extends back more than 150 years. Second, the role that computerized structural analysis has played in the design of existing structures is outlined. Finally, the computer capabilities themselves are tied to the programs written for structural analysis. Particular attention is given to the development and use of interactive computer graphics.

Chapters 2–5 represent closely allied subject matter related primarily to the direct stiffness method. Chapter 2 serves to define terminology, coordinate systems, and the most fundamental notions of structural behavior, but it also contains two developments of great generality. The first is the basic character of elemental relations in the form of stiffness and flexibility and their transformability from one to another and even to alternative formats. The second is the fundamental idea of direct stiffness analysis, described here by means of the simplest structural element.

A more formal treatment is given to direct stiffness analysis in Chapter 3 and, consequently, it is possible to examine more closely the implications for large-scale practical computation. The latter include considerations such as the characteristics of the algebraic equations that are to be formed and solved. In Chapters 4 and 5 the remaining tools needed for the linear elastic stiffness analysis of complete frames are established. The stiffness matrix of a rather general space-frame element is formulated and then applied, in illustrative examples, to a variety of specific situations.

Two comments on Chapters 2–5 are appropriate. The first is that, except for some editing of the text and modification of the examples and problems, they are identical to the same chapters of the first edition. The second is that this initial development of the subject features the basic physical conditions of structural analysis: equilibrium, compatibility, and the mechanical properties of the material. Reliance on work and energy concepts is kept to a minimum. Our feeling is that it is preferable to help the

---

[1]The name MASTAN2 is an acronym for *Ma*trix *St*ructural *An*alysis, 2nd Edition. It was developed by the authors for educational purposes only. It has no relationship to any existing commercial structural or finite element analysis program.

[2]MATLAB® is a registered trademark of The MathWorks Inc., 3 Apple Hill Drive, Natick, MA 01760-2098.

student become attuned to the overall approach to computerized structural analysis in this way, before proceeding to the more powerful and versatile—but more mathematical—virtual work–based methods.

We do introduce the formulation of matrix structural analysis on the basis of virtual work concepts in Chapters 6 and 7. A treatment of this type is necessary to give maximum scope to various aspects of practical design analysis of frameworks, such as tapered members and distributed loads, as well as to the study of nonlinear analysis and the finite element approach. The theoretical groundwork of the virtual work principle is laid in Chapter 6. Both virtual displacement and virtual force concepts are covered, but far greater attention is given to the former because of their role in stiffness formulations. Chapter 7 examines the implementation of the virtual work principle in linear elastic matrix structural analysis.

Chapters 8–10 comprise an introduction to the nonlinear matrix analysis of frameworks. The approach is one of using selected structural systems and types of material as the basis for explaining how theory can be translated into procedures useful in the design office. In the interest of focusing on sources and end results in these chapters, we defer to Chapter 12 the treatment of methods and details of nonlinear equation solution and to appendices the coverage of some aspects of the basic structural mechanics and behavior

An overview of sources of structural nonlinearity, levels of analysis used in dealing with them, and a system for casting the required equations in matrix stiffness method form is presented in Chapter 8. In the course of doing this the notions of the geometric stiffness matrix and a plastic reduction matrix are introduced. Also, the range of conditions to be considered is illustrated by the solution of some elementary problems by traditional methods. In Chapter 9, the groundwork prepared in the previous chapter is used as the base for the development and application of a matrix method for the nonlinear elastic (geometric nonlinear) analysis of both planar and space frames. The calculation of the elastic critical load of ideal systems is also treated. Chapter 10 employs the same base, plus the concepts of plastic hinge theory, for the material and geometric analysis of planar systems. Included is the determination of the inelastic critical load of ideal systems by an adaptation of tangent modulus theory.

Some of the more popular methods of solution of linear algebraic equations are studied in Chapter 11, and the same is done for nonlinear equations in Chapter 12. Equation solving is more in the realm of applied mathematics, numerical analysis, or computer science than in structural engineering. Nevertheless, since primary responsibility for the entire analysis generally belongs to structural engineers, they should have more than superficial knowledge of the important aspects of equation solving. Operations peculiar to the handling of equations in MASTAN2 are explained in detail. Otherwise, these chapters are introductions to the subject that summarize useful methods, mention pitfalls that may be encountered, and give guidance to more specialized sources.

In the interest of developing basic stiffness methods of analysis in the earlier chapters, we put off consideration of some procedures essential to the efficient and realistic solution of practical design analysis problems. Chapter 13 is devoted to the exposition of certain of these practices, including condensation of analysis equations prior to solution by elimination of specified unknowns, substructuring, the imposition of constraints, connection stiffness and finite joint size, the exploitation of symmetry and antisymmetry, and procedures for the economical reanalysis of structures when in an iterative design sequence. Attention is also given to various types of coordinate systems that are alternative to global coordinates, such as the local coordinate systems that are convenient for sloping supports.

In Appendix A, principles of structural mechanics and geometry fundamental to solving the problem of equilibrium of a deformed structure are reviewed and then employed in the derivation of terms that were incorporated without proof in the geometric stiffness matrices of Chapter 9. There are several reasons for placing this material in an appendix rather than in the main body of the text. Throughout the text we adhered to standard matrix notation. Concise treatment of the appended material requires use of some elementary concepts of tensors and indicial notation. It also requires the use of some less obvious definitions of stress and strain, such as second Piola-Kirchoff stresses and Green-Lagrange strains, and an excursion into the analysis of finite rotations. These changes in style and alteration of the analytical base are not abstruse, but we believe their introduction in the earlier chapters would have been an unnecessary intrusion on the main agrument. Nevertheless, although it is material that is not essential to an initial study, it is fundamental to a true understanding of the processes of nonlinear analysis and to their extension beyond the limits explored in the main text.

The problem of nonuniform torsion in three-dimensional frameworks is also addressed in this appendix but for a somewhat different reason. Practically, it is a key element in the resistance of many systems to the destabilizing effects of combined torsion and flexure. Failure to treat the subject would be a serious omission. But considerable theoretical and experimental research is needed to reduce its analysis to an established procedure. The approach described has a sound basis. It is worthy of application, but it must be viewed in the context of an area still very much under development.

Appendix B contains a discussion of the problem of distinguishing between the effects of rigid body motion and displacements resulting from the deformation of a body. It is a presented in an appendix rather than in the main body of the text because it is more of a critique rather than the presentation of methods, something that was done in Chapter 12.

The contents outlined here represent subject matter on the scale of two three-credit-hour, one-semester courses, with considerable selectivity of coverage and latitude in arrangement available to the instructor. For example, the first seven chapters plus selected portions of Chapters 11 and 13 might serve for an introductory junior- or senior-level course. At the same level, but for classes of students with a previous, limited introduction to the subject, the first five chapters could serve for review and reinforcement of understanding of basic concepts, with formal instruction starting with the virtual work approach (Chapters 6 and 7). The second level course, at the elective senior or introductory graduate level, might feature the virtual work approach to both linear and nonlinear elastic analysis, material nonlinear analysis, and stability studies (Chapters 6–10, Chapter 12, additional portions of Chapter 13, and reference to the appendices). The program MASTAN2 can be a valuable adjunct at all levels: for demonstration, for review of the text's numerous examples, for problem solving, as the base for analysis and design studies, and in support of individual programming. An attractive application is to have students develop their own programs, either as alternatives to the analysis routines in MASTAN2 or as additions to the program's existing routines.

Our co-author, Professor R. H. Gallagher, died in September 1997, while this edition was in preparation. The idea for the original book was his, and he was active in the planning of this edition and in the revision of several chapters. His contributions are vital to the work. The remaining authors prepared the material on nonlinear analysis and we bear sole responsibility for any of its shortcomings.

William McGuire
Ronald D. Ziemian

Acknowledgments

We are indebted to a number of people for their help in the preparation of this edition. Our work would not be complete without an acknowledgment of this debt and a particular offering of thanks to the following:

To Professors J. F. Abel and G. G. Deierlein for using drafts as classroom notes and for their wise advice throughout.

To Professors F. C. Filippou, J. F. Hajjar, E. M. Lui, and G. H. Paulino for their thorough, constructive reviews of an early draft and to Professor K. M. Mosalam for the benefit of his experience in using it in class.

To the men whose Cornell graduate research contributed to the development of material in the book: J. L. Gross, C. I. Pesquera, M. Gattass, J. G. Orbison, Y. B. Yang, S. I. Hilmy, J. L. Castañer, S. N. Sutharshana, D. W. White, C. S. Chen, C. Chrysostomou, and M. R. Attalla.

To Ms. N. Bulock and her colleagues at the MathWorks, Inc. and Professor J. Maneval of Bucknell University for their MATLAB support in developing MASTAN2.

To Ms. P. Welzel for her secretarial assistance in preparing the manuscript for publication.

Contents

Symbols

In matrix structural analysis, many physical quantities and mathematical operations must be represented symbolically. Preparation of the equations of analysis in a form suitable for computer solution requires that all symbols used be defined in a rigid fashion amenable to numerical interpretation. On the other hand, the development of these equations, with stress upon their physical significance, is often accomplished through the use of simple, less formal symbols–symbols that vary with the principle under discussion and have a clear physical connotation in the case at hand. In the interest of generality and uniformity, we shall use some basic symbols to denote certain quantities throughout the text. But the precise interpretation of any of these symbols must be obtained from the local context in which it is used, and in which it will be explained.

In general, we use the letter P to designate applied direct forces and P_m to designate applied moments. Similarly, R and R_m are used for direct and moment reactions. At their ends and within elements, we generally denote direct forces by F and moments by M. All of these symbols may carry clarifying subscripts and superscripts to indicate direction, point of application, or member to which the symbol applies. The symbols may appear in either single component or vector form. The symbols u, v, and w will designate translational displacements in the x, y, and z directions and θ_x, θ_y and θ_z rotational displacements about these axes. Generally, the letter k will refer to a stiffness quantity and d to a flexibility quantity.

Matrices are denoted by a boldface letter within the symbols [ ] (for a square or rectangular matrix), { } (for a column vector), $\lfloor\ \rfloor$ (for a row vector), and $\lceil\ \rfloor$ for a diagonal matrix. Matrix operations are represented by $[\ ]^{-1}$ as the inverse, $[\ ]^{\mathrm{T}}$ as the transpose, $|\ |$ as a determinant, and $\|\ \|$ as a norm.

As a further guide, the following is a list of the principal symbols used in the text. As indicated above, most of these may contain clarifying or modifying subscripts, or supplementary marks (overbars $\bar{\ }$, hat symbols ^, etc.) that will be defined in context. The same applies to the individual components of matrices and vectors listed below.

A	Area
B	Bimoment
C_w	Warping constant
[**d**]	Element flexibility matrix
[**D**]	Global flexibility matrix, diagonal matrix
E, [**E**]	Elastic modulus, matrix of elastic constants
e	Normal strain
e	Infinitesimal strain tensor
F	Normal or shearing force
{**F**}	Vector of element nodal forces

G	Shear modulus
$\{\mathbf{G}\}$	Gradient vector
$[\mathbf{G}]$	Constraint equation coefficients
HBW	Half-bandwidth
I	Moment of inertia
I_p	Polar moment of inertia
$[\mathbf{I}]$	Identity matrix
J	St. Venant torsion constant
k	Spring stiffness, rotational spring stiffness
$[\mathbf{k}]$	Element stiffness matrix
$[\mathbf{K}]$	Global stiffness matrix
L	Length
$[\mathbf{L}]$	Lower triangular matrix
l, m, n	Direction cosines
m	Ratio of bending moment to plastic moment
M	Bending or twisting moment
M_p	Plastic moment
$\lfloor \mathbf{N} \rfloor$	Vector of element shape functions
n	Number of degrees of freedom, number of nodes
$[\mathbf{0}], \{\mathbf{0}\}$	Null matrix and vector
$\{\mathbf{P}\}$	Vector of global nodal forces
P	Axial force
P_y	Axial yield or squash load
p	Ratio of axial force to squash load, number of elements
$\{\mathbf{Q}\}$	Vector of element force distribution functions
q	Distributed load intensity
$\{\mathbf{R}\}$	Vector of reaction forces, unbalanced load vector
$\{\mathbf{r}\}$	Residual vector
r	Radius of gyration
S	Current stiffness parameter
$\{\mathbf{s}\}$	Basis vector
T	Temperature change from stress-free state, torsional moment
$\mathbf{T}$	Cauchy stress tensor
$\tilde{\mathbf{T}}$	Second Piola-Kirchoff stress tensor
$[\mathbf{T}]$	Force transformation matrix
U, U^*	Strain energy and complementary strain energy
$[\mathbf{U}]$	Upper triangular matrix
u, v, w	Displacement components
V	Potential energy of applied loads
Vol	Volume
W, W^*	Work and complementary work
x, y, z	Cartesian coordinates
Z	Plastic section modulus

Greek Symbols

α	Coefficient of thermal expansion, factor defined in context
α, β, δ	Direction angles
β	Rate of twist, relaxation factor
$[\boldsymbol{\Gamma}]$, $[\gamma]$	Transformation matrix
γ	Shear strain
$\{\boldsymbol{\Delta}\}$	Vector of nodal point displacements
Δ	Displacement
δ	Virtual quantity, variation
ε	Green Lagrange strain tensor
ε_a	Absolute percent relative error
θ	Angular displacement
κ	Curvature, condition number
λ	Eigenvalue, load ratio, plastic strain factor, effective length factor
ν	Poisson's ratio
ξ	Nondimensional coordinate
Π	Total potential energy
π	3.1416...
ρ	Radius of curvature
τ	Shear stress, fraction of load increment
σ	Normal stress
σ_y	Yield stress
$\{\mathbf{Y}\}$	Eigenvector
Φ	Yield surface function
$[\boldsymbol{\Phi}]$	Static equilibrium matrix
ϕ	Angle of measure
ζ	Acceptable percent error tolerance
ω	Eigenvalue

In addition to the above literal and matrix symbolism, we shall use the following graphic symbols wherever it is desired to indicate or to stress some particular characteristic of a force or structure.

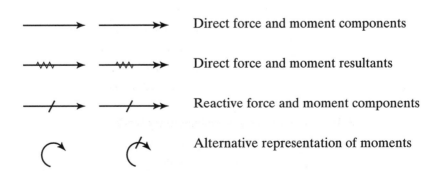

Direct force and moment components

Direct force and moment resultants

Reactive force and moment components

Alternative representation of moments

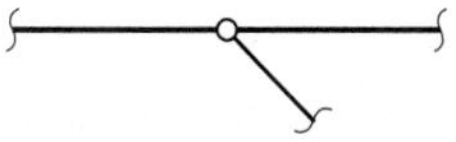 Pinned joint

 Rigid joint

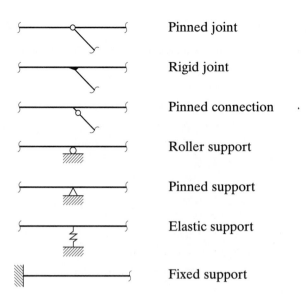

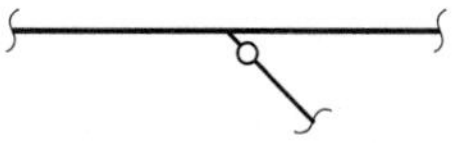 Pinned connection

 Roller support

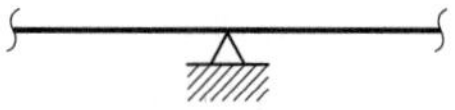 Pinned support

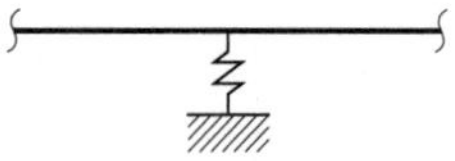 Elastic support

 Fixed support

Chapter 1

Introduction

One of the responsibilities of the structural design engineer is to devise arrangements and proportions of members that can withstand, economically and efficiently, the conditions anticipated during the lifetime of a structure. A central aspect of this function is the calculation of the distribution of forces within the structure and the displaced state of the system. Our objective is to describe modern methods for performing these calculations in the particular case of framed structures. The number of structures that are actually simple frameworks represents only a part of those whose idealization in the form of a framework is acceptable for the purposes of analysis. Buildings of various types, portions of aerospace and ship structures, and radio telescopes and the like can often be idealized as frameworks.

In design, both *serviceability limit states* and *strength limit states* should be considered. A serviceability limit state is one in which the structure would become unfit for normal service because of excessive deformation or vibration, or problems of maintenance or durability. Generally, linear elastic analysis gives a good picture of the actual response to service loads. A strength limit state is one in which the structure would become unsafe. Except for sudden brittle or fatigue crack propagation, the attainment of a strength limit is generally the result of increasingly nonlinear elastic or inelastic response that culminates in structural instability. Most often, the internal load distribution at a strength limit state is calculated by a linear method. In proportioning members, the analytical results are modified in an empirical or judgmental way to account for the effects of nonlinearity. Whereas this practice is common, the availability and application of practicable methods of nonlinear analysis is steadily increasing. One of the aims of this book is to provide an introduction to typical methods of this type.

Fundamentally, the behavior of all types of structures—frameworks, plates, shells, or solids—is described by means of differential equations. In practice, the writing of differential equations for framed structures is rarely necessary. It has been long established that such structures may be treated as assemblages of one-dimensional members. Exact or approximate solutions to the ordinary differential equations for each member are well-known. These solutions can be cast in the form of relationships between the forces and the displacements at the ends of the member. Proper combinations of these relationships with the equations of equilibrium and compatibility at the joints and supports yields a system of algebraic equations that describes the behavior of the structure.

Structures consisting of two- or three-dimensional components—plates, membranes, shells, solids—are more complicated in that rarely do exact solutions exist for the applicable partial differential equations. One approach to obtaining practical, numerical solutions is the *finite element method*. The basic concept of the method is that a *continuum* (the total structure) can be modeled analytically by its subdivision into regions (the *finite elements*), in each of which the behavior is described by a set of assumed functions representing the stresses or displacements in that region. This permits the problem formulation to be altered to one of the establishment of a system of algebraic equations.

The practical, numerical solution of problems in structural analysis thus is seen to involve the formation and solutions of systems—sometimes very large systems—of algebraic equations. Also, it should be fairly clear that a member of a framed structure is simply one example of a more broadly defined family of finite elements.

Viewed in this way, structural analysis may be broken down into five parts:

1. ***Basic mechanics.*** The fundamental relationships of stress and strain, compatibility, and equilibrium.
2. ***Finite element mechanics.*** The exact or approximate solution of the differential equations of the element.
3. ***Equation formulation.*** The establishment of the governing algebraic equations of the system.
4. ***Equation solution.*** Computational methods and algorithms.
5. ***Solution interpretation.*** The presentation of results in a form useful in design.

This book deals chiefly with parts 3, 4, and 5. Specifically, it is on *matrix structural analysis*. This is the approach to these parts that currently seems to be most suitable for automation of the equation formulation process and for taking advantage of the powerful capabilities of the computer in solving large-order systems of equations. An understanding of basic structural mechanics and basic matrix algebra is presumed. Only that segment of the finite element method relating to framed structures will be included, other aspects being left to texts specializing in the fundamentals of finite element mechanics and procedures (e.g., Refs. 1.1 and 1.2). Computational methods and algorithms will be discussed in Chapters 11 and 12, but more comprehensive coverage can be found in books on numerical analysis. (e.g., Ref. 1.3).

An appreciation of the approach to structural analysis we are taking requires some understanding of the history of this and related subjects. The following brief review may help.

1.1 A BRIEF HISTORY OF STRUCTURAL ANALYSIS

Although it was immediately preceded by the great accomplishments of the school of French elasticians, such as Navier and St. Venant, the period from 1850 to 1875 is a logical starting point for our review. The concept of framework analysis emerged during this period, through the efforts of Maxwell (Ref. 1.4), Castigliano (Ref. 1.5), and Mohr (Ref. 1.6), among others. At the same time, the concepts of matrices were being introduced and defined by Sylvester, Hamilton, and Cayley (Ref. 1.7). These concepts are the foundations of matrix structural analysis, which did not take form until nearly 80 years later.

An excellent chronicle of developments in structural mechanics in the period 1875 to 1920 is found in Timoshenko's *History of Strength of Materials* (Ref. 1.8). Very little progress was made in this period in the development of theory and the analytical techniques subsidiary to matrix structural analysis. To a great extent, this was due to practical limitations on the solvability of algebraic equations with more than a few unknowns. For the structures of primary interest in that period—trusses and frames—an analysis approach based on member forces as unknowns was almost universally employed.

Around 1920, as a result of the efforts of Maney (Ref. 1.9) in the United States and Ostenfeld (Ref. 1.10) in Denmark, the basic ideas of a truss and framework analysis approach based on displacement parameters as unknowns took form. These ideas represent the forerunners of matrix structural analysis concepts in vogue today. Severe limitations on the size of the problems that could be handled by either force or displacement unknowns nevertheless prevailed until 1932 when Hardy Cross introduced the method of moment distribution (Ref. 1.11). This method made feasible the solution of structural analysis problems that were an order of magnitude more complex than the most sophisticated problems treatable by the prior approaches. Moment distribution became the staple of structural frame analysis practice for the next 25 years. Today it remains of value as a ready method for solving small problems and as an aid in visualizing behavior, but it is primitive in power compared to computer methods and has been superseded by them in the solution of large problems.

Association of the mathematical concepts of matrix theory with the engineering concepts of structural analysis appeared in the 1930s, chiefly in the work of Frazer, Duncan, and Collar (Ref. 1.12). The liaison developed erratically through the 1940s, but there was no motivation for a firm union until digital computers were born in the early 1950s. Individuals who foresaw the impact of computers on both theory and practice then undertook the codification of the well-established framework analysis procedures in the format best suited to the computer, the matrix format. Two noteworthy developments were the publications of Argyris and Kelsey (Ref. 1.13) and Turner, Clough, Martin, and Topp (Ref. 1.14). These papers wedded the concepts of framework analysis and continuum analysis and cast the resulting procedures in a matrix format. They were a strong influence on developments in subsequent years. An important factor in the acceptance of computerized structural analysis in the civil engineering profession has been the series of ASCE conferences on electronic computation, especially the first three (Refs. 1.15–1.17).

The desirability of accounting for nonlinear elastic and inelastic behavior was recognized from the earliest days of the development of structural analysis, and significant contributions were made in the nineteenth century. By 1888 a practical theory for including the effect of elastic displacements in the analysis of suspension bridges had been evolved by Melan (Ref. 1.18), and in 1899 Goodman presented a clear explanation of the plastic hinge concept (Ref. 1.19). In the early years of the twentieth century, Steinman published an English translation of Melan's deflection theory (Ref. 1.20) and, as reviewed in Reference 1.21, contributions to plastic analysis and design were made by Kazinczy, Maier-Liebnitz, and others. But it was not until the latter half of this century that reasonably general, practicable methods for including both geometric and material nonlinearity in analysis became a reality.

Among the seminal events of this period was Turner, Dill, Martin, and Melosh's 1960 publication of a geometric stiffness matrix for the uniform axial force member (Ref. 1.22). On the material behavior side, a dramatic thrust occurred between the end of World War II and 1965. Most of the significant developments of that period are reported in References 1.21, 1.23, and 1.24. Since then, there has been a steadily increasing flow of contributions to the theory and application of nonlinear methods of analysis and design. Most of them have been incremental, but their cumulative effect has been profound.

1.2 COMPUTER PROGRAMS

When the structural engineer leaves the university and enters the design office, it is probable that he or she will encounter computational hardware and software of considerable sophistication. The available programs may require only the simplest data in order to bring about solutions for the problems at hand. With such tools available, it is natural for the student of engineering to question the utility, or even the necessity, of what may appear to be a study of already-programmed procedures. The answer to this question is in the responsibilities of the design engineer. The decisions that must be made by this individual, extending from the analytical description of the structure, thence to the evaluation of the correctness of the numerical analysis, and through to the interpretation of the output, are not amenable to computerization. The structural engineer must understand the principles of analysis. Computerization has relieved the burdens of the rote operations in design analysis, but it cannot relieve the engineer of responsibility for the designed structure.

Throughout this book, therefore, illustrative numerical examples and problems are used. Most of them are simple. The aim is to avoid obscuring the point or principle under study in a cloud of numerical computations. But in the process the advantage of matrix notation and the matrix approach may not become immediately apparent when it is clear that the algebraic equations for these elementary structures could have been compiled in a less formal, and perhaps more physically obvious, way. Such is not the case for most structures encountered in practice, however. They are larger, more complicated, and generally require more extensive calculations. It is the efficient analysis of such systems that we are mainly concerned with. Although adopting the matrix approach may not influence the way we think about structural behavior, one of its main virtues is that mentioned earlier: It provides the best format presently known for analyzing structures by computer.

1.2.1 Computational Flow and General Purpose Programs

The equations of the matrix/finite element approach are of a form so generally applicable that it is possible in theory to write a single computer program that will solve an almost limitless variety of problems in structural mechanics. Many commercially available general-purpose programs attempt to obtain this objective, although usually on a restricted scale. The advantage of general-purpose programs is not merely this capability but the unity afforded in the instruction of prospective users, input and output data interpretation procedures, and documentation.

The four components in the flowchart of Figure 1.1 are common to virtually all general-purpose, finite element analysis programs. As a minimum, the *input* phase should require of the user no information beyond that relating to the material of construction, geometric description of the finite element representation (including support conditions), and the conditions of loading. The more sophisticated general-purpose programs facilitate this input process through such features as prestored material-property schedules and graphical displays of the finite element idealization so that errors in input may be detected prior to performance of calculations.

In the phase comprised of the library of finite elements is the coded formulative process for the individual elements. Most general-purpose programs contain a variety of elements, samples of which are illustrated in Figure 1.2. The only element we are concerned with here is the framework element. Texts such as References 1.1 and 1.2 describe the use of other elements in analytically modeling a continuum. The *element library* phase of the general-purpose program receives the stored input data and establishes the element algebraic relationships by application of the relevant coded formulative process. This phase of the program also includes all operations necessary to position the element algebraic relationships for connection to the neighboring elements as well as the connection process itself. The latter operations produce the algebraic equations that characterize the response of structure.

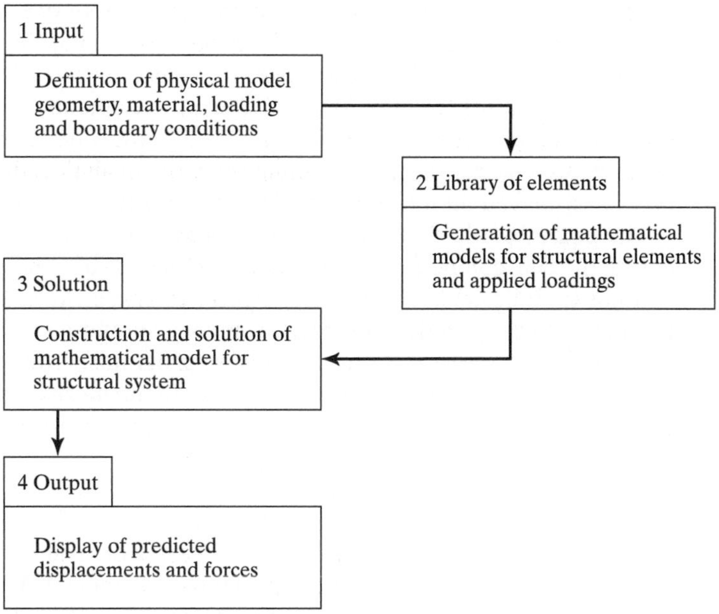

Figure 1.1 Structural analysis computational flow.

The *solution* phase operates on the equations of the problem formed in the prior phase. In the case of a linear static analysis program this may mean no more than the single solution of a set of linear algebraic equations for a known right-hand side. Nonlinear static analysis normally requires a series of incremental linear analyses. Solutions for dynamic response may require very extensive computations over a time-history of applied loads. In still other cases, particularly where the number of unknowns is very large, it may be advantageous to divide the total structure into several *substructures*. A finite element idealization of each substructure is made and analyzed, and the results are properly recombined into a solution for the full structure. Included in the solution phase are the back-substitution operations needed to obtain all the desired aspects of the solution.

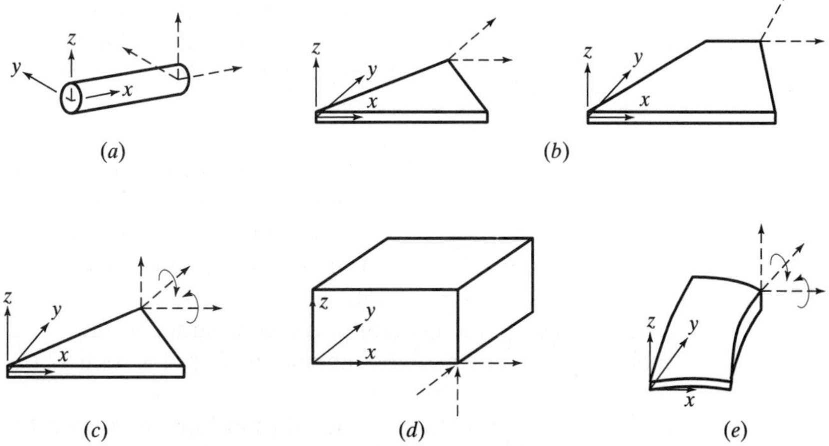

Figure 1.2 Sample finite elements. (*a*) Framework member. (*b*) Plane stress. (*c*) Flat plate bending. (*d*) Solid element. (*e*) Curved thin shell.

The *output* phase presents the analyst with a numerical or pictorial record of the solution upon which engineers can base decisions regarding the proportioning of the structure and other design questions. The numerical record is commonly a printed list of forces or stresses at points or within the finite elements, of displacements of specified points, and of other desired information. Pictorially, data such as moment diagrams, contour plots of the principal stresses, or plots of the displaced state of the structure may be displayed on the computer screen and printed as permanent records.

The phases of the general-purpose program have one feature in common: the *modularity* of their component aspects. Insofar as possible, the procedures treated in this text are formulated in a manner and form consistent with those found in widely distributed general-purpose programs. But they should also be of use to the student in preparing more limited or special-purpose programs. Often, programs for a particular purpose can be designed to treat that problem more efficiently than the general-purpose programs can.

1.2.2 The Program MASTAN2

A CD containing a structural analysis computer program entitled MASTAN2 is bundled with this textbook. As shown in Figures 1.3 and 1.4, MASTAN2 is an interactive graphics program that provides preprocessing, analysis, and postprocessing capabilities. Preprocessing options include definition of structural geometry, support conditions, applied loads, and element properties. The analysis routines provide the user the opportunity to perform first- or second-order elastic or inelastic analyses of two- or three-dimensional frames and trusses subjected to static loads. Post-

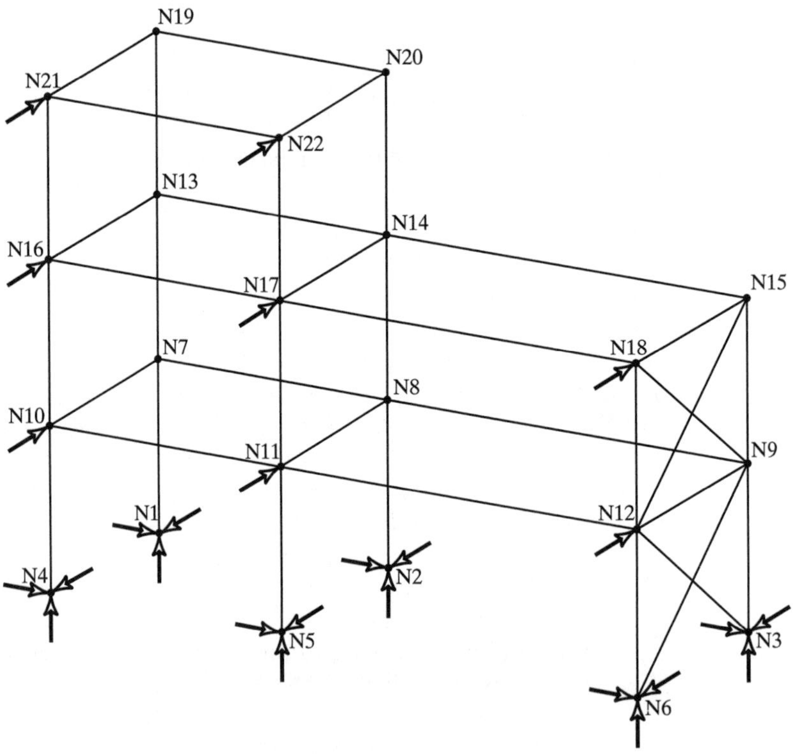

Figure 1.3 Preprocessing capabilities of MASTAN2.

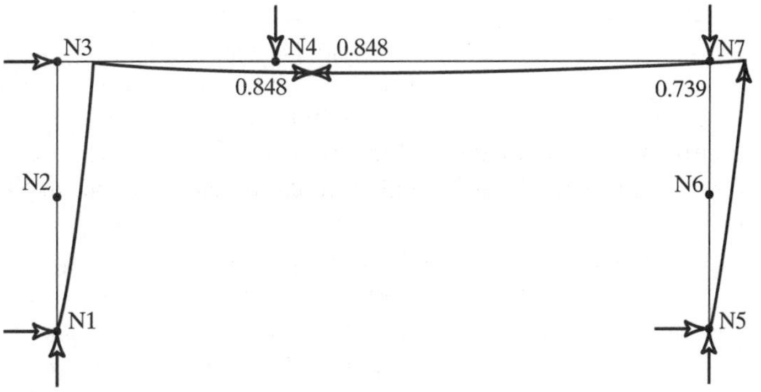

(*a*) Deformed shape and plastic hinge locations

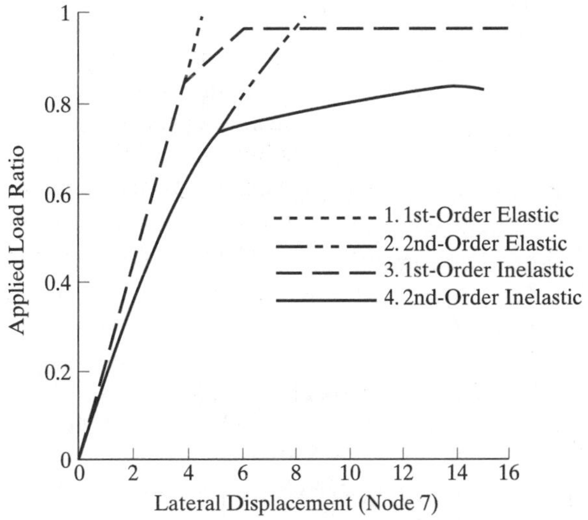

(*b*) Response curves

Figure 1.4 Postprocessing capabilities of MASTAN2.

processing capabilities include the interpretation of structural behavior through deformation and force diagrams, printed output, and facilities for plotting response curves.

In many ways, MASTAN2 is similar to today's commercially available software in functionality. The number of pre- and post-processing options, however, has been limited to minimize the amount of time needed for a user to become proficient with it. Many of the theoretical and numerical formulations presented in the context of the following chapters provide a basis for the program's linear and nonlinear analysis routines. In this regard, the reader is strongly encouraged to use this software as a tool for demonstration, reviewing examples, solving problems, and perhaps performing analysis and design studies. MASTAN2 has been written in modular format. Therefore, the reader is also provided the opportunity to develop and implement additional or alternative analysis routines directly within the program. Finally, it should be noted that MASTAN2 will execute on any computing platform where MATLAB is available.

REFERENCES

1.1 R. H. Gallagher, *Finite Element Analysis: Fundamentals*, Prentice-Hall, Englewood Cliffs, N.J., 1975.

1.2 K.-J. Bathe, *Finite Element Procedures*, Prentice-Hall, Englewood Cliffs, N.J., 1996.

1.3 G. H. Golub and C. F. Van Loan, *Matrix Computations*, Third Edition, The Johns Hopkins University Press, Baltimore, MD., 1996.

1.4 J. C. Maxwell, "On the Calculations of the Equilibrium and Stiffness of Frames," *Phil. Mag. (4), 27*, 294 (1864).

1.5 A. Castigliano, *Theorie de l'Equilibre des Systemes Elastiques*, Turin, 1879 (English translation, Dover Publications, New York, 1966).

1.6 O. Mohr, "Beitrag zur Theorie der Holz-und Eisen Konstruktionen," *Zeit des Architekten und Ingenieur Verienes zu Hannover*, 1868.

1.7 A. Cayley, "A Memoir on the Theory of Matrices," *Phil. Trans., 148*, 17–37 (1857).

1.8 S. Timoshenko, *History of Strength of Materials*, McGraw-Hill, New York, 1953.

1.9 G. B. Maney, *Studies in Engineering—No. 1*, University of Minnesota, Minneapolis, 1915.

1.10 A. Ostenfeld, *Die Deformationsmethode*, Springer-Verlag OHG, Berlin, 1926.

1.11 H. Cross, "Analysis of Continuous Frames by Distributing Fixed-End Moments," *Trans. ASCE, 96*, 1–10 (1932).

1.12 R. A. Frazer, W. J. Duncan, and A. R. Collar, *Elementary Matrices and Some Applications to Dynamics and Differential Equations*, Cambridge University Press, London, 1938.

1.13 J. H. Argyris, and S. Kelsey, *Energy Theorems and Structural Analysis*, Butterworth Scientific Publications, London, 1960.

1.14 M. J. Turner, R. Clough, H. Martin, and L. Topp, "Stiffness and Deflection Analysis of Complex Structures," *J. Aero. Sci., (23), 9*, 805–823 (Sept. 1956).

1.15 *Proceedings of the Conference on Electronic Computation.* Held in Kansas City, Mo., 1958, ASCE, N.Y.

1.16 *Proceedings of the 2nd Conference on Electronic Computation.* Held in Pittsburgh, Pa., 1960, ASCE, N.Y.

1.17 *Proceedings of the Third Conference on Electronic Computation.* Held in Boulder, CO., 1963. *Proc. ASCE, J. of the Struct. Div., 89*, No. ST4 (Aug. 1963).

1.18 J. Melan, *Theorie der eisernen Bogenbrücken und derHängebrücken, Handbuch der Ingenieurwissenschaften*, Leipzig, 1888.

1.19 J. Goodman, *Mechanics Applied to Engineering*, Longmans–Green, London, 1899.

1.20 J. Melan, Theory of Arches and Suspension Bridges, translation by D. B. Steinman, M. C. Clark Pub., Chicago, 1913.

1.21 *Plastic Design in Steel, A Guide and Commentary*, Manual No. 41, 2nd edition ASCE, New York, 1971.

1.22 M. J. Turner, E. H. Dill, H. C. Martin, and R. J. Melosh, "Large Deflections of Structures Subjected to Heating and External Loads" *J. Aero/Space Sciences, 27*, February, 1960.

1.23 J. F. Baker, M. R. Horne, and J. Heyman, *The Steel Skeleton, Vol. 2*, Cambridge University Press, Cambridge, 1956.

1.24 *Plastic Design of Multi-Story Steel Frames, Vols. 1 and 2*, Department of Civil Engineering, Lehigh University, Bethlehem, PA., 1965.

Chapter 2

Definitions and Concepts

This and the following three chapters contain the fundamentals of the displacement method of matrix structural analysis. Particularly useful components of the flexibility method are also treated. The terms *displacement method* and *flexibility method* refer to general approaches to analysis. In their simplest forms, both approaches can be reduced to sets of formal rules and procedures. However, a particular form of the displacement approach—the *direct stiffness method*—is dominant in general structural analysis and is therefore emphasized here. Use of the procedures developed in these chapters can give one a certain facility in analyzing simple structures, but further study is needed to appreciate the scope and power of these methods and their underlying principles of structural behavior. To promote such understanding, we later explore some of the most important ramifications of the fundamentals developed in Chapters 2–5.

We begin the present chapter with an explanation of the concept of degrees of freedom and then follow with a description of the principal coordinate systems and sign conventions to be used throughout the book. Idealization of framework structures for the purpose of analysis is examined next. We then proceed to the definition of influence coefficients: relationships between forces and displacements. Global stiffness equations for axial force members are developed and their use is illustrated in simple examples.

2.1 DEGREES OF FREEDOM

In this chapter we are concerned with the overall behavior of the elements of a structure as defined by the displacements of the structure's joints under the action of forces applied at the joints. The study of the relative displacements of points within individual members—the *strains*—and the distribution of forces per unit area—the *stresses*—is the subject of later chapters.

Displacement components required for the definition of the behavior of typical structures are illustrated in Figure 2.1. In the pin-jointed plane truss (Figure 2.1*a*), the members are stretched or compressed by the applied load. The net effect is the displaced structure shown, to exaggerated scale, by the solid lines. Except at supports or joints that are otherwise constrained, the movement of each joint can be described by two *translational displacement components*, such as u and v in the figure. As we'll demonstrate later, once the u and v components of all joints have been determined, the entire response—the reactions and member forces as well as the resultant displacements—can be defined. For a pin-jointed space truss, three components, such as u, v, and w in Figure 2.1*b*, are required at each joint.

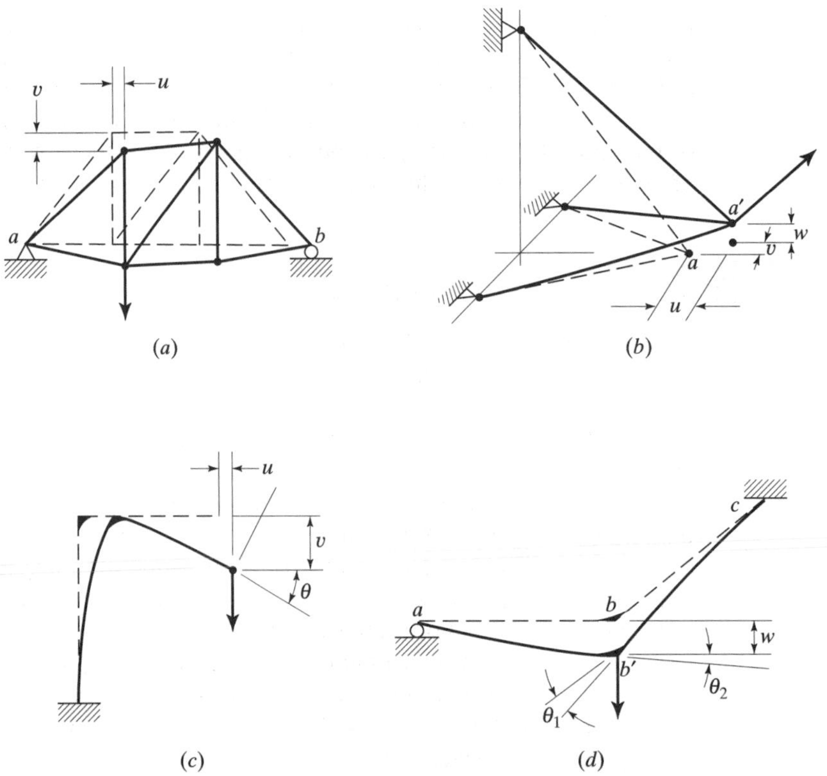

Figure 2.1 Joint displacements. (*a*) Pin-jointed plane truss. (*b*) Pin-jointed space truss. (*c*) Plane frame (in-plane loading). (*d*) Plane frame (out-of-plane loading).

In beams and rigid framed structures the elements are bent and perhaps twisted, in addition to being stretched or compressed. Figure 2.1*c* shows the *rotational displacement component*, θ, which, along with the translational components u and v, is necessary for the complete definition of the displacement of a joint in a plane frame. In a frame loaded normal to its plane, as in Figure 2.1*d*, two rotational components are needed at a joint. As shown, θ_1 is the common slope at the b end of member bc and twist at the b end of member ab. Likewise, θ_2 represents both a slope in ab and twist in bc. By an obvious extension, three translational and three rotational components will be required at a joint in a rigidly connected space frame.

Each displacement component illustrated in Figure 2.1 is a *degree of freedom*. In principle, the number of degrees of freedom of a system is the number of displacement components or coordinates needed to define its position in space at any time under any loading. Viewed in this way, all of the structures in Figure 2.1 have an infinite number of degrees of freedom since each one of the members in each structure is capable of deforming in an infinite number of modes if suitably excited. In dynamic analysis it is normally acceptable to avoid this complexity by lumping masses at selected points and considering only a finite number of degrees of freedom. In static analysis, in which displacement under a given loading will be in a single, forced mode, further simplification is possible through pre-analysis of each element and reduction of its behavior to a function of certain degrees of freedom at the element ends. Also, it is known that, for many structures, static response is more strongly dependent upon some types of deformation than upon others. In rigid frames, for example, displacements resulting from flexure are generally orders of magnitude larger than those due to uniform axial straining. In such cases it may be possible to neglect some degrees of freedom without noticeable loss of accuracy. Therefore, in practice, the number of degrees of freedom is not unique but is, instead, a function of the way in which the real structure has been idealized for analysis. Clearly, in this process the judgment and experience of the analyst come into play, but in any structure there will be a minimum number of degrees of freedom that must be considered to arrive at an acceptable result.

2.2 COORDINATE SYSTEMS AND CONDITIONS OF ANALYSIS

We shall most often work with a right-hand set of orthogonal axes identified by the symbols x, y, and z as shown in Figure 2.2. These axes remain fixed throughout the deformation of the structure, and displacements of points on the structure are referred to them. Consider a particle located at point g when the structure is in the unloaded, undeformed state. A force vector with components F_{xg}, F_{yg}, F_{zg} is applied to this particle. Under the action of this force the particle displaces to the point h. The translational displacement components of the particle are: $u_g = x_h - x_g$, $v_g = y_h - y_g$, and $w_g = z_h - z_g$. These displacements could also be shown as vector components at point g, just as the force components are. Positive values of the force and displacement components correspond to the positive sense of the coordinate axes.

Except where otherwise noted, we are limiting our attention to linear elastic behavior. That is, we are assuming that deformations are small, that material properties such as the modulus of elasticity remain constant during loading, and that the structure is nowhere stressed beyond its elastic limit. In Chapters 8–10 we shall discuss the consequences of these assumptions, and we'll develop several methods for incorporating sources of nonlinearity in the analysis. But as long as we restrict our consideration to linear elastic action, the components of the force vector may be considered to remain unchanged as the particle moves from g to h. Also, the mechanical aspects of this behavior, such as the work done by the forces F_{xg}, F_{yg}, F_{zg} acting through the displacements u_g, v_g, w_g, are not dependent upon the path taken to point h. Limiting consideration to linear behavior also implies that all equations will be formulated with respect to the geometry of the original, undeformed structure. Thus, the effect of joint displacements on these equations is not considered and the principle of superposition applies. Under this principle, the response of a structure to the application of a system of forces is identical to the summation of the responses of the same structure to the separate application of every force of the system.

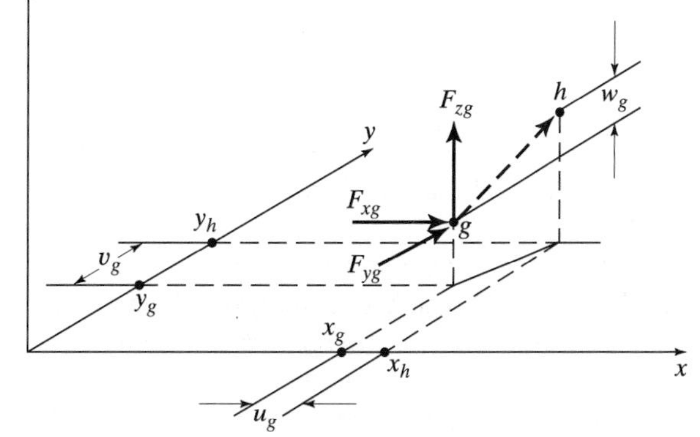

Figure 2.2 Displacement from point g to point h.

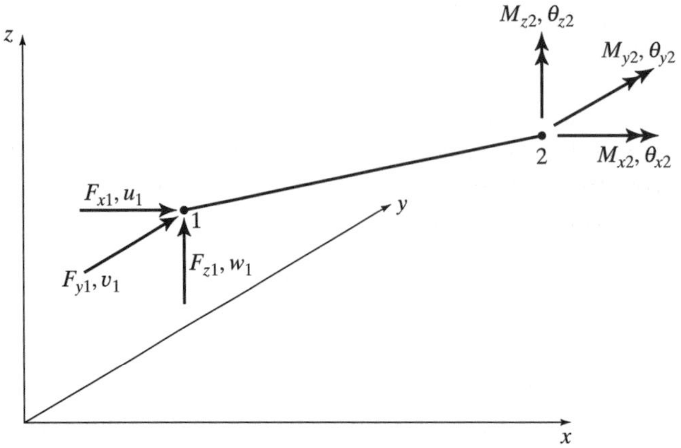

Figure 2.3 Forces, moments, and corresponding displacements.

When it is necessary to consider moments and rotational displacements, it may be done as shown at the 2 end of member 1-2 in Figure 2.3.[1] Generalizing the term *force* to include moments as well as linearly directed forces, the force components corresponding to the rotational displacement are the moments M_{x2}, M_{y2}, and M_{z2}.[2] Positive rotations and positive moments are defined in terms of the right-hand rule. In frame analysis the simplifying assumption is generally made that a line drawn normal to the elastic line of the beam or frame element remains normal to that line as the structure deforms under load. A measure of displacement at the joints of such structures is the rotation, θ, of the normal with respect to the undeformed state. As thus described in the coordinate system of the figure, the rotational displacement components for the point 2 may be represented by

$$\theta_{x2} = \left.\frac{\partial w}{\partial y}\right|_2 \qquad \theta_{y2} = -\left.\frac{\partial w}{\partial x}\right|_2 \qquad \theta_{z2} = \left.\frac{\partial v}{\partial x}\right|_2 \tag{2.1}$$

The right-hand rule accounts for the negative sign in the definition of θ_{y2}, since a positive displacement increment w gives a negative rotation θ_{y2}.[3]

As already indicated, the overall description of the behavior of a structure is accomplished through the medium of force and displacement components at designated points, commonly called *node points*, or just *nodes*. They are generally the physical joints of the structure, since these are the points of connection of the elements of the total, or *global*, analytical model. There are cases, as in the interior of members of varying cross section, in which the nodes are arbitrarily selected and do not have a

---

[1]The picture is simplified in the interest of focusing on these effects. Direct forces and moments may of course be present at both ends, as in Figure 2.7.

[2]The term *action* is also used in the same generalized sense. We'll use it, rather than *force*, on the occasions in which it is more descriptive of the case at hand.

[3]Equation 2.1 is based on the analysis of infinitesimal strains. We shall use it consistently. An equally valid alternative is

$$\theta_{x2} = -\left.\frac{\partial v}{\partial z}\right|_2 \qquad \theta_{y2} = \left.\frac{\partial u}{\partial z}\right|_2 \qquad \theta_{z2} = -\left.\frac{\partial u}{\partial y}\right|_2 \tag{2.1a}$$

The relationships between the two equations and the limitations of both are discussed in Section 4.6.2.

physical significance. Nevertheless, it is common to use the terms *node* and *joint* synonymously.

Similarly, the terms *member* and *element* are used interchangeably, but frequently, the former will connote a complete, physical component of a structure whereas the latter may refer to a segment of that component. The distinction should always be clear from the context.

The forces and displacements at the nodes of a given element form column vectors which we designate as {**F**} and {**Δ**}. (Braces, { }, denote column vectors.) For the element in Figure 2.3, for example, with direct forces at point 1 and moments at point 2,

$$\{\mathbf{F}\} = \lfloor F_{x1} \quad F_{y1} \quad F_{z1} \quad M_{x2} \quad M_{y2} \quad M_{z2} \rfloor^{\mathrm{T}}$$
$$\{\boldsymbol{\Delta}\} = \lfloor u_1 \quad v_1 \quad w_1 \quad \theta_{x2} \quad \theta_{y2} \quad \theta_{z2} \rfloor^{\mathrm{T}}$$

where, as we shall customarily do to save space, we have listed the contents of these vectors in transposed row vector form. An individual entry, Δ_i, in the vector of joint displacements, $\{\boldsymbol{\Delta}\} = \lfloor \Delta_1 \ldots \Delta_i \ldots \Delta_n \rfloor^{\mathrm{T}}$, is termed the *ith degree of freedom*. For the case of direct forces and moments at both ends, {**F**} and {**Δ**} would be 12×1 vectors.

The term *release* is often used in framework computer analysis programs. In a simplified sense, this term, which relates to support conditions, implies that all of the degrees of freedom at a joint are initially fixed and that certain components of displacement at the joint are then recognized as being actually free from constraint, that is, they are "released." In the roller support condition of point *b* of Figure 2.1*a*, for example, the horizontal displacement is a release. There is considerably more to the subject of releases. They will be discussed further and more precise definitions will be given in Section 13.5.

The first step in the formation of the force and displacement vectors is the definition of the nodal points and their location with respect to a coordinate system. We distinguish between *global* and *local* coordinate systems.[4] The global system is established for the complete structure. The local (*member* or *element*) axes are fixed to the respective elements and, since the members are in general differently oriented within a structure, these axes may differ from element to element. This is illustrated in Figure 2.4. When different types of axes are being compared or when they appear in the same portion of the text, the local axes will be identified by primes, as in the figure. No primes are used when local axes alone are used.

We use local axes in almost all of our formulations of element equations. Our convention for relating them to the elements and for numbering the joints of an element is illustrated in Figure 2.4*b*. The local x' axis is directed along the axis of the member. The joint at or closest to the origin of local coordinates is designated as joint 1 and the other node as joint 2. This convention is carried over into space structures, but in these it will be necessary, additionally, to define the orientation of the local y' and z' axes with respect to certain directions in a cross section of the element (see Chapter 5).

Global axes figure prominently in the development of the *global equations* (the equations of the complete structure) in Sections 3.1–3.3. Given the elastic properties of an element or structure in terms of local coordinates, the transformation of forces and displacements from these directions to the global directions can be constructed easily. The transformation procedures could be introduced at this point. However, in the interest of proceeding to overall analysis as quickly as possible, we postpone the subject of coordinate transformation to Section 5.1.

---

[4]A third type of coordinate system, *joint coordinates*, will be discussed in Section 13.4.

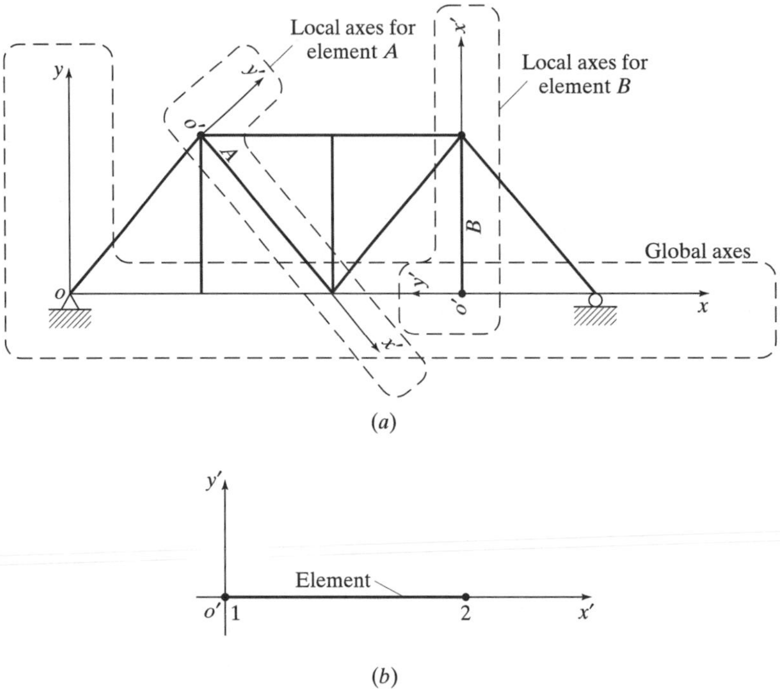

Figure 2.4 Coordinate axes and joint numbering. (*a*) Types of coordinate axes. (*b*) Joint numbering scheme in local coordinates.

2.3 STRUCTURE IDEALIZATION

To permit analysis, the actual structure must be idealized. Members, which have width and depth as well as length, are conventionally reduced to line elements. Their resistance to deformation is represented by material properties such as Young's modulus (E) and Poisson's ratio (ν), and by geometrical properties of the cross section, such as area (A), moment of inertia (I), and torsional constant (J). The behavior of the connections, that is, whether pinned, semirigid, rigid, yielding, etc., must be stipulated. How these idealizing decisions are made, or the necessary properties determined, is extremely important and involves considerable judgment. In the interest of focusing on the techniques of analysis, it will be assumed that the line diagrams and member properties used are valid idealizations of the real structure.

The concepts of idealization and analysis can be described with respect to the simple truss in Figure 2.5, in which the common assumption of pinned joints is implied. For the purpose of forming a *mathematical model* the state of stress in the members is represented by forces at the element ends. The corresponding displacements of these nodes—the degrees of freedom—are employed in the characterization of the displaced state of the element. The individual truss elements are isolated, and these forces and displacements are identified symbolically, as for the typical member in Figure 2.5*b*. Then, using the principles of elasticity and the laws of equilibrium, relations are formed between the joint forces and displacements. The truss is next reconstructed analytically by examining the equilibrium of member forces at each joint. By summing up the truss element forces in each direction at each joint in each coordinate direction and equating the result to the corresponding applied loads, the conditions of equilibrium within the truss are fully accounted for. The joining of truss members at the joints also ensures that the truss displaces as a structural entity without any discontinuities in the pattern

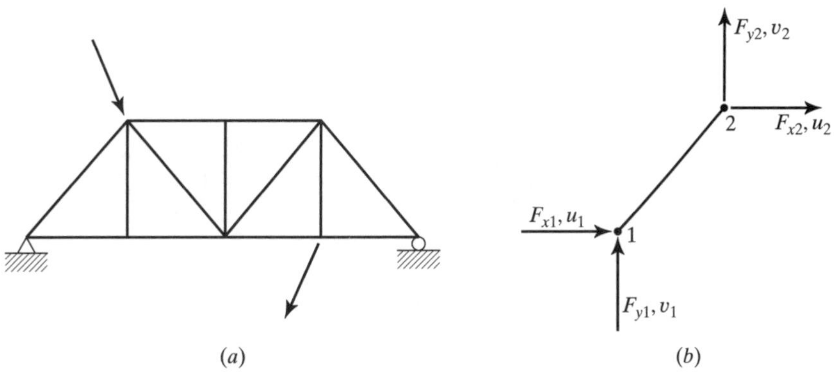

Figure 2.5 Idealized truss. (*a*) Truss. (*b*) Typical truss member.

of displacement, that is, that the conditions of internal compatibility are also accounted for. The solution is *exact* within the confines of linear elastic analysis, the assumption of pinned joints, and any other idealizing assumption made in converting the actual structure to a line diagram.

Although it hardly seems necessary to apply such a formal description as the foregoing to such a simple truss, the point has been to illustrate general concepts. They will apply to the most complex framed structure.

Framed structures are almost limitless in variety. A few examples of idealized frameworks are shown in Figure 2.6. All can be broken down into line elements, and all are

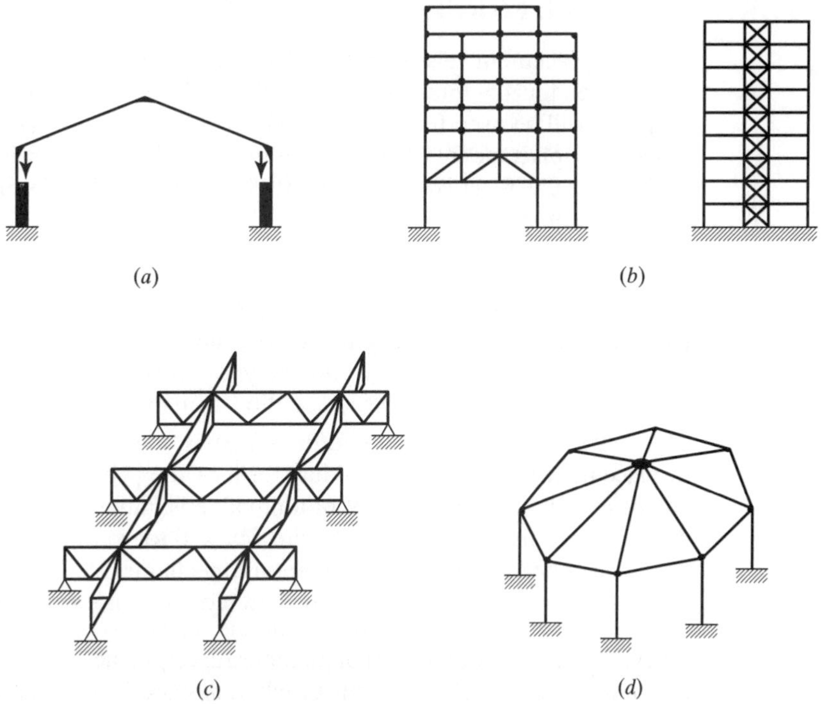

Figure 2.6 Typical framed structures. (*a*) Rigidly jointed plane frame. (*b*) Multistory frames—rigidly jointed and trussed. (*c*) Trussed space frame. (*d*) Rigidly jointed space frame.

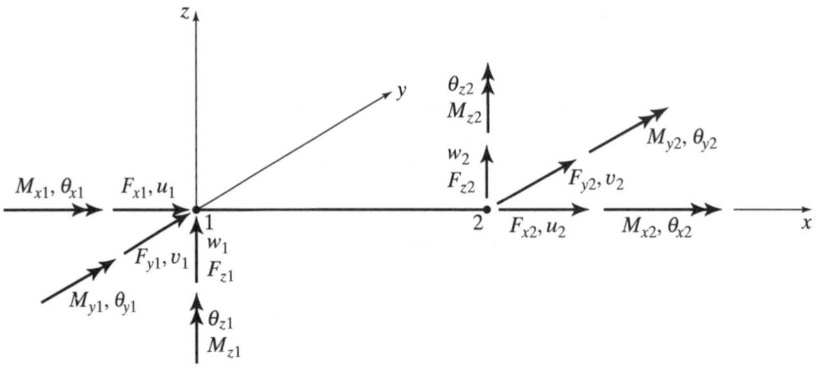

Figure 2.7 Framework element.

within the purview of this text. In Chapters 2–5 we treat only cases in which the component elements are *prismatic*, that is, straight and of uniform cross section from node to node. Nonprismatic elements are covered in Chapter 7.

The complete framework element has 12 nodal degrees of freedom and 12 nodal force components, as indicated in Figure 2.7.[5] Usually, however, members of the real structure are joined in ways such that, in the idealization, a number of these force components and degrees of freedom may be disregarded. For example, for the purpose of analysis, we may often assume that trusses are pin jointed. In beams, torsional moments may often be disregarded. Other reasons for neglecting certain effects have been cited in Section 2.1.

2.4 AXIAL FORCE ELEMENT: FORCE-DISPLACEMENT RELATIONSHIPS

Force-displacement relationships for the complete framework element will be developed in Section 4.5. First, however, the simple *axial force* element (also called the *truss* element) will be used for further demonstration of concepts and definitions.

Force-displacement equations, the relationships between joint forces and joint displacements are most commonly written in either of two forms: stiffness equations or flexibility equations.[6]

2.4.1 Element Stiffness Equations

Element stiffness equations are linear algebraic equations of the form

$$\{\mathbf{F}\} = [\mathbf{k}]\{\boldsymbol{\Delta}\} \tag{2.2}$$

The matrix $[\mathbf{k}]$ is the element stiffness matrix, and $\{\mathbf{F}\}$ and $\{\boldsymbol{\Delta}\}$ are element force and displacement vectors. An individual term of the $[\mathbf{k}]$ matrix, k_{ij}, is an *element stiffness coefficient.* If a displacement Δ_j of unit value is imposed and all other degrees of freedom are held fixed against movement ($\Delta_k = 0$ for $k \neq j$) the force F_i is equal in value to k_{ij}. The spring constant of elementary mechanics is a stiffness coefficient.

Consider a simple truss member directed along a local axis x and having a known

---

[5]This element is "complete" in the sense that it is capable of accounting for axial force, shear in two directions, biaxial bending, and *uniform (St. Venant) torsion.* See Section 7.4 for discussion of nonuniform torsion and the additional two degrees of freedom needed to account for *warping torsion.*

[6]A description of a third form, mixed force-displacement equations, may be found in the first edition of this text. Because of its limited use, it has not been included in this edition.

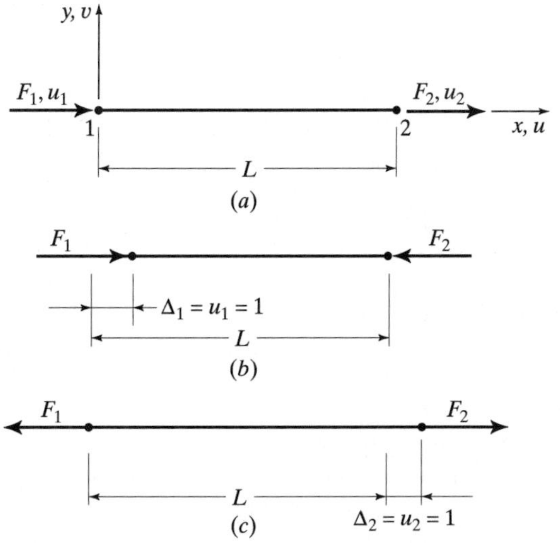

Figure 2.8 Axial force element—member on x axis. (a) Axial force element. (b) Unit displacement at 1. (c) Unit displacement at 2.

area A, and Young's modulus E (Figure 2.8a). Since it is assumed to be pin-ended and therefore capable of resisting only an axial force, it is a two-degree-of-freedom member. In Figure 2.8b a unit displacement in degree-of-freedom 1 is imposed with the other degree of freedom held fixed ($\Delta_1 = u_1 = 1$, $\Delta_2 = u_2 = 0$). The column of joint forces for this case is equal to the column of stiffness coefficients corresponding to Δ_1, or

$$\{\mathbf{F}\} = \{\mathbf{k}_{i1}\} \qquad (i = 1, 2)$$

where

$$\{\mathbf{F}\} = \lfloor F_1 \quad F_2 \rfloor^{\mathrm{T}}$$
$$\{\mathbf{k}_{i1}\} = \lfloor k_{11} \quad k_{21} \rfloor^{\mathrm{T}}$$

Clearly, $F_1 = k_{11}$ is the force required to impose a unit value of Δ_1, and $F_2 = k_{21}$ is the reactive force. Thus the column of stiffness coefficients $\{\mathbf{k}_{i1}\}$ represents a system of forces in equilibrium. The same interpretation can be placed on the second column of the element stiffness matrix (see Figure 2.8c). In Figures 2.8b and 2.8c the forces F_1 and F_2 are shown in the directions that correspond to the imposed positive displacements. In analysis, unknown forces are generally assumed to act in their positively defined directions, and stiffness coefficient signs are established accordingly. Thus the correct direction of any force can be determined from the sign of the analytical result.

To write the stiffness equations for this member, we have, from Figure 2.8b and the basic mechanics of deformable bodies:

$$u_1 = \frac{F_1 L}{EA}$$

or since we wish an expression for the end forces

$$F_1 = \frac{EA}{L} u_1$$

and, by equilibrium

$$F_2 = -F_1 = -\frac{EA}{L} u_1$$

In the same way, from Figure 2.8*c*

$$F_2 = -F_1 = \frac{EA}{L} u_2$$

Collected in matrix form, these equations are

$$\begin{Bmatrix} F_1 \\ F_2 \end{Bmatrix} = \begin{bmatrix} k_{11} & k_{12} \\ k_{21} & k_{22} \end{bmatrix} \begin{Bmatrix} u_1 \\ u_2 \end{Bmatrix} = \frac{EA}{L} \begin{bmatrix} 1 & -1 \\ -1 & 1 \end{bmatrix} \begin{Bmatrix} u_1 \\ u_2 \end{Bmatrix} \tag{2.3}$$

Note that the pertinent equilibrium equation is $\Sigma F_x = 0$, and that this results in a zero sum of the terms in each column of the stiffness matrix. The matrix is also symmetrical. This is a consequence of the necessary reciprocity of displacements, the laws of which will be discussed in Section 4.3.

A complete set of stiffness equations for an element, such as Equation 2.3, associates all of the effective node-point forces with all of the degrees of freedom. Mathematically, the stiffness matrix is singular. The reason is that, physically, the degrees of freedom include *rigid-body-motion* terms. For example, in defining the first column in this stiffness matrix, the member was restrained at node 2 by suppressing the displacement at that point (Figure 2.8*b*). If it had not been, any attempt to impose a displacement at node 1 would have been unresisted. The member could respond as an undeformed body and translate to the right an unlimited amount under the action of any nonzero value of the force F_1. Likewise, the displacement at node 1 was suppressed in defining the second column of the stiffness matrix. The suppressed displacements associated with any such support conditions are sets of displacements associated with rigid body motion. Such degrees of freedom, when extracted, enable a more concise description of the element stiffness properties (see Section 4.4).

2.4.2 Element Flexibility Equations

Element flexibility equations express, for elements supported in a stable manner, the joint displacements, $\{\mathbf{\Delta}_f\}$, as a function of the joint forces, $\{\mathbf{F}_f\}$:

$$\{\mathbf{\Delta}_f\} = [\mathbf{d}]\{\mathbf{F}_f\} \tag{2.4}$$

where $[\mathbf{d}]$ is the element flexibility matrix. An *element flexibility coefficient*, d_{ij}, is the value of the displacement Δ_i caused by a unit value of the force F_j. The subscript f on the force and displacement vectors refers to the degrees of freedom that are free to displace. It is used to emphasize the fact that these vectors exclude components related to the support conditions, that is, that the constrained degrees of freedom are absent from Equation 2.4.

Flexibility relationships can be written only for elements supported in a stable manner because rigid body motion of undetermined magnitude would otherwise result from application of applied forces. They can be applied in as many ways as there are stable and statically determinate support conditions. In contrast to the stiffness matrix of Equation 2.2, the flexibility matrix is not unique. The difference is that, in compiling the element stiffness matrix, displacements are specified (symbolically), and therefore rigid body motion presents no problem of definition.

Flexibility relationships can also be derived for elements supported in a statically indeterminate manner but they would not be of general use since, with certain exceptions, they could not be combined with other elements in the representation of a complex structure (see Section 4.4.1).

For the axial force member, the flexibility relationships assume the simple forms associated either with a pinned support at node 2 (Figure 2.8*b*):

$$\{u_1\} = \left[\frac{L}{EA}\right]\{F_1\}$$

or a pinned support at node 1 (Figure 2.8*c*):

$$\{u_2\} = \left[\frac{L}{EA}\right]\{F_2\}$$

When the complete element framework element is considered in Section 4.5 more complex flexibility equations will be found.

2.5 AXIAL FORCE ELEMENT—GLOBAL STIFFNESS EQUATIONS

Normally, the simplest way to form the *global* stiffness equations for the analysis of a structure is to start with element stiffness equations in convenient local coordinates—relationships such as Equation 2.3—and then to transform them to global axes by methods to be developed in Section 5.1. Mathematical transformations are not essential, however. Element stiffness equations in arbitrary coordinates—either local or global—can be developed from basic principles. Consider the axial force member in Figure 2.9, which is identical to the member in Figure 2.8. It has known properties and lies in an x-y plane, but in this case it makes an angle ϕ—which may be different from zero—with the x axis. Defining the degrees of freedom as displacement components parallel to the coordinate axes, there are in this instance four relevant quantities (u_1, v_1, u_2, v_2) rather than the two (u_1, u_2) that are sufficient for the member oriented as in Figure 2.8. Correspondingly, the resultant force acting on each end,

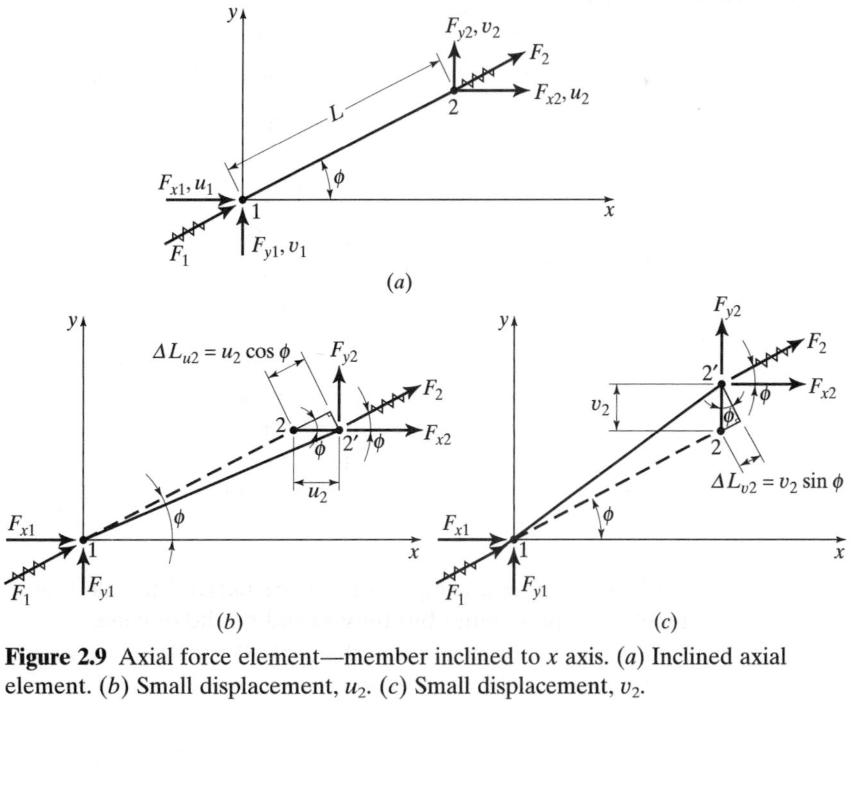

Figure 2.9 Axial force element—member inclined to x axis. (*a*) Inclined axial element. (*b*) Small displacement, u_2. (*c*) Small displacement, v_2.

which must be directed along the bar, may be resolved into its components in the coordinate directions, as shown in Figure 2.9*a*. The element stiffness matrix must now be of order 4 × 4.

To determine the column of stiffness coefficients relating the force components to u_2, impose a small displacement in the x direction at node 2 and hold all the other degrees of freedom fixed against displacement (Figure 2.9*b*). Making the usual assumptions of small displacement theory, the lengthening of the bar, ΔL_{u2}, is

$$\Delta L_{u2} = u_2 \cdot \cos\phi$$

The resultant force in the bar, F_2, is

$$F_2 = \frac{EA}{L}\Delta L_{u2} = \frac{EA\cos\phi}{L}\cdot u_2$$

All of the force components are related to F_2 by equilibrium. Following our assumption that the components of the force vector remain unchanged as node 2 moves from 2 to 2′, we have

$$F_{x2} = -F_{x1} = F_2\cos\phi = \frac{EA}{L}\cos^2\phi\cdot u_2$$

$$F_{y2} = -F_{y1} = F_2\sin\phi = \frac{EA}{L}\sin\phi\cos\phi\cdot u_2$$

Likewise, for a small displacement in the y direction at node 2 (Figure 2.9*c*), it follows that

$$F_{x2} = -F_{x1} = F_2\cos\phi = \frac{EA}{L}\sin\phi\cos\phi\cdot v_2$$

$$F_{y2} = -F_{y1} = F_2\sin\phi = \frac{EA}{L}\sin^2\phi\cdot v_2$$

Corresponding displacements can be imposed at node 1 and the results compiled in the element stiffness equations:

$$\begin{Bmatrix} F_{x1} \\ F_{y1} \\ F_{x2} \\ F_{y2} \end{Bmatrix} = \frac{EA}{L}\begin{bmatrix} \cos^2\phi & \sin\phi\cos\phi & -\cos^2\phi & -\sin\phi\cos\phi \\ \sin\phi\cos\phi & \sin^2\phi & -\sin\phi\cos\phi & -\sin^2\phi \\ -\cos^2\phi & -\sin\phi\cos\phi & \cos^2\phi & \sin\phi\cos\phi \\ -\sin\phi\cos\phi & -\sin^2\phi & \sin\phi\cos\phi & \sin^2\phi \end{bmatrix}\begin{Bmatrix} u_1 \\ v_1 \\ u_2 \\ v_2 \end{Bmatrix} \tag{2.5}$$

Equation 2.5 is a more general form of the axial member stiffness equation than Equation 2.3. It yields the element stiffness equations in arbitrary coordinates directly. For $\phi = 0$, Equation 2.5 reduces to Equation 2.3 after the deletion of null vectors, as of course it should.

2.6 EXAMPLES

The following examples illustrate the application of the concepts discussed in this chapter. Although not in themselves examples of matrix structural analysis, they contain previews of procedures that are part of formal matrix analysis methods.

Example 2.1 shows how member stiffness equations may be used to evaluate displacements in cases in which internal forces have already been calculated. The requirement of compatibility of displacements is satisfied through the simultaneous solution of two equations. Note that superscripts are used for defining unambiguously the element on which the force component acts.

Example 2.2 is an elementary application of the displacement method of analysis for forces and displacements.

In Example 2.3 stiffness equations are used to determine directly the force needed to obtain a desired displacement.

Example 2.4 is an elementary application of flexibility equations to the solution of a thermal loading problem.

Example 2.5 illustrates that structural equations may sometimes be used advantageously in the solution of small displacement kinematic problems.

In Example 2.6 some of the many different ways in which element stiffness equations are combined to form force-displacement relationships are illustrated.

EXAMPLE 2.1

A statically determinate truss is subjected to the load and resulting bar forces shown. What is the displacement of a? $E = 200{,}000$ MPa.

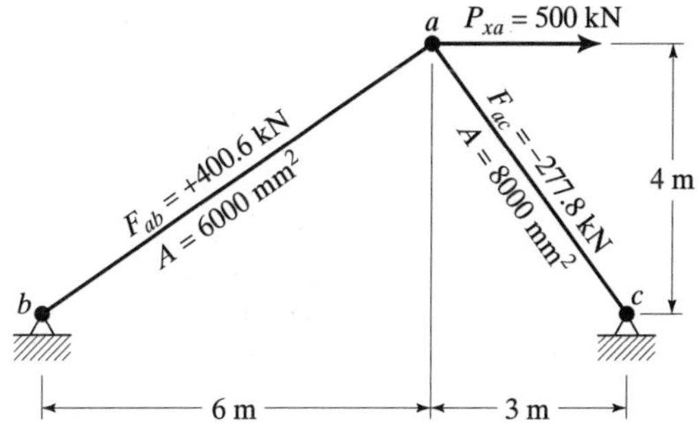

Consider ab:

$$\left(\frac{EA}{L}\right)_{ab} = \frac{200 \times 6 \times 10^3}{\sqrt{6^2 + 4^2} \times 10^3} = 166.4 \text{ kN/mm}$$

$$\phi_{ab} = \tan^{-1}\left(\frac{-4}{-6}\right) = 213.69°$$

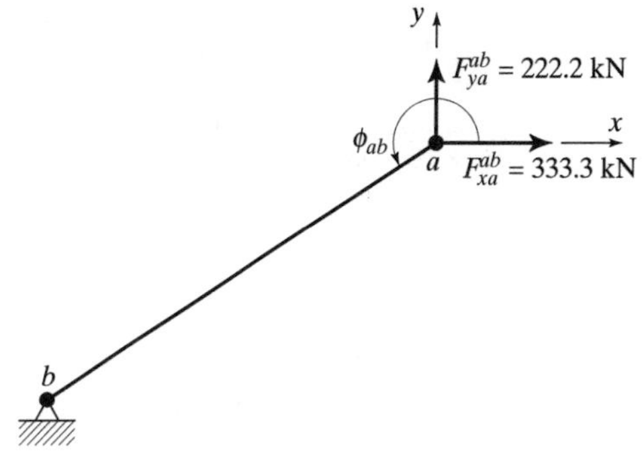

From the first part of Equation 2.5 (with $u_b = v_b = 0$),

$$F^{ab}_{xa} = 166.4(\cos^2 \phi_{ab} \cdot u_a + \sin \phi_{ab} \cos \phi_{ab} \cdot v_a)$$

$$333.3 = 166.4(0.6923u_a + 0.4615v_a) \qquad \text{(a)}$$

Consider ac:

$$\left(\frac{EA}{L}\right)_{ac} = \frac{200 \times 8 \times 10^3}{5 \times 10^3} = 320.0 \text{ kN/mm}$$

$$\phi_{ac} = \tan^{-1}\left(-\frac{4}{3}\right) = 306.87°$$

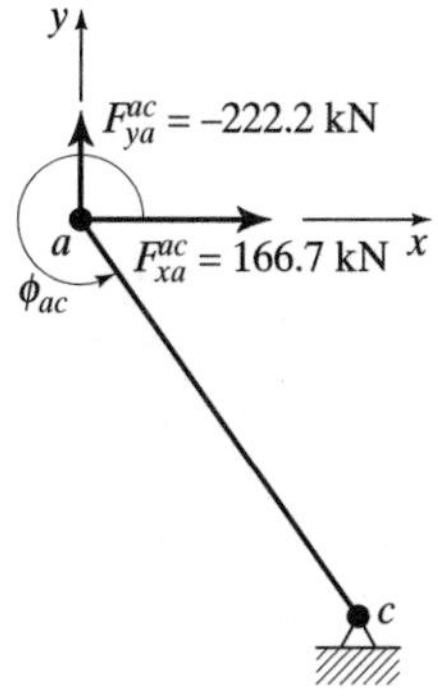

From the first part of Equation 2.5 (with $u_c = v_c = 0$),

$$F^{ac}_{xa} = 320.0(\cos^2 \phi_{ac} \cdot u_a + \sin \phi_{ac} \cos \phi_{ac} \cdot v_a)$$

$$166.7 = 320.0(0.3600u_a - 0.4800v_a) \qquad \text{(b)}$$

Solve Equations a and b simultaneously:

$$
\begin{aligned}
0.6923u_a + 0.4615v_a &= 2.003 \quad \text{(a)}\\
0.3600u_a - 0.4800v_a &= 0.5209 \quad \text{(b)}\\
\hline
u_a &= 2.41 \text{ mm} \rightarrow\\
v_a &= 0.72 \text{ mm} \uparrow\\
\overline{aa}' &= 2.52 \text{ mm} \nearrow
\end{aligned}
$$

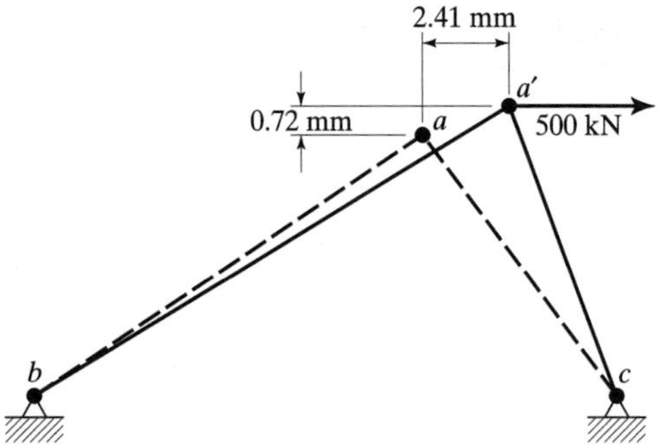

EXAMPLE 2.2

A statically indeterminate truss is subjected to the load shown. What are the bar forces and the displacement of a? $E = 200{,}000$ MPa.

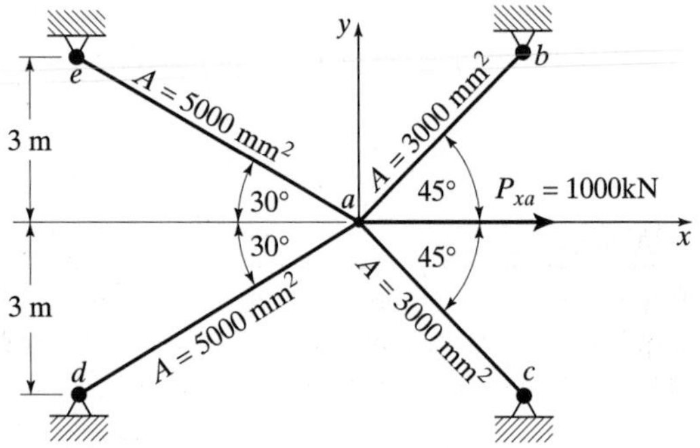

By symmetry, joint a must displace horizontally. Therefore, there is only one unknown degree of freedom, u_a From the first part of Equation 2.5, using the nomenclature indicated, and denoting q as the typical support point

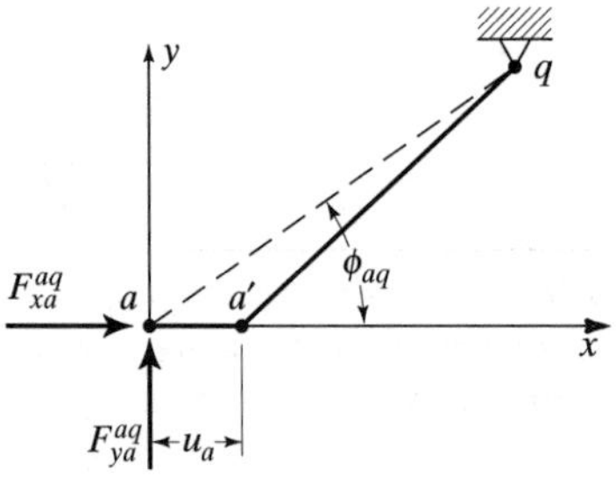

$$F_{xa}^{aq} = \left(\frac{EA}{L}\right)_{aq} \cdot (\cos\phi_{aq})^2 u_a$$

thus, for $q = b$

$$F_{xa}^{ab} = \frac{200 \times 3 \times 10^3 (\cos 45°)^2}{3\sqrt{2} \times 10^3} u_a = 70.71u_a \text{ kN}$$

$$F_{xa}^{ac} = \frac{200 \times 3 \times 10^3 (\cos 315°)^2}{3\sqrt{2} \times 10^3} u_a = 70.71u_a \text{ kN}$$

$$F_{xa}^{ad} = \frac{200 \times 5 \times 10^3 (\cos 210°)^2}{6 \times 10^3} u_a = 125.00u_a \text{ kN}$$

$$F_{xa}^{ae} = \frac{200 \times 5 \times 10^3 (\cos 150°)^2}{6 \times 10^3} u_a = 125.00u_a \text{ kN}$$

Write the equation of horizontal equilibrium of joint a with u_a as an unknown:

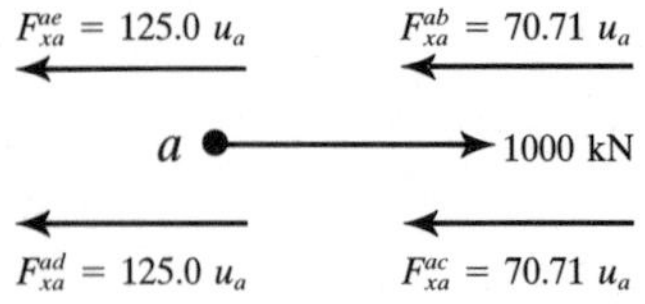

$$\Sigma H_a = 1000 - 2(70.71 + 125.0)u_a = 0$$

Solving for u_a,

$$u_a = 2.55 \text{ mm}$$

Calculate the bar forces using the member stiffness equations:

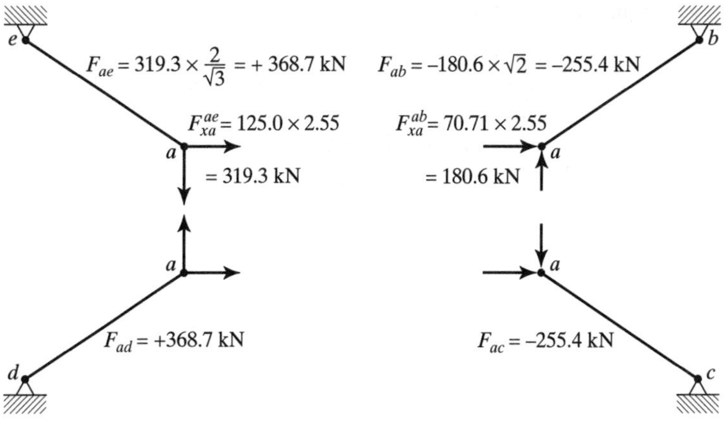

EXAMPLE 2.3

What is the magnitude and direction of the force P at point a required to displace that point vertically downward 5 mm without any horizontal displacement? E = 200,000 MPa.

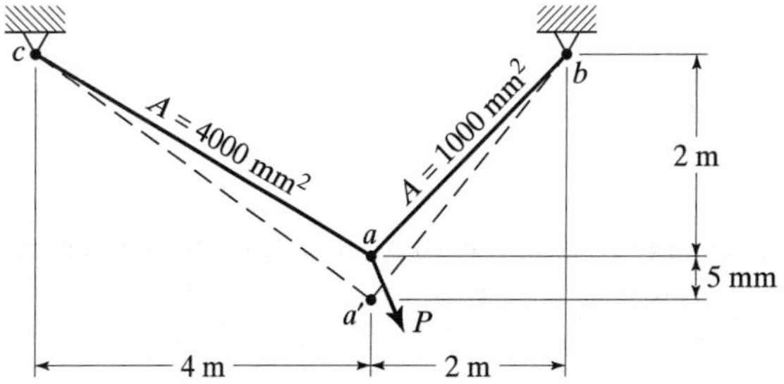

All degrees of freedom, except v_a, equal zero. $v_a = -5$ mm. From the first and second parts of Equation 2.5, using the nomenclature indicated, with q a typical support point

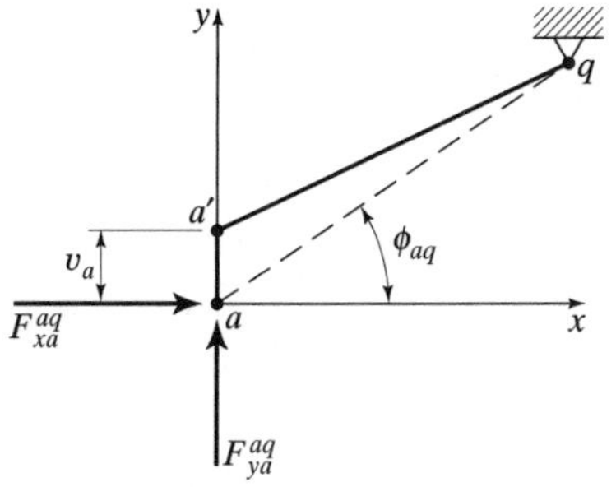

$$F_{xa}^{aq} = \left(\frac{EA}{L}\right)_{aq} \cdot (\sin \phi_{aq} \cos \phi_{aq}) \cdot v_a$$

$$F_{ya}^{aq} = \left(\frac{EA}{L}\right)_{aq} \cdot (\sin^2 \phi_{aq}) \cdot v_a$$

For ab, $\phi_{ab} = 45°$:

$$F_{xa}^{ab} = \frac{200 \times 1000}{2\sqrt{2} \times 10^3}(\sin 45° \cos 45°)(-5) = -176.8 \text{ kN}$$

$$F_{ya}^{ab} = 70.71(\sin^2 45°)(-5) = -176.8 \text{ kN}$$

For ac, $\phi_{ac} = \tan^{-1}(2/-4) = 153.43°$:

$$F_{xa}^{ac} = \frac{200 \times 4000}{\sqrt{2^2 + 4^2} \times 10^3}(\sin 153.43° \cos 153.43°)(-5) = 357.8 \text{ kN}$$

$$F_{ya}^{ac} = 178.89(\sin^2 153.43°)(-5) = -178.9 \text{ kN}$$

By equilibrium, the required force is the vector sum of these components:

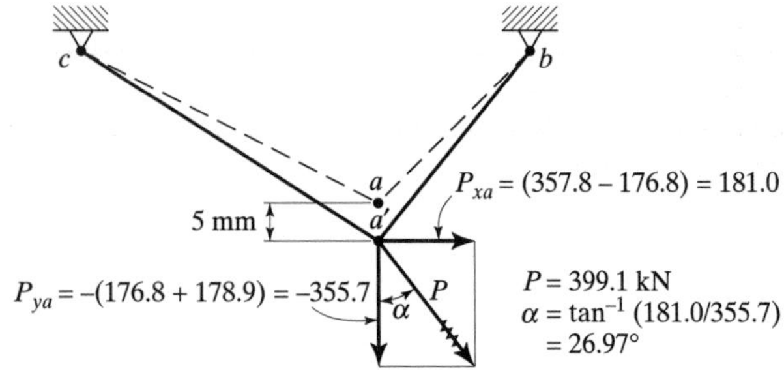

EXAMPLE 2.4

Two bars ab and bc are pinned together as shown. Bar bc is cooled 40°C. Determine the displacement of b and the force in the bars. Thermal expansion coefficient $\alpha = 1.17 \times 10^{-5}$ mm/mm°C. $E = 200{,}000$ MPa.

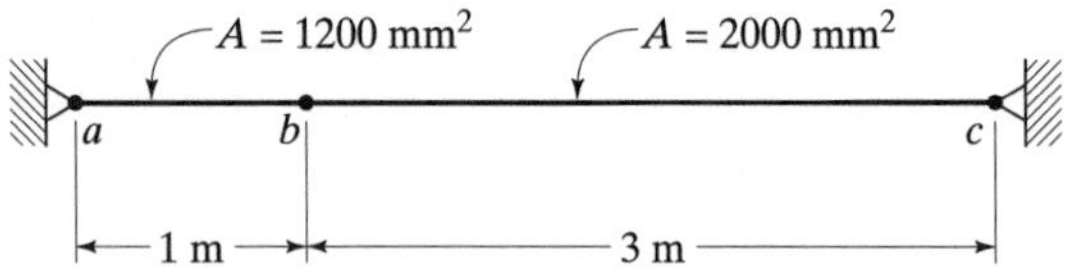

Assume the pin at b is disconnected and calculate the gap due to free thermal contraction of bar bc due to a temperature change, $T = 40$°C:

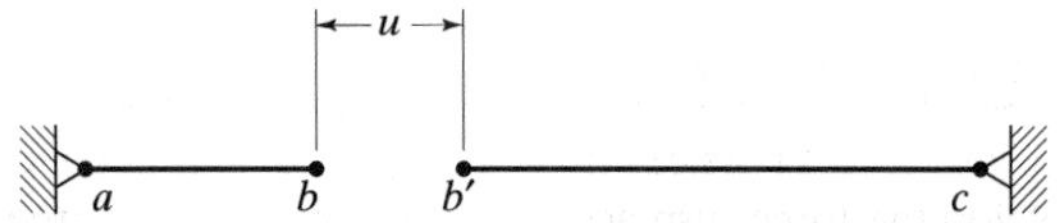

$$u = \alpha L_{bc} T$$
$$= 1.17 \times 10^{-5} \times 3 \times 10^3 \times 40 = 1.404 \text{ mm}$$

Apply equal tensile forces in each bar to close the gap:

$$u = u_{ab} + u_{bc}$$

Using flexibility relationships for u_{ab} and u_{bc},

$$u = F\left[\left(\frac{L}{EA}\right)_{ab} + \left(\frac{L}{EA}\right)_{bc}\right]$$

$$1.404 = F\left[\frac{1 \times 10^3}{200 \times 1.2 \times 10^3} + \frac{3 \times 10^3}{200 \times 2 \times 10^3}\right]$$

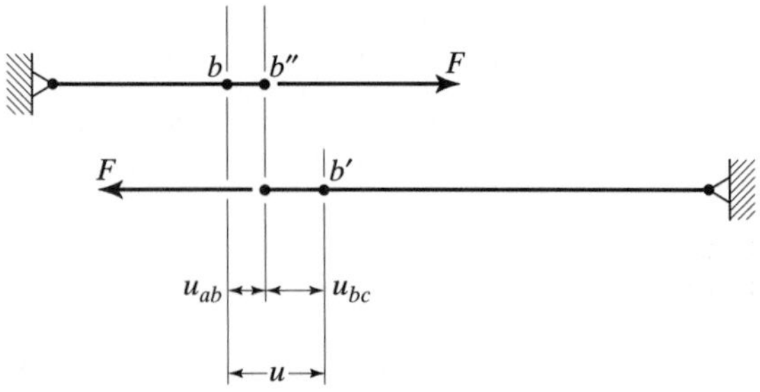

Solving for F and using the flexibility relationships for u_{ab} and u_{bc},

$$F = 120.3 \text{ kN}$$
$$u_{ab} = 0.501 \text{ mm}$$
$$u_{bc} = 0.903 \text{ mm}$$

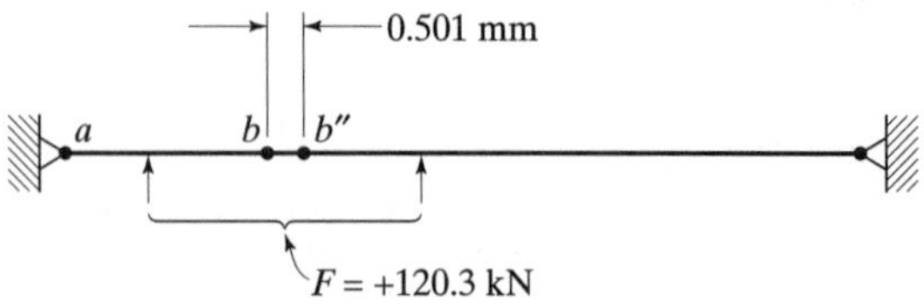

EXAMPLE 2.5

A weightless bar that is 5 m long rests against a slope as shown. If the point a is moved 10 mm to the right as indicated, how far will point b move up the slope?

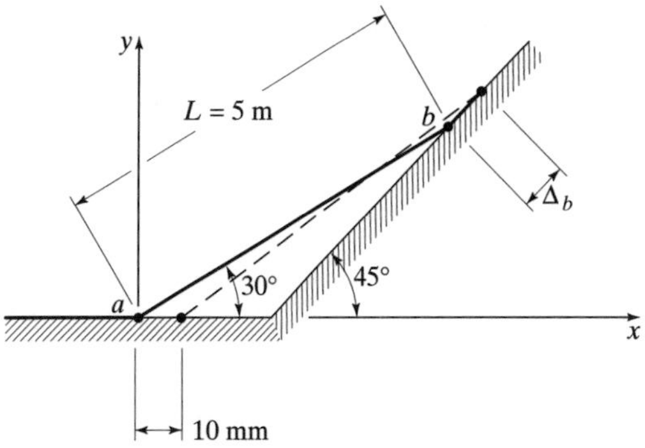

No forces are required; the problem is one of kinematics. From the first part of Equation 2.5, with the above notation,

$$F_{xa} = \frac{EA}{L}(\cos^2\phi \cdot u_a + \sin\phi\cos\phi \cdot v_a - \cos^2\phi \cdot u_b - \sin\phi\cos\phi \cdot v_b)$$

But

$$F_{xa} = 0 \qquad \phi = 30° \qquad u_a = 10 \text{ mm} \qquad v_a = 0 \qquad u_b = v_b = \Delta_b/\sqrt{2}$$

Thus

$$0 = \frac{EA}{L}[10\cos^2 30° - (\cos^2 30° + \sin 30°\cos 30°)\Delta_b/\sqrt{2}]$$

Solving for Δ_b,

$$\Delta_b = 8.97 \text{ mm}$$

EXAMPLE 2.6

Two straight bars of different properties are pinned together as shown. Develop the force-displacement relationships for three prescribed loading and support conditions.

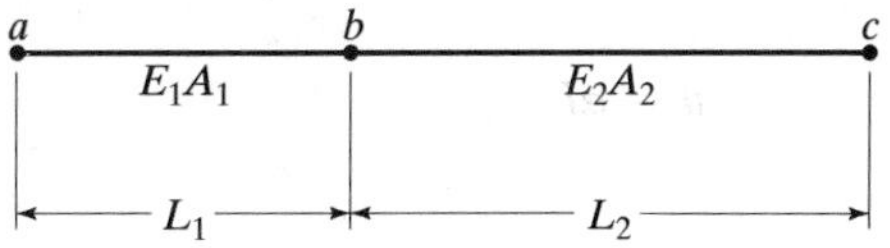

First write the element stiffness equations using Equation 2.3:

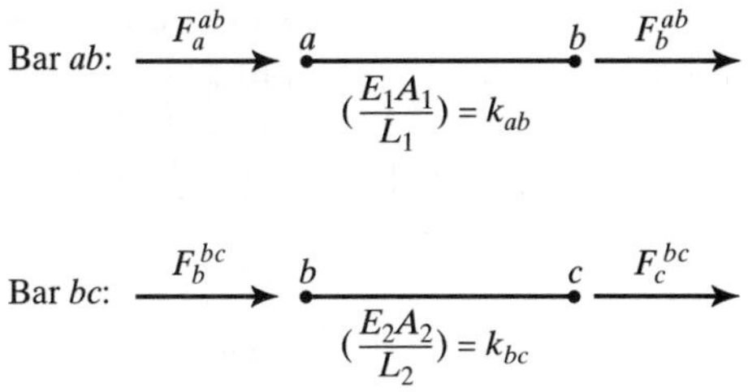

$$\begin{Bmatrix} F_a^{ab} \\ F_b^{ab} \end{Bmatrix} = k_{ab}\begin{bmatrix} 1 & -1 \\ -1 & 1 \end{bmatrix}\begin{Bmatrix} u_a \\ u_b \end{Bmatrix} \qquad \text{(a)}$$

$$\begin{Bmatrix} F_b^{bc} \\ F_c^{bc} \end{Bmatrix} = k_{bc}\begin{bmatrix} 1 & -1 \\ -1 & 1 \end{bmatrix}\begin{Bmatrix} u_b \\ u_c \end{Bmatrix} \qquad \text{(b)}$$

Case I:
From (a),

$$F_b^{ab} = P = k_{ab} \cdot u_b$$

From (b),

$$F_c^{bc} = P = k_{bc}(-u_b) + k_{bc}(u_c)$$

Eliminating u_b,

$$P = -\frac{k_{bc}}{k_{ab}}P + k_{bc}u_c$$

Boundary condition: $u_a = 0$
Equilibrium requirement:
$F_b^{ab} = F_c^{bc} = P$

The two force-displacement relationships are:

$$P = k_{ab}u_b$$

$$P = \frac{k_{ab} \cdot k_{bc}}{k_{ab} + k_{bc}} u_c$$

Case II:
From (a),

$$P = k_{ab}u_b$$

From (b),

$$0 = k_{bc}(-u_b + u_c), \qquad \text{or} \qquad u_c = u_b$$

The two force-displacement relationships are:

$$P = k_{ab}u_b$$

$$P = k_{ab}u_c$$

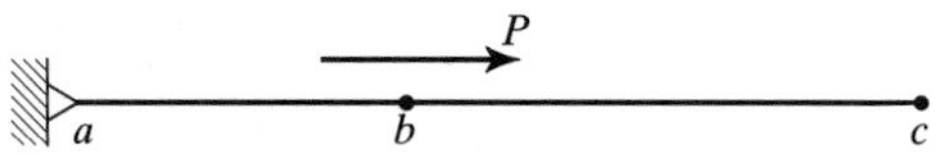

Boundary condition: $u_a = 0$
Equilibrium requirements:
$F_b^{ab} = P, F_c^{bc} = 0$

Case III:
From (a),

$$F_b^{ab} = k_{ab}u_b$$

From (b),

$$F_b^{bc} = k_{bc}u_b$$

From equilibrium,

$$P = F_b^{ab} + F_b^{bc} = k_{ab}u_b + k_{bc}u_b$$

The force-displacement relationship is:

$$P = (k_{ab} + k_{bc})u_b$$

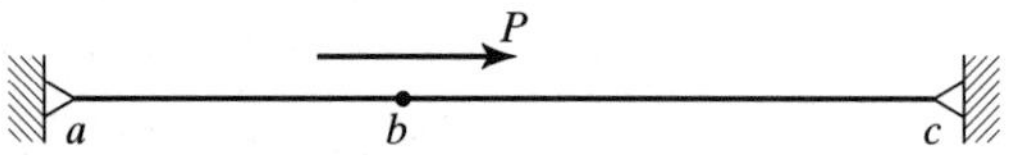

Boundary condition: $u_a = u_c = 0$
Equilibrium requirement
at node b:

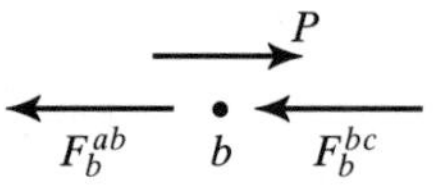

2.7 PROBLEMS

In the following problems all of the numerical calculations can be done with a pocket calculator, but it is suggested that if the reader is not familiar with the use of computer programs, the results of sample problems be checked with a program such as MASTAN2.

2.1 In each case shown, the vertical component of reaction at b is $2P/3$. Using stiffness equations for the support elements, calculate the displacement at b. Each supporting link has the same A, E, and L. Consider the beam itself undeformable, that is, capable of rigid body motion but not internal straining.

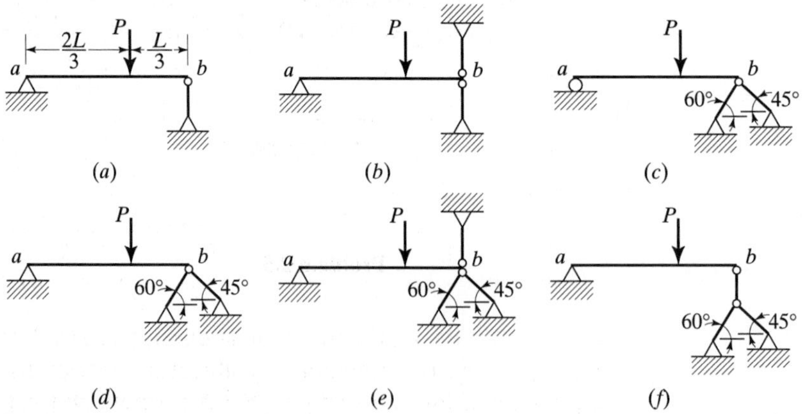

Problem 2.1

2.2 For the system shown, E is constant. Calculate (a) the horizontal force at b required to produce a displacement $u_b = 0.001L$, (b) the displacement at c under this force, and (c) the reactions at a and d.

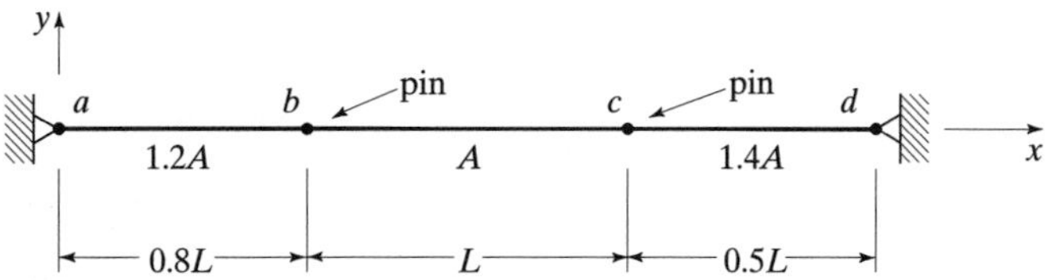

Problem 2.2

2.3 The system and the displacement u_b are the same as in Problem 2.2, but the horizontal force is applied at c. Calculate its magnitude, the displacement at c, and the reactions.

2.4 The truss shown is the same as in Example 2.1 except that the area of bar ab is a variable. $E = 200{,}000$ MPa. (a) For what value of A_{ab} is $v_a = 0$? Is this a function of the magnitude of the applied load? (b) Is it possible to obtain zero u_a by varying A_{ab}? (c) Suppose the applied load were vertical rather than horizontal, would it then be possible to obtain zero u_a by varying A_{ab}? If so, for what value of A_{ab}? What would be the corresponding v_a?

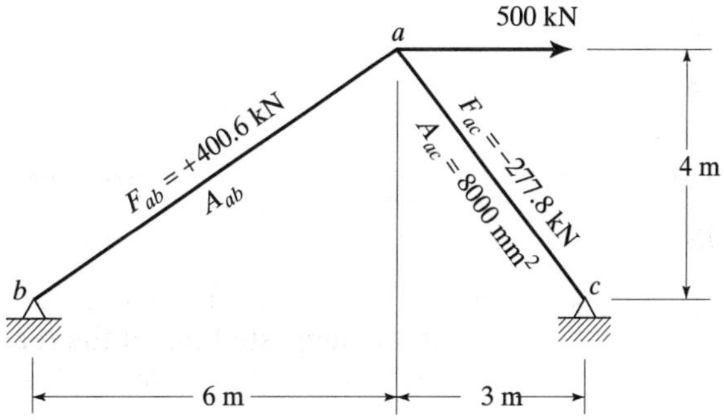

Problem 2.4

2.5 The two-bar axial system is the same as in Example 2.6 except that external axial forces are shown acting at each of the three nodes. Write the stiffness equations for the *assembled* system: $\{\mathbf{P}\} = [\mathbf{K}]\{\mathbf{\Delta}\}$, where $\{\mathbf{P}\} = \lfloor P_a \quad P_b \quad P_c \rfloor^{\mathrm{T}}$ and $\{\mathbf{\Delta}\} = \lfloor u_a \quad u_b \quad u_c \rfloor^{\mathrm{T}}$. Compare elements of $[\mathbf{K}]$ with values obtained in Example 2.6.

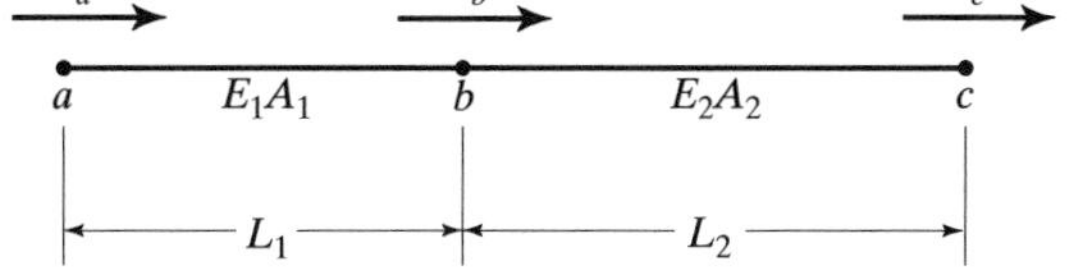

Problem 2.5

2.6 Using the matrix equation assembled in Problem 2.5 and assuming $P_b = 0$, eliminate u_b from the equations by algebraic operations and develop condensed matrix equations of the form $\{\mathbf{P}_r\} = [\mathbf{K}_r]\{\mathbf{\Delta}_r\}$, where $\{\mathbf{P}_r\} = \lfloor P_a \quad P_c \rfloor^{\mathrm{T}}$ and $\{\mathbf{\Delta}_r\} = \lfloor u_a \quad u_c \rfloor^{\mathrm{T}}$. Use the condensed equations to solve Case 1 of Example 2.6.

2.7 What is the magnitude and direction of the force P required to obtain displacement components $u_a = 2$ mm and $v_a = -3$ mm? Cross-sectional areas (mm$^2 \times 10^3$) are indicated on each bar. $E = 200{,}000$ MPa.

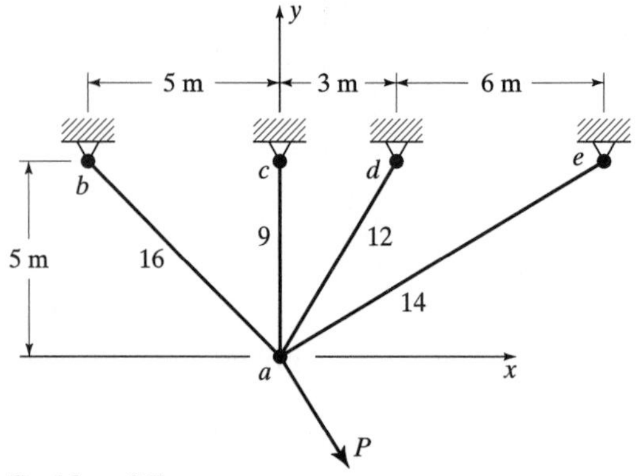

Problem 2.7

2.8 Using flexibility equations for the supporting links, calculate the displacement at b for part a of Problem 2.1. If the flexibility equations are used to calculate the displacement at b in part b of Problem 2.1, what additional condition must be invoked? Calculate the displacement at b in part b. As in Problem 2.1 consider the beam itself undeformable.

2.9 Comparing the flexibility and stiffness approaches to part a of Problem 2.1, at what points were the requirement of joint equilibrium and joint displacement compatibility invoked in the respective solutions?

2.10 Compare the two systems shown: a structural system subjected to an applied load P and a resistor network carrying a specified current I. Demonstrate that there is a mathematical analogy between force and current, flexibility and resistance, and displacement and voltage. Assume that the structural arrangement is such that the two parallel bars must elongate equally.

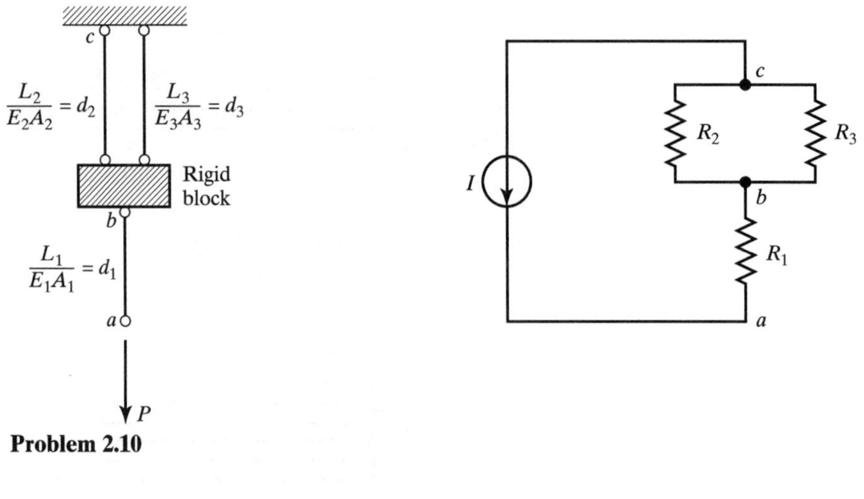

Problem 2.10

2.11 Compare the structural system shown with a capacitor network subjected to a battery voltage V. Demonstrate that there is a mathematical analogy between force and charge, stiffness and capacitance, and displacement and voltage. Make the same assumption regarding the elongation of the parallel bars as in Problem 2.10.

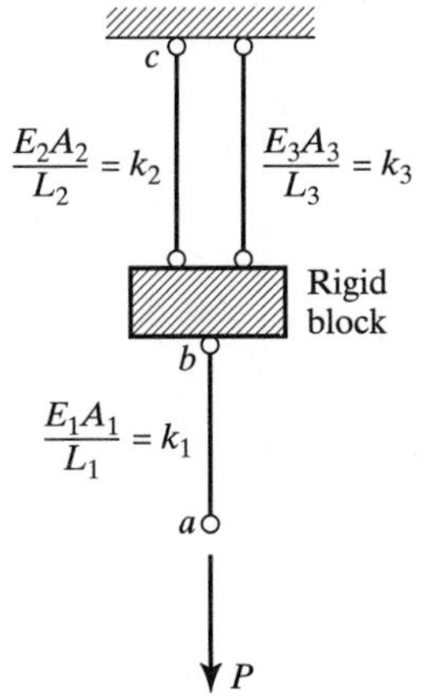

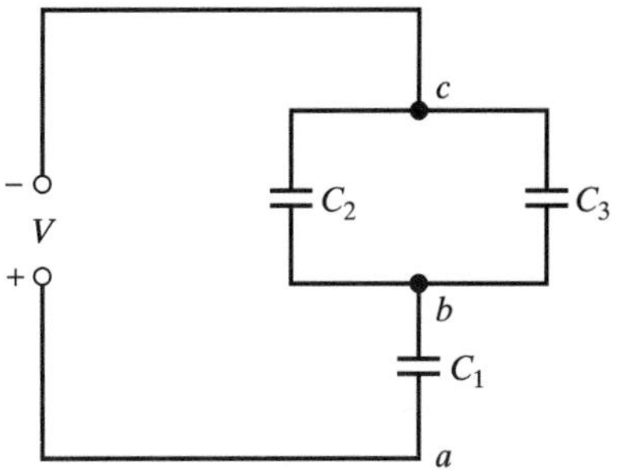

Problem 2.11

2.12 A steel bar having a cross-sectional area of 200 mm^2 is secured at its ends to an aluminum cylinder having a cross-sectional area of 350 mm^2. The cylinder is heated 50°C while the temperature of the bar remains unchanged. Calculate the changes in lengths and the internal forces in bar and cylinder. E_s = 200,000 MPa, E_{al} = 70,000 MPa, $\alpha_s = 1.17 \times 10^{-5}$ mm/mm°C, $\alpha_{al} = 2.31 \times 10^{-5}$ mm/mm°C.

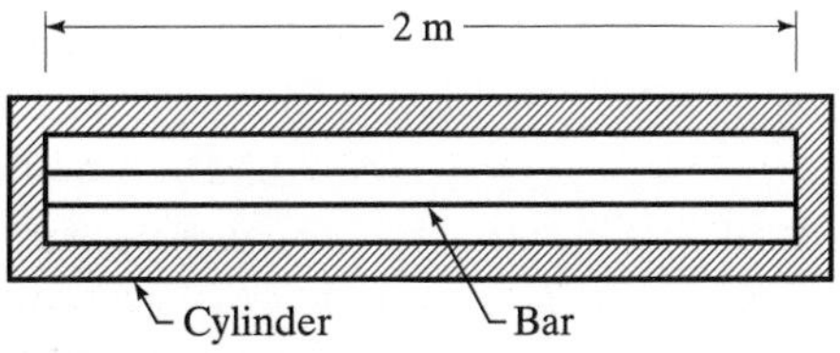

Problem 2.12

2.13 The composite member of Problem 2.12 is subjected to an external tension T as shown; calculate the changes in lengths and the internal forces in bar and cylinder.

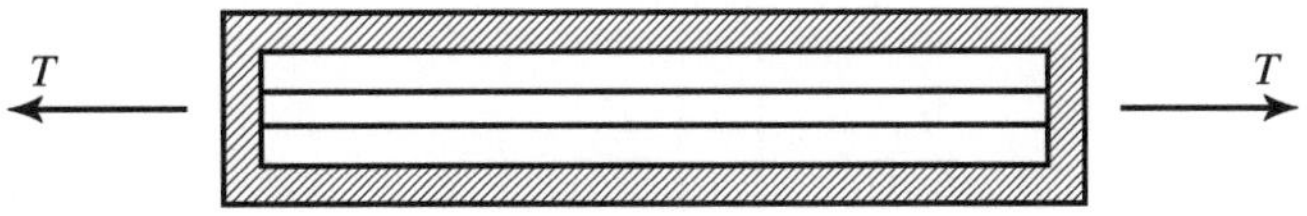

Problem 2.13

2.14 Two steel cylinders are clamped together by a steel bolt as shown. The bolt is loaded to an initial tension of 200 kN by tightening the nut. Plot a diagram of the force in the bolt versus T, an external load applied to the composite system after bolt tightening. A_{bolt} = 500 mm^2, $A_{cylinder}$ = 4800 mm^2, E = 200,000 MPa.

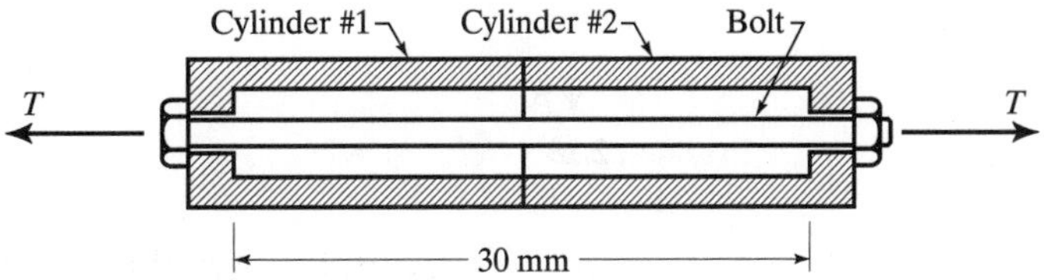

Problem 2.14

2.15 Use flexibility relationships, geometrical considerations, and equilibrium equations to calculate the displacement at a and the bar forces. $E = 200{,}000$ MPa.

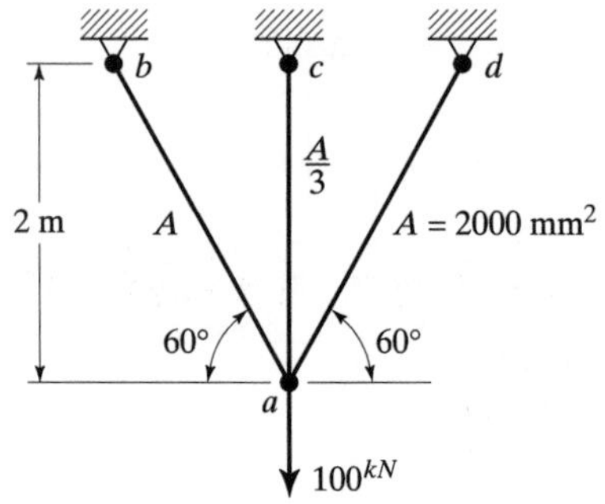

Problem 2.15

Chapter 3

Formation of the Global Analysis Equations

In this chapter we discuss the fundamentals of the direct stiffness approach to the formation of the equations of analysis, equations in which joint displacements play the role of unknowns. This approach requires only the notion and the algebraic form of the element stiffness matrix. And it merely involves the application of the conditions of equilibrium and continuity of displacement at the joints of the analytical model. The immediate objective of the chapter is to furnish readers with a means of forming the equations of analysis and to help them see their purpose by illustrating their solution and the interpretation of results in simple cases. Only axial force members are considered. Nevertheless, the text and the examples illustrate the essence of matrix structural analysis, and the ideas contained in them are part of the foundation on which the whole system is built.

In extending the scope and power of structural analysis, computer-oriented methods such as the direct stiffness method require an order in the preparation of a problem for analysis that was not always essential in the older, classical methods. The same skill, understanding, and imagination are needed, but it is also necessary to establish and obey a formal routine in the statement of the problem and in its solution.

3.1 DIRECT STIFFNESS METHOD—THE BASIC EQUATIONS

A complete set of force-displacement relationships for a framework element with n degrees of freedom is, from Equation 2.2,

$$\begin{array}{ccccccccccc} F_1 & = & k_{11}\Delta_1 & + & k_{12}\Delta_2 & + \cdots + & k_{1j}\Delta_j & + \cdots + & k_{1n}\Delta_n \\ \vdots & & \vdots & & \vdots & & \vdots & & \vdots \\ F_i & = & k_{i1}\Delta_1 & + & k_{i2}\Delta_2 & + \cdots + & k_{ij}\Delta_j & + \cdots + & k_{in}\Delta_n \\ \vdots & & \vdots & & \vdots & & \vdots & & \vdots \\ F_n & = & k_{n1}\Delta_1 & + & k_{n2}\Delta_2 & + \cdots + & k_{nj}\Delta_j & + \cdots + & k_{nn}\Delta_n \end{array} \tag{3.1}$$

In writing Equation 3.1 it is assumed that the degrees of freedom refer to the global axes of the structure. This is accomplished either by writing the equations directly in global coordinates as in Figure 2.9 and Equation 2.5, or by applying mathematical coordinate transformations to force-displacement relationships that have been written in local coordinates, as in Equation 2.3.[1]

The numbers 1, i, n identify the degrees of freedom at the joints of the element, and in this example they correspond to some convenient *global* numbering system for these joints. The global numbering system is completely independent of the local element numbering system. There are no inherent restrictions on the ways of designating global joints but, in any problem, there will be preferred systems. The order of degree-of-freedom numbering can have a strong effect on the efficiency of the equation solving process (see Section 11.4).

When the global analysis equations are formed using the direct stiffness method, all of the degrees of freedom appear in each of the rows of Equation 3.1: The element does not have a defined support condition. Thus, as shown in Section 2.4.1, the equations include rigid-body-motion terms.

Once the element force-displacement relationships have been numerically evaluated for all members of the structure, application of the *direct stiffness method* consists of their combination in an algebraic form that satisfies the requirements of static equilibrium and joint compatibility at all of the junction points of the assembled model. To illustrate the procedure, consider the formation of the force-displacement equations of the point q in the global x direction of the analytical model of the portion of a pin-jointed truss shown in Figure 3.1. The quantities in the x direction at this point are designated by the subscript i. The truss bars shown all lie in the x-y plane. For convenience, we have identified the degrees of freedom at each joint, but the only force that has been identified is the externally applied load, P_i, that acts in the x direction at point q.

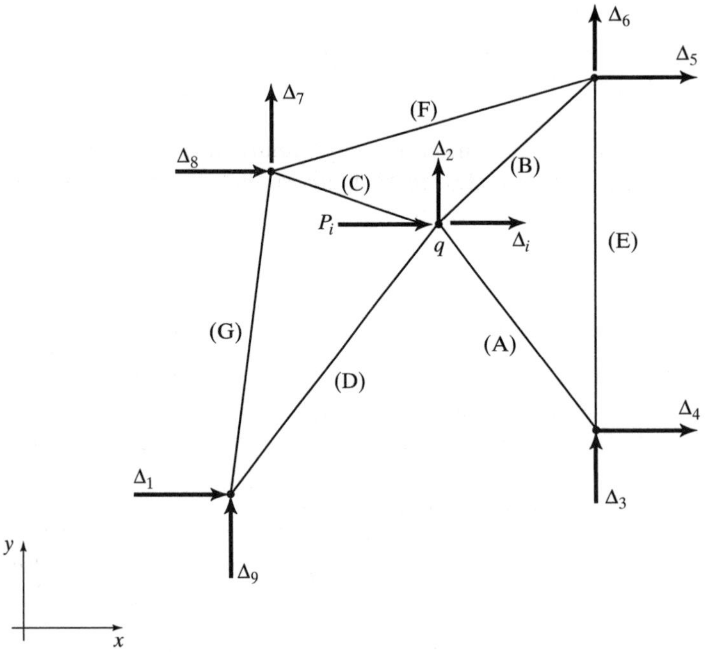

Figure 3.1 Representative interior joint of a plane truss.

[1]We defer formal development of the transformation equations to Section 5.1. For the axial force member the transformation matrix is rectangular (see Section 5.1.4) and in Example 5.1 it is used to relate the stiffness matrix of Equation 2.3 to that of Equation 2.5. Even without the benefit of the mathematics that precedes it, perusal of that example at this point will provide insight into the coordinate transformation process.

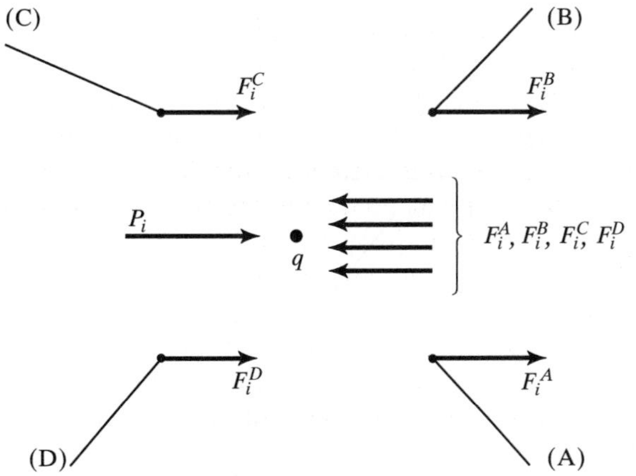

Figure 3.2 Study of equilibrium in the P_i-Δ_i directions.

For junction point equilibrium, the applied load must be equal to the sum of the internal forces acting on the bars meeting at that point.[2] To emphasize this operation, in Figure 3.2 we show the bars separated from the joint. From the condition of x-direction equilibrium,

$$P_i = F_i^A + F_i^B + F_i^C + F_i^D \tag{3.2}$$

where F_i^A is the global, x-direction internal force component on bar A, and so on. The force-displacement equations for the elements, each of the form of Equation 3.1, yield expressions for $F_i^A \ldots F_i^D$ in terms of the corresponding element degrees of freedom $\Delta_i^A \ldots \Delta_9^D$. Substitution of such expressions into Equation 3.2 results in:

$$\begin{aligned} P_i = {} & (k_{ii}^A \Delta_i^A + k_{i2}^A \Delta_2^A + k_{i3}^A \Delta_3^A + k_{i4}^A \Delta_4^A) \\ & + (k_{ii}^B \Delta_i^B + k_{i2}^B \Delta_2^B + k_{i5}^B \Delta_5^B + k_{i6}^B \Delta_6^B) \\ & + (k_{ii}^C \Delta_i^C + k_{i2}^C \Delta_2^C + k_{i7}^C \Delta_7^C + k_{i8}^C \Delta_8^C) \\ & + (k_{ii}^D \Delta_i^D + k_{i1}^D \Delta_1^D + k_{i2}^D \Delta_2^D + k_{i9}^D \Delta_9^D) \end{aligned} \tag{3.3}$$

Since, by the condition of compatibility of displacement, Δ_i is the same for bars A, B, C, and D in each degree of freedom, that is, $\Delta_i^A = \Delta_i^B = \Delta_i^C = \Delta_i^D = \Delta_i$, it follows that

$$\begin{aligned} P_i = {} & (k_{ii}^A + k_{ii}^B + k_{ii}^C + k_{ii}^D)\Delta_i + k_{i1}^D \Delta_1 \\ & + (k_{i2}^A + k_{i2}^B + k_{i2}^C + k_{i2}^D)\Delta_2 + k_{i3}^A \Delta_3 + \cdots + k_{i9}^D \Delta_9 \end{aligned} \tag{3.4}$$

or

$$P_i = K_{ii}\Delta_i + K_{i1}\Delta_1 + K_{i2}\Delta_2 + K_{i3}\Delta_3 + \cdots + K_{i9}\Delta_9 \tag{3.5}$$

This is the final form of the desired equations. The capitalized terms K_{ii}, K_{i1}, K_{i2}, . . . , K_{i9}, are *global stiffness coefficients* and Equation 3.5 is a *global stiffness equation.*

Note that each of the bars meeting at the indicated junction points possesses stiffness coefficients with common subscripts, e.g., k_{ii}^A, k_{ii}^B, k_{ii}^C, k_{ii}^D. When the subscripts of the coefficients of two or more different elements are identical, the elements have a degree of freedom in common, designated by the second subscript, and such coefficients are added to form one coefficient of the stiffness equation for the force represented by the first subscript.

Consider now what happens when an additional member, designated as H, frames into joint q (Figure 3.3). In Equation 3.2 this merely means that a force F_i^H is added

[2] A feature of the nomenclature is that internal (or element) forces are identified by the symbol F and external forces by the symbol P. Appropriate subscripts or superscripts are assigned in each case.

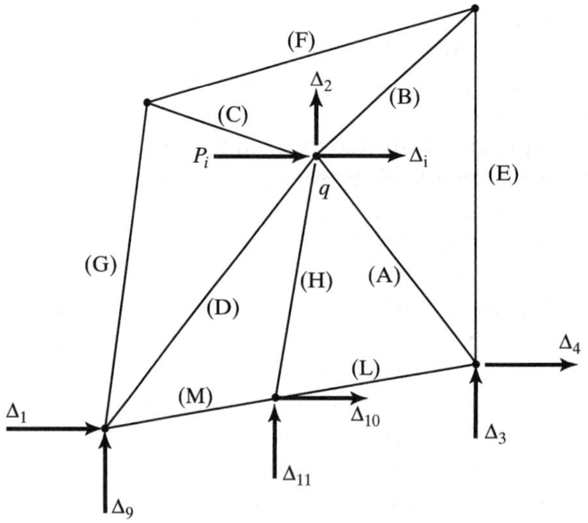

Figure 3.3 Effect of adding a member to a joint.

(Figure 3.4) and Equation 3.4 is supplemented by stiffness terms k_{ii}^H, k_{i2}^H, k_{i10}^H, k_{i11}^H. A force F_2^H would be added in the y-direction equilibrium equation at point q, and similarly for forces F_{10}^H and F_{11}^H at the opposite end of the member. There are no other contributions of member H to joint equilibrium equations. Thus it is a simple matter to revise the stiffness equations to include another member, and the effects are quite local.

Note, however, that after the equations have been solved for the displacements as functions of the applied loads, no simple modifications can be made to the resulting equations to account precisely for the presence of the new member. When a new member is added, the precise solution requires starting with stiffness equations that account for this member. But there are simple procedures for obtaining approximate solutions when structural changes are made. These will be taken up in Section 13.7.

In summary, joint equilibrium is satisfied through the formation of equations such as Equation 3.2, and joint displacement compatibility is satisfied by equating degrees of freedom, as in going from Equations 3.3 to 3.4. Therefore, if the element force-displacement relationships of Equation 3.1 satisfy the proper laws of materials (the

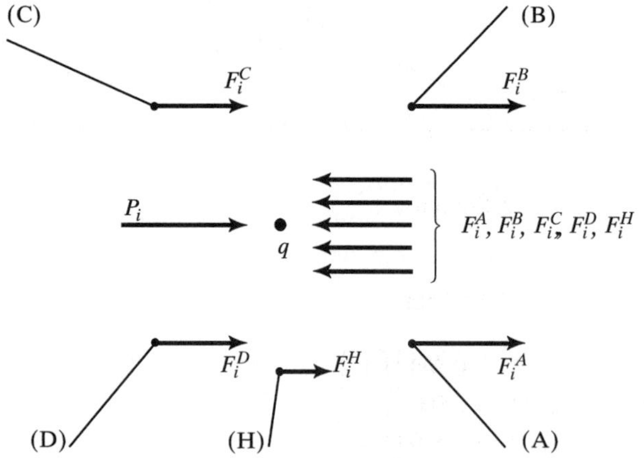

Figure 3.4 Equilibrium in the P_i-Δ_i direction, member added.

constitutive relationships) and the requirements of equilibrium within the elements, and if the proper supports (*boundary conditions*) are assigned to the assembled structure, then all the requirements for an exact solution—*within the limits of linear elastic theory*—are satisfied.

These concepts are illustrated in Examples 3.1–3.3.

In Example 3.1 the global stiffness equations for an unsupported truss are obtained from element stiffness equations, and it is demonstrated that the resulting stiffness matrix contains rigid-body-motion terms. Note that the axes used in forming the member stiffness equations are parallel to, but not coincident with, the overall global axes. There is no need for coincidence; Equation 2.5 yields the correct results without further transformation, provided that the member and overall global axes are parallel. This point will be discussed in Section 5.1.

Example 3.2 is an application of the equations of Example 3.1 to an adequately supported system. The ways in which various stiffness equations are used to determine displacements, reactions, and bar-force components are illustrated. Note that multipliers of known zero displacements have no effect on the results and thus may be eliminated from consideration.

Example 3.3 demonstrates the simplicity of modifying the stiffness matrix to account for structural changes. In the determination of the displacements, the inversion of the stiffness matrix is indicated symbolically but, for brevity, the actual calculations are not reproduced.

EXAMPLE 3.1

For the system shown:

1. Write the member force-displacement relationships in global coordinates.
2. Assemble the global stiffness equations.
3. Show that the global stiffness equations contain rigid-body-motion terms. $E = 200{,}000$ MPa.

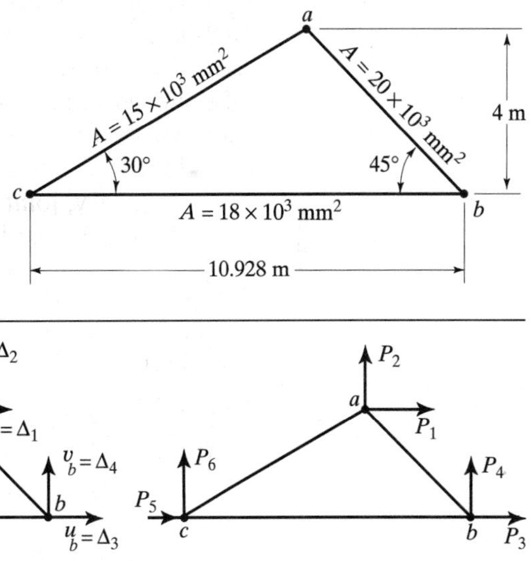

Define the coordinates, degrees of freedom, and external forces as follows:

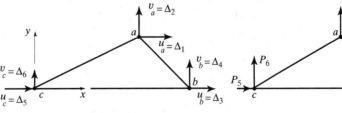

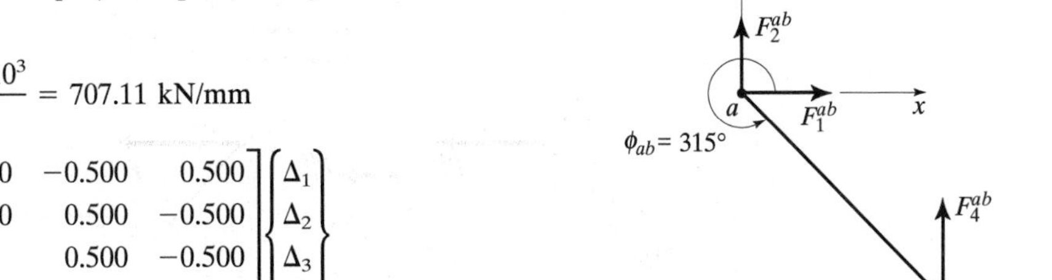

1. Member force-displacement relationships (see Equation 2.5):
Member ab

$$\left(\frac{EA}{L}\right)_{ab} = \frac{200 \times 20 \times 10^3}{4\sqrt{2} \times 10^3} = 707.11 \text{ kN/mm}$$

$$\begin{Bmatrix} F_1^{ab} \\ F_2^{ab} \\ F_3^{ab} \\ F_4^{ab} \end{Bmatrix} = 707.11 \begin{bmatrix} 0.500 & -0.500 & -0.500 & 0.500 \\ & 0.500 & 0.500 & -0.500 \\ & & 0.500 & -0.500 \\ \text{Sym.} & & & 0.500 \end{bmatrix} \begin{Bmatrix} \Delta_1 \\ \Delta_2 \\ \Delta_3 \\ \Delta_4 \end{Bmatrix}$$

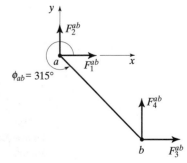

Member *bc*

$$\left(\frac{EA}{L}\right)_{bc} = \frac{200 \times 18 \times 10^3}{10.928 \times 10^3} = 329.43 \text{ kN/mm}$$

$$\begin{Bmatrix} F_3^{bc} \\ F_4^{bc} \\ F_5^{bc} \\ F_6^{bc} \end{Bmatrix} = 329.43 \begin{bmatrix} 1.000 & 0 & -1.000 & 0 \\ & 0 & 0 & 0 \\ & & 1.000 & 0 \\ \text{Sym.} & & & 0 \end{bmatrix} \begin{Bmatrix} \Delta_3 \\ \Delta_4 \\ \Delta_5 \\ \Delta_6 \end{Bmatrix}$$

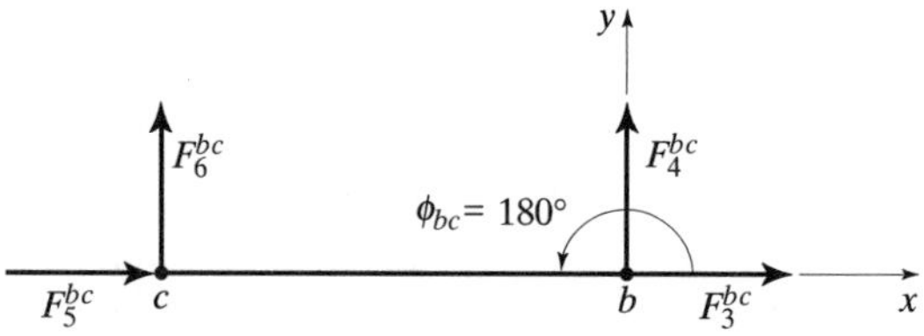

Member *ac*

$$\left(\frac{EA}{L}\right)_{ac} = \frac{200 \times 15 \times 10^3}{8 \times 10^3} = 375.00 \text{ kN/mm}$$

$$\begin{Bmatrix} F_1^{ac} \\ F_2^{ac} \\ F_5^{ac} \\ F_6^{ac} \end{Bmatrix} = 375.00 \begin{bmatrix} 0.750 & 0.433 & -0.750 & -0.433 \\ & 0.250 & -0.433 & -0.250 \\ & & 0.750 & 0.433 \\ \text{Sym.} & & & 0.250 \end{bmatrix} \begin{Bmatrix} \Delta_1 \\ \Delta_2 \\ \Delta_5 \\ \Delta_6 \end{Bmatrix}$$

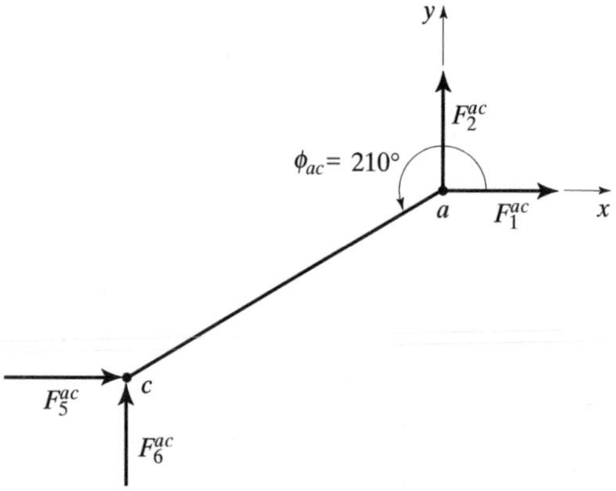

2. Global stiffness equations in matrix form (see Equation 3.5):

$$\begin{Bmatrix} P_1 \\ P_2 \\ P_3 \\ P_4 \\ P_5 \\ P_6 \end{Bmatrix} = 10^2 \begin{bmatrix} 6.348 & -1.912 & -3.536 & 3.536 & -2.812 & -1.624 \\ & 4.473 & 3.536 & -3.536 & -1.624 & -0.938 \\ & & 6.830 & -3.536 & -3.294 & 0 \\ & \text{Sym.} & & 3.536 & 0 & 0 \\ & & & & 6.107 & 1.624 \\ & & & & & 0.938 \end{bmatrix} \begin{Bmatrix} \Delta_1 \\ \Delta_2 \\ \Delta_3 \\ \Delta_4 \\ \Delta_5 \\ \Delta_6 \end{Bmatrix}$$

3. Rigid body motion. Adding rows 1 and 3 of the global stiffness matrix yields the vector:

$$\lfloor 2.812 \quad 1.624 \quad 3.294 \quad 0 \quad -6.107 \quad -1.624 \rfloor$$

which is the negative of row 5. Therefore, there is linear dependence, the determinant is zero, and the matrix is singular. This is a signal that, under an arbitrary load, the displacements are indefinite; that is, there may be rigid body motion.

EXAMPLE 3.2

The truss of Example 3.1 is supported and loaded as shown.

1. Calculate the displacements at *a* and *b*.
2. Calculate the reactions.
3. Calculate the bar forces. Use equations of Example 3.1.

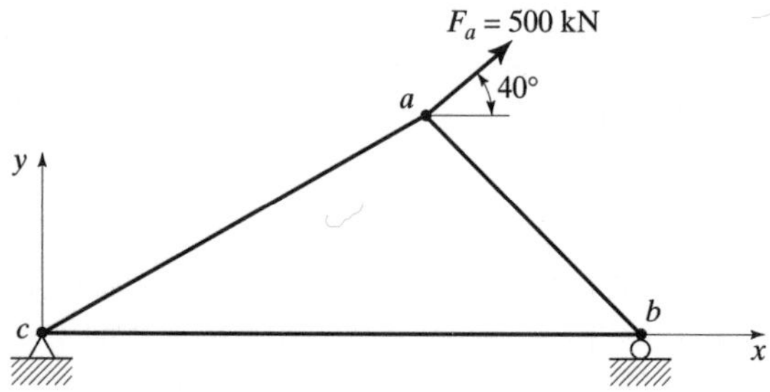

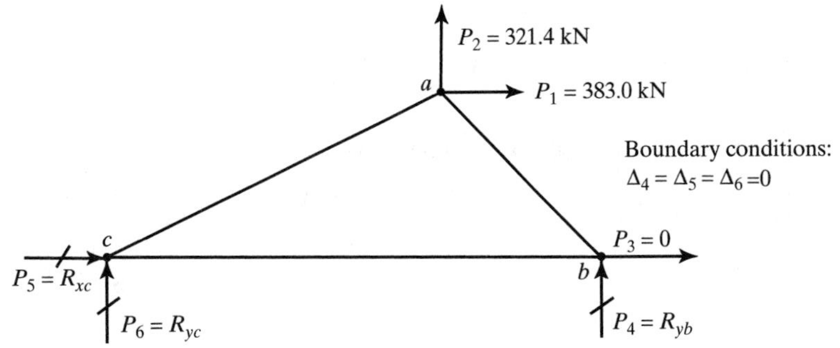

1. Displacements. The upper three global stiffness equations can be written as follows:

$$\begin{Bmatrix} 383.0 \\ 321.4 \\ 0 \end{Bmatrix} = 10^2 \begin{bmatrix} 6.348 & -1.912 & -3.536 \\ & 4.473 & 3.536 \\ \text{Sym.} & & 6.830 \end{bmatrix} \begin{Bmatrix} \Delta_1 \\ \Delta_2 \\ \Delta_3 \end{Bmatrix} + 10^2 \begin{bmatrix} 3.536 & -2.812 & -1.624 \\ -3.536 & -1.624 & -0.938 \\ -3.536 & -3.294 & 0 \end{bmatrix} \begin{Bmatrix} 0 \\ 0 \\ 0 \end{Bmatrix}$$

Inverting the first matrix and solving for the displacements yields

$$\lfloor \Delta_1 \quad \Delta_2 \quad \Delta_3 \rfloor = \lfloor 0.871 \quad 1.244 \quad -0.193 \rfloor \text{ mm}$$

2. Reactions. The lower three stiffness equations now yield the reactions:

$$\begin{Bmatrix} R_{yb} \\ R_{xc} \\ R_{yc} \end{Bmatrix} = 10^2 \overset{\Delta_1 \qquad\quad \Delta_2 \qquad\quad \Delta_3}{\begin{bmatrix} 3.536 & -3.536 & -3.536 \\ -2.812 & -1.624 & -3.294 \\ -1.624 & -0.938 & 0 \end{bmatrix}} \begin{Bmatrix} 0.871 \\ 1.244 \\ -0.193 \end{Bmatrix}$$

$$+ 10^2 \overset{\Delta_4 \qquad\quad \Delta_5 \qquad\quad \Delta_6}{\begin{bmatrix} 3.536 & 0 & 0 \\ & 6.107 & 1.624 \\ \text{Sym.} & & 0.938 \end{bmatrix}} \begin{Bmatrix} 0 \\ 0 \\ 0 \end{Bmatrix} = \begin{Bmatrix} -63.6 \\ -383.4 \\ -258.1 \end{Bmatrix} \text{ kN}$$

3. Bar forces. The bar forces may now be obtained from the member stiffness equations:
Member *ab*

$$\begin{Bmatrix} F_1^{ab} \\ F_2^{ab} \end{Bmatrix} = 707.11 \overset{\Delta_1 \qquad\quad \Delta_2 \qquad\quad \Delta_3}{\begin{bmatrix} 0.500 & -0.500 & -0.500 \\ -0.500 & 0.500 & 0.500 \end{bmatrix}} \begin{Bmatrix} 0.871 \\ 1.244 \\ -0.193 \end{Bmatrix} = \begin{Bmatrix} -63.6 \\ 63.6 \end{Bmatrix} \text{ kN}$$

$$F_{ab} = F_2^{ab} \cdot \sqrt{2} = +90.0 \text{ kN (tension)}$$

Member *bc*

$$\begin{Bmatrix} F_3^{bc} \\ F_4^{bc} \end{Bmatrix} = 329.43 \overset{\Delta_3}{\begin{bmatrix} 1.00 \\ 0 \end{bmatrix}} \{-0.193\} = \begin{Bmatrix} -63.6 \\ 0 \end{Bmatrix} \text{ kN}$$

$$F_{bc} = F_3^{bc} = -63.6 \text{ kN (compression)}$$

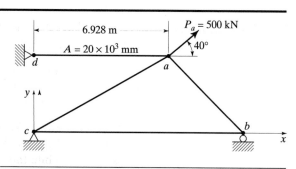

Member *ac*

$$\begin{Bmatrix} F_1^{ac} \\ F_2^{ac} \end{Bmatrix} = 375.00 \begin{array}{c} \begin{array}{cc} \Delta_1 & \Delta_2 \end{array} \\ \begin{bmatrix} 0.750 & 0.433 \\ 0.433 & 0.250 \end{bmatrix} \end{array} \begin{Bmatrix} 0.871 \\ 1.244 \end{Bmatrix} = \begin{Bmatrix} 447.0 \\ 258.0 \end{Bmatrix} \text{ kN}$$

$$F_{ac} = F_1^{ac} \cdot \frac{2}{\sqrt{3}} = +516.2 \text{ kN (tension)}$$

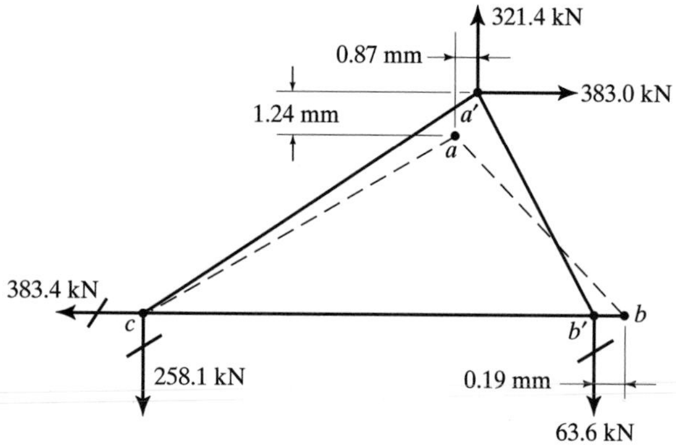

EXAMPLE 3.3

The truss shown is the same as in Example 3.2 except for the addition of the horizontal tie *ad*.

1. Calculate the displacements at *a* and *b*.
2. Calculate the reactions.

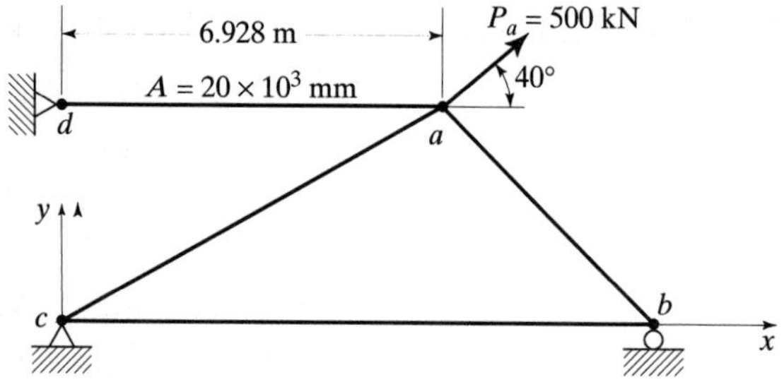

Member *ad*. Member force-displacement relationships at node *a* (see Equation 2.5):

$$F_1^{ad} = \frac{200 \times 20 \times 10^3}{6.928 \times 10^3}(1.000)\ \Delta_1 = 577.37\Delta_1 \text{ kN}$$

$$F_2^{ad} = 0$$

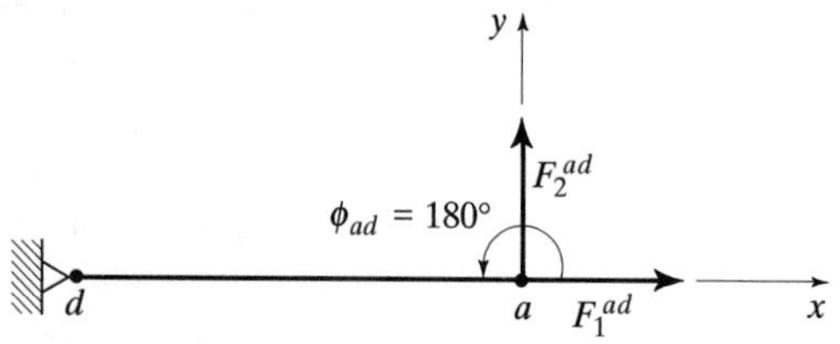

1. Displacements. No nonzero degrees of freedom have been added. The stiffness coefficient of *ad* can be added to the stiffness matrix of Example 3.2, resulting in the following:

$$\begin{Bmatrix} \Delta_1 \\ \Delta_2 \\ \Delta_3 \end{Bmatrix} = 10^{-2} \begin{bmatrix} \begin{pmatrix} 6.348 \\ +5.774 \end{pmatrix} & -1.912 & -3.536 \\ & 4.473 & 3.536 \\ \text{Sym.} & & 6.830 \end{bmatrix}^{-1} \begin{Bmatrix} 383.0 \\ 321.4 \\ 0 \end{Bmatrix}$$

$$= \begin{Bmatrix} 0.383 \\ 1.228 \\ -0.437 \end{Bmatrix} \text{ mm}$$

2. Reactions. Reactions may be obtained by adding an equation for R_{xd}, the horizontal reaction at d, to the equations of Example 3.2:

$$\begin{Bmatrix} R_{yb} \\ R_{xc} \\ R_{yc} \\ R_{xd} \end{Bmatrix} = 10^2 \begin{bmatrix} \Delta_1 & \Delta_2 & \Delta_3 \\ 3.536 & -3.536 & -3.536 \\ -2.812 & -1.624 & -3.294 \\ -1.624 & -0.938 & 0 \\ -5.774 & 0 & 0 \end{bmatrix} \begin{Bmatrix} 0.383 \\ 1.228 \\ -0.437 \end{Bmatrix} = \begin{Bmatrix} -144.3 \\ -163.2 \\ -177.4 \\ -221.1 \end{Bmatrix} \text{kN}$$

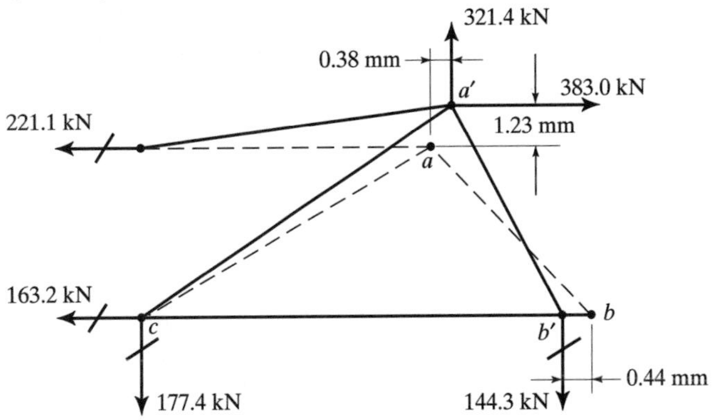

3.2 DIRECT STIFFNESS METHOD—THE GENERAL PROCEDURE

The previous discussion and examples suggest the following *automatic* approach to calculating the applied load versus displacement equations for the complete structure. Parts of its mathematical basis that have not been covered already are best explained after outlining the procedure:

1. Each element stiffness coefficient, after numerical evaluation, is assigned a double subscript (k_{ij}). The first subscript, i, designates the force for which the equation is written, while the second, j, designates the associated degree of freedom.
2. Provision is made for a square matrix whose size is equal to the number of degrees of freedom in the complete system, with the possibility that each force will be related to every displacement in the system. Each term in this array is identified by two subscripts. The first subscript (row) pertains to the force equation, the second (column) to the degree of freedom in question. The array is illustrated in Figure 3.5 for a structure with a total of n degrees of freedom. A search is then made through the member stiffness coefficients. When a coefficient is reached whose designating first subscript is 1, it is placed in row 1 in the column headed by its second subscript. For example, k_{13} is positioned as illustrated in the figure.
3. The procedure of step 2 is continued until all elements have been searched. Each time a coefficient is placed in a location where a value has already been placed, it is added to the latter. At the completion of this step all terms in row 1 have achieved their final value. Hence, for the ith degree of freedom,

$$K_{1i} = \Sigma k_{1i}$$

where the summation extends over all the members meeting at degree of freedom i.
4. The process of steps 2 and 3 is repeated for all other rows in order. The result is a complete set of coefficients of the stiffness equations for the entire structure

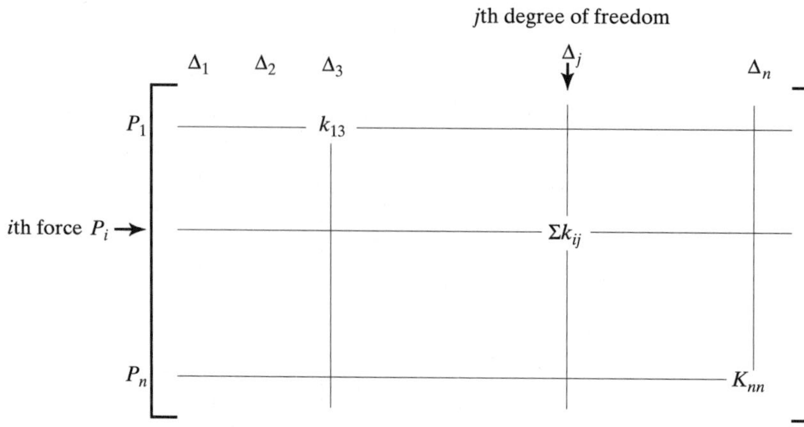

Figure 3.5 Formation of a global stiffness coefficient k_{ij} and insertion of k_{13}.

(the *global stiffness equations*) but with no recognition of the support conditions as yet. This process, by which the global equations are obtained from the element equations, is referred to as *assembly*.

5. The support conditions are accounted for by noting which displacements are zero and then removing from the equations the columns of stiffness coefficients multiplying these degrees of freedom.[3] This is a useful way of recognizing the fact shown in Equation 3.1 (and in Example 3.2) that multipliers of zero displacements have no effect on the force system. The immediate result, however, is to have more equations than unknowns. The surplus equations are those that pertain to the external forces at the support points, that is, the *reactions*. The rows representing these equations are now removed and saved for subsequent evaluation. Thus it will be found that an equal number of rows and columns have been extracted from the array, leaving a square, nonsingular matrix.[4]

6. The set of equations remaining after step 5 is solved for the remaining degrees of freedom. The internal forces acting on the ends of the elements are determined by back substitution of the solved degrees of freedom in the element force-displacement equations. These quantities may require a further transformation from global to local coordinates and finally a transformation into stresses.

The logic of the foregoing process will now be reviewed using a matrix formulation. It is assumed that the rationale for steps 1–4 is clear from the previous discussions of Equations 3.1–3.5 and that these steps have been performed, that is, that the global stiffness equations have been compiled in the form:

$$\{\mathbf{P}\} = [\mathbf{K}]\{\mathbf{\Delta}\} \tag{3.6}$$

Now we assume that the support degrees of freedom $\{\mathbf{\Delta}_s\}$ can be grouped together. After reordering rows and columns to separate quantities pertaining to the sup-

[3]The treatment of nonzero supports that follow a prescribed behavioral rule or are otherwise constrained is covered in Section 13.3.

[4]An alternative to this procedure is to take cognizance of the support conditions at the outset, that is, to form the element stiffness matrices only with respect to the unsupported degrees of freedom. Steps 2–4 then lead to the *reduced* stiffness matrix and step 5 is eliminated. This approach was implicit in Example 3.3, in which it was found possible to form the necessary stiffness matrix without adding terms associated with the degrees of freedom at support *d*.

ports from the remainder, we partition Equation 3.6 in a conformable fashion to yield:

$$\begin{Bmatrix} \mathbf{P}_f \\ \hline \mathbf{P}_s \end{Bmatrix} = \left[\begin{array}{c|c} \mathbf{K}_{ff} & \mathbf{K}_{fs} \\ \hline \mathbf{K}_{sf} & \mathbf{K}_{ss} \end{array}\right] \begin{Bmatrix} \boldsymbol{\Delta}_f \\ \hline \boldsymbol{\Delta}_s \end{Bmatrix} \tag{3.7}$$

where all quantities pertinent to the supports are assigned the subscript s and those relating to the remaining degrees of freedom have the subscript f. (Note the separation made in Example 3.2 where, in addition to using the above partitioning, the degrees of freedom were conveniently numbered in advance. Assigning the last sequence of numbers to the support degrees of freedom made subsequent renumbering unnecessary. Such a numbering scheme may be advantageous, but it is not essential.)

Expanding Equation 3.7 and noting that $\{\boldsymbol{\Delta}_s\} = 0$, we have

$$\{\mathbf{P}_f\} = [\mathbf{K}_{ff}]\{\boldsymbol{\Delta}_f\} \tag{3.8a}$$

$$\{\mathbf{P}_s\} = [\mathbf{K}_{sf}]\{\boldsymbol{\Delta}_f\} \tag{3.8b}$$

The general solution to Equation 3.8a is obtained symbolically by

$$\{\boldsymbol{\Delta}_f\} = [\mathbf{K}_{ff}]^{-1}\{\mathbf{P}_f\} = [\mathbf{D}]\{\mathbf{P}_f\} \tag{3.9}$$

where the matrix $[\mathbf{D}]$ is the set of global flexibility (displacement) coefficients.

We have emphasized that the operation of matrix inversion is symbolic. In practice, the process is normally one of equation solving by a method of the type described in Chapter 11. Also, the grouping of terms and partitioning indicated in Equation 3.7 was adopted for the sake of logic and clarity in presentation. In practice, it may be neither necessary nor convenient.

When $\{\boldsymbol{\Delta}_f\}$, the vector of displacements at all unsupported nodes, has been found from Equation 3.9, the support reactions $\{\mathbf{P}_s\}$ are found by substitution of the result in Equation 3.8b, yielding

$$\{\mathbf{P}_s\} = [\mathbf{K}_{sf}][\mathbf{D}]\{\mathbf{P}_f\} \tag{3.10}$$

To obtain the internal force distribution in the ith element, one may multiply the calculated degrees of freedom for that element, designated by $\{\boldsymbol{\Delta}^i\}$, by the element stiffness matrix $[\mathbf{k}^i]$, resulting in the numerical evaluation of the joint force components $\{F^i\}$. Thus

$$\{\mathbf{F}^i\} = [\mathbf{k}^i]\{\boldsymbol{\Delta}^i\} \tag{3.11}$$

These forces (direct forces and moments) will be vector components in the directions of the global coordinate axes. Separate operations are required to transform them into: (1) resultant forces, (2) components oriented with respect to the member axes (normal force, shear, bending moment, and torque), or (3) unit stresses. These transformations are often made part of the basic analysis, but occasionally it may be preferable to leave them as separate tasks for the analyst or designer.

The application of the general procedure is illustrated in Examples 3.4–3.6. The structures used are elementary pin-jointed trusses, but the examples contain most of the features found in the analysis of complex systems.

Example 3.4 is similar to Example 3.1 except that the calculations are arranged in a more formal fashion to follow the procedures outlined in this section. The details of a basic system for forming the global stiffness equations should be clear from this example.

Example 3.5 is similar to Example 3.2 except that the resultant bar-force calculations now follow a clear matrix formulation. The five displacements were calculated by computer, but the results may be verified easily by substitution in the stiffness equations.

In Example 3.6 it is shown that an increase in the number of constraints reduces the labor of solving for the displacements even though it increases the static redundancy of the system. Questions of indeterminacy are introduced in Section 3.4

EXAMPLE 3.4

For the system shown:

1. Write the force-displacement relationships in global coordinates.
2. Assemble the global stiffness equations.
3. Show that the stiffness equations contain rigid-body-motion terms.
   $E = 200{,}000$ MPa.

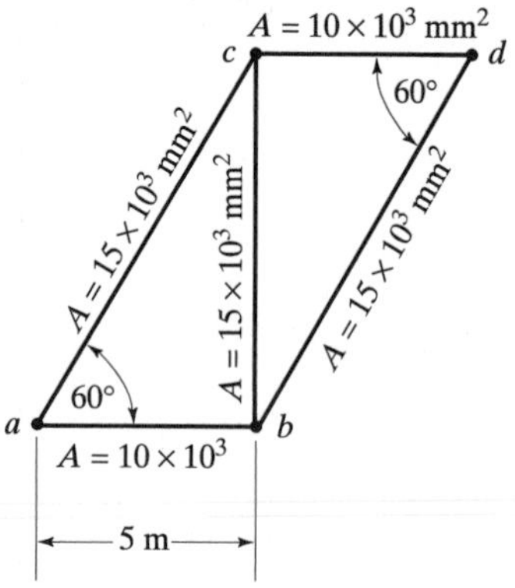

Define the coordinates, degrees of freedom, and external forces as follows:

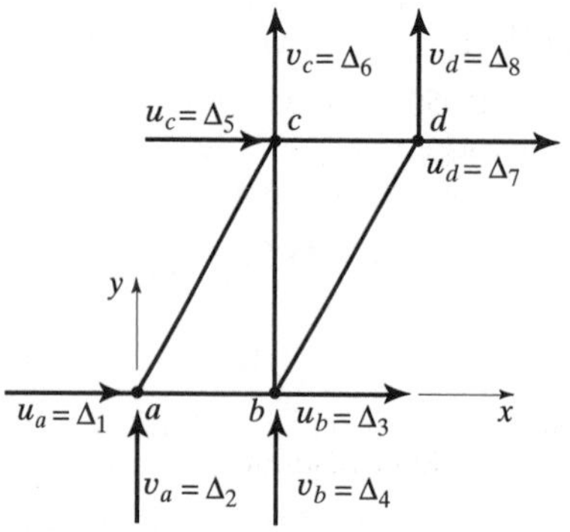

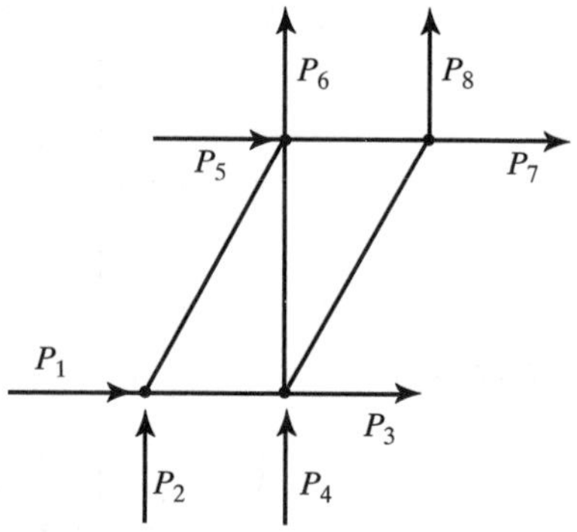

1. Member force-displacement relationships (see Equation 2.5):
   Member *ab*

$$\left(\frac{EA}{L}\right)_{ab} = \frac{200 \times 10 \times 10^3}{5 \times 10^3} = 400 \text{ kN/mm}$$

$$\begin{Bmatrix} F_1^{ab} \\ F_2^{ab} \\ F_3^{ab} \\ F_4^{ab} \end{Bmatrix} = \begin{bmatrix} k_{11} & k_{12} & k_{13} & k_{14} \\ k_{21} & k_{22} & k_{23} & k_{24} \\ k_{31} & k_{32} & k_{33} & k_{34} \\ k_{41} & k_{42} & k_{43} & k_{44} \end{bmatrix} \begin{Bmatrix} \Delta_1 \\ \Delta_2 \\ \Delta_3 \\ \Delta_4 \end{Bmatrix} = 400.00 \begin{bmatrix} 1.000 & 0 & -1.000 & 0 \\ & 0 & 0 & 0 \\ & & 1.000 & 0 \\ \text{Sym.} & & & 0 \end{bmatrix} \begin{Bmatrix} \Delta_1 \\ \Delta_2 \\ \Delta_3 \\ \Delta_4 \end{Bmatrix}$$

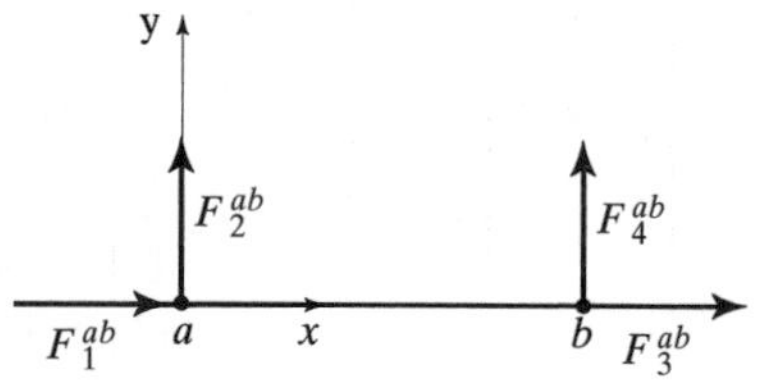

Member *cd*

$$\left(\frac{EA}{L}\right)_{cd} = 400 \text{ kN/mm}$$

$$\begin{Bmatrix} F_5^{cd} \\ F_6^{cd} \\ F_7^{cd} \\ F_8^{cd} \end{Bmatrix} = \begin{bmatrix} k_{55} & k_{56} & k_{57} & k_{58} \\ k_{65} & k_{66} & k_{67} & k_{68} \\ k_{75} & k_{76} & k_{77} & k_{78} \\ k_{85} & k_{86} & k_{87} & k_{88} \end{bmatrix} \begin{Bmatrix} \Delta_5 \\ \Delta_6 \\ \Delta_7 \\ \Delta_8 \end{Bmatrix} = 400.00 \begin{bmatrix} 1.000 & 0 & -1.000 & 0 \\ & 0 & 0 & 0 \\ & & 1.000 & 0 \\ \text{Sym.} & & & 0 \end{bmatrix} \begin{Bmatrix} \Delta_5 \\ \Delta_6 \\ \Delta_7 \\ \Delta_8 \end{Bmatrix}$$

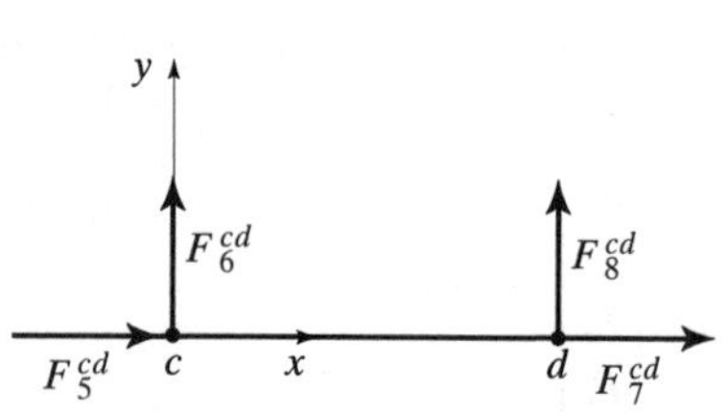

Member *ac*

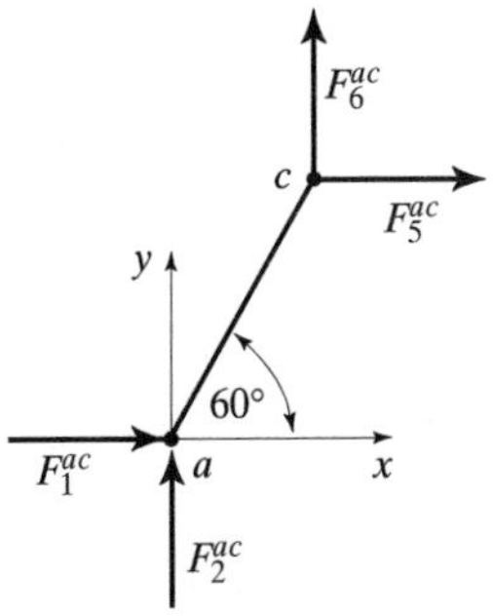

$$\left(\frac{EA}{L}\right)_{ac} = \frac{200 \times 15 \times 10^3}{10 \times 10^3} = 300 \text{ kN/mm}$$

$$\begin{Bmatrix} F_1^{ac} \\ F_2^{ac} \\ F_5^{ac} \\ F_6^{ac} \end{Bmatrix} = \begin{bmatrix} k_{11} & k_{12} & k_{15} & k_{16} \\ k_{21} & k_{22} & k_{25} & k_{26} \\ k_{51} & k_{52} & k_{55} & k_{56} \\ k_{61} & k_{62} & k_{65} & k_{66} \end{bmatrix} \begin{Bmatrix} \Delta_1 \\ \Delta_2 \\ \Delta_5 \\ \Delta_6 \end{Bmatrix}$$

$$= 300.00 \begin{bmatrix} 0.250 & 0.433 & -0.250 & -0.433 \\ & 0.750 & -4.333 & -0.750 \\ & & 0.250 & 0.433 \\ \text{Sym.} & & & 0.750 \end{bmatrix} \begin{Bmatrix} \Delta_1 \\ \Delta_2 \\ \Delta_5 \\ \Delta_6 \end{Bmatrix}$$

Member *bd*

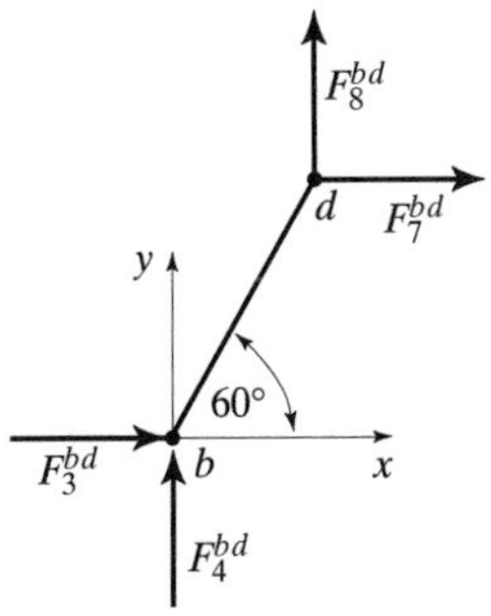

$$\left(\frac{EA}{L}\right)_{bd} = 300 \text{ kN/mm}$$

$$\begin{Bmatrix} F_3^{bd} \\ F_4^{bd} \\ F_7^{bd} \\ F_8^{bd} \end{Bmatrix} = \begin{bmatrix} k_{33} & k_{34} & k_{37} & k_{38} \\ k_{43} & k_{44} & k_{47} & k_{48} \\ k_{73} & k_{74} & k_{77} & k_{78} \\ k_{83} & k_{84} & k_{87} & k_{88} \end{bmatrix} \begin{Bmatrix} \Delta_3 \\ \Delta_4 \\ \Delta_7 \\ \Delta_8 \end{Bmatrix}$$

$$= 300.00 \begin{bmatrix} 0.250 & 0.433 & -0.250 & -0.433 \\ & 0.750 & -0.433 & -0.750 \\ & & 0.250 & 0.433 \\ \text{Sym.} & & & 0.750 \end{bmatrix} \begin{Bmatrix} \Delta_3 \\ \Delta_4 \\ \Delta_7 \\ \Delta_8 \end{Bmatrix}$$

Member *bc*

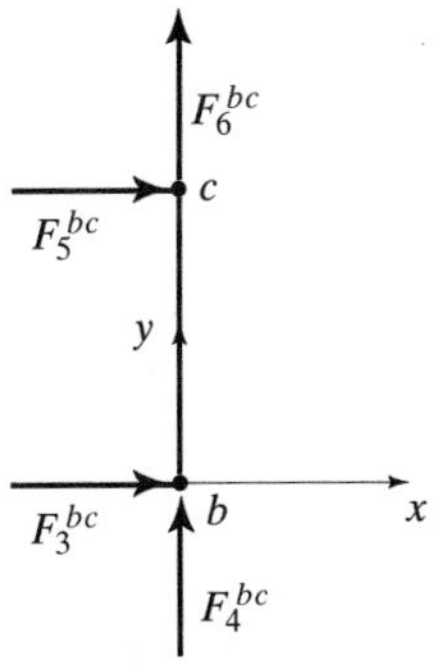

$$\left(\frac{EA}{L}\right)_{bc} = \frac{200 \times 15 \times 10^3}{5\sqrt{3} \times 10^3} = 346.41 \text{ kN/mm}$$

$$\begin{Bmatrix} F_3^{bc} \\ F_4^{bc} \\ F_5^{bc} \\ F_6^{bc} \end{Bmatrix} = \begin{bmatrix} k_{33} & k_{34} & k_{35} & k_{36} \\ k_{43} & k_{44} & k_{45} & k_{46} \\ k_{53} & k_{54} & k_{55} & k_{56} \\ k_{63} & k_{64} & k_{65} & k_{66} \end{bmatrix} \begin{Bmatrix} \Delta_3 \\ \Delta_4 \\ \Delta_5 \\ \Delta_6 \end{Bmatrix} = 346.41 \begin{bmatrix} 0 & 0 & 0 & 0 \\ & 1.000 & 0 & -1.000 \\ & & 0 & 0 \\ \text{Sym.} & & & 1.000 \end{bmatrix} \begin{Bmatrix} \Delta_3 \\ \Delta_4 \\ \Delta_5 \\ \Delta_6 \end{Bmatrix}$$

2. Global stiffness equations in matrix form (see Equations 3.5 and 3.6):

$$\begin{Bmatrix} P_1 \\ P_2 \\ P_3 \\ P_4 \\ P_5 \\ P_6 \\ P_7 \\ P_8 \end{Bmatrix} = 10^2 \begin{bmatrix} 4.750 & 1.299 & -4.000 & 0 & -0.750 & -1.299 & 0 & 0 \\ & 2.250 & 0 & 0 & -1.299 & -2.250 & 0 & 0 \\ & & 4.750 & 1.299 & 0 & 0 & -0.750 & -1.299 \\ & & & 5.714 & 0 & -3.464 & -1.299 & -2.250 \\ & & & & 4.750 & 1.299 & -4.000 & 0 \\ & & & & & 5.714 & 0 & 0 \\ & & & & & & 4.750 & 1.299 \\ & & \text{Sym.} & & & & & 2.250 \end{bmatrix} \begin{Bmatrix} \Delta_1 \\ \Delta_2 \\ \Delta_3 \\ \Delta_4 \\ \Delta_5 \\ \Delta_6 \\ \Delta_7 \\ \Delta_8 \end{Bmatrix}$$

3. Rigid body motion. Adding rows 1, 5, and 7 of the global stiffness matrix yields the vector:

$$\lfloor 4.000 \quad 0 \quad -4.750 \quad -1.299 \quad 0 \quad 0 \quad 0.750 \quad 1.299 \rfloor$$

which is the negative of row 3. Therefore, the matrix is singular. The displacements are indefinite—there may be rigid body motion.

EXAMPLE 3.5

The truss of Example 3.4 is supported and loaded as shown:

1. Calculate the displacements at *b*, *c*, and *d*.
2. Calculate the reactions.
3. Calculate the bar forces.
   Use equations of Example 3.4.

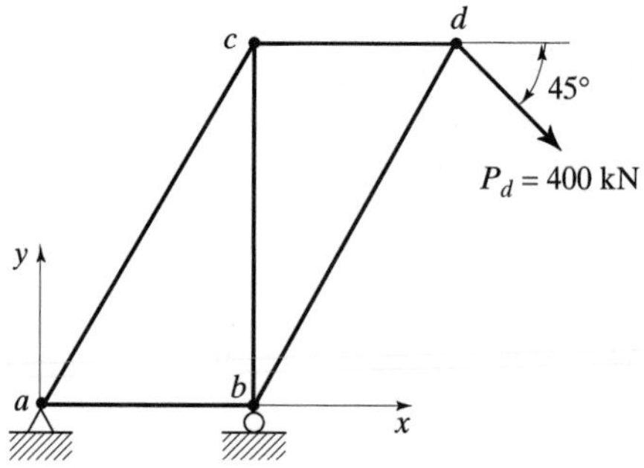

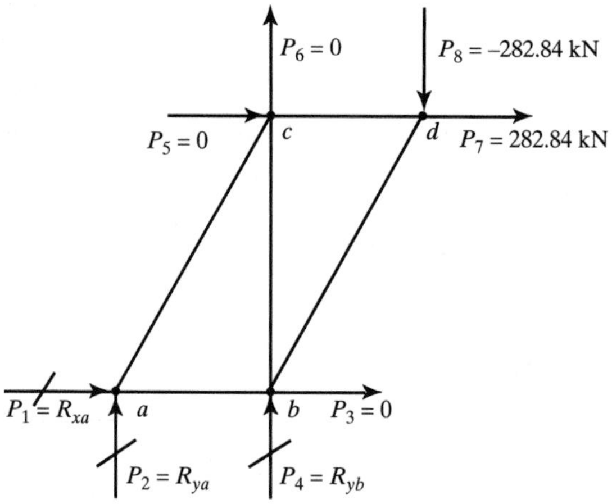

Boundary conditions: $\Delta_1 = \Delta_2 = \Delta_4 = 0$

1. Displacements. Remove columns and rows 1, 2, and 4 from the stiffness equations, leaving

$$\begin{Bmatrix} P_3 \\ P_5 \\ P_6 \\ P_7 \\ P_8 \end{Bmatrix} = \begin{Bmatrix} 0 \\ 0 \\ 0 \\ 282.84 \\ -282.84 \end{Bmatrix} = 10^2 \begin{bmatrix} 4.750 & 0 & 0 & -0.750 & -1.299 \\ & 4.750 & 1.299 & -4.000 & 0 \\ & & 5.714 & 0 & 0 \\ & \text{Sym.} & & 4.750 & 1.299 \\ & & & & 2.250 \end{bmatrix} \begin{Bmatrix} \Delta_3 \\ \Delta_5 \\ \Delta_6 \\ \Delta_7 \\ \Delta_8 \end{Bmatrix}$$

Solving for $\{\boldsymbol{\Delta}\}$ on a computer yields the following results, which may be checked by substitution in the above equations:

$$\lfloor \boldsymbol{\Delta} \rfloor = \lfloor \Delta_3 \quad \Delta_5 \quad \Delta_6 \quad \Delta_7 \quad \Delta_8 \rfloor = \lfloor -0.407 \quad 9.809 \quad -2.232 \quad 10.926 \quad -7.801 \rfloor \text{ mm}$$

2. Reactions. The remaining stiffness equations (rows 1, 2, and 4) are used as follows:

$$\begin{Bmatrix} P_1 \\ P_2 \\ P_4 \end{Bmatrix} = \begin{Bmatrix} R_{xa} \\ R_{ya} \\ R_{yb} \end{Bmatrix} = 10^2 \overset{\begin{matrix} \Delta_3 & \Delta_5 & \Delta_6 & \Delta_7 & \Delta_8 \end{matrix}}{\begin{bmatrix} -4.000 & -0.750 & -1.299 & 0 & 0 \\ 0 & -1.299 & -2.250 & 0 & 0 \\ 1.299 & 0 & -3.464 & -1.299 & -2.250 \end{bmatrix}} \begin{Bmatrix} -0.407 \\ 9.809 \\ -2.232 \\ 10.926 \\ -7.801 \end{Bmatrix}$$

$$= \begin{Bmatrix} -282.9 \\ -772.0 \\ 1056.2 \end{Bmatrix} \text{ kN}$$

3. Bar forces. Develop a formula for calculating bar forces from displacements: From equilibrium at the 2 end of a general member 1–2, the bar force F_{12} is

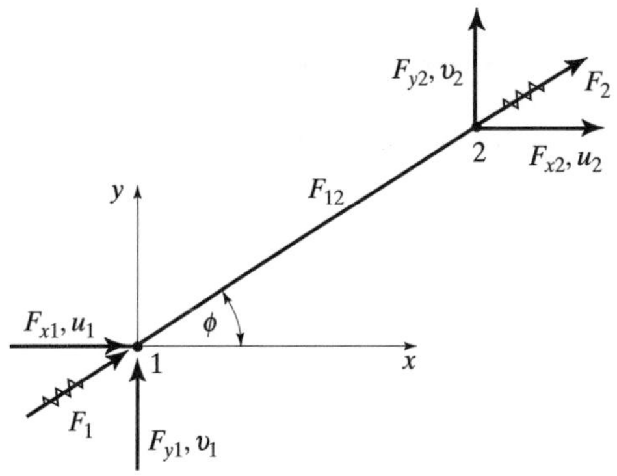

$$F_{12} = F_2 = F_{x2} \cos\phi + F_{y2} \cos(90 - \phi)$$

In matrix form this is

$$F_{12} = \lfloor \cos\phi \quad \sin\phi \rfloor \begin{Bmatrix} F_{x2} \\ F_{y2} \end{Bmatrix} \tag{a}$$

The member stiffness equations are, from Equation 3.11,

$$\begin{Bmatrix} F_{x1} \\ F_{y1} \\ F_{x2} \\ F_{y2} \end{Bmatrix} = \begin{bmatrix} k_{11} & k_{12} & k_{13} & k_{14} \\ k_{21} & k_{22} & k_{23} & k_{24} \\ k_{31} & k_{32} & k_{33} & k_{34} \\ k_{41} & k_{42} & k_{43} & k_{44} \end{bmatrix} \begin{Bmatrix} u_1 \\ v_1 \\ u_2 \\ v_2 \end{Bmatrix} \tag{b}$$

Substituting the last two of Equations b in Equation a gives the desired formula:

$$F_{12} = \lfloor \cos\phi \quad \sin\phi \rfloor \begin{bmatrix} k_{31} & k_{32} & k_{33} & k_{34} \\ k_{41} & k_{42} & k_{43} & k_{44} \end{bmatrix} \begin{Bmatrix} u_1 \\ v_1 \\ u_2 \\ v_2 \end{Bmatrix} \tag{c}$$

Member *ab* $\quad \phi = 0°$

$$F_{ab} = 400.0 \lfloor 1 \quad 0 \rfloor \overset{\begin{matrix} \Delta_1 & \Delta_2 & \Delta_3 & \Delta_4 \end{matrix}}{\begin{bmatrix} -1 & 0 & 1 & 0 \\ 0 & 0 & 0 & 0 \end{bmatrix}} \begin{Bmatrix} 0 \\ 0 \\ -0.407 \\ 0 \end{Bmatrix} = -162.8 \text{ kN}$$

Member *cd* $\quad \phi = 0°$

$$F_{cd} = 400.0 \lfloor 1 \quad 0 \rfloor \overset{\begin{matrix} \Delta_5 & \Delta_6 & \Delta_7 & \Delta_8 \end{matrix}}{\begin{bmatrix} -1 & 0 & 1 & 0 \\ 0 & 0 & 0 & 0 \end{bmatrix}} \begin{Bmatrix} 9.809 \\ -2.232 \\ 10.926 \\ -7.801 \end{Bmatrix} = +446.8 \text{ kN}$$

Member *ac* $\quad \phi = 60°$

$$F_{ac} = 300.0 \lfloor 0.500 \quad 0.866 \rfloor \overset{\begin{matrix} \Delta_1 & \Delta_2 & \Delta_5 & \Delta_6 \end{matrix}}{\begin{bmatrix} -0.250 & -0.433 & 0.250 & 0.433 \\ -0.433 & -0.750 & 0.433 & 0.750 \end{bmatrix}} \begin{Bmatrix} 0 \\ 0 \\ 9.809 \\ -2.232 \end{Bmatrix} = +891.4 \text{ kN}$$

Member bd $\quad \phi = 60°$

$$F_{bd} = 300.0\lfloor 0.500 \quad 0.866 \rfloor \begin{bmatrix} \Delta_3 & \Delta_4 & \Delta_7 & \Delta_8 \\ -0.250 & -0.433 & 0.250 & 0.433 \\ -0.433 & -0.750 & 0.433 & 0.750 \end{bmatrix} \begin{Bmatrix} -0.407 \\ 0 \\ 10.926 \\ -7.801 \end{Bmatrix} = -326.8 \text{ kN}$$

Member bc $\quad \phi = 90°$

$$F_{bc} = 346.41\lfloor 0 \quad 1 \rfloor \begin{bmatrix} \Delta_3 & \Delta_4 & \Delta_5 & \Delta_6 \\ 0 & 0 & 0 & 0 \\ 0 & -1 & 0 & 1 \end{bmatrix} \begin{Bmatrix} -0.407 \\ 0 \\ 9.809 \\ -2.232 \end{Bmatrix} = -773.2 \text{ kN}$$

EXAMPLE 3.6

The truss shown is the same as in Example 3.5 except for the addition of horizontal constraints at b and c. Calculate the displacements at c and d.

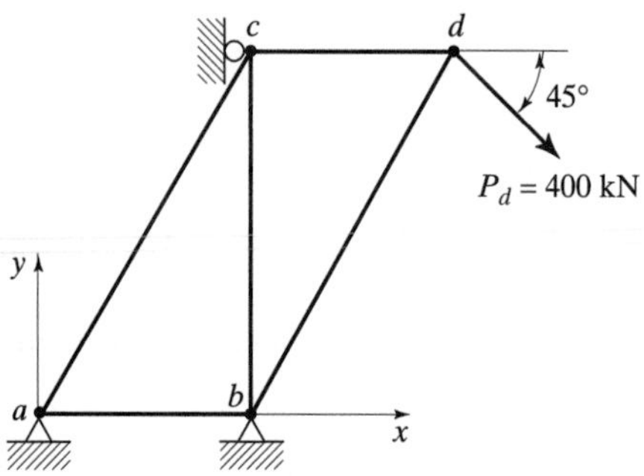

Remove columns and rows 1 to 5 from the stiffness equations, leaving

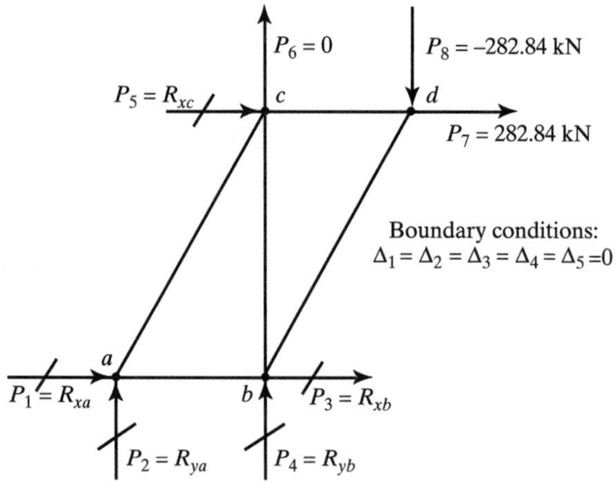

$$\begin{Bmatrix} P_6 \\ P_7 \\ P_8 \end{Bmatrix} = \begin{Bmatrix} 0 \\ 282.84 \\ -282.84 \end{Bmatrix}$$

$$= 10^2 \begin{bmatrix} 5.714 & 0 & 0 \\ & 4.750 & 1.299 \\ \text{Sym.} & & 2.250 \end{bmatrix} \begin{Bmatrix} \Delta_6 \\ \Delta_7 \\ \Delta_8 \end{Bmatrix}$$

Solving for $\{\boldsymbol{\Delta}\}$,

$$\begin{Bmatrix} \Delta_6 \\ \Delta_7 \\ \Delta_8 \end{Bmatrix} = 10^{-2} \begin{bmatrix} 0.175 & 0 & 0 \\ & 0.250 & -0.144 \\ \text{Sym.} & & 0.528 \end{bmatrix} \begin{Bmatrix} 0 \\ 282.84 \\ -282.84 \end{Bmatrix} = \begin{Bmatrix} 0 \\ 1.114 \\ -1.901 \end{Bmatrix} \text{mm}$$

3.3 SOME FEATURES OF THE STIFFNESS EQUATIONS

As has been shown, both element and global stiffness matrices are almost always symmetrical.[5] The conditions under which a set of stiffness coefficients for a structure

[5]Identification of the occurrence and the treatment of nonsymmetrical stiffness matrices are beyond the scope of this text. See Reference 3.1 for discussion of typical cases.

possesses the property of symmetry will be discussed in Section 4.3. Practically, this means that only the main diagonal terms and terms to one side of the main diagonal need to be stored in a computer program (see Section 11.4).

Also, the stiffness (equilibrium) equation for a given degree of freedom is influenced by the degrees of freedom associated with the elements connecting to that degree of freedom. The members shown in Figure 3.1 or in Example 3.4, could comprise but a small region of what is actually a very large structure. Bars such as E, F, and G of Figure 3.1, and any others that might exist beyond these, have no effect on Equation 3.5. In other words, the nonzero terms in a given row of a stiffness matrix consist only of the main diagonal and the terms corresponding to degrees of freedom at that joint and at other joints on the elements meeting at that joint. All other terms in the row are zero. When there are many degrees of freedom in the complete structure, the stiffness matrix may contain relatively few nonzero terms, in which case it is characterized as *sparse* or *weakly populated.*

Clearly, in the solution phase of the analysis it is advantageous to cluster all nonzero terms as close to the main diagonal as possible (see Figure 3.6), thereby isolating the zero terms and facilitating their removal in the solution process. This can be done by numbering the degrees of freedom in such a way that the columnar distance of the term most remote from the main diagonal term in each row is minimized, that is, by minimizing the *bandwidth.*

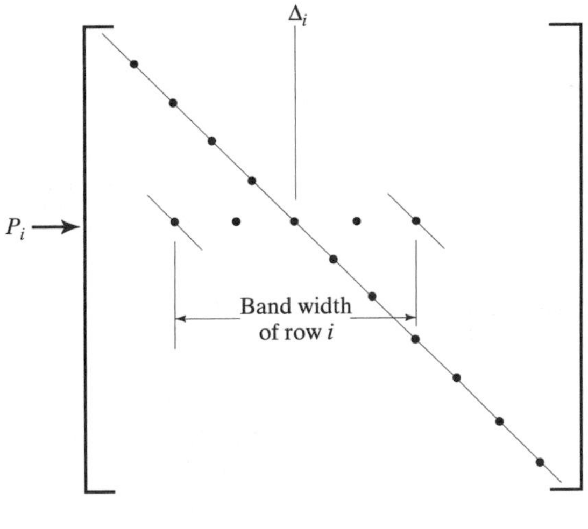

Figure 3.6 Typical final form of row of global stiffness matrix.

Bandwidth minimization is but one strategy for achieving efficiency in the equation-solving phase. Whatever the approach employed, in large-scale applications it is essential to the economy of the solution process that account be taken of the symmetry and sparseness of the stiffness matrix. Even the very elementary problem in Example 3.5 required the inversion of a 5 × 5 matrix. Real structures are often orders of magnitude larger than this. Equation-solving algorithms will be discussed in Chapter 11.

3.4 INDETERMINACY

Up to this point, the conventional concept of *static indeterminacy* has not been mentioned. Cognizance of static indeterminacy is in fact unnecessary in the direct stiffness approach. The structures in Examples 3.2 and 3.5 are statically determinate and those in Examples 3.3 and 3.6 are indeterminate, yet all were analyzed by identical procedures. The displacement approach rests on a comparable concept that may be called *kinematic indeterminacy*. Both can be clarified by comparing definitions of the two.

In the early, classical methods of analysis, a flexibility approach was normally employed. In that approach, one first determines the number of equations, in addition to those of equilibrium, that are required for analysis. The number of additional equations is the degree of *static indeterminacy*, that is, the number of redundant forces. If the conventional procedure of cutting the structure to reduce it to an equivalent statically determinate one is used, the additional equations are developed by employing the elastic characteristics of the system and applying the requirement of restoring the continuity that was destroyed when the system was cut. The unknowns in the analysis are the redundant forces at the cut sections or removed supports.

Kinematic indeterminacy refers to the number of displacements that are required to define the response of the structure. The degree of kinematic redundancy is equal to the number of degrees of freedom that must be conceptually constrained to reduce the system to one in which all joint displacements are zero or have predetermined values.[6] In the displacement approach the system is first reduced to this kinematically determinate condition by considering all unspecified degrees of freedom equal to zero. The number of such degrees of freedom is the number of equations that are required for analysis, that is, the number of rows of the $[\mathbf{K}_{ff}]$ submatrix of Equation 3.7. These equations are developed by employing the elastic characteristics of the system and applying the requirement of restoring the equilibrium that was impaired when the conceptual restraints were employed to reduce the system to an equivalent kinematically determinate one. Mathematically, the restoration of equilibrium was expressed in Equations 3.2–3.5. The unknowns in the analysis are the degrees of freedom.

From the preceding, two observations may be made:

1. The displacement approach appears more automatic since, in it, all of the unknown quantities are first reduced to zero to produce the "determinate" counterpart, whereas in the flexibility approach selectivity may be involved in designating "redundants."[7]

2. Since the labor of solution is largely a function of the number of unknowns, the flexibility approach would appear to be advantageous in structures in which the number of force redundants is less than the number of unknown degrees of freedom, and the displacement approach in all other cases. This view, however, fails to account for differences in effort required to form the equations to be solved in the respective approaches.[8]

The comparative examples in Figure 3.7 indicate that, except in small structures, the number of redundant forces is usually considerably greater than the degree of kinematic redundancy. In the figure, all relevant degrees of freedom are included. In eval-

---

[6]Support settlement problems in which estimated displacements of the supports are specified in the problem are cases in which displacements have predetermined values.

[7]Procedures for the automatic selection of redundant forces and the formulation of the stiffness and flexibility methods along parallel lines may be found in the first edition of this book.

[8]Note that in Example 3.5, the system is statically determinate but five times kinematically indeterminate, whereas in Example 3.6 a similar system is two times statically indeterminate and three times kinematically indeterminate. A flexibility approach to Example 3.5 would be very simple, whereas in Example 3.6 it would be relatively time-consuming. Conversely, the displacement solution of Example 3.6 is simpler than it is in Example 3.5. But, regardless of the differences in number of unknowns within each approach, in computerized frame analysis the displacement method is at least as efficacious, except for the determination of forces in statically determinate systems.

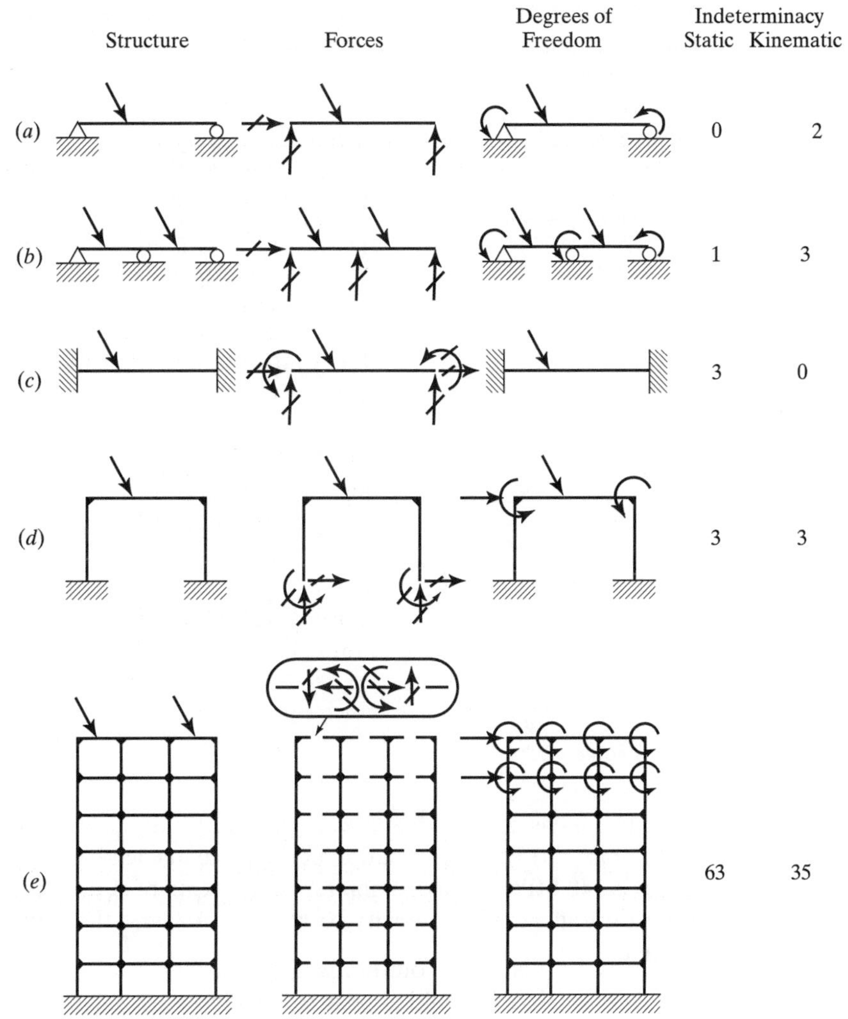

Figure 3.7 Static and kinematic indeterminacy.

uating kinematic indeterminacy, we make the common assumption that axial deformation may be neglected in analyzing flexurally loaded frames. As an example of the consequence of this, one need consider only one translational degree of freedom at each floor of the multi-story frame of Figure 3.7*e*.[9]

3.5 PROBLEMS

It is suggested that, in several of the following problems, particularly 3.6 and 3.15, a computer program such as MASTAN2 be used. In all cases, the computed results should be verified by manually checking key equilibrium and/or compatibility conditions.

[9]In tall buildings, the axial deformation of columns should be taken into account. This would add additional degrees of freedom to the analysis.

3.1 Compute the displacements, reactions, and bar forces for the trusses shown. Cross-sectional areas ($mm^2 \times 10^3$) are shown on each bar. $E = 200{,}000$ MPa.

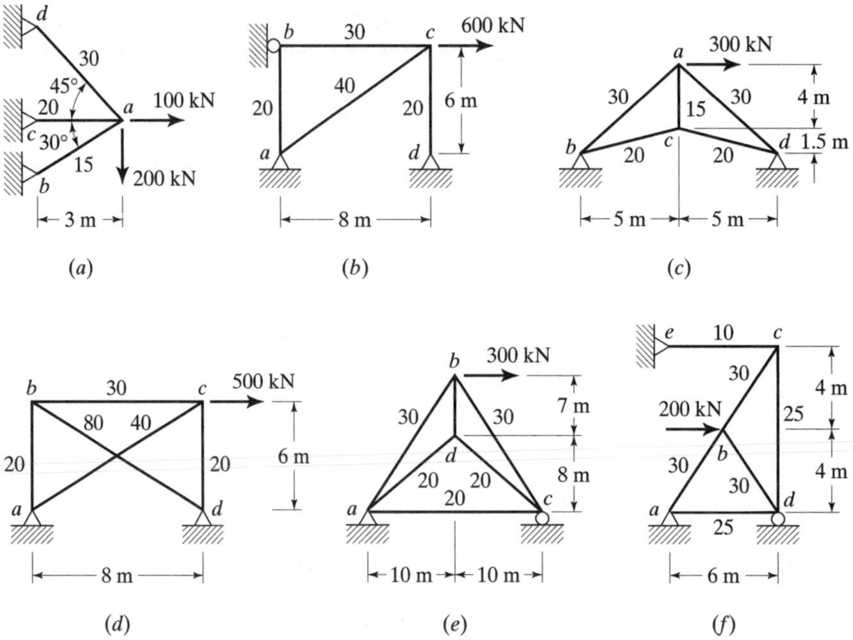

Problem 3.1

3.2 The areas of bars *ab* and *ac* in $mm^2 \times 10^3$ are as indicated. A varies from 0 to 40 $\times$ 10^3 mm^2. (a) Plot the variation in the force in each bar versus A. (b) If the allowable stress is 140 MPa, for approximately what value of A does the system have the highest ratio of load-carrying capacity to weight? Assume the same material is used for all three bars.

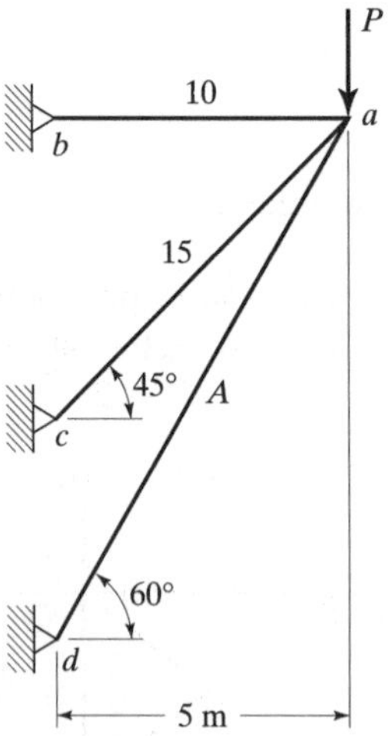

Problem 3.2

3.3 Repeat Problem 3.2 but, in addition to varying A, vary the angle of inclination of bar *ad* from 30 to 60° by 10° increments. What are the approximate values of A and the angle of inclination of *ad* that yield the highest ratio of load-carrying capacity to weight?

3.4 (a) Assemble the global stiffness equations for the truss shown. E and A of all bars are the same. (b) Assuming there will be no force applied at e, eliminate the corresponding degrees of freedom from the global equations by proper algebraic operations and develop a condensed matrix equation containing only the forces and degrees of freedom at c and d. (c) Calculate the reactions and bar forces for a 200-kN horizontal force acting to the right at d.

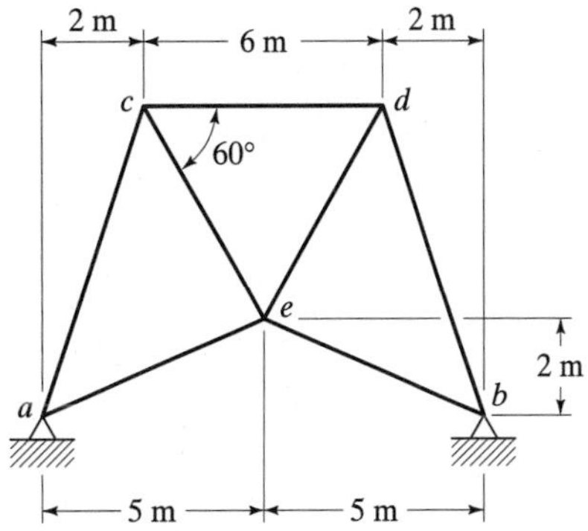

Problem 3.4

3.5 (a) Assemble the global stiffness equations for the truss in part a of the figure. (b) Modify the above to include bars *db* and *bf* as in part b. (c) Modify the above to include bars *gh* and *ij* as in part c. (d) Discuss the manner in which the changes may or may not affect the labor of solution. E and A of all bars are the same.

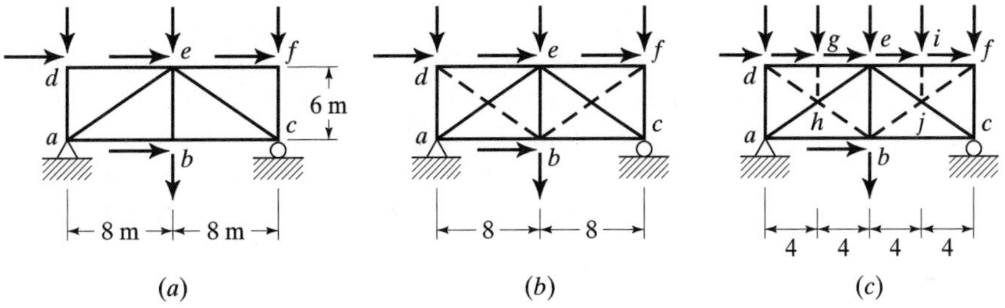

Problem 3.5

3.6 Use a computer program to calculate the displacements, reactions, and bar forces of the truss shown. $A = 40 \times 10^3$ mm^2 for all chord members and 25×10^3 mm^2 for all web members. $E = 200{,}000$ MPa.

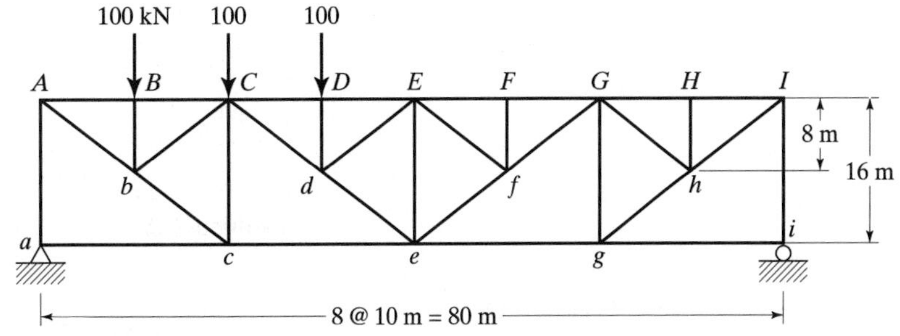

Problem 3.6

3.7 Assemble the $[\mathbf{K}_{ff}]$ matrices for the two trusses shown. Test to see whether each one is singular or not. What is the significance of your results? A and E are the same for all bars.

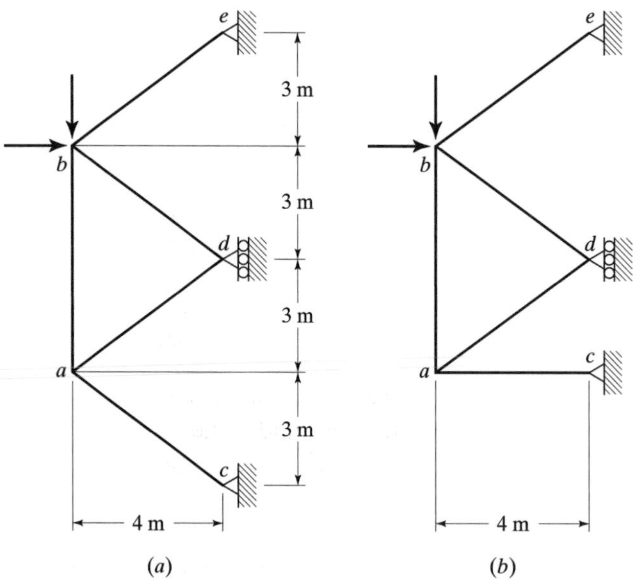

Problem 3.7

3.8 The symmetrical truss shown is made of three rods whose compressive resistance may be considered negligible. In its externally unloaded condition (part a of the figure) it is pretensioned so that the force in *ab* is 50 kN. $E = 200{,}000$ MPa. (a) Calculate the pretensioning force in *ac* and *ad* by statics. (b) Plot the vertical displacement at *a* versus the applied load P. Assume linear elastic behavior.

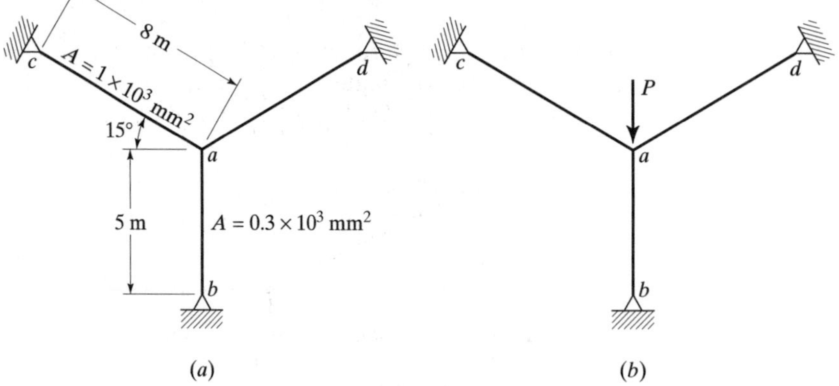

Problem 3.8

3.9 The truss shown is the same as in Example 3.4 but it is now supported as shown. Assume that the support at b settles vertically 10 mm. (a) Are any forces developed in the members? (b) Calculate the displacements of joint c and d. (c) Sketch the displaced structure.

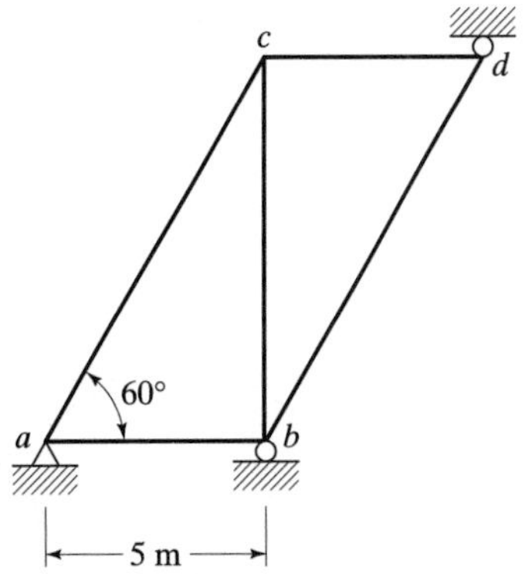

Problem 3.9

3.10 A and E are the same for all bars. (a) Assemble the $[\mathbf{K}_{ff}]$ matrix. (b) Sketch the displaced structure under the loads P. (c) Use considerations of symmetry to reduce the order of the $[\mathbf{K}_{ff}]$ matrix. (d) Calculate the displacements at b and c.

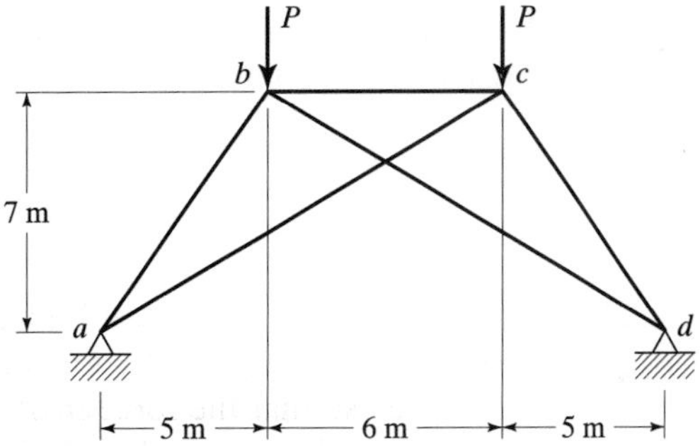

Problem 3.10

3.11 Consider the structure of Problem 3.10, carrying—instead of the loads P—two horizontal loads H acting to the right at b and c. Sketch the displaced structure, use considerations of antisymmetry to reduce the order of $[\mathbf{K}_{ff}]$, and calculate the displacement at b and c.

3.12 Three bars of equal A and E are arranged as shown. (a) Assemble the $[\mathbf{K}_{ff}]$ matrix and test for singularity. (b) Can a useful solution be obtained when $P_b = P_c$? Demonstrate. (c) What happens when $P_c > P_b$?

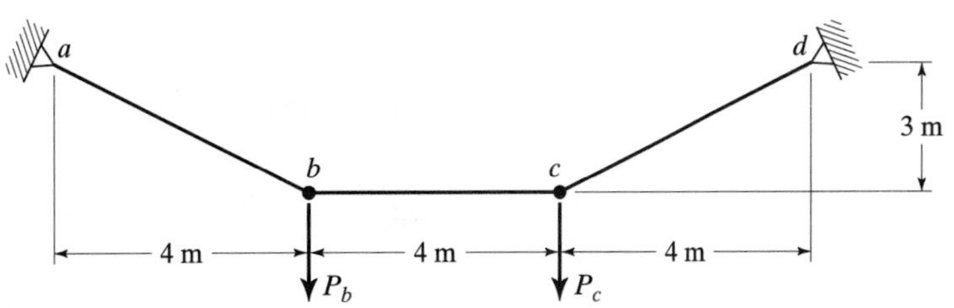

Problem 3.12

3.13 Compute the displacements, reactions, and bar forces for the system shown. $A = 2 \times 10^3$ mm^2 and $E = 200{,}000$ MPa for each member.

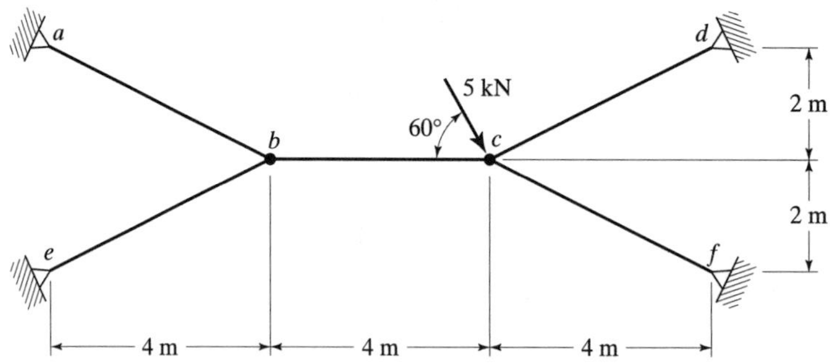

Problem 3.13

3.14 If the system shown were modeled as an analogous capacitor network, what would be the relative values of the required capacitances (see Problem 2.11)? Consider vertical displacement only. $E = 200{,}000$ MPa.

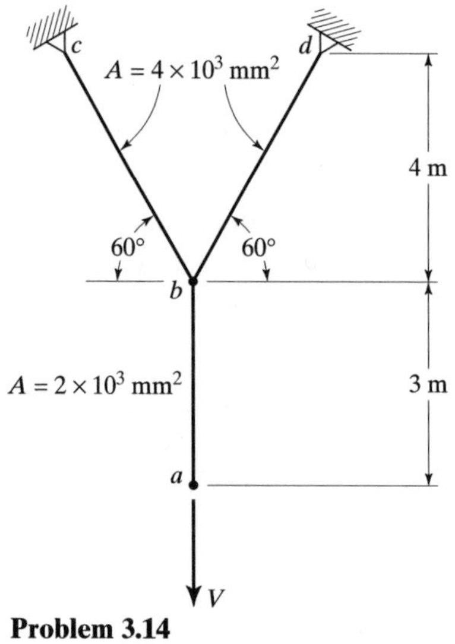

Problem 3.14

3.15 Use a computer program to calculate the displacements, reactions, and bar forces of the trusses shown. $A = 40 \times 10^3$ mm^2 for all chord members and 25×10^3 mm^2 for all web members. $E = 200{,}000$ MPa.

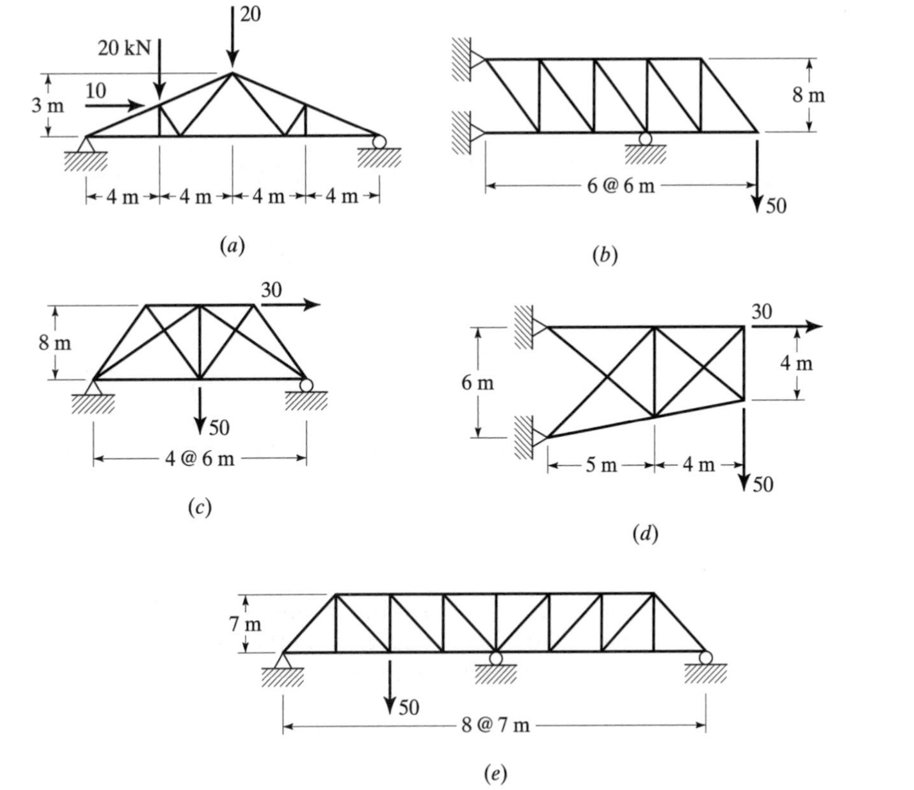

Problem 3.15

REFERENCE

3.1 K.-J. Bathe, *Finite Element Procedures*, Prentice Hall, Englewood Cliffs, N.J., 1996.

Chapter 4

Stiffness Analysis of Frames—I

In this and the following chapter we develop the remaining tools needed for the linear elastic stiffness analysis of complete frames. Here we are using the term *frames* in a generic sense to include structures such as pin-jointed planar and space trusses, rigid-jointed planar and space frames, and combinations of these forms; in short, frameworks of any shape, jointing scheme, and loading. By the end of Chapter 5 we can demonstrate the solution of complex, realistic problems. Nevertheless, there will still remain much more to be covered in later chapters—in fundamental areas of structural mechanics and behavior, in nonlinear analysis, and in methods of application such as the techniques of equation solution.

Our main purpose in the present chapter is the development and application of the stiffness matrix of the prismatic, bisymmetrical, 12-degree-of-freedom member (see Figure 2.7). This will be done in local coordinates. Formation of this matrix requires an understanding of the material stress-strain relationships, and it is facilitated by the use of energy concepts and the reciprocal theorem. These are reviewed to the extent useful to the development of the stiffness matrix. The subjects of coordinate transformation and the effects of loads applied between nodal points, self straining, and temperature change will be introduced in Chapter 5.

4.1 STRESS-STRAIN RELATIONSHIPS

Theories of structural behavior must incorporate laws relating stresses to strains. Application of any theory therefore requires knowledge of physical properties of the material used. As mentioned in Chapter 2, in these early chapters we are limiting our attention to linear, elastic structures. We also assume that the material used is homogeneous and isotropic. It follows from these restrictions that there are three properties of interest, only two of which are independent. The three properties are:

- E Young's modulus, the ratio of normal stress to the corresponding strain in a uniaxially stressed element.
- G The modulus of rigidity (shear modulus), the ratio of shearing stress to shearing strain.
- ν Poisson's ratio, the ratio of transverse strain to axial strain in a uniaxially stressed element.

Figure 4.1 shows a small element of unit length, width, and thickness stressed in two different ways: uniaxial tension of intensity σ (Figure 4.1*a*) and pure shear of intensity τ (Figure 4.1*b*). For convenience the unit thickness is shown to reduced scale. From

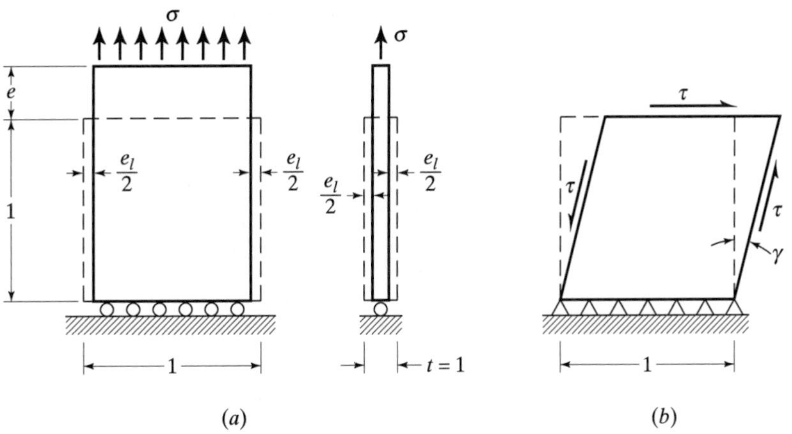

Figure 4.1 Basic deformations.

the figures and the above definitions, we have the following equations relating strain to stress:

$$e = \frac{\sigma}{E} \tag{4.1a}$$

$$e_l = -\nu e = -\frac{\nu\sigma}{E} \tag{4.1b}$$

$$\gamma = \frac{\tau}{G} \tag{4.1c}$$

The minus sign of e_l signifies that the material contracts.

By equilibrium, the state of pure shear in the element of Figure 4.1*b* is equal to a state of equal tensile and compressive stresses in the diagonal directions of the figure. Analysis of the small elastic strains in the element leads to the following relationship between the modulus of rigidity and the other two properties (see Section 6 of Reference 4.1):

$$G = \frac{E}{2(1 + \nu)} \tag{4.2}$$

Equation 4.2 is the most commonly used form of the dependency relationship, but any of the three properties of the material could be made the dependent variable if desired.

The most widely used civil engineering structural materials, steel and concrete, have uniaxial stress-strain diagrams of the types shown in Figure 4.2. Mild steel yields with a pronounced permanent elongation at a stress σ_{ym}, whereas high-strength steel yields gradually, requiring an arbitrary definition of its yield stress, σ_{yh}, such as the commonly used 0.2% offset criterion (Figure 4.2*a*). Yield stresses of structural steel used in general practice vary from less than 250 MPa (36 ksi) to more than 700 MPa (100 ksi), with 345 MPa (50 ksi) yield a commonly specified value. At stresses below the yield, steel stressed either in tension or compression comes close to satisfying the criterion of ideal elasticity. The Young's modulus and Poisson's ratio of steel are always close to 200,000 MPa (29,000 ksi) and 0.3, respectively.

The ultimate compressive strength, $\sigma_u = f'_c$, of normal weight concrete is usually in the 21 to 55 MPa (3 to 8 ksi) range (Figure 4.2*b*). The most common ranges for cast-in-place buildings have been from 21 to 41 MPa (3 to 6 ksi), but with a trend toward

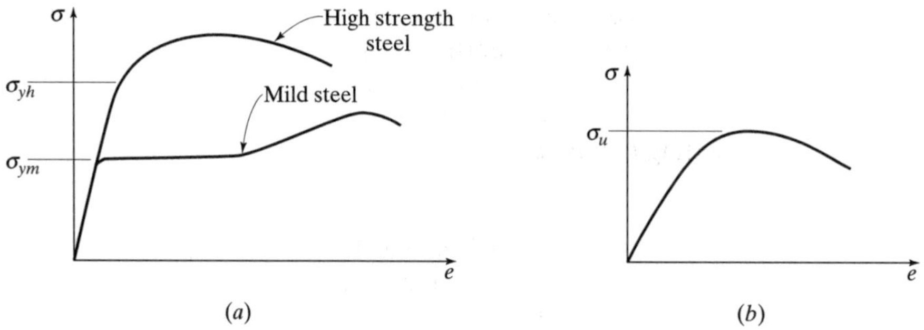

Figure 4.2 Typical σ-e diagrams. (*a*) Steel. (*b*) Concrete (compression).

using higher strength values, for example 69 MPa (10 ksi) in the vertical components of high-rise buildings. The mechanical properties of concrete are less predictable than those of steel, but under short-duration compressive stress not greater than $f_c'/2$, its behavior is reasonably linear. For normal weight concrete, the values $E_c = 4700\sqrt{f_c'}$ MPa ($1800\sqrt{f_c'}$ ksi, with $\sqrt{f_c'}$ expressed in ksi) and $\nu = 0.15$ are often used in the analysis of concrete structures. Concrete creeps–that is, its strain increases–at a diminishing rate, under constant compressive stress. Steel doesn't creep at ordinary temperatures. Both materials fall short of the ideals of homogeneity and isotropy, concrete more so than steel. In most analyses these shortcomings and the complications they cause are ignored. But there are many practical problems in which doing this can compromise the validity of the results.

4.2 WORK AND ENERGY

The work, dW, of a force F acting through a change in displacement $d\Delta$ in the direction of that force is the product $F\,d\Delta$. Over a total displacement Δ_1 the total work W is $\int_0^{\Delta_1} F\,d\Delta$ (Figure 4.3*a*).

In this text we are concerned with *static behavior*, that is, with forces that are applied to a structure in a gradual manner (inertia forces due to dynamic behavior are assumed

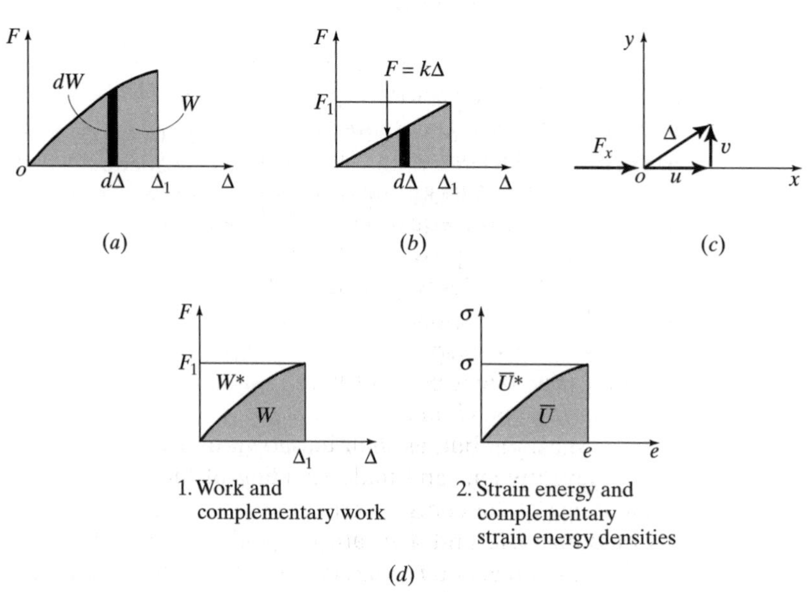

Figure 4.3 Force-displacement relationship.

negligible). Also, as long as we limit our attention to linear behavior, displacements and forces are proportional by definition. Thus a single force varies linearly with displacement from zero to its final intensity F_1. This relationship can be written as $F = k\Delta$, where k is a constant (Figure 4.3*b*). In this case, $W = k\int_0^{\Delta_1} \Delta\, d\Delta = k(\Delta_1)^2/2$. Also, letting $k = F_1/\Delta_1$

$$W = \tfrac{1}{2}F_1\Delta_1$$

Both the force and the displacement at a point are representable as vectors **F** and **Δ** and, in the symbolism of vector algebra, the work of the force is then represented by the dot product,

$$W = \tfrac{1}{2}\mathbf{F}\cdot\boldsymbol{\Delta}$$

We deal with matrix symbolism, however, where the dot product has the form

$$W = \tfrac{1}{2}\lfloor \mathbf{F} \rfloor\{\boldsymbol{\Delta}\} \tag{4.3}$$

In the illustration of Figure 4.3*c*, where $F_y = 0$, this product is given by

$$W = \tfrac{1}{2}\lfloor F_x \quad 0 \rfloor \begin{Bmatrix} u \\ v \end{Bmatrix} = \tfrac{1}{2}F_x u$$

The above considerations, which are couched in terms of a force acting on one element, are readily generalized to the case of a system of forces acting on a structure and the displacements of the points of application of these forces. In such cases the force and displacement vectors are of the form:

$$\lfloor \mathbf{P} \rfloor = \lfloor P_{x1} \quad P_{y1} \quad P_{z1} \quad P_{x2} \cdots P_{zn} \rfloor$$
$$\{\boldsymbol{\Delta}\} = \lfloor u_1 \quad v_1 \quad w_1 \quad u_2 \cdots w_n \rfloor^{\mathrm{T}}$$

and Equation 4.3, with $\lfloor \mathbf{F} \rfloor = \lfloor \mathbf{P} \rfloor$, gives the work of the system of forces $\lfloor \mathbf{P} \rfloor$.

Equation 4.3 can be transformed into expressions exclusively in terms of either the forces or the displacements by substitution of the stiffness (Equation 2.2) or the flexibilities (Equation 2.4). Therefore, from Equations 4.3 and 2.2,

$$W = \tfrac{1}{2}\lfloor \boldsymbol{\Delta} \rfloor[\mathbf{k}]\{\boldsymbol{\Delta}\} = U \tag{4.4a}$$

and, Equations 4.3 and 2.4,

$$W^* = \tfrac{1}{2}\lfloor \mathbf{F}_f \rfloor[\mathbf{d}]\{\mathbf{F}_f\} = U^* \tag{4.4b}$$

in which $\{\mathbf{F}_f\}$ is the vector of forces not related to the support conditions.

In Equations 4.4, U describes the *strain energy* of deformation corresponding to the work W, W^* the *complementary work* done by the forces acting on the element, and U^* the *complementary strain energy* of deformation corresponding to W^*. From the development of the equations it is obvious that, for linear elastic behavior, $W^* = W$ and thus $U^* = U$, provided there is no initial strain (see Sections 5.3 and 7.5). But Figure 4.3*d* illustrates why we have defined the terms of Equation 4.4b as we have just done. When the response to a force is nonlinear, W^* is unequal to W, as indicated in Figure 4.3*d*1. Correspondingly, the strain energy and complementary strain energy densities (the values per unit volume, $\overline{U}$ and $\overline{U}^*$ in Figure 4.3*d*2) and thus their integrated resultants, U and U^*, are also unequal. Complementary energy has no direct physical meaning but, as shall be shown in Section 6.5, it is a concept that is the key to the development and understanding of the *principle of virtual forces*, one of the foundations of the energy methods of structural analysis.

Equations 4.4a and 4.4b are of quadratic form, that is, their expansion yields U and U^* as homogeneous, quadratic polynomials in the parameters $\lfloor \boldsymbol{\Delta} \rfloor$ and $\lfloor \mathbf{F}_f \rfloor$. These equations were developed in terms of work done on an element. It is clear, however, that they apply to entire systems as well, once appropriate substitutions are made for the definitions of forces, displacements, stiffnesses, and flexibilities, as in writing the system force and displacement vectors earlier.

Example 4.1 illustrates the necessary equality of work and energy calculated in alternative ways.

EXAMPLE 4.1

For the truss of Example 3.2 demonstrate the equivalence of:

1. Work calculated by Equation 4.3.
2. Work calculated by Equation 4.4a.
3. Strain energy calculated using a basic definition.
   Use results of Examples 3.1 and 3.2.

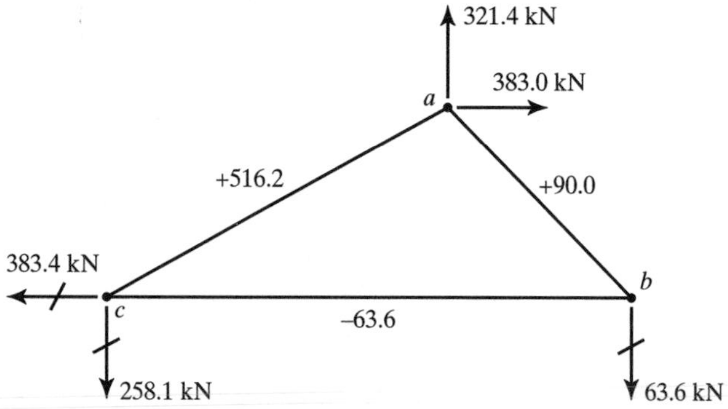

1. Equation 4.3, redefining terms with respect to the whole structure, is

$$W = \tfrac{1}{2}\lfloor \mathbf{P} \rfloor\{\boldsymbol{\Delta}\} = \tfrac{1}{2}\lfloor 383.4 \quad 321.4 \quad 0 \rfloor \begin{Bmatrix} 0.871 \\ 1.244 \\ -0.193 \end{Bmatrix} = 367.0 \text{ J}$$

2. Equation 4.4a, redefining terms with respect to the whole structure, is

$$W = \tfrac{1}{2}\lfloor \boldsymbol{\Delta} \rfloor[\mathbf{K}_{ff}]\{\boldsymbol{\Delta}\}$$

$$= \frac{10^2}{2}\lfloor 0.871 \quad 1.244 \quad -0.193 \rfloor \begin{bmatrix} 6.348 & -1.912 & -3.536 \\ & 4.473 & 3.536 \\ \text{Sym.} & & 6.830 \end{bmatrix} \begin{Bmatrix} 0.871 \\ 1.244 \\ -0.193 \end{Bmatrix} = 367.0 \text{ J}$$

3. Strain energy. Designating a bar force by the symbol F and summing over all members, the strain energy of this pin-jointed truss is

$$U = \sum \frac{F^2L}{2EA} = \frac{1}{2}\left[\frac{(90.0)^2}{707.11} + \frac{(-63.6)^2}{329.43} + \frac{(516.2)^2}{375.00}\right] = 367.0 \text{ J}$$

4.3 RECIPROCITY

As has been noted, flexibility and stiffness coefficients for linear elastic behavior have the property of reciprocity ($d_{ij} = d_{ji}$ and $k_{ij} = k_{ji}$). This property is important from the standpoint of computational efficiency and it is useful in checking formulated or computed coefficients. To prove reciprocity and thereby define its scope and limitations, consider the work done on the supported structure in Figure 4.4 as load F_1 is first applied, followed by load F_2.[1] We designate the total amount of work as W_I. From Figure 4.4*a*, for gradual application of F_1,

$$W_{I_1} = \tfrac{1}{2}(\Delta_1)_1F_1 = \tfrac{1}{2}(d_{11}F_1)F_1 \tag{4.5}$$

[1]Even though we view the system of Figure 4.4 as a complete structure, we retain our single element notation. Similarly, in the discussion to follow we retain the element notation in the reference to Equation 3.7, which was written for complete systems. Justification for this is given in the next section.

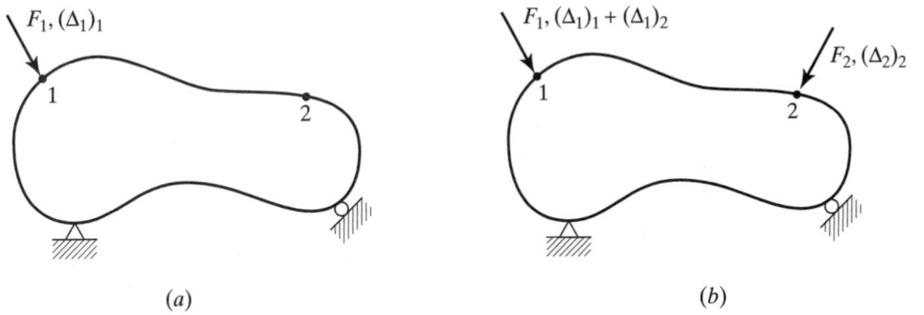

Figure 4.4 Sequential application of loads. (*a*) Force F_1 applied at 1. (*b*) Force F_2 applied at 2.

where the subscript on W_I and on (Δ_1) means *due to force* 1. Now applying F_2 with F_1 held constant, and using similar designations, the increment of work is (Figure 4.4*b*)

$$\begin{aligned} W_{I_2} &= \tfrac{1}{2}(\Delta_2)_2 F_2 + (\Delta_1)_2 F_1 \\ &= \tfrac{1}{2}(d_{22}F_2)F_2 + (d_{12}F_2)F_1 \end{aligned} \tag{4.6}$$

The factor 1/2 is not applied to the second term of Equation 4.6 because F_1 exists, in full force, all during the time F_2 is gradually applied. The total work, W_I is

$$W_I = W_{I_1} + W_{I_2} = \tfrac{1}{2}d_{11}(F_1)^2 + \tfrac{1}{2}d_{22}(F_2)^2 + d_{12}F_2F_1 \tag{4.7}$$

Reversing the sequence of application of forces and recomputing the components of work, we have for initial application of F_2 (the work in this second sequence is identified as W_{II})

$$W_{II_2} = \tfrac{1}{2}(\Delta_2)_2 F_2 = \tfrac{1}{2}d_{22}(F_2)^2 \tag{4.5a}$$

followed by the application of F_1:

$$W_{II_1} = \tfrac{1}{2}(\Delta_1)_1 F_1 + (\Delta_2)_1 F_2 = \tfrac{1}{2}d_{11}(F_1)^2 + d_{21}F_1F_2 \tag{4.6a}$$

so that

$$W_{II} = W_{II_1} + W_{II_2} = \tfrac{1}{2}d_{22}(F_2)^2 + \tfrac{1}{2}d_{11}(F_1)^2 + d_{21}F_1F_2 \tag{4.7a}$$

Since, for a linear system, the sequence of application of loads is immaterial from the standpoint of the work performed, we can equate the two expressions for W and, after canceling terms,

$$d_{21} = d_{12}$$

And, generally,

$$d_{ij} = d_{ji} \tag{4.8}$$

This algebraic statement is known as *Maxwell's reciprocal theorem.*

As shall be demonstrated formally in the next section, the stiffness matrix for a given set of force conditions (the $\mathbf{k}_{ff}$ matrix in Equation 3.7) is the inverse of the flexibility matrix of the system when supported in the same way. Since it has just been demonstrated that the flexibility matrix is symmetric and since it is known that the inverse of a symmetric matrix is symmetric, it follows that the stiffness matrix is symmetric, or

$$k_{ij} = k_{ji} \tag{4.9}$$

Maxwell's reciprocal theorem is usually defined as a special case of *Betti's law*, which states that the work done by a system of forces $\{\mathbf{P}_1\}$ acting through the displacements $\{\boldsymbol{\Delta}_1\}_2$ that are caused by the system of forces $\{\mathbf{P}_2\}$, is equal to the work done by the forces $\{\mathbf{P}_2\}$, acting through displacements $\{\boldsymbol{\Delta}_2\}_1$ that correspond to $\{\mathbf{P}_1\}$, that is,

$$\lfloor \boldsymbol{\Delta}_1 \rfloor_2 \{\mathbf{P}_1\} = \lfloor \boldsymbol{\Delta}_2 \rfloor_1 \{\mathbf{P}_2\} \tag{4.10}$$

4.4 FLEXIBILITY-STIFFNESS TRANSFORMATIONS

Given one form of element force-displacement relationship, it is possible to obtain the alternate forms through simple operations. These transformations have many uses. Since they rest only on general mathematical operations and the principles of equilibrium and energy, they apply not only to simple elements but also to complete systems, the members of which undergo compatible deformation. They are developed here because they are particularly useful in forming the framework element stiffness matrix.

4.4.1 Stiffness-to-Flexibility Transformation

To illustrate the transformation of stiffness to flexibility, we use a three-bar system of the type analyzed in Examples 3.1 and 3.2, but for simplicity, we treat it as a single element by using the notation of Chapter 2 (see Figure 4.5). This is permissible for the reason given in the preceding paragraph. The bar assemblage may be thought of as one component of a larger structure. To comply with the requirement that an element must be supported in a stable, statically determinate manner in order to obtain a useful flexibility relationship (see Section 2.4.2), we support it as in Figure 4.5*b*.

Quantities related to the supports will be assigned the subscript s and those related to the remaining degrees of freedom the subscript f. Thus the element stiffness matrix, Equation 2.2, may be partitioned as follows (compare with Equations 3.6 and 3.7):

$$\begin{Bmatrix} \mathbf{F}_f \\ \hline \mathbf{F}_s \end{Bmatrix} = \left[\begin{array}{c|c} \mathbf{k}_{ff} & \mathbf{k}_{fs} \\ \hline \mathbf{k}_{sf} & \mathbf{k}_{ss} \end{array}\right] \begin{Bmatrix} \boldsymbol{\Delta}_f \\ \hline \boldsymbol{\Delta}_s \end{Bmatrix} \tag{4.11}$$

where, for the condition shown in Figure 4.5*b*, each of the submatrices ($[\mathbf{k}_{ff}]$, etc.) is a 3×3 matrix and

$$\begin{aligned} \{\mathbf{F}_f\} &= \lfloor F_{x1} \quad F_{y1} \quad F_{x2} \rfloor^{\mathrm{T}} & \{\mathbf{F}_s\} &= \lfloor F_{y2} \quad F_{x3} \quad F_{y3} \rfloor^{\mathrm{T}} \\ \{\boldsymbol{\Delta}_f\} &= \lfloor u_1 \quad v_1 \quad u_2 \rfloor^{\mathrm{T}} & \{\boldsymbol{\Delta}_s\} &= \lfloor v_2 \quad u_3 \quad v_3 \rfloor^{\mathrm{T}} \end{aligned} \tag{4.12}$$

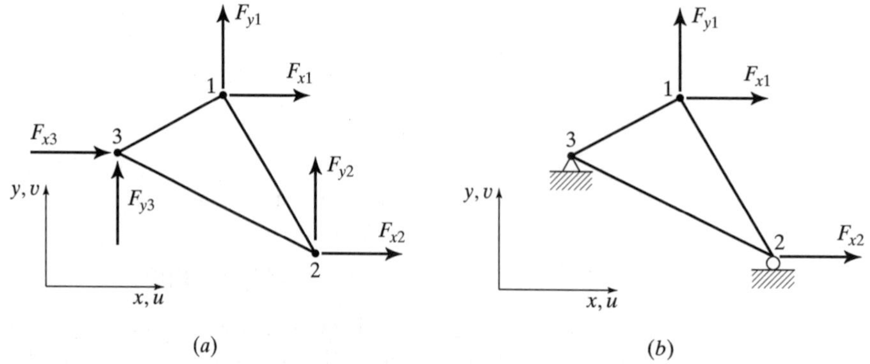

Figure 4.5 Three-bar element. (*a*) Unsupported element. (*b*) Supported element.

Now, since $\{\boldsymbol{\Delta}_s\} = \{\mathbf{0}\}$ from the nature of the supports,

$$\begin{Bmatrix} \mathbf{F}_f \\ \hline \mathbf{F}_s \end{Bmatrix} = \begin{bmatrix} \mathbf{k}_{ff} \\ \hline \mathbf{k}_{sf} \end{bmatrix} \{\boldsymbol{\Delta}_f\} \tag{4.13}$$

The equations above the partition line are an independent set of equations relating the external forces $\{\mathbf{F}_f\}$ to the corresponding node point displacements (compare Example 3.2). Solution of these by inversion of $[\mathbf{k}_{ff}]$ yields

$$\{\boldsymbol{\Delta}_f\} = [\mathbf{d}]\{\mathbf{F}_f\} \tag{4.14}$$

where

$$[\mathbf{d}] = [\mathbf{k}_{ff}]^{-1} \tag{4.15}$$

The matrix $[\mathbf{d}]$ is the desired set of element flexibility coefficients. Thus the flexibilities are derived from the stiffnesses by merely defining a stable, statically determinate support system, removing from the stiffness matrix the rows and columns corresponding to the support components, and inverting the remainder.

Of the two conditions on the supports, it was pointed out in Section 2.4.2 that the first—stability—is required to prevent unrestricted rigid body motion under the application of applied forces. Regarding the second—determinacy—it can be shown that when an element is supported in a statically determinate manner, the complementary strain energy has a value that is independent of the specific set of support conditions that has been chosen (Example 4.6 to follow). If a statically indeterminate support condition is used, then strains that are dependent upon the chosen supports are introduced into the element. The complementary strain energy is then different for each support condition. Since the global analysis equations can be obtained through a summation of the element strain energies these must, in general, be independent of local element support conditions.

4.4.2 Flexibility-to-Stiffness Transformation

To reverse the above process, that is, to develop the complete stiffness matrix from a given flexibility matrix, we begin by inverting the flexibility matrix. But then we have to add something since, obviously, inversion of the flexibility matrix yields no direct information on stiffness coefficients related to degrees of freedom that have been equated to zero in defining the support system.

We start by inverting a given flexibility matrix:

$$\{\mathbf{F}_f\} = [\mathbf{d}]^{-1}\{\boldsymbol{\Delta}_f\} = [\mathbf{k}_{ff}]\{\boldsymbol{\Delta}_f\} \tag{4.16}$$

or

$$[\mathbf{k}_{ff}] = [\mathbf{d}]^{-1} \tag{4.17}$$

Now, it will be recalled that each column of a complete stiffness matrix represents a system of forces in equilibrium (Section 2.4.1). Also, the forces in the "s" portion of each column are the statically determinate support forces. Hence we can use the equations of equilibrium to construct equations of the form:

$$\{\mathbf{F}_s^j\} = [\boldsymbol{\Phi}]\{\mathbf{F}_f^j\} \tag{4.18}$$

where the superscript j identifies the column in question of the stiffness matrix and $[\boldsymbol{\Phi}]$ is an equilibrium matrix, or matrix of coefficients of the equations of static equilibrium. Construction of equilibrium matrices is a simple process which will be illustrated in

Example 4.2 and other, later, examples. It follows that any set of forces $\{\mathbf{F}_f\}$, stemming from one *or a combination* of columns of the element stiffness matrix, are related to the forces $\{\mathbf{F}_s\}$ in the lower portion of the matrix by a relationship of the above form, that is

$$\{\mathbf{F}_s\} = [\mathbf{\Phi}]\{\mathbf{F}_f\} \tag{4.19}$$

By substitution of Equation 4.16 in 4.19,

$$\{\mathbf{F}_s\} = [\mathbf{\Phi}][\mathbf{d}]^{-1}\{\mathbf{\Delta}_f\} \tag{4.20}$$

or, from Equation 4.13,

$$\{\mathbf{F}_s\} = [\mathbf{k}_{sf}]\{\mathbf{\Delta}_f\} \tag{4.21}$$

in which

$$[\mathbf{k}_{sf}] = [\mathbf{\Phi}][\mathbf{d}]^{-1} \tag{4.22}$$

In Equations 4.17 and 4.22 we have the necessary relationships for all terms in the columns that premultiply $\{\mathbf{\Delta}_f\}$ in Equation 4.11.

Next we seek the submatrices that premultiply $\{\mathbf{\Delta}_s\}$ in Equation 4.11; that is, $[\mathbf{k}_{fs}]$ and $[\mathbf{k}_{ss}]$. To establish $[\mathbf{k}_{fs}]$, we simply invoke the reciprocal theorem (Equation 4.9), which in matrix form may be written as

$$[\mathbf{k}_{fs}] = [\mathbf{k}_{sf}]^{\mathrm{T}} = [\mathbf{d}]^{-1}[\mathbf{\Phi}]^{\mathrm{T}} \tag{4.23}$$

(Note that the transpose of a product of matrices equals the reversed product of the transposes of the respective matrices. Also, since $[\mathbf{d}]^{-1}$ is symmetric, its transpose is itself).

In forming $[\mathbf{k}_{ss}]$, the lower portion of the columns that premultiply $\{\mathbf{\Delta}_s\}$, we recall from Equations 4.18 and 4.19 that such terms are simply given by the premultiplication of the corresponding top portion of the stiffness matrix by the equilibrium matrix $[\mathbf{\Phi}]$. Hence

$$[\mathbf{k}_{ss}] = [\mathbf{\Phi}][\mathbf{k}_{fs}] = [\mathbf{\Phi}][\mathbf{d}]^{-1}[\mathbf{\Phi}]^{\mathrm{T}} \tag{4.24}$$

Thus, from Equations 4.11, 4.17, 4.22, 4.23, and 4.24, the constructed stiffness matrix assumes the form:

$$[\mathbf{k}] = \left[\begin{array}{c|c} [\mathbf{d}]^{-1} & [\mathbf{d}]^{-1}[\mathbf{\Phi}]^{\mathrm{T}} \\ \hline [\mathbf{\Phi}][\mathbf{d}]^{-1} & [\mathbf{\Phi}][\mathbf{d}]^{-1}[\mathbf{\Phi}]^{\mathrm{T}} \end{array}\right] \tag{4.25}$$

In summary it is seen that the stiffness matrix is constructed from the inverse of the flexibility matrix $[\mathbf{d}]$ and a matrix that derives from the element static equilibrium relationships—the equilibrium matrix $[\mathbf{\Phi}]$. The property of symmetry was invoked in constructing $[\mathbf{k}_{fs}]$ from $[\mathbf{k}_{sf}]$. The $[\mathbf{k}_{ss}]$ portion of the result is given by a matrix triple product in which the premultiplier of the central matrix is equal to the transpose of the postmultiplying matrix. This triple product, known as a *congruent transformation*, produces a symmetric matrix when the central matrix is symmetric; thus $[\mathbf{k}_{ss}]$ is assured to be symmetric. Equation 4.25 is a general formula for transformation from flexibility to a stiffness form that includes rigid-body-motion degrees of freedom. The number, s, of support forces is dictated by the requirements of stable, statically determinate support, but there is no limit on the number, f, of external forces, that is, the order of the flexibility matrix.

The use of Equation 4.25 is illustrated in Example 4.2. Included as part of the example is an illustration of obtaining flexibility matrices from stiffness coefficients.

EXAMPLE 4.2

Using Equation 4.25, develop the stiffness matrix for the axial force member in arbitrary coordinates (Equation 2.5). E, A, and L are constant.

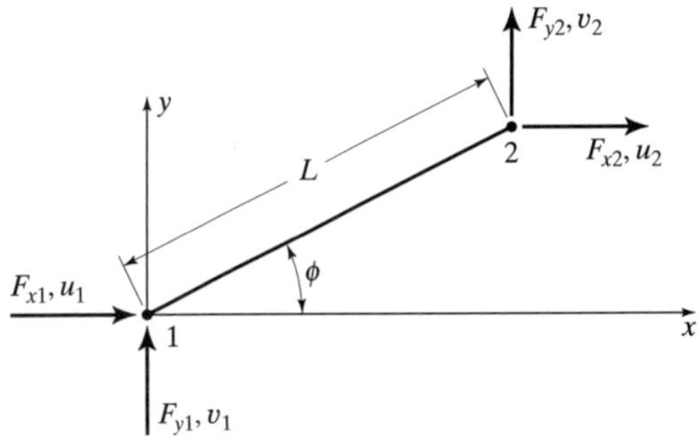

Support member as shown and write equations for $[\mathbf{\Phi}]$ and $[\mathbf{d}]$:

$$\{\mathbf{F}_f\} = \{\mathbf{F}_{x2}\} \qquad \{\mathbf{F}_s\} = \lfloor F_{x1} \quad F_{y1} \quad F_{y2} \rfloor^{\mathrm{T}}$$
$$\{\mathbf{\Delta}_f\} = \{\mathbf{u}_2\} \qquad \{\mathbf{\Delta}_s\} = \lfloor u_1 \quad v_1 \quad v_2 \rfloor^{\mathrm{T}}$$

Thus

$$\{\mathbf{F}_s\} = [\mathbf{\Phi}]\{\mathbf{F}_f\}$$

or

$$\begin{Bmatrix} F_{x1} \\ F_{y1} \\ F_{y2} \end{Bmatrix} = \begin{bmatrix} -1 \\ -\tan\phi \\ \tan\phi \end{bmatrix} \{F_{x2}\}$$

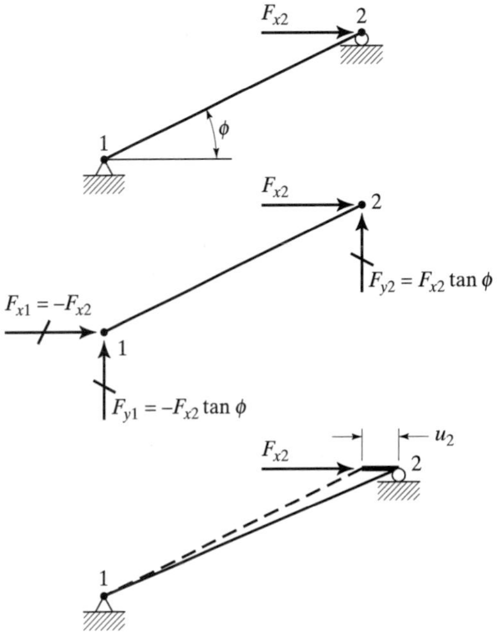

Applying F_{x2}, the only displacement is u_2. From Figure 2.9 and accompanying equations,

$$F_{x2} = \frac{EA}{L} \cos^2\phi \cdot u_2$$

Thus

$$[\mathbf{d}] = \left[\frac{L}{EA} \sec^2\phi \right]$$

Apply Equations 4.17, 4.23, 4.24:

$$[\mathbf{k}_{ff}] = [\mathbf{d}]^{-1} = \frac{EA}{L} [\cos^2\phi] \tag{4.17}$$

$$\begin{aligned} [\mathbf{k}_{fs}] &= [\mathbf{k}_{sf}]^{\mathrm{T}} = [\mathbf{d}]^{-1}[\mathbf{\Phi}]^{\mathrm{T}} \\ &= \frac{EA}{L} \lfloor -\cos^2\phi \quad -\sin\phi\cos\phi \quad \sin\phi\cos\phi \rfloor \end{aligned} \tag{4.23}$$

$$\begin{aligned} [\mathbf{k}_{ss}] &= [\mathbf{\Phi}][\mathbf{d}]^{-1}[\mathbf{\Phi}]^{\mathrm{T}} \\ &= \frac{EA}{L} \begin{bmatrix} \cos^2\phi & \sin\phi\cos\phi & -\sin\phi\cos\phi \\ \sin\phi\cos\phi & \sin^2\phi & -\sin^2\phi \\ -\sin\phi\cos\phi & -\sin^2\phi & \sin^2\phi \end{bmatrix} \end{aligned} \tag{4.24}$$

Collecting terms as in Equation 4.11 yields

$$\begin{Bmatrix} F_{x2} \\ \hline F_{x1} \\ F_{y1} \\ F_{y2} \end{Bmatrix} = \frac{EA}{L} \left[\begin{array}{c|ccc} \cos^2\phi & -\cos^2\phi & -\sin\phi\cos\phi & \sin\phi\cos\phi \\ \hline -\cos^2\phi & \cos^2\phi & \sin\phi\cos\phi & -\sin\phi\cos\phi \\ -\sin\phi\cos\phi & \sin\phi\cos\phi & \sin^2\phi & -\sin^2\phi \\ \sin\phi\cos\phi & -\sin\phi\cos\phi & -\sin^2\phi & \sin^2\phi \end{array} \right] \begin{Bmatrix} u_2 \\ \hline u_1 \\ v_1 \\ v_2 \end{Bmatrix}$$

which is identical to Equation 2.5 after reordering rows and columns.

4.5 THE FRAMEWORK ELEMENT STIFFNESS MATRIX

We now consider a 12-degree-of-freedom element that is straight, prismatic, and symmetrical about both principal cross-sectional axes (bisymmetrical). For such a section, the shear center coincides with the centroid. We shall consider displacements resulting from uniform axial deformation, flexural deformation, and twisting deformation. We will neglect displacements resulting from transverse shearing deformations and those resulting from the out-of-plane (longitudinal) warping of a cross section that torsional forces may cause. These geometrical limitations and disregarded deformations limit the applicability of the resulting equations to some extent, but they are sufficient for the analysis of the majority of the frameworks encountered in structural engineering practice. Cases in which consideration of transverse shear and torsional warping may be necessary or desirable are discussed briefly in Section 4.6. Methods for including them in the analysis are presented in Sections 7.6 and 7.4, respectively.

The type of member studied, its orientation with respect to a local coordinate system, and the nomenclature used is illustrated in Figure 4.6. A wide flanged member is shown, but the analysis applies to any bisymmetrical section. The local x axis coincides with the centroidal axis of the element. For illustrative purposes the local y axis is assumed to define the minor principal axis of a cross section and the local z axis its major principal axis. Positive signs are in the directions indicated. The right-hand rule is used for moments and rotational displacements. The restrictions described earlier leave us with an element that is sufficient for most practical purposes and, moreover, they greatly simplify the development of the element stiffness matrix in that they yield an analytical problem in which a number of effects are *uncoupled*. By this we mean a situation in which a particular force vector causes a displacement only in the same vector direction. For example, the axial forces F_{x1} and F_{x2} in Figure 4.6 can only cause axial deformation, and hence only the axial displacements u_1 and u_2 in the element. Also, the torques M_{x1} and M_{x2} only cause torsional deformations and hence only twisting displacements θ_{x1} and θ_{x2}. This means that the stiffness coefficient relating any of these force components to a degree of freedom to which it is not coupled must be zero. From this, plus the principle of reciprocity, it follows that bending moments and transverse shearing forces are coupled neither to axial nor to torsional effects in a single element. Furthermore, since the section is bisymmetrical, it is known from elementary strength of materials that bending moments and shearing forces in one principal plane cause no deformation outside of this plane; therefore, they are not coupled to rotational or translational displacement in the other principal plane. Consequently, our problem devolves into the consideration of four separates cases, with the assembly of

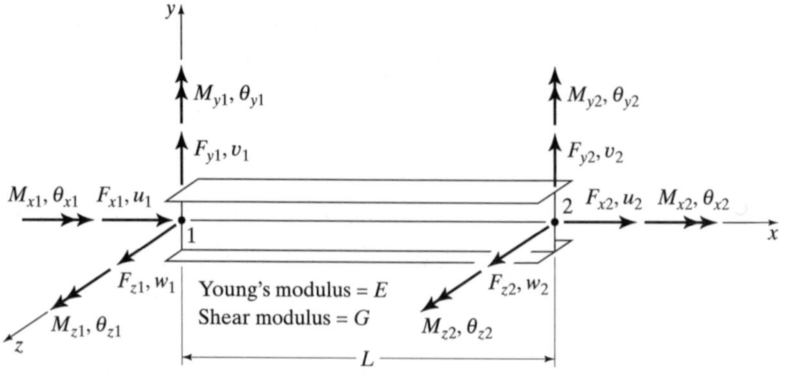

Figure 4.6 Bisymmetrical framework element.

the results in a 12×12 stiffness matrix having zero coefficients for all of the uncoupled forces and displacements.

The four cases are:

1. An axial force member.
2. A pure torsional member.
3. A beam bent about one principal axis.
4. A beam bent about the other principal axis.

4.5.1 Axial Force Member

The stiffness matrix for the axial force member in local coordinates was developed in Section 2.4.1 (Equation 2.3). It may also be written using Equation 2.5 or the results of Example 4.2. Nevertheless, it will be redone here as a further illustration of the flexibility-stiffness transformations of Section 4.4. Consider the member in Figure 4.7 in which the nomenclature of Figure 4.6 has been retained and a stable, statically determinate support system has been prescribed. The displacement at point 2 under the applied force $\{\mathbf{F}_f\} = F_{x2}$ is

$$u_2 = \int_0^L e\,dx = \int_0^L \frac{\sigma\,dx}{E} = \int_0^L \frac{F_{x2}\,dx}{EA} = \frac{F_{x2}L}{EA}$$

Hence, $[\mathbf{d}] = L/EA$ or, from Equation 4.17,

$$[\mathbf{k}_{ff}] = \frac{EA}{L}$$

By equilibrium, the reactive force $\{\mathbf{F}_s\} = F_{x1} = -F_{x2}$. And, from Equation 4.18,

$$[\boldsymbol{\Phi}] = -1$$

Thus, from Equation 4.25, in which the equilibrium relationships are invoked

$$[\mathbf{k}] = \frac{EA}{L}\begin{bmatrix} 1 & -1 \\ -1 & 1 \end{bmatrix} \tag{4.26a}$$

and

$$\begin{Bmatrix} F_{x1} \\ F_{x2} \end{Bmatrix} = \frac{EA}{L}\begin{bmatrix} 1 & -1 \\ -1 & 1 \end{bmatrix}\begin{Bmatrix} u_1 \\ u_2 \end{Bmatrix} \tag{4.26b}$$

4.5.2 Pure Torsional Member

Mathematically, the simple torsional member (a shaft) is identical to the axial force member since, by comparison of Figures 4.7 and 4.8, we see that in the two cases the

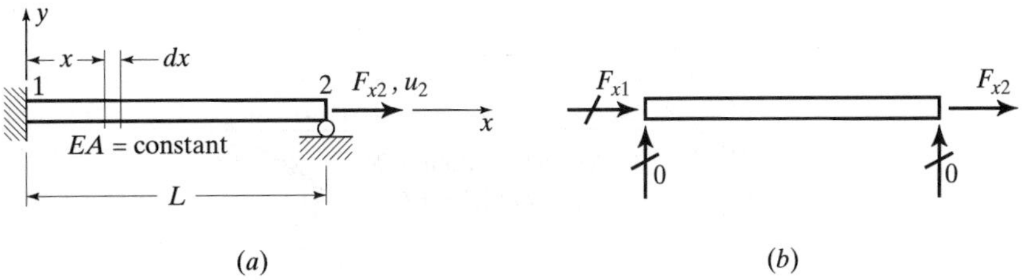

Figure 4.7 Axial force element. (*a*) Support system. (*b*) Free-body diagram.

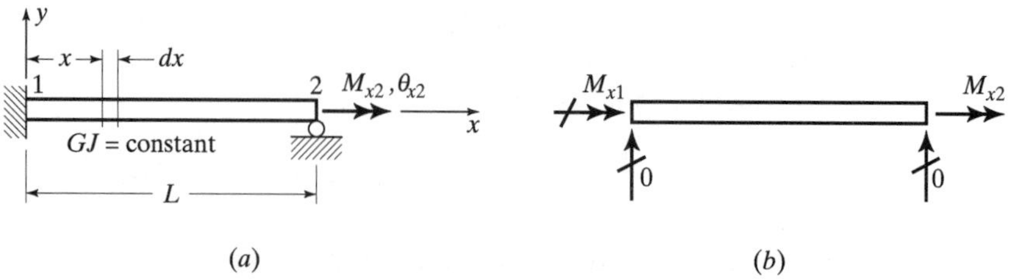

Figure 4.8 Pure torsion element. (*a*) Support system. (*b*) Free-body diagram.

forces and displacements can be represented by sets of vectors of the same type. The physical difference is that whereas the axial member stretches or shortens uniformly, the shaft twists uniformly. From elementary strength of materials we know that, for a shaft subjected to a pure torque M_{x2}, the rate of twist, β, that is, the change in rotation of the cross section about the longitudinal axis per unit of length along that axis, can be expressed as

$$\beta = \frac{M_{x2}}{GJ}$$

where M_{x2} is the torque at the section considered, G is the modulus of rigidity, and J is the *torsional constant*—a geometric property of the cross section. The dimensions of J are length units to the fourth power. For the special case of the circular, cylindrical shaft, it is equal to the polar moment of inertia of the cross section. Its value in other cases will be illustrated in later examples.

The rate of twist is the measure of the torsional strain. If it is integrated along the full length of the member portrayed in Figure 4.8, the total rotational displacement at point 2, θ_{x2}, is obtained:

$$\theta_{x2} = \int_0^L \beta\, dx = \int_0^L \frac{M_{x2}\, dx}{GJ} = \frac{M_{x2}L}{GJ}$$

Thus, by comparison with the axial force member, $[\mathbf{d}] = L/GJ$, $[\mathbf{\Phi}] = -1$, and it follows that

$$[\mathbf{k}] = \frac{GJ}{L}\begin{bmatrix} 1 & -1 \\ -1 & 1 \end{bmatrix} \tag{4.27a}$$

and

$$\begin{Bmatrix} M_{x1} \\ M_{x2} \end{Bmatrix} = \frac{GJ}{L}\begin{bmatrix} 1 & -1 \\ -1 & 1 \end{bmatrix}\begin{Bmatrix} \theta_{x1} \\ \theta_{x2} \end{Bmatrix} \tag{4.27b}$$

4.5.3 Beam Bent About Its z Axis

Since we are considering a bisymmetrical member, we only need to treat bending about one axis in detail. Relationships for bending about the other follow readily by replacement of the relevant subscripts and adjustment of algebraic signs. But since this section also introduces the basic force-displacement relationships for the flexural member, it is appropriate to discuss certain fundamental aspects of beam flexure and to define conditions and terminology that will be used here and at a later time.

The stresses and strains at any cross section caused by bending about the z axis are directed along the x axis of the member. They vary linearly with respect to the y axis and are constant in the z direction for a fixed value of y. In elementary mechanics it is shown that the strain, e_x is given by

$$e_x = -\frac{y}{\rho} = -y\frac{d^2v}{dx^2} \tag{4.28}$$

where ρ, the radius of curvature, is approximated by $1/(d^2v/dx^2)$.[2] Also, $\sigma_x = Ee_x$, so that

$$\sigma_x = -Ey\frac{d^2v}{dx^2} \tag{4.29}$$

The direct stress σ_x has a zero stress resultant ($\int_A \sigma_x\,dA = 0$), but the moment about the z axis produced by σ_x is nonzero and equal to M_z that is,

$$M_z = -\int_A \sigma_x y\,dA \tag{4.30}$$

In the analysis of bending, one customarily works with stress resultants such as bending moments and associated transverse shears, rather than with unit stresses. For this reason, in subsequent portions of the text, we frequently refer to the moments as the "stresses" of bending, their relationship to the actual stresses being given by Equation 4.30.

One further analogy will prove useful in later work. Substitution of Equation 4.29 into Equation 4.30 gives

$$M_z = \int_A E\frac{d^2v}{dx^2}y^2\,dA \tag{4.31a}$$

and, since E is constant and d^2v/dx^2 is not a function of the cross-sectional coordinate y,

$$M_z = E\frac{d^2v}{dx^2}\int_A y^2\,dA = EI_z\frac{d^2v}{dx^2} \tag{4.31b}$$

where $I_z = \int_A y^2\,dA$ is the moment of inertia of the section about the z axis. If, in Equation 4.31b, M_z is regarded as the "stress" of bending, then d^2v/dx^2 is correspondingly the "strain" of bending. The "elastic coefficient" connecting "stress" to "strain" is no longer just E; it is now EI_z. The factor I_z results from the integration of behavior on the cross section, just as M_z is the integrated bending effect of the stresses σ_x.

The development of the stiffness matrix of a beam element is somewhat more difficult than that for the axial and torsional elements. The stress is related to the strain by a second-order, rather than first-order, differential equation (Equation 4.31b). Integration produces two constants of integration that must satisfy the specified support conditions. Also, several possibilities exist for the stable, statically determinate support that is needed for the generation of a flexibility matrix. The most obvious choices are: (1) to place simple supports at each end (the simple beam), and (2) to fix either end, leaving the other free (the cantilever). Different degrees of freedom are suppressed in each case and, consequently, different flexibility matrices will be obtained. But the

[2]The negative signs in Equation 4.28 are occasioned by the fact that we are assuming tensile strain to be positive. Likewise, in Equation 4.30 to follow, we are assuming positive bending moments to be those that cause positive curvature.

resulting complete stiffness matrix must be the same, since it includes all degrees of freedom. We will use the cantilever system shown in Figure 4.9.[3]

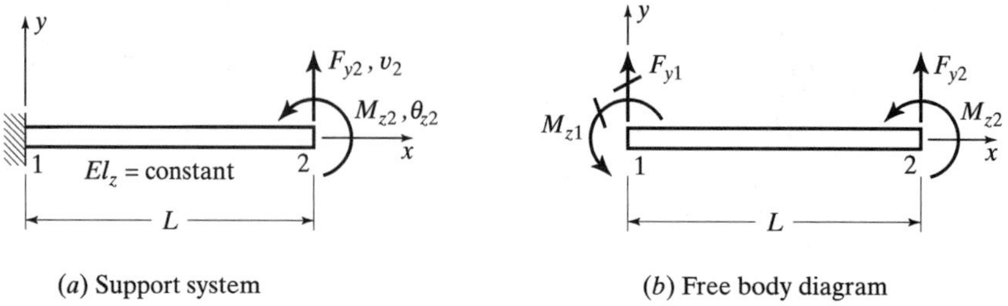

Figure 4.9 Beam bent about z axis. (a) Support system. (b) Free-body diagram.

The flexibility matrix sought is the one that satisfies Equation 4.14, which in this case reads

$$\begin{Bmatrix} v_2 \\ \theta_{z2} \end{Bmatrix} = [\mathbf{d}]\begin{Bmatrix} F_{y2} \\ M_{z2} \end{Bmatrix}$$

The terms in [**d**] are developed in Example 4.3. The result is

$$[\mathbf{d}] = \frac{L}{EI_z}\begin{bmatrix} \frac{L^2}{3} & \frac{L}{2} \\ \frac{L}{2} & 1 \end{bmatrix}$$

The remainder of the process of forming [**k**] is presented in Example 4.4. Thus, after reordering the rows and columns of that example, we have

$$\begin{Bmatrix} F_{y1} \\ M_{z1} \\ \hdashline F_{y2} \\ M_{z2} \end{Bmatrix} = \frac{EI_z}{L}\left[\begin{array}{cc:cc} \frac{12}{L^2} & \frac{6}{L} & -\frac{12}{L^2} & \frac{6}{L} \\ \frac{6}{L} & 4 & -\frac{6}{L} & 2 \\ \hdashline -\frac{12}{L^2} & -\frac{6}{L} & \frac{12}{L^2} & -\frac{6}{L} \\ \frac{6}{L} & 2 & -\frac{6}{L} & 4 \end{array}\right]\begin{Bmatrix} v_1 \\ \theta_{z1} \\ \hdashline v_2 \\ \theta_{z2} \end{Bmatrix} \tag{4.32}$$

EXAMPLE 4.3

Develop the flexibility matrix for a prismatic beam supported as a cantilever.

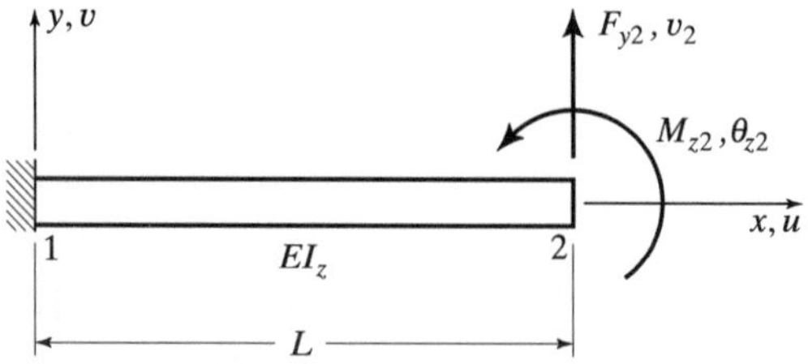

[3]In this case the choice of support conditions is immaterial; it's a matter of personal preference. But there are cases in which certain choices are preferable from the standpoint of integration and satisfaction of support conditions.

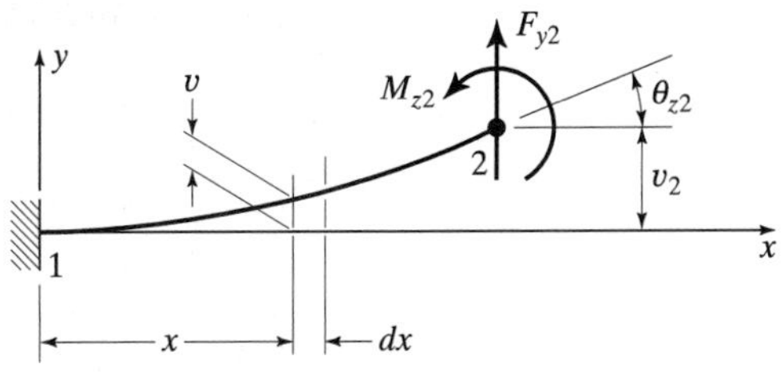

From the differential equation of the elastic curve, Equation 4.31b

$$\frac{d^2v}{dx^2} = \frac{M_z}{EI_z} = \frac{1}{EI_z}[F_{y2}(L - x) + M_{z2}]$$

Integrating,

$$\frac{dv}{dx} = \frac{1}{EI_z}\left[F_{y2}\left(Lx - \frac{x^2}{2}\right) + M_{z2}x\right] + C_1$$

and

$$v = \frac{1}{EI_z}\left[F_{y2}\left(\frac{Lx^2}{2} - \frac{x^3}{6}\right) + \frac{M_{z2}x^2}{2}\right] + C_1x + C_2$$

Boundary conditions:

$$\left.\frac{dv}{dx}\right|_{x=0} = 0 \quad \therefore \quad C_1 = 0 \qquad \text{and} \qquad v|_{x=0} = 0 \quad \therefore \quad C_2 = 0$$

Elastic curve equations:

$$v = \frac{F_{y2} \cdot x^2}{2EI_z}\left(L - \frac{x}{3}\right) + \frac{M_{z2}x^2}{2EI_z} \qquad \text{and}$$

$$\frac{dv}{dx} = \frac{F_{y2} \cdot x}{EI_z}\left(L - \frac{x}{2}\right) + \frac{M_{z2} \cdot x}{EI_z}$$

Determine flexibility influence coefficients:

$$v_2 = v|_{x=L} = \frac{F_{y2}L^3}{3EI_z} + \frac{M_{z2}L^2}{2EI_z} \qquad \text{and} \qquad \theta_{z2} = \left.\frac{dv}{dx}\right|_{x=L} = \frac{F_{y2}L^2}{2EI_z} + \frac{M_{z2}L}{EI_z}$$

Assemble equations in matrix form:

$$\begin{Bmatrix} v_2 \\ \theta_{z2} \end{Bmatrix} = \frac{L}{EI_z}\begin{bmatrix} \frac{L^2}{3} & \frac{L}{2} \\ \frac{L}{2} & 1 \end{bmatrix}\begin{Bmatrix} F_{y2} \\ M_{z2} \end{Bmatrix}$$

EXAMPLE 4.4

Using Equation 4.25 and the results of Example 4.3, develop the stiffness matrix for the prismatic beam shown.

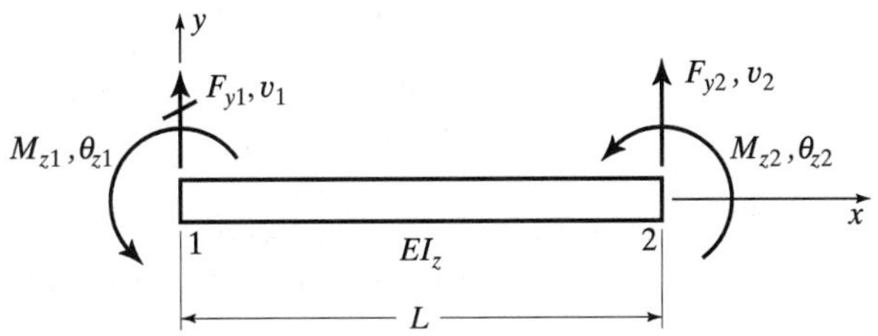

Write equations for $[\mathbf{\Phi}]$ and $[\mathbf{d}]^{-1}$:

$$\begin{Bmatrix} F_{y1} \\ M_{z1} \end{Bmatrix} = \begin{bmatrix} -1 & 0 \\ -L & -1 \end{bmatrix}\begin{Bmatrix} F_{y2} \\ M_{z2} \end{Bmatrix} \qquad \text{Thus} \qquad [\mathbf{\Phi}] = -\begin{bmatrix} 1 & 0 \\ L & 1 \end{bmatrix}$$

$$[\mathbf{d}]^{-1} = \frac{EI_z}{L}\begin{bmatrix} \frac{L^2}{3} & \frac{L}{2} \\ \frac{L}{2} & 1 \end{bmatrix}^{-1} = \frac{EI_z}{L}\begin{bmatrix} \frac{12}{L^2} & -\frac{6}{L} \\ -\frac{6}{L} & 4 \end{bmatrix}$$

Apply Equation 4.25

$$[\mathbf{k}] = \left[\begin{array}{c|c} [\mathbf{d}]^{-1} & [\mathbf{d}]^{-1}[\boldsymbol{\Phi}]^{\mathrm{T}} \\ \hline [\boldsymbol{\Phi}][\mathbf{d}]^{-1} & [\boldsymbol{\Phi}][\mathbf{d}]^{-1}[\boldsymbol{\Phi}]^{\mathrm{T}} \end{array}\right]$$

$$[\mathbf{d}]^{-1}[\boldsymbol{\Phi}]^{\mathrm{T}} = \frac{EI_z}{L}\begin{bmatrix} \frac{12}{L^2} & -\frac{6}{L} \\ -\frac{6}{L} & 4 \end{bmatrix}\begin{bmatrix} -1 & -L \\ 0 & -1 \end{bmatrix} = \frac{EI_z}{L}\begin{bmatrix} -\frac{12}{L^2} & -\frac{6}{L} \\ \frac{6}{L} & 2 \end{bmatrix}$$

$$[\boldsymbol{\Phi}][\mathbf{d}]^{-1}[\boldsymbol{\Phi}]^{\mathrm{T}} = \frac{EI_z}{L}\begin{bmatrix} -1 & 0 \\ -L & -1 \end{bmatrix}\begin{bmatrix} -\frac{12}{L^2} & -\frac{6}{L} \\ \frac{6}{L} & 2 \end{bmatrix} = \frac{EI_z}{L}\begin{bmatrix} \frac{12}{L^2} & \frac{6}{L} \\ \frac{6}{L} & 4 \end{bmatrix}$$

Collecting terms,

$$[\mathbf{k}] = \frac{EI_z}{L}\left[\begin{array}{cc|cc} \frac{12}{L^2} & -\frac{6}{L} & -\frac{12}{L^2} & -\frac{6}{L} \\ -\frac{6}{L} & 4 & \frac{6}{L} & 2 \\ \hline -\frac{12}{L^2} & \frac{6}{L} & \frac{12}{L^2} & \frac{6}{L} \\ -\frac{6}{L} & 2 & \frac{6}{L} & 4 \end{array}\right] \quad (\text{columns: } v_2,\ \theta_{z2},\ v_1,\ \theta_{z1})$$

4.5.4 Beam Bent About Its *y* Axis

The development of the stiffness matrix for the beam bent about its other principal axis is identical to the previous case and need not be repeated. There is, however, a complication in signs that is impossible to avoid if we are to employ uniform notation. The source of the problem may be seen by comparing Figures 4.10 and 4.9*b*. In each case the member is viewed from the positive end of the coordinate axis that is normal to the plane of the paper, and the forces and degrees of freedom are shown in their positive directions. All moments and rotations are positive counterclockwise on the ends of the members. In the case of the beam bent about its z axis, however, direct forces and translational degrees of freedom are positive upward, whereas in the other case they are positive downward. Therefore, the conversion of flexibility and stiffness matrices developed for the case in Figure 4.9 to the case in Figure 4.10 requires a change in the sign of all influence coefficients relating direct forces to rotations and,

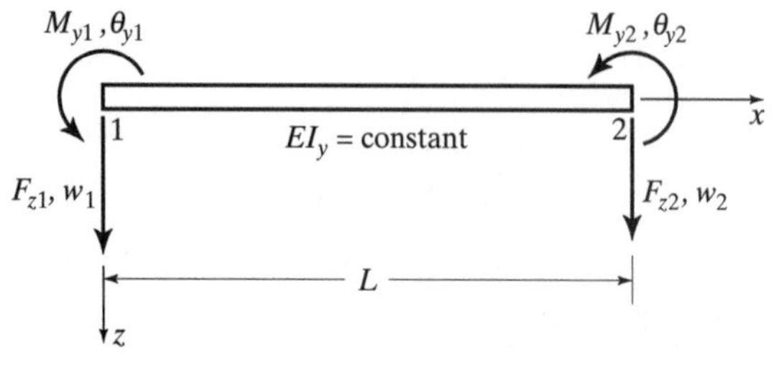

Figure 4.10 Beam bent about y axis.

for reciprocity, those relating moments to translational displacements. Thus we have, for the beam bent about its y axis,

$$\begin{Bmatrix} F_{z1} \\ M_{y1} \\ \hline F_{z2} \\ M_{y2} \end{Bmatrix} = \frac{EI_y}{L} \left[\begin{array}{cc|cc} \frac{12}{L^2} & -\frac{6}{L} & -\frac{12}{L^2} & -\frac{6}{L} \\ -\frac{6}{L} & 4 & \frac{6}{L} & 2 \\ \hline -\frac{12}{L^2} & \frac{6}{L} & \frac{12}{L^2} & \frac{6}{L} \\ -\frac{6}{L} & 2 & \frac{6}{L} & 4 \end{array}\right] \begin{Bmatrix} w_1 \\ \theta_{y1} \\ \hline w_2 \\ \theta_{y2} \end{Bmatrix} \tag{4.33}$$

It is left as an exercise for the reader to demonstrate the validity of the sign changes made.

4.5.5 Complete Element Stiffness Matrix

All that remains of this development is to assemble the foregoing results in the complete stiffness equation for the 12-degree-of-freedom, bisymmetrical member. Since for small displacements the axial force effects, torsion, and bending about each axis are uncoupled, the influence coefficients relating these effects are zero. Therefore, taking Equations 4.26b, 4.27b, 4.32, and 4.33, suitably reordering rows and columns, and letting $G = E/2(1 + \nu)$ (Equation 4.2), we arrive at the result shown as Equation 4.34 in Figure 4.11.

$$\begin{Bmatrix} F_{x1} \\ F_{y1} \\ F_{z1} \\ M_{x1} \\ M_{y1} \\ M_{z1} \\ \hline F_{x2} \\ F_{y2} \\ F_{z2} \\ M_{x2} \\ M_{y2} \\ M_{z2} \end{Bmatrix} = E \left[\begin{array}{cccccc|cccccc}
\frac{A}{L} & 0 & 0 & 0 & 0 & 0 & -\frac{A}{L} & 0 & 0 & 0 & 0 & 0 \\
0 & \frac{12I_z}{L^3} & 0 & 0 & 0 & \frac{6I_z}{L^2} & 0 & -\frac{12I_z}{L^3} & 0 & 0 & 0 & \frac{6I_z}{L^2} \\
0 & 0 & \frac{12I_y}{L^3} & 0 & -\frac{6I_y}{L^2} & 0 & 0 & 0 & -\frac{12I_y}{L^3} & 0 & -\frac{6I_y}{L^2} & 0 \\
0 & 0 & 0 & \frac{J}{2(1+\nu)L} & 0 & 0 & 0 & 0 & 0 & -\frac{J}{2(1+\nu)L} & 0 & 0 \\
0 & 0 & -\frac{6I_y}{L^2} & 0 & \frac{4I_y}{L} & 0 & 0 & 0 & \frac{6I_y}{L^2} & 0 & \frac{2I_y}{L} & 0 \\
0 & \frac{6I_z}{L^2} & 0 & 0 & 0 & \frac{4I_z}{L} & 0 & -\frac{6I_z}{L^2} & 0 & 0 & 0 & \frac{2I_z}{L} \\
\hline
-\frac{A}{L} & 0 & 0 & 0 & 0 & 0 & \frac{A}{L} & 0 & 0 & 0 & 0 & 0 \\
0 & -\frac{12I_z}{L^3} & 0 & 0 & 0 & -\frac{6I_z}{L^2} & 0 & \frac{12I_z}{L^3} & 0 & 0 & 0 & -\frac{6I_z}{L^2} \\
0 & 0 & -\frac{12I_y}{L^3} & 0 & \frac{6I_y}{L^2} & 0 & 0 & 0 & \frac{12I_y}{L^3} & 0 & \frac{6I_y}{L^2} & 0 \\
0 & 0 & 0 & -\frac{J}{2(1+\nu)L} & 0 & 0 & 0 & 0 & 0 & \frac{J}{2(1+\nu)L} & 0 & 0 \\
0 & 0 & -\frac{6I_y}{L^2} & 0 & \frac{2I_y}{L} & 0 & 0 & 0 & \frac{6I_y}{L^2} & 0 & \frac{4I_y}{L} & 0 \\
0 & \frac{6I_z}{L^2} & 0 & 0 & 0 & \frac{2I_z}{L} & 0 & -\frac{6I_z}{L^2} & 0 & 0 & 0 & \frac{4I_z}{L}
\end{array}\right] \begin{Bmatrix} u_1 \\ v_1 \\ w_1 \\ \theta_{x1} \\ \theta_{y1} \\ \theta_{z1} \\ \hline u_2 \\ v_2 \\ w_2 \\ \theta_{x2} \\ \theta_{y2} \\ \theta_{z2} \end{Bmatrix} \tag{4.34}$$

Figure 4.11 Bisymmetrical framework element stiffness matrix.

4.6 A COMMENTARY ON DEFORMATIONS AND DISPLACEMENT VARIABLES

The deformations neglected in the development of Equation 4.34 (Figure 4.11) and the particular variables used in the definition of displacements and their derivatives deserve further explanation at this point.

4.6.1 Neglected Deformations

As mentioned earlier, in developing Equation 4.34 deformations due to transverse shearing stresses and torsional warping were neglected. These effects are illustrated in the sketches of Figures 4.12 and 4.13.

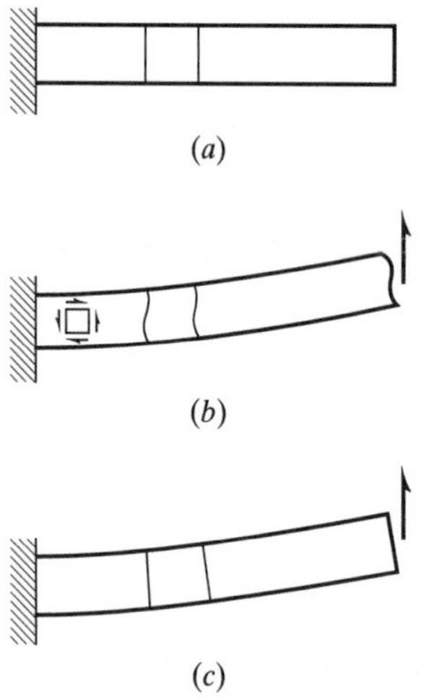

Figure 4.12 Warping caused by transverse shear.

When a beam is flexed, the transverse shear stresses in the web cause cross sections to warp in the longitudinal direction as in Figure 4.12*b*. We have neglected this, employing instead the conventional beam theory, which assumes that plane sections remain plane in flexure (Figure 4.12*c* and Example 4.3). Except in beams that have large depth-to-span ratios or local regions of high shear-to-moment ratios, the additional displacement due to web shear strain is of little consequence. Approximate procedures for including it are presented in Section 7.6.

Similarly, when all but circular cross-sectional shafts are twisted, cross sections warp longitudinally as shown in Figure 4.13*b*. In shafts of closed cross section, such as cylin-

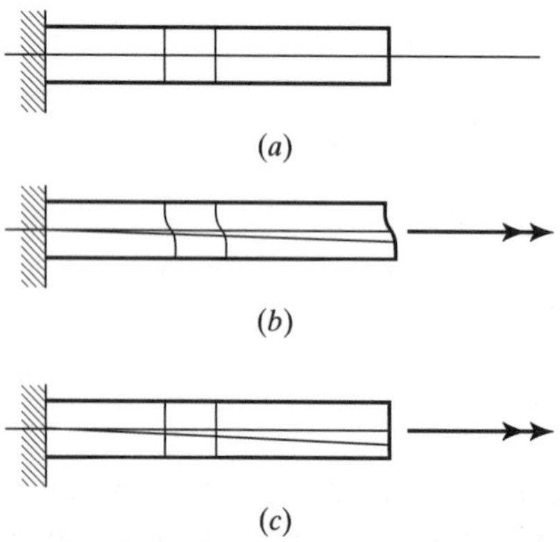

Figure 4.13 Warping caused by torsion.

drical or rectangular tubes, or rectangular reinforced concrete beams, the increased flexibility due to this type of deformation is generally small and may be neglected (Figure 4.13*c*). In members of open cross section, such as steel wide-flange or light-gage beams, resistance to out-of-plane warping, which may arise either as a result of fixing certain cross sections or varying applied torque along the member, may alter substantially the manner in which the member and the entire framework respond to torque. When torsional effects are small, as they are in many civil engineering frames, the assumption of a constant rate of twist used in the development of Equation 4.27 is adequate. When torque is significant, the warping resistance effect should be included. A procedure for doing this is presented in Section 7.4.

4.6.2 Displacement Variables

In Figure 4.6 and throughout Section 4.5 the rotational degrees of freedom were listed as angles, but in the development of the flexural stiffness relationships derivatives of lateral displacement were used. Only in the analysis of torsion was an angular measurement (the rate of twist) used. The connection between components of rotation and lateral displacement derivatives was explained in Section 2.2, in which it was noted that in conventional frame analysis it is generally assumed that a line drawn normal to the elastic line of a frame element remains normal to that line as the structure deforms under load. In that case the rotational components at a node such as the 2 end of the element of Figure 4.6 can be represented as in Equation 2.1, which is reproduced here for reference:

$$\theta_{x2} = \left.\frac{\partial w}{\partial y}\right|_2 \qquad \theta_{y2} = -\left.\frac{\partial w}{\partial x}\right|_2 \qquad \theta_{z2} = \left.\frac{\partial v}{\partial x}\right|_2 \tag{2.1}$$

Not only are lateral displacements and their derivatives at the core of elementary beam theory but, as shall be shown in Chapter 7, they are indispensable in the "shape function" approach to contemporary finite element analysis. For this reason, it is desirable to summarize the background and limitations of Equation 2.1.

During the general homogeneous deformation of a body, any element in it may be strained, translated, and rotated. If the strains are infinitesimally small, the unit elongations in three orthogonal directions and the engineering shearing strains related to the same directions are (see Sections 81–83 of Reference 4.1):

$$\begin{aligned} e_x &= \frac{\partial u}{\partial x} & e_y &= \frac{\partial v}{\partial y} & e_z &= \frac{\partial w}{\partial z} \\ \gamma_{xy} &= \left(\frac{\partial u}{\partial y} + \frac{\partial v}{\partial x}\right) & \gamma_{xz} &= \left(\frac{\partial u}{\partial z} + \frac{\partial w}{\partial x}\right) & \gamma_{yz} &= \left(\frac{\partial v}{\partial z} + \frac{\partial w}{\partial y}\right) \end{aligned} \tag{4.35a}$$

Also, the components of rotation about the principal axes of strain are

$$\theta_x = \frac{1}{2}\left(\frac{\partial w}{\partial y} - \frac{\partial v}{\partial z}\right) \qquad \theta_y = \frac{1}{2}\left(\frac{\partial u}{\partial z} - \frac{\partial w}{\partial x}\right) \qquad \theta_z = \frac{1}{2}\left(\frac{\partial v}{\partial x} - \frac{\partial u}{\partial y}\right) \tag{4.35b}$$

By definition, the shearing strains in the principal planes of strain are zero. In that case, $\partial u/\partial y = -(\partial v/\partial x)$, $\partial u/\partial z = -(\partial w/\partial x)$, and $\partial v/\partial z = -(\partial w/\partial y)$ from Equation 4.35a, and, from Equation 4.35b,

$$\theta_x = \frac{\partial w}{\partial y} = -\frac{\partial v}{\partial z} \qquad \theta_y = -\frac{\partial w}{\partial x} = \frac{\partial u}{\partial z} \qquad \theta_z = \frac{\partial v}{\partial x} = -\frac{\partial u}{\partial y} \tag{4.35c}$$

Thus it is seen that the representation of rotational displacement components by displacement derivatives as in Equation 2.1 or 2.1a is based on the analysis of infinitesimal strains and that only when they are infinitesimal is it rigorously correct. But in metallic and similarly behaving constructional materials the actual strains are small. For example, unit extensions are generally of the order of a few thousandths of an inch per inch. For this reason, conventional linear elastic analysis, which is based in part on the relationships of Equation 2.1, can give a good picture of the response to service loads of most structures. In nonlinear analysis we include finite strains that are the source of higher-order terms neglected in the infinitesimal strain theory. These will be identified and treated in Chapter 9 and Appendix A.

4.7 EXAMPLES

The following examples illustrate the applications of many of the concepts and results developed here. The parts, but not the whole, of Equation 4.34 are used. Use of the entire set of coefficients of Equation 4.34 occurs in space frame analysis. Discussion of this is postponed until the next chapter, where the coordinate transformations needed for the assembly of the global stiffness equations of complete space frames are developed. We should note that in these examples and throughout the book we will follow the convention of plotting bending moments on the tension side of the member.

Example 4.5 demonstrates that flexibility matrices may be generated readily by techniques of classical analysis, such as the conjugate beam method.

Example 4.6 verifies a statement made in Section 4.4 regarding the equality of complementary energy under different statically determinate support systems.

In Example 4.7 it is shown that the beam stiffness matrix may be obtained directly from the conventional slope deflection equations.

Example 4.8 is a preparatory example similar to Example 3.4. The matrices developed here will be used in a number of the examples that follow in this and later chapters.

In Example 4.9 the results of the previous example are applied to a simple case. To illustrate some of the possibilities of matrix solutions, this five-degree-of-freedom system is solved by partitioning. Solution techniques will be discussed in Chapter 11.

Example 4.10 is a further application of the same basic stiffness equations, and in Example 4.11, the same system is investigated for the effect of an assumed support settlement. In Example 4.12 a bracket is added to the two-element beam, and a simple torsional effect is investigated.

Example 4.13 is a preview of frame analysis problems. Here the emphasis is on the physical interpretation of the stiffness influence coefficients. Note that although in Example 4.13 the nodes are lettered a, b, and c as in Example 4.8 and the member sections correspond, the orientation of member bc with respect to the coordinate system is different than it is in Example 4.8. This ad hoc arrangement should cause no difficulty in interpretation of the influence coefficients. But after transformation procedures are introduced in the next chapter, problems of this sort will be solved in a more automatic way.

In Example 4.14 some of the versatility of matrix procedures in combining special elements is introduced. Later chapters contain additional examples of this powerful aspect of the matrix formulation.

Example 4.15 is an application of matrix analysis to the classical problem of a beam on a continuous elastic foundation in which it is assumed that the stiffness characteristics of the supporting medium are known. As they should be, the results obtained are in almost precise agreement with those of the classical solution. Problems of this type first became apparent in attempts to analyze the forces in railway rails. The same analytical problems are encountered in systems such as cylindrical shells and structural grillages (References 4.3 and 4.4), so the matrix approach can be used to advantage in addressing those as well.

EXAMPLE 4.5

Develop the flexibility matrix for a prismatic beam on simple supports.

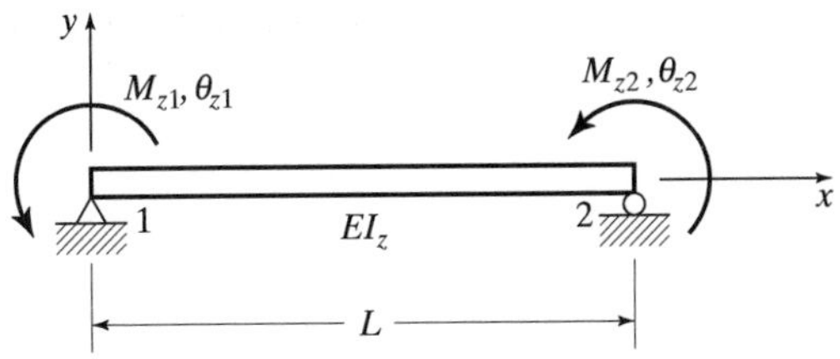

Determine flexibility influence coefficients for M_{z1} and M_{z2} using the conjugate beams and elastic weights shown:

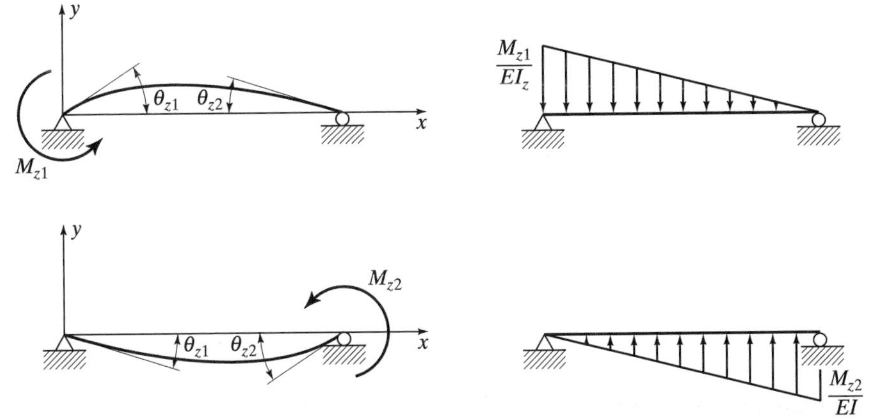

$$\theta_{z1} = \frac{M_{z1}L}{3EI_z} \qquad \theta_{z2} = -\frac{M_{z1}L}{6EI_z}$$

$$\theta_{z1} = -\frac{M_{z2}L}{6EI_z} \qquad \theta_{z2} = \frac{M_{z2}L}{3EI_z}$$

Assemble equations in matrix form:

$$\begin{Bmatrix} \theta_{z1} \\ \theta_{z2} \end{Bmatrix} = \frac{L}{6EI_z}\begin{bmatrix} 2 & -1 \\ -1 & 2 \end{bmatrix}\begin{Bmatrix} M_{z1} \\ M_{z2} \end{Bmatrix}$$

EXAMPLE 4.6

Show that the alternative forms of the beam flexibility matrix in Examples 4.3 and 4.5 yield the same complementary energy.

From Example 4.3, using Equation 4.4b,

$$U_1^* = \tfrac{1}{2}\lfloor \mathbf{F}_f \rfloor [\mathbf{d}]\{\mathbf{F}_f\} = \tfrac{1}{2}\lfloor F_{y2} \quad M_{z2} \rfloor \left(\frac{L}{EI_z}\right)\begin{bmatrix} \frac{L^2}{3} & \frac{L}{2} \\ \frac{L}{2} & 1 \end{bmatrix}\begin{Bmatrix} F_{y2} \\ M_{z2} \end{Bmatrix}$$

$$U_1^* = \frac{L}{2EI_z}\left(M_{z2}^2 + L \cdot M_{z2} \cdot F_{y2} + \frac{L^2 F_{y2}^2}{3}\right) \qquad \text{(a)}$$

From Example 4.5,

$$U_2^* = \tfrac{1}{2}\lfloor \mathbf{F}_f \rfloor [\mathbf{d}]\{\mathbf{F}_f\} = \tfrac{1}{2}\lfloor M_{z1} \quad M_{z2} \rfloor \left(\frac{L}{6EI_z}\right)\begin{bmatrix} 2 & -1 \\ -1 & 2 \end{bmatrix}\begin{Bmatrix} M_{z1} \\ M_{z2} \end{Bmatrix}$$

$$U_2^* = \frac{L}{6EI_z}\left(M_{z1}^2 - M_{z1}M_{z2} + M_{z2}^2\right) \qquad \text{(b)}$$

But, by equilibrium

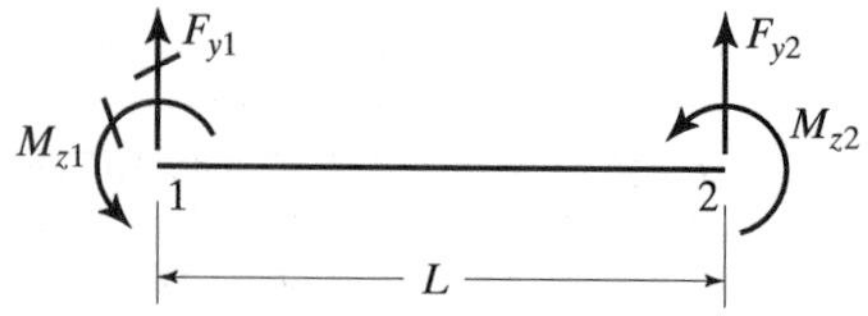

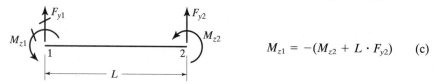

$$M_{z1} = -(M_{z2} + L \cdot F_{y2}) \qquad \text{(c)}$$

Substituting Equation c in Equation b,

$$U_2^* = \frac{L}{2EI_z}\left(M_{z2}^2 + L \cdot M_{z2} \cdot F_{y2} + \frac{L^2 F_{y2}^2}{3}\right) = U_1^* \quad \text{QED}$$

EXAMPLE 4.7

Using the "slope-deflection" equation, write the stiffness matrix for a prismatic beam element.

The straight element 1-2 is displaced by the forces shown. There are no loads between nodal points. The "slope-deflection" equations for the member are as follows (see Ref. 4.2)

$$M_{z1} = \frac{EI_z}{L}\left(4\theta_{z1} + 2\theta_{z2} - \frac{6\delta}{L}\right) \quad \text{and} \quad M_{z2} = \frac{EI_z}{L}\left(2\theta_{z1} + 4\theta_{z2} - \frac{6\delta}{L}\right)$$

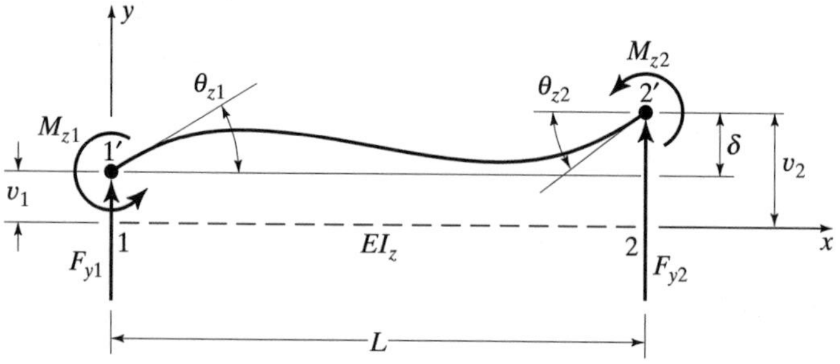

Letting $\delta = v_2 - v_1$,

$$M_{z1} = \frac{EI_z}{L}\left(4\theta_{z1} + 2\theta_{z2} + \frac{6v_1}{L} - \frac{6v_2}{L}\right) \quad \text{and} \quad M_{z2} = \frac{EI_z}{L}\left(2\theta_{z1} + 4\theta_{z2} + \frac{6v_1}{L} - \frac{6v_2}{L}\right)$$

By equilibrium,

$$F_{y1} = \frac{M_{z1} + M_{z2}}{L} = \frac{EI_z}{L}\left(\frac{6\theta_{z1}}{L} + \frac{6\theta_{z2}}{L} + \frac{12v_1}{L^2} - \frac{12v_2}{L^2}\right)$$

and

$$F_{y2} = -\left(\frac{M_{z1} + M_{z2}}{L}\right) = \frac{EI_z}{L}\left(-\frac{6\theta_{z1}}{L} - \frac{6\theta_{z2}}{L} - \frac{12v_1}{L^2} + \frac{12v_2}{L^2}\right)$$

Assemble equations in matrix form:

$$\begin{Bmatrix} F_{y1} \\ M_{z1} \\ F_{y2} \\ M_{z2} \end{Bmatrix} = \frac{EI_z}{L}\begin{bmatrix} \frac{12}{L^2} & \frac{6}{L} & -\frac{12}{L^2} & \frac{6}{L} \\ & 4 & -\frac{6}{L} & 2 \\ & & \frac{12}{L^2} & -\frac{6}{L} \\ \text{Sym.} & & & 4 \end{bmatrix}\begin{Bmatrix} v_1 \\ \theta_{z1} \\ v_2 \\ \theta_{z2} \end{Bmatrix}$$

which is identical to Equation 4.32.

EXAMPLE 4.8

For the system shown:

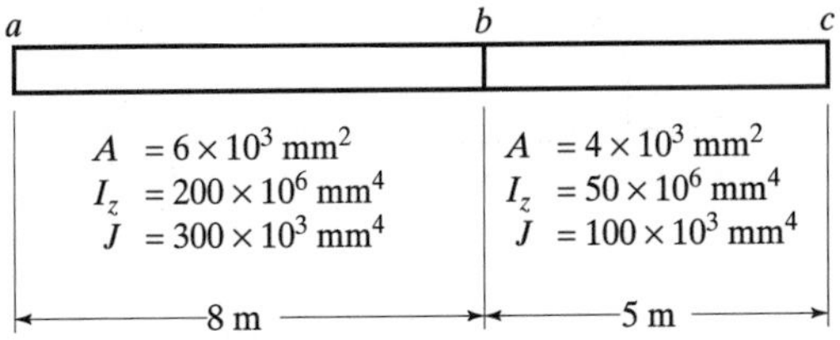

1. Write the element stiffness matrices—assume no bending normal to the plane of the paper.
2. Assemble the global stiffness equations.
   $E = 200{,}000$ MPa and $\nu = 0.3$.

Define the coordinates, degrees of freedom, and external forces as follows:

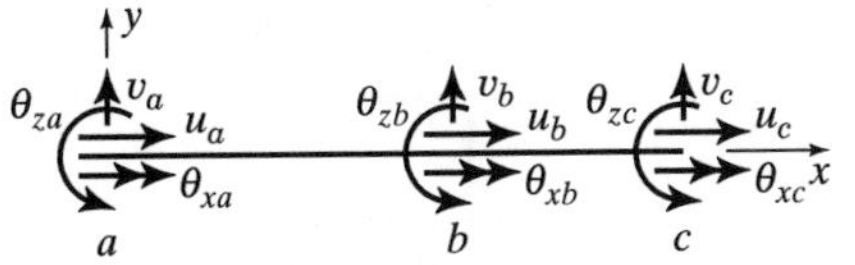

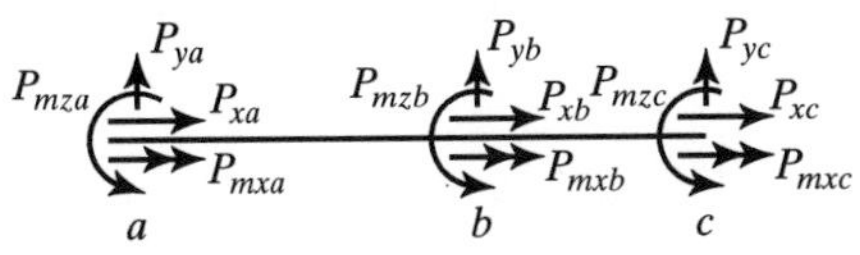

1. ***Element stiffness equations.*** Use Equation 4.34, omitting out-of-plane shear and bending degrees of freedom, that is, w_1, θ_{y1}, w_2, *and* θ_{y2}.

Member *ab*

$$
\begin{Bmatrix} F_{xa}^{ab} \\ F_{ya}^{ab} \\ M_{xa}^{ab} \\ M_{za}^{ab} \\ F_{xb}^{ab} \\ F_{yb}^{ab} \\ M_{xb}^{ab} \\ M_{zb}^{ab} \end{Bmatrix} = 200 \left[\begin{array}{cccc|cccc}
0.750 & 0 & 0 & 0 & -0.750 & 0 & 0 & 0 \\
 & 0.00469 & 0 & 18.750 & 0 & -0.00469 & 0 & 18.750 \\
 & & 14.423 & 0 & 0 & 0 & -14.423 & 0 \\
 & & & 1.0 \times 10^5 & 0 & -18.750 & 0 & 0.5 \times 10^5 \\
\hline
 & & & & 0.750 & 0 & 0 & 0 \\
 & \text{Sym} & & & & 0.00469 & 0 & -18.750 \\
 & & & & & & 14.423 & 0 \\
 & & & & & & & 1.0 \times 10^5
\end{array}\right] \begin{Bmatrix} u_a \\ v_a \\ \theta_{xa} \\ \theta_{za} \\ u_b \\ v_b \\ \theta_{xb} \\ \theta_{zb} \end{Bmatrix}
$$

Member *bc*

$$
\begin{Bmatrix} F_{xb}^{bc} \\ F_{yb}^{bc} \\ M_{xb}^{bc} \\ M_{zb}^{bc} \\ F_{xc}^{bc} \\ F_{yc}^{bc} \\ M_{xc}^{bc} \\ M_{zc}^{bc} \end{Bmatrix} = 200 \left[\begin{array}{cccc|cccc}
0.800 & 0 & 0 & 0 & -0.800 & 0 & 0 & 0 \\
 & 0.00480 & 0 & 12.000 & 0 & -0.00480 & 0 & 12.000 \\
 & & 7.692 & 0 & 0 & 0 & -7.692 & 0 \\
 & & & 0.4 \times 10^5 & 0 & -12.000 & 0 & 0.2 \times 10^5 \\
\hline
 & & & & 0.800 & 0 & 0 & 0 \\
 & \text{Sym} & & & & 0.00480 & 0 & -12.000 \\
 & & & & & & 7.692 & 0 \\
 & & & & & & & 0.4 \times 10^5
\end{array}\right] \begin{Bmatrix} u_b \\ v_b \\ \theta_{xb} \\ \theta_{zb} \\ u_c \\ v_c \\ \theta_{xc} \\ \theta_{zc} \end{Bmatrix}
$$

2. ***Global stiffness equations***

$$
\begin{Bmatrix} P_{xa} \\ P_{ya} \\ P_{mxa} \\ P_{mza} \\ P_{xb} \\ P_{yb} \\ P_{mxb} \\ P_{mzb} \\ P_{xc} \\ P_{yc} \\ P_{mxc} \\ P_{mzc} \end{Bmatrix} = 200 \left[\begin{array}{cccc|cccc|cccc}
0.750 & 0 & 0 & 0 & -0.750 & 0 & 0 & 0 & 0 & 0 & 0 & 0 \\
 & 0.00469 & 0 & 18.750 & 0 & -0.00469 & 0 & 18.750 & 0 & 0 & 0 & 0 \\
 & & 14.423 & 0 & 0 & 0 & -14.423 & 0 & 0 & 0 & 0 & 0 \\
 & & & 1.0 \times 10^5 & 0 & -18.750 & 0 & 0.5 \times 10^5 & 0 & 0 & 0 & 0 \\
\hline
 & & & & 1.550 & 0 & 0 & 0 & -0.800 & 0 & 0 & 0 \\
 & & & & & 0.00949 & 0 & -6.750 & 0 & -0.00480 & 0 & 12.000 \\
 & & & & & & 22.115 & 0 & 0 & 0 & -7.692 & 0 \\
 & & & & & & & 1.4 \times 10^5 & 0 & -12.000 & 0 & 0.2 \times 10^5 \\
\hline
 & & & & & & & & 0.800 & 0 & 0 & 0 \\
 & \text{Sym} & & & & & & & & 0.00480 & 0 & -12.000 \\
 & & & & & & & & & & 7.692 & 0 \\
 & & & & & & & & & & & 0.4 \times 10^5
\end{array}\right] \begin{Bmatrix} u_a \\ v_a \\ \theta_{xa} \\ \theta_{za} \\ u_b \\ v_b \\ \theta_{xb} \\ \theta_{zb} \\ u_c \\ v_c \\ \theta_{xc} \\ \theta_{zc} \end{Bmatrix}
$$

EXAMPLE 4.9

The beam of Example 4.8 is supported and loaded as shown.

1. Calculate the displacements at b and c.
2. Calculate the reactions.

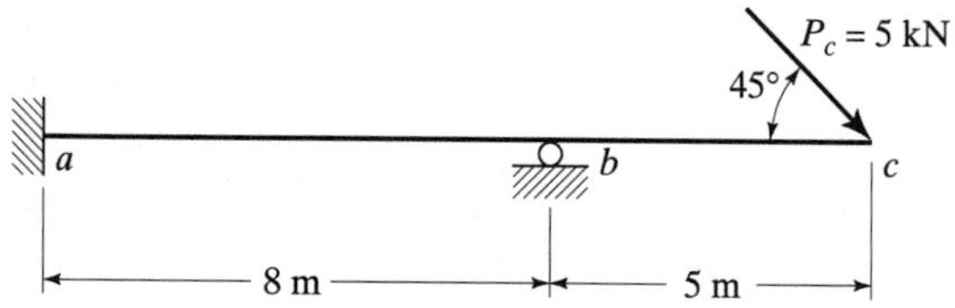

Boundary conditions, $u_a = v_a = \theta_{xa} = \theta_{za} = v_b = 0$
Also, $\theta_{xb} = \theta_{xc} = 0$ because of absence of torque.

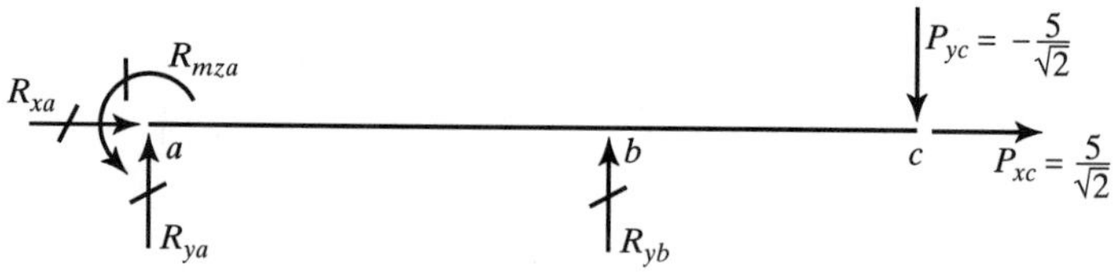

1. ***Displacements.*** Remove the u_a, v_a, θ_{xa}, θ_{za}, v_b, θ_{xb}, θ_{xc} rows and columns from the stiffness equations leaving

$$\begin{Bmatrix} P_{xb} \\ P_{mzb} \\ P_{xc} \\ P_{yc} \\ P_{mzc} \end{Bmatrix} = \begin{Bmatrix} 0 \\ 0 \\ 5/\sqrt{2} \\ -5/\sqrt{2} \\ 0 \end{Bmatrix} = 200 \begin{bmatrix} 1.500 & 0 & -0.800 & 0 & 0 \\ & 1.4 \times 10^5 & 0 & -12.000 & 0.2 \times 10^5 \\ & & 0.800 & 0 & 0 \\ & \text{Sym.} & & 0.00480 & -12.000 \\ & & & & 0.4 \times 10^5 \end{bmatrix} \begin{Bmatrix} u_b \\ \theta_{zb} \\ u_c \\ v_c \\ \theta_{zc} \end{Bmatrix}$$

As an alternative to solving on a computer as in Example 3.5, solve by operating on partitioned matrices after reordering; thus

$$\begin{Bmatrix} P_{xb} \\ P_{xc} \\ \hline P_{yc} \\ \hline P_{mzb} \\ P_{mzc} \end{Bmatrix} = \begin{Bmatrix} 0 \\ 5/\sqrt{2} \\ \hline -5/\sqrt{2} \\ \hline 0 \\ 0 \end{Bmatrix} = 200 \left[\begin{array}{cc|c|cc} 1.550 & -0.800 & 0 & 0 & 0 \\ -0.800 & 0.800 & 0 & 0 & 0 \\ \hline 0 & 0 & 0.00480 & -12.000 & -12.000 \\ \hline 0 & 0 & -12.000 & 1.4 \times 10^5 & 0.2 \times 10^5 \\ 0 & 0 & -12.000 & 0.2 \times 10^5 & 0.4 \times 10^5 \end{array}\right] \begin{Bmatrix} u_b \\ u_c \\ \hline v_c \\ \hline \theta_{zb} \\ \theta_{zc} \end{Bmatrix}$$

Expanding the top partition and solving for $\lfloor u_b \quad u_c \rfloor^{\mathrm{T}}$ yields

$$\begin{Bmatrix} u_b \\ u_c \end{Bmatrix} = 8.33 \times 10^{-3} \begin{bmatrix} 0.800 & 0.800 \\ 0.800 & 1.550 \end{bmatrix} \begin{Bmatrix} 0 \\ 5/\sqrt{2} \end{Bmatrix} = \begin{Bmatrix} 0.024 \\ 0.046 \end{Bmatrix} \text{ mm}$$

Expanding the bottom partition and solving for $\lfloor \theta_{zb} \quad \theta_{zc} \rfloor^{\mathrm{T}}$, in terms of v_c

$$\begin{Bmatrix} \theta_{zb} \\ \theta_{zc} \end{Bmatrix} = 1.923 \times 10^{-5} \begin{bmatrix} 0.4 & -0.2 \\ -0.2 & 1.4 \end{bmatrix} \begin{Bmatrix} 12.000 \\ 12.000 \end{Bmatrix} v_c = 23.077 \times 10^{-5} \begin{Bmatrix} 0.2 \\ 1.2 \end{Bmatrix} v_c$$

Expanding the middle partition and using the above expression for $\lfloor \theta_{zb} \quad \theta_{zc} \rfloor^{\mathrm{T}}$,

$$\frac{-5}{\sqrt{2}} = 200 \left[0.00480 v_c - 23.077 \times 10^{-5} \lfloor 12.000 \quad 12.000 \rfloor \begin{Bmatrix} 0.2 \\ 1.2 \end{Bmatrix} v_c \right] \quad \text{or } v_c = -19.15 \text{ mm}$$

Substituting v_c in the expression for $\lfloor \theta_{zb} \quad \theta_{zc} \rfloor^{\mathrm{T}}$ yields

$$\begin{Bmatrix} \theta_{zb} \\ \theta_{zc} \end{Bmatrix} = \begin{Bmatrix} -0.00088 \\ -0.00530 \end{Bmatrix} \text{ rad}$$

2. ***Reactions.*** The remaining rows of the stiffness equations are used as follows:

$$\begin{Bmatrix} R_{xa} \\ R_{ya} \\ R_{mza} \\ R_{yb} \end{Bmatrix} = 200 \begin{bmatrix} -0.750 & 0 & 0 & 0 & 0 \\ 0 & 0 & 0 & 18.750 & 0 \\ 0 & 0 & 0 & 0.5 \times 10^5 & 0 \\ 0 & 0 & -0.00480 & -6.750 & 12.000 \end{bmatrix} \begin{Bmatrix} 0.024 \\ 0.046 \\ -19.15 \\ -0.00088 \\ -0.00530 \end{Bmatrix}$$

(columns: u_b, u_c, v_c, θ_{zb}, θ_{zc})

$$= \begin{Bmatrix} -3.60 \text{ kN} \\ -3.30 \text{ kN} \\ -8.80 \text{ kNm} \\ 6.85 \text{ kN} \end{Bmatrix}$$

Moment diagram

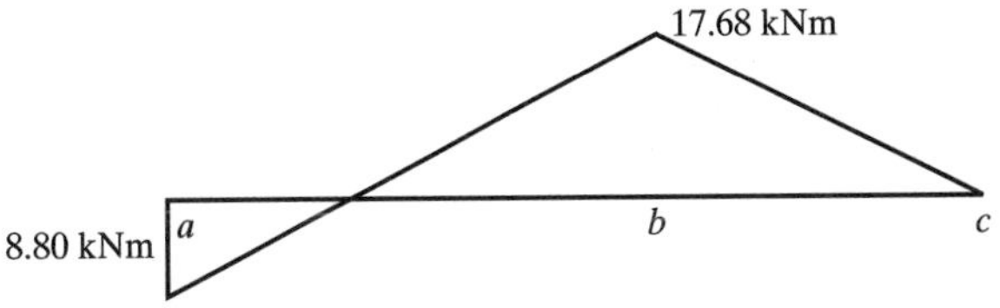

Deflected structure

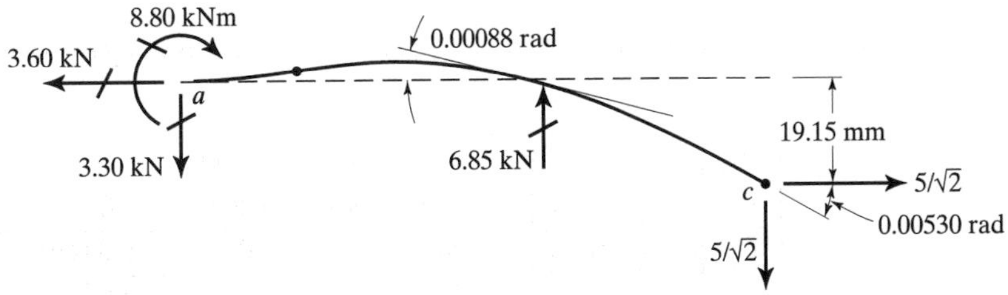

EXAMPLE 4.10

The beam of Example 4.8 is supported and loaded as shown.

1. Calculate the rotation at b.
2. Calculate the rotation and bending moments.

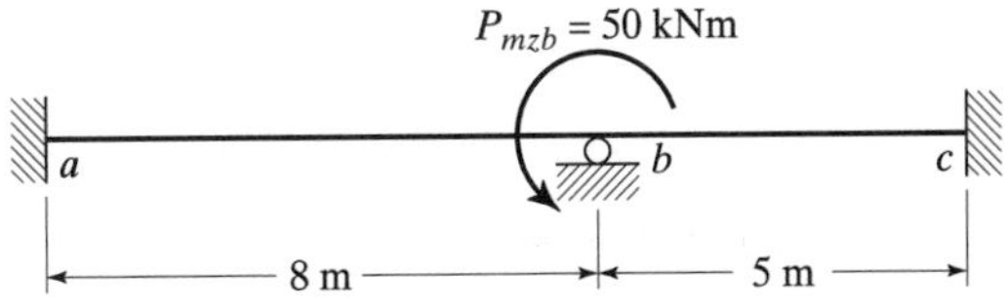

Boundary conditions: $u_a = v_a = \theta_{xa} = \theta_{za} = v_b = u_c = \theta_{xc} = \theta_{zc} = 0$

Also, $u_b = \theta_{xb} = 0$ because of absence of axial force and torque.

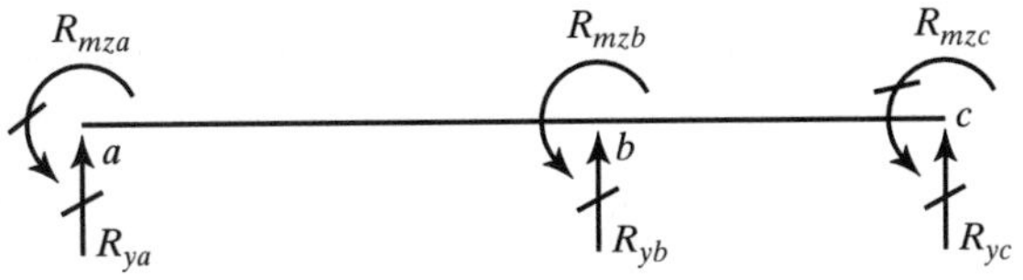

1. ***Displacements.*** Remove all except the θ_{zb} row and column from the stiffness equations, leaving

$$\{P_{mzb}\} = \{50 \times 10^3\} = 200[1.4 \times 10^5]\theta_{zb} \quad \text{or} \quad \theta_{zb} = 0.001786 \text{ rad}$$

2. *Reactions.* The remaining rows of the stiffness equations yield

$$\begin{Bmatrix} R_{ya} \\ R_{mza} \\ R_{yb} \\ R_{yc} \\ R_{mzc} \end{Bmatrix} = 200 \begin{Bmatrix} 18.750 \\ 0.5 \times 10^5 \\ -6.750 \\ -12.000 \\ 0.2 \times 10^5 \end{Bmatrix} \theta_{zb} = \begin{Bmatrix} 6.70 \text{ kN} \\ 17.86 \text{ kNm} \\ -2.41 \text{ kN} \\ -4.29 \text{ kN} \\ 7.14 \text{ kNm} \end{Bmatrix}$$

Obtain M_{zb}^{ab} and M_{zc}^{bc} from the member stiffness equations, Example 4.8:

$$M_{zb}^{ab} = 200[1 \times 10^5]\theta_{zb} = 35.72 \text{ kNm} \qquad M_{zc}^{bc} = 200[0.4 \times 10^5]\theta_{zb} = 14.28 \text{ kNm}$$

Moment diagram

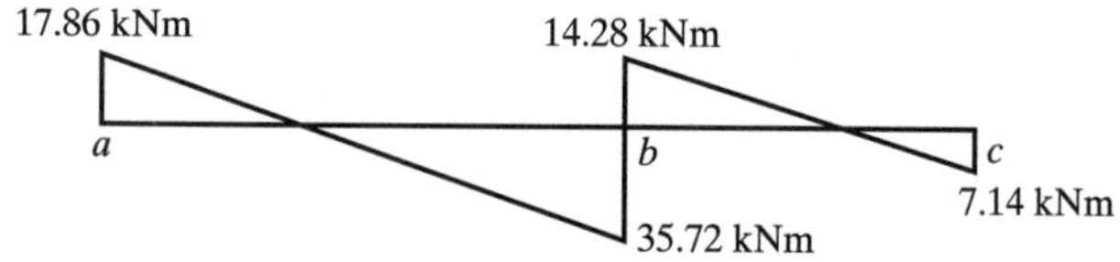

Deflected structure

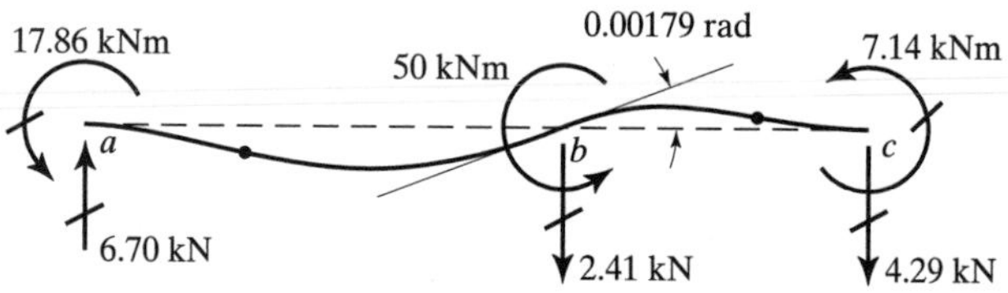

EXAMPLE 4.11

The beam of Example 4.8 is supported as shown. The support at b settles 20 mm, carrying the beam with it.

1. Calculate the rotation at b.
2. Calculate the reactions and bending moments.

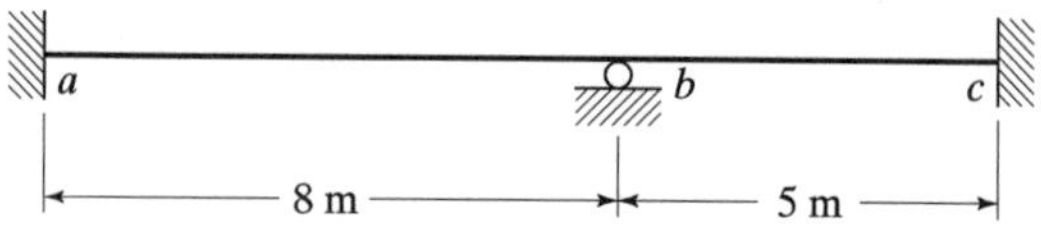

Boundary conditions: $u_a = v_a = \theta_{xa} = \theta_{za} = u_c = v_c = \theta_{xc} = \theta_{zc} = 0$.
Also $u_b = \theta_{xb} = 0$ because of absence of axial force and torque.

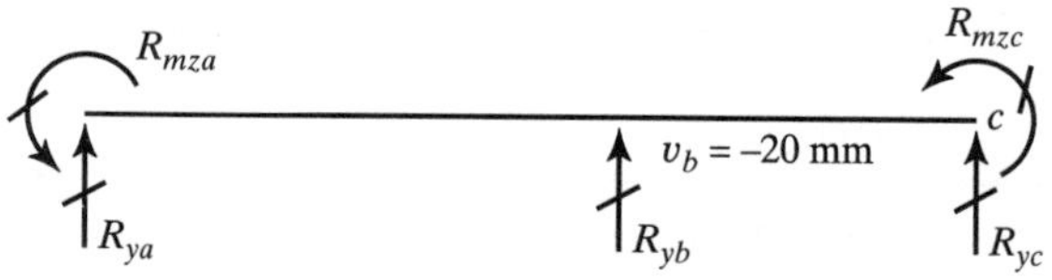

1. *Displacements.* Remove all except the v_b and θ_{zb} rows and columns from the stiffness equations leaving

$$\begin{Bmatrix} R_{yb} \\ P_{mzb} \end{Bmatrix} = \begin{Bmatrix} R_{yb} \\ 0 \end{Bmatrix} = 200 \begin{bmatrix} 0.00949 & -6.750 \\ -6.750 & 1.4 \times 10^5 \end{bmatrix} \begin{Bmatrix} -20 \\ \theta_{zb} \end{Bmatrix}$$

Solving the lower equation for θ_{zb},

$$\theta_{zb} = -0.0009643 \text{ rad}$$

2. *Reactions.* The remaining rows of the stiffness equations yield

$$\begin{Bmatrix} R_{ya} \\ R_{mza} \\ R_{yb} \\ R_{yc} \\ R_{mzc} \end{Bmatrix} = 200 \begin{bmatrix} \overset{v_b}{-0.00469} & \overset{\theta_{zb}}{18.750} \\ -18.750 & 0.5 \times 10^5 \\ 0.00949 & -6.750 \\ -0.00480 & -12.000 \\ 12.000 & 0.2 \times 10^5 \end{bmatrix} \begin{Bmatrix} -20.0 \\ -0.0009643 \end{Bmatrix} = \begin{Bmatrix} 15.14 \text{ kN} \\ 65.36 \text{ kNm} \\ -36.66 \text{ kN} \\ 21.51 \text{ kN} \\ -51.86 \text{ kNm} \end{Bmatrix}$$

Obtain M_{zb}^{ab} and M_{zb}^{bc} from the member stiffness equations, Example 4.8:

$$M_{zb}^{ab} = 200[-18.750v_b + 1.0 \times 10^5\theta_{zb}] = 55.71 \text{ kNm}$$
$$M_{zb}^{bc} = 200[12.00v_b + 0.4 \times 10^5\theta_{zb}] = -55.71 \text{ kNm}$$

Moment diagram

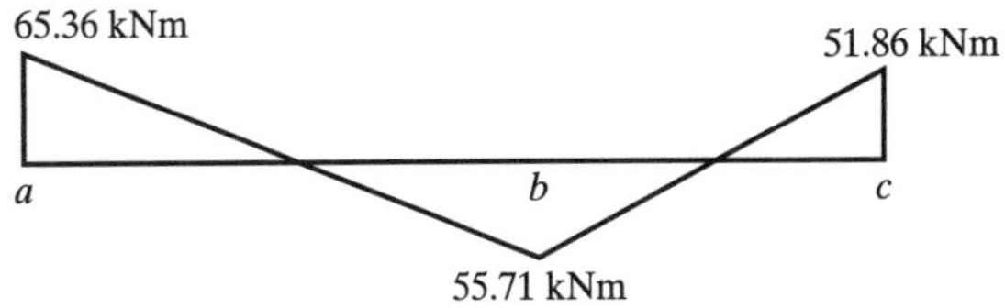

Deflected structure

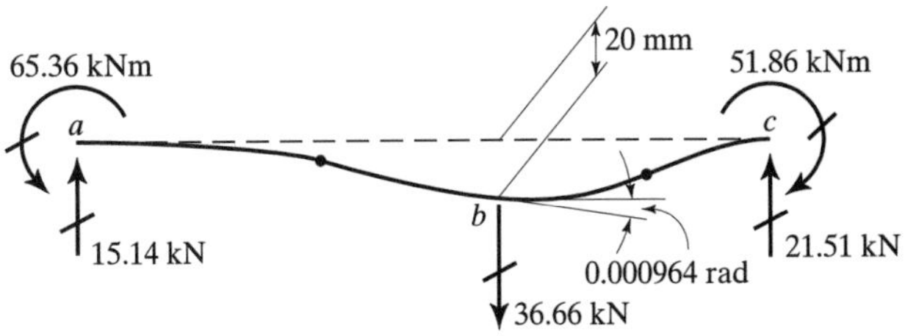

EXAMPLE 4.12

The beam of Example 4.8 is supported as shown and loaded by a 1 kN force applied to a rigid bracket projecting from the beam at b.

1. Calculate the displacement at b.
2. Calculate the reactions, torques, and bending moments.

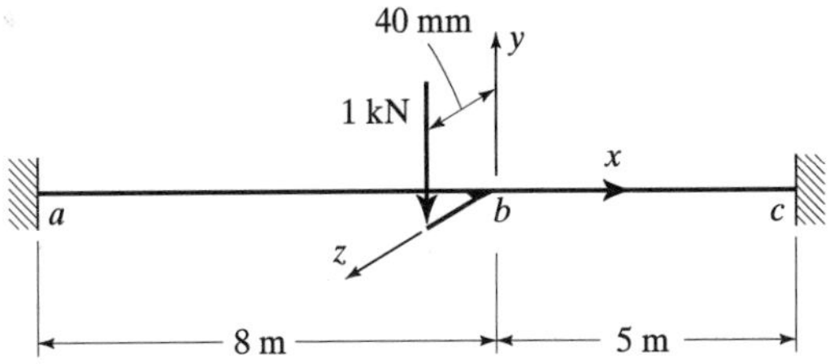

Boundary conditions: $u_a = v_a = \theta_{xa} = \theta_{za} = u_c = v_c = \theta_{xc} = \theta_{zc} = 0$.
Also, $u_b = 0$ because of absence of axial force.

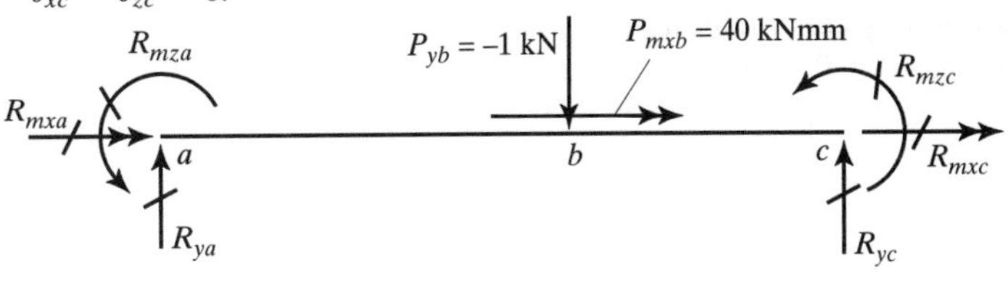

1. ***Displacements.*** Remove all except the v_b, θ_{xb}, and θ_{zb} rows and columns from the stiffness equations leaving

$$\begin{Bmatrix} P_{yb} \\ P_{mxb} \\ P_{mzb} \end{Bmatrix} = \begin{Bmatrix} -1 \\ 40 \\ 0 \end{Bmatrix} = 200 \begin{bmatrix} 0.00949 & 0 & -6.750 \\ & 22.115 & 0 \\ \text{Sym.} & & 1.4 \times 10^5 \end{bmatrix} \begin{Bmatrix} v_b \\ \theta_{xb} \\ \theta_{zb} \end{Bmatrix}$$

Solving the second equation for θ_{xb},

$$\theta_{xb} = 0.009044 \text{ rad}$$

Solving the first and third equations for $\lfloor v_b \quad \theta_{zb} \rfloor^T$,

$$\begin{Bmatrix} v_b \\ \theta_{zb} \end{Bmatrix} = 3.897 \times 10^{-6} \begin{bmatrix} 1.4 \times 10^5 & 6.750 \\ 6.750 & 0.00949 \end{bmatrix} \begin{Bmatrix} -1 \\ 0 \end{Bmatrix} = \begin{Bmatrix} -0.545 \text{ mm} \\ -0.0000263 \text{ rad} \end{Bmatrix}$$

2. *Reactions.* The remaining rows of the stiffness matrix yield

$$
\begin{Bmatrix} R_{ya} \\ R_{mxa} \\ R_{mza} \\ R_{yc} \\ R_{mxc} \\ R_{mzc} \end{Bmatrix} = 200
\begin{bmatrix}
\overset{v_b}{-0.00469} & \overset{\theta_{xb}}{0} & \overset{\theta_{zb}}{18.750} \\
0 & -14.423 & 0 \\
-18.750 & 0 & 0.5 \times 10^5 \\
-0.00480 & 0 & -12.000 \\
0 & -7.692 & 0 \\
12.000 & 0 & 0.2 \times 10^5
\end{bmatrix}
\begin{Bmatrix} -0.545 \\ 0.009044 \\ -0.0000263 \end{Bmatrix} =
\begin{Bmatrix} 0.423 \text{ kN} \\ -26.1 \text{ kNmm} \\ 1.781 \text{ kNm} \\ 0.586 \text{ kN} \\ -13.9 \text{ kNmm} \\ -1.413 \text{ kNm} \end{Bmatrix}
$$

Torque diagram

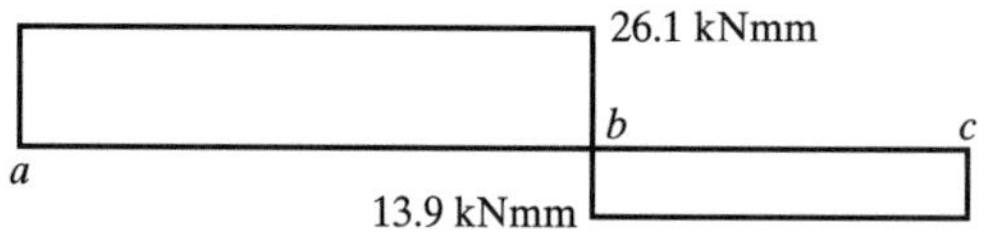

Moment diagram

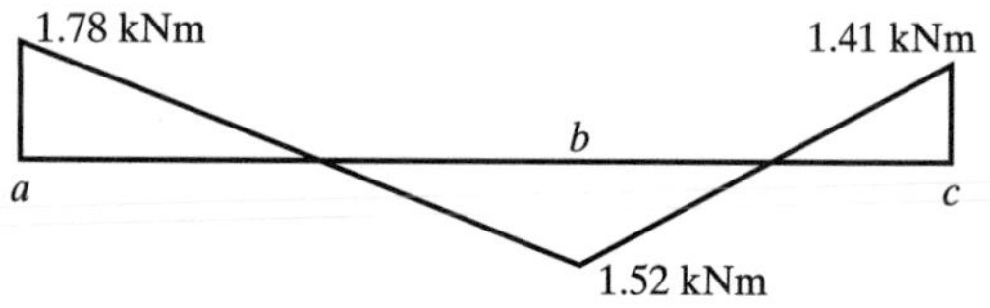

Deflected structure

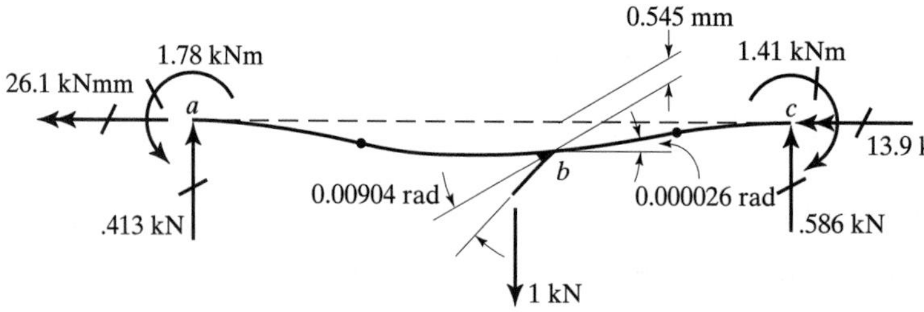

EXAMPLE 4.13

The rigid frame shown is made of the elements of Example 4.8.

1. Using the influence coefficients of Example 4.8 assemble the stiffness equations relating the forces applied at b to the degrees of freedom at that joint.
2. Calculate the displacement at b.

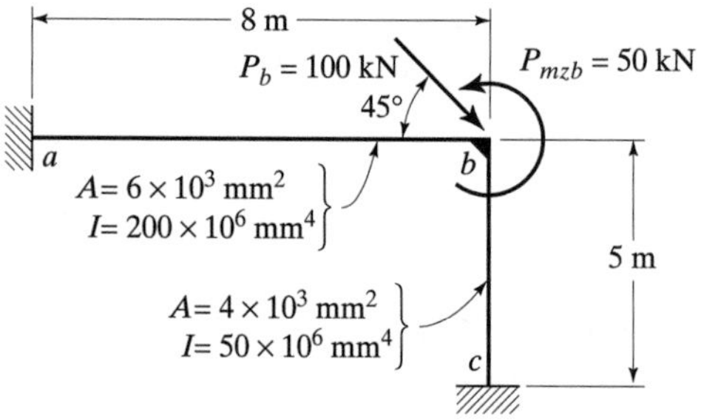

Boundary conditions:

$$u_a = v_a = \theta_{za} = u_c = v_c = \theta_{zc} = 0$$

Nonzero degrees of freedom:

$$u_b,\ v_b \text{ and } \theta_{zb}$$

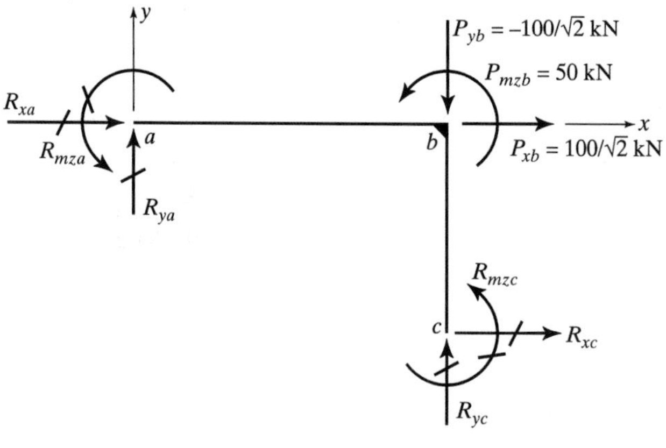

1. ***Stiffness equations.*** Select coefficients from Example 4.8 on the basis of physical behavior. Apply a horizontal displacement, u_b:

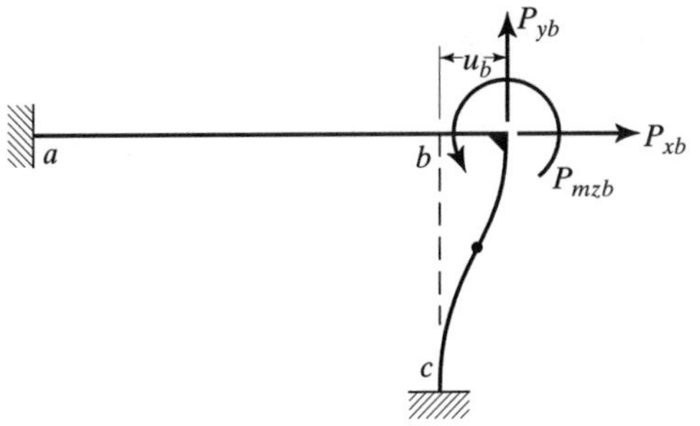

$$
\begin{aligned}
P_{xb} &= 200(0.750 + 0.00480)u_b = 200 \times 0.7548u_b \\
P_{yb} &= 200(0 + 0)u_b = 0 \\
P_{mzb} &= 200(0 + 12.000)u_b = 200 \times 12.000u_b
\end{aligned}
$$

Apply a vertical displacement, v_b:

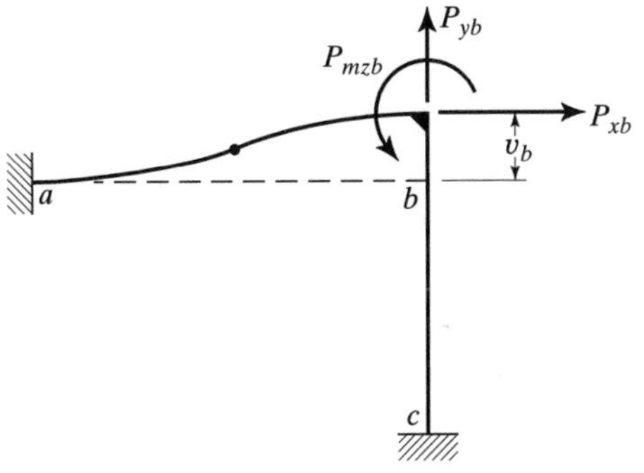

$$
\begin{aligned}
P_{xb} &= 200(0 + 0)v_b = 0 \\
P_{yb} &= 200(0.00469 + 0.800)v_b = 200 \times 0.80469v_b \\
P_{mzb} &= 200(-18.750 + 0)v_b = -200 \times 18.75v_b
\end{aligned}
$$

Apply a rotation, θ_{zb}:

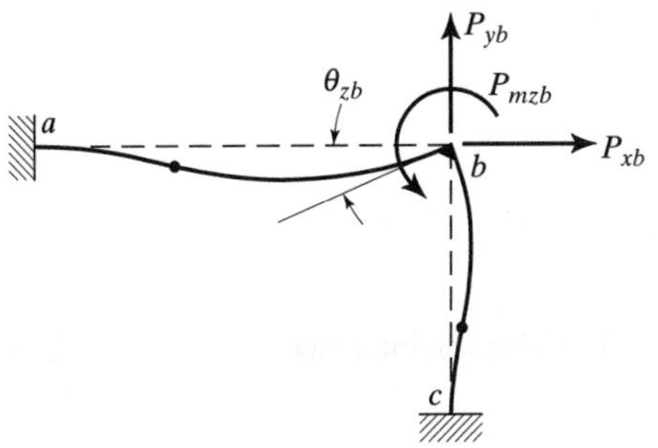

$$
\begin{aligned}
P_{xb} &= 200(0 + 12.000)\theta_{zb} = 200 \times 12.00\theta_{zb} \\
P_{yb} &= 200(-18.750 + 0)\theta_{zb} = -200 \times 18.75\theta_{zb} \\
P_{mzb} &= 200(1.0 \times 10^5 + 0.4 \times 10^5)\theta_{zb} = 200 \times 1.4 \times 10^5\theta_{zb}
\end{aligned}
$$

Assemble equations in matrix form and apply joint loads:

$$
\begin{Bmatrix} P_{xb} \\ P_{yb} \\ P_{mzb} \end{Bmatrix} = \begin{Bmatrix} 100/\sqrt{2} \\ -100/\sqrt{2} \\ 50 \times 10^3 \end{Bmatrix} = 200 \begin{bmatrix} 0.7548 & 0 & 12.000 \\ & 0.8047 & -18.750 \\ \text{Sym.} & & 1.4 \times 10^5 \end{bmatrix} \begin{Bmatrix} u_b \\ v_b \\ \theta_{zb} \end{Bmatrix}
$$

2. ***Displacements.*** Solving the stiffness equations yields

$$
\begin{Bmatrix} u_b \\ v_b \\ \theta_{zb} \end{Bmatrix} = \begin{Bmatrix} 0.4414 \text{ mm} \\ -0.3998 \text{ mm} \\ 0.00169 \text{ rad} \end{Bmatrix}
$$

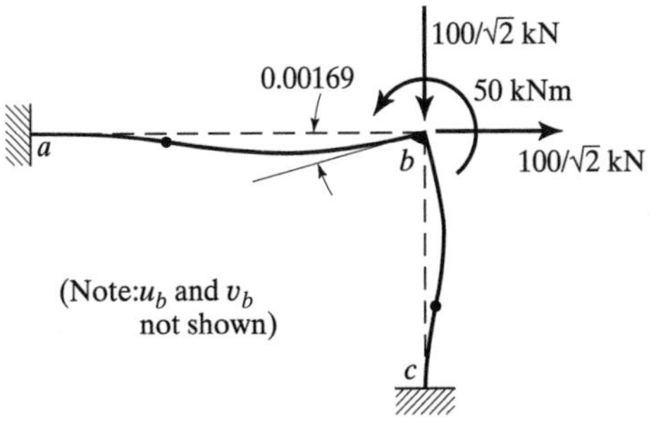

EXAMPLE 4.14

The two beam elements shown are joined by a hinge at 0. Develop the stiffness matrix for the combined member.

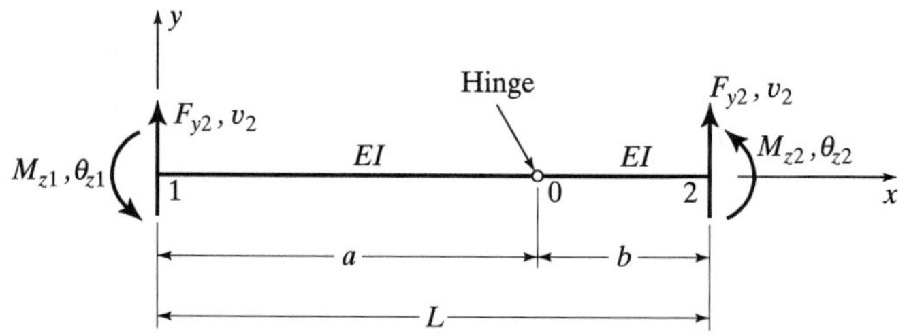

Write separate stiffness equations for each element using Equation 4.32. Account for:

1. Equilibrium conditions.
   (a) Zero bending moment at hinge.
   (b) Equal shear on two sides of hinge.
2. Compatibility condition; equal transverse displacement at hinge.
3. Special conditions; slope discontinuity at hinge.

Equations. Incorporating the above conditions, we have

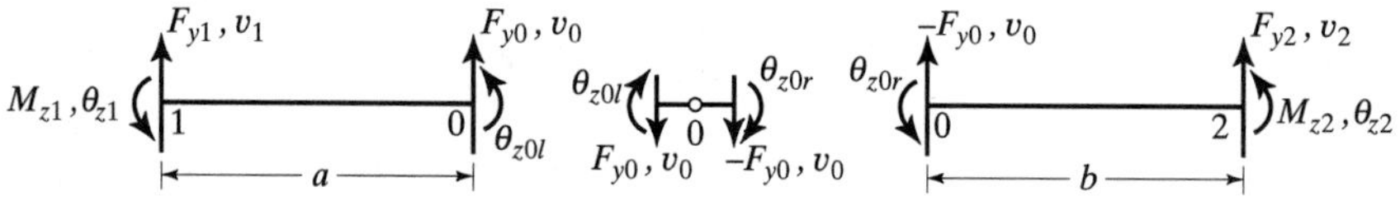

$$\begin{Bmatrix} F_{y1} \\ M_{z1} \\ F_{y0} \\ 0 \end{Bmatrix} = EI \begin{bmatrix} \frac{12}{a^3} & \frac{6}{a^2} & -\frac{12}{a^3} & \frac{6}{a^2} \\ & \frac{4}{a} & -\frac{6}{a^2} & \frac{2}{a} \\ & & \frac{12}{a^3} & -\frac{6}{a_2} \\ \text{Sym.} & & & \frac{4}{a} \end{bmatrix} \begin{Bmatrix} v_1 \\ \theta_1 \\ v_0 \\ \theta_{z0l} \end{Bmatrix} \qquad \begin{Bmatrix} -F_{yo} \\ 0 \\ F_{y2} \\ M_{z2} \end{Bmatrix} = EI \begin{bmatrix} \frac{12}{b^3} & \frac{6}{b^2} & -\frac{12}{b^3} & \frac{6}{b^2} \\ & \frac{4}{b} & -\frac{6}{b^2} & \frac{2}{b} \\ & & \frac{12}{b^3} & -\frac{6}{b^2} \\ \text{Sym.} & & & \frac{4}{b} \end{bmatrix} \begin{Bmatrix} v_0 \\ \theta_{z0r} \\ v_2 \\ \theta_{z2} \end{Bmatrix}$$

Assemble two sets in a single matrix format, ordered and partitioned as follows:

$$\begin{Bmatrix} F_{y1} \\ M_{z1} \\ F_{y2} \\ M_{z2} \\ \hline 0 \\ 0 \\ 0 \end{Bmatrix} = EI \left[\begin{array}{cccc|ccc} \frac{12}{a^3} & \frac{6}{a^2} & 0 & 0 & -\frac{12}{a^3} & \frac{6}{a^2} & 0 \\ \frac{6}{a^2} & \frac{4}{a} & 0 & 0 & -\frac{6}{a^2} & \frac{2}{a} & 0 \\ 0 & 0 & \frac{12}{b^3} & -\frac{6}{b^2} & -\frac{12}{b^3} & 0 & -\frac{6}{b^2} \\ 0 & 0 & -\frac{6}{b^2} & \frac{4}{b} & \frac{6}{b^2} & 0 & \frac{2}{b} \\ \hline -\frac{12}{a^3} & -\frac{6}{a^2} & -\frac{12}{b^3} & \frac{6}{b^2} & 12\left(\frac{1}{a^3}+\frac{1}{b^3}\right) & -\frac{6}{a^2} & \frac{6}{b^2} \\ \frac{6}{a^2} & \frac{2}{a} & 0 & 0 & -\frac{6}{a^2} & \frac{4}{a} & 0 \\ 0 & 0 & -\frac{6}{b^2} & \frac{2}{b} & \frac{6}{b^2} & 0 & \frac{4}{b} \end{array}\right] \begin{Bmatrix} v_1 \\ \theta_{z1} \\ v_2 \\ \theta_{z2} \\ \hline v_0 \\ \theta_{z0l} \\ \theta_{z0r} \end{Bmatrix}$$

Eliminate degrees of freedom at hinge, $\lfloor \Delta_o \rfloor^T = \lfloor v_0 \quad \theta_{z0l} \quad \theta_{z0r} \rfloor^T$, by operating on partitioned equations. Do symbolically. Submatrices are as indicated above.

$$\begin{Bmatrix} \mathbf{F}_e \\ \mathbf{0} \end{Bmatrix} \quad \left[\begin{array}{c|c} \mathbf{K}_{ee} & \mathbf{K}_{eo} \\ \hline \mathbf{K}_{oe} & \mathbf{K}_{oo} \end{array}\right] \begin{Bmatrix} \boldsymbol{\Delta}_e \\ \boldsymbol{\Delta}_o \end{Bmatrix}$$

Expand lower partition and solve for $\{\boldsymbol{\Delta}_o\}$:

$$\{\boldsymbol{\Delta}_o\} = -[\mathbf{K}_{oo}]^{-1}[\mathbf{K}_{oe}]\{\boldsymbol{\Delta}_e\} \tag{a}$$

Expand upper partition and substitute Equation a for $\{\boldsymbol{\Delta}_o\}$:

$$\{\mathbf{F}_e\} = [[\mathbf{K}_{ee}] - [\mathbf{K}_{eo}][\mathbf{K}_{oo}]^{-1}[\mathbf{K}_{oe}]]\{\boldsymbol{\Delta}_e\}$$

Thus the stiffness matrix $[\mathbf{K}]$ of the combined member is

$$[\mathbf{K}] = [[\mathbf{K}_{ee}] - [\mathbf{K}_{eo}][\mathbf{K}_{oo}]^{-1}[\mathbf{K}_{oe}]]$$

Carrying out the indicated operations yields (Do as *Problem 4.5*)

$$\{\mathbf{F}_e\} = [\mathbf{K}]\{\boldsymbol{\Delta}_e\}$$

$$\begin{Bmatrix} F_{y1} \\ M_{z1} \\ F_{y2} \\ M_{z2} \end{Bmatrix} = \frac{3EI}{(a^3 + b^3)} \begin{bmatrix} 1 & a & -1 & b \\ & a^2 & -a & ab \\ & & 1 & -b \\ \text{Sym.} & & & b^2 \end{bmatrix} \begin{Bmatrix} v_1 \\ \theta_{z1} \\ v_2 \\ \theta_{z2} \end{Bmatrix}$$

EXAMPLE 4.15

Member ab is a beam on an elastic foundation $I_{ab} = 128.5$ in^4. Foundation modulus $k = 1.5$ k/in^2. $E = 29{,}000$ ksi.

Determine the deflection and the bending moment in the beam at c by:

1. Elastic theory.
2. The stiffness method.

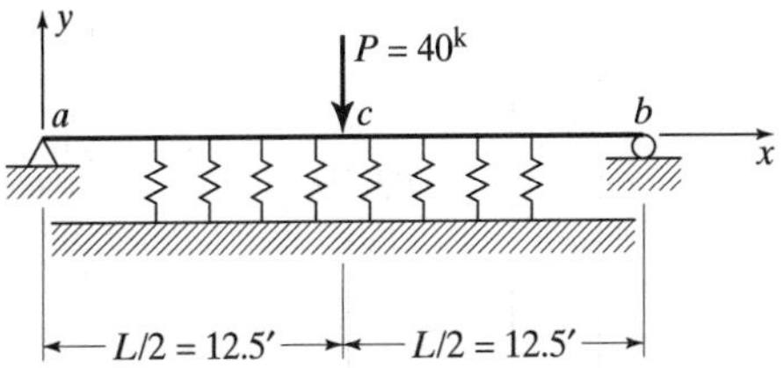

1. Assuming linear elastic behavior of the beam and the supporting foundation it may be shown that (see Ref. 4.3):

$$v_c = \frac{P\beta}{2k}\frac{\sinh \beta L - \sin \beta L}{\cosh \beta L + \cos \beta L} \quad \text{(a)} \qquad \text{and} \qquad M_c = \frac{P}{4\beta}\frac{\sinh \beta L + \sin \beta L}{\cosh \beta L + \cos \beta L} \quad \text{(b)}$$

in which

$$\beta = \sqrt[4]{\frac{k}{4EI_{ab}}} = \sqrt[4]{\frac{1.5}{4 \times 29{,}000 \times 128.5}} = 0.01781\,\frac{1}{\text{in}}$$

From (a) $v_c = 0.237$ in. from (b) $M_c = 554$ in. kips

2. Model as a beam supported on 19 equally spaced axial members of $L = 15$ in.

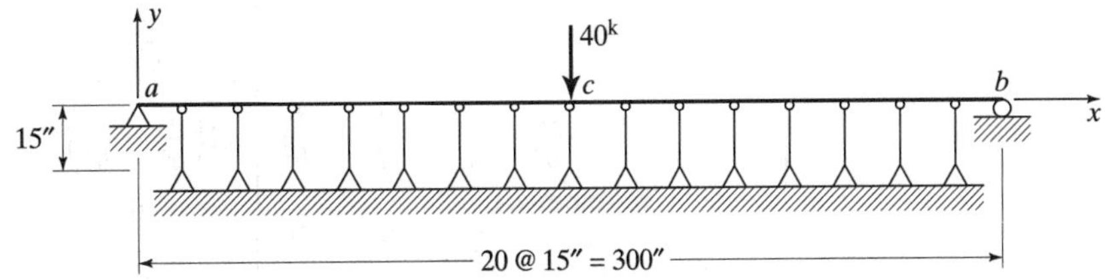

k per support $= 1.5 \times 15 = 22.5$ k/in $\qquad A_{\text{sup.}} = \dfrac{22.5 \times 15}{29{,}000} = 0.0116$ in^2

Deflected structure:

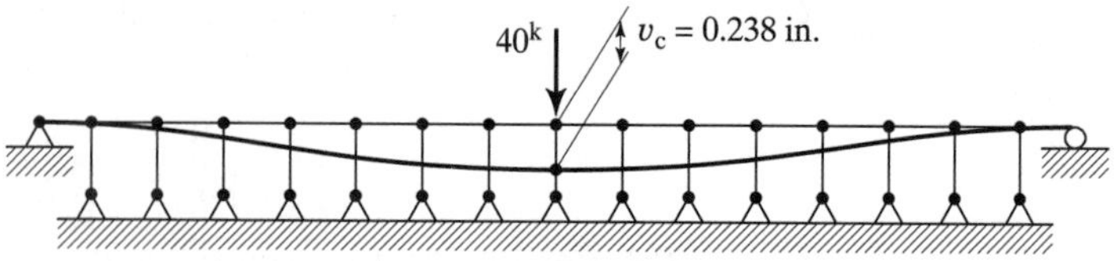

Moment diagram:

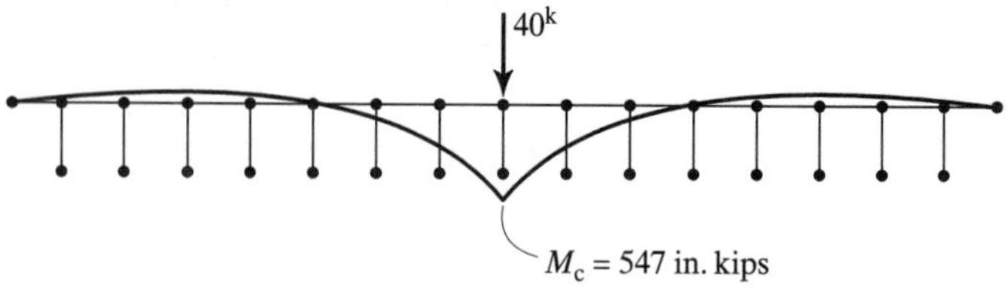

4.8 PROBLEMS

As in Chapter 3, it is suggested that, in several of the following problems, a computer program such as MASTAN2 be used and that the results be manually checked.

4.1 Assemble the global stiffness equations for the beam shown. Neglect axial deformation. Compute the displacements, reactions, and internal forces for the loading and support conditions indicated.

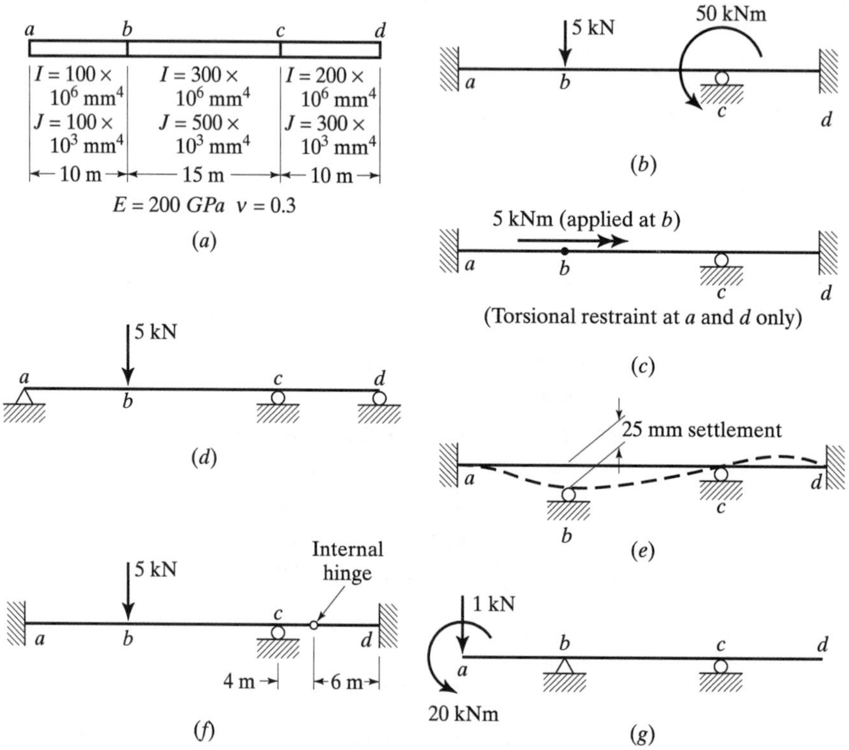

Problem 4.1

4.2 Develop the stiffness matrix for the beam in Figure 4.10, that is, verify Equation 4.33.

4.3 One way to support a beam in a stable, statically determinate fashion is as indicated (rotation at the left end is prevented). (a) Verify that the flexibility matrix is

$$\begin{Bmatrix} v_1 \\ \theta_{z2} \end{Bmatrix} = \frac{L}{6EI_z} \begin{bmatrix} 2L^2 & -3L \\ -3L & 6 \end{bmatrix} \begin{Bmatrix} F_{y1} \\ M_{z2} \end{Bmatrix}$$

(b) Verify that the complementary strain energy equals that of the cantilever and simply supported element (see Example 4.6). (c) Develop the beam stiffness matrix using this flexibility matrix and verify Equation 4.32.

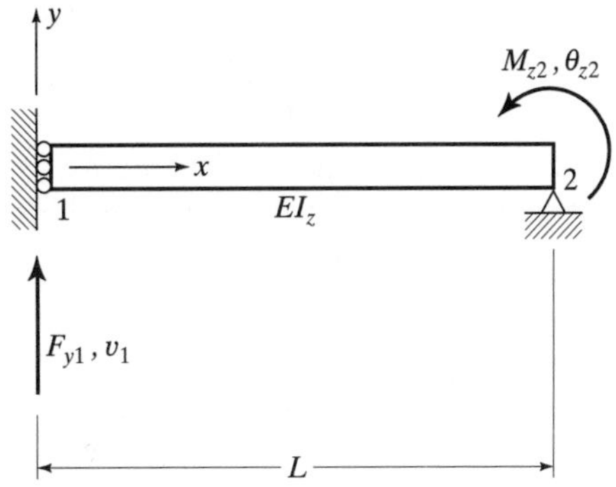

Problem 4.3

4.4 Compute the displacements, reactions, and internal forces for the beams shown in the figure. Insert nodes as required. E and I are constant.

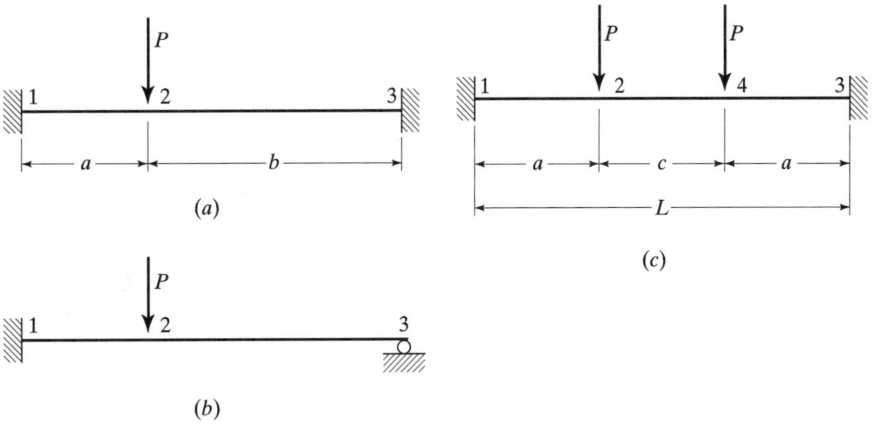

Problem 4.4

4.5 Carry out all of the algebraic operations needed to obtain the final combined member stiffness matrix of Example 4.14.

4.6 (a) By algebraic operations on Equation 4.32 develop a condensed stiffness matrix for a beam element in which the bending moment at one end is always zero (part a of the figure below). (b) Use this stiffness matrix in the analysis of the beam in part b.

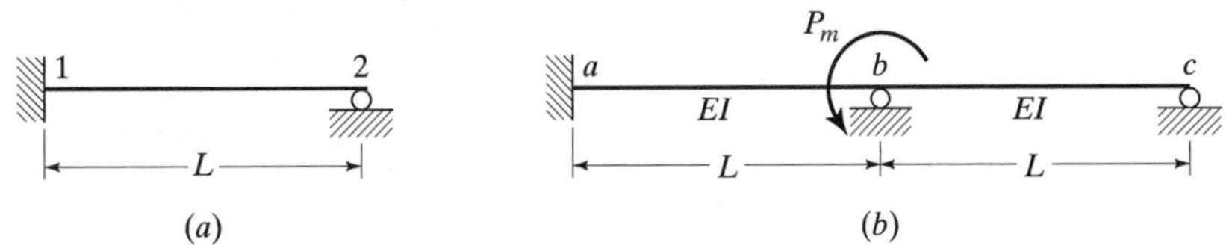

Problem 4.6

4.7 Compare the stiffness method with the method of moment distribution.

4.8 Compute the displacements, reactions, and internal forces for the systems shown. $E = 200{,}000$ MPa.

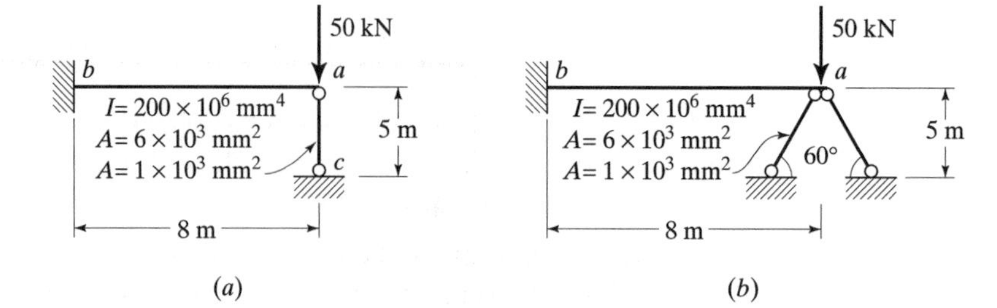

Problem 4.8

4.9 Compute the displacements, reactions, and internal forces for the systems shown. For the beams $I = 700 \times 10^6$ mm^4, $A = 15 \times 10^3$ mm^2. For the struts, $A = 10 \times 10^3$ mm^2, $E = 200{,}000$ MPa.

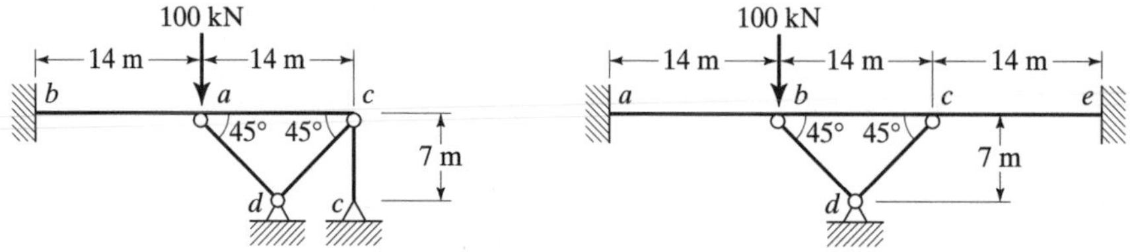

Problem 4.9

4.10 Compute the displacements, reactions, and internal forces for the system shown. For the beam $I = 500 \times 10^6$ mm^4, $A = 12 \times 10^3$ mm^2. For the struts, $A = 8 \times 10^3$ mm^2, $E = 200{,}000$ MPa.

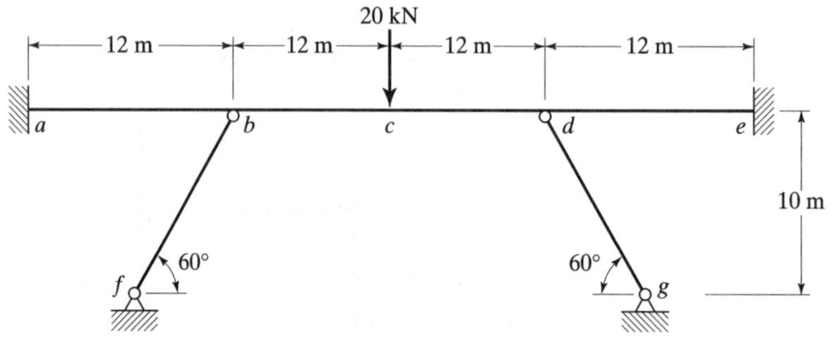

Problem 4.10

4.11 Same as Problem 4.10 but place an internal hinge halfway between a and b.

4.12 Compute the displacements, reactions, and internal forces for the systems shown. $E = 200{,}000$ MPa.

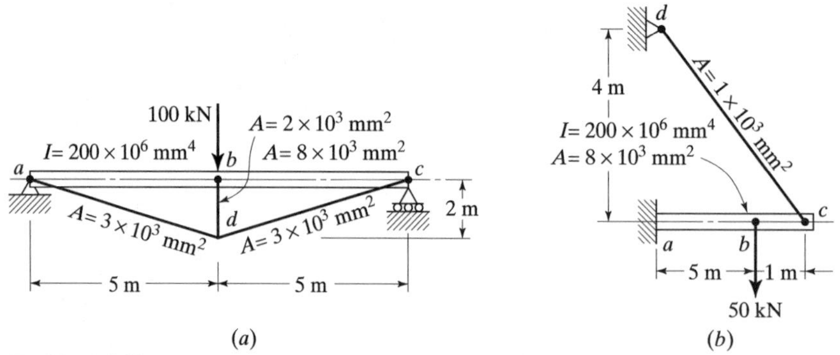

Problem 4.12

4.13 Plot the bending moment at a versus I_1/I_2 as I_1/I_2 varies from 1 to 10.

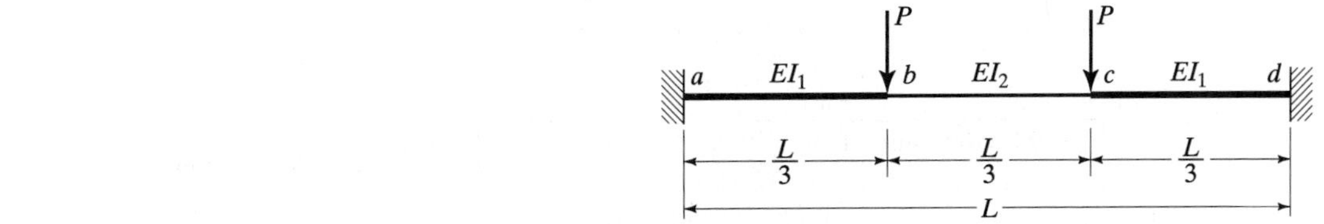

Problem 4.13

4.14 Plot the ratio of the bending moment at a to the bending moment at c versus I_1/I_2 as I_1/I_2 varies from 1 to 10. Assume A is infinite (neglect axial deformation of the strut).

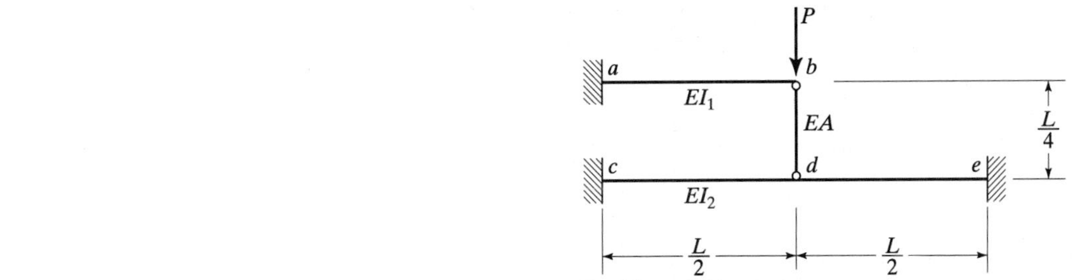

Problem 4.14

4.15 Compute the displacements, reactions, and internal forces for the system shown. $E = 200{,}000$ MPa.

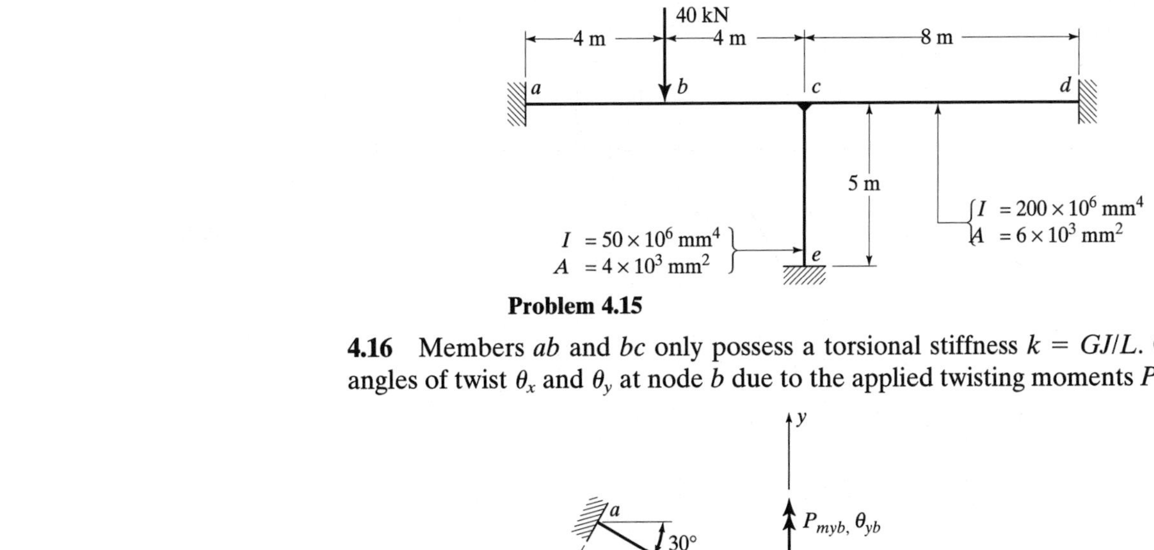

Problem 4.15

4.16 Members ab and bc only possess a torsional stiffness $k = GJ/L$. Calculate the angles of twist θ_x and θ_y at node b due to the applied twisting moments P_{mxb} and P_{myb}.

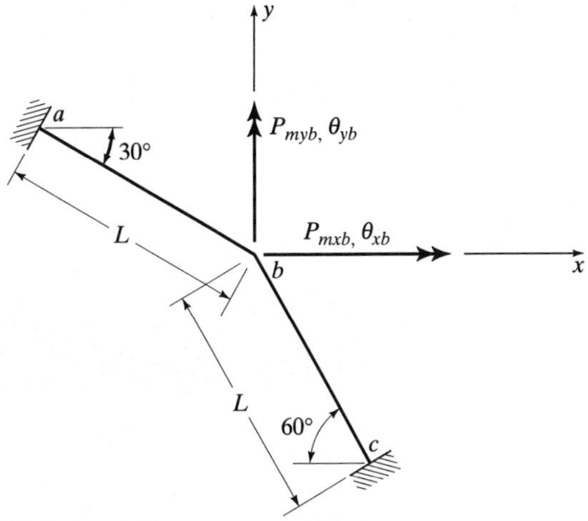

Problem 4.16

REFERENCES

4.1 S. Timoshenko and J. N. Goodier, *Theory of Elasticity*, Third Edition, McGraw-Hill, New York, 1970.

4.2 C. H. Norris, J. B. Wilbur, and S. Utku, *Elementary Structural Analysis*, Third Edition, McGraw-Hill, New York, 1976.

4.3 S. Timoshenko, *Strength of Materials, Part II, Advanced Theory and Problems*, Third Edition, Van Nostrand, Princeton, N.J., 1956.

4.4 J. P. Den Hartog, *Advanced Strength of Materials*, McGraw-Hill, New York, 1952.

Chapter 5

Stiffness Analysis of Frames—II

In this chapter, with the introduction of methods of performing coordinate transformations and for treating loads applied between nodal points, self-straining, and temperature change, we complete the development of the basic tools required for the linear elastic analysis of a framework of any shape, connection type, and loading.

The transformation of degrees-of-freedom, forces, and force-displacement relationships from one coordinate system to another can be treated as an exercise in analytical geometry. It is important, however, to keep in mind the physical problem being addressed, that is, to remember that merely changing the frame of reference can in no way alter the actual response of the structure.

The concept of fixed-end forces will be introduced in dealing with distributed loads, self-straining, and thermal effects. Understanding this notion requires that one visualize the application and subsequent removal of temporary, fictitious restraints on the degrees-of-freedom and grant to such constraints the ability to resist force. It is a concept that is at the heart of the displacement approach to analysis.

5.1 COORDINATE TRANSFORMATIONS

We have already discussed the reasons for employing a number of coordinate systems in the analysis of a particular structure and, as in Figure 2.4, we have indicated several logical choices. The problem can be illustrated further by Figure 5.1, a tripodal space frame with fixed supports and a rigid joint at *a*. The structure is shown placed in a convenient global coordinate system with axes x, y, and z. A certain natural action is exerted on the *a* end of member *ab* as the system deforms under the applied loads. This action can be portrayed graphically or its components can be listed in vector fashion. There are many ways to do this, but the action itself is immutable.

To illustrate some possibilities we first consider the case in which the resultant action on *ab* is a single force and not a couple, that is, it is the bound vector F_{ab} with an origin at point q (Figure 5.1*b*). The position of q with respect to *a* can be stated in terms of the radial distance ρ, the azimuth angle α, and the elevation angle β. Similarly, we could list the azimuth and elevation angles, γ and δ, to the positive direction of the force. The six quantities α, β, ρ, γ, δ, and the magnitude F_{ab} of the resultant force completely define the natural action on the *a* end of this member. This is physically clear, and it would be a useful way to define the action if *ab* were an isolated member and F_{ab} a load applied to a fixed external loading point. But it doesn't lend itself to the representation of element stiffness, and therefore it will not be considered in this connection.

Two other options are illustrated in Figures 5.1*c* and 5.1*d*. In the former a convenient

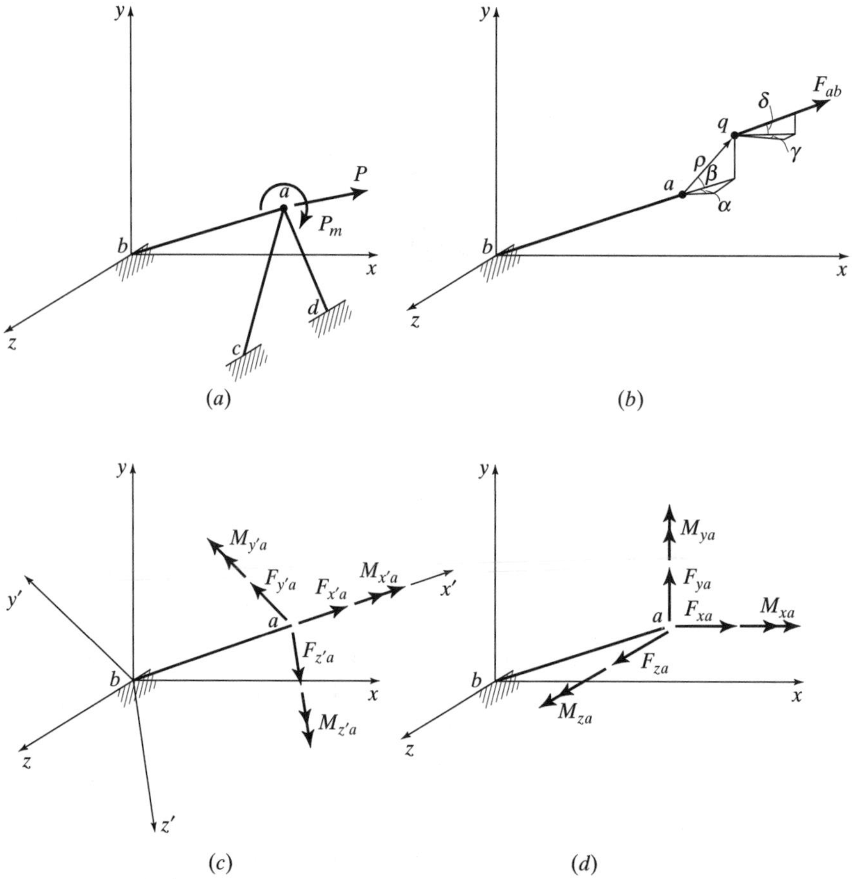

Figure 5.1 Alternative representations of force. (*a*) Structure. (*b*) Resultant force. (*c*) Local coordinates. (*d*) Global coordinates.

set of local coordinates is established and the action is described in terms of the six quantities $F_{x'a}$, $F_{y'a}$, $F_{z'a}$, $M_{x'a}$, $M_{y'a}$, and $M_{z'a}$, that is, direct force and moment components acting parallel to the local axes. In the latter we view the same action as the six quantities F_{xa}, F_{ya}, F_{za}, M_{xa}, M_{ya}, M_{za}, direct force and moment components acting parallel to the global axes.

The following vectors are thus alternative representations of the same quantity:

$$\{\mathbf{F}'\} = \lfloor F_{x'a} \quad F_{y'a} \quad F_{z'a} \quad M_{x'a} \quad M_{y'a} \quad M_{z'a} \rfloor^{\mathrm{T}}$$
$$\{\mathbf{F}\} = \lfloor F_{xa} \quad F_{ya} \quad F_{za} \quad M_{xa} \quad M_{ya} \quad M_{za} \rfloor^{\mathrm{T}}$$

Having either of these, the other can be obtained by a suitable transformation. We use both repeatedly since we are interested in transforming force and displacement vectors and, by logical extension, stiffness matrices, from local to global coordinates and vice versa.

5.1.1 Transformation Matrices

Since, as shown in Figure 2.7, we can portray displacements (translations and rotations) by line vectors in exactly the same way as we portray forces (direct forces and mo-

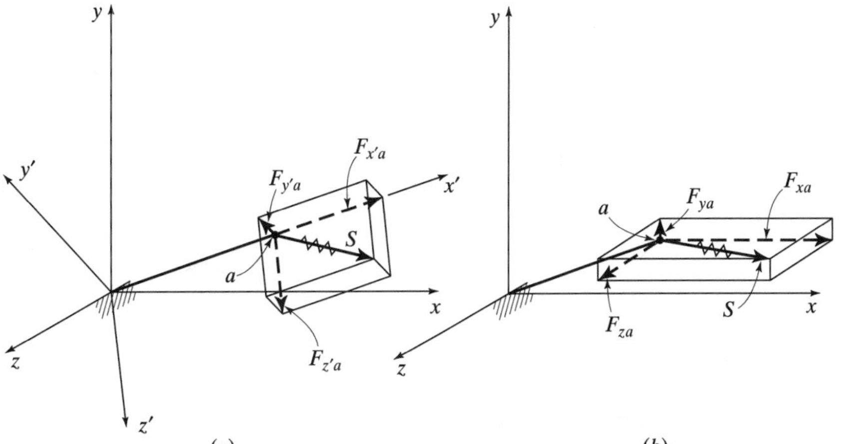

Figure 5.2 Direct force at *a*. (*a*) Local coordinates. (*b*) Global coordinates.

ments), the rules for transformation of displacements and forces are identical.[1] Furthermore, as shown below, the necessary transformations may be accomplished through rotation matrices that consist of direction cosines. This is the result of two things. First, the equations that comprise the set of global analysis equations involve equilibrium and compatibility at particular joints (see Figure 3.2 and Equations 3.3–3.5) and, in these, we simply need to transform between vectors passing through the joint: rotation, but not translation, is required. Second, we are using orthogonal coordinate systems.

For the reasons given above, the direct force components in Figures 5.1*c* and 5.1*d* have the same resultant. In the same way, the resultants of the two sets of moment components are equal. Consequently, in developing transformations linking local coordinates to global coordinates, one may treat direct forces and moments separately. Rules developed for direct forces also apply to moments, since they are each represented by sets of orthogonal vectors.

In Figure 5.2, the direct force components in the two systems are compared. The resultant direct force (*S* in the Figure), can be resolved into either its local components $F_{x'a}$, $F_{y'a}$, and $F_{z'a}$ (Figure 5.2*a*) or its global components F_{xa}, F_{ya}, and F_{za} (Figure 5.2*b*). It follows that each of the global components may in turn be resolved into components in the local directions, and vice versa. The easiest way to express one set of components in terms of the other is through direction cosines. Using the nomenclature for direction cosines defined in Figure 5.3, we have, for the local components.

$$
\begin{aligned}
F_{x'a} &= F_{xa}\cos\alpha_{x'} + F_{ya}\cos\beta_{x'} + F_{za}\cos\delta_{x'} \\
F_{y'a} &= F_{xa}\cos\alpha_{y'} + F_{ya}\cos\beta_{y'} + F_{za}\cos\delta_{y'} \\
F_{z'a} &= F_{xa}\cos\alpha_{z'} + F_{ya}\cos\beta_{z'} + F_{za}\cos\delta_{z'}
\end{aligned}
$$

[1]This statement is valid for linear analysis. However, as shown in texts such as Reference 5.1, finite rotations of a rigid body do not obey the law of vector addition. The finite rotation effect, which requires special consideration in three-dimensional nonlinear analysis, is studied in Section A.4 of the Appendix.

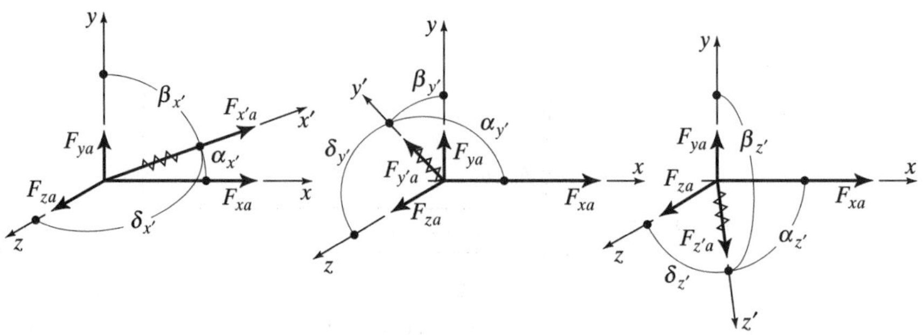

Figure 5.3 Direction angles.

Using, respectively, the symbols l, m, and n, with appropriate subscripts, to designate the cosines of the nine direction angles between the global x, y, and z axes and the subscripted local axes, we have

$$\begin{aligned} F_{x'a} &= l_{x'}F_{xa} + m_{x'}F_{ya} + n_{x'}F_{za} \\ F_{y'a} &= l_{y'}F_{xa} + m_{y'}F_{ya} + n_{y'}F_{za} \\ F_{z'a} &= l_{z'}F_{xa} + m_{z'}F_{ya} + n_{z'}F_{za} \end{aligned} \tag{5.1}$$

In matrix notation this becomes

$$\begin{Bmatrix} F_{x'a} \\ F_{y'a} \\ F_{z'a} \end{Bmatrix} = \begin{bmatrix} l_{x'} & m_{x'} & n_{x'} \\ l_{y'} & m_{y'} & n_{y'} \\ l_{z'} & m_{z'} & n_{z'} \end{bmatrix} \begin{Bmatrix} F_{xa} \\ F_{ya} \\ F_{za} \end{Bmatrix} \tag{5.2}$$

or, in short form,

$$\{\mathbf{F}_{F'}\} = [\boldsymbol{\gamma}]\{\mathbf{F}_F\} \tag{5.3}$$

The matrix $[\boldsymbol{\gamma}]$ of the nine direction cosines relating the local coordinate system to the global one is a *rotation matrix*.

Recalling that the sum of the squares of the direction cosines for any axis is unity, we have

$$\begin{aligned} l_{x'}^2 + m_{x'}^2 + n_{x'}^2 &= 1 \\ l_{y'}^2 + m_{y'}^2 + n_{y'}^2 &= 1 \\ l_{z'}^2 + m_{z'}^2 + n_{z'}^2 &= 1 \end{aligned} \tag{5.4}$$

The length of any vector is the square root of the sum of the squares of its components. Thus each row of $[\boldsymbol{\gamma}]$ has unit length. Also, for any two orthogonal axes such as x' and y', x' and z', and z' and y', the sum of the products of the corresponding direction cosines (the scalar product) is zero:

$$\begin{aligned} l_{x'}l_{y'} + m_{x'}m_{y'} + n_{x'}n_{y'} &= 0 \\ l_{y'}l_{z'} + m_{y'}m_{z'} + n_{y'}n_{z'} &= 0 \\ l_{z'}l_{x'} + m_{z'}m_{x'} + n_{z'}n_{x'} &= 0 \end{aligned} \tag{5.5}$$

Together, Equations 5.4 and 5.5 state that the $[\boldsymbol{\gamma}]$ matrix consists of rows of orthogonal unit vectors. Such a matrix is called an *orthogonal matrix*. It is a square matrix having the distinguishing property of an inverse equal to its transpose:

$$[\boldsymbol{\gamma}]^{-1} = [\boldsymbol{\gamma}]^{\mathrm{T}} \tag{5.6}$$

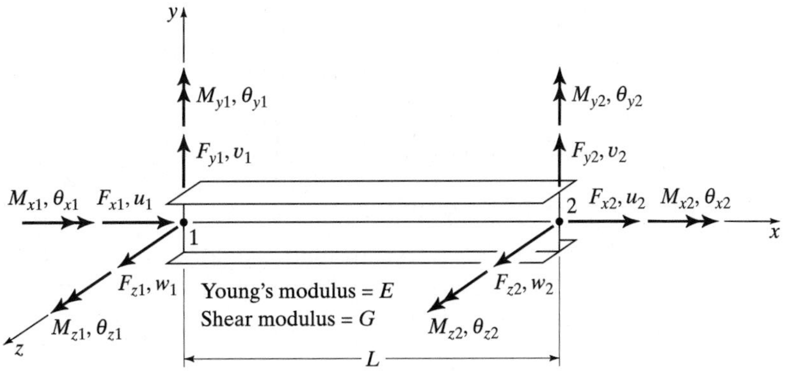

Figure 4.6 Bisymmetrical framework element.

This property may be verified readily by the identity $[\boldsymbol{\gamma}][\boldsymbol{\gamma}]^{-1} = [\mathbf{I}] = [\boldsymbol{\gamma}][\boldsymbol{\gamma}]^{\mathrm{T}}$. It will be shown later (Equation 5.14) that use of the orthogonality of the $[\boldsymbol{\gamma}]$ matrix simplifies considerably the labor of transforming coordinates.

Recognizing again that direct forces and moments transform independently, and that all that has been said about forces applied to one end of a member applies equally well to the other end, we can use Equation 5.3 directly in compiling the force transformation equation for the 12-degree-of-freedom framework element. Referring to Figure 4.6 (repeated here) for general nomenclature, and using primes to designate local coordinates, this equation is

$$\begin{Bmatrix} F_{x'1} \\ F_{y'1} \\ F_{z'1} \\ M_{x'1} \\ M_{y'1} \\ M_{z'1} \\ F_{x'2} \\ F_{y'2} \\ F_{z'2} \\ M_{x'2} \\ M_{y'2} \\ M_{z'2} \end{Bmatrix} = \begin{bmatrix} [\boldsymbol{\gamma}] & \mathbf{0} & \mathbf{0} & \mathbf{0} \\ \mathbf{0} & [\boldsymbol{\gamma}] & \mathbf{0} & \mathbf{0} \\ \mathbf{0} & \mathbf{0} & [\boldsymbol{\gamma}] & \mathbf{0} \\ \mathbf{0} & \mathbf{0} & \mathbf{0} & [\boldsymbol{\gamma}] \end{bmatrix} \begin{Bmatrix} F_{x1} \\ F_{y1} \\ F_{z1} \\ M_{x1} \\ M_{y1} \\ M_{z1} \\ F_{x2} \\ F_{y2} \\ F_{z2} \\ M_{x2} \\ M_{y2} \\ M_{z2} \end{Bmatrix} \tag{5.7}$$

or, in shorthand

$$\{\mathbf{F}'\} = [\boldsymbol{\Gamma}]\{\mathbf{F}\} \tag{5.8}$$

where

$$[\boldsymbol{\Gamma}] = \begin{bmatrix} [\boldsymbol{\gamma}] & & & \\ & [\boldsymbol{\gamma}] & & \\ & & [\boldsymbol{\gamma}] & \\ & & & [\boldsymbol{\gamma}] \end{bmatrix} \tag{5.9}$$

($\lceil\ \rfloor$ symbolizes a diagonal matrix in the text). Just as $[\boldsymbol{\gamma}]$ is orthogonal, so too is $[\boldsymbol{\Gamma}]$, or

$$[\boldsymbol{\Gamma}]^{-1} = [\boldsymbol{\Gamma}]^{\mathrm{T}} \tag{5.10}$$

Also, considering that the 12 degrees of freedom of the member in Figure 4.6 are portrayed by vectors just as the forces are, we may use the same transformation for these, or

$$\{\mathbf{\Delta}'\} = [\mathbf{\Gamma}]\{\mathbf{\Delta}\} \tag{5.11}$$

5.1.2 Transformation of Degrees of Freedom

Having the necessary transformation matrix, the transformation equations for the member stiffness matrix follows directly. If we write Equation 2.2 in local coordinates,

$$\{\mathbf{F}'\} = [\mathbf{k}']\{\mathbf{\Delta}'\} \tag{5.12}$$

From Equation 5.11

$$\{\mathbf{F}'\} = [\mathbf{k}'][\mathbf{\Gamma}]\{\mathbf{\Delta}\} \tag{5.13a}$$

and, from Equation 5.8

$$[\mathbf{\Gamma}]\{\mathbf{F}\} = [\mathbf{k}'][\mathbf{\Gamma}]\{\mathbf{\Delta}\} \tag{5.13b}$$

or

$$\{\mathbf{F}\} = [\mathbf{\Gamma}]^{-1}[\mathbf{k}'][\mathbf{\Gamma}]\{\mathbf{\Delta}\} \tag{5.13c}$$

Using the property of orthogonality (Equation 5.10),

$$\{\mathbf{F}\} = [\mathbf{\Gamma}]^{\mathrm{T}}[\mathbf{k}'][\mathbf{\Gamma}]\{\mathbf{\Delta}\} \tag{5.14}$$

Thus

$$\{\mathbf{F}\} = [\mathbf{k}]\{\mathbf{\Delta}\} \tag{5.15}$$

where

$$[\mathbf{k}] = [\mathbf{\Gamma}]^{\mathrm{T}}[\mathbf{k}'][\mathbf{\Gamma}] \tag{5.16}$$

This is the equation most generally used for finding the member stiffness matrix in global coordinates once one has formulated it in local coordinates and knows the geometrical relationship between the two sets of axes.

5.1.3 Transformations and Energy

We have arrived at the necessary transformations in a direct and rigorous way, but it is useful to reconsider the subject from an energy approach. It will bring out certain principles that may not be apparent from the above and which will be used later.

The transformed degrees of freedom are not necessarily equal in number to the original degrees of freedom. For example, compare Figures 2.8 and 2.9. In the first figure the behavior of the axial force member is defined by two degrees of freedom, whereas in the second figure, in which the only difference is the coordinate system, four degrees of freedom are required. The relationships connecting the two displacement vectors may be written symbolically, as before

$$\{\mathbf{\Delta}'\} = [\mathbf{\Gamma}]\{\mathbf{\Delta}\} \tag{5.11}$$

where, now, $[\mathbf{\Gamma}]$ need not be the square matrix defined in Equation 5.7 and, if it isn't, the inverse, $[\mathbf{\Gamma}]^{-1}$ of Equation 5.13c, does not exist. To treat this condition we again suppose the equations to be transformed are

$$\{\mathbf{F}'\} = [\mathbf{k}']\{\mathbf{\Delta}'\} \tag{5.12}$$

Also, we assume that each force component F'_i of the vector $\{\mathbf{F}'\}$ produces the work

$$\tfrac{1}{2}F'_i\Delta'_i$$

during the displacement Δ'_i, with no work done under any other displacement components in $\{\boldsymbol{\Delta}'\}$. This is another consequence of employing components that act along orthogonal axes. Such force and displacement vectors are called *conjugate vectors*. Both sets, that is, $\{\boldsymbol{\Delta}'\}$, $\{\mathbf{F}'\}$ and $\{\boldsymbol{\Delta}\}$, $\{\mathbf{F}\}$ are taken to be conjugate sets of vectors. Changing the frame of reference cannot alter the amount of work done. Therefore, for the work to remain invariant under the imposed transformation, the following equality must be satisfied:

$$\tfrac{1}{2}\lfloor\mathbf{F}'\rfloor\{\boldsymbol{\Delta}'\} = \tfrac{1}{2}\lfloor\mathbf{F}\rfloor\{\boldsymbol{\Delta}\}$$

and, from Equation 5.11

$$\lfloor\mathbf{F}'\rfloor[\boldsymbol{\Gamma}]\{\boldsymbol{\Delta}\} = \lfloor\mathbf{F}\rfloor\{\boldsymbol{\Delta}\}$$

Thus

$$\lfloor\mathbf{F}'\rfloor[\boldsymbol{\Gamma}] = \lfloor\mathbf{F}\rfloor$$

or, by transposition,

$$\{\mathbf{F}\} = [\boldsymbol{\Gamma}]^{\mathrm{T}}\{\mathbf{F}'\} \tag{5.17}$$

Since Equation 5.11 was invoked in this derivation, the transformation of displacements presented in Equation 5.11 implies the transformation of forces given by Equation 5.17. The force and displacement transformations are termed *contragredient* under the stipulated conditions of conjugacy. The conclusion of immediate interest is that, if the force transformation is first defined as in Equation 5.17, then the displacement-transformation matrix of Equation 5.11 is given by the transpose of the force-transformation matrix. The principle of contragredience is of considerable importance when the displacement (or force) transformation is readily constructed from physical meaning, but the formation of the conjugate vector is not readily perceived. This occurs, for example, when the condensation of degrees of freedom is accomplished by means of a transformation process (see Section 13.3).

To follow this approach in developing element stiffness transformations, it is convenient to deal with strain energy and external work, quantities that were introduced in Section 4.2. Again we require that work remain invariant under the imposed transformation, permitting direct substitution of Equation 5.11 into Equations 4.3 and 4.4a. Thus, from Equation 4.3,

$$\begin{aligned} W &= \tfrac{1}{2}\lfloor\boldsymbol{\Delta}'\rfloor\{\mathbf{F}'\} = \tfrac{1}{2}\{\boldsymbol{\Delta}'\}^{\mathrm{T}}\{\mathbf{F}'\} \\ &= \tfrac{1}{2}\lfloor\boldsymbol{\Delta}\rfloor[\boldsymbol{\Gamma}]^{\mathrm{T}}\{\mathbf{F}'\} = \tfrac{1}{2}\lfloor\boldsymbol{\Delta}\rfloor\{\mathbf{F}\} \end{aligned} \tag{5.18a}$$

and, from Equation 4.4a,

$$\begin{aligned} W &= \tfrac{1}{2}\lfloor\boldsymbol{\Delta}'\rfloor[\mathbf{k}']\{\boldsymbol{\Delta}'\} = \tfrac{1}{2}\lfloor\boldsymbol{\Delta}\rfloor[\boldsymbol{\Gamma}]^{\mathrm{T}}[\mathbf{k}'][\boldsymbol{\Gamma}]\{\boldsymbol{\Delta}\} \\ &= \tfrac{1}{2}\lfloor\boldsymbol{\Delta}\rfloor[\mathbf{k}]\{\boldsymbol{\Delta}\} \end{aligned} \tag{5.18b}$$

Hence the transformed stiffness matrix is again given by

$$[\mathbf{k}] = [\boldsymbol{\Gamma}]^{\mathrm{T}}[\mathbf{k}'][\boldsymbol{\Gamma}] \tag{5.16}$$

The force vector is of course transformed by Equation 5.17. The transformation of [**k**′] into [**k**] by Equation 5.16 is a congruent transformation (see page 64). Therefore, if [**k**′] is symmetric, [**k**] will also be symmetric.

5.1.4 Rectangular Transformation Matrices

The main reason for introducing the energy approach at this juncture is to show that, under the condition of contragredience, it is not necessary to use the condition that the inverse of the transformation matrix equals its transpose. Transposition can be invoked directly. This allows convenient definition of nonsquare coordinate axis transformation matrices. As noted earlier, the stiffness matrix of the axial member in Figure 2.8 features the two axial displacements. The same element disposed in an x-y plane as in Figure 2.9 is described by four displacement components. Arranging the local x' axis along the member of Figure 2.9 and defining the direction cosines as we have here, the relationship between the local and global degrees of freedom becomes

$$\begin{Bmatrix} u_1' \\ u_2' \end{Bmatrix} = \begin{bmatrix} l_{x'} & m_{x'} & 0 & 0 \\ 0 & 0 & l_{x'} & m_{x'} \end{bmatrix} \begin{Bmatrix} u_1 \\ v_1 \\ u_2 \\ v_2 \end{Bmatrix}$$

The validity of this transformation may be verified by inspection, but it is done formally in Example 5.1 by deleting irrelevant rows and columns from the complete transformation matrix of Equation 5.7. In this example the same transformation matrix is used to demonstrate the equivalence of Equations 2.3 and 2.5, which were developed independently from basic physical principles in Chapter 2.

Additional demonstrations of basic transformations are contained in Examples 5.2 through 5.5.

Example 5.2 is similar to Example 5.1 in that it illustrates the reduction of the complete 12-degree-of-freedom transformation matrix to a special situation—in this case the planar framework member. The partitioning of the [**Γ**] matrix follows Equation 5.7.

In Example 5.3, the results of several previous examples are used to demonstrate the assembly of the global stiffness equations for a typical rigid frame. Axial deformation effects are included for completeness even though, in most practical civil engineering frames of this type, they may be neglected in linear analysis. The example contains merely a listing of representative matrices and not details of their manipulation. The primed degrees of freedom refer to the local coordinates indicated.

The small space truss analysis of Example 5.4 is illustrative of the application of the same transformation techniques to space structures.

In Example 5.5 the problem of determining the direction cosines of the principal axes of an oblique member is solved by viewing the total transformation as a series of rotations. Most general-purpose computer programs have some built-in method for transforming coordinates that require only simple descriptive input. Nevertheless, the analyst should understand the geometrical and analytical problems involved.

EXAMPLE 5.1

Verify Equation 2.5 using Equation 2.3 and the transformation Equations 5.7 and 5.16.

Review development of Equations 2.3 and 2.5.

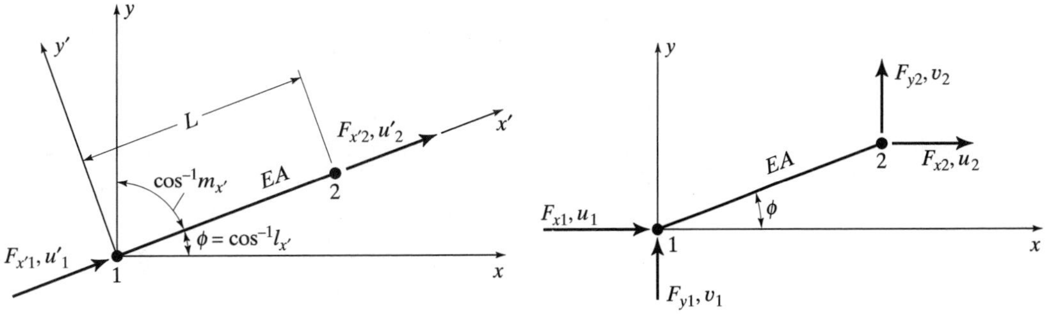

(*a*) (*b*)

Equation 2.3 was developed from a sketch similar to sketch *a* of the above figure. It may be written

$$\{\mathbf{F}'\} = [\mathbf{k}']\{\mathbf{\Delta}'\}$$

where $[\mathbf{k}']$ is a 2×2 matrix.
Equation 2.5 was developed from sketch *b*. It may be written

$$\{\mathbf{F}\} = [\mathbf{k}]\{\mathbf{\Delta}\}$$

where $[\mathbf{k}]$ is a 4×4 matrix.
Transform $[\mathbf{k}']$ to $[\mathbf{k}]$ using Equation 5.16. $[\mathbf{k}] = [\mathbf{\Gamma}]^T[\mathbf{k}'][\mathbf{\Gamma}]$, as follows: In Equation 5.7 delete all columns except those corresponding to F_{x1}, F_{y1}, F_{x2}, and F_{y2} and all rows except those corresponding to $F_{x'1}$ and $F_{x'2}$, leaving

$$\begin{Bmatrix} F_{x'1} \\ F_{x'2} \end{Bmatrix} = \begin{bmatrix} l_{x'} & m_{x'} & 0 & 0 \\ 0 & 0 & l_{x'} & m_{x'} \end{bmatrix} \begin{Bmatrix} F_{x1} \\ F_{y1} \\ F_{x2} \\ F_{y2} \end{Bmatrix}$$

Thus

$$[\mathbf{\Gamma}] = \begin{bmatrix} l_{x'} & m_{x'} & 0 & 0 \\ 0 & 0 & l_{x'} & m_{x'} \end{bmatrix}$$

which is the transformation matrix of this section.
From Equations 2.3, 5.16 and the above figure,

$$[\mathbf{k}] = \frac{EA}{L} \begin{bmatrix} \cos\phi & 0 \\ \sin\phi & 0 \\ 0 & \cos\phi \\ 0 & \sin\phi \end{bmatrix} \begin{bmatrix} 1 & -1 \\ -1 & 1 \end{bmatrix} \begin{bmatrix} \cos\phi & \sin\phi & 0 & 0 \\ 0 & 0 & \cos\phi & \sin\phi \end{bmatrix}$$

or

$$[\mathbf{k}] = \frac{EA}{L} \begin{bmatrix} \cos^2\phi & \sin\phi\cos\phi & -\cos^2\phi & -\sin\phi\cos\phi \\ & \sin^2\phi & -\sin\phi\cos\phi & -\sin^2\phi \\ & & \cos^2\phi & \sin\phi\cos\phi \\ \text{Sym.} & & & \sin^2\phi \end{bmatrix}$$

which is the stiffness matrix of Equation 2.5.

EXAMPLE 5.2

The member shown is part of a frame located and loaded in the x-y plane.

1. Reduce the general Equation 5.7 to the particular transformation equation for this case.
2. Demonstrate the orthogonality of the transformation matrix.

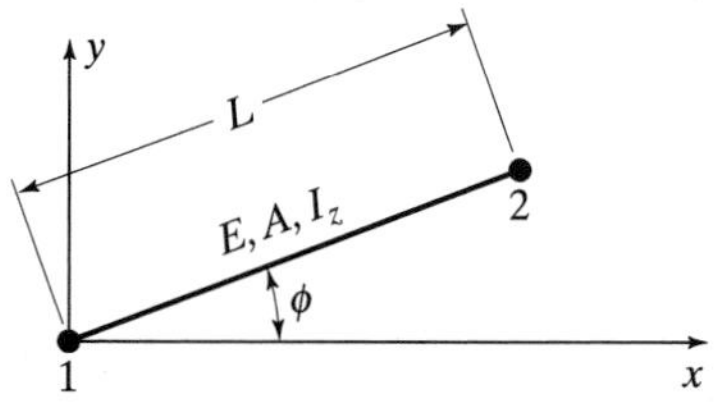

Define the local and global coordinates, degrees of freedom, and forces:

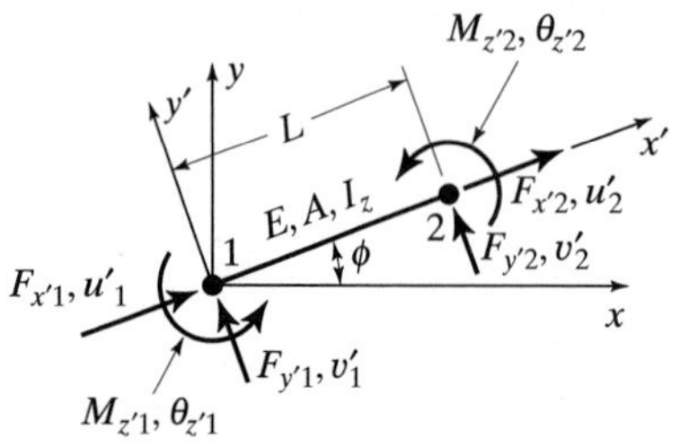

Local system

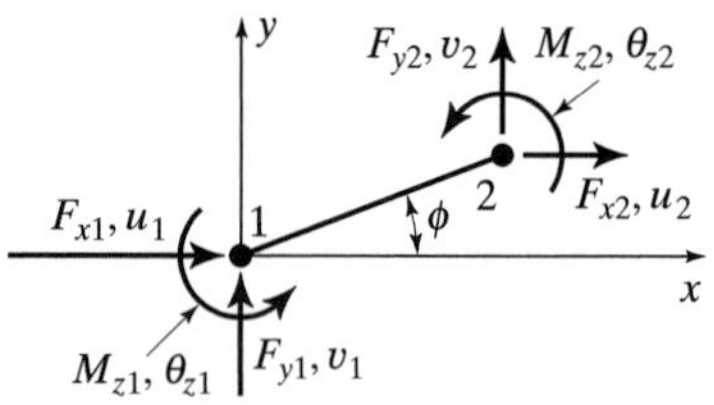

Global system

1. Deleting all irrelevant rows and columns from Equation 5.7 leaves

$$
\begin{Bmatrix} F_{x'1} \\ F_{y'1} \\ M_{z'1} \\ F_{x'2} \\ F_{y'2} \\ M_{z'2} \end{Bmatrix} =
\left[\begin{array}{cc|c|cc|c}
l_{x'} & m_{x'} & 0 & 0 & 0 & 0 \\
l_{y'} & m_{y'} & 0 & 0 & 0 & 0 \\ \hline
0 & 0 & n_{z'} & 0 & 0 & 0 \\ \hline
0 & 0 & 0 & l_{x'} & m_{x'} & 0 \\
0 & 0 & 0 & l_{y'} & m_{y'} & 0 \\ \hline
0 & 0 & 0 & 0 & 0 & n_{z'}
\end{array}\right]
\begin{Bmatrix} F_{x1} \\ F_{y1} \\ M_{z1} \\ F_{x2} \\ F_{y2} \\ M_{z2} \end{Bmatrix}
$$

or

$$
[\Gamma] = \left[\begin{array}{cc|c|cc|c}
\cos\phi & \sin\phi & 0 & 0 & 0 & 0 \\
-\sin\phi & \cos\phi & 0 & 0 & 0 & 0 \\ \hline
0 & 0 & 1 & 0 & 0 & 0 \\ \hline
0 & 0 & 0 & \cos\phi & \sin\phi & 0 \\
0 & 0 & 0 & -\sin\phi & \cos\phi & 0 \\ \hline
0 & 0 & 0 & 0 & 0 & 1
\end{array}\right]
$$

2. Orthogonality.
   It may be seen that $[\Gamma]^T[\Gamma] = [\mathbf{I}]$
   Thus $[\Gamma]^T = [\Gamma]^{-1}$

EXAMPLE 5.3

The rigid frame shown has $E = 200{,}000$ MPa and the following member properties: for ab, cd, ed, $A = 4 \times 10^3$ mm^2, $I = 50 \times 10^6$ mm^4; for bc, $A = 6 \times 10^3$ mm^2, $I = 200 \times 10^6$ mm^4. Using the results of Example 4.8 and the transformation matrix of Equation 5.2, develop the global stiffness equations for the structure, including flexural and axial deformations.

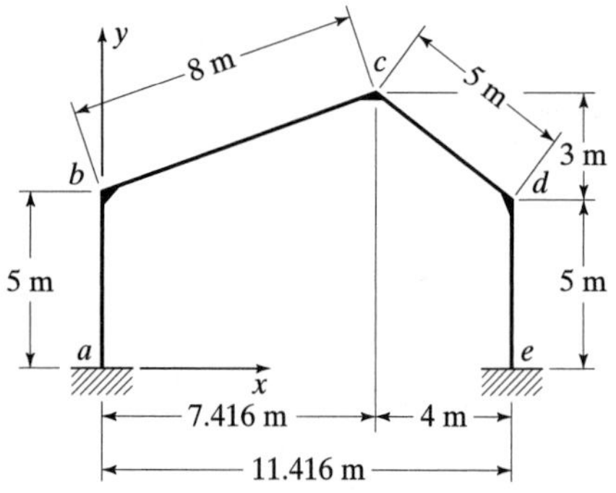

Consider member ab in detail; show $[\mathbf{k}]$ only for members bc, cd, and ed:

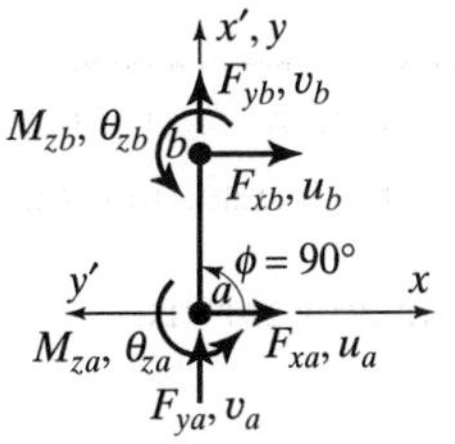

ab

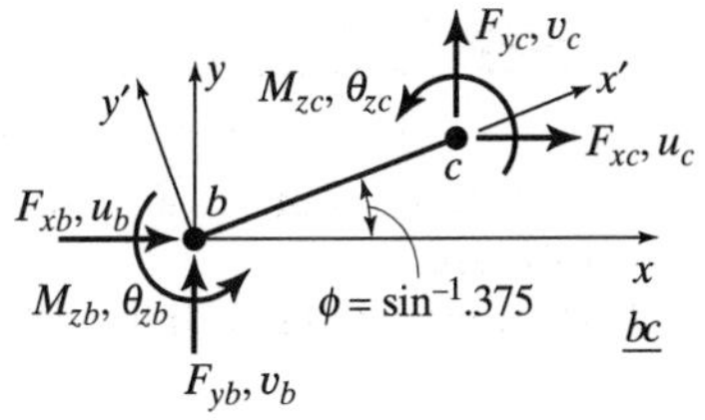

bc

Member ab

$[\mathbf{k}']$:

$$
200\begin{bmatrix}
0.800 & 0 & 0 & -0.800 & 0 & 0 \\
 & 0.0048 & 12.00 & 0 & -0.0048 & 12.00 \\
 & & 0.4\times10^5 & 0 & -12.00 & 0.2\times10^5 \\
 & & & 0.800 & 0 & 0 \\
 & \text{Sym.} & & & 0.0048 & -12.00 \\
 & & & & & 0.4\times10^5
\end{bmatrix}
\begin{matrix} u'_a & v'_a & \theta_{z'a} & u'_b & v'_b & \theta_{z'b} \end{matrix}
$$

$[\boldsymbol{\Gamma}]$:

$$
\left[\begin{array}{cc|c|cc|c}
0 & 1 & 0 & 0 & 0 & 0 \\
-1 & 0 & 0 & 0 & 0 & 0 \\ \hline
0 & 0 & 1 & 0 & 0 & 0 \\ \hline
0 & 0 & 0 & 0 & 1 & 0 \\
0 & 0 & 0 & -1 & 0 & 0 \\ \hline
0 & 0 & 0 & 0 & 0 & 1
\end{array}\right]
$$

$[\mathbf{k}]$:

$$
200\begin{bmatrix}
0.0048 & 0 & -12.00 & -0.0048 & 0 & -12.00 \\
 & 0.800 & 0 & 0 & -0.800 & 0 \\
 & & 0.4\times10^5 & 12.00 & 0 & 0.2\times10^5 \\
 & & & 0.0048 & 0 & 12.00 \\
 & \text{Sym.} & & & 0.800 & 0 \\
 & & & & & 0.4\times10^5
\end{bmatrix}
\begin{matrix} u_a & v_a & \theta_{za} & u_b & v_b & \theta_{zb} \end{matrix}
$$

Member bc

$[\mathbf{k}]$:

$$
200\begin{bmatrix}
0.6452 & 0.2591 & -7.0313 & -0.6452 & -0.2591 & -7.0313 \\
 & 0.1095 & 17.381 & -0.2591 & -0.1095 & 17.381 \\
 & & 1\times10^5 & 7.0313 & -17.381 & 0.5\times10^5 \\
 & & & 0.6452 & 0.2591 & 7.0313 \\
 & \text{Sym.} & & & 0.1095 & -17.381 \\
 & & & & & 1\times10^5
\end{bmatrix}
\begin{matrix} u_b & v_b & \theta_{zb} & u_c & v_c & \theta_{zc} \end{matrix}
$$

Member cd:

$[\mathbf{k}]$:

$$
200\begin{bmatrix}
0.5137 & -0.3817 & 7.200 & -0.5137 & 0.3817 & 7.200 \\
 & 0.2911 & 9.600 & 0.3817 & -0.2911 & 9.600 \\
 & & 0.4\times10^5 & -7.200 & -9.600 & 0.2\times10^5 \\
 & & & 0.5137 & -0.3817 & -7.200 \\
 & \text{Sym.} & & & 0.2911 & -9.600 \\
 & & & & & 0.4\times10^5
\end{bmatrix}
\begin{matrix} u_c & v_c & \theta_{zc} & u_d & v_d & \theta_{zd} \end{matrix}
$$

Member *ed*:

[k]:

$$
200\begin{bmatrix}
 & u_e & v_e & \theta_{ze} & u_d & v_d & \theta_{zd} \\
 & 0.0048 & 0 & -12.00 & -0.0048 & 0 & -12.00 \\
 & & 0.800 & 0 & 0 & -0.800 & 0 \\
 & & & 0.4 \times 10^5 & 12.00 & 0 & 0.2 \times 10^5 \\
 & & & & 0.0048 & 0 & 12.00 \\
 & & \text{Sym.} & & & 0.800 & 0 \\
 & & & & & & 0.4 \times 10^5
\end{bmatrix}
$$

Assembled global stiffness equation:

$\lfloor P_{xa}\ P_{ya}\ P_{mza} \mid P_{xb}\ P_{yb}\ P_{mzb} \mid P_{xc}\ P_{yc}\ P_{mzc} \mid P_{xd}\ P_{yd}\ P_{mzd} \mid P_{xe}\ P_{ye}\ P_{mze}\rfloor^T =$

200 ×

u_a	v_a	θ_{za}	u_b	v_b	θ_{zb}	u_c	v_c	θ_{zc}ft	u_d	v_d	θ_{zd}	u_e	v_e	θ_{ze}
0.0048	0	−12.00	−0.0048	0	−12.00									
	0.800	0	0	−0.800	0		**0**			**0**			**0**	
		0.4 × 10^5	12.00	0	0.2 × 10^5									
			0.6500	0.2591	4.969	−0.6452	−0.2591	−7.0313						
				0.9095	17.381	−0.2591	−0.1095	17.381		**0**			**0**	
					1.4 × 10^5	7.0313	−17.381	0.5 × 10^5						
						1.1589	−0.1226	14.231	−0.5137	0.3817	7.200			
				Sym			0.4006	−7.781	0.3817	−0.2911	9.600		**0**	
								1.4 × 10^5	−7.200	−9.600	0.2 × 10^5			
									0.5185	−0.3817	4.80	−0.0048	0	12.00
										1.0911	−9.60	0	−0.800	0
											0.8 × 10^5	−12.00	0	0.2 × 10^5
												0.00480	0	−12.00
													0.800	0
														0.4 × 10^5

$\{u_a\ v_a\ \theta_{za} \mid u_b\ v_b\ \theta_{zb} \mid u_c\ v_c\ \theta_{zc} \mid u_d\ v_d\ \theta_{zd} \mid u_e\ v_e\ \theta_{ze}\}$

EXAMPLE 5.4

A pin-jointed space truss is supported and loaded as shown. $E = 200{,}000$ MPa. Bar areas are:

$$A_{ab} = 20 \times 10^3\ \text{mm}^2$$
$$A_{ac} = 30 \times 10^3\ \text{mm}^2$$
$$A_{ad} = 40 \times 10^3\ \text{mm}^2$$
$$A_{ae} = 30 \times 10^3\ \text{mm}^2$$

1. Calculate the displacement at *a*.
2. Calculate the reactions.

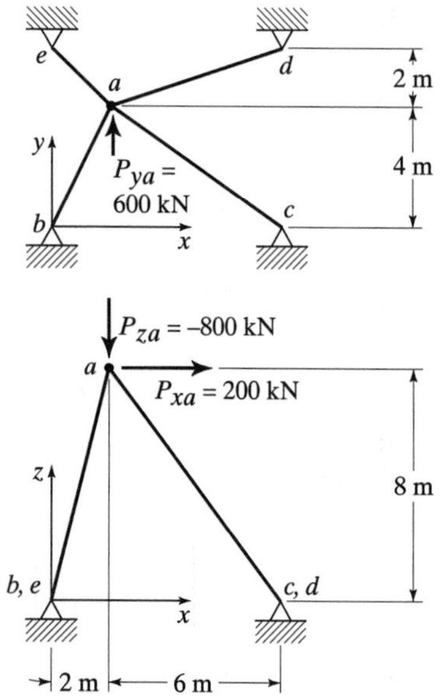

Develop the three-dimensional stiffness matrix for an axial force member. Stating Equation 4.34 in the local coordinates shown and eliminating all irrelevant degrees of freedom,

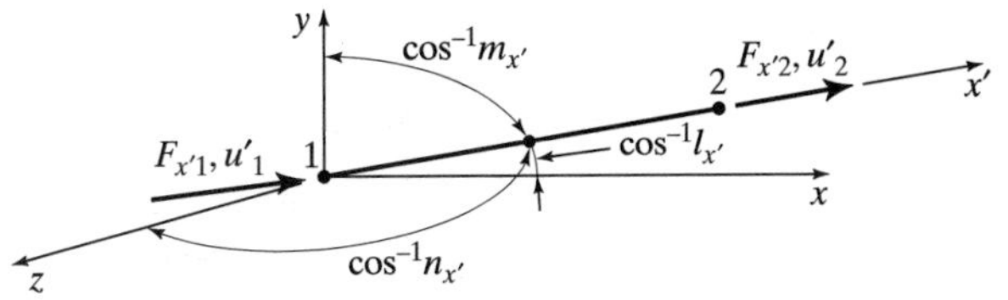

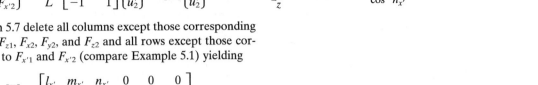

$$\begin{Bmatrix} F_{x'1} \\ F_{x'2} \end{Bmatrix} = \frac{EA}{L}\begin{bmatrix} 1 & -1 \\ -1 & 1 \end{bmatrix}\begin{Bmatrix} u'_1 \\ u'_2 \end{Bmatrix} = [\mathbf{k}']\begin{Bmatrix} u'_1 \\ u'_2 \end{Bmatrix}$$

In Equation 5.7 delete all columns except those corresponding to F_{x1}, F_{y1}, F_{z1}, F_{x2}, F_{y2}, and F_{z2} and all rows except those corresponding to $F_{x'1}$ and $F_{x'2}$ (compare Example 5.1) yielding

$$[\boldsymbol{\Gamma}] = \begin{bmatrix} l_{x'} & m_{x'} & n_{x'} & 0 & 0 & 0 \\ 0 & 0 & 0 & l_{x'} & m_{x'} & n_{x'} \end{bmatrix}$$

Using Equation 5.16, $[\mathbf{k}] = [\boldsymbol{\Gamma}]^{\mathrm{T}}[\mathbf{k}'][\boldsymbol{\Gamma}]$,

$$[\mathbf{k}] = \frac{EA}{L}\begin{array}{c} \begin{matrix} u_1 & v_1 & w_1 & u_2 & v_2 & w_2 \end{matrix} \\ \left[\begin{array}{ccc|ccc} l_{x'}^2 & l_{x'}m_{x'} & l_{x'}n_{x'} & -l_{x'}^2 & -l_{x'}m_{x'} & -l_{x'}n_{x'} \\ & m_{x'}^2 & m_{x'}n_{x'} & -l_{x'}m_{x'} & -m_{x'}^2 & -m_{x'}n_{x'} \\ & & n_{x'}^2 & -l_{x'}n_{x'} & -m_{x'}n_{x'} & -n_{x'}^2 \\ \hline & & & l_{x'}^2 & l_{x'}m_{x'} & l_{x'}n_{x'} \\ & \text{Sym} & & & m_{x'}^2 & m_{x'}n_{x'} \\ & & & & & n_{x'}^2 \end{array}\right] \end{array}$$

1. ***Displacements.*** Record direction cosines for each member. Locate local origin for each member at a.

Member	x	y	z	L	$l_{x'}$	$m_{x'}$	$n_{x'}$
ab	−2	−4	−8	9.165	−0.2182	−0.4364	−0.8729
ac	6	−4	−8	10.770	0.5571	−0.3714	−0.7428
ad	6	2	−8	10.198	0.5883	0.1961	−0.7845
ae	−2	2	−8	8.485	−0.2357	0.2357	−0.9428

Record direction cosine products multiplied by A/L. (Values of A/L are in mm.)

Member	(A/L)	×	$l_{x'}^2$	$m_{x'}^2$	$n_{x'}^2$	$l_{x'}m_{x'}$	$l_{x'}n_{x'}$	$m_{x'}n_{x'}$
ab	2.182		0.1034	0.4156	1.663	0.2078	0.4156	0.8312
ac	2.785		0.8643	0.3841	1.537	−0.5762	−1.1525	0.7683
ad	3.922		1.3574	0.1508	2.414	0.4524	−1.8101	−0.6034
ae	3.536		0.1964	0.1964	3.143	−0.1964	0.7858	−0.7858
	Σ		2.5215	1.1469	8.757	−0.1124	−1.7612	0.2103

Referring to the $[\mathbf{k}]$ matrix and adding stiffnesses in the usual way, the global stiffness equations for the nonzero degrees of freedom are

$$\begin{Bmatrix} P_{xa} \\ P_{ya} \\ P_{za} \end{Bmatrix} = \begin{Bmatrix} 200 \\ 600 \\ -800 \end{Bmatrix} = 200\begin{bmatrix} 2.522 & -0.1124 & -1.7612 \\ & 1.147 & 0.2103 \\ \text{Sym.} & & 8.757 \end{bmatrix}\begin{Bmatrix} u_a \\ v_a \\ w_a \end{Bmatrix}$$

Solving for the displacements,

$$\lfloor \boldsymbol{\Delta} \rfloor = \lfloor u_a \quad v_a \quad w_a \rfloor = \lfloor 0.1783 \quad 2.722 \quad -0.4863 \rfloor \text{ mm}$$

2. ***Reactions.*** Using the properties of the element stiffness matrices (see above table), the global components of the reactions are

$$
\begin{Bmatrix} R_{xb} \\ R_{yb} \\ R_{zb} \\ \hline R_{xc} \\ R_{yc} \\ R_{zc} \\ \hline R_{xd} \\ R_{yd} \\ R_{zd} \\ \hline R_{xe} \\ R_{ye} \\ R_{ze} \end{Bmatrix} = -200 \begin{bmatrix} \overset{u_a}{0.1034} & \overset{v_a}{0.2078} & \overset{w_a}{0.4156} \\ 0.2078 & 0.4156 & 0.8312 \\ 0.4156 & 0.8312 & 1.663 \\ \hline 0.8643 & -0.5762 & -1.1525 \\ -0.5762 & 0.3841 & 0.7683 \\ -1.1525 & 0.7683 & 1.537 \\ \hline 1.3574 & 0.4524 & -1.8101 \\ 0.4524 & 0.1508 & -0.6034 \\ -1.810 & -0.6034 & 2.414 \\ \hline 0.1964 & -0.1964 & 0.7858 \\ -0.1964 & 0.1964 & -0.7858 \\ 0.7858 & -0.7858 & 3.143 \end{bmatrix} \begin{Bmatrix} 0.1783 \\ 2.722 \\ -0.4863 \end{Bmatrix} = \begin{Bmatrix} -76.4 \\ -152.8 \\ -305.6 \\ \hline 170.8 \\ -113.8 \\ -227.7 \\ \hline -470.7 \\ -156.9 \\ 627.8 \\ \hline 176.3 \\ -176.3 \\ 705.5 \end{Bmatrix} \text{kN}
$$

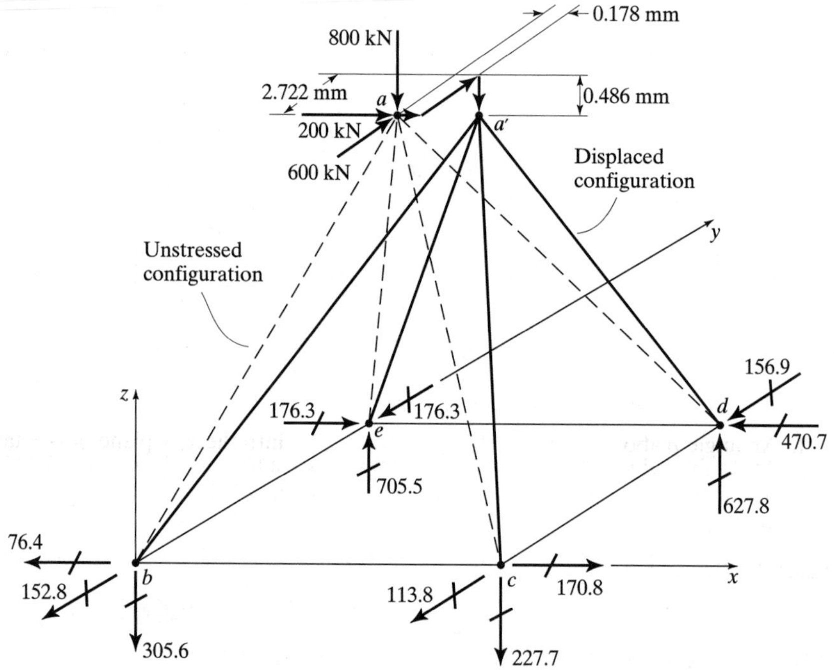

EXAMPLE 5.5

Find the direction cosines of the member shown. x and z are horizontal; y is vertical. The x' axis is along the member. The $x'y'$ plane makes a dihedral angle of 30° with a vertical plane through the member axis. Demonstrate that the matrix of direction cosines $[\gamma]$ can be generated by three successive rotations.

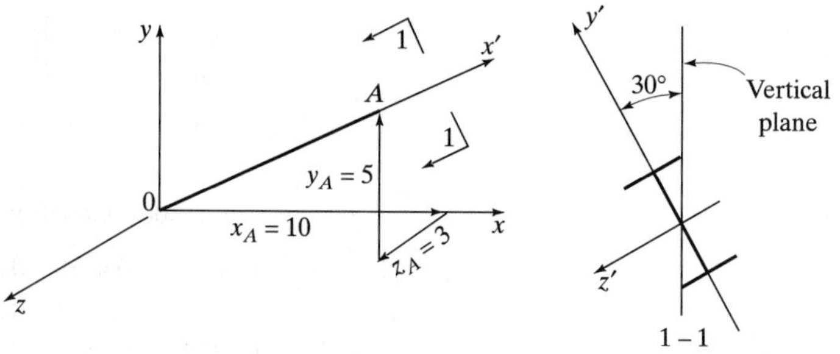

From the figure on the right

$$\overline{OA} = \sqrt{10^2 + 5^2 + 3^2} = 11.576$$

$$l_{x'} = 10/11.576 = 0.8639 \qquad \alpha_{x'} = \cos^{-1} l_{x'} = 30.248°$$

$$m_{x'} = 5/11.576 = 0.4319 \qquad \beta_{x'} = \cos^{-1} m_{x'} = 64.411°$$

$$n_{x'} = 3/11.576 = 0.2592 \qquad \delta_{x'} = \cos^{-1} n_{x'} = 74.930°$$

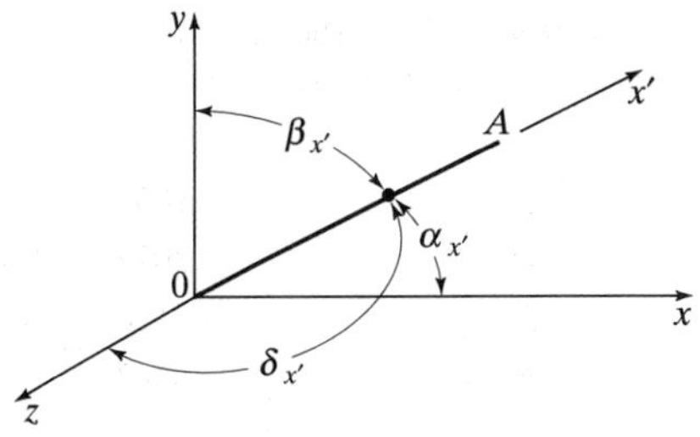

In the figure at the right, subscripts indicate the position of each axis at the *end* of each rotation. Following our sign convention, ρ is a negative angle as pictured.

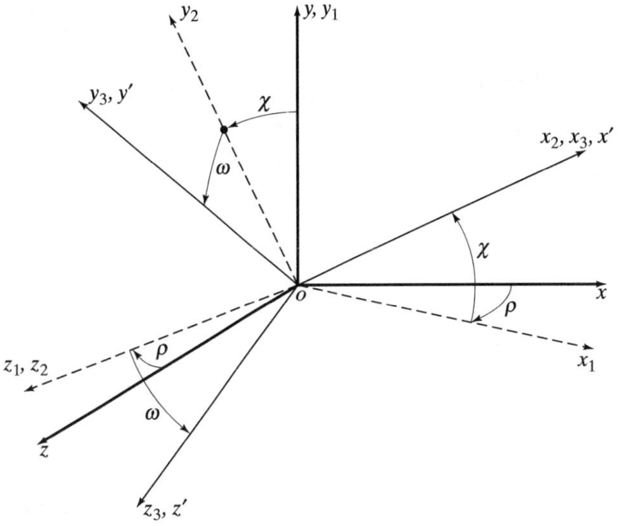

The successive rotations are as follows:

1. An angle ρ about the y axis, bringing the x' axis into the x_1-y plane. $\rho = -\tan^{-1}(3/10) = -16.699°$.
2. An angle χ about the z_1 axis, bringing the x axis into coincidence with the x' axis. $\chi = (90 - \beta_{x'}) = 25.589°$.
3. An angle ω about the x_2 axis, bringing the y axis into coincidence with the y' axis and the z axis into coincidence with the z' axis. $\omega = 30°$.

Symbolically we may write

$$\{\mathbf{x}'\} = [\gamma]\{\mathbf{x}\} = [\gamma_\omega][\gamma_\chi][\gamma_\rho]\{\mathbf{x}\}$$

where $\{\mathbf{x}\}$ and $\{\mathbf{x}'\}$ are coordinate vectors and $[\gamma_\rho]$, $[\gamma_\chi]$ and $[\gamma_\omega]$ are transformation matrices defining the three rotations described above. Using the definitions of Equation 5.2,

$$[\gamma] = \begin{bmatrix} 1 & 0 & 0 \\ 0 & \cos\omega & \sin\omega \\ 0 & -\sin\omega & \cos\omega \end{bmatrix} \begin{bmatrix} \cos\chi & \sin\chi & 0 \\ -\sin\chi & \cos\chi & 0 \\ 0 & 0 & 1 \end{bmatrix} \begin{bmatrix} \cos\rho & 0 & -\sin\rho \\ 0 & 1 & 0 \\ \sin\rho & 0 & \cos\rho \end{bmatrix}$$

$$= \begin{bmatrix} 1 & 0 & 0 \\ 0 & 0.8660 & 0.5000 \\ 0 & -0.5000 & 0.8660 \end{bmatrix} \begin{bmatrix} 0.9019 & 0.4319 & 0 \\ -0.4319 & 0.9019 & 0 \\ 0 & 0 & 1 \end{bmatrix} \begin{bmatrix} 0.9578 & 0 & 0.2873 \\ 0 & 1 & 0 \\ -0.2873 & 0 & 0.9578 \end{bmatrix}$$

$$= \begin{bmatrix} 0.8638 & 0.4319 & 0.2591 \\ -0.5019 & 0.7811 & 0.3714 \\ -0.0420 & -0.4510 & 0.8915 \end{bmatrix}$$

5.2 LOADS BETWEEN NODAL POINTS

Generally, structures must resist loads applied between joints or natural nodal points of the system. Three possibilities are shown in Figure 5.4. Loading may range from a few concentrated loads to an infinite variety of uniformly or nonuniformly distributed loads.

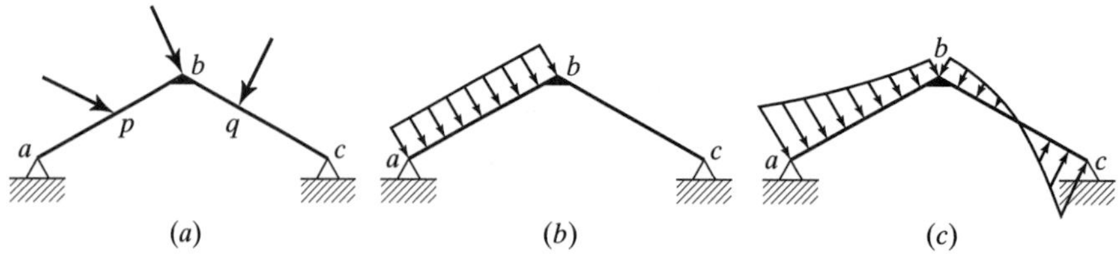

Figure 5.4 Loads between nodal points.

One way to treat intermediate concentrated loads is to insert artificial nodes, such as p and q in Figure 5.4*a*. The solution then proceeds as in many of the earlier examples. The degrees of freedom at the artificial nodes are added to the total inventory, and the necessary additional equations are written by considering the requirements of equilibrium at these nodes. The internal element forces on each side of nodes such as p and q must equilibrate the external loads applied at these points. Assigning the identical degree-of-freedom designation to the corresponding unknown displacement components at the two sides of the node ensures the satisfaction of compatibility. This procedure was used in Example 4.12 and its use was implied in several problems of Chapter 4.

Additional nodes may also be used when the load is distributed as in Figures 5.4*b* and 5.4*c*. Following a process selected or devised by the analyst, the distributed loads are "lumped" as concentrated loads at suitably selected arbitrary nodes, and the degrees of freedom at these and the actual joints are treated as the unknowns of the problem. There are various ways of approaching or obtaining equivalence between the lumped and the actual loading. In all cases the lumped loads must be statically equivalent to the distributed loads they replace. For example, the transverse load at the artificial joint must equal the total transverse component of the distributed load associated with that joint.

Use of artificial nodes may be the most expeditious route to a solution that is inherently approximate but satisfactory for all practical purposes. It is not a very elegant procedure, however, and the addition of the displacements at the arbitrary nodes to the unknown degrees of freedom may increase the labor of solution unduly. For these reasons, it is desirable to have recourse to a rigorous method that eliminates the need for "artificial" joints.

The approach that is of most general use with the displacement method is one that employs the related concepts of fictitious joint restraint, fixed end forces, and equivalent nodal loads. It will be explained by example. But in preparation for doing this, it is useful to recapitulate an approach we have already used by illustrating its application to the continuous beam of Figure 5.5*a*. The solution of this problem of a system loaded only at a natural node would start with the writing of Equation 3.6:

$$\{\mathbf{P}\} = [\mathbf{K}]\{\mathbf{\Delta}\} \tag{3.6}$$

The support and the remaining degrees of freedom would then be grouped as in Equation 3.7:

$$\begin{Bmatrix} \mathbf{P}_f \\ \hline \mathbf{P}_s \end{Bmatrix} = \left[\begin{array}{c|c} \mathbf{K}_{ff} & \mathbf{K}_{fs} \\ \hline \mathbf{K}_{sf} & \mathbf{K}_{ss} \end{array}\right] \begin{Bmatrix} \mathbf{\Delta}_f \\ \hline \mathbf{\Delta}_s \end{Bmatrix} \tag{3.7}$$

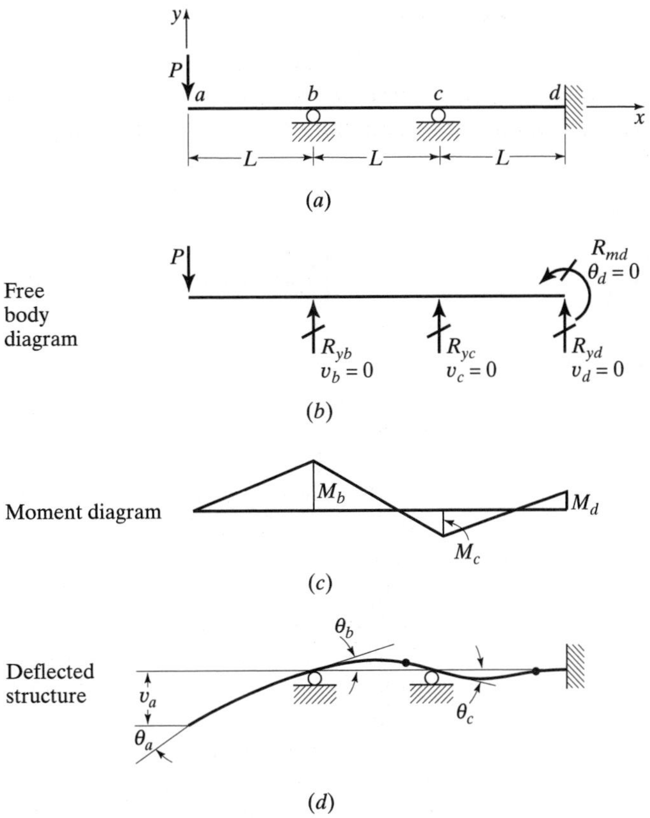

Figure 5.5 Continuous beam—nodal-point load.

For the particular loading, coordinate designations, and support conditions shown, this becomes

$$\begin{Bmatrix} -P \\ 0 \\ 0 \\ 0 \\ \hdashline R_{yb} \\ R_{yc} \\ R_{yd} \\ R_{md} \end{Bmatrix} = \left[\begin{array}{c:c} \mathbf{K}_{ff} & \mathbf{K}_{fs} \\ \hdashline \mathbf{K}_{sf} & \mathbf{K}_{ss} \end{array}\right] \begin{Bmatrix} v_a \\ \theta_a \\ \theta_b \\ \theta_c \\ \hdashline 0 \\ 0 \\ 0 \\ 0 \end{Bmatrix} \tag{5.19}$$

Solution for the unknown displacements, reactions, and internal forces proceeds in the usual way. The moment diagram and the deflected structure are drawn in Figures 5.5*c* and 5.5*d*.

Now consider the same structure but with a uniformly distributed load of intensity q in the center span, as in Figure 5.6. This loading, which is between nodal points, will be treated in two stages and the results summed, as indicated in the sketches of parts *a*, *b*, and *c* of the figure.

First, as in Figure 5.6*a*, grant the existence of fictitious external constraints capable of reducing the nodal degrees of freedom to zero (clamping the joints). The constraining forces, which in this case consist of two direct forces and two moments, are shown in their positive sense under our sign convention. It should be clear that P_{mc}^F must

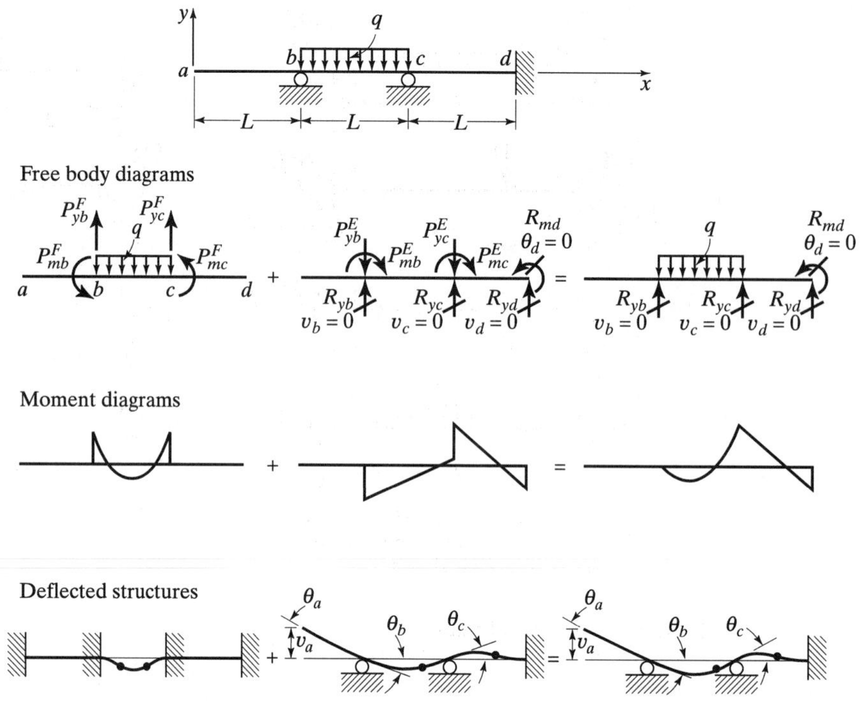

(*a*) System fully constrained + (*b*) Removal of artificial constraints = (*c*) Resultant system

Figure 5.6 Continuous beam—intermediate load.

actually be negative in order to constrain the rotation at point c. It should also be accepted that these forces are completely independent of the system—they are not being supplied by the real beam *or* its real supports. We presume that the solution for the fixed-end forces is known. In the case shown, the absolute values of the end reactions and fixing moments are $qL/2$ and $qL^2/12$, respectively. Solutions for some other cases are summarized in Table 5.1. Knowing the fixed-end forces, the internal forces and deformations corresponding to the assumed constraints can be calculated, as portrayed in the bending moment diagram and deflected structure of Figure 5.6*a*.

Now it is necessary to remove the fictitious constraints by applying to the nodes loads that are equal and opposite to the fixed-end forces and permitting the system to deform under the action of these likewise fictitious loads. The reversed fixed-end forces are called *equivalent nodal loads*. They are indicated in Figure 5.6*b* of the example structure by use of the symbol P^E and the appropriate subscripts. Solving in the usual way for the displacements, reactions, and internal forces caused by these loads results in the bending moment diagram and deflected structure of Figure 5.6*b*.

Last, the total solution is obtained by summing the two parts. In arriving at the free body diagram of Figure 5.6*c*, the two sets of fictitious forces have canceled each other and we are left with the real system in which all the requirements of equilibrium and compatibility are satisfied. The bending moment diagram and deflected structure of this part of the figure are the final, correct ones.

There are several ways to represent this physical process algebraically. We must first recognize that the internal forces and displacements of the fixed-end problem (Figure 5.6*a*) must be obtained by some means not detailed here, and reserved for addition to the solution of the nodal displacement problem (Figure 5.6*b*). One way to formu-

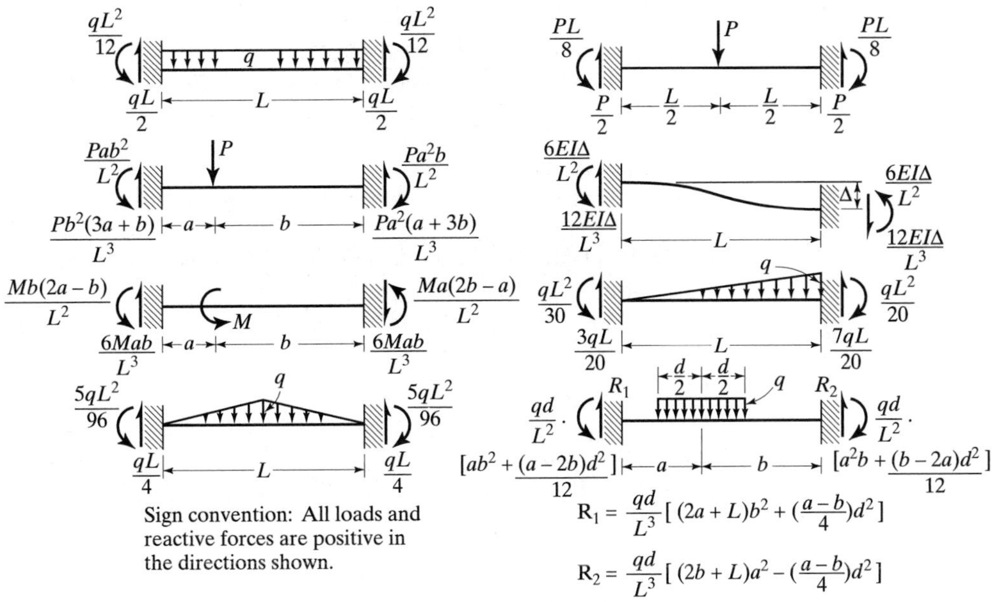

Table 5.1 Fixed-end forces.

late the solution to the displacement problem is after the fashion of Equation 3.6, *but with the addition of the vector of fixed-end forces to the right-hand side of the equation*; thus

$$\{\mathbf{P}\} = [\mathbf{K}]\{\boldsymbol{\Delta}\} + \{\mathbf{P}^F\} \tag{3.6a}$$

Physically, this states that, in the absence of any nodal displacement, i.e., $\{\boldsymbol{\Delta}\} = \mathbf{0}$, $\{\mathbf{P}\}$ would be equal to the vector of fixed end forces, $\{\mathbf{P}^F\}$. Conceptually, this formation is useful because it helps to keep clear the distinction between any real nodal loads and reaction components of $\{\mathbf{P}\}$ and the fictitious components that comprise $\{\mathbf{P}^F\}$.

The support and the remaining degrees of freedom of Equation 3.6a may be grouped as before:

$$\left\{\begin{array}{c}\mathbf{P}_f \\ \hline \mathbf{P}_s\end{array}\right\} = \left[\begin{array}{c|c}\mathbf{K}_{ff} & \mathbf{K}_{fs} \\ \hline \mathbf{K}_{sf} & \mathbf{K}_{ss}\end{array}\right]\left\{\begin{array}{c}\boldsymbol{\Delta}_f \\ \hline \boldsymbol{\Delta}_s\end{array}\right\} + \left\{\begin{array}{c}\mathbf{P}_f^F \\ \hline \mathbf{P}_s^F\end{array}\right\} \tag{3.7a}$$

For the particular loading and coordinate designations in Figure 5.6, the above may be written as (compare with Equation 5.19):

$$\left\{\begin{array}{c}0\\0\\0\\0\\ \hline R_{yb}\\R_{yc}\\R_{yd}\\R_{md}\end{array}\right\} = \left[\begin{array}{c|c}\mathbf{K}_{ff} & \mathbf{K}_{fs} \\ \hline \mathbf{K}_{sf} & \mathbf{K}_{ss}\end{array}\right]\left\{\begin{array}{c}v_a\\\theta_a\\\theta_b\\\theta_c\\ \hline 0\\0\\0\\0\end{array}\right\} + \left\{\begin{array}{c}0\\0\\qL^2/12\\-qL^2/12\\ \hline qL/2\\qL/2\\0\\0\end{array}\right\} \tag{5.20}$$

Transference of the $\{\mathbf{P}^F\}$ vector in any of the above formulations (Equations 3.6a, 3.7a, or 5.20) *to the left-hand side of the equation* is the algebraic equivalent of applying the reversed fixed-end forces as nodal loads in Figure 5.6*b*. Hence

$$\{\mathbf{P}\} - \{\mathbf{P}^F\} = \{\mathbf{P}\} + \{\mathbf{P}^E\} = [\mathbf{K}]\{\boldsymbol{\Delta}\} \tag{5.21}$$

where $\{\mathbf{P}^E\}$ is the *vector of equivalent nodal loads* defined in the discussion of Figure 5.6*b*.

For the illustrative example of Figure 5.6, the statement of the displacement problem is therefore

$$\left\{\begin{array}{c} 0 \\ 0 \\ -qL^2/12 \\ +qL^2/12 \\ \hline R_{yb} - qL/2 \\ R_{yc} - qL/2 \\ R_{yd} \\ R_{md} \end{array}\right\} = \left[\begin{array}{c|c} \mathbf{K}_{ff} & \mathbf{K}_{fs} \\ \hline \mathbf{K}_{sf} & \mathbf{K}_{ss} \end{array}\right] \left\{\begin{array}{c} v_a \\ \theta_a \\ \theta_b \\ \theta_c \\ \hline 0 \\ 0 \\ 0 \\ 0 \end{array}\right\} \tag{5.22}$$

Solution for the unknown nodal displacements, the real reactions (R_{yb}, R_{yc}, R_{yd}, R_{md}), and the internal forces proceeds in the usual way, but with appropriate accounting for the loads between nodes. Thus, from Equation 3.7a, for $\{\boldsymbol{\Delta}_s\} = 0$, we have

$$\{\boldsymbol{\Delta}_f\} = [\mathbf{K}_{ff}]^{-1}\{\mathbf{P}_f - \mathbf{P}_f^F\} \tag{3.7b}$$

and

$$\{\mathbf{P}_s\} = [\mathbf{K}_{sf}]\{\boldsymbol{\Delta}_f\} + \{\mathbf{P}_s^F\} \tag{3.7c}$$

In determining forces and displacements within the loaded members, one must remember to add the results of Equation 5.22 to the solution of the fixed end problem. Formally, this part of the problem can be symbolized by augmenting the element Equations 3.11 in the same way that Equations 3.6 and 3.7 were modified to obtain the above global equations. Thus

$$\{\mathbf{F}\} = [\mathbf{k}]\{\boldsymbol{\Delta}\} + \{\mathbf{F}^F\} \tag{3.11a}$$

These concepts are illustrated in Examples 5.6–5.9. We note that Equations 3.6a, 3.7a, 5.20, and 5.21 are cast in a global analysis form. In accordance with our convention of designating forces acting on the ends of elements, we use the symbols $\{\mathbf{F}^F\}$ and $\{\mathbf{F}^E\}$ to designate, respectively, vectors of fixed end forces and equivalent nodal loads formulated on an element basis. Components of these vectors will be symbolized by F^F, M^F, F^E, and M^E, with appropriate subscripts. As shown in the examples, joint equilibrium requires that element fixed-end forces appear in the global equations with magnitudes and signs unchanged.

Example 5.6 is a straightforward beam analysis problem. In Example 5.7, a simple planar rigid frame is studied. In forming the stiffness equations, only those parts needed at each stage are written. Axial deformation effects are included for illustration. The example shows that there are no restrictions against combining real and equivalent nodal forces, but it must be done in algebraically proper fashion.

Example 5.8 illustrates the analysis of a plane frame loaded normal to its plane. Such a structure is often called a *grid*. The primed degrees of freedom refer to the local coordinates indicated. The partitioning of the [Γ] matrix follows Equation 5.7. As in Example 5.3, much of the detail of matrix formulation and manipulation is omitted.

In Example 5.9, two analyses of a portal frame are simplified considerably by using two artifices commonly employed in classical frame analysis: (1) suppressing (equating to zero) displacements that are known in advance to be very small with respect to the remainder, and (2) taking advantage of symmetry and antisymmetry to combine coefficients of related degrees of freedom. The concepts of symmetry and antisymmetry are discussed further in Section 13.6.

Although the devices used in Example 5.9 are used correctly and effectively, a word of caution regarding the employment of presumed shortcuts in a stiffness analysis is in order. Unless proper procedures are followed, significant contributions to the stiffness of the structure may be overlooked or incorrectly represented in the analytical model. In some cases solution of the stiffness equations will prove impossible but, in others, plausible but unrealistic results may be obtained. Problems 5.6 and 5.7 relate to some of the consequences of improper modeling.

EXAMPLE 5.6

The beam of Example 4.8 is supported and loaded as shown.

1. Calculate the displacements at *a* and *b*.
2. Calculate the reactions and bending moments.

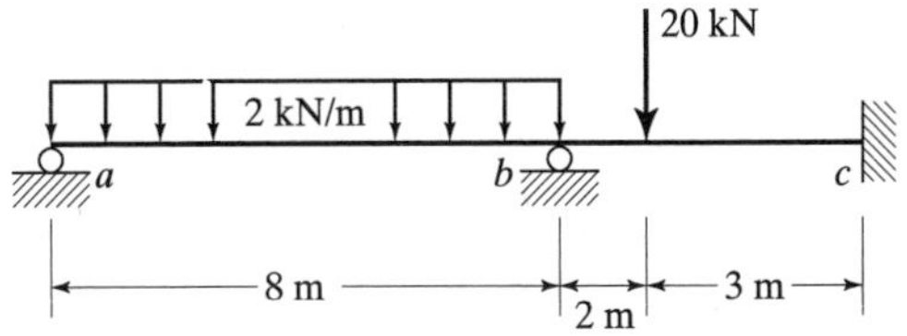

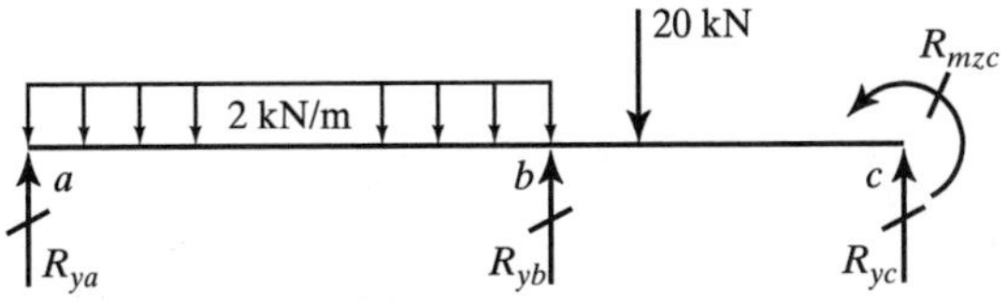

Fixed end forces:

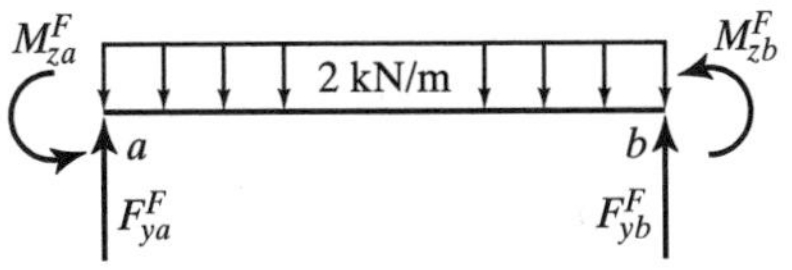

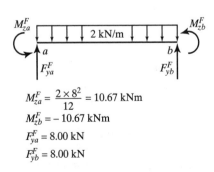

$$M_{za}^F = \frac{2 \times 8^2}{12} = 10.67 \text{ kNm}$$
$$M_{zb}^F = -10.67 \text{ kNm}$$
$$F_{ya}^F = 8.00 \text{ kN}$$
$$F_{yb}^F = 8.00 \text{ kN}$$

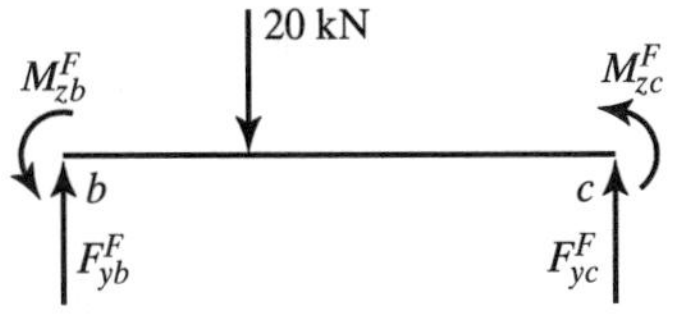

$$M_{zb}^F = \frac{20 \times 2 \times 3^2}{5^2} = 14.40 \text{ kNm}$$
$$M_{zc}^F = \frac{-20 \times 2^2 \times 3}{5^2} = -9.60 \text{ kNm}$$
$$F_{yb}^F = \frac{3 \times 20}{5} + \left(\frac{14.4 - 9.6}{5}\right) = 12.96 \text{ kN}$$
$$F_{yc}^F = 7.04 \text{ kN}$$

1. ***Displacements.*** Use Equation 3.6a and the relevant stiffness equations of Example 4.8.

$$\begin{Bmatrix} R_{ya} \\ P_{mza} \\ R_{yb} \\ P_{mzb} \\ R_{yc} \\ R_{mzc} \end{Bmatrix} = 200 \begin{bmatrix} 0.00469 & 18.75 & -0.00469 & 18.75 & 0 & 0 \\ & 1 \times 10^5 & -18.75 & 0.5 \times 10^5 & 0 & 0 \\ & & 0.00949 & -6.75 & -0.00480 & 12.00 \\ & \text{Sym.} & & 1.4 \times 10^5 & -12.00 & 0.2 \times 10^5 \\ & & & & 0.00480 & -12.00 \\ & & & & & 0.4 \times 10^5 \end{bmatrix} \begin{Bmatrix} v_a \\ \theta_{za} \\ v_b \\ \theta_{zb} \\ v_c \\ \theta_{zc} \end{Bmatrix} + \begin{Bmatrix} 8.00 \\ 10.67 \times 10^3 \\ 20.96 \\ 3.73 \times 10^3 \\ 7.04 \\ -9.60 \times 10^3 \end{Bmatrix}$$

Let $P_{mza} = P_{mzb} = 0$; $v_a = v_b = v_c = \theta_{zc} = 0$; reorder and partition:

$$\begin{Bmatrix} 0 \\ 0 \\ \hline R_{ya} \\ R_{yb} \\ R_{yc} \\ R_{mzc} \end{Bmatrix} = 200 \left[\begin{array}{cc|cccc} \theta_{za} & \theta_{zb} & v_a & v_b & v_c & \theta_{zc} \\ 1 \times 10^5 & 0.5 \times 10^5 & 18.75 & -18.75 & 0 & 0 \\ & 1.4 \times 10^5 & 18.75 & -6.75 & -12.00 & 0.2 \times 10^5 \\ \hline & & 0.00469 & -0.00469 & 0 & 0 \\ & \text{Sym} & & 0.00949 & -0.00480 & 12.00 \\ & & & & 0.00480 & -12.00 \\ & & & & & 0.4 \times 10^5 \end{array}\right] \begin{Bmatrix} \theta_{za} \\ \theta_{zb} \\ \hline 0 \\ 0 \\ 0 \\ 0 \end{Bmatrix} + \begin{Bmatrix} 10.67 \times 10^3 \\ 3.73 \times 10^3 \\ \hline 8.00 \\ 20.96 \\ 7.04 \\ -9.60 \times 10^3 \end{Bmatrix}$$

Expand the upper partition and solve for $\lfloor \theta_{za} \quad \theta_{zb} \rfloor^T$:

$$\begin{Bmatrix} \theta_{za} \\ \theta_{zb} \end{Bmatrix} = \frac{10^{-2}}{200} \begin{bmatrix} 1 & 0.5 \\ 0.5 & 1.4 \end{bmatrix}^{-1} \begin{Bmatrix} -10.67 \\ -3.73 \end{Bmatrix} = \begin{Bmatrix} -5.684 \times 10^{-4} \\ 0.698 \times 10^{-4} \end{Bmatrix} \text{rad}$$

2. *Reactions.* From the lower partition.

$$\begin{Bmatrix} R_{ya} \\ R_{yb} \\ R_{yc} \\ R_{mzc} \end{Bmatrix} = 2 \times 10^{-2} \begin{bmatrix} \theta_{za} & \theta_{zb} \\ 18.75 & 18.75 \\ -18.75 & -6.75 \\ 0 & -12.00 \\ 0 & 0.2 \times 10^5 \end{bmatrix} \begin{Bmatrix} -5.684 \\ 0.698 \end{Bmatrix} + \begin{Bmatrix} 8.00 \\ 20.96 \\ 7.04 \\ -9.60 \times 10^3 \end{Bmatrix}$$

$$= \begin{Bmatrix} 6.13 \text{ kN} \\ 23.00 \text{ kN} \\ 6.87 \text{ kN} \\ -9.32 \times 10^3 \text{ kNmm} \end{Bmatrix}$$

Bending moments. Use the member stiffness equations of Example 4.8.

$$\begin{Bmatrix} M_{za}^{ab} \\ M_{zb}^{ab} \\ M_{zb}^{bc} \end{Bmatrix} = 2 \times 10^{-2} \begin{bmatrix} \theta_{za} & \theta_{zb} \\ 1 \times 10^5 & 0.5 \times 10^5 \\ 0.5 \times 10^5 & 1 \times 10^5 \\ 0 & 0.4 \times 10^5 \end{bmatrix} \begin{Bmatrix} -5.684 \\ 0.698 \end{Bmatrix} + \begin{Bmatrix} 10.67 \times 10^3 \\ -10.67 \times 10^3 \\ 14.40 \times 10^3 \end{Bmatrix}$$

$$= \begin{Bmatrix} 0 \\ -14.96 \times 10^3 \\ 14.96 \times 10^3 \end{Bmatrix} \text{kNmm}$$

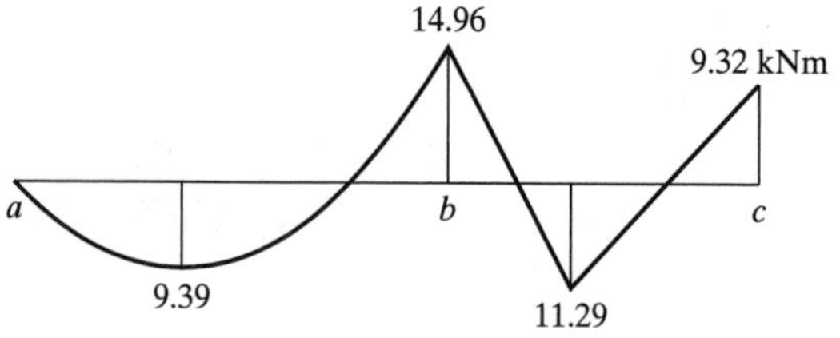

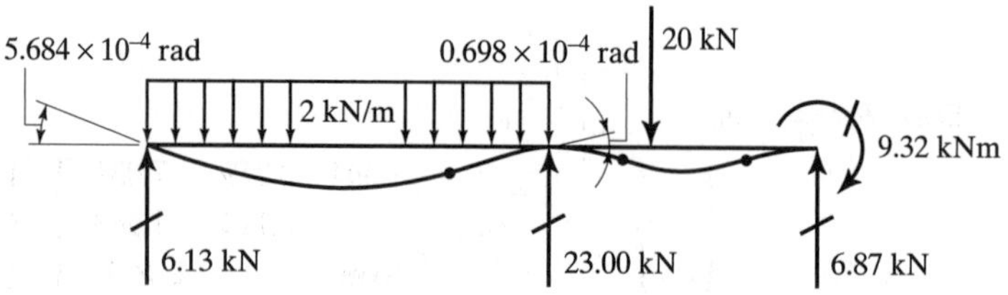

EXAMPLE 5.7

The rigid frame shown is made of elements studied in Examples 4.8 and 5.3.

1. Calculate the displacement at b. Include flexural and axial deformation effects.
2. Calculate the reactions.

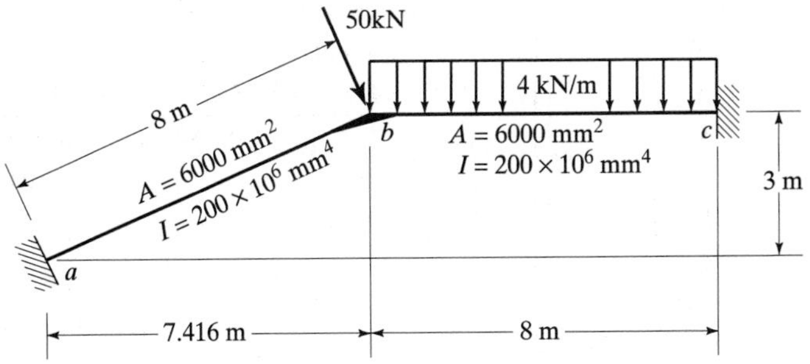

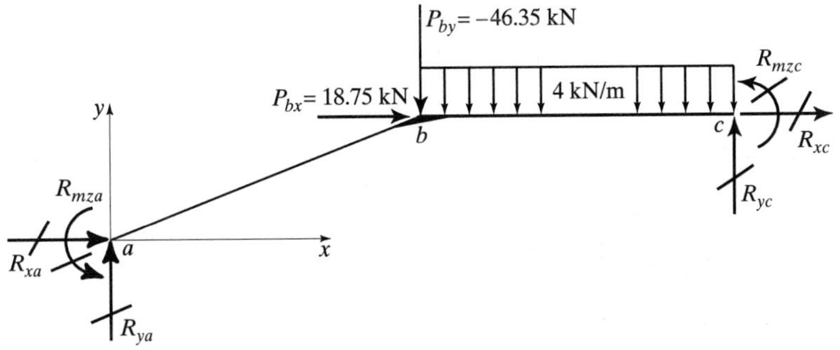

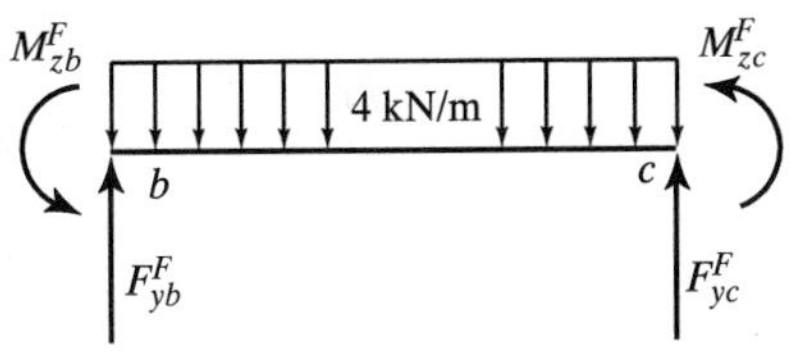

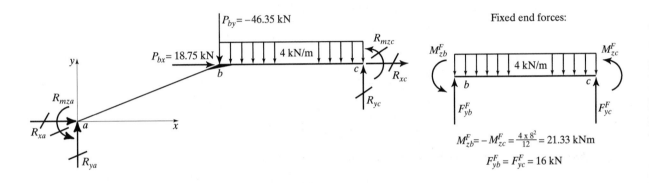

1. ***Displacements.*** Use Equation 3.8a and stiffness equations for the nonzero degrees of freedom drawn from Examples 4.8 and 5.3.

$$\begin{Bmatrix} P_{xb} \\ P_{yb} \\ P_{mzb} \end{Bmatrix} = \begin{Bmatrix} 18.75 \\ -46.35 \\ 0 \end{Bmatrix}$$

$$= 200\begin{bmatrix} (0.645 + 0.750) & (0.259 + 0) & (7.031 + 0) \\ & (0.109 + 0.00469) & (-17.382 + 18.75) \\ \text{Sym.} & & (1 + 1)10^5 \end{bmatrix}\begin{Bmatrix} u_b \\ v_b \\ \theta_{zb} \end{Bmatrix} + \begin{Bmatrix} 0 \\ 16 \\ 21.33 \times 10^3 \end{Bmatrix}$$

Solve for $\lfloor u_b \quad v_b \quad \theta_{zb} \rfloor^T$.

$$\begin{Bmatrix} u_b \\ v_b \\ \theta_{zb} \end{Bmatrix} = 0.5 \times 10^{-2}\begin{bmatrix} 1.394 & 0.259 & 7.031 \\ & 0.114 & 1.368 \\ \text{Sym.} & & 2 \times 10^5 \end{bmatrix}^{-1}\begin{Bmatrix} 18.75 \\ -62.35 \\ -21.33 \times 10^3 \end{Bmatrix} = \begin{Bmatrix} 0.09982 \text{ mm} \\ -4.996 \text{ mm} \\ -0.000534 \text{ rad} \end{Bmatrix}$$

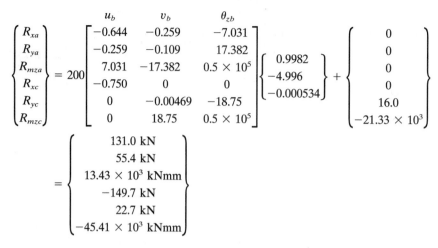

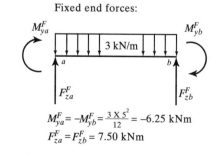

2. ***Reactions.*** Use member stiffness equations from Examples 4.8 and 5.3.

$$
\begin{Bmatrix} R_{xa} \\ R_{ya} \\ R_{mza} \\ R_{xc} \\ R_{yc} \\ R_{mzc} \end{Bmatrix}
= 200
\begin{array}{c}
\begin{array}{ccc} \quad u_b \quad & \quad v_b \quad & \quad \theta_{zb} \quad \end{array} \\
\begin{bmatrix}
-0.644 & -0.259 & -7.031 \\
-0.259 & -0.109 & 17.382 \\
7.031 & -17.382 & 0.5 \times 10^5 \\
-0.750 & 0 & 0 \\
0 & -0.00469 & -18.75 \\
0 & 18.75 & 0.5 \times 10^5
\end{bmatrix}
\end{array}
\begin{Bmatrix} 0.9982 \\ -4.996 \\ -0.000534 \end{Bmatrix}
+
\begin{Bmatrix} 0 \\ 0 \\ 0 \\ 0 \\ 16.0 \\ -21.33 \times 10^3 \end{Bmatrix}
$$

$$
= \begin{Bmatrix} 131.0 \text{ kN} \\ 55.4 \text{ kN} \\ 13.43 \times 10^3 \text{ kNmm} \\ -149.7 \text{ kN} \\ 22.7 \text{ kN} \\ -45.41 \times 10^3 \text{ kNmm} \end{Bmatrix}
$$

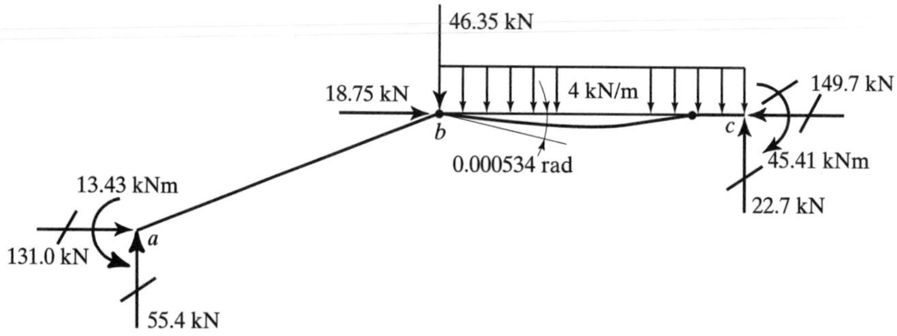

EXAMPLE 5.8

The rigid frame (grid) shown lies in a horizontal plane. The loads act vertically. Member *ab* has the properties of member *bc* of Example 4.8, and member *bc* has the properties of member *ab* of that example.

1. Calculate the displacement at *b*.
2. Calculate the reactions.

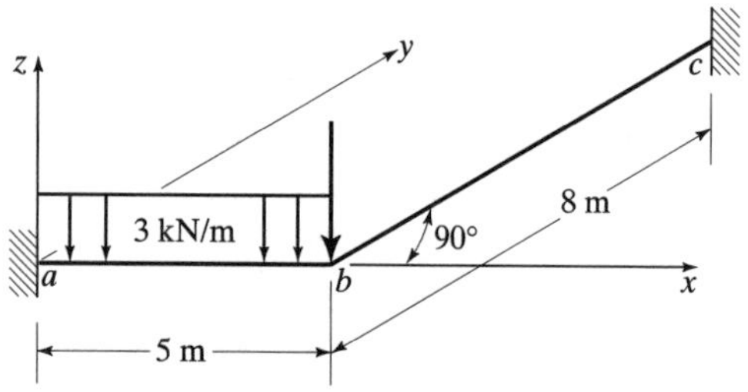

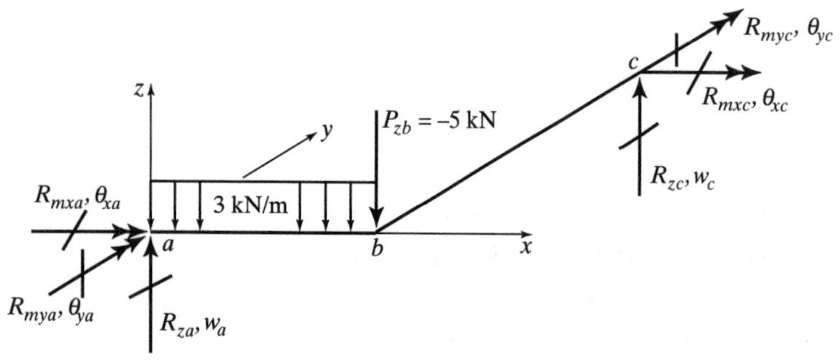

Fixed end forces:

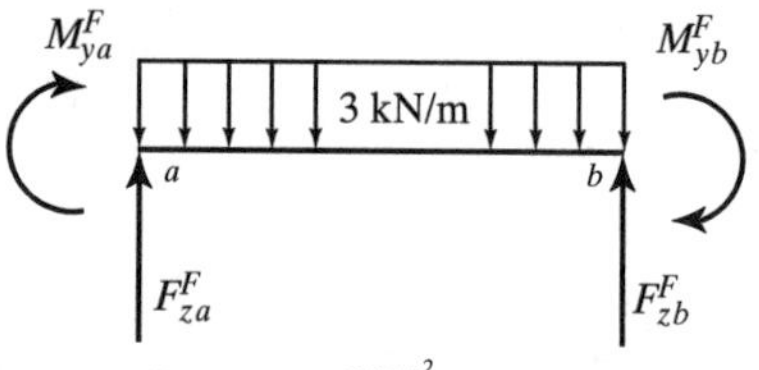

$M^F_{ya} = -M^F_{yb} = \frac{3 \times 5^2}{12} = -6.25$ kNm

$F^F_{za} = F^F_{zb} = 7.50$ kNm

1. ***Displacements.*** Following the style of Example 5.3 and using the results of Example 4.8, develop the member stiffness matrices in global coordinates.

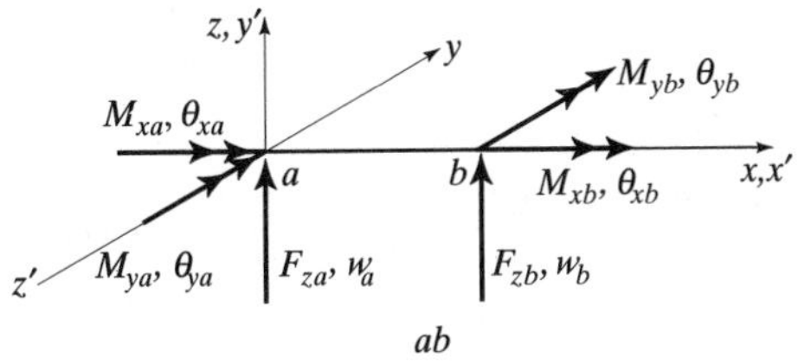

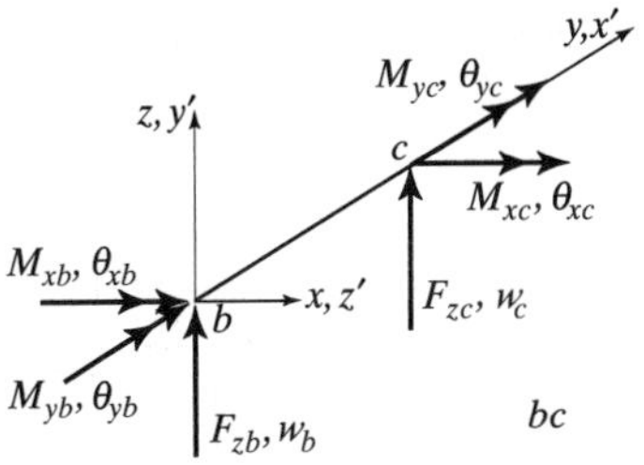

Member *ab*

[**k'**]:

$$200\begin{array}{c} \\ \end{array}\begin{bmatrix} v'_a & \theta_{x'a} & \theta_{z'a} & v'_b & \theta_{x'b} & \theta_{z'b} \\ \hline 0.0048 & 0 & 12.00 & -0.0048 & 0 & 12.00 \\ & 7.692 & 0 & 0 & -7.692 & 0 \\ & & 0.4 \times 10^5 & -12.00 & 0 & 0.2 \times 10^5 \\ & & & 0.0048 & 0 & -12.00 \\ & & \text{Sym.} & & 7.692 & 0 \\ & & & & & 0.4 \times 10^5 \end{bmatrix}$$

[**Γ**]:

$$\begin{bmatrix} 1 & 0 & 0 & 0 & 0 & 0 \\ 0 & 1 & 0 & 0 & 0 & 0 \\ 0 & 0 & -1 & 0 & 0 & 0 \\ 0 & 0 & 0 & 1 & 0 & 0 \\ 0 & 0 & 0 & 0 & 1 & 0 \\ 0 & 0 & 0 & 0 & 0 & -1 \end{bmatrix}$$

[**k**]:

$$200\begin{bmatrix} w_a & \theta_{xa} & \theta_{ya} & w_b & \theta_{xb} & \theta_{yb} \\ \hline 0.0048 & 0 & -12.00 & -0.0048 & 0 & -12.00 \\ & 7.692 & 0 & 0 & -7.692 & 0 \\ & & 0.4 \times 10^5 & 12.00 & 0 & 0.2 \times 10^5 \\ & & & 0.0048 & 0 & 12.00 \\ & & \text{Sym.} & & 7.692 & 0 \\ & & & & & 0.4 \times 10^5 \end{bmatrix}$$

Member *bc*

[**k'**]:

$$200\begin{bmatrix} v'_b & \theta_{x'b} & \theta_{z'b} & v'_c & \theta_{x'c} & \theta_{z'c} \\ \hline 0.00469 & 0 & 18.75 & -0.00469 & 0 & 18.75 \\ & 14.423 & 0 & 0 & -14.423 & 0 \\ & & 1 \times 10^5 & -18.75 & 0 & 0.5 \times 10^5 \\ & & & 0.00469 & 0 & -18.75 \\ & \text{Sym.} & & & 14.423 & 0 \\ & & & & & 1 \times 10^5 \end{bmatrix}$$

[**Γ**]:

$$\begin{bmatrix} 1 & 0 & 0 & 0 & 0 & 0 \\ 0 & 0 & 1 & 0 & 0 & 0 \\ 0 & 1 & 0 & 0 & 0 & 0 \\ 0 & 0 & 0 & 1 & 0 & 0 \\ 0 & 0 & 0 & 0 & 0 & 1 \\ 0 & 0 & 0 & 0 & 1 & 0 \end{bmatrix}$$

[**k**]:

$$200\begin{bmatrix} w_b & \theta_{xb} & \theta_{yb} & w_c & \theta_{xc} & \theta_{yc} \\ \hline 0.00469 & 18.75 & 0 & -0.00469 & 18.75 & 0 \\ & 1 \times 10^5 & 0 & -18.75 & 0.5 \times 10^5 & 0 \\ & & 14.423 & 0 & 0 & -14.423 \\ & & & 0.00469 & -18.75 & 0 \\ & \text{Sym.} & & & 1 \times 10^5 & 0 \\ & & & & & 14.423 \end{bmatrix}$$

Assemble global stiffness equations for the nonzero degrees of freedom: w_b, θ_{xb}, θ_{yb}:

$$\begin{Bmatrix} P_{zb} \\ P_{mxb} \\ P_{myb} \end{Bmatrix} = \begin{Bmatrix} -5.00 \\ 0 \\ 0 \end{Bmatrix} = 200 \begin{bmatrix} 0.00949 & 18.75 & 12.00 \\ & 1 \times 10^5 & 0 \\ \text{Sym.} & & 0.4 \times 10^5 \end{bmatrix} \begin{Bmatrix} w_b \\ \theta_{xb} \\ \theta_{yb} \end{Bmatrix} + \begin{Bmatrix} 7.50 \\ 0 \\ 6.25 \times 10^3 \end{Bmatrix}$$

Solve for $\lfloor w_b \quad \theta_{xb} \quad \theta_{yb} \rfloor^T$:

$$\begin{Bmatrix} w_b \\ \theta_{xb} \\ \theta_{yb} \end{Bmatrix} = 5.264 \times 10^{-8} \begin{bmatrix} 0.4 \times 10^8 & -7.5 \times 10^3 & -12 \times 10^3 \\ & 2.365 & 2.25 \\ \text{Sym.} & & 5.974 \end{bmatrix} \begin{Bmatrix} -12.50 \\ 0 \\ -6.25 \times 10^3 \end{Bmatrix}$$

$$= \begin{Bmatrix} -22.37 \text{ mm} \\ 4.195 \times 10^{-3} \text{ rad} \\ 5.931 \times 10^{-3} \text{ rad} \end{Bmatrix}$$

2. *Reactions.* Use member stiffness equations:

$$\begin{Bmatrix} R_{za} \\ R_{mxa} \\ R_{mya} \\ R_{zc} \\ R_{mxc} \\ R_{myc} \end{Bmatrix} = 200 \begin{bmatrix} w_b & \theta_{xb} & \theta_{yb} \\ -0.00480 & 0 & -12.00 \\ 0 & -7.692 & 0 \\ 12.00 & 0 & 0.2 \times 10^5 \\ -0.00469 & -18.75 & 0 \\ 18.75 & 0.5 \times 10^5 & 0 \\ 0 & 0 & -14.423 \end{bmatrix} \begin{Bmatrix} -22.37 \\ 4.195 \times 10^{-3} \\ 5.931 \times 10^{-3} \end{Bmatrix} + \begin{Bmatrix} 7.50 \\ 0 \\ -6.25 \times 10^3 \\ 0 \\ 0 \\ 0 \end{Bmatrix}$$

$$= \begin{Bmatrix} 14.74 \text{ kN} \\ -6.45 \text{ kNmm} \\ -36.21 \times 10^3 \text{ kNmm} \\ 5.25 \text{ kN} \\ -41.94 \times 10^3 \text{ kNmm} \\ -17.11 \text{ kNmm} \end{Bmatrix}$$

Torque diagram:

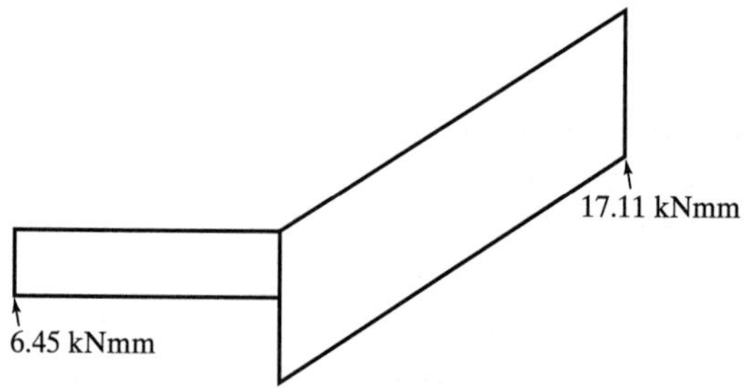

Moment diagram:

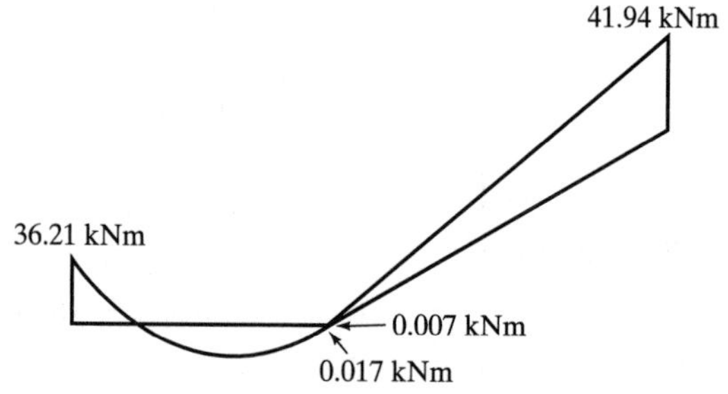

Deflected structure:

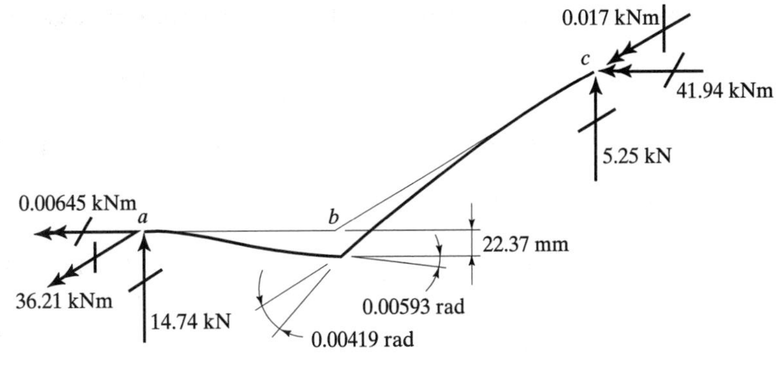

EXAMPLE 5.9

In the planar rigid frame shown, member ab corresponds to member ab of Example 5.3, member bc corresponds to member ab of Example 4.8, and member dc corresponds to member ed of Example 5.3. Neglect axial deformations.

1. Calculate the displacements under a uniformly distributed vertical load of 2 kN/m. Use symmetry to reduce number of unknowns.
2. Calculate the displacements under horizontal loads of 2.5 kN at b and c. Use antisymmetry to reduce the number of unknowns.

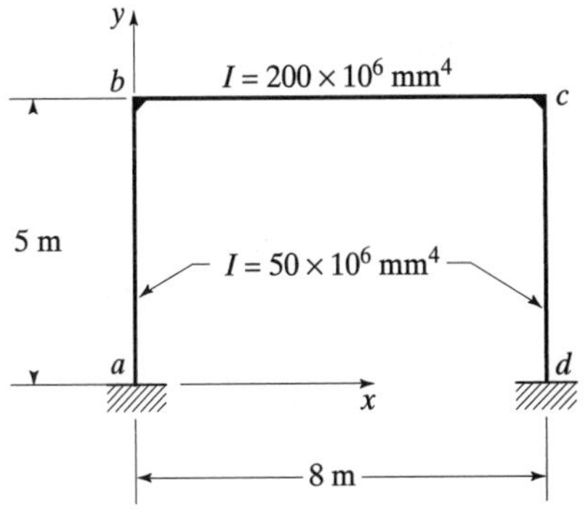

Neglecting axial deformations, $v_b = v_c = 0$. Assemble global stiffness equations for nonzero degrees of freedom. Use Examples 4.8 and 5.3.

$$\begin{Bmatrix} P_{xb} \\ P_{mb} \\ P_{xc} \\ P_{mc} \end{Bmatrix} = 200 \begin{bmatrix} 0.00480 & 12.000 & 0 & 0 \\ & 1.4 \times 10^5 & 0 & 0.5 \times 10^5 \\ \text{Sym.} & & 0.00480 & 12.000 \\ & & & 1.4 \times 10^5 \end{bmatrix} \begin{Bmatrix} u_b \\ \theta_b \\ u_c \\ \theta_c \end{Bmatrix} + \begin{Bmatrix} P_{xb}^F \\ P_{mb}^F \\ P_{xc}^F \\ P_{mc}^F \end{Bmatrix}$$

1. ***Uniform vertical load.*** By symmetry, $u_c = -u_b$ and $\theta_c = -\theta_b$. Considering both symmetry and zero axial displacements, $u_b = u_c = 0$. Stiffness equations may be written

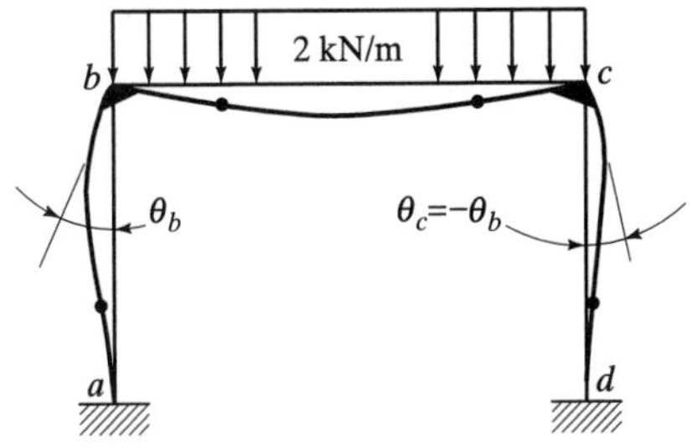

$$\begin{Bmatrix} 0 \\ 0 \\ 0 \\ 0 \end{Bmatrix} = 200 \begin{bmatrix} 0.00480 & 12.000 & 0 & 0 \\ & 1.4 \times 10^5 & 0 & 0.5 \times 10^5 \\ \text{Sym.} & & 0.00480 & 12.000 \\ & & & 1.4 \times 10^5 \end{bmatrix} \begin{Bmatrix} 0 \\ \theta_b \\ 0 \\ -\theta_b \end{Bmatrix}$$

$$+ \begin{Bmatrix} 0 \\ 10.67 \times 10^3 \\ 0 \\ -10.67 \times 10^3 \end{Bmatrix} \text{kNmm}$$

Expand second equation and solve for θ_b:

$$0 = 200(1.4 - 0.5)10^5 \theta_b + 10.67 \times 10^3$$

$$\theta_b = -0.000593 \text{ rad}$$

2. ***Horizontal load.*** Letting $u_c = u_b$ and noting that, by antisymmetry, $\theta_c = \theta_b$, stiffness equations may be written:

$$\begin{Bmatrix} P_{xb} \\ P_{mb} \\ P_{xc} \\ P_{mc} \end{Bmatrix} = \begin{Bmatrix} 2.5 \\ 0 \\ 2.5 \\ 0 \end{Bmatrix} = 200 \begin{bmatrix} 0.00480 & 12.000 & 0 & 0 \\ & 1.4 \times 10^5 & 0 & 0.5 \times 10^5 \\ \text{Sym.} & & 0.00480 & 12.000 \\ & & & 1.4 \times 10^5 \end{bmatrix} \begin{Bmatrix} u_b \\ \theta_b \\ u_b \\ \theta_b \end{Bmatrix}$$

Simplify the first two equations and solve for $\lfloor u_b \quad \theta_b \rfloor^{\mathrm{T}}$

$$\begin{Bmatrix} 2.5 \\ 0 \end{Bmatrix} = 200 \begin{bmatrix} 0.00480 & 12.000 \\ 12.000 & 1.9 \times 10^5 \end{bmatrix} \begin{Bmatrix} u_b \\ \theta_b \end{Bmatrix}$$

$$\begin{Bmatrix} u_b \\ \theta_b \end{Bmatrix} = \begin{Bmatrix} 3.09 \text{ mm} \\ -0.000195 \text{ rad} \end{Bmatrix}$$

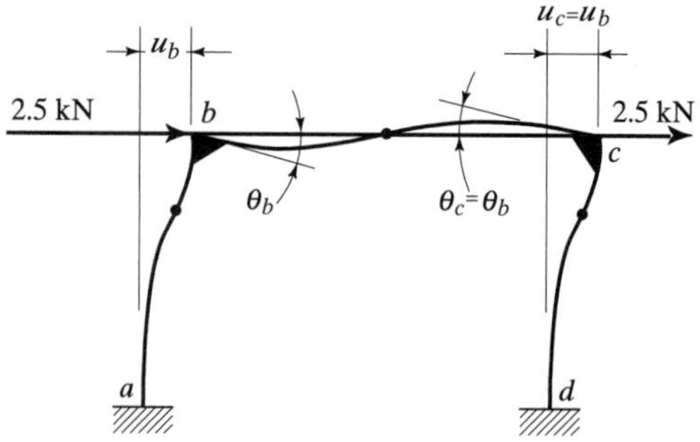

5.3 SELF-STRAINING—INITIAL AND THERMAL STRAIN CONDITIONS

A self-strained structure may be defined as any system that is internally strained and in a state of stress while at rest and sustaining no external load. Several examples are illustrated in Figure 5.7. A bar fabricated overly long and forced into place between fixed ends is one case (Figure 5.7*a*). To ensure tightness in the diagonal bracing of truss systems, some tension bars may be fabricated short and then drawn into position, producing self-straining of the type indicated in Figure 5.7*b*. If, for some reason, a beam is fabricated with an upward bow and then pulled into place, as in Figure 5.7*c*,

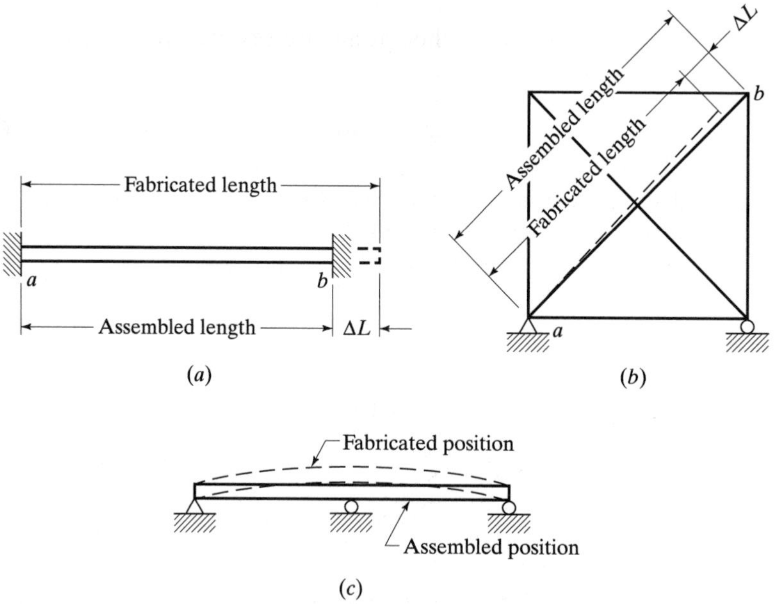

Figure 5.7 Self-strained structures. (*a*) Compressed fit. (*b*) Draw of tie rod. (*c*) Fabricated bow.

self-straining occurs. Shrink-fit problems, prestressed concrete girders, and cable-stayed bridges, in which the cables are adjusted to produce a bow in a girder that is fabricated straight, are other examples of self-strained structures. The analysis of all of these circumstances is within the scope of the approach to be described below. The state of residual stress that may exist in an unloaded member that has previously been loaded until part of the system has yielded is another example of self-straining, but it is not one that is treated in this chapter (Example 10.9 illustrates this type of self-straining).

Temperature changes with respect to the unloaded, stress-free state of a structure may also cause self-straining. If the natural expansion or contraction of a system under such temperature changes is resisted in any way, internal forces develop and the system may be self-strained. If the bar in Figure 5.8 is heated, it tries to expand an amount $\Delta L = \alpha LT$, where α is the coefficient of thermal expansion and T is the temperature change, presumed to be uniform in the illustration. If this expansion is prevented by fixed ends, an axial force is transmitted to the bar and it is correspondingly stressed.

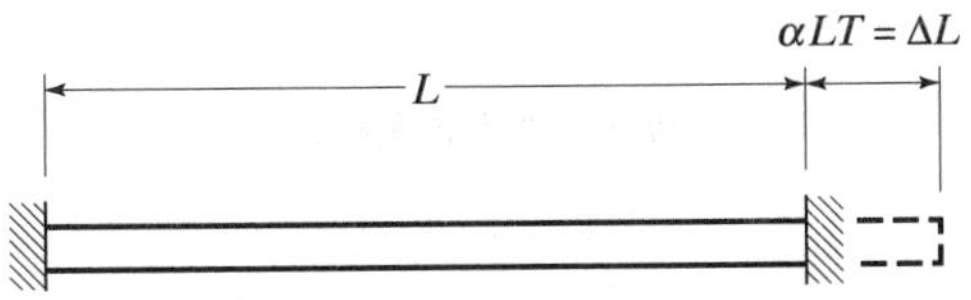

Figure 5.8 Temperature restraint.

All of the structures in Figures 5.7 and 5.8 have one thing in common, they are statically indeterminate. It is impossible for a statically determinate elastic structure to be self-strained. In the statically determinate, pin-jointed truss of Figure 5.9*a*, an imposed shortening or lengthening of any bar, from any cause, may be accommodated by a stress-free change of geometry. Likewise, the right end of the simply supported beam in Figure 5.9*b* can settle, or the end can expand or contract thermally, without any stress change. Although no stresses are produced in such cases, design requirements might prompt the calculation of the displacements that occur.

At this point it is important to recall that strain is a kinematic quantity. As defined by Love (Ref. 5.2, p. 32): "Whenever, owing to any cause, changes take place in the relative positions of the parts of a body the body is said to be *strained*." In almost all of the problems treated earlier, the causative agent was stress. But stress is not necessary. Strain may also result from temperature change, for example. As soon as we recognize temperature change as a source of strain, it becomes clear that we must

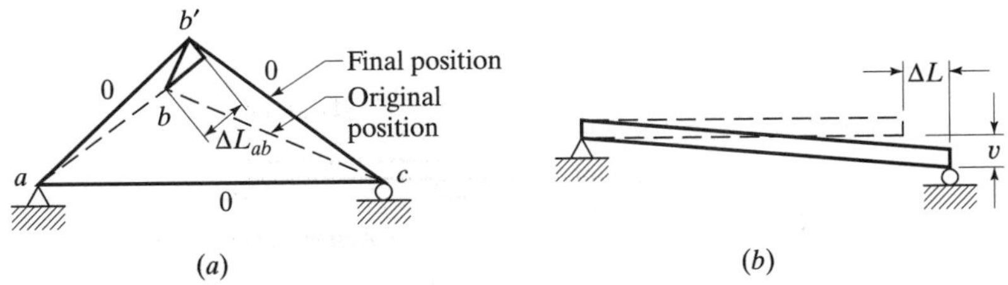

Figure 5.9 Displacements of statically determinate structure. (*a*) Simple truss—result of elongation of bar *ab*. (*b*) Simple beam—result of beam expansion and support settlement.

measure strain from some reference temperature and the geometry of the system at that temperature. From this it is a simple step to the recognition that *any* geometrical configuration of a system that is different from a defined *reference configuration* and that is not the result of rigid body motion must be the result of the strain, that is, changes in the relative positions of the parts of the body.

To illustrate these concepts, we define the assembled (solid line) configuration of each of the structures in Figures 5.7 and 5.8 as its reference configuration. Each can be viewed as having undergone a strain *prior* to assembly. In the members *ab* of Figures 5.7*a* and 5.7*b*, for example, the displacements ΔL can be regarded as having arisen because of unit strains $\Delta L/L$. Initial strains can likewise be regarded as the source of the curved fabricated state of the beam in Figure 5.7*c*. The thermal expansion of the member in Figure 5.8 is due to a unit thermal strain αT. The forces required for the assembly of these initially strained members cause further strains throughout the system. Hence, the reference configuration is internally strained and in a state of stress while sustaining no external load; that is, it is self-strained. The strains both prior and subsequent to assembly must be accounted for in the structural analysis.[2]

5.3.1 Initial Strain Problems

Structural analysis for the effects of self-straining can be accomplished through an approach that, in principle, is identical to the one developed earlier for loads applied between nodal points:

1. We first assume the presence of fictitious constraints at each of the degrees of freedom of the reference configuration. We then determine the magnitude of the forces required of these constraints to counteract the source of the initial strain and maintain compatibility in the reference state with no nodal point displacement, that is, in a fixed-end condition. For example, suppose one bar of an indeterminate truss is fabricated short a prescribed amount (Figure 5.7*b*). The fixed-end forces would be those needed to draw the two ends of the bar into their compatible, but undisplaced, position in the reference configuration and thus to permit them to be joined to the other members, so that, subsequently, they can function as regular force resisting elements of the assembled structure.
2. We next annul these fixed-end forces by applying forces equal and opposite to them as equivalent nodal loads acting on the assembled structure, and we calculate the displacements they cause.
3. The resultant displacements, reactions, and internal forces are, just as in the case of distributed loads, the sum of those found in the fixed-end analysis and the nodal displacement analysis (see again Figure 5.6).

The process is illustrated in Example 5.10. Bar *ad* is fabricated 3 mm too short. What are the consequences? As seen in the example, large forces and substantial displacements are developed, purely as the result of self-straining. Example 5.11, a support settlement problem, is an additional illustration. Here the fixed end forces are those caused by the prescribed settlement with all other degrees of freedom restrained. The coefficients needed for determining these fictitious forces may be taken directly from

---

[2]The view that initial strains are the origin of self-straining conditions in structures is important in the development of general procedures in matrix structural analysis and will be adopted in Section 7.5. Consistent with the approach we have taken in these early chapters, however, we are following a simple physical approach here.

the beam element stiffness matrix as shown in the example. Comparison of this example and Example 4.11 discloses that support settlement problems can be solved as either self-straining or specified joint-displacement problems.

EXAMPLE 5.10

The truss shown is the same as in Example 3.3 except that the bar *ad* is fabricated 3 mm too short.

1. Calculate the displacement at *a* and *b* after *ad* is joined to the remainder.
2. Calculate the reactions.
   $E = 200{,}000$ MPa.

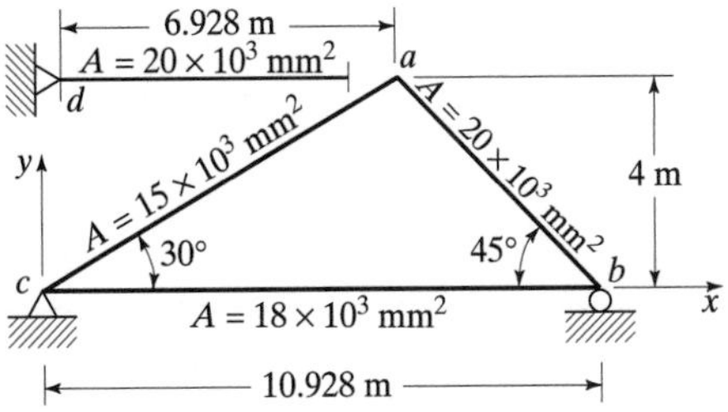

Assembled structure:

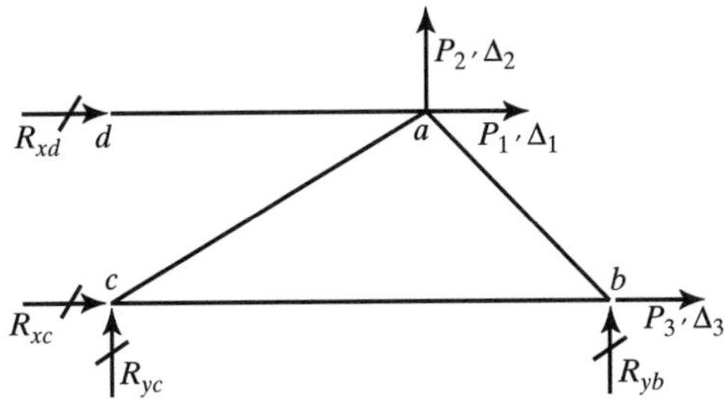

Fixed end forces (see Example 3.3):

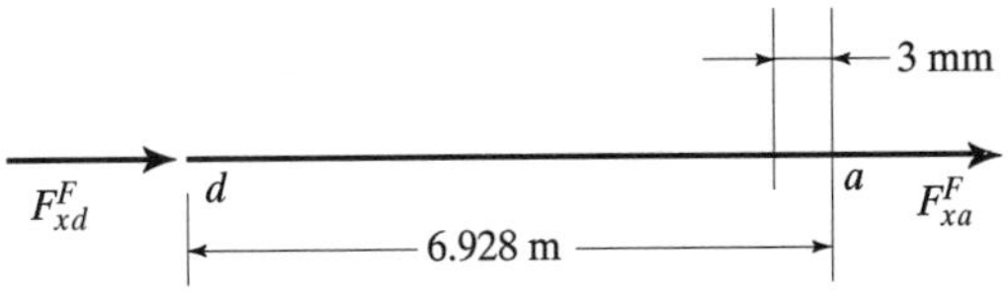

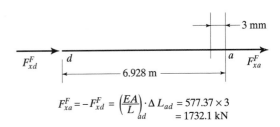

$$F^F_{xa} = -F^F_{xd} = \left(\frac{EA}{L}\right)_{ad} \cdot \Delta L_{ad} = 577.37 \times 3 = 1732.1 \text{ kN}$$

1. ***Displacements.*** Following Example 3.3 and Equation 3.8a

$$\begin{Bmatrix} P_1 \\ P_2 \\ P_3 \end{Bmatrix} = \begin{Bmatrix} 0 \\ 0 \\ 0 \end{Bmatrix} = 10^2 \begin{bmatrix} 12.122 & -1.912 & -3.536 \\ & 4.473 & 3.536 \\ \text{Sym.} & & 6.830 \end{bmatrix} \begin{Bmatrix} \Delta_1 \\ \Delta_2 \\ \Delta_3 \end{Bmatrix} + \begin{Bmatrix} 1732.1 \\ 0 \\ 0 \end{Bmatrix}$$

Solving for the displacements,

$$\lfloor \Delta_1 \quad \Delta_2 \quad \Delta_3 \rfloor = \lfloor -1.683 \quad -0.0518 \quad -0.8448 \rfloor \text{ mm}$$

2. *Reactions.* See Example 3.3.

$$\begin{Bmatrix} R_{yb} \\ R_{xc} \\ R_{yc} \\ R_{xd} \end{Bmatrix} = 10^2 \begin{bmatrix} \overset{\Delta_1}{3.536} & \overset{\Delta_2}{-3.536} & \overset{\Delta_3}{-3.536} \\ -2.812 & -1.624 & -3.294 \\ -1.624 & -0.938 & 0 \\ -5.774 & 0 & 0 \end{bmatrix} \begin{Bmatrix} -1.683 \\ -0.0518 \\ -0.8448 \end{Bmatrix} + \begin{Bmatrix} 0 \\ 0 \\ 0 \\ -1732.1 \end{Bmatrix}$$

$$= \begin{Bmatrix} -278.1 \\ 760.0 \\ 278.1 \\ -760.2 \end{Bmatrix} \text{kN}$$

EXAMPLE 5.11

The beam of Example 4.8 is supported as shown. The support at b settles 15 mm, carrying the beam with it.

1. Calculate the displacements at a and b.
2. Calculate the reactions.

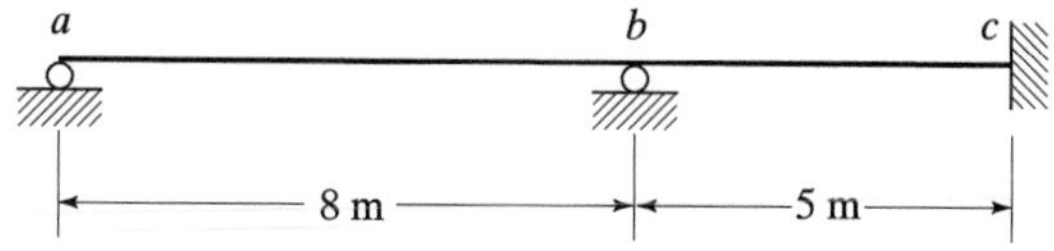

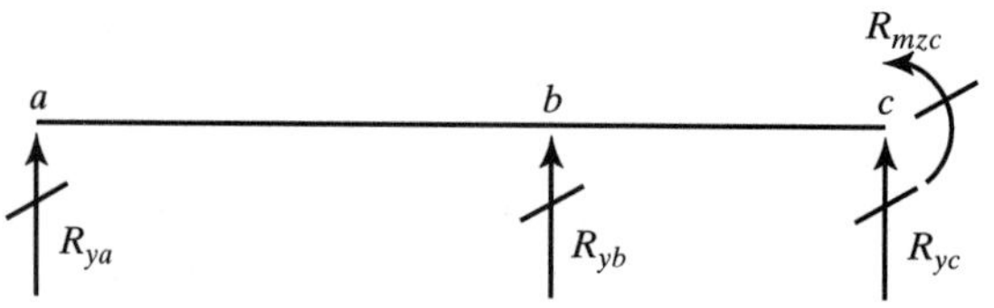

Fixed-end forces–Impose a 15-mm vertical displacement at b with no other joint displacement. Using Example 4.8 or Table 5.1, the fixed-end forces are as follows:

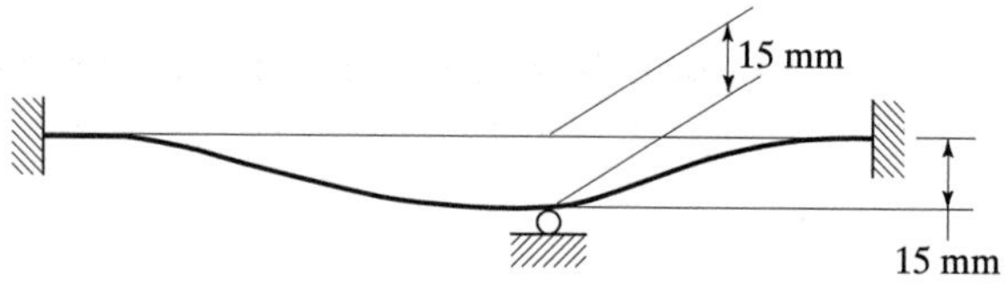

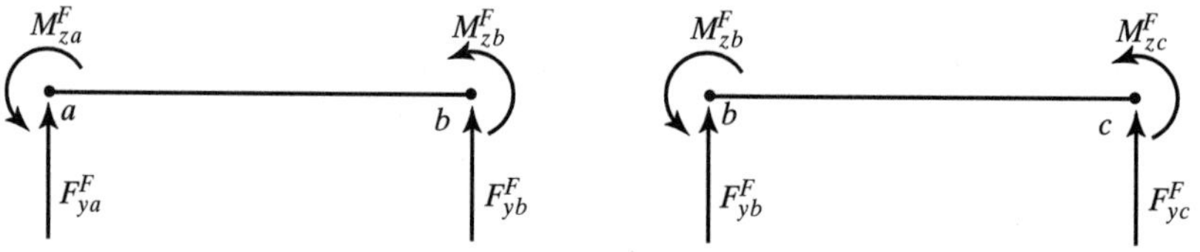

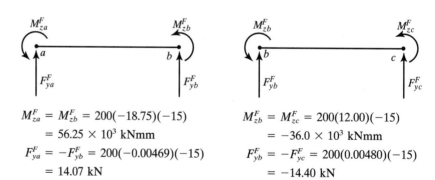

$$M^F_{za} = M^F_{zb} = 200(-18.75)(-15) = 56.25 \times 10^3 \text{ kNmm}$$

$$F^F_{ya} = -F^F_{yb} = 200(-0.00469)(-15) = 14.07 \text{ kN}$$

$$M^F_{zb} = M^F_{zc} = 200(12.00)(-15) = -36.0 \times 10^3 \text{ kNmm}$$

$$F^F_{yb} = -F^F_{yc} = 200(0.00480)(-15) = -14.40 \text{ kN}$$

1. *Displacements.* Use the stiffness equations of Example 5.6 (which is the same structure) and allow no further variation in v_b.

$$\begin{Bmatrix} P_{mza} \\ P_{mzb} \end{Bmatrix} = \begin{Bmatrix} 0 \\ 0 \end{Bmatrix} = 200 \times 10^5 \begin{bmatrix} 1 & 0.5 \\ 0.5 & 1.4 \end{bmatrix} \begin{Bmatrix} \theta_{za} \\ \theta_{zb} \end{Bmatrix} + \begin{Bmatrix} 56.25 \times 10^3 \\ 20.25 \times 10^3 \end{Bmatrix}$$

Solving for the displacements,

$$\lfloor \theta_{za} \quad \theta_{zb} \rfloor = \lfloor -0.002984 \quad 0.000342 \rfloor \text{ rad}$$

2. ***Reactions.*** See Example 5.6.

$$
\begin{Bmatrix} R_{ya} \\ R_{yb} \\ R_{yc} \\ R_{mzc} \end{Bmatrix} = 2 \times 10^{-2} \begin{array}{c} \begin{array}{cc} \theta_{za} \qquad & \qquad \theta_{zb} \end{array} \\ \begin{bmatrix} 18.75 & 18.75 \\ -18.75 & -6.75 \\ 0 & -12.00 \\ 0 & 0.2 \times 10^5 \end{bmatrix} \end{array} \begin{Bmatrix} -29.84 \\ 3.42 \end{Bmatrix} + \begin{Bmatrix} 14.07 \\ -28.47 \\ 14.40 \\ -36.0 \times 10^3 \end{Bmatrix}
$$

$$
= \begin{Bmatrix} 4.16 \text{ kN} \\ -17.74 \text{ kN} \\ 13.58 \text{ kN} \\ -34.63 \times 10^3 \text{ kNmm} \end{Bmatrix}
$$

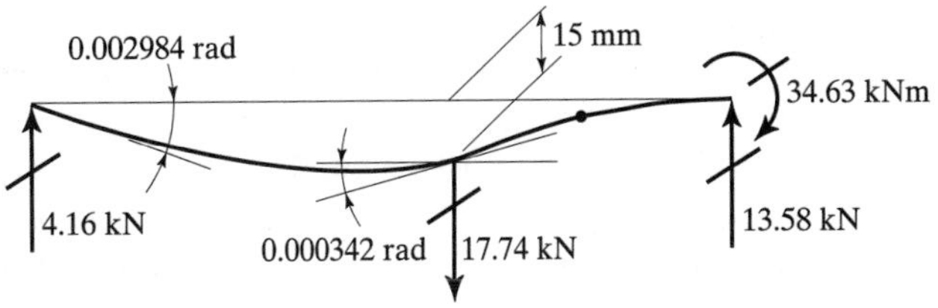

5.3.2 Thermal Strain Problems

To treat thermal problems in the way just described, a method for computing the fixed-end forces resulting from temperature change is needed. A general method is given in Section 7.5. Here, we develop the required relationships for two cases of general interest through simple reasoning.

5.3.2.1 Uniform Temperature Change in an Axial Member We consider a prismatic bar of area A, Young's modulus E, and thermal expansion coefficient α (Figure 5.10a).

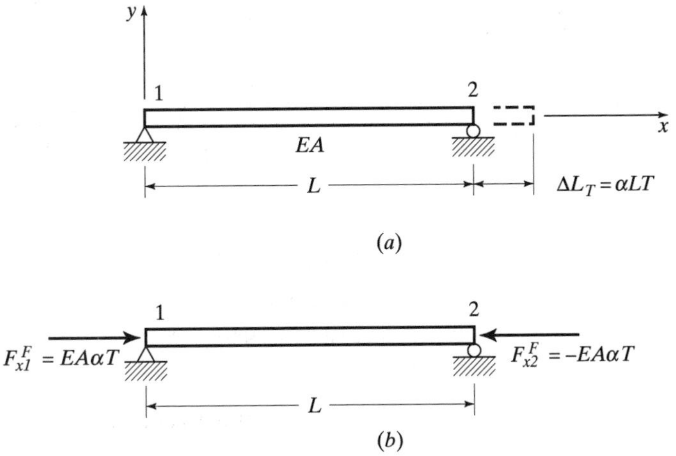

Figure 5.10 Uniform temperature expansion.

Under a uniform temperature increase T the bar would elongate an amount

$$
\Delta L = \alpha LT
$$

if permitted to expand freely. But this is the elongation that would be suppressed by fictitious nodal constraints. For the axial force member, the constraining force would be

$$F = \frac{EA}{L} \cdot \Delta L$$

Combining the two equations and following our usual local sign convention,

$$F_{x1}^F = -F_{x2}^F = EA\alpha T \tag{5.23}$$

Note that these forces are independent of the bar's length.

5.3.2.2 A Thermal Gradient Through the Depth of a Beam Here we consider a bisymmetrical beam of moment of inertia I, depth h, Young's modulus E, and expansion coefficient α (Figure 5.11*a*). The entire bottom surface is heated an amount T_l and the top surface an amount T_u, with a linear temperature gradient $\Delta T = (T_l - T_u)$ in between. If free to expand, the axis of the beam would elongate an amount $\alpha L(T_l + T_u)/2$ and, in addition, the beam would curve in a circular arc as in Figure 5.11*b*. The fixed-end forces necessary to suppress the longitudinal expansion can be calculated from Equation 5.23, with T in that formula taken as $(T_l + T_u)/2$. The fact that the curvature is circular needs to be proved and then we have to show what forces are necessary to constrain it. To do this, consider the thermally induced curvature of an element originally dx long when it is permitted to expand freely. The element is shown to enlarged scale in Figure 5.11c. The angle change between its two faces is

$$d\theta = \frac{\alpha(T_l - T_u)dx}{h}$$

or

$$\frac{d\theta}{dx} = \frac{\alpha(\Delta T)}{h} = \frac{\alpha(T_l - T_u)}{h}$$

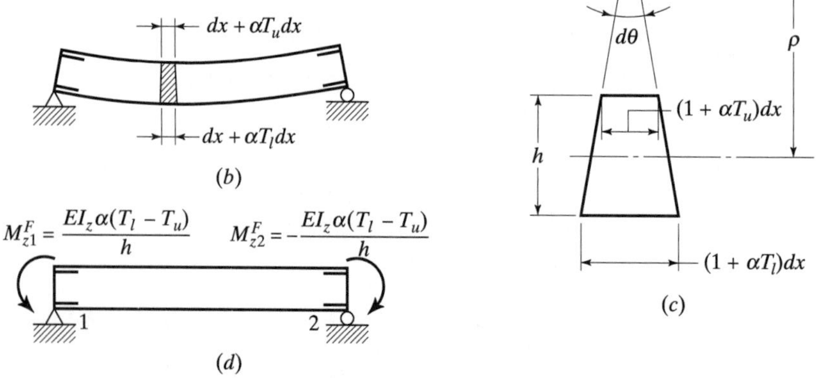

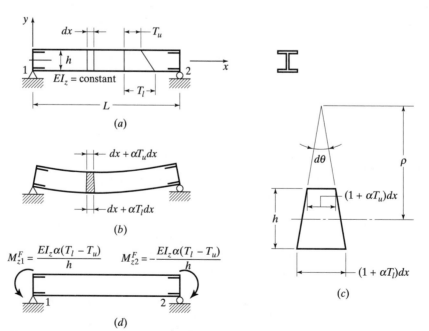

Figure 5.11 Thermal gradient through a beam.

where, here, ΔT is the difference between lower and upper surface temperature. But, neglecting second order effects, $d\theta/dx = 1/\rho$, the curvature of the section. Thus

$$\frac{1}{\rho} = \frac{\alpha(T_l - T_u)}{h}$$

The element selected is typical of conditions all along the span. Thus, since the curvature is constant, the beam is bent in a circular arc. To constrain this we need simply add end moments as in Figure 5.11*d*. From the differential equation of the elastic curve, Equation 4.31*b*,

$$\frac{1}{\rho} = \frac{d^2v}{dx^2} = \frac{M_z}{EI_z}$$

Therefore, combining the two equations for curvature, we have, with proper regard for the signs of the fixed-end moments (see Figure 5.11*d* in which the constraining moments are shown in their physically correct directions),

$$M_{z1}^F = -M_{z2}^F = \frac{EI_z\alpha(T_l - T_u)}{h} \tag{5.24}$$

Again, the fixed-end forces are independent of the length.

The methods of thermal analysis are illustrated in Examples 5.12 and 5.13.

Example 5.12 is an example of the thermal-force analysis of a small truss. The primed forces represent forces in the local coordinate directions, and the unprimed forces represent components in the global coordinate direction. For each member, the origin of local coordinates is taken at the *a* end.

EXAMPLE 5.12

In the pin-jointed truss shown, all bars are cooled 20°C. $\alpha = 1.2 \times 10^{-5}$ mm/mm°C. $E = 200{,}000$ MPa.

1. Calculate the displacement at *a*.
2. Calculate the reactions.
3. Reduce the thermal displacement under uniform temperature change to zero by suitably altering the areas of bars *ac* and *ad*.

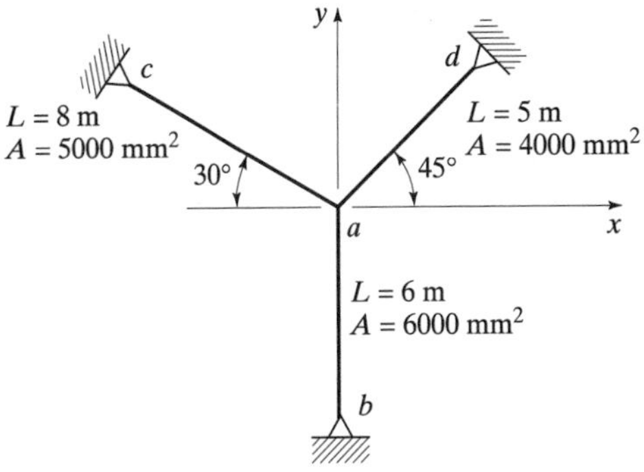

Fixed-end forces—Following Equation 5.23,

$$F_{xa}^F = F_{xb}^F = 0$$

$$F_{ya}^F = -F_{yb}^F = -F_{x'a}^F$$

$$= -200 \times 6 \times 10^3 \times 1.2 \times 10^{-5}(-20) = 288.0 \text{ kN}$$

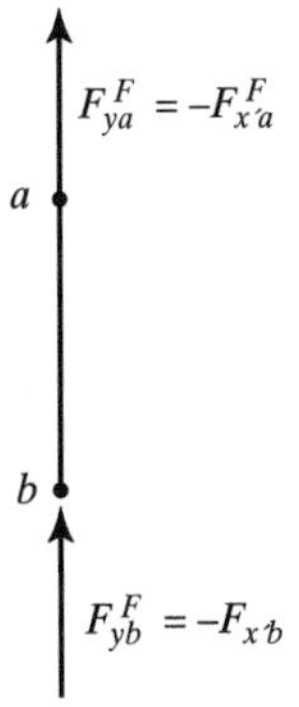

$$F^F_{xa} = -F^F_{xc} = -F^F_{x'a}\cos 30°$$
$$= -200 \times 5 \times 10^3 \times 1.2 \times 10^{-5}(-20)\cos 30° = 207.9 \text{ kN}$$
$$F^F_{ya} = -F^F_{yc} = F^F_{x'a}\sin 30°$$
$$= 200 \times 5 \times 10^3 \times 1.2 \times 10^{-5}(-20)\sin 30° = -120.0 \text{ kN}$$

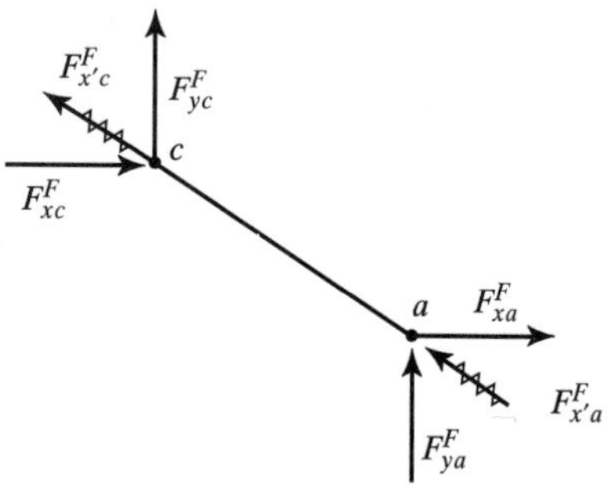

$$F^F_{xa} = F^F_{ya} = -F^F_{xd} = -F^F_{yd} = F^F_{x'a}\cos 45°$$
$$= 200 \times 4 \times 10^3 \times 1.2 \times 10^{-5}(-20)\cos 45° = -135.8 \text{ kN}$$

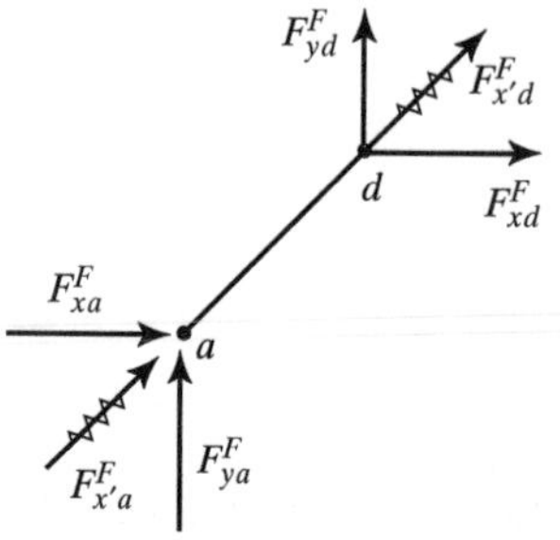

1. ***Displacements.*** Record direction angles, direction cosines, and products multiplied by A/L (see Example 5.4). A/L is in mm. See Figure 5.3 for definition of direction angles.

Member	$\alpha_{x'}$	$l_{x'}$	$m_{x'}$	(A/L)	$l^2_{x'}(A/L)$	$m^2_{x'}(A/L)$	$l_{x'}m_{x'}(A/L)$
ab	270°	0	−1	1.000	0	1.000	0
ac	150°	−0.8660	0.5000	0.6250	0.4688	0.1563	−0.2706
ad	45°	0.7071	0.7071	0.8000	0.4000	0.4000	0.4000
				Σ	0.8688	1.5563	0.1294

Stiffness equations for displacement at a (see Equation 3.6a):

$$\begin{Bmatrix} P_{xa} \\ P_{ya} \end{Bmatrix} = \begin{Bmatrix} 0 \\ 0 \end{Bmatrix} = 200\begin{bmatrix} 0.8688 & 0.1294 \\ 0.1294 & 1.5563 \end{bmatrix}\begin{Bmatrix} u_a \\ v_a \end{Bmatrix} + \begin{Bmatrix} (0 + 207.9 - 135.8) \\ (288.0 - 120.0 - 135.8) \end{Bmatrix}$$

Solving for the displacements,

$$\begin{Bmatrix} u_a \\ v_a \end{Bmatrix} = \begin{Bmatrix} -0.4045 \\ -0.0698 \end{Bmatrix} \text{ mm}$$

2. ***Reactions.*** Use negatives of member influence coefficients in the above table.

$$\begin{Bmatrix} R_{xb} \\ R_{yb} \\ R_{xc} \\ R_{yc} \\ R_{xd} \\ R_{yd} \end{Bmatrix} = 200\begin{bmatrix} u_a & v_a \\ 0 & 0 \\ 0 & -1.000 \\ -0.4688 & 0.2706 \\ 0.2706 & -0.1563 \\ -0.4000 & -0.4000 \\ -0.4000 & -0.4000 \end{bmatrix}\begin{Bmatrix} -0.4045 \\ -0.0698 \end{Bmatrix} + \begin{Bmatrix} 0 \\ -288.0 \\ -207.9 \\ 120.0 \\ 135.8 \\ 135.8 \end{Bmatrix} = \begin{Bmatrix} 0 \\ -274.0 \\ -173.8 \\ 100.2 \\ 173.8 \\ 173.8 \end{Bmatrix} \text{ kN}$$

3. ***Nullify displacements.*** Net, thermal, fixed-end forces must be zero. For uniform temperature change and fixed geometry (truss configuration), fixed-end forces are a function of the bar areas alone (see Equation 5.23). Thus we have

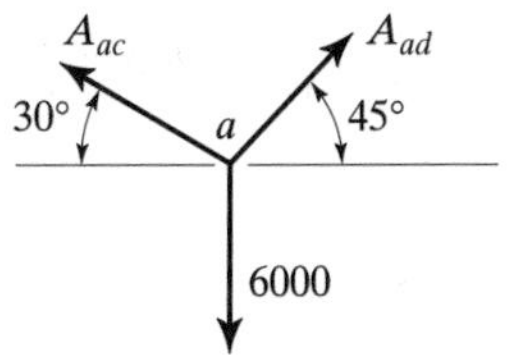

$$-A_{ac}\cos 30 + A_{ad}\cos 45 = 0$$
$$A_{ac}\sin 30 + A_{ad}\sin 45 = 6000$$

$$-0.8660\,A_{ac} + 0.7071\,A_{ad} = 0$$
$$0.5000\,A_{ac} + 0.7071\,A_{ad} = 6000$$

$$A_{ac} = 4393 \text{ mm}^2 \qquad A_{ad} = 5380 \text{ mm}^2$$

In Example 5.13 the differential heating of a continuous beam is studied. Although a point of inflection appears in the beam, the bending moment is positive throughout. Stressing may be thought of as occurring as the beam is strained from the position it would assume if the thermal deformation were permitted to occur in some unrestrained way (say, through removal of statically redundant supports).

EXAMPLE 5.13

The beam is the same as in Example 5.6. The depth of span *ab* is 400 mm and that of span *bc* is 200 mm. The bottom surface of both spans is heated 10°C and the top surface 20°C, with a uniform gradient in between. $\alpha = 1.2 \times 10^{-5}$ mm/mm°C.

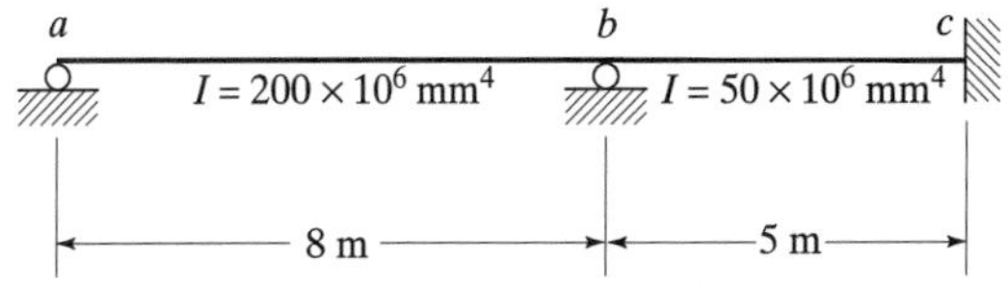

1. Calculate the displacements at *a* and *b*.
2. Calculate the reactions.

Fixed-end forces (see Equation 5.24):

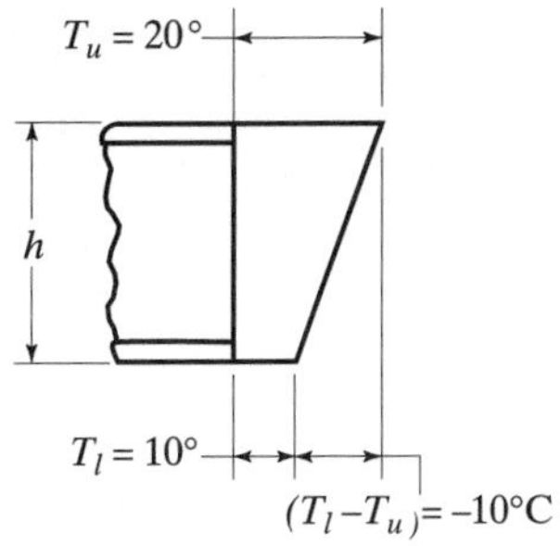

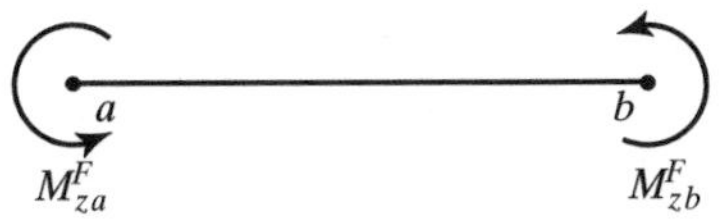

$$M^F_{za} = -M^F_{zb} = \frac{200 \times 200 \times 10^6 \times 1.2 \times 10^{-5}(-10)}{400} = -12.0 \times 10^3 \text{ kNmm}$$

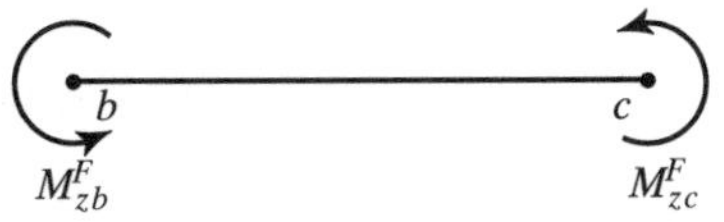

$$M^F_{zb} = -M^F_{zc} = \frac{200 \times 50 \times 10^6 \times 1.2 \times 10^{-5}(-10)}{200} = -6.0 \times 10^3 \text{ kNmm}$$

1. ***Displacements.*** Use the stiffness equations of Example 5.6:

$$\begin{Bmatrix} P_{mza} \\ P_{mzb} \end{Bmatrix} = \begin{Bmatrix} 0 \\ 0 \end{Bmatrix} = 200 \times 10^5 \begin{bmatrix} 1 & 0.5 \\ 0.5 & 1.4 \end{bmatrix} \begin{Bmatrix} \theta_{za} \\ \theta_{zb} \end{Bmatrix} + \begin{Bmatrix} -12.0 \times 10^3 \\ (12.0 - 6.0)10^3 \end{Bmatrix}$$

Solving for the displacements,

$$\lfloor \theta_{za} \quad \theta_{zb} \rfloor = \lfloor 0.0008609 \quad -0.0005217 \rfloor \text{ rad}$$

2. *Reactions.* See Example 5.6.

$$\begin{Bmatrix} R_{ya} \\ R_{yb} \\ R_{yc} \\ R_{mzc} \end{Bmatrix} = 200 \begin{bmatrix} \theta_{za} & \theta_{zb} \\ 18.75 & 18.75 \\ -18.75 & -6.75 \\ 0 & -12.00 \\ 0 & 0.2 \times 10^5 \end{bmatrix} \begin{Bmatrix} 0.0008609 \\ -0.0005217 \end{Bmatrix} + \begin{Bmatrix} 0 \\ 0 \\ 0 \\ 6 \times 10^3 \end{Bmatrix}$$

$$= \begin{Bmatrix} 1.27 \text{ kN} \\ -2.52 \text{ kN} \\ 1.25 \text{ kN} \\ 3.913 \times 10^3 \text{ kNmm} \end{Bmatrix}$$

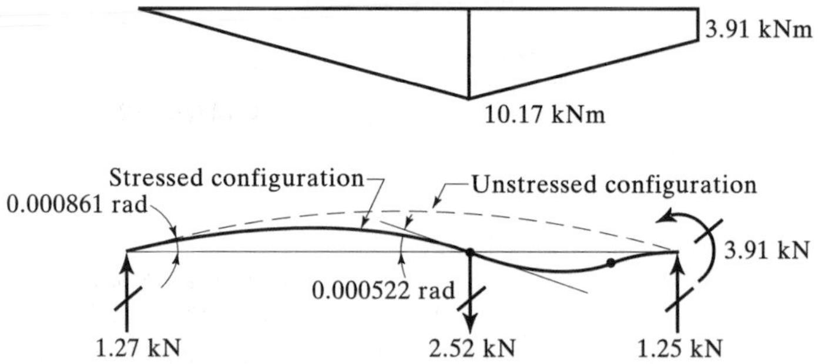

5.4 PROBLEMS

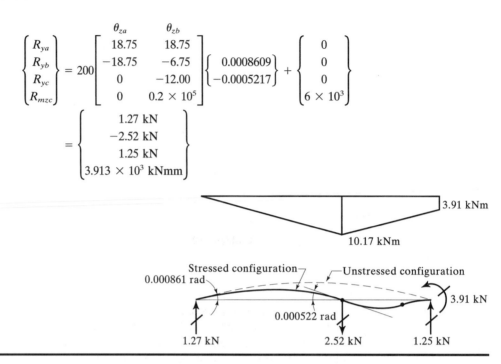

As in the previous chapters, it is suggested that, in several of the following problems, a computer program such as MASTAN2 be used and the results checked manually.

5.1 In the sketches below, the x and z axes are horizontal and the y axis is vertical. In part a of the figure the $x'y'$ plane makes a dihedral angle of 75° with the vertical plane through OA and in part b the $x'y'$ plane makes a dihedral angle of 90° with the vertical plane. Determine the direction cosines for the two elements OA.

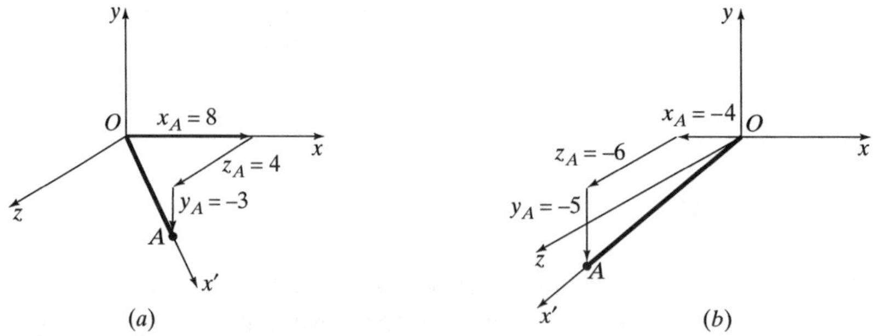

Problem 5.1

5.2 Two sets of parallel, orthogonal axes are shown. Six vector components act at O', the origin of one set. Develop the transformation matrix that relates six statically equivalent parallel components (not shown on the sketch) acting at O to those shown.

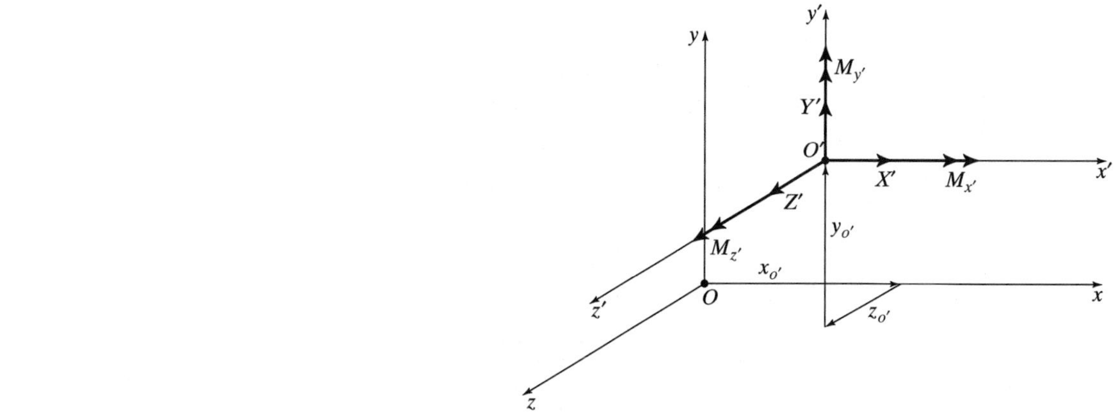

Problem 5.2

5.3 The two sets of axes shown lie in the same plane. Using the rotation matrix $[\gamma]$ and the results of Problem 5.2, develop a transformation matrix that makes P_x, P_y, and P_{mz} statically equivalent to $P_{x'}$, $P_{y'}$, and $P_{mz'}$. Show that it can be expressed as the product of a translation and a rotation matrix.

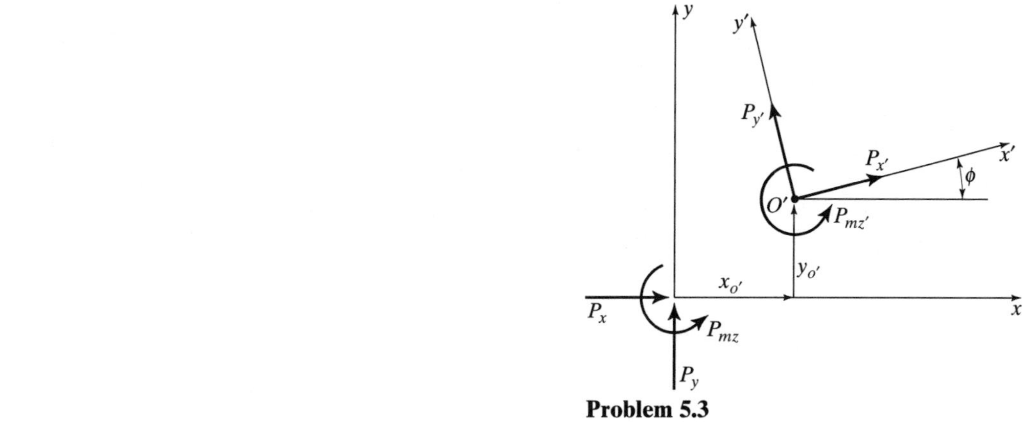

Problem 5.3

5.4 The member A is the same as in Example 5.5. Using the results of that example, plus an extension of the reasoning used in Problem 5.3, construct an equilibrium matrix $[\Phi]$ that relates reactive components acting at O parallel to the global axes (these components are not shown in the figure) to the forces acting at A that are expressed in local coordinates.

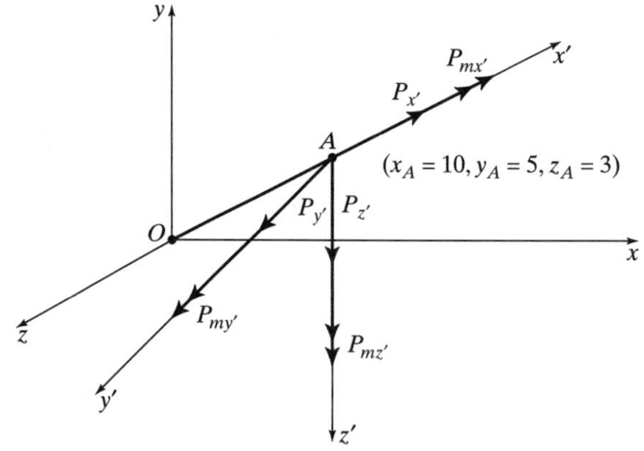

Problem 5.4

5.5 Compute the displacements, reactions, and internal forces for the systems shown. E = constant.

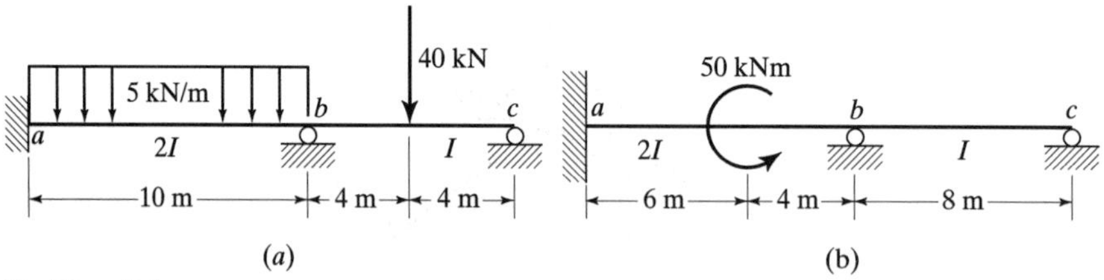

Problem 5.5

5.6 Assume E = constant and A = zero for all members; calculate u, v, and θ for joint a. Also calculate the distribution of internal forces. Why isn't the answer realistic? What is the shortcoming of the solution and how can it be corrected? What answers would one obtain to the same problem using moment distribution? What assumptions as to deformation are implicit in the moment distribution solution?

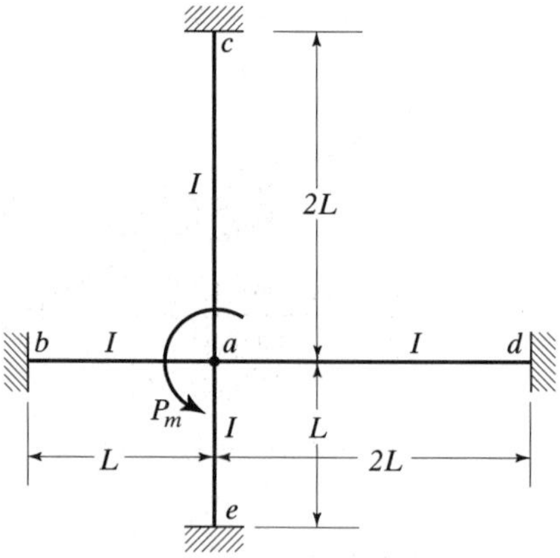

Problem 5.6

5.7 Assume E = constant and A = zero for both members. Assemble the stiffness equations required for the solution of u, v, and θ at joints b and c. Show that this solution fails in this case. What is the difficulty and how can it be corrected? Why was a stiffness solution (albeit an unrealistic one) obtainable in Problem 5.6 but not in this case?

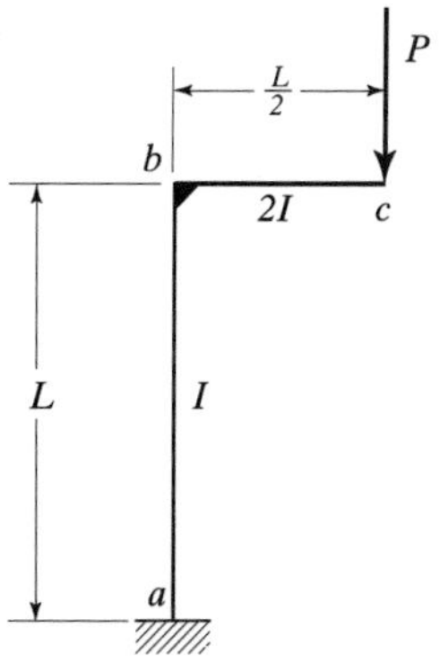

Problem 5.7

5.8 Compute the displacements, reactions, and internal forces for the systems shown. Neglect axial deformation (i.e., assume it to be zero) except in the structures for which areas are presented. $E = 200{,}000$ MPa.

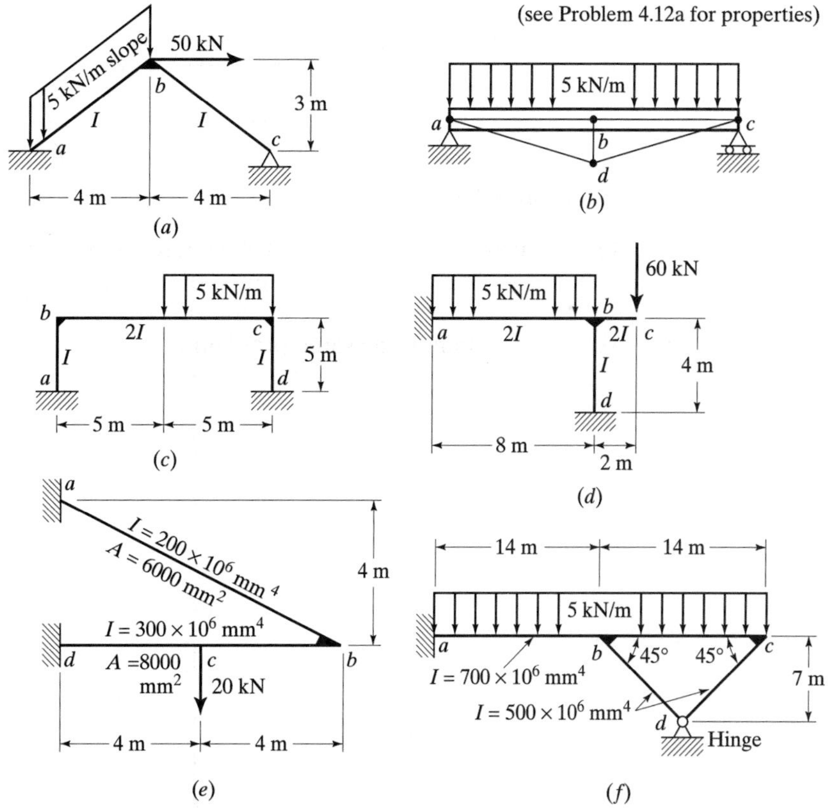

Problem 5.8

5.9 The structure shown is the same as in Example 5.3. Using a computer, calculate the displacements, reactions, and internal forces. Solve for (a) axial deformation included, and (b) axial deformation neglected. What difference does it make? Double the moment of inertia of the columns, solve, and compare results. Double the moment of inertia of the beams (with column I's at original value) and do likewise.

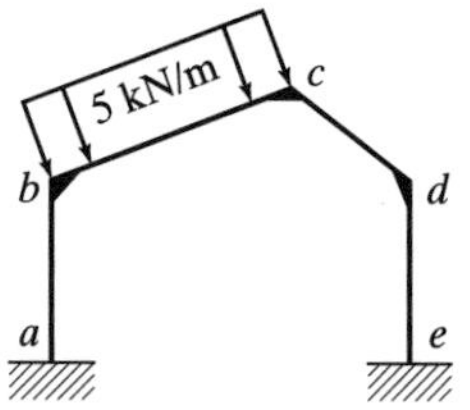

Problem 5.9

5.10 Compute the displacements, reactions, and internal forces for the systems shown.

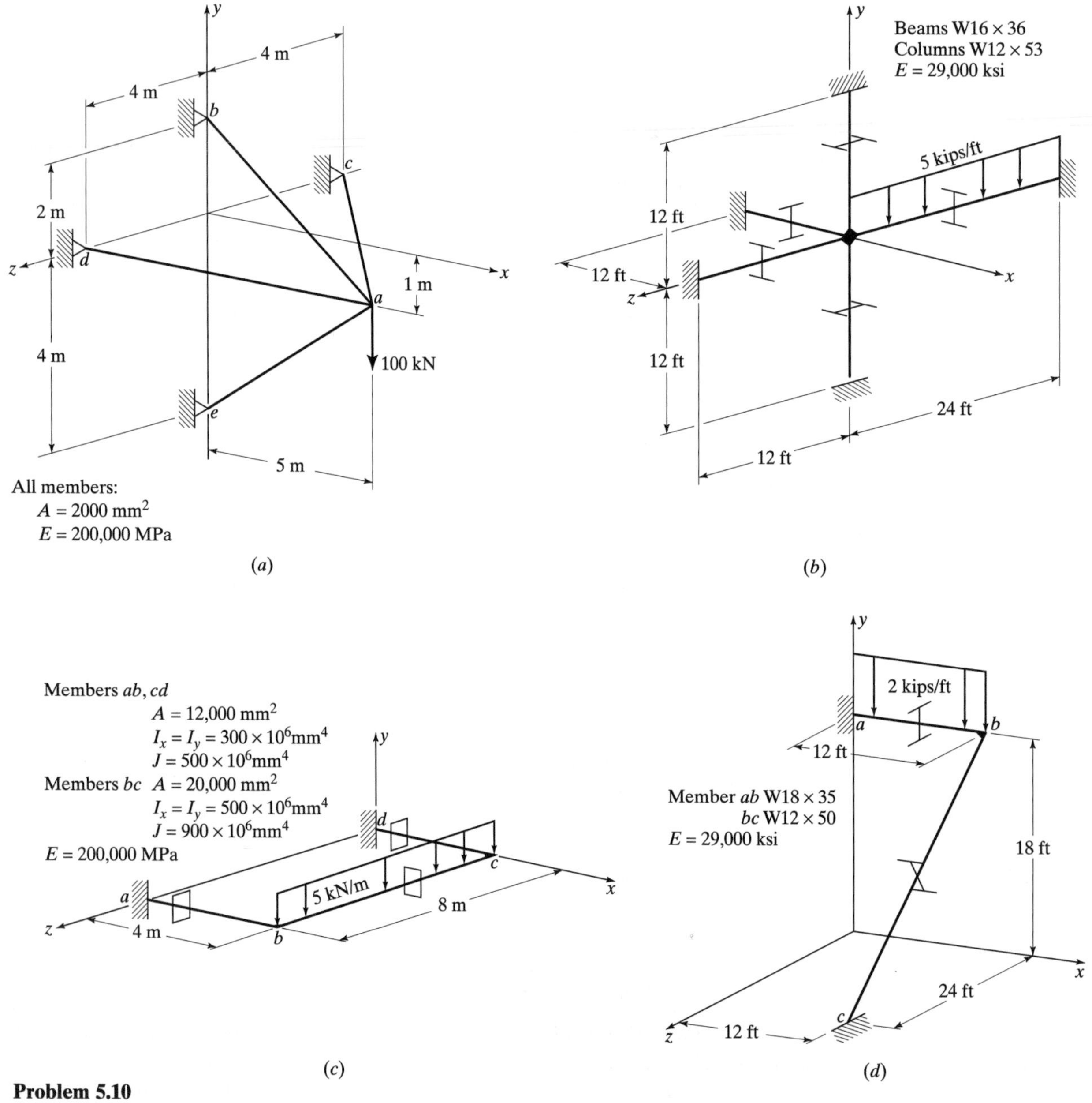

Problem 5.10

5.11 Bars *ac* and *ce* are heated 30°C. Compute the nodal-point displacements and the bar forces. $A = 6000\ \text{mm}^2$ all bars, $E = 200{,}000$ MPa, $\alpha = 1.2 \times 10^{-5}$ mm/mm°C.

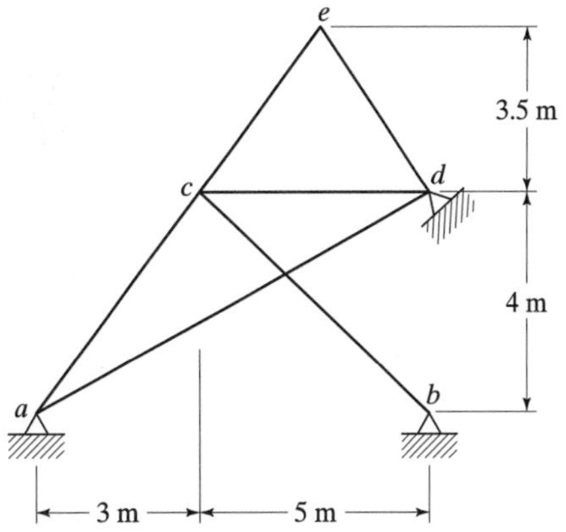

Problem 5.11

5.12 The depth of spans *ab* and *cd* is 200 mm and that of span *bc* is 400 mm. The bottom surface of span *bc* is heated 40°C and the top surface 15°C, with a uniform temperature gradient in between; the other spans remain unheated. $\alpha = 1.2 \times 10^{-5}$ mm/mm°C, $E = 200{,}000$ MPa. Compute the displacements, reactions, and internal forces.

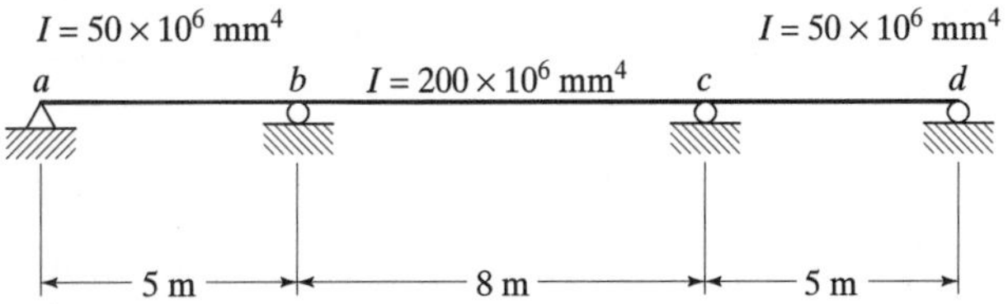

Problem 5.12

5.13 The problem is the same as the previous one except for the struts at *b* and *c*. These are heated 15°C in addition to the beam being heated as prescribed in Problem 5.12.

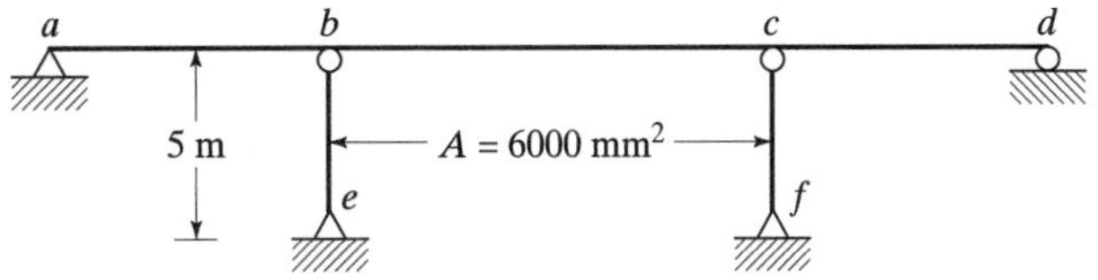

Problem 5.13

5.14 The beam shown is of constant section throughout its length and it is subjected to a depthwise thermal gradient ΔT all along its length. Calculate the displacements, reactions, and internal forces for (a) $L_2 = L_1$, and (b) $L_2 = 2L_1$. Compare results.

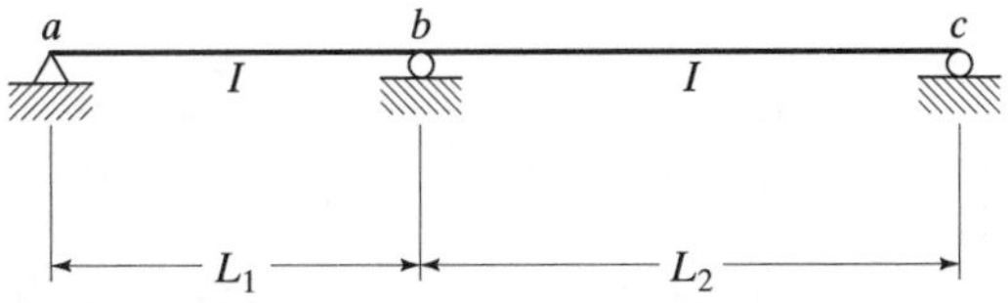

Problem 5.14

5.15 (a) It is desired to pretension the system in part *a* of the figure so that, when the geometrical configuration shown is obtained, the tension in *ab* is 50 kN. To what lengths should the individual bars be fabricated if this is to be attained? $E =$

200,000 MPa. (b) The problem is the same but, in this case, there is an additional bar *ae*, as shown in part *b*.

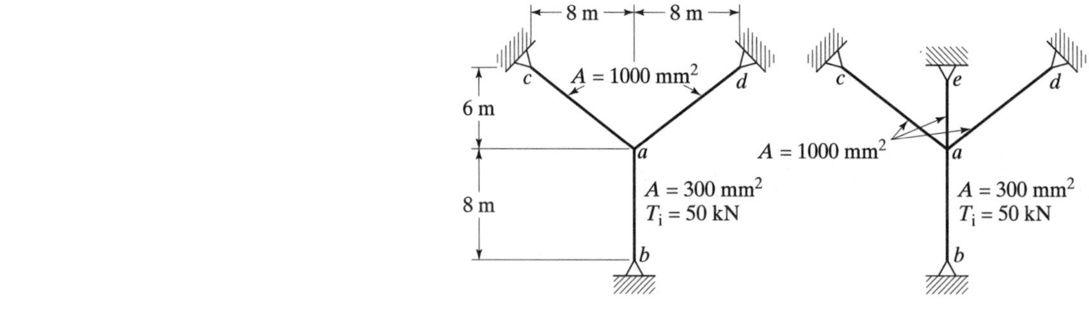

Problem 5.15

5.16 Two steel cylinders are clamped together by a hot driven rivet. The heads are formed on the rivet at a temperature 1000°C higher than the cylinder. The rivet cools against the resistance of the cylinder. Assume that the rivet is elastic throughout the cooling process. What are the forces in the system after the rivet has cooled 1000°C? $A_{\text{rivet}} = 500\ \text{mm}^2$, $A_{\text{cylinder}} = 4800\ \text{mm}^2$. $E = 200{,}000$ MPa, $\alpha = 1.2 \times 10^{-5}$ mm/m°C

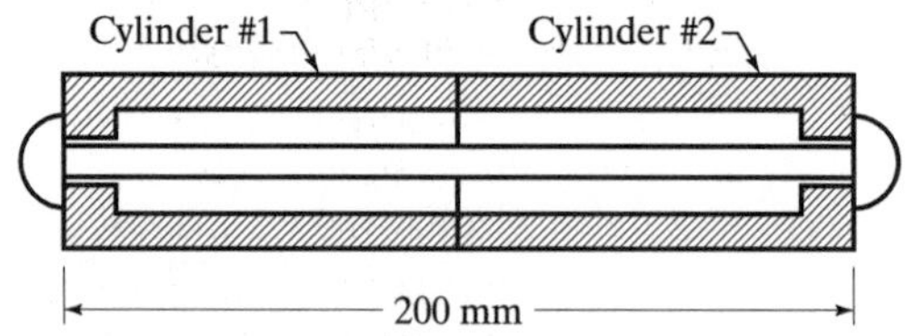

Problem 5.16

5.17 Given the axial force member stiffness equations in orthogonal coordinates x and y (Equation 2.5), develop the transformations necessary to transform these equations into the oblique coordinate system defined by the axes η and ξ and the included angle ψ. Compare the transformed equations.

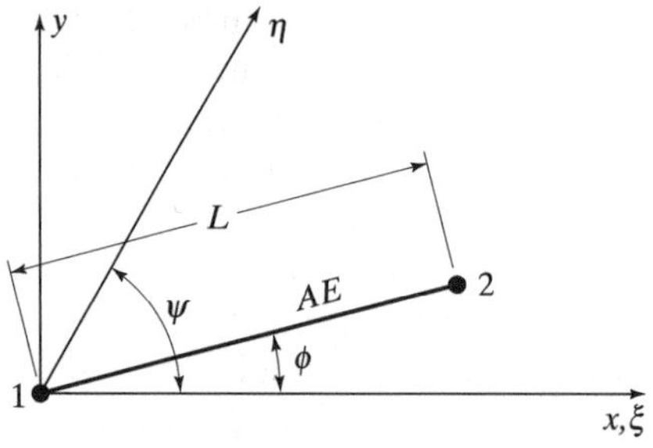

Problem 5.17

REFERENCES

5.1 F. P. Beer and E. R. Johnson, *Vector Mechanics for Engineers: Statics*, Sixth Edition, McGraw-Hill, New York, 1996.

5.2 A. E. H. Love, *A Treatise on the Mathematical Theory of Elasticity*, Fourth Edition, Dover Publications, New York, 1944.

Chapter 6

Virtual Work Principles

In prior chapters we established the relationships of framework analysis by dealing directly with the basic conditions of equilibrium and displacement continuity for elastic structures. That approach has considerable appeal because of its straightforward nature, but another method—virtual work—can be exploited for the same purpose as well as for the development of important analytical tools that are beyond the reach of the simple physical approach. *Virtual work* is a general term that includes two distinct principles, those of *virtual displacements* and *virtual forces*.

Virtual work concepts, in one form or another, will have been encountered by any student who has been exposed to elementary structural analysis. Because our purpose in dealing with such principles is to establish a basis for general framework analysis, it is necessary to take a closer look at theoretical questions than is usually the case in introductory structural analysis. In Chapter 7 we demonstrate the utility of these theories in matrix structural analysis. (References 6.1–6.3 contain further discussions of the theories and their application.)

This chapter begins with an examination of the concept of work in structural mechanics, a topic treated briefly in Section 4.2. The notion of a virtual, or imaginary, displacement is introduced and, by study of the response of a rigid body to this type of displacement, the most elementary form of the principle of virtual displacements is established. The usefulness of this form of the principle is limited to the calculation of support reactions and member forces in statically determinate structures.

The *principle of virtual displacements* is of value principally in the formulation of stiffness equations, which imply deformable behavior. Therefore, the concepts developed for rigid bodies are established for deformable bodies in Section 6.2. Detailed virtual work expressions are constructed for axial, torsional, and flexural elements in Section 6.3. Although we emphasize the theoretical foundations of virtual work, a proper appreciation of certain of its features and subtleties can be gained only through specific examples. Hence, Section 6.4 examines the principle of virtual displacements as applied to a few classical analysis problems.

The *principle of virtual forces* is the alternative, or *dual*, of the principle of virtual displacements and is the basis of the direct formulation of element flexibility relationships. Although the stiffness formulation is emphasized in computerized structural analysis, there are many cases in which the formulation of flexibility relationships and their transformation to stiffness terms as described in Section 4.4 is simpler than the direct formulation of stiffness relationships. The theoretical basis of the principle of virtual forces is therefore established in Section 6.5. We treat the principle in terms of its role in element flexibility formulation rather than as an exposition of its historical place in classical indeterminate structural analysis.

6.1 PRINCIPLE OF VIRTUAL DISPLACEMENTS—RIGID BODIES

To develop the principle of virtual displacements we must amplify the definition of work given previously in Section 4.2. There, it was stated that the work of a force F is

$$W = \int_0^{\Delta_0} F\, d\Delta \tag{6.1}$$

where Δ is the displacement of the point of application of the force in the direction of the force, with Δ_0 the final amplitude of displacement. This development will be limited to forces that are gradually applied, so that the displacement grows in proportion to the force as shown in Figure 6.1*a*. The figure refers to conditions of linear elastic behavior. This restriction is unnecessary in the development of virtual work principles in their more general form, but it is adopted here because we deal only with linear elastic structures in this and the next chapter.

For a given intensity of the load F_0 the amount of work produced, W_0, is given by the shaded area of Figure 6.1*a*. If a small increment of displacement, $d\Delta$, is imposed through the action of a small increment of force, dF, the change of work is

$$dW_0 = F_0(d\Delta) + 1/2(dF)(d\Delta) \tag{6.2}$$

If we discard the second term on the right side, based on order-of-magnitude considerations, we have the "first-order" change of work:

$$dW_0 = F_0(d\Delta) \tag{6.2a}$$

The same form of relationship is obtained if, as pictured in Figure 6.1*b*, it is imagined that the displacement increment occurs *without* the action of a corresponding force increment. Such an imaginary or hypothetical displacement is termed a *virtual displacement*, $\delta\Delta$, and the increment of the work of the existing force acting through the virtual displacement, $\delta W_0 = F_0(\delta\Delta)$, is known as *virtual work*. The virtual displacement, being imaginary and unrelated to any motivating force, need not be restricted to smallness but will, instead, be arbitrary in direction and undefined in magnitude. If the virtual displacement is not small, due account must be taken of this fact in the resulting formulation. Only small virtual displacements are treated here.

In the following we will use $\delta\Delta$ and the like to designate virtual displacements.[1]

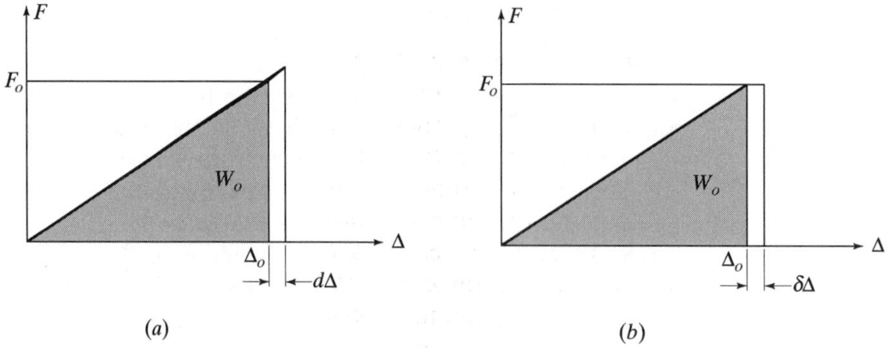

Figure 6.1 Force–displacement relationship.

[1]In the first edition of this text an overbar was placed on all virtual displacement symbols to distinguish them from real displacements. Overbars are omitted in this edition because we will use the δ symbol as an indicator of virtual quantities and not as a differential operator. Therefore, there should be no danger of misinterpretation.

Virtual work will be designated as δW, frequently with a subscript indicative of its source.

We first consider a single particle acted upon by a system of forces $F_1, \ldots, F_i, \ldots, F_s$, which are in equilibrium. Denoting the direction cosines between an arbitrary direction and the forces as $\lambda_1, \ldots, \lambda_i, \ldots, \lambda_s$, the condition of equilibrium of the forces is $\sum_{i=1}^{s} F_i\lambda_i = 0$.

Suppose the particle is subjected to a virtual displacement $\delta\Delta$ in an arbitrary direction. The virtual work is then

$$\delta W = \sum_{i=1}^{s} F_i \cdot (\delta\Delta) \tag{6.3}$$

where $\cdot$ denotes the dot product. Introducing the direction cosines that give the components of the forces F_i in the direction of $\delta\Delta$ this can be written as

$$\delta W = F_1\lambda_1(\delta\Delta) + \cdots F_i\lambda_i(\delta\Delta) + \cdots F_s\lambda_s(\delta\Delta) + \cdots = \left(\sum_{i}^{s} F_i\lambda_i\right)(\delta\Delta) \tag{6.3a}$$

and, since the bracketed terms that multiply $\delta\Delta$ are zero in accordance with the condition of force equilibrium, it follows that $\delta W = 0$. This is the virtual displacement principle, which can be expressed as follows:

For a particle subjected to a system of forces in equilibrium, the work due to a virtual displacement is zero.

It is of interest to ascertain whether the virtual displacement principle can be used to determine the converse, that is, if a system of forces acting on a particle is in equilibrium. This would appear to be difficult, since the virtual displacement is arbitrary, implying an infinity of directions in which the virtual work can be assessed. Any displacement, however, can be defined by components in the independent coordinate directions, such as three independent components in a general three-dimensional state. Thus one can establish the conditions of equilibrium in the latter case by writing virtual work equations for three independent virtual displacements. The converse of the above statement of the principle of virtual displacements can therefore be written:

A particle is in equilibrium under the action of a system of forces if the virtual work is zero for every independent virtual displacement.

The above statements apply also to a system of particles under the action of an equilibrated set of forces, since the virtual work equations can obviously be written for each and every particle. Consideration of such systems of particles, or rigid bodies, demands a more careful examination of the description of the virtual displacement.

Consider, for purposes of this examination, the beam pictured in Figure 6.2*a*. The member shown is nondeformable; it is supported in a statically determinate manner and is subjected to a concentrated applied load of P_{y3}. There would appear to be no scope for application of the virtual displacement principle, since no displacement seems possible. The support forces can replace the support constraints, however, as in Figure 6.2*b*. If the conditions of equilibrium are written and solved, these support forces can be calculated. Then the structure can be represented, as in Figure 6.2*b*, as a free body under the action of an equilibrated force system without any consideration of support conditions. The structures represented in Figures 6.2*a* and 6.2*b* are entirely equivalent from the point of view of structural analysis. Hence, a virtual displacement can be defined for Figure 6.2*b* as shown in Figures 6.2*c*. Since the beam is rigid, this virtual

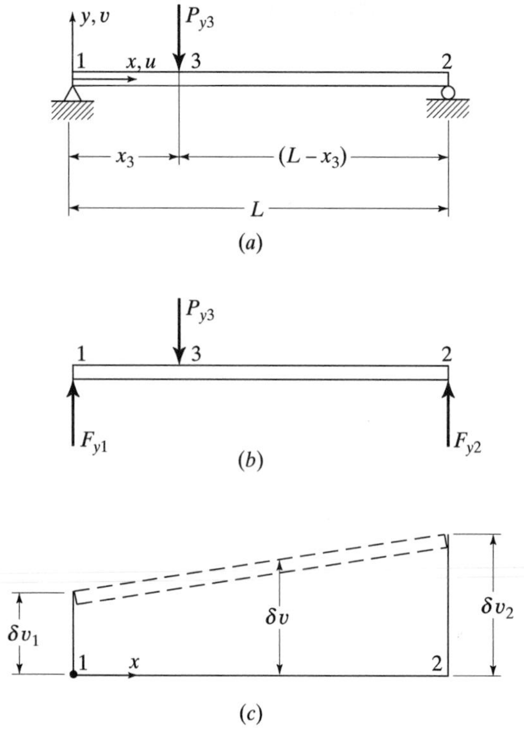

Figure 6.2 Rigid body analysis of simply supported beam by principle of virtual displacements: (*a*) Actual structure. (*b*) Structure with defined support reactions. (*c*) Virtual displacement.

displacement must represent rigid body motion, which is a straight line described by the equation:

$$\delta v = \left(1 - \frac{x}{L}\right)\delta v_1 + \frac{x}{L}\,\delta v_2 \tag{6.4}$$

The virtual displacement principle can now be written as

$$\delta W = F_{y1}\delta v_1 + F_{y2}\delta v_2 - P_{y3}\delta v_3 = 0 \tag{6.5}$$

and, from Equation 6.4 with $x = x_3$,

$$\begin{aligned}\delta W &= F_{y1}\delta v_1 + F_{y2}\delta v_2 - P_{y3}\left(1 - \frac{x_3}{L}\right)\delta v_1 - P_{y3}\frac{x_3}{L}\,\delta v_2 \\ &= \left[F_{y1} - P_{y3}\left(1 - \frac{x_3}{L}\right)\right]\delta v_1 + \left(F_{y2} - P_{y3}\frac{x_3}{L}\right)\delta v_2 = 0\end{aligned} \tag{6.5a}$$

Now since the virtual displacements are arbitrary, the terms δv_1 and δv_2 can take on any value, including zero. This means that the multipliers of δv_1 and δv_2 in Equation 6.5*a* must each equal zero. Thus we have

$$F_{y1} = P_{y3}\left(1 - \frac{x_3}{L}\right)$$

and

$$F_{y2} = \frac{P_{y3}x_3}{L}$$

Application of the principle of virtual displacement has produced the two relevant equations of equilibrium of this body. If we had introduced just one support force (F_{y2})

in place of the support constraint, then the admissible virtual displacement would lead, by means of the principle of virtual displacements, to only one of the above two equilibrium equations. On the other hand, if the virtual displacement state were generalized to include an axial component, then all three equilibrium equations of this rigid planar structure would be produced.

Example 6.1 demonstrates that the principle of virtual displacements can be employed to calculate directly the force in a member in the interior of a statically determinate structure. The force in the member is represented as acting on the joints at the ends of the member, but the virtual displacement takes place as if the constraining action of the member itself were absent.

EXAMPLE 6.1

Calculate the force in member 3-6 of the truss shown, using the principle of virtual work.

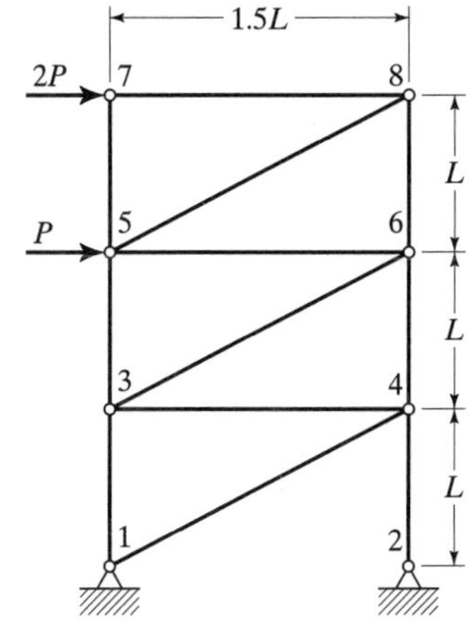

Replace member 3-6 by the equal and opposite forces $F_{3\text{-}6}$ acting on the joints to which the member is connected. Assume, for purposes of defining the virtual displaced state, that the constraining action of the member is absent. Since the member has been effectively removed through this step, the indicated virtual displacement can be enforced. By the principle of virtual displacements,

$$2P(\delta u_5) + P(\delta u_5) - F_{3\text{-}6}\left[\frac{1.5}{\sqrt{3.25}}(\delta u_5)\right] = 0$$

$$F_{3\text{-}6} = 2\sqrt{3.25}P$$

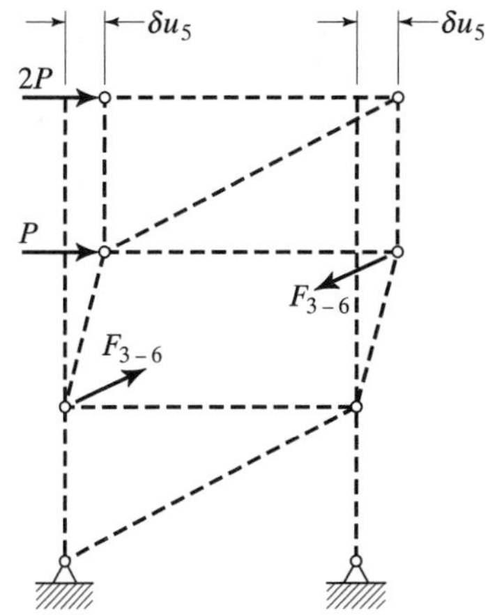

In summary, the principle of virtual displacements can be employed to establish the equilibrium equations whose solution gives reactions or statically determinate internal member forces. In either case the reactive or member force is represented, but the support or constraining action is removed to permit participation of the corresponding displacement component in the definition of the virtual displacement state.

The principle of virtual displacements, as an approach to the construction of rigid-body equilibrium equations, is of no direct value in matrix structural analysis. Element stiffness formulations include all of the element degrees of freedom and incorporate within themselves the element rigid-body equilibrium equations. Nevertheless, the examples given above illustrate the most fundamental application of the principle and emphasize its relationship to the conditions of equilibrium.

Our interest lies in deformable bodies rather than in rigid bodies because only by taking into account deformability are we able to calculate the response of statically indeterminate structures. Therefore, we devote the next section to the extension of the principle of virtual displacements to deformable bodies.

6.2 PRINCIPLE OF VIRTUAL DISPLACEMENTS—DEFORMABLE BODIES

Application of the principle of virtual displacements to deformable bodies requires that the total virtual work of a system be clearly defined, which means that distinctions must be made between external and internal work. The relationships that describe these terms are developed in the following.

To establish the desired relationships we examine the simplest arrangement of structural elements, a pair of axial members connected in series as shown in Figure 6.3*a*. The right end (joint 3) is fixed, and loads P_1, P_2 are applied to joints 1 and 2. The stiffnesses of the axial members are designated as k_1 and k_2.

Figure 6.3*b* shows the free body diagrams of the joints and elements. All forces are initially shown as acting in the direction of the positive x axis. The internal forces exerted by the joints on the members are designated as F_{12}, F_{21}, F_{23}, while primes are employed to distinguish the action of the internal forces on the joints. The force on a joint and the counterpart member force must be equal and opposite. Thus

$$
\begin{aligned}
F_{12} &= -F'_{12} \\
F_{21} &= -F'_{21} \\
F_{23} &= -F'_{23}
\end{aligned}
\tag{6.6}
$$

Also, the condition for equilibrium of element 1 is

$$F_{12} + F_{21} = 0 \tag{6.7}$$

or, in view of Equations 6.6,

$$F'_{12} = -F'_{21} \tag{6.7a}$$

The conditions of equilibrium of joints 1 and 2 are, respectively,

$$P_1 + F'_{12} = 0 \tag{6.8a}$$

$$P_2 + F'_{21} + F'_{23} = 0 \tag{6.8b}$$

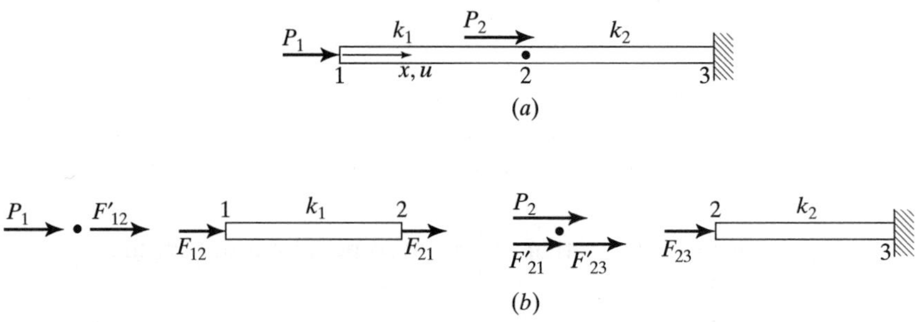

Figure 6.3 Two-element axial structure. (*a*) Actual structure. (*b*) Free-body diagrams of joints and elements.

Now consider the effect of virtual displacements δu_1 and δu_2 of the joints 1 and 2. Work is a scalar quantity so that the total virtual work of the system of joints is simply the sum of the virtual work of the component parts. In this case, therefore, the total virtual work is

$$\begin{aligned} \delta W &= (P_1 + F'_{12})\delta u_1 + (P_2 + F'_{21} + F'_{23})\delta u_2 \\ &= (P_1\delta u_1 + P_2\delta u_2) + (F'_{12}\delta u_1 + F'_{21}\delta u_2 + F'_{23}\delta u_2) \end{aligned} \tag{6.9}$$

The first term on the right side of Equation 6.9 represents the virtual work of the applied loads and is designated as δW_{ext}, that is

$$\delta W_{\text{ext}} = (P_1\delta u_1 + P_2\delta u_2) \tag{6.10}$$

The second term on the right side is the virtual work of the internal forces acting on the joints. By substitution of Equations 6.6, this term can be transformed into one expressed in terms of the forces exerted by the joints on the member, that is

$$(F'_{12}\delta u_1 + F'_{21}\delta u_2 + F'_{23}\delta u_2) = -(F_{12}\delta u_1 + F_{21}\delta u_2 + F_{23}\delta u_2) \tag{6.11}$$

The expression within the parentheses on the right represents the virtual work of the internal forces acting on the members, or the internal virtual work, δW_{int}.[2] Hence in the present case

$$\delta W_{\text{int}} = (F_{12}\delta u_1 + F_{21}\delta u_2 + F_{23}\delta u_2) \tag{6.12}$$

Furthermore, in view of Equations 6.9 through 6.12, for the conditions under study,

$$\delta W = \delta W_{\text{ext}} - \delta W_{\text{int}}$$

By application of the above procedure to any other type of deformable structure, it can be shown that this is a general expression for the total virtual work. To establish the properties possessed by this expression it is only necessary to return to the earlier statement of the total virtual work (Eq. 6.9) and substitute into the second term on the right (i.e., $-\delta W_{\text{int}}$) the equilibrium relationships given by Equations 6.8*a* and *b*. We then have

$$\delta W = (P_1\delta u_1 + P_2\delta u_2) + (-P_1\delta u_1 - P_2\delta u_2) = 0$$

or

$$\delta W = \delta W_{\text{ext}} - \delta W_{\text{int}} = 0 \tag{6.13}$$

This is the algebraic statement of the principle of virtual displacements for deformable structures. Again, although it has been established for a special case, its validity can be confirmed by application of the above procedure to any other type of structure. It is completely consistent with the statement of the virtual displacements principle of the previous section. It is now necessary to recognize that it encompasses a system of particles and the virtual work is the total virtual work, both internal and external. The verbal statement of Equation 6.13 is as follows:

> *For a deformable structure in equilibrium under the action of a system of applied forces, the external virtual work due to an admissible virtual displaced state is equal to the internal virtual work due to the same virtual displacements.*

Observe that the qualifying term, *admissible*, prefaces *virtual displaced state*. One would expect that there are some limitations on the choices that can be made of the

---

[2]This quantity may be viewed as energy stored within the member as a result of the virtual displacements. For this reason it is also called the *virtual strain energy*, δU, a designation that was used in the first edition of this text.

virtual displacements, and it is indeed the case. The term *admissible* is employed to emphasize this consideration. We define these limitations in the next section, where the method followed in using the virtual displacements principle is outlined, and virtual work expressions are given for the specific modes of structural action with which we are concerned.

6.3 VIRTUAL DISPLACEMENTS ANALYSIS PROCEDURE AND DETAILED EXPRESSIONS

6.3.1 General Procedure

Although the stiffness properties of the elements in Section 6.2 were defined at the outset, no use was made of them in the development of the principle of virtual displacements for deformable bodies. This circumstance can be used to advantage in certain situations, but the use to which the principle of virtual displacements will be put in this text requires the introduction of the structural stiffness properties. They are introduced to transform the internal virtual work from an expression written in terms of forces and virtual displacements into an expression in terms of displacements and virtual displacements.

As an example, in the structure of Figure 6.3*a*, we have the following relationships between the member stiffness and the joint forces and displacements:

$$F_{12} = k_1(u_1 - u_2)$$
$$F_{23} = k_2 u_2$$

Hence, by substitution in Equation 6.12,

$$\delta W_{\text{int}} = k_1(u_1 - u_2)\delta u_1 - k_1(u_1 - u_2)\delta u_2 + k_2 u_2 \delta u_2$$

This is an expression in terms of the displacements (u_1, u_2) and the virtual displacements (δu_1, δu_2).

It is tempting to refer to displacements such as u_1 and u_2 as *actual* displacements. All of the developments thus far have either explicitly dealt with or implied a procedure leading to the solution that is exact within the limits of linear elastic behavior. For these cases the displacements are indeed the actual displacements of the linear system. However, the principle of virtual displacements finds its most powerful application in the development of approximate solutions. In these, the state upon which the virtual displacement is imposed may itself be an approximation of the actual displaced state (see Section 6.4.2). Thus, rather than calling them actual displacements, we will use the adjective *real*—in its sense of something rooted in nature—to distinguish between tangible and virtual displacements.

The character of both the real and virtual displacements, and the limitations imposed upon them, deserve careful attention. As emphasized previously, the solution to a problem in structural mechanics is exact if it meets all relevant conditions of equilibrium and displacement continuity. The foregoing development of the principle of virtual displacements gives explicit attention only to the condition of equilibrium. The condition of displacement continuity, which requires that the displacements are continuous functions of the coordinates of the structure (x, y, z), must be met implicitly. That is, the real and virtual displacements must be of a form such that the displacement continuity conditions are satisfied from the outset.[3] *Admissible displaced states* are those that meet these conditions.

---

[3]It should be noted that this requirement still permits the virtual displacement of supports and joints as in Figure 6.2 and Example 6.1, provided that the supports are represented by support-reaction forces rather than by the actual constraints against displacement.

With the above in mind, an outline can be given of a general procedure of structural analysis based on the principle of virtual displacements. One begins with a description of the real displaced state by means of admissible functions that have undetermined multipliers. The virtual displaced state is similarly described, except that the chosen admissible functions have arbitrary multipliers (e.g., δu_1 and δu_2 in the development above). These functions are, in turn, used in the construction of the terms δW_{ext} and δW_{int}. Equating δW_{ext} and δW_{int} produces conditions that enable the evaluation of the undetermined multipliers of the real displaced state. These values are such that the conditions of equilibrium are satisfied. It will be found that the arbitrary multipliers may be canceled from the resulting expressions or that they may lead to conditions that can only be satisfied by certain relationships among the real quantities of the problem.

The tasks that remain before this general procedure can be applied to specific problems are the formulation of explicit expressions for the internal virtual work δW_{int} and the external virtual work δW_{ext} for the various modes of structural action.

6.3.2 Internal Virtual Work

The primary modes of structural action with which we are concerned are axial, torsional, and flexural behavior. In the following discussion, we develop relationships for δW_{int} for members of length L sustaining these modes of behavior.

Consider first an axial member, a differential segment of which is shown in Figure 6.4*a*. The stress σ_x that acts upon the segment throughout the virtual displacement is the stress corresponding to the real displacement. The virtual displacement of the left end of the segment is designated as δu. The virtual displacement will, in general, change from one point to the next along the segment so that at its right end it will be, to the first order in dx, $\delta u + [d(\delta u)/dx]dx$.

To demonstrate the relationship between virtual displacements and virtual strains we will first treat the differential segment as a free body. The real forces at its ends

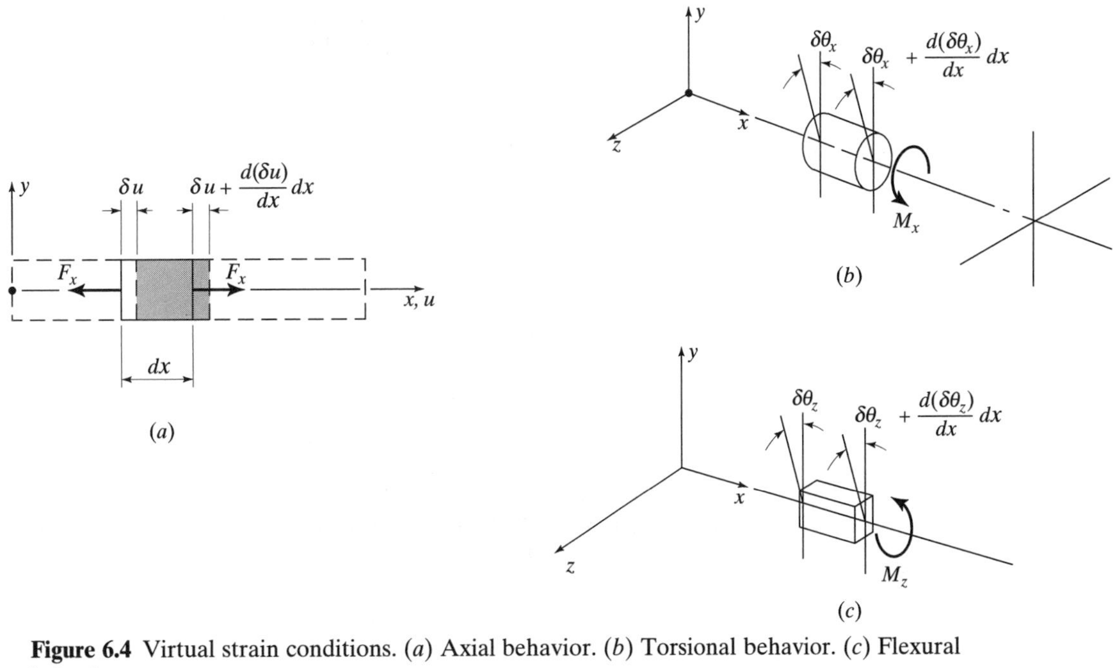

Figure 6.4 Virtual strain conditions. (*a*) Axial behavior. (*b*) Torsional behavior. (*c*) Flexural behavior.

are F_x, with a negative sign assigned to the force on the left face, since it acts in a direction opposite to the positive x direction. The virtual work of the force on the left face is therefore $-F_x\delta u$ and that of that force on the right face is $F_x\{\delta u + [d(\delta u)/dx]dx\}$. Thus for the segment as a free body, $\delta W_{\text{ext}} = -\delta u F_x + \{\delta u + [d(\delta u)/dx]dx\}F_x = [d(\delta u)/dx]F_x dx$ and, since $\delta W_{\text{ext}} = \delta W_{\text{int}}$ from Equation 6.13,

$$\delta W_{\text{int}} = \frac{d(\delta u)}{dx} F_x dx \tag{6.14a}$$

Now, axial strain is defined as the rate of change of the axial displacement with respect to the axial coordinate; that is, for linear behavior $e_x = du/dx$. In the present case we have a virtual axial displacement δu. Defining the virtual strain δe_x in an analogous manner, we have $\delta e_x = d(\delta u)/dx$. Also, $F_x = \sigma_x A$. Thus, Equation 6.14*a* can be written as

$$\delta W_{\text{int}} = \delta e_x \sigma_x A\, dx \tag{6.14b}$$

This relationship pertains to a differential segment dx. For a complete axial member of length L, it follows that the change in internal virtual work due to virtual displacement is

$$\delta W_{\text{int}} = \int_0^L \delta e_x \sigma_x A\, dx \tag{6.15a}$$

Finally, by use of Hooke's law ($\sigma_x = Ee_x$), this can be written entirely in terms of strain:

$$\delta W_{\text{int}} = \int_0^L \delta e_x EA e_x dx \tag{6.15b}$$

Since the real and virtual strains are the derivatives of the real and virtual displacements $[e_x = du/dx,\ \delta e_x = d(\delta u)/dx]$, the internal virtual work can also be written as a function only of the real displacements and virtual displacements:

$$\delta W_{\text{int}} = \int_0^L \left[\frac{d(\delta u)}{dx}\right] EA \frac{du}{dx}\, dx \tag{6.15c}$$

This means that the internal work can be calculated if expressions are available for the real and virtual displacements, as will be the case in the work to follow.

In the case of torsion, the "strain" β is the rate of change of rotation of the cross section about the longitudinal axis (see Section 4.5.2). If, in the differential segment of a torsion element, Figure 6.4*b*, the left face undergoes rotational displacement θ_x, the right face is displaced $\theta_x + (d\theta_x/dx)dx$, and the rate of change of rotational displacement is

$$\beta = \frac{1}{dx}\left[\left(\theta_x + \frac{d\theta_x}{dx}\, dx\right) - \theta_x\right] = \frac{d\theta_x}{dx} \tag{6.16}$$

Correspondingly, the torsional strain due to a virtual displacement in torsion is

$$\delta\beta = \frac{d(\delta\theta_x)}{dx} \tag{6.17}$$

The change in internal virtual work due to a virtual twisting displacement in the presence of a real twisting moment M_x is, in an element of length L,

$$\delta W_{\text{int}} = \int_0^L (\delta\beta)(M_x)dx = \int_0^L \frac{d(\delta\theta_x)}{dx} M_x dx \tag{6.18}$$

and, since $M_x = GJ(d\theta_x/dx)$, this can be written entirely in terms of displacement qualities:

$$\delta W_{\text{int}} = \int_0^L \frac{d(\delta\theta_x)}{dx} GJ \frac{d\theta_x}{dx} dx \tag{6.18a}$$

In examining beam flexure we consider a member whose axis is coincident with the x axis, has a cross section symmetric about the y axis, and is subjected to bending about the z axis. As explained in Section 4.5.3, the curvature $d^2v/dx^2 = \kappa_z$ can be regarded as the strain for bending behavior. Figure 6.4c, which describes the displacements of a differential segment of a flexural member, is further illustration of this point. The deformation of the segment is characterized by the rate of change of rotation θ_z with respect to the x coordinate. The rotation of the cross section is equal to the slope of the neutral axis, $\theta_z = dv/dx$.[4] The rate of change of θ_z with respect to x is, then

$$\frac{1}{dx}\left[\left(\theta_z + \frac{d\theta_z}{dx} dx\right) - \theta_z\right] = \frac{1}{dx}\left[\left(\frac{dv}{dx} + \frac{d^2v}{dx^2} dx\right) - \frac{dv}{dx}\right] = \frac{d^2v}{dx^2} = \kappa_z \tag{6.19}$$

It follows that the virtual bending strain $\delta\kappa_z$ is, for the virtual transverse displacement δv,

$$\delta\kappa_z = \frac{d^2(\delta v)}{dx^2} \tag{6.20}$$

It was also explained in Section 4.5.3 that the "stress" in flexure is the bending moment M_z. Therefore, in an element of length L, the internal virtual work due to virtual displacement of the flexural member is

$$\delta W_{\text{int}} = \int_0^L (\delta\kappa_z)(M_z)dx = \int_0^L \frac{d^2(\delta v)}{dx^2} M_z dx \tag{6.21}$$

Also, since the moment is related to the curvature by $M_z = EI_z(d^2v/dx^2)$ (see Eq. 4.31b), this can be written as

$$\delta W_{\text{int}} = \int_0^L \frac{d^2(\delta v)}{dx^2} EI_z \frac{d^2v}{dx^2} dx \tag{6.21a}$$

A comment is in order regarding the conditions of continuity in flexure: The transverse displacements, v and δv, must of course be continuous. The flexural deformation state is also characterized by the rotations θ_z and $\delta\theta_z$ which, as we have noted, are equal to the slopes of the elastic axis, dv/dx and $d(\delta v)/dx$. Hence, in order for the real and virtual displacement states to be admissible, they must also possess continuous first derivatives. Furthermore, the displacement boundary conditions in flexure in a given problem might involve not only v and δv but also dv/dx and $d(\delta v)/dx$.

It is useful in the work that follows to have available a general formula for δW_{int} applicable to any or all of the above cases and also to problems of structural mechanics wherein the state of stress might be two- or three-dimensional. A general designation of the state of stress at a point in a structure is in the form of a column vector $\{\boldsymbol{\sigma}\}$, which includes all components of stress that are present. For a three-dimensional state of stress, for example, we have $\{\boldsymbol{\sigma}\} = \lfloor \sigma_x \ \ \sigma_y \ \ \sigma_z \ \ \tau_{xy} \ \ \tau_{yz} \ \ \tau_{zx} \rfloor^{\mathrm{T}}$. Similarly, the virtual strains can be designated by a row vector $\lfloor \delta\mathbf{e} \rfloor$ which, for a three-dimensional state of strain is $\lfloor \delta e_x \ \ \delta e_y \ \ \delta e_z \ \ \delta\gamma_{xy} \ \ \delta\gamma_{yz} \ \ \delta\gamma_{zx} \rfloor$. Now, the internal virtual work per unit volume of a multiaxial stress state is given by $\delta e_x \cdot \sigma_x + \delta e_y \cdot \sigma_y + \ldots$, which can be

[4]As noted in Chapter 4, this statement implies neglect of deformation due to transverse shear.

written as $\lfloor \delta \mathbf{e} \rfloor \{\boldsymbol{\sigma}\}$. For a structure whose volume is symbolized by (vol), the internal virtual work is, therefore,

$$\delta W_{\text{int}} = \int_{\text{vol}} \lfloor \delta \mathbf{e} \rfloor \{\boldsymbol{\sigma}\} d(\text{vol}) \tag{6.22}$$

The symbolism of Equation 6.22 is consistent with each of the definitions given above for the cases of axial, torsional, and flexural behavior. For axial behavior, for example $\lfloor \delta \mathbf{e} \rfloor = \delta e_x$, $\{\boldsymbol{\sigma}\} = \sigma_x$, and $d(\text{vol}) = A\,dx$, yielding $\delta W_{\text{int}} = \int_0^L \delta e_x \cdot \sigma_x A\,dx$, as before.

The principle of virtual displacements in the formulation of stiffness equations requires that δW_{int} be expressed solely in terms of strain. For this purpose one can introduce the stress-strain relationships $\{\boldsymbol{\sigma}\} = [\mathbf{E}]\{\mathbf{e}\}$ into Equation 6.22, resulting in

$$\delta W_{\text{int}} = \int_{\text{vol}} \lfloor \delta \mathbf{e} \rfloor [\mathbf{E}] \{\mathbf{e}\} d(\text{vol}) \tag{6.23}$$

where $[\mathbf{E}]$ is the matrix of elastic constants. For framework elements $[\mathbf{E}]$ is merely the scalar value E, or in the case of torsion, $E/2(1 + \nu)$.

6.3.3 External Virtual Work

The calculation of the external virtual work δW_{ext} is a simple matter in the case of concentrated loads and can be represented symbolically as

$$\sum_{i=1}^{s} (\delta \Delta_i) P_i$$

where $\delta \Delta_i$ refers to the virtual displacement of the ith degree of freedom and P_i is the applied load in that degree of freedom, in a situation in which loads are applied to s degrees of freedom. For applied moments, of course, the virtual displacement is an angular displacement $\delta \theta_{xi}$ (or $\delta \theta_{yi}$ or $\delta \theta_{zi}$) and the applied loads are moments, P_{mx} (or P_{my} or P_{mz}).

Note especially those cases in which the virtual displacements include rigid body motion to enable the construction of equations for determination of support reactions. This was illustrated in Section 6.1 and will again be illustrated in the next section. In such cases, the reaction at the support that is released to enable rigid-body virtual displacement must be designated and included in the expression for δW_{ext}.

Distributed load cases also deserve attention. Denoting the distributed loads as q and combining with the virtual work of concentrated loads,

$$\delta W_{\text{ext}} = \int \delta \Delta \cdot q dx + \sum_{i=1}^{s} (\delta \Delta_i) P_i \tag{6.24}$$

where the limits of the integral are taken to define the portion of the member on which the load acts. Use of Equation 6.24 is illustrated below in Example 6.5 for a case of flexure. It is of central importance in the exploitation of virtual work concepts in the formulation of fixed end forces described in Section 7.5.

It was observed in Section 6.2 that work (and therefore energy) expressions produce scalar quantities. Thus, when an element is subjected to the combined action of axial load, torsion, and flexure, the change in internal virtual work due to virtual displacement is simply the sum of the δW_{int}'s for the respective actions. Correspondingly, the change in external virtual work is simply the sum of δW_{ext} for the component external actions. This point is brought out in Example 6.2.

EXAMPLE 6.2

A prismatic member of uniform cross section, with geometric properties A, J, I_y, and I_z is fixed at one end and subjected to an axial load (F_{x2}), a twisting moment (M_{x2}), shears (F_{y2} and F_{z2}), and bending moments (M_{y2} and M_{z2}) at the opposite end. The displacements that result from the application of the forces are as given below. The virtual displacements chosen are of the same general shape. Construct the expressions for the internal virtual work in the member due to axial, torsional, and simple flexural behavior.

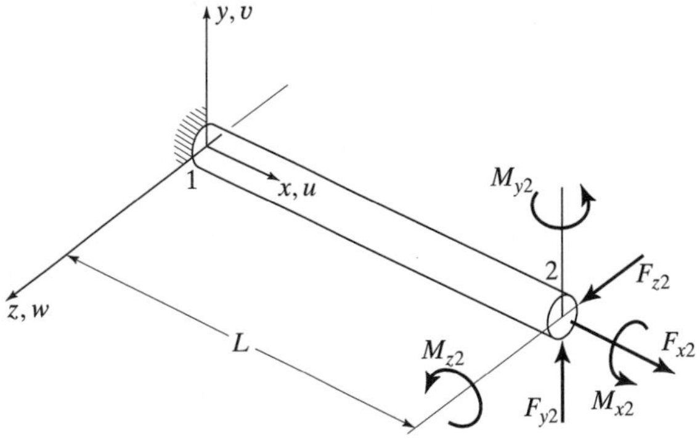

The real displacement state can be described by the following functions.[5]

$$u = \frac{x}{L}u_2 \qquad \theta_x = \frac{x}{L}\theta_{x2}$$

$$v = \left[3\left(\frac{x}{L}\right)^2 - 2\left(\frac{x}{L}\right)^3\right]v_2 + x\left[\left(\frac{x}{L}\right)^2 - \frac{x}{L}\right]\theta_{z2}$$

$$w = \left[3\left(\frac{x}{L}\right)^2 - 2\left(\frac{x}{L}\right)^3\right]w_2 - x\left[\left(\frac{x}{L}\right)^2 - \frac{x}{L}\right]\theta_{y2}$$

In accordance with the problem statement, the virtual displacements are written in the same form:

$$\delta u = \frac{x}{L}\delta u_2 \qquad \delta\theta_x = \frac{x}{L}\delta\theta_{x2}$$

$$\delta v = \left[3\left(\frac{x}{L}\right)^2 - 2\left(\frac{x}{L}\right)^3\right]\delta v_2 + x\left[\left(\frac{x}{L}\right)^2 - \frac{x}{L}\right]\delta\theta_{z2}$$

$$\delta w = \left[3\left(\frac{x}{L}\right)^2 - 2\left(\frac{x}{L}\right)^3\right]\delta w_2 - x\left[\left(\frac{x}{L}\right)^2 - \frac{x}{L}\right]\delta\theta_{y2}$$

In this case, the desired quantity is given by the sum of Equations 6.15c, 6.18a, and 6.21a, plus Equation 6.21a written with respect to bending about the y axis:

$$\delta W_{\text{int}} = \int_0^L \left[\frac{d(\delta u)}{dx}\right]EA\left(\frac{du}{dx}\right)dx + \int_0^L \left[\frac{d(\delta\theta_x)}{dx}\right]GJ\left(\frac{d\theta_x}{dx}\right)dx$$
$$+ \int_0^L \left[\frac{d^2(\delta v)}{dx^2}\right]EI_z\left(\frac{d^2v}{dx^2}\right)dx + \int_0^L \left[\frac{d^2(\delta w)}{dx^2}\right]EI_y\left(\frac{d^2w}{dx^2}\right)dx$$

Upon performance of the indicated differentiation on the given displacements (u, θ_x, v, w, δu, $\delta\theta_x$, δv, δw), substitution into this expression, and integration, we have

$$\delta W_{\text{int}} = \delta u_2\left(\frac{EA}{L}\right)u_2 + \delta\theta_{x2}\left(\frac{GJ}{L}\right)\theta_{x2} + \delta v_2\left(\frac{12EI_z}{L^3}\right)v_2 - \delta v_2\left(\frac{6EI_z}{L^2}\right)\theta_{z2}$$
$$-\delta\theta_{z2}\left(\frac{6EI_z}{L^2}\right)v_2 + \delta\theta_{z2}\left(\frac{4EI_z}{L}\right)\theta_{z2} + \delta w_2\left(\frac{12EI_y}{L^3}\right)w_2 + \delta w_2\left(\frac{6EI_y}{L^2}\right)\theta_{y2}$$
$$+ \delta\theta_{y2}\left(\frac{4EI_y}{L}\right)\theta_{y2} + \delta\theta_{y2}\left(\frac{6EI_y}{L^2}\right)w_2$$

[5]The derivation of these expressions is presented in Section 7.1.

6.4 CONSTRUCTION OF ANALYTICAL SOLUTIONS BY THE PRINCIPLE OF VIRTUAL DISPLACEMENTS

6.4.1 Exact Solutions

We now have available the general statement of the principle of virtual displacements (Eq. 6.13) and the ingredients of the terms that comprise the virtual work. The purpose of this section is to illustrate the use of the principle. The approach taken is one that seeks "analytical" rather than "numerical" solutions. The structures studied are treated as a whole, instead of being subdivided into segments or elements. The analytical approach enables us to identify key theoretical aspects of the principle of virtual displacements that require clarification or have escaped identification in the foregoing development of the principle.

Application of the principle of virtual displacements leads directly to expressions for loads in terms of displacements, that is, stiffness equations. These expressions are not in themselves solutions to problems, except in cases where the displacements are specified and the loads are to be calculated. Generally, when the expressions are obtained, they must then be solved to give the displacements in terms of the specified loads. The adaptation of the virtual displacements principle to the formulation of framework element stiffness equations and to the construction of global stiffness equations is given in the next chapter.

As a first example of a case in which an exact solution is possible, consider the uniform axial member of Figure 6.5*a*. The element is fixed at point 1 and subjected to a load F_{x2} at point 2 where the displacement is u_2. We know that under these circumstances the displacement varies linearly between points 1 and 2. Therefore, we adopt the form of the actual displacement to describe the real displacement u:

$$u = \frac{x}{L} u_2$$

This gives a strain:

$$e_x = \frac{du}{dx} = \frac{u_2}{L}$$

Now, the construction of the expressions for δW_{int} and δW_{ext} requires the definition of the virtual displacements and corresponding virtual strains. One convenient way to define these is to use the same form as that of the real displacements. (Alternative modes of description are discussed later.) Thus we choose $\delta u = (x/L)\delta u_2$, and $[d(\delta u)/dx] = \delta u_2/L$.

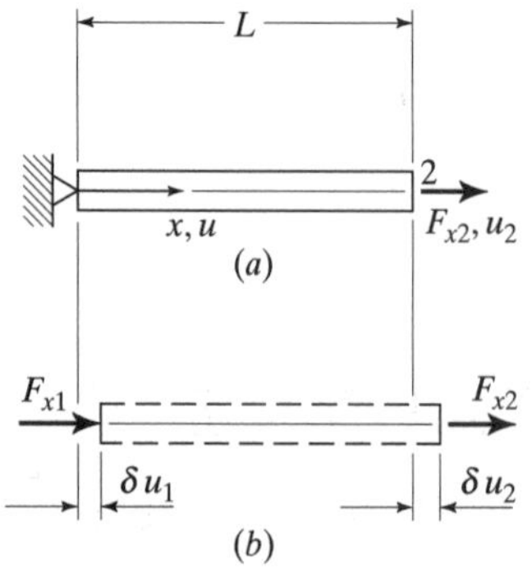

Figure 6.5 Uniform axial member. (*a*) Member. (*b*) Virtual displaced state.

Writing now the expression for δW_{int} for the axial member (Equation 6.15c)

$$\delta W_{\text{int}} = \int_0^L \left(\frac{d(\delta u)}{dx}\right) EA \frac{du}{dx}\, dx = \int_0^L \frac{\delta u_2}{L} EA \frac{u_2}{L}\, dx$$

and, since δu_2, u_2, A, E, and L are all constant within the member and $\int_0^L dx = L$

$$\delta W_{\text{int}} = (\delta u_2)\left(u_2 \frac{EA}{L}\right)$$

Also

$$\delta W_{\text{ext}} = (\delta u_2) F_{x2}$$

Equating δW_{int} and δW_{ext}, consistent with the principle of virtual displacements,

$$(\delta u_2) u_2 \left(\frac{EA}{L}\right) = (\delta u_2) F_{x2}$$

and, since δu_2 is arbitrary (it can take on any desired value), it follows that

$$u_2 \frac{EA}{L} = F_{x2}$$

This is a stiffness equation. Solving for the displacement in terms of the applied load,

$$u_2 = F_{x2} \frac{L}{EA}$$

which is the well-known exact result for this problem.

The virtual displacement met the same boundary condition (constraint of point 1) as the real displacement. Suppose the virtual displacement is chosen of a form that does not meet this condition. For example, if we choose

$$\delta u = \left(1 - \frac{x}{L}\right)\delta u_1 + \frac{x}{L}\,\delta u_2$$

we have a virtual displacement that still varies linearly but also involves displacement of point 1. The new virtual displacement parameter δu_1, which has been introduced, permits rigid-body virtual displacement. As Section 6.1 and its illustrations disclosed, such a virtual displacement of a support point requires replacement of the support designation by the support-reaction force designation F_{x1} (Fig. 6.5*b*). We now have

$$\delta W_{\text{ext}} = F_{x1}\delta u_1 + F_{x2}\delta u_2$$

Also, since now

$$\delta e_x = \frac{d\left[\left(1 - \frac{x}{L}\right)\delta u_1 + \left(\frac{x}{L}\right)\delta u_2\right]}{dx} = -\frac{\delta u_1}{L} + \frac{\delta u_2}{L}$$

we have

$$\delta W_{\text{int}} = \int_0^L \left(-\frac{\delta u_1}{L} + \frac{\delta u_2}{L}\right) EA \frac{u_2}{L}\, dx$$

$$= -(\delta u_1) u_2 \frac{EA}{L} + (\delta u_2) u_2 \frac{EA}{L}$$

Equating δW_{ext} and δW_{int} results in

$$F_{x1}\delta u_1 + F_{x2}\delta u_2 = -(\delta u_1)u_2 \frac{EA}{L} + (\delta u_2)u_2 \frac{EA}{L}$$

and, because of the arbitrary character of δu_1 and δu_2,

$$F_{x2} = u_2 \frac{EA}{L}$$

$$F_{x1} = -u_2 \frac{EA}{L} = -F_{x2}$$

The second equation is the equation of overall equilibrium of the element.

We find, therefore, that the virtual displacement is not required to meet the same support conditions as the total displacement. If it does not meet these conditions, however, the pertinent support forces must be introduced and will appear in the expression for δW_{ext}. Application of the principle of virtual displacements then establishes, for these terms, equations of equilibrium of the overall structure.

The real displacement of the above example was constrained by support conditions and the virtual displacement was not. The extra degree of freedom of the latter led to a separate statement of the rigid-body equilibrium conditions. In the formulation of element stiffness equations, as we observed in Section 2.4 and subsequently, it is desired that all joint displacements be represented (no support constraint) and that the rigid-body equilibrium conditions be contained within these stiffness equations. It can be surmised that the proper approach to take in such cases is to define unconstrained representations of both the real and virtual displacements.

Reference was made earlier to the notion of employing virtual displacement states of a shape different from the real displacement state. For example, consider in the above the use of a function that meets the same boundary condition as the real displacement but is of quadratic form:

$$\delta u = \left(\frac{x}{L}\right)^2 (\delta u_2)$$

then

$$\frac{d}{dx}(\delta u) = \frac{2x}{L^2}(\delta u_2)$$

In this case,

$$\delta W_{\text{int}} = \int_0^L \frac{2x}{L^2}(\delta u_2)EA\frac{u_2}{L}\,dx = 2\frac{EA}{L^3}(\delta u_2)u_2 \int_0^L x\,dx = (\delta u_2)\left(u_2 \frac{EA}{L}\right)$$

which is identical to the expression obtained previously.

Still another form for the virtual displacement for this case is

$$\delta u = (\delta u_2)\sin\frac{\pi x}{2L}$$

for which the strain is

$$\frac{d}{dx}(\delta u) = \frac{\pi}{2L}(\delta u_2)\cos\frac{\pi x}{2L}$$

In this case,

$$\begin{aligned}\delta W_{\text{int}} &= \int_0^L \frac{\pi}{2L}(\delta u_2)\cos\frac{\pi x}{2L}\cdot EA\frac{u_2}{L}\,dx\\ &= (\delta u_2)EA\frac{u_2}{L}\int_0^L \frac{\pi}{2L}\cos\frac{\pi x}{2L}\,dx\\ &= (\delta u_2)u_2\frac{EA}{L}\end{aligned}$$

which is again identical to the expression obtained when the virtual and real displacements are of the same form.

The above two forms of virtual displacement were applied to a real displacement that was of the form of the exact solution. Not only were the real displacements continuous but the stresses associated with them satisfied the condition of equilibrium. Although it is worthwhile to assign a physical meaning to virtual displacements, another view that can be taken is that they are simply arbitrary multipliers of the equilibrium equations. Reexamination of the development of Section 6.1 and 6.2 should confirm the latter view. Hence, if the chosen real displacements correspond to stresses that satisfy identically the conditions of equilibrium, any form of admissible virtual displacement will suffice to produce the exact solution.

6.4.2 Approximate Solutions and the Significance of the Chosen Virtual Displacements

Different forms of admissible virtual displacement generally produce different solutions when the real displacements are themselves approximations to the exact displacements. Recognition of this fact is important because the customary practical formulations of stiffness relationships for more complex phenomena in frameworks (e.g., taper, nonlinearity, instability, and dynamics) and for nearly all finite element representations are based on real displacements that are approximations.

This point can be illustrated quite simply by assigning a taper to the axial member of the above problem and solving for the end-point displacement u_2. The taper (Fig. 6.6) is given by $A = A_1(1 - x/2L)$. Because of it, the stress and strain vary with the axial coordinate, and the axial displacement is no longer accurately described by a linear function $u = xu_2/L$. The linear function gives constant stress, $\sigma_x = E du/dx = (E/L)u_2$, which violates the condition of equilibrium,[6] $dF_x/dx = d[\sigma_x A_1(1 - x/2L)]/dx = 0$. Nevertheless, to illustrate the significance of the principle of virtual displacements when the real displacements are approximate, we employ $u = xu_2/L$.

First, if we use a virtual displacement of the same form as the real displacement ($\delta u = x\delta u_2/L$) and consider that the area A of Equation 6.15c is now a variable, we have

$$\delta W_{\text{int}} = \int_0^L \left(\frac{\delta u_2}{L}\right)E\left(\frac{u_2}{L}\right)A_1\left(1 - \frac{x}{2L}\right)dx = (\delta u_2)(u_2)\frac{3EA_1}{4L}$$

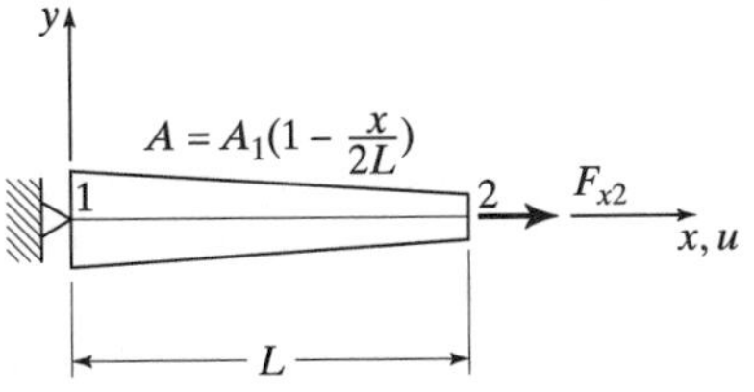

Figure 6.6 Tapered axial member.

[6]The equation of equilibrium for axial behavior is given in Section 6.5.1, Equation 6.25.

and since δW_{ext} is still $(\delta u_2)F_{x2}$, the principle of virtual displacement gives

$$\frac{3u_2EA_1}{4L} = F_{x2}$$

Next, we try $\delta u = (x/L)^2(\delta u_2)$ and retain $u = (x/L)u_2$. This results in $\delta W_{int} = 2(\delta u_2)(u_2)EA_1/3L$ and

$$\frac{2u_2EA_1}{3L} = F_{x2}$$

Finally, for $(\delta u) = (\delta u_2)\sin(\pi x/2L)$ and $u = (x/L)u_2$, $\delta W_{int} = (1 - 1/\pi)(\delta u_2)(u_2)EA_1/L$ and

$$\frac{0.6817u_2EA_1}{L} = F_{x2}$$

Clearly, three different solutions have been obtained for three different choices of virtual displacement. The exact solution is $0.721u_2EA_1/L = F_{x2}$ (see Section 7.3). None of the above represent the exact solution, but each is a valid approximate solution.

The explanation of why these different solutions are obtained is as follows. The statement of the principle of virtual work, given previously, implied that the solution would be correct for $\delta W_{int} = \delta W_{ext}$. What, in fact, occurs here is that both δW_{int} and δW_{ext} are approximate because of the approximate nature of the chosen real displacement function. The condition $\delta W_{int} = \delta W_{ext}$ is then *enforced*, providing a basis for the calculation of the undetermined parameter (u_2) of this function. Although the approximate real displacement function cannot give exact satisfaction of the equilibrium conditions at every point in the structure, it can be shown that the enforcement of the condition $\delta W_{int} = \delta W_{ext}$ results in a weighted-average satisfaction of the equilibrium conditions throughout the structure. If more terms are employed in the approximation of the displacement, the resulting weighted averages come closer to the correct solution, provided the proper conditions on a mathematical series are met.

If different solutions are obtained for different choices of virtual displacements for given approximate real displacement, what choice should be made? *The standard procedure is to adopt a virtual displacement of the same form as the real displacement.* This proves convenient in the integration of the resulting expression for δW_{int}. Furthermore, alternative choices may bring about unsymmetric forms of the resulting stiffness equations, in violation of the reciprocal theorem (Problem 7.12).

6.4.3 Examples

Examples 6.3–6.5 amplify the considerations established here. In Example 6.3 the internal virtual work established in Example 6.2 is supplemented by the virtual work of the applied loads, and the principle of virtual displacements is invoked to produce relationships between the applied loads and the displacements at their point of application. The example serves to demonstrate application of the principle to cases involving combined axial, torsional, and flexural behavior.

EXAMPLE 6.3

Construct, using the principle of virtual displacements, the expressions for the displacement of the tip (point 2) of the cantilever beam of Example 6.2.

To solve this problem we need expressions for δW_{int} and δW_{ext}. δW_{int} was established in Example 6.2. For δW_{ext}:

$$\delta W_{ext} = (\delta u_2)F_{x2} + (\delta v_2)F_{y2} + (\delta w_2)F_{z2} + (\delta\theta_{x2})M_{x2} + (\delta\theta_{y2})M_{y2} + (\delta\theta_{z2})M_{z2}$$

Equating this to the expression for δW_{int} found in Example 6.2, we have:
For independent δu_2 $(\delta v_2, \ldots \delta\theta_{z2} = 0)$

$$F_{x2} = \frac{EA}{L} u_2$$

For independent δv_2

$$F_{y2} = \frac{12EI_z}{L^3} v_2 - \frac{6EI_z}{L^2} \theta_{z2}$$

For independent δw_2

$$F_{z2} = \frac{12EI_y}{L^3} w_2 + \frac{6EI_y}{L^2} \theta_{y2}$$

For independent $\delta\theta_{x2}$

$$M_{x2} = \frac{GJ}{L} \theta_{x2}$$

For independent $\delta\theta_{y2}$

$$M_{y2} = \frac{6EI_y}{L^2} w_2 + \frac{4EI_y}{L} \theta_{y2}$$

For independent $\delta\theta_{z2}$

$$M_{z2} = -\frac{6EI_z}{L^2} v_2 + \frac{4EI_z}{L} \theta_{z2}$$

The conditions of equilibrium of internal and external forces can be established by replacement, in the virtual work expression, of the terms such as $E(du/dx)A$ by the corresponding forces (in this case, F_x). Overall equilibrium equations for this member can be established by including the support reactions (at point 1) in the analytical model and by imposition of displacements and virtual displacements in the form of rigid body motions.

Example 6.4 describes how the principle of virtual displacements can be used to advantage in the analysis of a statically indeterminate truss structure where all members are joined to a single point. The resulting pair of equations is coupled and must be solved simultaneously. Although in this simple case the solution is easily obtained, in general the virtual displacements principle is not advantageous in hand-calculation analysis of statically indeterminate trusses. This is because of the complexity of defining the external and internal work resulting from a system of arbitrary joint displacements.

EXAMPLE 6.4

Calculate the displacement of point 4 and the member forces through the application of the principle of virtual displacements.

$E = 200{,}000$ MPa. All areas: 15×10^4 mm^2.

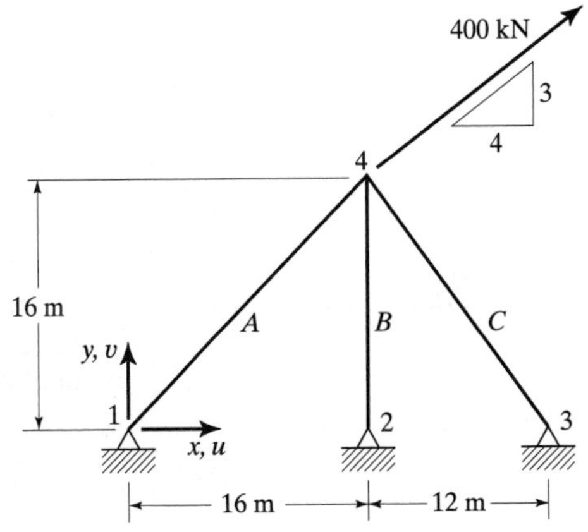

Form the internal virtual work due to virtual displacements δu_4 and δv_4. This first requires the calculation of the elongation of members A, B, and C due to the actual and virtual displacements of point 4.

For element A (see figure),

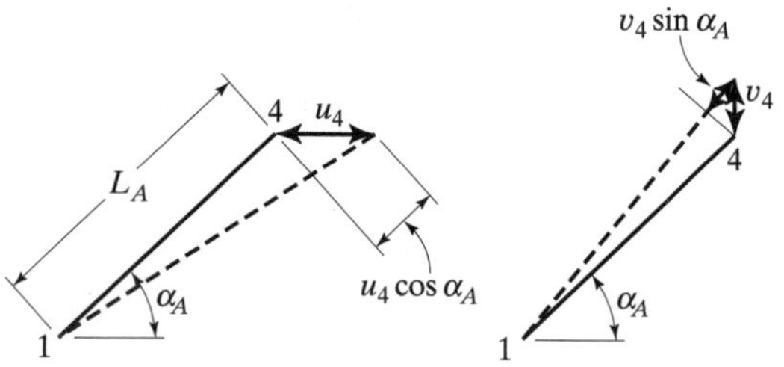

$$e_A = \frac{u_4 \cos \alpha_A}{L_A} + \frac{v_4 \sin \alpha_A}{L_A}$$
$$\delta e_A = \frac{\delta u_4}{L_A} \cos \alpha_A + \frac{\delta v_4}{L_A} \sin \alpha_A \tag{a}$$

For element B,

$$e_B = \frac{v_4}{L_B}, \qquad \delta e_B = \frac{\delta v_4}{L_B} \tag{b}$$

For element C (same sketch as element A),

$$e_C = \frac{u_4}{L_C} \cos \alpha_C + \frac{v_4}{L_C} \sin \alpha_C$$
$$\delta e_C = \frac{\delta u_4}{L_C} \cos \alpha_C + \frac{\delta v_4}{L_C} \sin \alpha_C \tag{c}$$

Thus

$$\begin{aligned}
\delta W_{\text{int}} &= \Sigma \int (\delta e \cdot \sigma) d(\text{vol.}) = \Sigma(\delta e \cdot \sigma \cdot \text{vol.}) \\
&= \delta e_A e_A EAL_A + \delta e_B e_B EAL_B + \delta e_C e_C EAL_C \\
&= \delta u_4 u_4 \frac{EA}{L_A} \cos^2 \alpha_A + \delta v_4 v_4 \frac{EA}{L_A} \sin^2 \alpha_A + \delta u_4 v_4 \frac{EA}{L_A} \sin \alpha_A \cos \alpha_A \\
&\quad + \delta v_4 u_4 \frac{EA}{L_A} \sin \alpha_A \cos \alpha_A + \delta v_4 v_4 \frac{EA}{L_B} + \delta u_4 u_4 \frac{EA}{L_C} \cos^2 \alpha_C \\
&\quad + \delta v_4 v_4 \frac{EA}{L_C} \sin^2 \alpha_C + \delta v_4 u_4 \frac{EA}{L_C} \cos \alpha_C \sin \alpha_C + \delta u_4 v_4 \frac{EA}{L_C} \cos \alpha_C \sin \alpha_C
\end{aligned}$$

With $L_A = 22.626$ m, $L_B = 16$ m, and $L_C = 20$ m,

$$\cos \alpha_A = \sin \alpha_A = 0.707 \qquad \cos \alpha_C = -0.6 \qquad \sin \alpha_C = 0.8$$

$$\begin{aligned}
\delta W_{\text{int}} &= \delta u_4 \left(\frac{0.5}{22.626} u_4 + \frac{0.5}{22.626} v_4 + \frac{0.36}{20} u_4 - \frac{0.48}{20} v_4 \right) EA \\
&\quad + \delta v_4 \left(\frac{0.5}{22.626} v_4 + \frac{0.5}{22.626} u_4 + \frac{v_4}{16} + \frac{0.64}{20} v_4 - \frac{0.48}{20} u_4 \right) EA
\end{aligned}$$

Form the virtual work due to applied loads:

$$\delta W_{\text{ext}} = \delta u_4 \cdot 320 + \delta v_4 \cdot 240$$

Equating δW_{int} and δW_{ext} using values for A and E, and noting the arbitrary character of δu_4 and δv_4 gives

$$\begin{aligned}
1203 u_4 - 57 v_4 &= 320 \\
-57 u_4 + 3498 v_4 &= 240
\end{aligned}$$

Solution of these yields $u_4 = 0.2695$ mm and $v_4 = 0.0730$ mm. The member forces are obtained by substituting u_4 and v_4 into Equations a, b, and c to give strains and multiplying these by EA. The final results are: $F_A = 321$ kN, $F_B = 136.9$ kN, and $F_C = -155$ kN.

Example 6.5 is a problem that involves only one form of behavior—bending—but displacements that are a function of two undetermined parameters. Because of the choice of trigonometric functions used to describe the approximate real displaced state and the virtual displaced state, the resulting equations are uncoupled, and the undetermined parameters can be solved for independently. This approach, which is based on a series representation of the displaced state, is known as the *Rayleigh-Ritz method*. Because it is approximate, the chosen displacement function violates the conditions of equilibrium.

EXAMPLE 6.5

Using the principle of virtual displacements and the following two-term trigonometric expansion in description of the transverse displacement, calculate the displacements at joints 2 and 3. Compare this solution with the exact answer (EI_z constant).

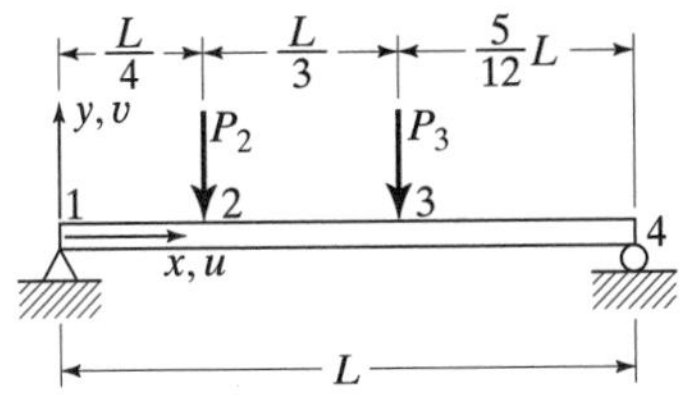

$$v = a_1 \sin\frac{\pi x}{L} + a_2 \sin\frac{2\pi x}{L}$$

$$\kappa_z = \frac{d^2 v}{dx^2} = -a_1\left(\frac{\pi}{L}\right)^2 \sin\frac{\pi x}{L} - a_2 4\left(\frac{\pi}{L}\right)^2 \sin\frac{2\pi x}{L}$$

$$\delta v = \delta a_1 \sin\frac{\pi x}{L} + \delta a_2 \sin\frac{2\pi x}{L}$$

$$\delta\kappa_z = \frac{d^2(\delta v)}{dx^2} = -\delta a_1\left(\frac{\pi}{L}\right)^2 \sin\frac{\pi x}{L} - \delta a_2 4\left(\frac{\pi}{L}\right)^2 \sin\frac{2\pi x}{L}$$

Form the internal virtual work

$$\delta W_{\text{int}} = \int_0^L \delta\kappa_z \cdot \kappa_z EI_z dx$$

$$= \int_0^L \left[\delta a_1\left(\frac{\pi}{L}\right)^2 \sin\frac{\pi x}{L} + \delta a_2 \cdot 4\left(\frac{\pi}{L}\right)^2 \sin\frac{2\pi x}{L}\right]$$

$$\times \left[a_1\left(\frac{\pi}{L}\right)^2 \sin\frac{\pi x}{L} + a_2 \cdot 4\left(\frac{\pi}{L}\right)^2 \sin\frac{2\pi x}{L}\right] EI_z dx$$

$$\delta W_{\text{int}} = \frac{EI_z\pi^4}{L^4}\int_0^L \left[\delta a_1 \cdot a_1 \sin^2\frac{\pi x}{L} + 16\delta a_2 \cdot a_2 \sin^2\frac{2\pi x}{L}\right.$$

$$\left. + 4(\delta a_1 \cdot a_2 + \delta a_2 \cdot a_1)\sin\frac{\pi x}{L}\sin\frac{2\pi x}{L}\right] dx$$

and since

$$\int_0^L \sin(\pi x/L)\cdot\sin(2\pi x/L)\,dx = 0$$

and

$$\int_0^L \sin^2\frac{\pi x}{L}\,dx = \int_0^L \sin^2\frac{2\pi x}{L}\,dx = \frac{L}{2}$$

$$\delta W_{\text{int}} = \frac{EI_z\pi^4}{2L^3}\delta a_1 \cdot a_1 + \frac{8EI_z\pi^4}{L^3}\delta a_2 \cdot a_2$$

Form the virtual work of the applied loads:

$$\delta W_{\text{ext}} = P_2\left(\delta a_1 \sin\frac{\pi}{4} + \delta a_2 \sin\frac{\pi}{2}\right) + P_3\left(\delta a_1 \sin\frac{7\pi}{12} + \delta a_2 \sin\frac{7\pi}{6}\right)$$

$$= \delta a_1(0.707P_2 + 0.966P_3) + \delta a_2(P_2 - 0.5P_3)$$

By the principle of virtual displacements,

$$\delta W_{int} = \delta W_{ext}$$

$$\frac{EI_z\pi^4}{2L^3}\delta a_1 \cdot a_1 + \frac{8EI_z\pi^4}{L_3}\delta a_2 \cdot a_2 = \delta a_1(0.707P_2 + 0.966P_3) + \delta a_2(P_2 - 0.5P_3)$$

And, because of the independent nature of δa_1 and δa_2,

$$a_1 = \frac{2L^3}{EI_z\pi^4}(0.707P_2 + 0.966P_3) \qquad a_2 = \frac{L_3}{8EI_z\pi^4}(P_2 - 0.5P_3)$$

so that

$$v = \frac{L^3}{EI_z\pi^4}\left[(1.414P_2 + 1.932P_3)\sin\frac{\pi x}{L} + (0.125P_2 - 0.0625P_3)\sin\frac{2\pi x}{L}\right]$$

$$v_2 = \frac{L^3}{EI_z\pi^4}(1.125P_2 + 1.303P_3) = \frac{L^3}{EI_z}(0.0116P_2 + 0.0134P_3)$$

$$\left[\text{versus } \frac{L^3}{EI_z}(0.0117P_2 + 0.0133P_3) \text{ exact}\right]$$

$$v_3 = \frac{L^3}{EI_z\pi^4}(1.303P_2 + 1.898P_3) = \frac{L^3}{EI_z}(0.0134P_2 + 0.0195P_3)$$

$$\left[\text{versus } \frac{L^3}{EI_z}(0.0133P_2 + 0.0197P_3) \text{ exact}\right]$$

It can be observed that in Examples 6.3 and 6.5 different approaches were taken to the definition of the displaced state. In Example 6.3 the displaced state is written directly in terms of the *physical displacements*, that is, the displacements $u_2, v_2, \ldots, \theta_{z2}$ of point 2. In Example 6.5, the displaced state is written in terms of the parameter a_1 and a_2, known as *generalized displacements* or *generalized parameters*. These parameters are carried through to a solution in terms of the applied loads, at which point they are substituted into the originally chosen expression for the displaced state. It would have been possible, alternatively, to fit the parameters a_1 and a_2 to the transverse displacements at two points on the beam (e.g., the center and a quarter point) and obtain an expression for the displacement at any point in terms of the displacements at these two points before performing the virtual work calculation.

In Chapter 7, where the virtual work principle is employed to formulate element stiffness equations, it will be necessary to work with the displaced states in terms of physical parameters, the element joint displacements. Thus, if the displaced state is first described in terms of generalized parameters, it must then be transformed to an expression in terms of physical parameters. A detailed examination of the procedure to be followed is presented at the start of that chapter, in Section 7.1.

6.5 PRINCIPLE OF VIRTUAL FORCES

6.5.1 Equations of Equilibrium

Stresses and strains are the *dual* parameters of structural analysis. They are connected to each other by the elastic constants. In the principle of virtual displacements the internal virtual work resulted from virtual strains and real stresses. It is logical to expect, and it is indeed the case, that there is a dual work principle in which the internal virtual work is due to the action of virtual stresses on real strains. This dual principle is the *principle of virtual forces*.

In the principle of virtual displacements we examined the changes in external and internal work that took place on account of a perturbation, or virtual change, of a displacement state about a real state of stress and strain. This process gave equations for the loads in terms of the displacements, that is, stiffness equations. We now examine an imaginary, or virtual, state of applied loading and its associated internal stress system. The loading and its associated internal stress system are designated the *virtual force system* in what follows. Application of this system in conjunction with the principle of virtual forces leads to expressions for displacements in terms of forces, that is, flexibility equations.

The conditions associated with a displacement state are the conditions of displacement continuity. Consequently, the virtual displacements were required to meet these conditions, consistent with the chosen description of support conditions. The conditions associated with a stress state are those of equilibrium; therefore, the fundamental requirement on virtual force systems is that they meet the relevant conditions of equilibrium. Thus, preliminary to the definition of the principle of virtual forces, we must first discuss the relationships for the force resultants of axial, torsional, and flexural behavior, especially those representing the conditions of equilibrium.

Consider first the axial member, a differential segment of which is shown in Figure 6.7*a*. For generality, the member is assumed to be of nonuniform cross section, with area A, subjected to a distributed axial load q, expressed in units of force per unit length.

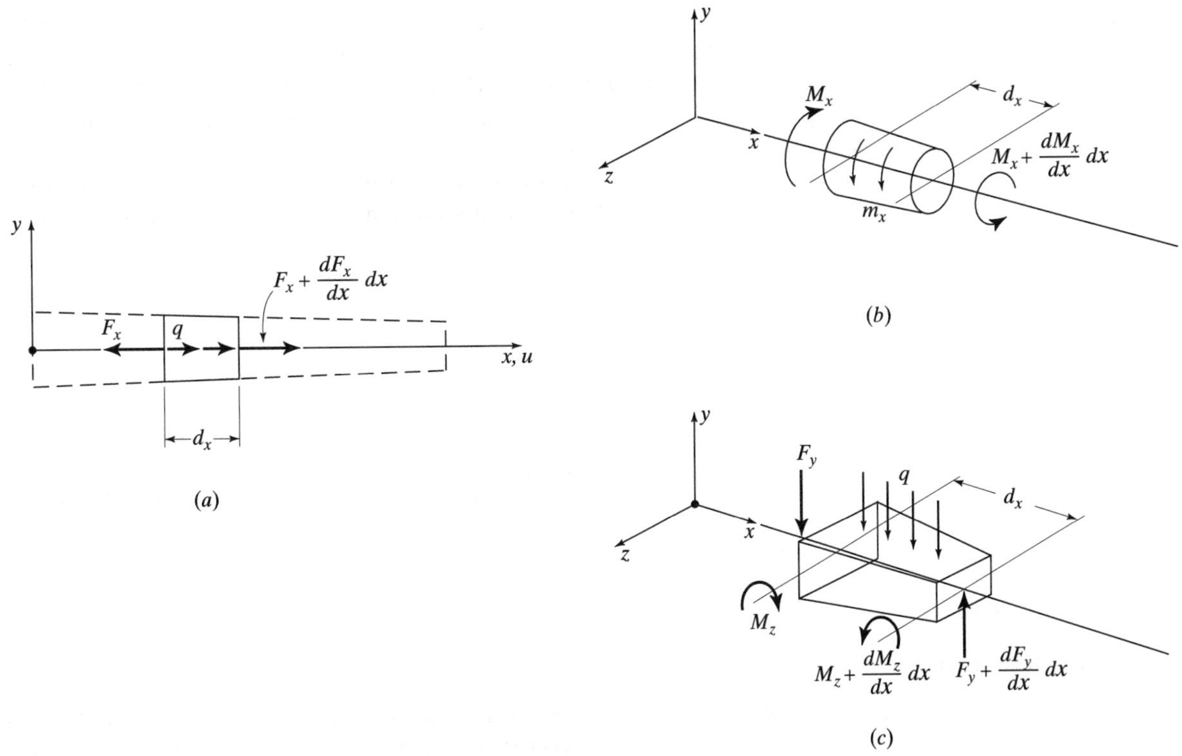

Figure 6.7 Forces acting on differential elements of axial, torsional, and flexural members. (*a*) Axial behavior (*b*) Torsional behavior. (*c*) Flexural behavior.

The internal force on the left face is F_x. (F_x is positive when tensile.) Because of the presence of the applied load q, this force will change along the length and at the right face it is $F_x + (dF_x/dx)dx$. The condition of equilibrium in the horizontal direction can now be written as

$$\sum F_x = \left(F_x + \frac{dF_x}{dx}dx\right) - F_x + q \cdot dx = 0$$

or

$$\frac{dF_x}{dx} + q = 0 \tag{6.25}$$

The more fundamental measure of structural action on a cross section is stress, which is related to the force by $F_x = A\sigma_x$.

Similar relationships hold for the case of torsional behavior (see Figure 6.7*b*). The left face of the illustrated differential element is subjected to a twisting moment M_x and, because of the presence of a distributed applied twisting moment m_x, the moment on the right face is $M_x + (dM_x/dx)dx$. The condition of equilibrium of moments about the x axis gives

$$\sum M_x = \left(M_x + \frac{dM_x}{dx}dx\right) - M_x + m_x dx = 0$$

or

$$\frac{dM_x}{dx} + m_x = 0 \tag{6.26}$$

Again, there is a more fundamental measure of structural action, which in this case is the shearing stress $\tau = M_x r/J$, where J is the torsion constant and r is the radius from the x axis to the point at which τ is evaluated.

In examining beam flexure we consider, as before, a differential element (Figure 6.7*c*) that has a cross section that is symmetric about the y axis. A moment M_z and a shear F_y act on the element's left face. A distributed transverse load q acts on the element in the y direction. Because of the action of an applied distributed load, q, the shear at the right face is $F_y + (dF_y/dx)dx$. The moment at the right face is $M_z + (dM_z/dx)dx$. From y-direction force equilibrium,

$$\sum F_y = \left(F_y + \frac{dF_y}{dx}dx\right) - F_y - qdx = 0$$

so that

$$\frac{dF_y}{dx} - q = 0 \tag{6.27}$$

Also, from the equilibrium of moments about the z axis at the right end of the element (the higher-order effect of q is disregarded),

$$\sum M_z = \left(M_z + \frac{dM_z}{dx}dx\right) - M_z - F_y \cdot dx = 0$$

or

$$\frac{dM_z}{dx} = F_y \tag{6.28}$$

Finally, by substitution of (6.28) into (6.27),

$$\frac{d^2M_z}{dx^2} = q \tag{6.29}$$

Equations 6.27 and 6.28 are the basic equations of equilibrium for beam bending. When the internal bending moment satisfies Equation 6.29, both Equations 6.27 and 6.28 are satisfied.

The stresses due to the shear and bending moment are, respectively, $\tau_{xy} = F_yQ_z/I_zb$ and $\sigma_x = M_zy/I_z$. In the equation for τ_{xy}, b is the width of the element at the point where the shear stress is calculated and Q_z is the first moment of area beyond that point, taken with respect to the neutral axis.

6.5.2 Characteristics of Virtual Force Systems

We can now proceed to the definition of the characteristics of virtual force systems. The fundamental requirement on any such system is that it meet the relevant conditions of equilibrium. These are represented, externally, by the overall equations of equilibrium (e.g., $\Sigma F_x = 0$, etc.) and internally by Equations 6.25, 6.26, or 6.29, as appropriate.

Consider, in illustration of this requirement, the imposition of a virtual force δF_{y1} to the left end of the cantilever beam of Figure 6.8*a*. The appropriate virtual force system, in terms of internal shears and moments, is simply that given by elementary statics and shown in Figures 6.8*b* and 6.8*c*. If the bending moment distributions were, however, assumed to be of the form $\delta M_z = (x^2/L)\delta F_{y1}$ rather than the correct form $x(\delta F_{y1})$, then the relevant equilibrium condition (Equation 6.28) gives for δF_{y1}, the virtual shear at any section:

$$\delta F_y = \frac{d\delta M_z}{dx} = \frac{2x}{L}\,\delta F_{y1}$$

which is not consistent with the correct relationship $\delta F_y = \delta F_{y1}$. This assumed virtual force system would not be an acceptable virtual force system.

Generally, and in particular for statically indeterminate structures, there is more

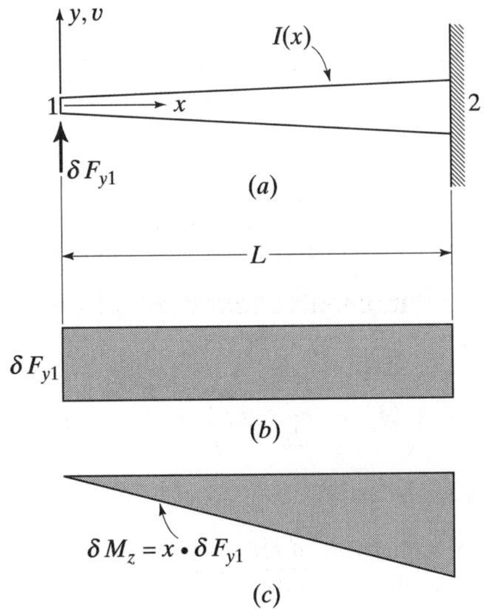

Figure 6.8 Virtual force systems for cantilever beam. (*a*) Structure. (*b*) Shear. (*c*) Bending moment.

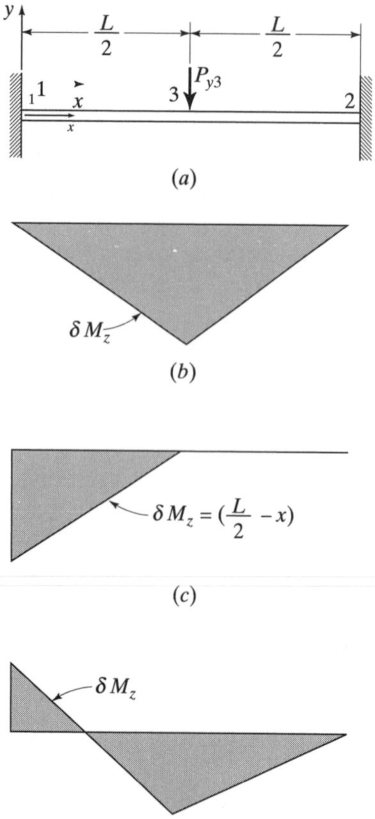

Figure 6.9 Fixed beam under concentrated load—alternative virtual force systems. (*a*) Structure. (*b*) Moment diagram—simple support conditions. (*c*) Moment diagram—cantilever support conditions. (*d*) Moment diagram—fixed support at left and hinge at right.

than one acceptable choice of a virtual force system. Figure 6.9 illustrates this point. The given analysis condition (Figure 6.9*a*) is a fixed end beam under the action of a concentrated load P_{y3}. The moment diagrams of Figures 6.9*b*–6.9*d* are, respectively, the solutions for the beam with simple supports, cantilever support, and fixity at the left end and a hinge at the right end, each under the action of the applied load P_{y3}. It is easily verified that application of the equilibrium conditions (Equations 6.27 and 6.28) gives a statically consistent force-distribution relationship for each support system. Hence, in the problem at hand, any one of these is acceptable as a virtual force system for a virtual force δP_{y3}.

Despite the wide number of choices available in the definition of virtual force systems for statically indeterminate structures, practical considerations favor the use of systems calculated for a statically determinate form of the same structure, such as the simple beam (Fig. 6.9*b*) of the problem of Figure 6.9*a*. The use of internal force distributions corresponding to the statically indeterminate forms of the structure are generally not practical. They are certainly not of use in the development of computer-oriented analysis tools since, most often, it is precisely this information that is being sought through the solution.

6.5.3 Formulation of the Virtual Forces Principle

It was stated at the beginning of this section that the principle of virtual forces stems from the investigation of the changes in work and energy that occur on account of the imposition of a system of virtual forces on the state achieved under the real applied

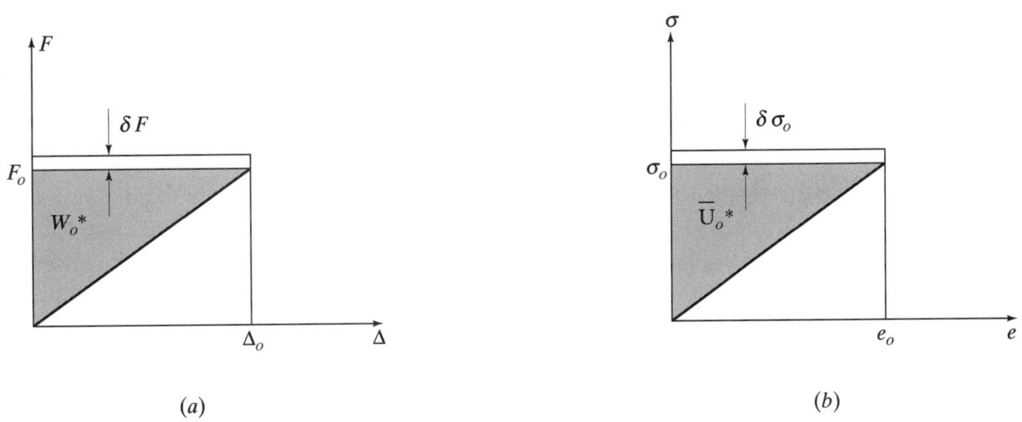

Figure 6.10 Complementary external and internal virtual work.

loads. The virtual force system produces *external complementary virtual work*, δW^*_{ext}, and *internal complementary virtual work*, δW^*_{int}.[7]

The term "complementary" can be explained by reference to Figure 6.10*a*, a plot of the relationship between a force F and its corresponding displacement Δ, for linear elastic behavior. As explained in Section 4.2, Figure 4.3, and again in Section 6.1, Figure 6.1, the area between the F-Δ line and the Δ axis is the external work. The shaded area between the F-Δ line and the F axis is *complementary* to the external work and is designated as the external complementary work. Thus, if the intensity of the force is incremented by an amount δF, the virtual force, then the external complementary virtual work δW^*_{ext} is $(\delta F)(\Delta_0)$, where Δ_0 is the intensity of the displacement at the total load F_0.

A general expression for the internal complementary virtual work δW^*_{int} can be established by reference to Figure 6.10*b*, a plot of stress versus strain at a representative point in a structure. In analogy to the description of external virtual load-displacement circumstances, the action of the virtual stress $\delta\sigma_0$ (caused by the external virtual load) on the real state of strain e_0 results in a complementary virtual strain energy per unit of volume equal to $\lfloor \delta\boldsymbol{\sigma}_0 \rfloor \{\mathbf{e}_0\}$. The internal complementary virtual work in the complete body of volume (vol) is, therefore

$$\delta W^*_{\text{int}} = \int_{\text{vol}} \lfloor \delta\boldsymbol{\sigma}_0 \rfloor \{\mathbf{e}_0\} d(\text{vol}) \tag{6.30}$$

Because we wish to deal with force states rather than displacement states, the relationship between stress and strain is invoked to replace $\{\mathbf{e}_0\}$ by stress $\{\boldsymbol{\sigma}_0\}$. This has been designated as $\{\boldsymbol{\sigma}_0\} = [\mathbf{E}]\{\mathbf{e}_0\}$ for the general state of stress and, since we require strain in terms of stress, we write it as $\{\mathbf{e}_0\} = [\mathbf{E}]^{-1}\{\boldsymbol{\sigma}_0\}$. Hence, Equation 6.30 becomes

$$\delta W^*_{\text{int}} = \int_{\text{vol}} \lfloor \delta\boldsymbol{\sigma}_0 \rfloor [\mathbf{E}]^{-1} \{\boldsymbol{\sigma}_0\} d(\text{vol}) \tag{6.31}$$

We are now in a position to state the *principles of virtual forces*, which is simply a condition for the balance of internal and external complementary virtual work for the condition of an imposed virtual force system:

$$\delta W^*_{\text{ext}} = \delta W^*_{\text{int}} \tag{6.32}$$

[7]This quantity is also called *complementary virtual strain energy*, δU^*, a designation that was used in the first edition of this text.

It has already been emphasized that the real and virtual force states must each satisfy the conditions of equilibrium. Equation 6.32 represents the conditions of compatibility. When the real force state corresponds to a state of strain which meets the conditions of compatibility, the solution is exact.

The solutions to be established by use of the principle of virtual forces are all exact (within the limits of the assumption of linear behavior) and we use Equation 6.32 to achieve the evaluation of the parameters that characterize such solutions. This is done in the next section and again in Section 7.6. Therefore, with exact solutions in mind, we can give the following verbal statement of the principle of virtual forces (see Ref. 6.1).

The strains and displacements in a deformable system are compatible and consistent with the constraints if and only if the external complementary virtual work is equal to the internal complementary virtual work for every system of virtual forces and stresses that satisfies the conditions of equilibrium.

If the real force state does not correspond to a deformational state that exactly satisfies compatibility, then Equation 6.32 can be used to enforce an approximate satisfaction of the conditions of compatibility. This is analogous to the state of affairs in the virtual displacements approach, when the real displacement corresponds to a stress state that does not satisfy equilibrium, and enforcement of the virtual displacement condition (Eq. 6.13) results in an approximate satisfaction of equilibrium. In this text, however, we are not concerned with approximate solutions by means of the principle of virtual forces.

To make use of this principle it is necessary to establish specific expressions for the respective modes of structural behavior that are of interest in framework analysis. In so doing, the subscript 0 is removed from all terms. Also, the stresses are replaced by the respective stress resultant terms.

For axial members (Fig. 6.7*a*), we have $\{\boldsymbol{\sigma}\} = F_x/A, \lfloor \delta\sigma \rfloor = \delta F_x/A, [\mathbf{E}]^{-1} = 1/E$, and $d(\text{vol}) = A\,dx$. Therefore, Equation 6.31 becomes

$$\delta W^*_{\text{int}} = \int_0^L \delta F_x \cdot \frac{F_x}{EA} \cdot dx \tag{6.33}$$

For the flexural member (Fig. 6.7*c*), we have $\sigma = \sigma_x = M_z \cdot y/I_z, \lfloor \delta\sigma \rfloor = \delta\sigma_x = \delta M_z \cdot y/I_z$, and $[\mathbf{E}]^{-1} = 1/E$. Thus, Equation 6.31 becomes

$$\begin{aligned} \delta W^*_{\text{int}} &= \int_{\text{vol}} \left(\delta M_z \cdot \frac{y}{I_z}\right)\left(M_z \cdot \frac{y}{I_z}\right) \frac{d(\text{vol})}{E} \\ &= \int_0^L \frac{1}{E}\left(\frac{\delta M_z}{I_z}\right)\left(\frac{M_z}{I_z}\right)\left(\int_A y^2 dA\right)dx = \int_0^L \delta M_z \cdot \frac{M_z}{EI_z}\,dx \end{aligned} \tag{6.34}$$

Through similar reasoning we find that, for torsion,

$$\delta W^*_{\text{int}} = \int_0^L \delta M_x \cdot \frac{M_x}{GJ}\,dx \tag{6.35}$$

6.5.4 Construction of Analytical Solutions by Virtual Forces Principle

As in Section 6.4, where we explored the practical implications of the principle of virtual displacements through the medium of analytical solutions, we now examine solutions by the principle of virtual forces. The solutions to be obtained are for displacements due to the action of specified forces.

To use the principle of virtual forces we must have expressions for the real and virtual internal force systems. The total internal force system is that resulting from the given applied loads. The virtual internal force system results from the application of a virtual load that corresponds to the desired displacements.

The simplest example concerns the calculation of displacement u_2 of the uniform axial member (Fig. 6.5*a*) due to the end load F_{x2}. The real internal force system is, from statics, $F_x = F_{x2}$. The virtual force is chosen in correspondence with the desired displacement and therefore is δF_{x2}, resulting in an internal virtual force system $\delta F_x = \delta F_{x2}$. We have, from Equation 6.33,

$$\delta W^*_{\text{int}} = \int_0^L \frac{\delta F_{x2} \cdot F_{x2}}{EA}\, dx = \delta F_{x2} \cdot F_{x2} \frac{L}{EA}$$

The external complementary virtual work is simply $\delta F_{x2} \cdot u_2$. Thus, from the principle of virtual forces

$$\delta F_{x2} \cdot u_2 = \delta F_{x2} \cdot F_{x2} \frac{L}{EA}$$

or

$$u_2 = F_{x2} \frac{L}{EA}$$

which is the correct solution.

Consider next the uniform flexural member, Figure 6.11, simply supported at either end and subjected to the central load, $-P_c$. We seek to calculate the displacement under this load so that the virtual force is $-\delta P_c$. The internal force system, characterized by the moments M_z is (Fig. 6.11*b*)

$$M_z = P_c \frac{x}{2} \qquad \left(0 \le x \le \frac{L}{2}\right)$$

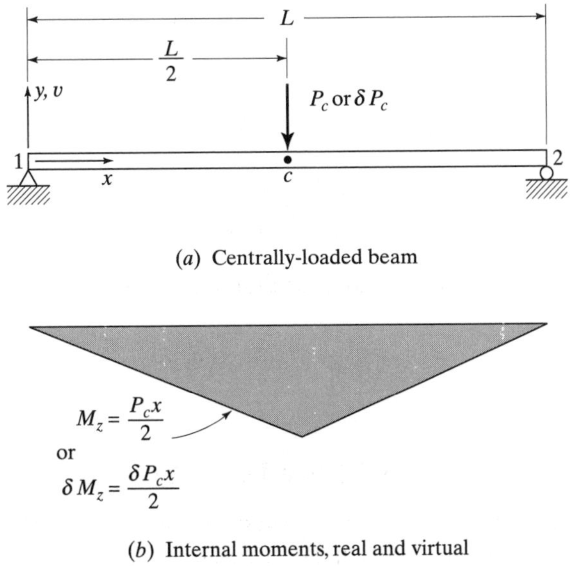

Figure 6.11 Simply supported beam. Total and virtual force systems. (*a*) Structure. (*b*) Moment diagram for simple support conditions.

and we choose the virtual internal force system to be of the same form:

$$\delta M_z = (\delta P_c)\frac{x}{2} \qquad \left(0 \le x \le \frac{L}{2}\right)$$

Since these are derived from statical analysis, they meet the conditions of equilibrium. We have, from Equation 6.34 and invoking symmetry about c,

$$\delta W^*_{\text{int}} = \frac{2}{EI_z}\int_0^{L/2}\left(\delta P_c \cdot \frac{x}{2}\right)\left(P_c \cdot \frac{x}{2}\right)dx = \delta P_c \cdot P_c \cdot \frac{L^3}{48EI_z}$$

The external complementary virtual work is $\delta W^*_{\text{ext}} = -\delta P_c \cdot v_c$, and the principle of virtual forces gives

$$-\delta P_c \cdot v_c = \delta P_c \cdot P_c \cdot \frac{L^3}{48EI_z}$$

or, upon cancellation of δP_c from both sides,

$$v_c = -\frac{P_c L^3}{48EI_z}$$

In the case described above we placed a single virtual external load δP_i at the point i at which a displacement solution was desired. (In flexure, when an angular displacement is sought, the virtual force is a moment.) Since the virtual displacement is cancelled from both sides of the equation $\delta W^*_{\text{ext}} = \delta W^*_{\text{int}}$, its value is immaterial. Therefore, it is customary to set δP_i equal to unity, and for this reason the principle of virtual forces, in practical application, is generally termed the *unit load method*. The procedure of setting the virtual external load to unity (with due account being taken of the sign of the virtual force) is followed in Examples 6.6–6.8.

EXAMPLE 6.6

Using the principle of virtual forces, calculate the displacements of points 2 and 3 of the beam in Example 6.5.

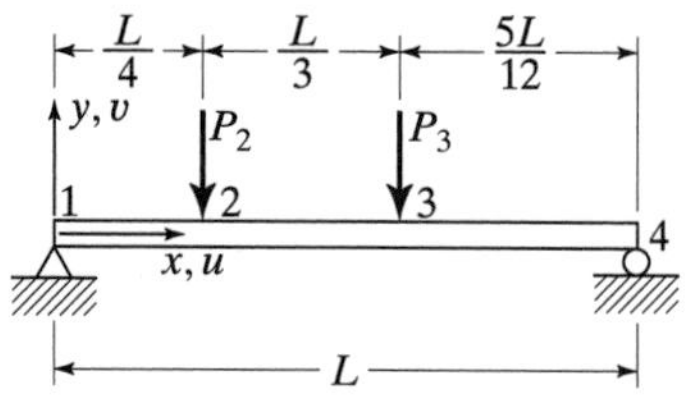

The internal moments due to the applied loads are

$$M_z = \frac{x}{12}(9P_2 + 5P_3) \qquad \left(0 \le x \le \frac{L}{4}\right)$$

$$M_z = \frac{1}{4}(L - x)P_2 + \frac{5}{12}P_3 x \qquad \left(\frac{L}{4} \le x \le \frac{7L}{12}\right)$$

$$M_z = \frac{P_2}{4}(L - x) + \frac{7P_3}{12}(L - x) \qquad \left(\frac{7L}{12} \le x \le L\right)$$

Two systems of virtual forces are needed for calculation of v_2 and v_3, respectively. For v_2 we place a virtual external load $\delta P_2 = -1$ at point 2. The corresponding virtual moments are then

$$\begin{aligned}\delta M_z &= \frac{3x}{4} \qquad \left(0 \le x \le \frac{L}{4}\right)\\ &= \frac{L-x}{4} \qquad \left(\frac{L}{4} \le x \le L\right)\end{aligned}$$

In this case, the internal complementary virtual work

$$\begin{aligned}W^*_{\text{int}} &= \frac{1}{EI_z}\left\{\int_0^{L/4}\left(\frac{3x}{4}\right)\left(\frac{x}{12}\right)(9P_2 + 5P_3)dx\right.\\ &\quad + \int_{L/4}^{7L/12}\left(\frac{L-x}{4}\right)\left(\frac{1}{4}\right)\left[(L-x)P_2 + \frac{5}{3}P_3x\right]dx\\ &\quad \left.+ \int_{7L/12}^{L}\left(\frac{L-x}{4}\right)\left(\frac{L-x}{4}\right)\left(P_2 + \frac{7}{3}P_3\right)dx\right\}\\ &= \frac{L^3}{EI_z}(0.0117P_2 + 0.0133P_3)\end{aligned}$$

Also, the external complementary work is $\delta W^*_{\text{ext}} = (\delta P_2) \cdot v_2 = -1 \cdot v_2$. Equating δW^*_{int} and δW^*_{ext} we have

$$v_2 = -\frac{L^3}{EI_z}(0.0117P_2 + 0.0133P_3)$$

Similarly, by placing a virtual force $\delta P_3 = -1$ at point 3 and again pursuing the virtual forces procedure, we find

$$v_3 = -\frac{L^3}{EI_z}(0.0133P_2 + 0.0197P_3)$$

Just as the virtual displaced state in the principle of virtual displacements need not correspond to the real displaced state, the virtual force state need not correspond to the real force state in the application of the principle of virtual forces. This circumstance often proves useful in the analytical determination of the displacements of statically indeterminate structures. For example, if a solution is available for the internal force distribution due to the applied loads, then a statically determinate virtual force system can be employed in calculation of displacements by means of the principle of virtual forces.

This is illustrated in Example 6.7. The exact solution for the desired displacement is obtained even through the chosen virtual stress field does not correspond to a displacement field that meets the displacement continuity conditions of the problem. (The displaced state of the simply supported beam is in violation of the fixed-support condition at point 2.)

Example 6.8 demonstrates that care must be exercised, however, in the choice of virtual force fields so that the results obtained have practical value. The logical choice of a virtual internal force distribution is the force associated with the actual cantilever support (virtual force system 1). This leads to the conventional solution $v_c = 5P_1L^3/48EI_z$. Virtual force system (2), however, which is associated with simple support, gives the solution $v_1/2 - v_c = P_1L^3/16EI_z$. Although this is a valid solution, its form is not of direct practical value. It requires a separate determination of v_c and this cannot be accomplished using simple support conditions for the virtual force system.

EXAMPLE 6.7

Calculate the displacement v_c at the center of the beam shown, using the principle of virtual forces.

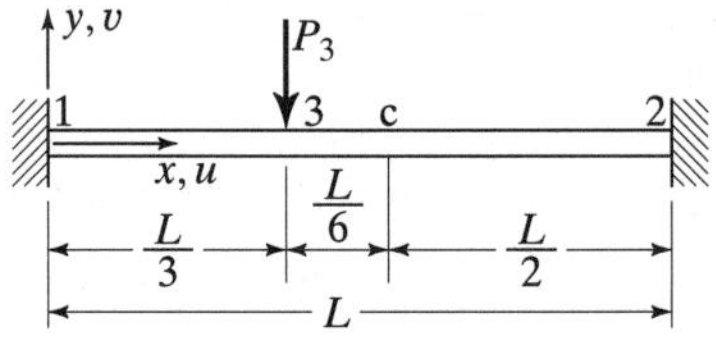

The beam is statically indeterminate, and the determination of the internal force distribution is, in itself, a problem for solution by the principle of virtual forces, using redundant force concepts. Since the problem seeks only a value of displacement, it will be assumed that the indeterminate structural analysis has already been performed, giving the following relationships for the internal moments in terms of the applied loads P_3.

$$M_z = \frac{4}{27} P_3(5x - L) \qquad \left(0 \le x \le \frac{L}{3}\right)$$

$$M_z = \frac{P_3}{27}(5L - 7x) \qquad \left(\frac{L}{3} \le x \le L\right)$$

In order to obtain the displacement v_c, a virtual force $\delta P_c = -1$ is placed at point c. Since the associated internal moments need only satisfy the conditions of equilibrium, we calculate them as if the beam were simply supported at both ends. This gives

$$\delta M_z = \frac{x}{2} \qquad \left(0 \le x \le \frac{L}{2}\right)$$

$$= \frac{L - x}{2} \qquad \left(\frac{L}{2} \le x \le L\right)$$

Thus

$$\delta W^*_{\text{int}} = \frac{1}{EI_z}\left[\int_0^{L/3}\left(\frac{x}{2}\right)(5x - L)\frac{4P_3}{27}\,dx + \int_{L/3}^{L/2}\left(\frac{x}{2}\right)(5L - 7x)\frac{P_3}{27}\,dx + \int_{L/2}^{L}\frac{L - x}{2}(5L - 7x)\frac{P_3}{27}\,dx\right] = \frac{5P_3L^3}{1296EI_z}$$

Also,

$$\delta W^*_{\text{ext}} = -1 \cdot v_c$$

Hence, by $\delta W^*_{\text{int}} = \delta W^*_{\text{ext}}$

$$v_c = -\frac{5P_3L^3}{1296EI_z}$$

EXAMPLE 6.8

Calculate the displacement at point c of the cantilever beam shown, using the virtual forces approach with the following systems of virtual forces: (1) a system based on the actual support condition and (2) a system based on simple support of the beam.

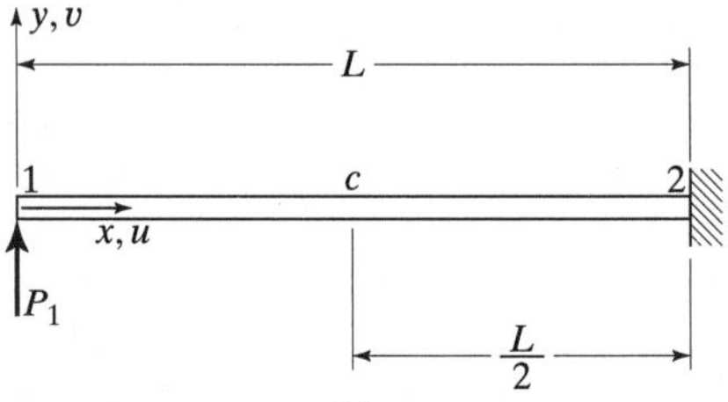

(a)

The actual internal force system, defined by M_z, is shown in part b of the figure. Proceeding to the solution for virtual force system (1), we place a unit load at point c and obtain the internal virtual force system shown in part c of the figure. Thus

$$\delta W^*_{\text{ext}} = 1 \cdot v_c$$

$$\delta W^*_{\text{int}} = \frac{1}{EI_z}\int_{L/2}^{L}\left(x - \frac{L}{2}\right)(P_1 x)dx = \frac{5P_1L^3}{48EI_z}$$

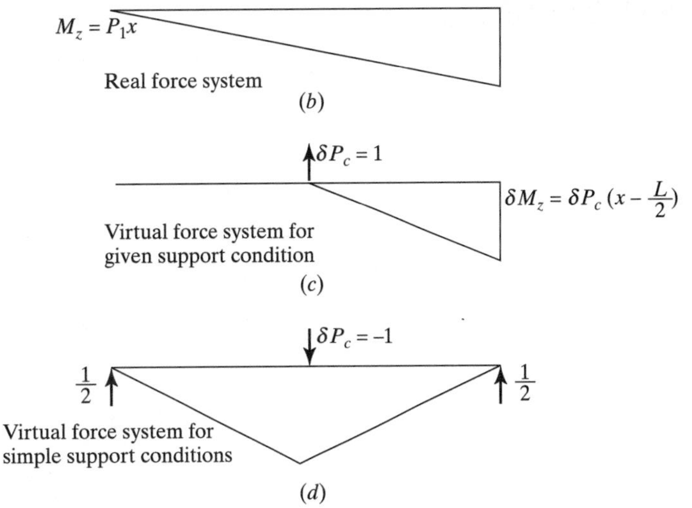

Hence, by $\delta W^*_{\text{ext}} = \delta W^*_{\text{int}}$

$$v_c = \frac{5P_1L^3}{48EI_z}$$

The virtual force system (2) gives the internal virtual force system shown in part d of the figure. In this case, we have

$$\delta W^*_{\text{ext}} = (\tfrac{1}{2})v_1 - (1)v_c$$

also

$$\delta W^*_{\text{int}} = \frac{1}{EI_z}\int_{0}^{L/2}\left(\frac{x}{2}\right)(P_1 x)dx + \frac{1}{EI_z}\int_{L/2}^{L}\frac{L - x}{2}(P_1 x)dx$$
$$= \frac{P_1L^3}{16EI_z}$$

Hence, by $\delta W^*_{\text{ext}} = \delta W^*_{\text{int}}$

$$\frac{v_1}{2} - v_c = \frac{P_1L^3}{16EI_z}$$

Our intention in introducing the principle of virtual forces has been to establish a basis for the formulation of element flexibility equations, a topic which will be taken up in detail in Section 7.6. In such cases the element will always be a statically determinate structure, and the virtual forces will be chosen in the same form as the real forces. As mentioned in the introduction to this chapter, our principal interest in flexibility relationships is in their value in the formation of element stiffness equations through simple flexibility-stiffness transformations. The role of the principle of virtual forces in statically indeterminate structural analysis, where it is most prominent in classical methods, has only been mentioned briefly. To use the principle effectively in large-scale, computerized indeterminate structural analysis it is necessary to have a more organized scheme than the one that is customarily adopted in classical treatments of the principle. The essential components of one such scheme may be found in Chapters 6 and 7 of the first edition of this text.

6.6 PROBLEMS

6.1 Calculate, by use of virtual work, the force in member 4-6.

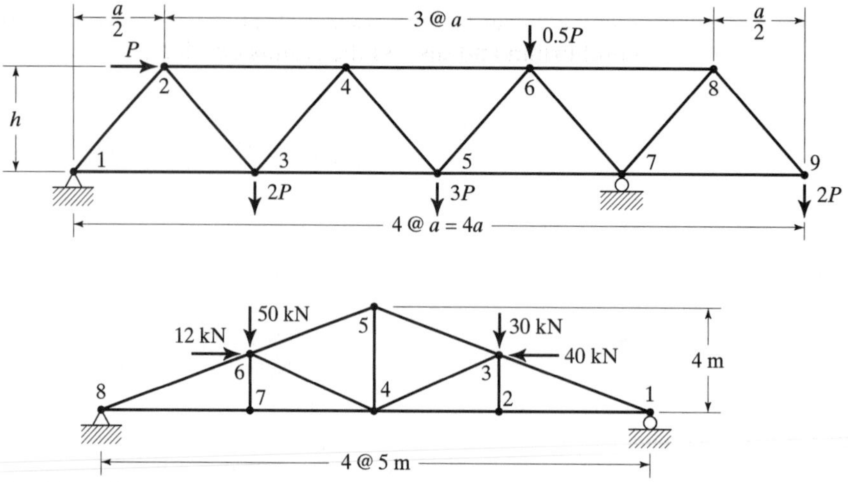

Problem 6.1

6.2 Calculate, by virtual work, the reaction at point 5.

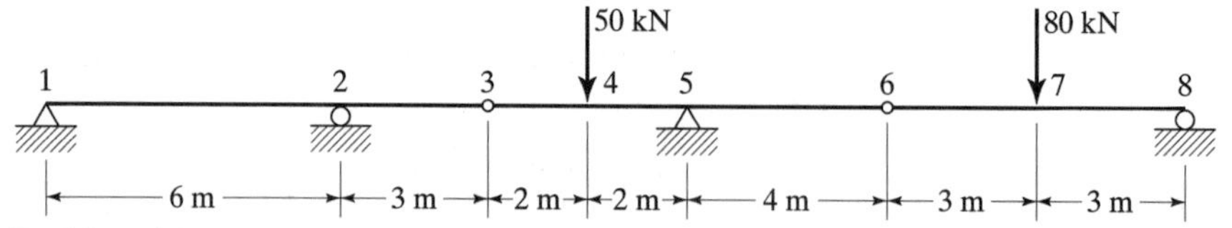

Problem 6.2

6.3 Calculate, by virtual work, the reactions at points 1 and 2 of Problem 6.2.

6.4 Establish expressions for the total strain energy, U, of combined axial, torsional, and flexural behavior (Example 6.2) by treating δW_{int} as a differential and assuming linear behavior.

6.5 A nonlinear stress-strain law is defined as $\sigma = E_0(\varepsilon + 100\varepsilon^2)$ where E_0 is the modulus of elasticity at $\varepsilon = 0$. Develop an expression for the internal virtual work for an axial member composed of this material.

6.6 Calculate the displacement at point A, using the principle of virtual displacements. Use $v = v_A \sin \pi x/L$ and a similar form for δv.

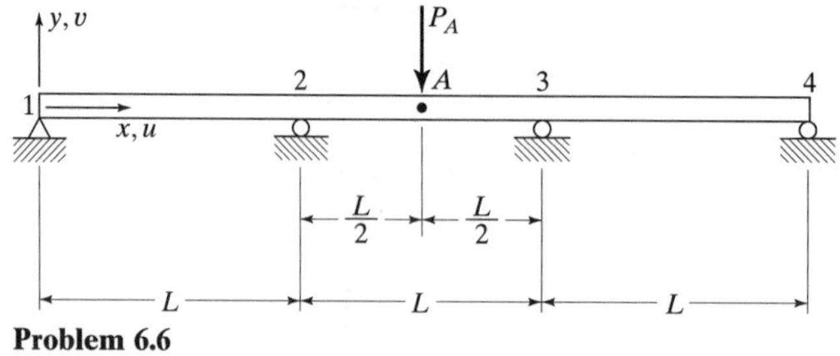

Problem 6.6

6.7 Solve Problem 6.6 by means of the following expansion for v and a similar expansion for δv.

$$v = a_1 \sin \frac{\pi x}{L} + a_2 \sin \frac{3\pi x}{L}$$

6.8 Calculate, by means of the principle of virtual displacements, the deflection of the end point of the cantilever beam shown. Use the following approximation of the transverse displacement (EI constant). (First fit the coefficients a_1 and a_2 to the displacements at points 2 and 3, v_2 and v_3). Check for satisfaction of the conditions of equilibrium and discuss the results obtained.

$$v = \left(1 - \cos\frac{\pi x}{2L}\right)a_1 + \left(1 - \cos\frac{3\pi x}{2L}\right)a_2$$

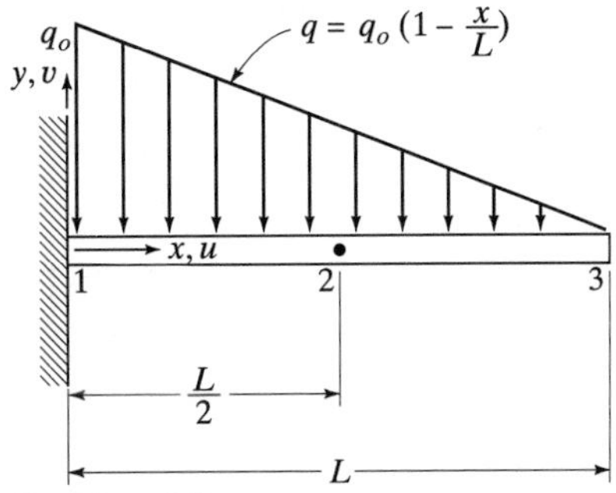

Problem 6.8

6.9 Solve Problem 6.8, using $v = a_1x^2 + a_2x^3$. (Fit the coefficients a_1 and a_2 to the displacements at point 3, v_3 and θ_{x3}.)

6.10 Calculate, by means of the principle of virtual displacements, the displacement at point A, using $v = x(x - L)a_1 + x^2(x - L)a_2$ and a similar form for δv_1 (EI constant).

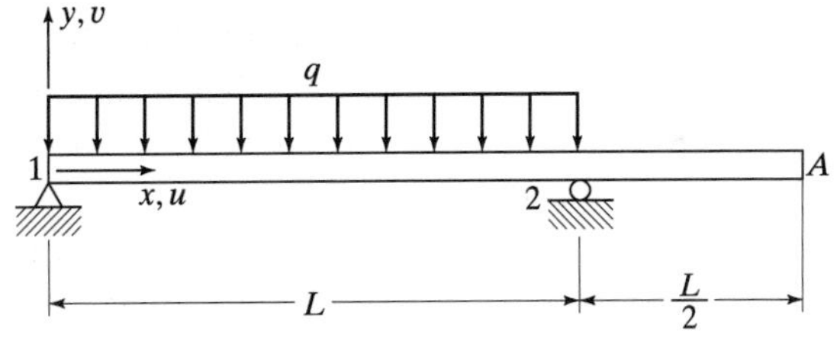

Problem 6.10

6.11 Using the principle of virtual displacements, solve for the displacements v_1 and θ_{z1} of the cantilever beam and for the support reactions F_{y2} and M_{z2}.
Use

$$v = \frac{1}{2}\left(1 + \cos\frac{\pi x}{L}\right)v_1, \qquad \delta v = \frac{1}{2}\left(1 + \cos\frac{\pi x}{L}\right)\delta v_1$$

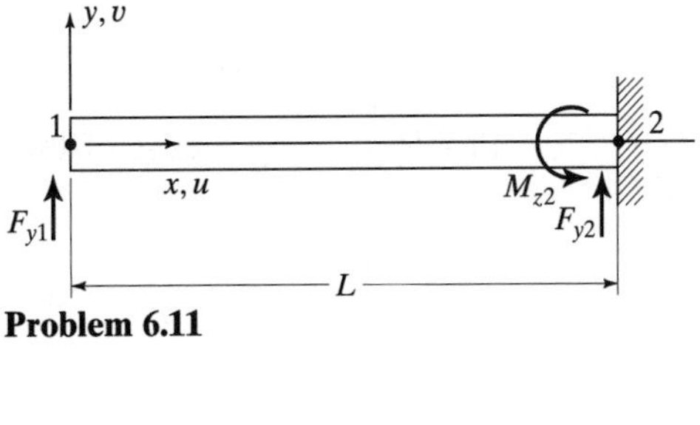

Problem 6.11

6.12 Determine, by means of the principle of virtual displacements, the approximate value of the transverse displacement at point 2. Use $v = -\sin(\pi x/10)v_2$ in description of the displaced shape of the beam. $E = 200{,}000$ MPa.

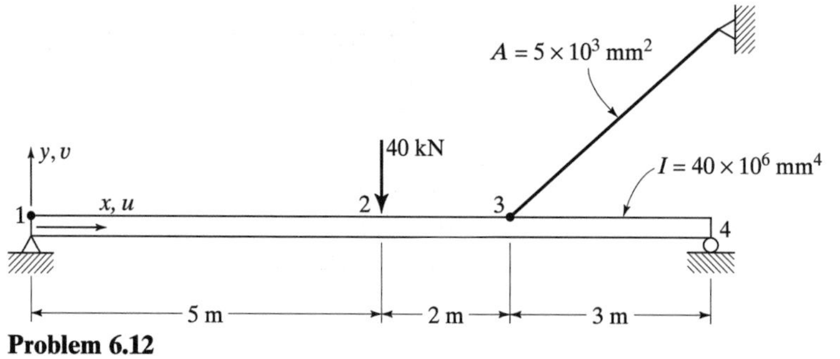

Problem 6.12

6.13 Calculate, by means of the principle of virtual displacements, the displacement of point 5. All member areas are $A = 10 \times 10^4$ mm², $E = 200{,}000$ MPa.

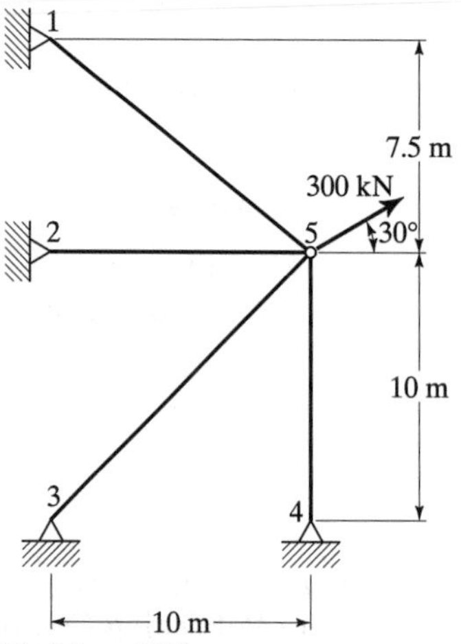

Problem 6.13

6.14 Calculate, by virtual work, the transverse displacement of point 3. $E = 200{,}000$ MPa, $I = 100 \times 10^6$ mm⁴.

$$v = a_1 \sin \frac{\pi x}{12} + a_2 \sin \frac{\pi x}{4}$$

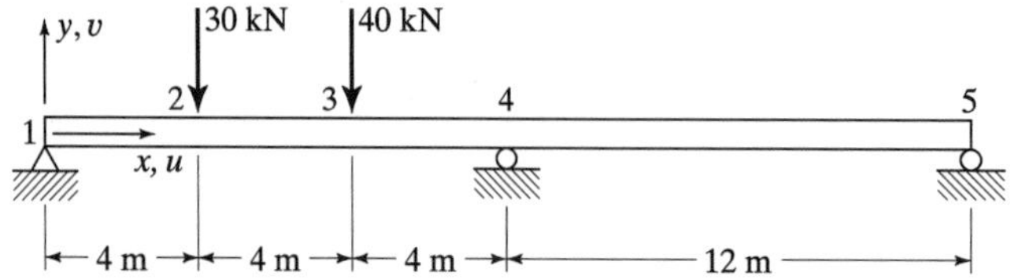

Problem 6.14

6.15 Solve for the displacement at point 3 by use of the principle of virtual forces. The exact solution for the moments in the beam, due to the applied loads is,

$$\begin{aligned} M_z &= -45 + 22.5x \qquad (0 \le x \le 6) \\ &= 270 - 30x \qquad (6 \le x \le 9) \end{aligned}$$

Use for the virtual force distribution a statically determinate bending moment distribution. $E = 200{,}000$ MPa.

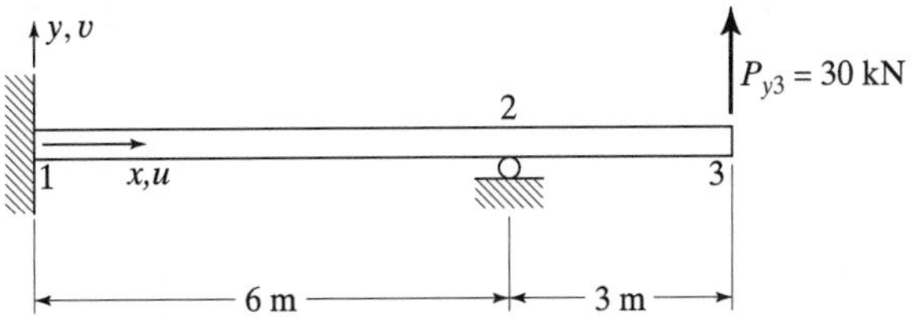

Problem 6.15

6.16 Derive expressions for the angular and transverse displacement, θ_{z2} and v_2 by use of the principle of virtual forces. The member is of rectangular cross section and has unit width.

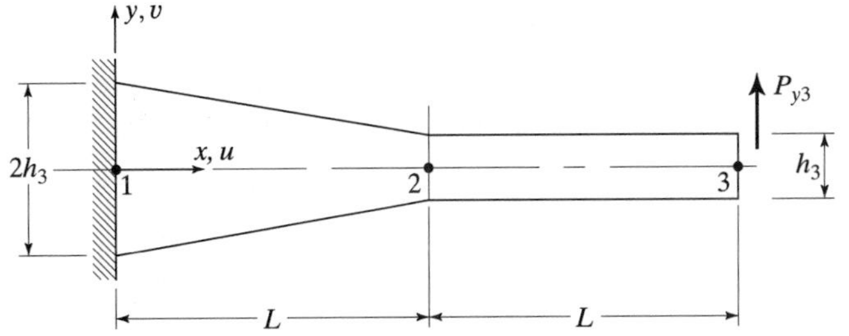

Problem 6.16

6.17 Calculate the displacement at the center ($x = L/2$) of the structure of Example 6.5, using the principle of virtual forces.

6.18 Calculate the displacement θ_{zb} of Example 4.10, using the solution of that example for the real force system and a statically determinate virtual force system.

REFERENCES

6.1 J. T. Oden and E. A. Ripperger, *Mechanics of Elastic Structures*, 2nd Edition, Hemisphere Publishing Co., N.Y., 1981.

6.2 J. S. Przemieniecki, *Theory of Matrix Structural Analysis*, McGraw-Hill, N.Y., 1968.

6.3 J. H. Argyris and S. Kelsey, *Energy Theorems and Structural Analysis*, Butterworth Scientific Publications, London, 1960.

Chapter 7

Virtual Work Principles in Framework Analysis

In this chapter we apply the principles of virtual work established in Chapter 6 to the construction of the algebraic relationships of framework matrix structural analysis. A development that is entirely parallel to the direct formulation procedure given earlier can be accomplished through virtual work. To include such parallel developments is desirable since it equips the reader to cope more easily with continuing developments in this field, which might be based on either direct formulation procedures or on virtual work concepts.

The principles of virtual work are applied herein principally for their facility in the formulation of approximate solutions. This facility has no identifiable counterpart in the direct formulation procedures. For example, an expeditious and relatively accurate approximate formulation can be established for tapered members, and a quite general approach can be made to the treatment of distributed loads. Also, it is the approach we shall use in the formulation of the equations of geometric nonlinear and elastic critical load analysis in Chapter 9.

The starting point of this chapter, Section 7.1, delineates the manner of describing the displaced state of framework elements. When the resulting element formulation is intended to be approximate, the approximation is nearly always in the form of a description of the displaced state, or "shape function," of the elements.

With a means of description of the displaced state of elements in hand, we proceed in Section 7.2 to the formulation of the standard stiffness relationship of constant-section framework elements, using the principle of virtual displacements. Relationships already established in Chapters 2–5 are reformulated in this manner. The material presented merely serves to illustrate the details of the procedure. The next section, however, takes up the formulation of approximate stiffness equations for tapered elements, an entirely new topic. Similarly, nonuniform torsion, the condition in which torque is resisted by a combination of the simple torsional shearing stresses studied in Section 4.5.2 and resistance to out-of-plane warping, is treated in Section 7.4. Then, on the basis of virtual displacement considerations, a completely general approach to the treatment of loads applied between the end points of framework elements is presented. This approach encompasses the extremes of point loads and distributed loads and confirms the validity of expressions derived in Section 5.2.

The element flexibility relationships employed up to this point were those for the uniform axial member and beam element and were established in Section 4.5 through direct reasoning. Practical structures, however, may contain tapered or curved elements or elements of such form and behavior that flexibility equations are not readily derivable through that route. Nevertheless, a general approach to the formulation of element flexibility relationships is available on the basis of the principle of virtual forces presented in Section 6.5. In Section 7.6 it is applied to the formulation of element flexibility equations. Included is the treatment of transverse shearing deformation in beams.

7.1 DESCRIPTION OF THE DISPLACED STATE OF ELEMENTS

7.1.1 Definition of the Shape Function Mode of Description

In linear elastic analysis the only ingredients needed for the construction of element stiffness equations by the virtual displacements approach are:

1. The elastic constants relating the stresses and strains of the material.
2. Descriptions of the real and virtual displaced states of the element.
3. The relevant differential relationships between strain and displacement.

The elastic constants are known from laboratory testing. The differential relationships between strain and displacement are basic relationships in structural mechanics and have already been defined for the cases of interest. It has been pointed out (Section 6.4) that the virtual displaced state is logically chosen to be of the same shape as the real displaced state. Thus, to form element stiffness equations by means of the virtual displacement approach, the only preliminary task that remains is the description of the displaced shape of the element, which is then employed for both the virtual and real displaced states.

The displaced state of the simplest structural elements, such as those studied previously, can be found by solution of the differential equations that govern the behavior of these elements (e.g., Equation 4.31b in the case of bending). To use the virtual displacement approach would appear to be redundant in part, since, as demonstrated in Section 4.5, the element force-displacement equations can be constructed directly from these solutions. It will be shown in Sections 7.3 and 7.5, however, that a description of the displaced state of an element that is exact for such conditions can be extremely useful in the construction, by the virtual displacement principle, of terms that account for nonuniform cross-sectional properties and distributed loads.

Alternatively, it is possible to begin with an assumption as to the displaced state of the element that is not obtained from a solution to a differential equation. In the case of the simplest form of an element it might, fortuitously, correspond to the exact displaced state. Generally, however, the assumption made will be an approximation.

Whatever the approach taken to the definition of the displaced state of an element, our objective is to produce an algebraic expression in terms of all of the displacements at the element node points, that is, an expression of the form

$$\begin{aligned} \Delta &= N_1\Delta_1 + N_2\Delta_2 + \cdots N_i\Delta_i + \cdots N_n\Delta_n \\ &= \sum_{i=1}^{n} N_i\Delta_i = \lfloor \mathbf{N} \rfloor \{\boldsymbol{\Delta}\} \end{aligned} \tag{7.1}$$

where Δ is the displacement component in question (such as u in the case of an axial member, or v in the case of the beam), Δ_i is the ith degree of freedom of the element, N_i is the "shape function" corresponding to Δ_i, and n is the total number of degrees of freedom at the node points of the element. We elaborate on the physical meaning of a shape function below; for the present, it is simply identified as a function of the element coordinates.

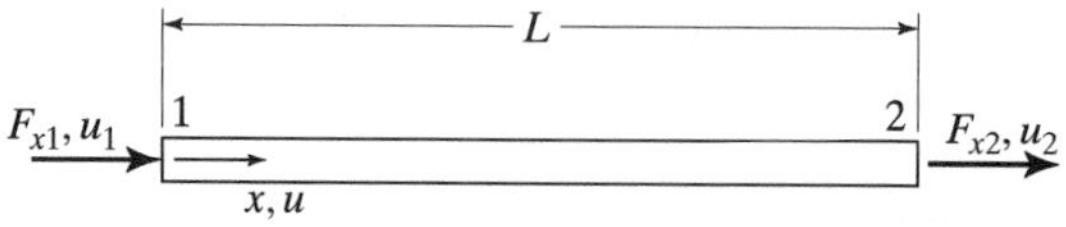

Figure 7.1 Axial force element.

7.1.2 Formulation of Shape Functions

Consider, for example, the axial member (Fig. 7.1). Here, $\Delta = u$, $\Delta_1 = u_1$ and $\Delta_2 = u_2$. Hence, the form of the displacement expression is

$$u = N_1u_1 + N_2u_2 \tag{7.2}$$

The pure torsion member (Figure 7.2) is described by end-point displacements that are the angular displacements θ_{x1} and θ_{x2}. The angular displacement at any point between nodes 1 and 2, θ_x, can be expressed as

$$\theta_x = N_1\theta_{x1} + N_2\theta_{x2} \tag{7.3}$$

Another example is the flexural element (Fig. 7.3), where $\Delta = v$, $\Delta_1 = v_1$, $\Delta_2 = v_2$, $\Delta_3 = \theta_{z1}$, and $\Delta_4 = \theta_{z2}$. In this case the displacement would be described by

$$v = N_1v_1 + N_2v_2 + N_3\theta_{z1} + N_4\theta_{z2} \tag{7.4}$$

The derivation of the general frame element stiffness matrix in Chapter 4 was based upon the superposition of simple axial and torsional behavior and flexure about two perpendicular axes. Consequently, the axial, torsional, and flexural displacement functions cited above are sufficient for the derivation of the same relationships through application of virtual work concepts. It now remains to establish the algebraic form and physical significance of the shape functions (N_i) for these cases. We begin with the simplest case, that of the axial member of uniform cross-sectional area A.

For this element it is known that the axial strain is constant. Since $e_x = du/dx$, u is a linear expression in x, the most general statement of which is

$$u = a_1 + a_2x \tag{7.5}$$

The terms a_1 and a_2 are constants that, as yet, have no assigned physical meaning. We also observed that two constants are needed to match the two node-point displacements, u_1 and u_2. Proceeding to the evaluation of Equation 7.5 at the node points we have, at point 1, where $x = 0$,

$$u = u_1 = a_1$$

Also, at point 2, where $x = L$ and with the above result,

$$u_2 = u_1 + a_2L$$

so that

$$a_2 = (u_2 - u_1)/L$$

By substitution of these expressions for a_1 and a_2 into Equation 7.5

$$u = \left(1 - \frac{x}{L}\right)u_1 + \frac{x}{L}u_2 \tag{7.6}$$

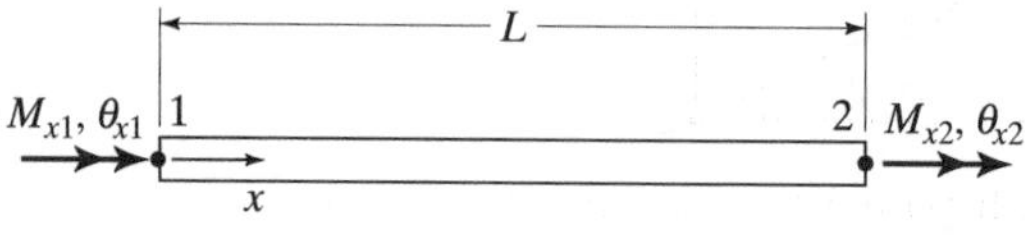

Figure 7.2 Pure torsion element.

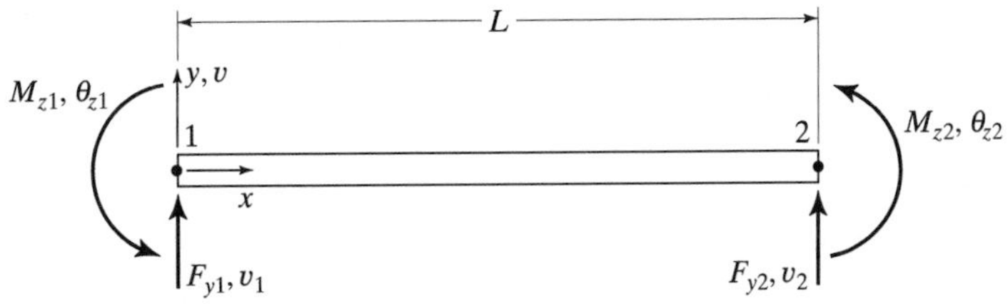

Figure 7.3 Flexural element.

Comparison of Equations 7.2 and 7.6 discloses that $N_1 = (1 - (x/L))$, $N_2 = x/L$. It is instructive to examine these shape functions more closely for the sake of establishing properties applicable to all element-shape functions. First, the shape functions, which for this case are nondimensional, have unit value when evaluated at the nodal point to which they refer and are zero when evaluated at all other nodal points. This must be the case for $u = u_1$ when Equation 7.6 is evaluated at point 1 ($x = 0$) and $u = u_2$ at point 2 ($x = L$).

Figure 7.4 shows the variation of N_1 and N_2 as a function of the element coordinate x. Each plot describes the displaced shape of the element when the element is given a unit displacement at the node point corresponding to N_i (point 1 in the case of N_1, point 2 in the case of N_2), and the displacements at all other degrees of freedom are held to zero. It describes the displaced shape of the element under these circumstances, hence the designation *shape function* for N_i.

Definition of the displaced state of the pure torsional element (Fig. 7.2) follows along the same lines as for the axial member. The angular displacement (θ_x) is given in terms of two node-point displacements (Eq. 7.3), and so it is logical to employ a linear description of the angular displacement at any point along the element:

$$\theta_x = a_1 + a_2 x$$

Clearly, the process of evaluation of a_1 and a_2 results in

$$\theta_x = \left(1 - \frac{x}{L}\right)\theta_{x1} + \frac{x}{L}\,\theta_{x2} \tag{7.7}$$

so that $N_1 = (1 - (x/L))$, $N_2 = x/L$ as in the case of the axial member.

Construction of the shape functions for flexural behavior involves more extensive algebra than the above cases but follows along the same lines. The element (Fig. 7.3) is described in terms of four joint displacements, v_1, v_2, θ_{z1}, θ_{z2}. Note, as emphasized in prior chapters, that in flexure the node-point angular displacements are the derivatives of the transverse displacements, that is $\theta_{z1} = dv/dx|_1$, $\theta_{z2} = dv/dx|_2$.

A polynomial expression has been used in the case of the axial and torsional elements, and it is logical to adopt a polynomial expression for the flexural element. Since there are now four joint displacements, the polynomial in question is cubic:

$$v = a_1 + a_2 x + a_3 x^2 + a_4 x^3 \tag{7.8}$$

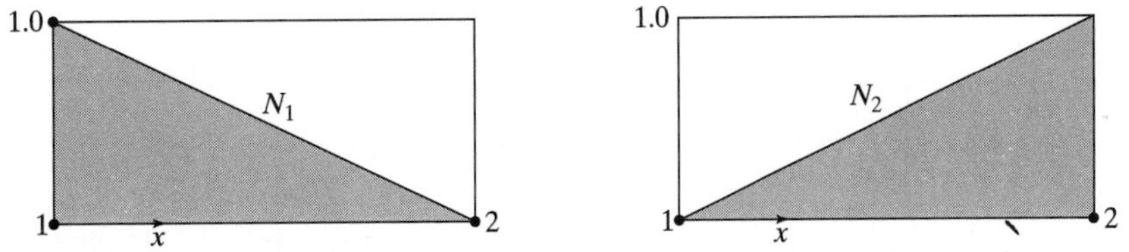

Figure 7.4 Shape functions for uniform axial element.

Evaluation of v, at point 1, where $x = 0$,

$$v_1 = a_1 \tag{7.8a}$$

At point 2, where $x = L$,

$$v_2 = a_1 + a_2L + a_3L^2 + a_4L^3 \tag{7.8b}$$

To evaluate θ_{z1} and θ_{z2}, we need an expression for slope:

$$\theta_z = \frac{dv}{dx} = a_2 + 2a_3x + 3a_4x^2$$

Hence, at point 1,

$$\theta_{z1} = a_2 \tag{7.8c}$$

and at point 2,

$$\theta_{z2} = a_2 + 2a_3L + 3a_4L^2 \tag{7.8d}$$

Equations 7.8a–7.8d are four equations in four unknowns, which can be solved to yield

$$a_1 = v_1$$
$$a_2 = \theta_{z1}$$
$$a_3 = \frac{1}{L^2}(-3v_1 + 3v_2 - 2\theta_{z1}L - \theta_{z2}L)$$

and

$$a_4 = \frac{1}{L^3}(2v_1 - 2v_2 + \theta_{z1}L + \theta_{z2}L)$$

Substitution of these into Equation 7.8 and collection of terms gives

$$\begin{aligned} v = &\left[1 - 3\left(\frac{x}{L}\right)^2 + 2\left(\frac{x}{L}\right)^3\right]v_1 + \left[3\left(\frac{x}{L}\right)^2 - 2\left(\frac{x}{L}\right)^3\right]v_2 \\ &+ x\left(1 - \frac{x}{L}\right)^2\theta_{z1} + x\left[\left(\frac{x}{L}\right)^2 - \frac{x}{L}\right]\theta_{z2} \end{aligned} \tag{7.9}$$

Thus, by comparison with Equation 7.4,

$$N_1 = 1 - 3\left(\frac{x}{L}\right)^2 + 2\left(\frac{x}{L}\right)^3 \qquad N_3 = x\left(1 - \frac{x}{L}\right)^2$$

$$N_2 = 3\left(\frac{x}{L}\right)^2 - 2\left(\frac{x}{L}\right)^3 \qquad N_4 = x\left[\left(\frac{x}{L}\right)^2 - \frac{x}{L}\right]$$

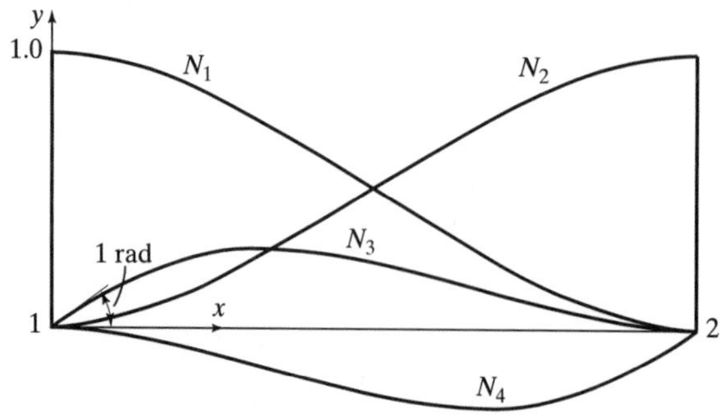

Figure 7.5 Shape functions for flexure of uniform beam element.

These shape functions are plotted in Figure 7.5. It is instructive for the reader to sketch these on the basis of the definition of shape functions given above. The process described above for the axial, torsional, and flexural elements is illustrated further in Examples 7.1 and 7.2.

EXAMPLE 7.1

It is desired that a stiffness matrix be formed, using the virtual displacement principle for an axial member with three node points. Construct the shape function on which the formulation will be based.

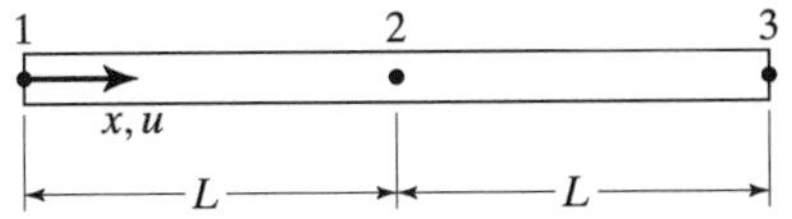

Since the shape function will involve three joint displacements u_1, u_2, and u_3, it is appropriate to choose a quadratic (three-term) polynomial as the basis for the shape function. Thus

$$u = a_1 + a_2 x + a_3 x^2$$

By evaluation at $x = 0$, $x = L$, and $x = 2L$,

$$u_1 = a_1$$
$$u_2 = a_1 + a_2 L + a_3 L^2$$
$$u_3 = a_1 + 2a_2 L + 4a_3 L^2$$

Solving,

$$a_1 = u_1 \qquad a_2 = \frac{1}{2L}(-3u_1 + 4u_2 - u_3) \qquad a_3 = \frac{1}{2L^2}(u_1 - 2u_2 + u_3)$$

Substitution of these into the original expression for u gives

$$u = \left(1 - \frac{3}{2}\frac{x}{L} + \frac{x^2}{2L^2}\right)u_1 + \left(\frac{2x}{L} - \frac{x^2}{L^2}\right)u_2 - \left(\frac{x}{2L} - \frac{x^2}{2L^2}\right)u_3$$

This expression can be checked to confirm that $N_i = 1$ at x_i and $N_i = 0$ at points other than i.

EXAMPLE 7.2

The illustrated tapered beam is of rectangular cross section, of width b, and tapering depth $h_1(1 + x/L)$. Construct an expression for the displaced shape that satisfies exactly the governing differential equation

$$\frac{d^2}{dx^2}\left(EI\frac{d^2v}{dx^2}\right) = 0$$

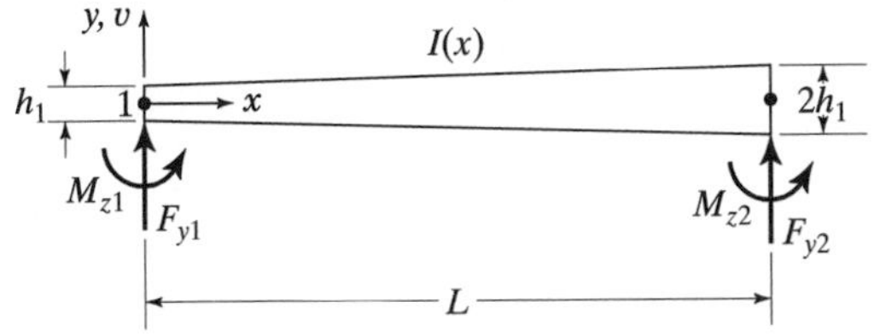

The moment of inertia is

$$I = \frac{bh_1^3}{12}\left(1 + \frac{x}{L}\right)^3$$

Thus

$$\frac{Ebh_1^3}{12}\frac{d^2}{dx^2}\left[\left(1 + \frac{x}{L}\right)^3 \frac{d^2v}{dx^2}\right] = 0$$

or

$$\frac{d^2}{dx^2}\left[\left(1 + \frac{x}{L}\right)^3 \frac{d^2v}{dx^2}\right] = 0$$

Integrating this expression twice,

$$\left(1+\frac{x}{L}\right)^3 \frac{d^2v}{dx^2} = C_1 + C_2x$$

where C_1 and C_2 are constants of integration. Rearranging,

$$\frac{d^2v}{dx^2} = \left(1+\frac{x}{L}\right)^{-3} C_1 + x\left(1+\frac{x}{L}\right)^{-3} C_2$$

By integration,

$$\frac{dv}{dx} = -\frac{L}{2}\frac{C_1}{\left(1+\frac{x}{L}\right)^2} + \left[\frac{L^2}{2}\frac{1}{\left(1+\frac{x}{L}\right)^2} - \frac{1}{\left(1+\frac{x}{L}\right)}\right]C_2 + C_3$$

Integrating again,

$$v = \frac{L^2}{2}\frac{C_1}{\left(1+\frac{x}{L}\right)} - L^3\left[\frac{1}{2\left(1+\frac{x}{L}\right)} + \ln\left(1+\frac{x}{L}\right)\right]C_2 + C_3x + C_4$$

This expression is evaluated at the end points ($v_1 = v$ at $x = 0$, $v_2 = v$ at $x = L$), differentiated and evaluated at the end points ($dv/dx = \theta_{z1}$ at $x = 0$, $dv/dx = \theta_{z2}$ at $x = L$), and the resulting equations are solved for C_1, C_2, C_3, C_4. Substitution of the solution for $C_1, \ldots, C_4$ into the equation for v gives the desired expression in terms of shape functions (Problem 7.4).

7.1.3 Characteristics of Shape Functions

Some observations should be made regarding the foregoing. First, although polynomial expressions were the appropriate bases of the shape functions for simple axial and flexural members (Eqs. 7.5 and 7.8 and Ex. 7.1), a logarithmic expression gave the exact basis for the tapered beam shape function (Ex. 7.2). Other types of functions, such as trigonometric expressions, also find application. Nevertheless, polynomial functions are prevalent and are used both when they represent the exact displaced shape and in approximation of logarithmic, trigonometric, or other functions. Examples of the latter are given in Section 7.3.

Second, the above shape functions (Eq. 7.6 and 7.7 and the results of Examples 7.1 and 7.2) each include representation of both rigid body motion and elastic deformational behavior. These two classes of behavior are seen more clearly in the underlying polynomial expression. For example, in the case of the axial member (Eq. 7.5) the term a_1 portrays rigid body motion and a_2x embodies the strain. In the flexural member (Eq. 7.8), these behaviors are characterized by a_1, a_2x and a_3x^2, a_4x^3, respectively.

Third, since shape functions are usually nondimensional multipliers of the joint displacements, it may be convenient to express them in terms of nondimensional coordinates. For the axial member, for example, one can introduce the coordinate $\xi = x/L$, which then has a range of 0 to 1 from one end of the element to the other. Equation 7.6 can then be written

$$u = (1 - \xi)u_1 + \xi u_2$$

In the case of flexural members, there is a mixture of translational and angular displacements (Eq. 7.9). The multipliers of the angular displacement must have the units of a length dimension. Thus, N_3 can be written as $x(1 - x/L)^2 = x(1 - \xi)^2$, which mixes the x and ξ coordinates. To avoid this one can express the angular displacements as $\theta_1 L$ and $\theta_2 L$, in which case $N_3 = (x/L)(1 - x/L)^2 = \xi(1 - \xi)^2$. This mode of expression of the angular displacement is advantageous in scaling a system of equations in attempts to improve solution accuracy (see Section 11.6.4).

The approach to the construction of element shape functions taken in this section can be summarized as follows. The displaced state of the element is first described in terms of parameters a_i, one for each element joint displacement. The choice of the displaced state may be based upon an intuitive understanding of the behavior of the elements, solution of the governing differential equation (Ex. 7.2), or simply through the choice of a polynomial or other expansion containing as many terms as there are element degrees of freedom. The expression for the displaced state is evaluated at each degree of freedom Δ_i, producing equations that are sufficient in number to determine a_i in terms of Δ_i. The multipliers of the joint displacements Δ_i in the resulting expression for the displaced state are the shape functions.

There is a more elegant way of producing expressions for the displaced state of elements, directly in the desired form $\Delta = \lfloor \mathbf{N} \rfloor \{\boldsymbol{\Delta}\}$. This is through the use of interpolation formulas that fit an expression for a variable through specified points at which the variable is known. The interpolation approach is detailed in Reference 7.1. The route of first selecting an expression in terms of parameters a_i and then eliminating these parameters has been adopted here for simplicity and because it makes clear the shape of displacement upon which each term of $\lfloor \mathbf{N} \rfloor$ is based.

7.2 VIRTUAL DISPLACEMENTS IN THE FORMULATION OF ELEMENT STIFFNESS EQUATIONS

7.2.1 Construction of Expressions for Real and Virtual Displacements

Having established expressions for the displacement of any point on an element in terms of the element node-point displacements, we now employ these expressions in the principle of virtual displacements to produce element stiffness equations.

The expression for internal work is written in terms of strains. To obtain these, the displacements must be differentiated in the manner appropriate to the type of structural action under study. The relevant strain-displacement equations were discussed in Section 6.3.2. For an axial member, for example, the displacement is u and the strain $e_x = du/dx$. For torsion, the displacement is the rotation about the x axis, θ_x, and the strain β is the rate of change of this rotation with respect to the axial coordinate, that is, $\beta = d\theta_x/dx$. For beams, the displacement is denoted as v and the "strain" is the curvature $\kappa_z = d^2v/dx^2$.

In general, then, the strain (e) is obtained by differentiation of Equation 7.1 and we symbolize this operation by

$$e = \Delta' = \lfloor \mathbf{N}' \rfloor \{\boldsymbol{\Delta}\} \tag{7.10}$$

where the prime denotes the appropriate differentiation with respect to the spatial coordinate. This may require first derivatives (as in the case of axial and torsional elements) or second derivatives (as in bending). Also, we note from Equations 7.6–7.9 that only the shape functions $\lfloor \mathbf{N} \rfloor$ are functions of the spatial coordinates and are therefore the only terms subject to differentiation. Since the joint displacements are specific values and not functions of the coordinates, they are not subject to differentiation.

To illustrate Equation 7.10 we first examine the axial member where, from Equation 7.6,

$$u = \left\lfloor \left(1 - \frac{x}{L}\right) \quad \frac{x}{L} \right\rfloor \begin{Bmatrix} u_1 \\ u_2 \end{Bmatrix}$$

so that

$$e_x = \frac{du}{dx} = \left\lfloor -\frac{1}{L} \quad \frac{1}{L} \right\rfloor \begin{Bmatrix} u_1 \\ u_2 \end{Bmatrix} \tag{7.11a}$$

Similarly, for torsion, from Equation 7.7,

$$\beta = \frac{d\theta_x}{dx} = \left\lfloor -\frac{1}{L} \quad \frac{1}{L} \right\rfloor \begin{Bmatrix} \theta_{x1} \\ \theta_{x2} \end{Bmatrix} \tag{7.12a}$$

For the beam element, from Equation 7.9 (with "strain," $e = \kappa_z = d^2v/dx^2$),

$$\kappa_z = \frac{d^2v}{dx^2} = \left\lfloor \frac{6}{L^2}\left(1 - \frac{2x}{L}\right) \middle| \frac{2}{L}\left(\frac{3x}{L} - 1\right) \middle| \frac{6}{L^2}\left(\frac{2x}{L} - 1\right) \middle| \frac{2}{L}\left(\frac{3x}{L} - 2\right) \right\rfloor \begin{Bmatrix} v_2 \\ \theta_{z2} \\ v_1 \\ \theta_{z1} \end{Bmatrix} \tag{7.13a}$$

(The order of terms in the displacement vector has been changed to conform to the order employed in the stiffness matrix derivation of Example 4.4).

To form the complete statement of internal virtual work, it is necessary to have expressions for the virtual strains. The approach taken is to adopt the same form of displaced state for the virtual displacements as for the real displacements. Thus, paraphrasing Equation 7.1,

$$\delta\Delta = \lfloor \mathbf{N} \rfloor \{\delta\boldsymbol{\Delta}\} \tag{7.14}$$

where $\{\delta\boldsymbol{\Delta}\}$ denotes the element node-point virtual displacements. The virtual strain, δe, is then

$$\delta e = \lfloor \mathbf{N}' \rfloor \{\delta\boldsymbol{\Delta}\} \tag{7.15}$$

The specific forms of this expression for axial, torsional, and bending behaviors are

$$\delta e_x = \left\lfloor -\frac{1}{L} \quad \frac{1}{L} \right\rfloor \begin{Bmatrix} \delta u_1 \\ \delta u_2 \end{Bmatrix} \tag{7.11b}$$

$$\delta\beta = \left\lfloor -\frac{1}{L} \quad \frac{1}{L} \right\rfloor \begin{Bmatrix} \delta\theta_{x1} \\ \delta\theta_{x2} \end{Bmatrix} \tag{7.12b}$$

$$\delta\kappa_z = \left\lfloor \frac{6}{L^2}\left(1 - \frac{2x}{L}\right) \middle| \frac{2}{L}\left(\frac{3x}{L} - 1\right) \middle| \frac{6}{L^2}\left(\frac{2x}{L} - 1\right) \middle| \frac{2}{L}\left(\frac{3x}{L} - 2\right) \right\rfloor \begin{Bmatrix} \delta v_2 \\ \delta\theta_{z2} \\ \delta v_1 \\ \delta\theta_{z1} \end{Bmatrix} \tag{7.13b}$$

7.2.2 Virtual Displacements Formula for an Element Stiffness Matrix

The expression of the principle of virtual displacements can now be written for an individual element, treating it as if it were a structure in isolation. The basic statement of this principle was established in Section 6.2 as

$$\delta W = \delta W_{\text{ext}} - \delta W_{\text{int}} = 0 \tag{6.13}$$

Also, it was shown in Section 6.3 that

$$\delta W_{\text{int}} = \int_{\text{vol}} \lfloor \delta\mathbf{e} \rfloor [\mathbf{E}] \{\mathbf{e}\}\, d(\text{vol}) \tag{6.23}$$

where $\{\mathbf{e}\}$ and $\lfloor \delta\mathbf{e} \rfloor$ are the real and virtual strains that are characteristic of the mode of behavior under study, for example $\{\mathbf{e}\} = e_x$ for axial behavior. $[\mathbf{E}]$ is the relevant elastic constant. The element is loaded by the joint forces $F_1, \ldots, F_i, \ldots, F_n$. (In Section 7.5 we take up the matter of distributed loads and loads applied at points other than the element node points). Thus, $\delta W_{ext} = \Sigma_{i=1}^{n} \delta\Delta_i F_i$, and the virtual work principle becomes

$$\int_{vol} \lfloor \delta\mathbf{e} \rfloor [\mathbf{E}]\{\mathbf{e}\}\, d(\text{vol}) = \sum_{i=1}^{n} \delta\Delta_i F_i = \lfloor \delta\mathbf{\Delta} \rfloor \{\mathbf{F}\}$$

By substitution of Equation 7.10 and 7.15 we have, for an element described by n degrees of freedom,[1]

$$\lfloor \delta\mathbf{\Delta} \rfloor \left[\int_{vol} \{\mathbf{N}'\}[\mathbf{E}]\lfloor \mathbf{N}' \rfloor\, d(\text{vol}) \right] \{\mathbf{\Delta}\} = \lfloor \delta\mathbf{\Delta} \rfloor \{\mathbf{F}\} \tag{7.16a}$$

or

$$\lfloor \delta\mathbf{\Delta} \rfloor [\mathbf{k}]\{\mathbf{\Delta}\} = \lfloor \delta\mathbf{\Delta} \rfloor \{\mathbf{F}\} \tag{7.16b}$$

where

$$[\mathbf{k}] = \left[\int_{vol} \{\mathbf{N}'\}[\mathbf{E}]\lfloor \mathbf{N}' \rfloor\, d(\text{vol}) \right] \tag{7.17}$$

Under the condition that the virtual displacements are completely arbitrary,

$$[\mathbf{k}]\{\mathbf{\Delta}\} = \{\mathbf{F}\} \tag{7.18}$$

Equation 7.17 is the general expression for an element stiffness matrix, derived through use of the virtual displacements principle. Equation 7.18 represents the set of element stiffness equations to which it refers.

7.2.3 Application to Standard Axial, Torsional, and Flexural Elements

For an axial member of uniform section (Fig. 7.1), $[\mathbf{E}] = E$, $d(\text{vol}) = A\, dx$, and the expressions for real and virtual strains are given by Equations 7.11a and 7.11b. Thus, by substitution into Equation 7.17,

$$[\mathbf{k}] = \left[\int_0^L \begin{Bmatrix} -\dfrac{1}{L} \\ \dfrac{1}{L} \end{Bmatrix} E \left\lfloor -\dfrac{1}{L} \quad \dfrac{1}{L} \right\rfloor A\, dx \right] = \frac{EA}{L} \begin{bmatrix} 1 & -1 \\ -1 & 1 \end{bmatrix}$$

which is the already established result for the axial force member.

For the torsional element $[\mathbf{E}] = G$, $d(\text{vol})$ is replaced by $J\, dx$, and the real and virtual strains are defined by Equations 7.12a and 7.12b. Again, using these in Equation 7.17,

$$[\mathbf{k}] = \left[\int_0^L \begin{Bmatrix} -\dfrac{1}{L} \\ \dfrac{1}{L} \end{Bmatrix} G \left\lfloor -\dfrac{1}{L} \quad \dfrac{1}{L} \right\rfloor J\, dx \right] = \frac{GJ}{L} \begin{bmatrix} 1 & -1 \\ -1 & 1 \end{bmatrix}$$

This was derived, in a different manner, in Section 4.5 (Eq. 4.27a).

---

[1]Generally, $[\mathbf{N}]$ and $[\mathbf{N}']$ are matrices. Since in this chapter we deal only with separate modes of behavior, for each of which $\{\mathbf{e}\}$ has a single component, they are vectors here.

For the beam element, $[\mathbf{E}] = E$ and, after integration across the depth, $d(\text{vol})$ leads to $I_z\,dx$. The real and virtual strain expressions are given by Equations 7.13a and 7.13b. After substitution in Equation 7.17,

$$[\mathbf{k}] = \left[\int_{\text{vol}} \begin{bmatrix} \frac{6}{L^2}\left(1 - \frac{2x}{L}\right) \\ \frac{2}{L}\left(\frac{3x}{L} - 1\right) \\ \frac{6}{L^2}\left(\frac{2x}{L} - 1\right) \\ \frac{2}{L}\left(\frac{3x}{L} - 2\right) \end{bmatrix} EI_z \left\lfloor \frac{6}{L^2}\left(1 - \frac{2x}{L}\right) \middle| \frac{2}{L}\left(\frac{3x}{L} - 1\right) \middle| \frac{6}{L^2}\left(\frac{2x}{L} - 1\right) \middle| \frac{2}{L}\left(\frac{3x}{L} - 2\right) \right\rfloor dx\right]$$

$$[\mathbf{k}] = \frac{EI_z}{L}\begin{array}{c} \begin{matrix} v_2 & \theta_{z2} & v_1 & \theta_{z1} \end{matrix} \\ \begin{bmatrix} 12/L^2 & -6/L & -12/L^2 & -6/L \\ & 4 & 6/L & 2 \\ \text{Sym.} & & 12/L^2 & 6/L \\ & & & 4 \end{bmatrix} \end{array}$$

In this case, the external virtual work is $\lfloor \delta v_2 \;\; \delta\theta_{z2} \;\; \delta v_1 \;\; \delta\theta_{z1} \rfloor \cdot \lfloor F_{y2} \;\; M_{z2} \;\; F_{y1} \;\; M_{z1} \rfloor^{\mathrm{T}}$. The result is the one obtained by alternative means in Example 4.4 (see also Eq. 4.32).

7.3 NONUNIFORM ELEMENTS

Practical structures may contain members of nonuniform cross section, as shown in Figure 7.6*a*. It is, of course, possible to represent such members by a series of uniform-section elements, as shown in Figure 7.6*b*. But this may require an inordinate number of unknown displacement components to yield a solution wherein the error due to the approximation is within acceptable limits. The formulation of exact stiffness equations for a nonuniform element is an alternative, but, as Examples 7.2 and 7.9 disclose, this can be rather complicated. A convenient approach, described below, uses the principle of virtual displacements. It enables the construction of exact or approximate stiffness relationships for nonuniform members.

The expression for the element stiffness matrix (Eq. 7.17), derived from the principle

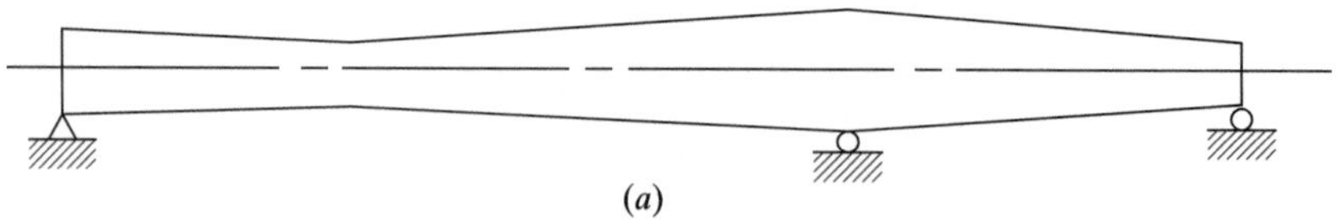

(*a*)

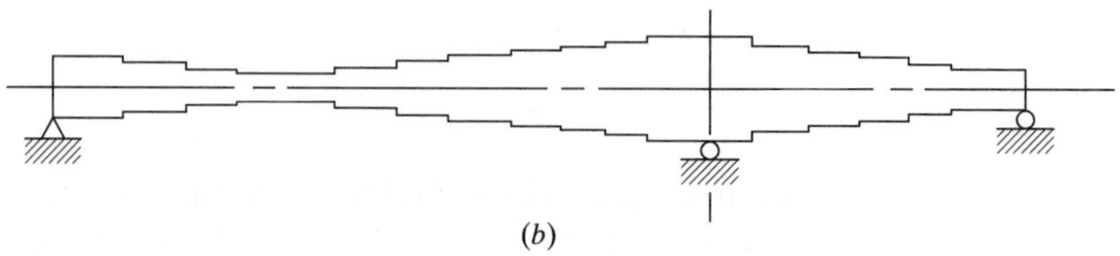

(*b*)

Figure 7.6 Continuous beam with nonuniform cross section. (*a*) Actual structure. (*b*) Stepped representation.

of virtual displacements, requires the derivatives of the displacement shape functions $\lfloor \mathbf{N}' \rfloor$ and a description of the element geometry as embodied in d(vol). These representations of element behavior and geometry can be either exact or approximate. (See Section 6.4 for the requirements of an exact solution).

To illustrate the approximate procedure we examine the case of a linearly tapering axial member whose cross-sectional area at any point is given by

$$A = A_1\left(1 - \frac{rx}{L}\right)$$

where A_1 is the cross-sectional area at point 1, and r is a parameter that can be adjusted to give a specific ratio of end areas. This is similar to the description of taper as used for the element of Figure 6.6. Using the shape functions for the uniform axial member $[N_1 = (1 - x/L), N_2 = x/L]$, the expression for the approximate stiffness matrix is

$$[\mathbf{k}] = \left[\int_0^L \begin{Bmatrix} -\dfrac{1}{L} \\ \dfrac{1}{L} \end{Bmatrix} E \left\lfloor -\frac{1}{L} \quad \frac{1}{L} \right\rfloor A_1\left(1 - \frac{rx}{L}\right) dx \right]$$

and, upon integration

$$[\mathbf{k}] = \frac{EA_1}{L}\left(1 - \frac{r}{2}\right)\begin{bmatrix} 1 & -1 \\ -1 & 1 \end{bmatrix}$$

The result, in this case, is a stiffness matrix that is of the same form as the stiffness matrix for a uniform axial force element with "effective area" $A = A_1(1 - r/2)$. (This is due to the circumstance of constant strain arising from the chosen approximate displaced state. It does not occur for flexural elements where the strain due to the chosen displaced state, in the form of the curvature, usually varies over the length of the element.)

In this case the exact displaced shape can be found through solution of the differential equation for axial behavior. This equation has not been developed previously, but it is established quite easily. The condition of equilibrium for a differential segment of an axial member is $d(A\sigma_x)/dx = 0$ (see Section 6.5). The strain-displacement equation is $e_x = du/dx$ and, by Hooke's law, $\sigma_x = Ee_x$. Combination of these relationships gives

$$\frac{d}{dx}\left(EA\frac{du}{dx}\right) = 0 \tag{7.19}$$

In the present case, with $A = A_1(1 - rx/L)$, the solution of this equation for boundary conditions $u = u_1$ at $x = 0$ and $u = u_2$ at $x = L$ is

$$u = \left[1 - \frac{ln(1 - rx/L)}{ln(1 - r)}\right]u_1 + \left[\frac{ln(1 - rx/L)}{ln(1 - r)}\right]u_2$$

Using this in the formula for the stiffness matrix (Equation 7.17), we obtain

$$[\mathbf{k}] = -\frac{EA_1}{L}\cdot\frac{r}{ln(1 - r)}\begin{bmatrix} 1 & -1 \\ -1 & 1 \end{bmatrix}$$

For $r = 1/2$ this gives a value of $0.7213EA_1/L$ for each stiffness coefficient, while the approximate formulation gives $0.75EA_1/L$, an error of 4%.

Example 7.3 illustrates the approximate approach for the case of the tapered cantilever beam.

EXAMPLE 7.3

Determine the deflection (v) at point 1 of the member of Example 7.2 due to F_{y1} using an approximate formulation, based on the shape functions of a uniform beam (Equation 7.9). ($M_{z1} = 0$). The member is supported as a cantilever at point 2.

Since $v_2 = \theta_{z2} = 0$, the relevant stiffness equations are

$$\begin{Bmatrix} F_{y1} \\ 0 \end{Bmatrix} = \begin{bmatrix} k_{11} & k_{12} \\ k_{21} & k_{22} \end{bmatrix} \begin{Bmatrix} v_1 \\ \theta_{z1} \end{Bmatrix}$$

with $I = (bh_1^3/12)(1 + x/L)^3$ and using the curvature shape functions of Equation 7.13a,

$$k_{11} = \int_0^L \left[\frac{6}{L^2}\left(\frac{2x}{L} - 1\right)\right]^2 \frac{Eh_1{}^3b}{12}\left(1 + \frac{x}{L}\right)^3 dx = \frac{81}{20} Eh_1{}^3 \frac{b}{L^3}$$

$$k_{12} = k_{21} = \int_0^L \frac{6}{L^2}\left(\frac{2x}{L} - 1\right) \times \frac{2}{L}\left(\frac{3x}{L} - 2\right) \frac{Eh_1{}^3b}{12}\left(1 + \frac{x}{L}\right)^3 dx = \frac{29Eh_1{}^3b}{20L^2}$$

$$k_{22} = \int_0^L \left[\frac{2}{L}\left(\frac{3x}{L} - 2\right)\right]^2 \frac{Eh_1{}^3b}{12}\left(1 + \frac{x}{L}\right)^3 dx = \frac{3Eh_1{}^3b}{4L}$$

Thus, upon evaluation of the terms of the stiffness matrix,

$$\begin{Bmatrix} F_{y1} \\ 0 \end{Bmatrix} = \frac{Eh_1{}^3b}{20L^3} \begin{bmatrix} 81 & 29L \\ 29L & 15L^2 \end{bmatrix} \begin{Bmatrix} v_1 \\ \theta_{z1} \end{Bmatrix}$$

The solution of these equations gives

$$v_1 = 0.06684 \frac{F_{y1}L^3}{EI_1}$$

This compares with the exact solution $0.06815F_{y1}L^3/EI_1$ (1.92% error), obtained as a solution to Example 7.9

The solution errors in Example 7.3 and in the above tapered axial member are due to the failure of the chosen polynomial displacement state to satisfy the condition of equilibrium. For the axial member, this condition is represented by Equation 7.19, which gives, for $u = (1 - rx/L)u_1 + (rx/L)u_2$,

$$A_1 \frac{d}{dx}\left(1 - \frac{rx}{L}\right)E \frac{d}{dx}\left[\left(1 - \frac{rx}{L}\right)u_1 + \frac{rx}{L} u_2\right] = EA_1 \frac{r^2}{L^2}(u_1 - u_2)$$

The term on the right is zero only when r equals zero. A similar exercise for the shape function used in Example 7.3 will disclose that it does not meet the equilibrium condition for flexural behavior (Problem 7.9). As the length of these elements becomes smaller, because of the refinement of the analysis grid, the error in equilibrium becomes smaller and the exact solution is approached.

7.4 NONUNIFORM TORSION

The stiffness matrix for an element in pure torsion was developed from simple strength of materials principles in Section 4.5.2 and by the virtual displacement approach in Section 7.2.3 of this chapter. It was pointed out in Section 4.6 that whereas that matrix is adequate for the analysis of many systems in which torsional effects are small, it is limited by the fact that it neglects resistance to cross-sectional out-of-plane warping (Figure 4.13). Warping resistance can be significant in the response of many frames consisting of members of open cross section, such as the wide flange steel shape. Indeed, it can be the dominant factor in their resistance to torsion and combined torsional-flexural effects if warping deformation is restrained in any way. The relevant elastic theory is developed in References 7.2 and 7.3, and its application is illustrated in Reference 7.4.

7.4.1 An Element Stiffness Matrix

In the following the principle of virtual displacements is used to develop an element stiffness matrix that accounts for warping restraint in bisymmetrical wide flange shapes. Elementary engineering theory is used in formulating the governing differential equation, but the result is the same as that obtained by the more general methods of the cited references. The numerical solution of the equation is approximate, but it will be found to be in good agreement with analytical solutions of representative problems.

A wide flange member subjected to an applied torque at the right end is shown in Figure 7.7. If there is no restraint against longitudinal displacement of the fibers, its rate of twist is constant, all cross sections are in a state of pure shear, and the stiffness matrix of Equation 4.27 and Section 7.2.3 applies. But the cross sections will warp out of plane uniformly, and as the member twists, its flanges will displace laterally as indicated in Figure 7.7*a*, in which twisting but not warping is prevented at the left end. If, however, longitudinal displacement is restrained in any way, the rate of twist is no longer constant, and both normal and shearing stresses can exist on cross sections. Cross-sectional out-of-plane warping can still occur, but it will vary along the length. As a consequence, the flanges will bend as they displace laterally as in Figure 7.7*b*, in which both twisting and warping are prevented at the left end. In this condition, known as *nonuniform torsion*, internal resistance to the applied torque becomes more complex, but it can be analyzed by the introduction of an additional degree of freedom, the rate of twist θ'_x, and its conjugate variable, a self equilibrating action B, called the *bimoment.*

Without undue simplification, the wide flange member can be viewed as one in which the applied torque is resisted by a combination of the continuous flow of shear stresses on a section and the resistance of the flanges to lateral bending (Figure 7.8). The former

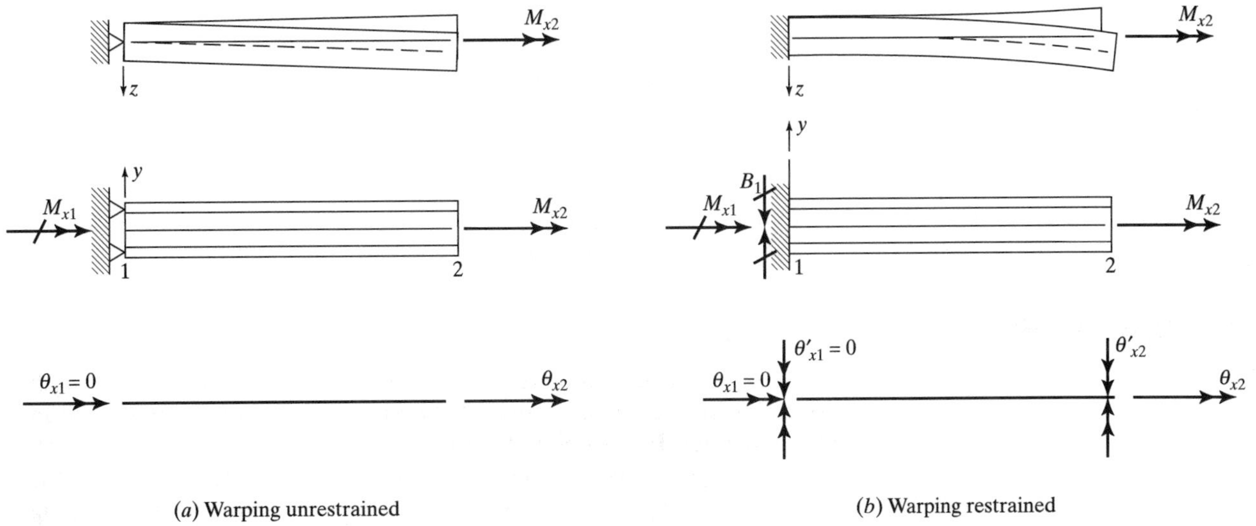

Figure 7.7 Torsion member.

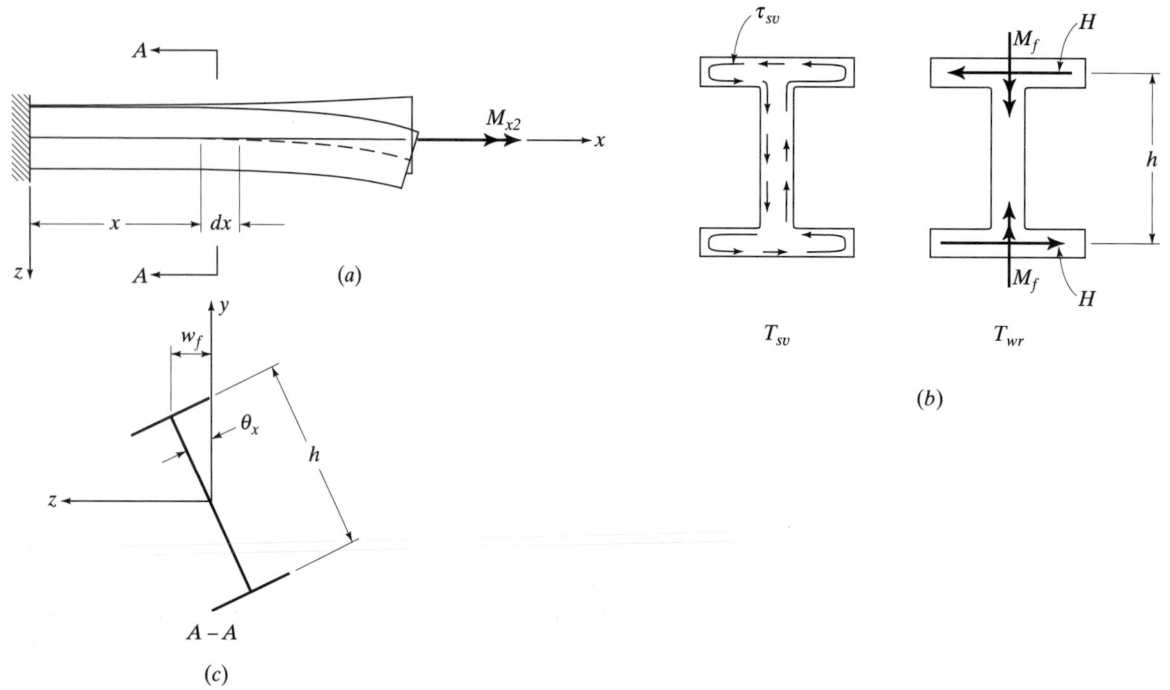

Figure 7.8 Nonuniform torsion.

is commonly called *St. Venant torque*, after the developer of the theory of pure torsion, and the latter *warping restraint torque* or just *warping torque*. The normal stresses due to flange bending must be self equilibrating since the net action on the cross section is a torque. But, as seen in Figure 7.8*b*, the product of the flange shears associated with transverse bending and the effective depth of the section constitutes a twist resisting couple. For a small angle of twist the quantities of interest can be expressed as follows:

From Section 4.5.2, noting that $\beta = \theta_x'$, the St. Venant torque is

$$T_{sv} = GJ\theta_x' \tag{7.20}$$

In the development of the warping restraint torque, we will apply the conventional moment-curvature relationship to the flanges. Noting that the flange moment of inertia is approximately half that of the section, and letting $w_f = \theta_x h/2$ (Figure 7.8*c*)

$$M_f h = EI_f\left(\frac{d^2 w_f}{dx^2}\right)h = \frac{EI_y h^2}{4}\left(\frac{d^2\theta_x}{dx^2}\right)$$

Defining the self-equilibrating action $M_f h$ as the bimoment, B, and designating the section property $I_y h^2/4$ as a *warping constant*, C_w, of the wide flange section, we have

$$B = EC_w\theta_x'' \tag{7.21}$$

Values of the warping constant for other sections may be found in the cited references. Now, noting that $T_{wr} = Hh = -(dM_f/dx)h = -dB/dx$

$$T_{wr} = -EC_w\theta_x''' \tag{7.22}$$

Thus, since the total torsional resistance is the sum of the St. Venant and warping resistance torques we have, from Equations 7.20 and 7.22:

$$M_x = GJ\theta_x' - EC_w\theta_x''' \tag{7.23}$$

In employing virtual displacements to formulate the element stiffness equations for a member of this type, the internal and external virtual work expressions developed in Section 6.3 may be used for the virtual work associated with St. Venant torsion. For warping torque we have, using Figures 7.7*b* and 7.8 and the above relationships

$$\delta W_{\text{ext}} = 2M_{f1}\, \delta\left(\frac{dw_f}{dx}\right)_{x=0} = B_1\, \delta\theta'_{x1} \tag{7.24}$$

in which M_{f1} is the flange moment at end 1, and

$$\delta W_{\text{int}} = 2\int_0^L \delta\kappa_f M_f\, dx = 2\int_0^L \delta\left(\frac{d^2 w_f}{dx^2}\right)\frac{EC_w}{h}\,\theta''_x\, dx$$

which can be reduced to

$$\delta W_{\text{int}} = \int_0^L \delta\theta''_x EC_w \theta''_x\, dx \tag{7.25}$$

Thus, for a torsional element with warping constraint at both ends, the external virtual work of nonuniform torsion is $\delta W_{\text{ext}} = \lfloor \delta\theta_{x1} \quad \delta\theta_{x2} \quad \delta\theta'_{x1} \quad \delta\theta'_{x2} \rfloor \lfloor M_{x1} \quad M_{x2} \quad B_1 \quad B_2 \rfloor^{\text{T}}$.

From Equation 7.17 and the internal virtual work equations for St. Venant and warping torsion, it follows that the stiffness matrix for such a four degree of freedom element can be expressed as

$$[\mathbf{k}] = \left[\int_0^L \{\mathbf{N}'\} G \lfloor \mathbf{N}' \rfloor J\, dx + \int_0^L \{\mathbf{N}''\} E \lfloor \mathbf{N}'' \rfloor C_w\, dx\right] \tag{7.26}$$

For a torsional member in which warping is unrestrained, the rate of twist is constant. Therefore, it was correct to employ a linear description of the twist in the development of the shape function used in the earlier analysis (see Equation 7.7). This is not the case in nonuniform torsion, however. In this the shape functions must be approximate. In the following, both the real and the virtual angle of twist are represented by a cubic polynomial. Thus

$$\begin{aligned} \theta_x &= a_1 + a_2 x + a_3 x^2 + a_4 x^3 \\ \theta'_x &= a_2 + 2a_3 x + 3a_4 x^2 \end{aligned} \tag{7.27}$$

Noting the identity of these expressions and those of Equation 7.8 for the deflection and slope of a beam element, it is seen that $\lfloor \mathbf{N} \rfloor = \lfloor N_1 \quad N_2 \quad N_3 \quad N_4 \rfloor$, in which the shape functions are those of Equation 7.9. Substituting the derivatives of these in Equation 7.26 and integrating results in

$$[\mathbf{k}] = \left[GJ \begin{array}{c} \begin{array}{cccc} \theta_{x1} & \theta_{x2} & \theta'_{x1} & \theta'_{x2} \end{array} \\ \begin{bmatrix} \dfrac{6}{5L} & -\dfrac{6}{5L} & \dfrac{1}{10} & \dfrac{1}{10} \\ & \dfrac{6}{5L} & -\dfrac{1}{10} & -\dfrac{1}{10} \\ \text{Sym.} & & \dfrac{2L}{15} & -\dfrac{L}{30} \\ & & & \dfrac{2L}{15} \end{bmatrix} \end{array} + \frac{EC_w}{L} \begin{array}{c} \begin{array}{cccc} \theta_{x1} & \theta_{x2} & \theta'_{x1} & \theta'_{x2} \end{array} \\ \begin{bmatrix} \dfrac{12}{L^2} & -\dfrac{12}{L^2} & \dfrac{6}{L} & \dfrac{6}{L} \\ & \dfrac{12}{L^2} & -\dfrac{6}{L} & -\dfrac{6}{L} \\ \text{Sym.} & & 4 & 2 \\ & & & 4 \end{bmatrix} \end{array} \right] \tag{7.28}$$

The previously developed 12-degree-of-freedom stiffness matrix of Equation 4.34 can be modified and augmented by the elements of [**k**] to form a 14-degree-of-freedom stiffness matrix for the linear elastic analysis of members subjected to combined axial force, bending, and nonuniform torsion. The result is Equation A.55 of Appendix A in which the St. Venant torsional terms of Equation 4.34 have been replaced by the above, and columns and rows related to the rate of twist and bimoment have been added. For our present purpose—the linear analysis of members under torsion alone—the four-degree-of-freedom stiffness matrix of Equation 7.28 is sufficient. The essential stiffness equation may be written in a form that emphasizes the relative roles of St. Venant and warping torsion. Thus, letting

$$\frac{EC_w}{GJL^2} = \alpha \tag{7.29}$$

and combining terms in [**k**], we have

$$\begin{Bmatrix} M_{x1} \\ M_{x2} \\ B_1 \\ B_2 \end{Bmatrix} = \frac{GJ}{L} \begin{bmatrix} \left(\frac{6}{5} + 12\alpha\right) & -\left(\frac{6}{5} + 12\alpha\right) & L\left(\frac{1}{10} + 6\alpha\right) & L\left(\frac{1}{10} + 6\alpha\right) \\ & \left(\frac{6}{5} + 12\alpha\right) & -L\left(\frac{1}{10} + 6\alpha\right) & -L\left(\frac{1}{10} + 6\alpha\right) \\ & \text{Sym.} & L^2\left(\frac{2}{15} + 4\alpha\right) & L^2\left(-\frac{1}{30} + 2\alpha\right) \\ & & & L^2\left(\frac{2}{15} + 4\alpha\right) \end{bmatrix} \begin{Bmatrix} \theta_{x1} \\ \theta_{x2} \\ \theta'_{x1} \\ \theta'_{x2} \end{Bmatrix} \tag{7.30}$$

7.4.2 Application and Examples

Judicious treatment of torsion in frame members requires consideration of the following:

1. ***The relative magnitude of the warping restraint and St. Venant effects.*** As seen from the parameter α of Equations 7.29 and 7.30, for a given material the significance of warping restraint increases with the ratio C_w/JL^2 and, for a given cross section, the effect of warping torsion decreases with increase in beam length. But also, since the bimoment is self-equilibrating, the effect of a given source of restraint is, in principle, a localized one. For deep, relatively thin-walled wide flange sections, the warping constant is so much larger than the torsional constant that warping restraint such as fixing the flanges at the end of member may have a major influence on the torsional response of spans of practical length. It may still be significant for shallower, stockier sections. But for closed sections, such as the rectangular tube, it is generally inconsequential. Warping resistance of open sections in which the component elements meet at a point, such as angle and cruciform sections, is also negligible.

2. ***The boundary conditions.*** In principle, torsional and flexural support conditions are independent of each other. For example, the torsionally simple supports of Figure 7.7*a* will provide flexural fixity about the section's major axis if the two idealized hinges which constitute the support are restrained in the x direction. On the other hand, the torsionally fixed support of Figure 7.7*b* could be flexurally simple with respect to rotation about the section's minor axis. This would be the case if the member were welded to a plate thick enough to prevent warping of the member but that itself would be supported by hinges that would permit it to rotate about the member's minor axis.

Additional questions arise in the definition of torsional boundary conditions at the junction of noncolinear members: for example, at a beam-to-column joint. For these reasons, we shall consider only the two extremes of torsional support: warping-free and warping-fixed at the ends of assemblages of colinear elements. A scheme for accommodating noncolinear elements and conditions of partial warping restraint is presented in Reference 7.6, and other possibilities for defining boundary conditions are described in Reference 7.7.

3. ***Solution accuracy.*** It should be noted that the underlying assumption of a cubic polynomial for the angle of twist does not satisfy a requirement of internal equilibrium. For the member considered, in which there is no load between the nodal points, dM_x/dx must be equal to zero. Substitution of the derivatives of the cubic polynomial of Equation 7.27 into Equation 7.23 reveals that this condition is not satisfied. Also, the accuracy of the terms of the solution obtained through the use of Equation 7.30 decreases as the order of the derivative to which they are related increases. Thus the calculated distribution of the bimoment and the warping torque may not be as reliable as the nodal forces, the angle of twist, and the St. Venant torque. But Equation 7.30 is efficient in computation and it can produce results in good agreement with known analytical solutions (Refs. 7.5 and 7.6). Further, the use of higher order approximating polynomials may lead to other, more serious, computational problems (Ref. 7.8). For these reasons, use of the cubic polynomial can be recommended. Methods for overcoming some of its shortcomings are incorporated in the following examples.

In Example 7.4, Equation 7.30 is applied to the analysis of a shaft using a single element idealization of the member. The distribution of twist is determined from the calculated end effects and the cubic shape function. The derivative of this function is then used in the calculation of the St. Venant torque (Eq. 7.20). To avoid reliance upon the higher derivative of Equation 7.22, the warping torque is determined as the difference between the total and the St. Venant torque.

Example 7.5 illustrates the influence of the relative values of α. For the case in which warping is restrained there is little difference in the twisting of the three sections. But comparison of the warping fixed and warping free response shows that St. Venant torsional resistance dominates in the shallowest section, whereas warping resistance does in the deepest one. Related to this is the fact that the effect of torsionally fixing the left end decays most rapidly in the shallowest section. In this example, a four element idealization of the members was used. Member subdivision of this sort is another way to improve the accuracy of an approximate solution. In this case, however, there is very little difference between the results obtained by it and by those from the single element idealization of Example 7.4. Both are excellent approximations of the analytical solution of the differential equation (see Ref. 7.4).

EXAMPLE 7.4

The member shown is fixed against warping and twisting at the left end and free at the right end.

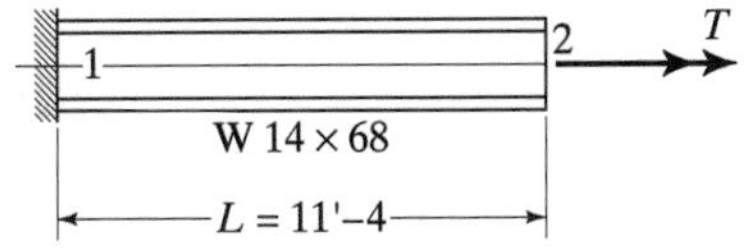

1. Determine the reactions.
2. Plot the angle of twist and the apportionment of T between St. Venant and warping torque.

1. Section properties: $J = 3.02$ in.^{4}, $C_w = 5380$ in.^{6}, $E = 29{,}000$ ksi, $\nu = 0.3$

$$\alpha = (2.6 \times 5380)/(3.02 \times 136^2) = 0.250$$

From Equation 7.30

$$\begin{Bmatrix} T \\ 0 \end{Bmatrix} = \frac{GJ}{L}\begin{bmatrix} 4.200 & -1.600L \\ -1.600L & 1.133L^2 \end{bmatrix}\begin{Bmatrix} \theta_{x2} \\ \theta'_{x2} \end{Bmatrix}$$

Thus

$$\theta_{x2} = 0.515\,\frac{TL}{GJ}, \quad \theta'_{x2} = 0.727\,\frac{T}{GJ}$$

and

$$\begin{Bmatrix} M_{x1} \\ B_1 \end{Bmatrix} = T\begin{bmatrix} -4.200 & 1.600L \\ -1.600L & 0.467L^2 \end{bmatrix}\begin{Bmatrix} 0.515 \\ \dfrac{0.727}{L} \end{Bmatrix} = \begin{Bmatrix} -T \\ -0.484TL \end{Bmatrix}$$

Reactions:

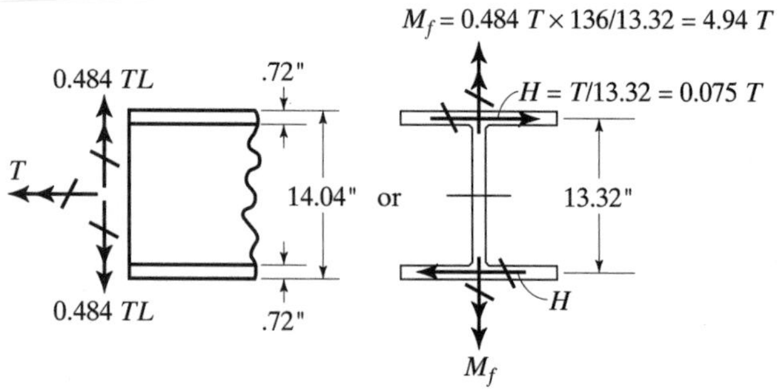

2. Twist and torque distribution:
From Equation 7.9

$$\theta_x = \left[3\left(\frac{x}{L}\right)^2 - 2\left(\frac{x}{L}\right)^3\right]\theta_{x2} + \left[\left(\frac{x}{L}\right)^3 - \left(\frac{x}{L}\right)^2\right]L\theta'_{x2}$$

$$\theta'_x = \left[6\left(\frac{x}{L}\right) - 6\left(\frac{x}{L}\right)^2\right]\frac{\theta_{x2}}{L} + \left[3\left(\frac{x}{L}\right)^2 - 2\left(\frac{x}{L}\right)\right]\theta'_{x2}$$

and Equation 7.20:

Angle of twist

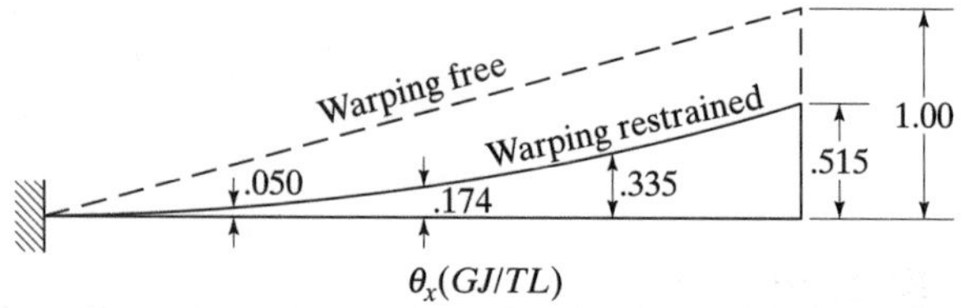

Torque distribution

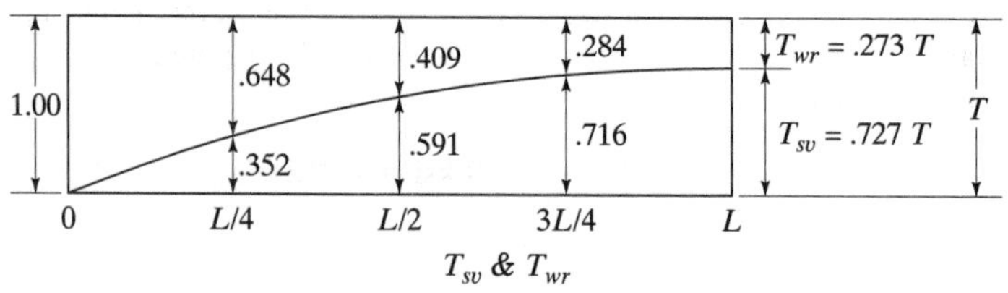

EXAMPLE 7.5

Compare the torsional response of three shafts of essentially equal weight:

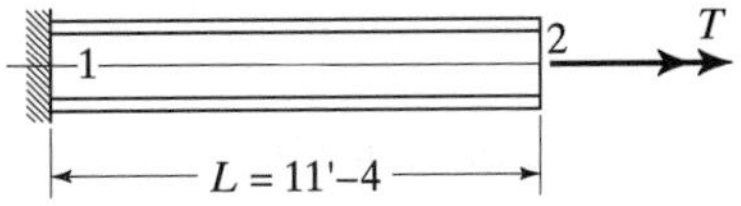

1. W8 × 67
2. W14 × 68
3. W24 × 68

$E = 29{,}000$ ksi, $\nu = 0.3$

Section properties:

	W8 × 67	W14 × 68	W24 × 68
J	5.06 in^4	3.02 in.^{4}	1.87 in.^{4}
C_w	1440 in.^{6}	5380 in.^{6}	9430 in.^{6}
α	0.040	0.250	0.709

Using Equation 7.30 and four element discretization:

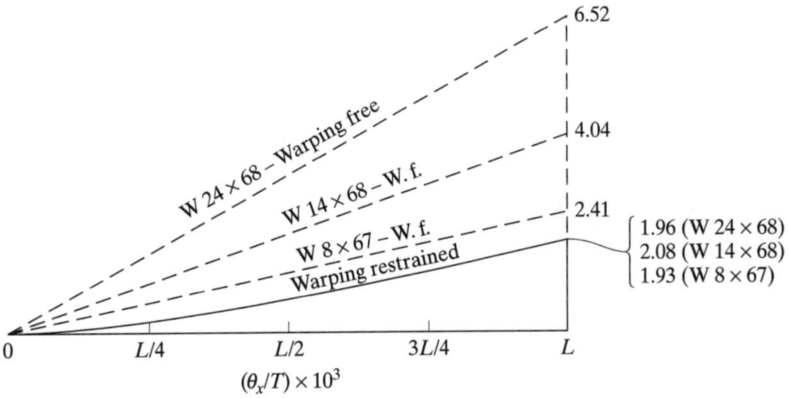

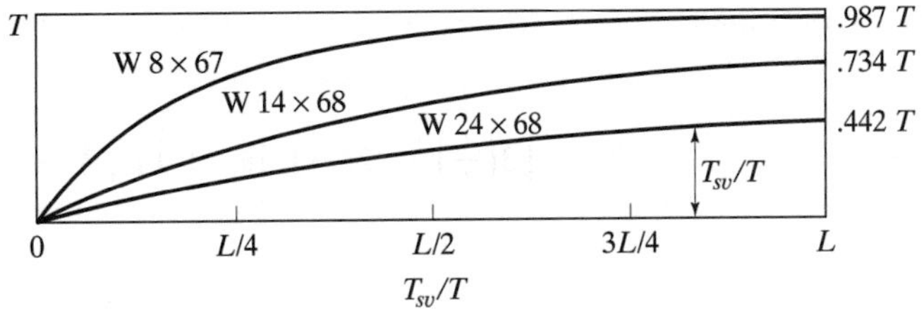

Example 7.6 is another case for which analytical results may be found in Reference 7.4. Again, torsionally fixing the ends has a profound effect on the twisting of the member.

EXAMPLE 7.6

A torque, T, is applied at the quarter point of the member shown. Compare its torsional response for warping free and warping fixed conditions at both ends.

$E = 29{,}000$ ksi, $\nu = 0.3$

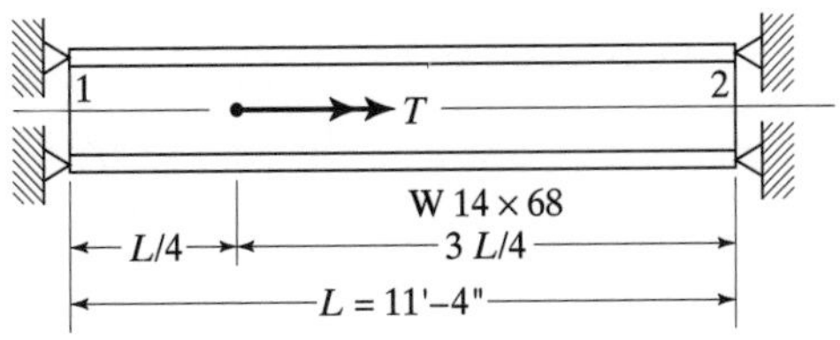

Use Equation 7.30 and four element discretization:

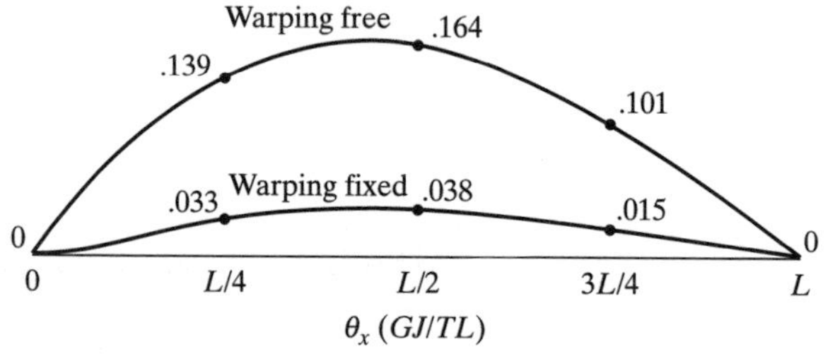

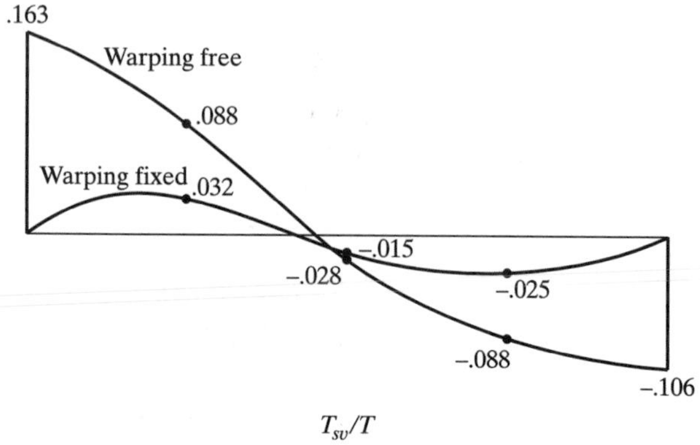

7.5 LOADS BETWEEN NODAL POINTS AND INITIAL STRAIN EFFECTS—A GENERAL APPROACH

The principle of virtual displacements, when intermediate loads act on members or when initial strains are present, gives a particularly convenient transformation of these loads and initial strains into equivalent joint forces. Such transformations were developed earlier (Sections 5.2 and 5.3) for simple cases. The formulas of Sections 5.2 and 5.3 can be duplicated with ease by means of the virtual work approach. The significant attribute of the virtual work approach, however, is that it is able to deal with rather complicated conditions in a unified and convenient manner.

We examine first the treatment of intermediate loads on members. Consider, for purposes of the development, a framework member subjected to a distributed load $\mathbf{q}$, oriented arbitrarily with respect to the x, y, z axes. Consider also the virtual displaced state of the element, designated by $\delta\boldsymbol{\Delta}$. For axial behavior, for example, $\delta\boldsymbol{\Delta} = \delta u$. The external virtual work of the load $\mathbf{q}$ in a differential length of the element dx is the dot product $\delta\boldsymbol{\Delta}\cdot\mathbf{q}\,dx$; and for the total length of the element, it is (from Equation 6.24)

$$\delta W_{\text{ext}} = \int_0^L \delta\boldsymbol{\Delta}\cdot\mathbf{q}\,dx \tag{7.31}$$

Adopting now the description of $\delta\boldsymbol{\Delta}$ in terms of the joint virtual displacements (Equation 7.14), we have

$$\delta W_{\text{ext}} = \lfloor\delta\boldsymbol{\Delta}\rfloor\left\{\int_0^L \{\mathbf{N}\}\cdot\mathbf{q}\,dx\right\} = \lfloor\delta\boldsymbol{\Delta}\rfloor\{\mathbf{F}^E\} \tag{7.31a}$$

so that

$$\left\{\int_0^L \{\mathbf{N}\} \cdot \mathbf{q}\, dx\right\} = \{\mathbf{F}^E\} \tag{7.32}$$

This is a general formula for the effective nodal loads $\{\mathbf{F}^E\}$. Clearly, the joint forces $\{\mathbf{F}^E\}$ produce the same virtual work as the distributed loads. For this reason it is customary to describe such joint forces, when calculated by use of Equation 7.32, as "work-equivalent" loads.

A specific case of the above is the uniformly loaded axial member, where $\mathbf{q} = q_x$ and

$$\{\mathbf{N}\} = \begin{Bmatrix} 1 - \dfrac{x}{L} \\ \dfrac{x}{L} \end{Bmatrix}$$

Hence, by Equation 7.32

$$\{\mathbf{F}^E\} = \int_0^L \begin{Bmatrix} \left(1 - \dfrac{x}{L}\right) \\ \dfrac{x}{L} \end{Bmatrix} q_x\, dx = \frac{q_x}{2}\begin{Bmatrix} 1 \\ 1 \end{Bmatrix}$$

In this case the equivalent nodal loads are those that one would obtain by simple apportionment of the total load to the two joints.

The more significant conditions are encountered in the case of flexural members, where $\delta\boldsymbol{\Delta} = \delta v$ and the distributed loads (q_y) act in the y direction (Fig. 7.9). Also, concentrated loads $P_3, \ldots, P_i, \ldots, P_r$ may be present. Thus the external virtual work is, using Equation 7.31 for the distributed load portion,

$$\delta W_{\text{ext}} = \int_0^L \delta v \cdot q_y\, dx + \sum_{i=3}^{r} \delta v_i \cdot P_i$$

where δv symbolizes the virtual transverse displacement at all points along the beam, and δv_i is the virtual transverse displacement at the point of application of P_i. Therefore, δv_i denotes the evaluation of δv at point i.

As in previous developments, we now assume that the virtual displaced state of the beam element can be described in terms of the four joint displacements δv_1, δv_2, $\delta\theta_{z1}$, and $\delta\theta_{z2}$, that is, in the form of Equation 7.13b.

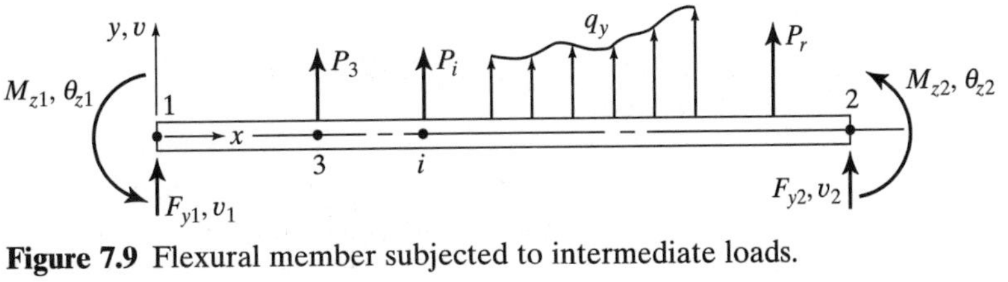

Figure 7.9 Flexural member subjected to intermediate loads.

Substitution of Equation 7.13b into the expression for the virtual work of the applied loads results in

$$\delta W_{\text{ext}} = \lfloor \delta v_1 \ \delta v_2 \ \delta\theta_{z1} \ \delta\theta_{z2} \rfloor \left\{ \begin{Bmatrix} \int N_1 q_y \, dx \\ \int N_2 q_y \, dx \\ \int N_3 q_y \, dx \\ \int N_4 q_y \, dx \end{Bmatrix} + \begin{Bmatrix} \sum N_1^i P_i \\ \sum N_2^i P_i \\ \sum N_3^i P_i \\ \sum N_4^i P_i \end{Bmatrix} \right\} \tag{7.33}$$

where the summations extend from $i = 3$ to $i = r$, and $N_1^i, \ldots, N_4^i$ designate the shape functions evaluated at each point i. This can be written more concisely as

$$\delta W_{\text{ext}} = \lfloor \delta\mathbf{\Delta} \rfloor \{\{\mathbf{F}^d\} + \{\mathbf{F}^c\}\} = \lfloor \delta\mathbf{\Delta} \rfloor \{\mathbf{F}^E\} \tag{7.33a}$$

where $\{\mathbf{F}^E\}$, as in Section 5.2, is the vector of effective nodal forces corresponding to the distributed loads q_y and the within-span forces P_i. Thus

$$\{\mathbf{F}^E\} = \{\mathbf{F}^d\} + \{\mathbf{F}^c\} \tag{7.34}$$

and

$$\{\mathbf{F}^d\} = \begin{Bmatrix} \int N_1 q_y \, dx \\ \int N_2 q_y \, dx \\ \int N_3 q_y \, dx \\ \int N_4 q_y \, dx \end{Bmatrix} \quad (7.34\text{a}) \qquad \{\mathbf{F}^c\} = \begin{Bmatrix} \sum N_1^i P_i \\ \sum N_2^i P_i \\ \sum N_3^i P_i \\ \sum N_4^i P_i \end{Bmatrix} \quad (7.34\text{b})$$

If Equation 7.33a is used in the principle of virtual work (Eq. 6.13), as a supplement to the δW_{ext} of the nodal forces ($\lfloor \delta\mathbf{\Delta} \rfloor \{F\}$), the element stiffness Equation 7.18 is extended to[2]

$$[\mathbf{k}]\{\mathbf{\Delta}\} = \{\mathbf{F}\} + \{\mathbf{F}^E\} \tag{7.18a}$$

Initial strain effects, which have been discussed in Section 5.3 bear a similarity to distributed load effects in that they are represented by vectors of equivalent joint forces. They derive, however, from the internal virtual work rather than from the external virtual work.

Initial strain effects are basically represented in the stress-strain law. Consider, for purposes of explanation, direct strain behavior in the x direction. When initial strain is present, the total strain is the sum of the strain due to stress, σ_x/E, and the initial strain, e_x^i, that is

$$e_x = \frac{\sigma_x}{E} + e_x^i$$

Solving this expression for stress, we obtain

$$\sigma_x = Ee_x - Ee_x^i \tag{7.35}$$

[2]In Equation 5.21, $\{\mathbf{P}^E\}$ was used to designate equivalent nodal loads whereas in Equation 7.18a $\{\mathbf{F}^E\}$ is used for the same quantities. The only difference is one of symbolism: Equation 5.21 was developed in the context of a global analysis example and Equation 7.18a from the analysis of element joint forces.

and generally,

$$\{\sigma\} = [\mathbf{E}]\{\mathbf{e}\} - [\mathbf{E}]\{\mathbf{e}^i\} \tag{7.35a}$$

Consequently, the internal virtual work, given previously by Equation 6.23, is now

$$\begin{aligned}\delta W_{\text{int}} &= \int_{\text{vol}} \lfloor \delta\mathbf{e} \rfloor \{\boldsymbol{\sigma}\}\, d(\text{vol}) \\ &= \int_{\text{vol}} \lfloor \delta\mathbf{e} \rfloor ([\mathbf{E}]\{\mathbf{e}\} - [\mathbf{E}]\{\mathbf{e}^i\})\, d(\text{vol}) \\ &= \int_{\text{vol}} \lfloor \delta\mathbf{e} \rfloor [\mathbf{E}]\{\mathbf{e}\}\, d(\text{vol}) - \int_{\text{vol}} \lfloor \delta\mathbf{e} \rfloor [\mathbf{E}]\{\mathbf{e}^i\}\, d(\text{vol})\end{aligned} \tag{6.23a}$$

Now, the first integral on the right side is the same integral as established earlier for conditions where initial strains are absent and is the basis for the element stiffness equations (see Section 6.3, Equation 6.23). The second integral leads to a vector of "initial forces" $\{\mathbf{F}^i\}$ by substitution of the expression for the virtual strains in terms of the joint displacements, given by Equation 7.15:

$$\begin{aligned}\int_{\text{vol}} \delta\mathbf{e}[\mathbf{E}]\{\mathbf{e}^i\}\, d(\text{vol}) &= \lfloor \delta\boldsymbol{\Delta} \rfloor \int_{\text{vol}} \{\mathbf{N}'\}[\mathbf{E}]\{\mathbf{e}^i\}\, d(\text{vol}) \\ &= \lfloor \delta\boldsymbol{\Delta} \rfloor \{\mathbf{F}^i\}\end{aligned} \tag{7.36}$$

where

$$\{\mathbf{F}^i\} = \int_{\text{vol}} \{\mathbf{N}'\}[\mathbf{E}]\{\mathbf{e}^i\}\, d(\text{vol}) \tag{7.37}$$

Thus, upon substitution of the results of Equation 7.36 into Equation 6.23a, and application of the principle of virtual displacements, we have

$$\{\mathbf{F}\} = [\mathbf{k}]\{\boldsymbol{\Delta}\} - \{\mathbf{F}^i\} \tag{7.18b}$$

For an axial member, for example, $\{\mathbf{N}'\} = \dfrac{1}{L}\begin{Bmatrix} -1 \\ 1 \end{Bmatrix}$ and $[\mathbf{E}] = E$. Suppose the initial strain is due to a uniform temperature change T. Then $\{\mathbf{e}^i\} = \alpha T$ and the initial forces are

$$\{\mathbf{F}^i\} = \int_0^L \frac{1}{L}\begin{Bmatrix} -1 \\ 1 \end{Bmatrix} E\alpha TA\, dx = EA\alpha T\begin{Bmatrix} -1 \\ 1 \end{Bmatrix}$$

This is the result obtained in Section 5.3, Equation 5.23, in a more intuitive manner. The present development, however, is based on Equation 7.37, which permits a more general formulation. For example, initial strains that are a function of the coordinates of the element can be handled.

We consider, in the case of bending, a member with a bisymmetrical cross section and a linear variation of initial strain across the depth (h) of the element. These strains have a value of ($\Delta e^i/2$) at the top and bottom of the section and zero value at the neutral axis. It was shown in Section 5.3 that for this case the curvature due to initial strain is $\kappa^i = e^i/h$. Adopting, in the formula for initial strains, Equation 7.35, the same analogies adopted previously for the flexural element stiffness matrix, we have

$$\{\mathbf{F}^i\} = \int_0^L \{\mathbf{N}'\} EI\kappa^i\, dx \tag{7.38}$$

$\{\mathbf{N}'\}$ now represents the second derivatives of the shape functions for transverse displacement, such as those given in Equation 7.13a for the simple flexural element.

Example 7.7 illustrates the application of the work-equivalent load approach to cases that were studied by an alternative approach in Chapter 5. Example 7.8 illustrates the calculation of the initial forces for a linear temperature change across the depth of a flexural element where the change varies along the length of the member.

EXAMPLE 7.7

Verify, by use of the virtual work approach, the effective nodal loads $\{\mathbf{F}^E\}$ for the beam element and loadings shown. Results for these cases are given in Table 5.1.

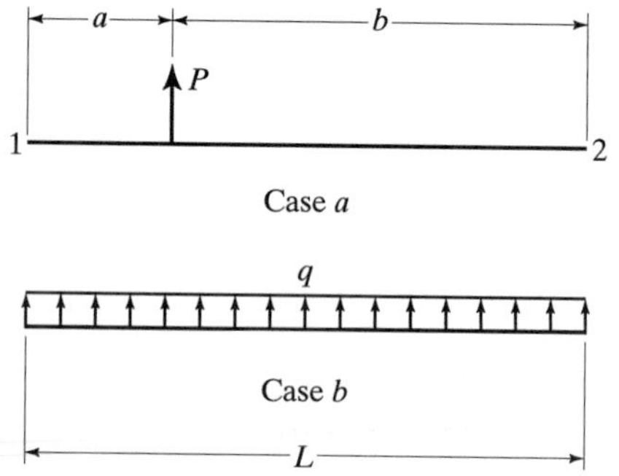

For Case a, using the shape functions of Equation 7.9.

$$\{\mathbf{F}^E\} = \begin{Bmatrix} \left[1 - 3\left(\frac{x}{L}\right)^2 + 2\left(\frac{x}{L}\right)^3\right]\Big|_{x=a} P \\ \left(\frac{x}{L}\right)^2\left(3 - \frac{2x}{L}\right)\Big|_{x=a} P \\ x\left(1 - \frac{x}{L}\right)^2\Big|_{x=a} P \\ \frac{x^2}{L}\left(\frac{x}{L} - 1\right)\Big|_{x=a} P \end{Bmatrix} = \begin{Bmatrix} \frac{b^2}{L^3}(3a + b)P \\ \frac{a^2}{L^3}(a + 3b)P \\ \frac{ab^2}{L^2} P \\ -\frac{a^2 b}{L^2} P \end{Bmatrix} = \begin{Bmatrix} F_{y1}^E \\ F_{y2}^E \\ M_{z1}^E \\ M_{z2}^E \end{Bmatrix}$$

For Case b,

$$\begin{Bmatrix} \int_0^L \left[1 - 3\left(\frac{x}{L}\right)^2 + 2\left(\frac{x}{L}\right)^3\right] q\,dx \\ \int_0^L \left[3\left(\frac{x}{L}\right)^2 - 2\left(\frac{x}{L}\right)^3\right] q\,dx \\ \int_0^L x\left(1 - \frac{x}{L}\right)^2 q\,dx \\ \int_0^L x\left(\frac{x^2}{L^2} - \frac{x}{L}\right) q\,dx \end{Bmatrix} = \begin{Bmatrix} \frac{qL}{2} \\ \frac{qL}{2} \\ \frac{qL^2}{12} \\ -\frac{qL^2}{12} \end{Bmatrix} = \begin{Bmatrix} F_{y1}^E \\ F_{y2}^E \\ M_{z1}^E \\ M_{z2}^E \end{Bmatrix}$$

EXAMPLE 7.8

The simple flexural element (Figure 7.3), which in this case is of bisymmetrical cross section and of depth h, sustains a temperature difference across the depth given by $\Delta T = (1 - x/L)(\Delta T_1) + (x/L)(\Delta T_2)$. (The neutral axis temperature change is zero.) Calculate the relevant vector of initial forces.

It is shown, in Section 5.3, that the initial curvature in this case is $\kappa^i = \alpha(\Delta T)/h$. Hence, by Equation 7.38 and using Equation 7.13a for $\{\mathbf{N}'\}$, we have

$$\{\mathbf{F}^i\} = \frac{EI\alpha}{h}\begin{Bmatrix} \int_0^L \frac{6}{L^2}\left(\frac{2x}{L} - 1\right)\left[\left(1 - \frac{x}{L}\right)(\Delta T_1) + \frac{x}{L}(\Delta T_2)\right] dx \\ \int_0^L -\frac{6}{L^2}\left(\frac{2x}{L} - 1\right)\left[\left(1 - \frac{x}{L}\right)(\Delta T_1) + \frac{x}{L}(\Delta T_2)\right] dx \\ \int_0^L \frac{2}{L}\left(\frac{3x}{L} - 2\right)\left[\left(1 - \frac{x}{L}\right)(\Delta T_1) + \frac{x}{L}(\Delta T_2)\right] dx \\ \int_0^L \frac{2}{L}\left(\frac{3x}{L} - 1\right)\left[\left(1 - \frac{x}{L}\right)(\Delta T_1) + \frac{x}{L}(\Delta T_2)\right] dx \end{Bmatrix} = \frac{EI\alpha}{hL}\begin{Bmatrix} \Delta T_2 - \Delta T_1 \\ \Delta T_1 - \Delta T_2 \\ -L(\Delta T)_1 \\ L(\Delta T)_2 \end{Bmatrix}$$

(The ordering of terms in the vector is the same as in Example 7.7)

7.6 VIRTUAL FORCES IN THE FORMULATION OF ELEMENT FORCE-DISPLACEMENT EQUATIONS

7.6.1 Construction of Element Equations by the Principle of Virtual Forces

We have established that the virtual displacement principle is the basis for the direct formulation of element stiffness equations. The dual principle of virtual forces is, conversely, the basis of the direct formulation of element flexibility equations.

Element flexibility equations are of considerable value in practice since they can be transformed into element stiffness equations by inversion and by supplemental matrix operations detailed in Section 4.4. There are many conditions under which the formulation of element flexibility equations is straightforward, while the direct formulation of the corresponding element stiffness is extremely difficult. Such conditions include the formulation of the beam element for combined transverse shear and flexural deformations, tapered members, and elements with curved or irregular axes.

The principle of virtual forces was stated in Section 6.5 as follows

$$\delta W^*_{\text{ext}} = \delta W^*_{\text{int}} \tag{6.32}$$

where

$$\delta W^*_{\text{int}} = \int_{\text{vol}} \lfloor \delta\boldsymbol{\sigma} \rfloor [\mathbf{E}]^{-1}\{\boldsymbol{\sigma}\}\, d(\text{vol}) \tag{6.31}$$

The real ($\boldsymbol{\sigma}$) and the virtual ($\delta\boldsymbol{\sigma}$) stresses are, for a given mode of structural behavior, single components of stress: uniaxial direct stress for axial or flexural behavior and shear stress for simple torsion. $[\mathbf{E}]^{-1}$ denotes the inverse of the relevant elastic constant. For the particular case of element flexibility formulation, where the joint displacements are $\Delta_1, \ldots, \Delta_i, \ldots, \Delta_f$ and the virtual joint forces are $\delta F_1, \ldots, \delta F_i, \ldots, \delta F_f$, δW^*_{ext} becomes

$$\delta W^*_{\text{ext}} = \sum_{i=1}^{f} (\delta F_i)(\Delta_i) = \lfloor \delta \mathbf{F}_f \rfloor \{\boldsymbol{\Delta}\} \tag{7.39}$$

We have changed the symbolism from $\{\mathbf{P}_f\}$ to $\{\mathbf{F}_f\}$ because the latter is reserved for designation of element joint forces. The subscripts f emphasize that the virtual joint forces and associated displacements refer to the points that are free to displace on elements that are supported in a stable, statically determinate manner. Thus, the number of node points, f, represented above is s fewer than for the stiffness formulation, where s designates the number of rigid body motions. For flexure, the displacements include angular displacements θ_i for which the corresponding virtual forces are the moments δM_i.

In the following we direct our attention to the complementary internal virtual work δW^*_{int}, since this is the source of coefficients f_{ij} of the element flexibility relationships. To evaluate δW^*_{int} for this purpose, expressions must be constructed that give the virtual and real stresses, $\lfloor \delta\boldsymbol{\sigma} \rfloor$ and $\{\boldsymbol{\sigma}\}$ in terms of joint real and virtual forces, $\{\mathbf{F}_f\}$ and $\{\delta\mathbf{F}_f\}$. An intermediate step, however, is the transformation of the stresses into the stress resultants of the structural action in question. This step was detailed in Section 6.5 and gives, for axial behavior (Fig. 7.10*a*),

$$\delta W^*_{\text{int }a} = \int_0^L \delta F_x \left(\frac{F_x}{EA}\right) dx \tag{6.33}$$

where δF_x and F_x are the virtual and real internal axial forces. For a pure torsional element oriented as in Figure 7.10*b*,

$$\delta W^*_{\text{int }t} = \int_0^L \delta M_x \left(\frac{M_x}{GJ}\right) dx \tag{6.35}$$

where δM_x and M_x are the virtual and real torsional moments (about the x axis), and J is the torsional constant. For a flexural element with coordinate axes as shown in Figure 7.10*c*,

$$\delta W^*_{\text{int }b} = \int_0^L \delta M_z \left(\frac{M_z}{EI_z}\right) dx \tag{6.34}$$

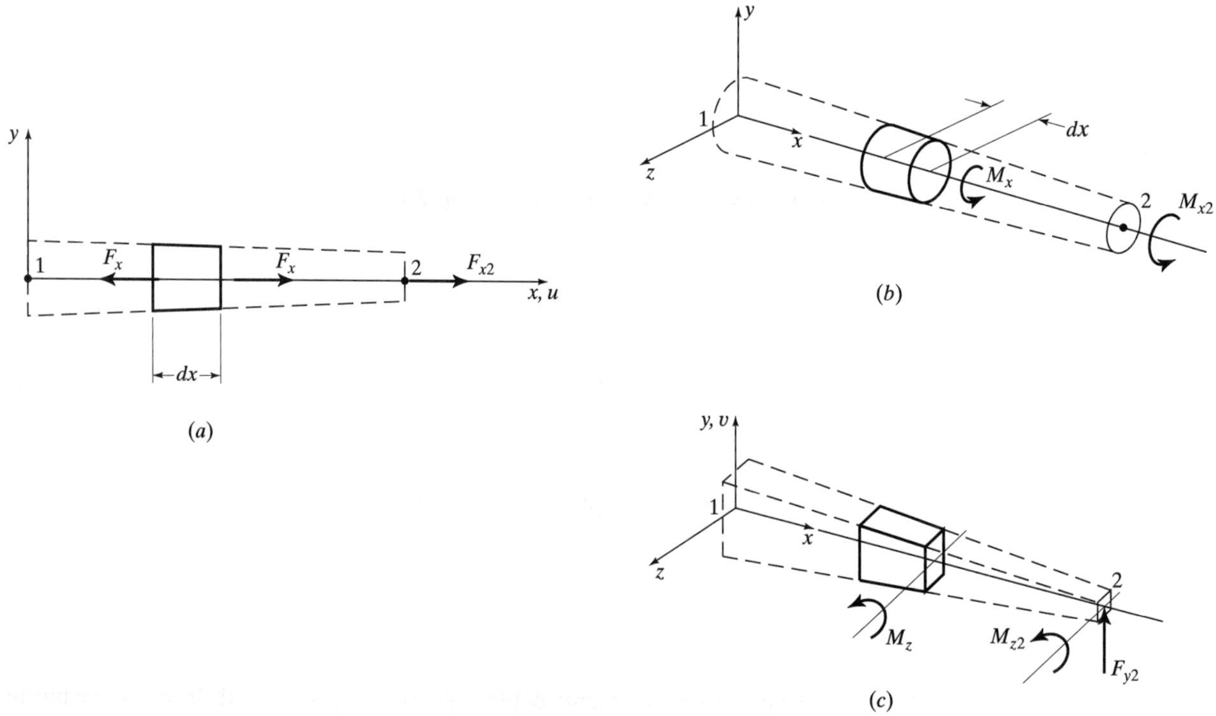

Figure 7.10 Internal force resultants for axial, torsional, and flexural elements (virtual forces δF_{x2}, δM_{x2}, δF_{y2}, δM_{z2} corresponding to $F_{x2}, \ldots, M_{z2}$ are not shown). (*a*) Axial force element. (*b*) Torsional element. (*c*) Flexural element (*x*-*y* plane).

where δM_z and M_z are the virtual and real internal moments at an arbitrary point within the element.

The approach taken in the principle of virtual displacements, where the displacement at an arbitrary point within the element is related to the joint displacements by means of shape functions, suggests the use of analogous functions that relate the internal forces to the joint forces in the principle of virtual forces. Symbolically, this can be written as

$$F_x(\text{or } M_x \text{ or } M_z) = \lfloor \mathbf{Q} \rfloor \{\mathbf{F}_f\} \tag{7.40}$$

for axial force members, or torsional or flexural members, as the case may be. $\lfloor \mathbf{Q} \rfloor$, the vector of functions of the force distribution must be chosen in such a way that the conditions of equilibrium are satisfied, a consideration that was emphasized in Section 6.5. It will be shown below that the choice of $\lfloor \mathbf{Q} \rfloor$ to meet such conditions presents no difficulty.

It is both convenient and appropriate to choose the same shapes of virtual and real force systems. The equilibrium conditions, cited above, are then sought only once, the integration of δW^*_{int} is simplified, and the resulting flexibility matrix is guaranteed to be symmetric. The chosen form of the virtual force system is, consequently,

$$\delta F_x(\text{or } \delta M_x \text{ or } \delta M_z) = \lfloor \mathbf{Q} \rfloor \{\delta \mathbf{F}_f\} \tag{7.41}$$

Substitution of Equations 7.40 and 7.41 into δW^*_{int} yields

$$\delta W^*_{\text{int}} = \lfloor \delta \mathbf{F}_f \rfloor [\mathbf{d}] \{\mathbf{F}_f\} \tag{7.42}$$

where, for the axial force member (Eq. 6.33, Fig. 7.10*a*),

$$[\mathbf{d}] = \left[\int_0^L \{\mathbf{Q}\} \frac{1}{EA} \lfloor \mathbf{Q} \rfloor \, dx \right] \tag{7.43a}$$

$$\{\mathbf{F}_f\} = F_{x2} \qquad \{\delta \mathbf{F}_f\} = \delta F_{x2} \tag{7.43b,c}$$

and for the torsional element (Eq. 6.35, Fig. 7.10*b*),

$$[\mathbf{d}] = \left[\int_0^L \{\mathbf{Q}\} \frac{1}{GJ} \lfloor \mathbf{Q} \rfloor \, dx \right] \tag{7.44a}$$

$$\{\mathbf{F}_f\} = M_{x2} \qquad \{\delta \mathbf{F}_f\} = \delta M_{x2} \tag{7.44b,c}$$

and, for the flexural element (Equation 6.34, Figure 7.10*c*),

$$[\mathbf{d}] = \left[\int_0^L \{\mathbf{Q}\} \frac{1}{EI} \lfloor \mathbf{Q} \rfloor \, dx \right] \tag{7.45a}$$

$$\{\mathbf{F}_f\} = \begin{Bmatrix} F_{y2} \\ M_{z2} \end{Bmatrix} \qquad \{\delta \mathbf{F}_f\} = \begin{Bmatrix} \delta F_{y2} \\ \delta M_{z2} \end{Bmatrix} \tag{7.45b,c}$$

Note that each of these elements could have been supported differently, resulting in different specific forms for $\{\mathbf{F}_f\}$ and $\{\delta \mathbf{F}_f\}$. Note also that they are portrayed as having nonuniform cross sections. It is demonstrated in the following that the principle of virtual forces enables a very expeditious treatment of nonuniform elements.

A convenient illustration of the use of the principle of virtual forces in the formulation of element flexibility relationships is given by the tapered axial member (Fig. 6.6). We choose $A = A_1(1 - rx/L)$. The relationship between the joint force and internal force state is $F_x = F_{x2}$. Hence, $\lfloor \mathbf{Q} \rfloor$ in this case is merely unity. The flexibility matrix is a single term and is obtained by the substitution of the foregoing into Equation 7.43a, resulting in

$$[\mathbf{d}] = d_{22} = \frac{1}{EA_1}\int_0^L \frac{dx}{(1 - rx/L)} = -\frac{L}{EA_1 r}\, ln(1 - r)$$

This is the exact solution for the flexibility coefficient.

To establish the stiffness matrix for this element we now use the procedure of Section 4.4, Equation 4.25. Here, $\{\mathbf{F}_s\} = F_{x1} = -F_{x2}$, so that $[\mathbf{\Phi}] = -1$. Hence, by Equation 4.25,

$$[\mathbf{k}] = \begin{bmatrix} d_{22}{}^{-1} & -d_{22}{}^{-1} \\ -d_{22}{}^{-1} & d_{22}{}^{-1} \end{bmatrix} = -\frac{EA_1}{L}\frac{r}{ln(1 - r)}\begin{bmatrix} 1 & -1 \\ -1 & 1 \end{bmatrix}$$

which is the same as the result obtained in Section 7.3.

It is important to observe why this exact solution was obtained with ease, whereas the solution by use of the virtual displacements principle (Section 6.4) was considerably more difficult to obtain. The virtual and real forces required by the virtual forces principle are given by the equation of statics; the variations that may be present in the section or material properties (A, I_z, E, etc.) do not affect these equations. Exact displacement functions for elements with such variations can only be obtained through solution of differential equations with varying coefficients and are generally of rather complex form.

Still another simple illustration is given by the construction of the flexibility matrix for the flexural element, Figure 7.10c, but with constant section properties. From statics, we can write

$$M_z = \lfloor (L - x) \quad 1 \rfloor \begin{Bmatrix} F_{y2} \\ M_{z2} \end{Bmatrix}$$

so that $\lfloor \mathbf{Q} \rfloor = \lfloor (L - x) \quad 1 \rfloor$. Then,

$$\delta M_z = \lfloor (L - x) \quad 1 \rfloor \begin{Bmatrix} \delta F_{y2} \\ \delta M_{z2} \end{Bmatrix}$$

With the above relationships Equation 7.45a becomes, for an element of length L,

$$[\mathbf{d}] = \frac{1}{EI_z}\left[\int_0^L \begin{Bmatrix} (L - x) \\ 1 \end{Bmatrix} \lfloor (L - x) \quad 1 \rfloor \, dx\right] = \frac{L}{EI_z}\begin{bmatrix} \dfrac{L^2}{3} & \dfrac{L}{2} \\ \dfrac{L}{2} & 1 \end{bmatrix} \tag{7.46}$$

This matrix was derived in Example 4.3 by solution of the governing differential equation of beam bending.

Example 7.9 is a further illustration of the above. It deals with the tapered beam element for which exact shape functions were established in Example 7.2 and an approximate stiffness matrix and a solution for the tip displacement were obtained in Example 7.3. With virtual forces the exact flexibility matrix is established quite easily.

The exact stiffness matrix can be obtained by inversion of this flexibility matrix and use of Equation 4.25. The same result will be obtained by use of the exact displacement functions in the virtual displacements principle, Equation 7.17.

EXAMPLE 7.9

Formulate the flexibility matrix for the tapered cantilever beam of Example 7.2 using the principle of virtual forces. Calculate the displacement v_1 due to the force F_{y1} and compare with the solution of Example 7.3.

In this case the relationship between the internal real and virtual moments and the joint real and virtual forces are

$$M_z = \lfloor x \quad -1 \rfloor \begin{Bmatrix} F_{y1} \\ M_{z1} \end{Bmatrix}, \qquad \delta M_z = \lfloor x \quad -1 \rfloor \begin{Bmatrix} \delta F_{y1} \\ \delta M_{z1} \end{Bmatrix}$$

so that $\lfloor \mathbf{Q} \rfloor = \lfloor x \quad -1 \rfloor$.

From Equation 7.45a,

$$[\mathbf{d}] = \left[\int_0^L \begin{Bmatrix} x \\ -1 \end{Bmatrix} \lfloor x \quad -1 \rfloor \frac{12}{Ebh_1^3\left(1 + \dfrac{x}{L}\right)^3} dx \right]$$

$$= \frac{12L}{Eh_1^3 b} \begin{bmatrix} (ln\ 2 - 5/8)L^2 & \dfrac{L}{8} \\ \dfrac{L}{8} & \dfrac{1}{4} \end{bmatrix}$$

For $M_{z1} = 0$, this gives $v_1 = 0.06815F_{y1}L^3/EI_1$. The solution for Example 7.3 is $v_1 = 0.06684F_{y1}L^3/EI_1$.

7.6.2 Further Applications—Shearing Deformations and Curved Elements

The principle of virtual forces is a highly advantageous approach in situations in which more than one mode of action may require representation within the element and in the case of "line elements." The first is exemplified by the combination of flexure and transverse shear deformation, or the combination of axial, torsional, and flexural behavior. In such cases the component modes of structural action each produce a contribution to the virtual change in complementary internal work δW^*_{int}. Therefore, for a bisymmetrical element of length L, in the presence of axial, torsional, and flexural deformations,

$$\delta W^*_{\text{int}} = \int_0^L \delta F_x \left(\frac{F_x}{EA} \right) dx + \int_0^L \delta M_z \left(\frac{M_z}{EI_z} \right) dx + \int_0^L \delta M_x \left(\frac{M_x}{GJ} \right) dx + \int_0^L \delta M_y \left(\frac{M_y}{EI_y} \right) dx$$

For line elements (framework elements, but especially complicated elements of the kind illustrated in Fig. 7.11) one need only fix one end of the member, say, point 2 for the case shown. Expressions for the real stresses and virtual stresses at any point in the member in terms of the free-end real and virtual forces (the forces at point 1)

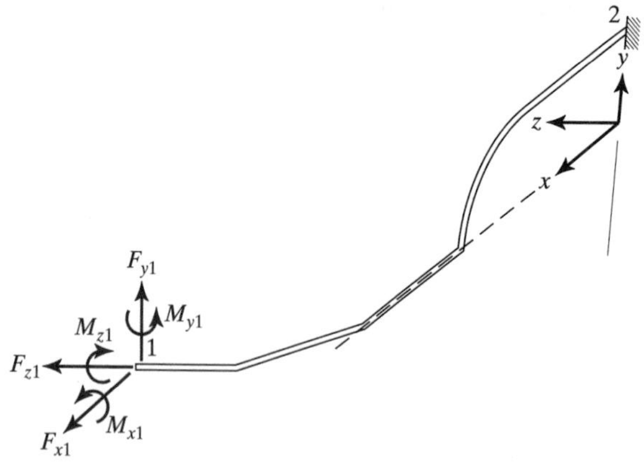

Figure 7.11 Nonprismatic framework element.

can then be constructed straightforwardly from statics. These relationships are of the form of Equations 7.40 and 7.41, where $\{\mathbf{F}_f\}$ and $\{\delta\mathbf{F}_f\}$ are the real and virtual forces at point 1.

These features of the virtual forces approach can be illustrated by its application to two problems: (1) the effect of shearing deformation on the deflection of a beam, and (2) the analysis of a ring beam.

7.6.2.1 Shearing Deformation of a Beam The flexural element of Figure 7.3 will be the basis for analyzing the effect of shear. We define a stable, statically determinate support condition by fixing the left end (point 1 in Figure 7.3) and use the principle of virtual forces to develop the relevant flexibility matrix. The counterpart stiffness matrix is then constructed by the procedure of Section 4.4.

In dealing with transverse shear deformation we adopt the simplified notion of an "equivalent shear area," A_s. This area, multiplied by the shearing stress (τ_{na}) at the neutral axis gives the total shear force on the cross section. Thus there is assumed to be a uniform shear stress $\tau_{na} = F_{y2}/A_s$ on the cross section. The corresponding neutral axis shear strain is $\gamma = F_{y2}/A_sG$ and in the expression for the component of complementary internal virtual work due to transverse shear, $\delta W^*_{\text{int } s}$, we have $d(\text{vol}) = A_s\, dx$, $\sigma = \tau_{na}$, $\delta\sigma = \delta\tau_{na}$, $\mathbf{E}^{-1} = 1/G$, $\{\mathbf{F}_f\} = F_{y2}$, $\{\delta\mathbf{F}_f\} = \delta F_{y2}$, $\lfloor \mathbf{Q} \rfloor = 1$. Hence

$$\delta W^*_{\text{int } s} = \int_0^L \left(\frac{\delta F_{y2}}{A_s}\right)\frac{1}{G}\left(\frac{F_{y2}}{A_s}\right)A_s\, dx = \frac{\delta F_{y2}\cdot F_{y2}L}{A_sG}$$

To this must be added the complementary internal virtual work due to bending, denoted here by $\delta W^*_{\text{int } b}$. The flexibility terms for $\delta W^*_{\text{int } b}$ have already been established and are given by Equation 7.46. Therefore,

$$\delta W_{\text{int}} = \delta W^*_{\text{int } b} + \delta W^*_{\text{int } s} = \lfloor \delta F_{y2} \quad \delta M_{z2} \rfloor \left[\begin{array}{c:c} \dfrac{L^3}{3EI_z} + \dfrac{L}{A_sG} & \dfrac{L^2}{2EI_z} \\ \hdashline \dfrac{L^2}{2EI_z} & \dfrac{L}{EI_z} \end{array}\right] \begin{Bmatrix} F_{y2} \\ M_{z2} \end{Bmatrix} \tag{7.47}$$

The desired flexibility matrix $[\mathbf{d}]$ is the central matrix of this product. The equilibrium matrix $[\mathbf{\Phi}]$ needed for the transformation of this matrix into a stiffness matrix is unaffected by the presence of shearing deformation and is thus the same as that presented in Example 4.4. Using that matrix and the above matrix [**d**] in Equation 4.25 yields, with $EI_z/A_sG = \eta$,

$$\begin{Bmatrix} F_{y2} \\ M_{z2} \\ F_{y1} \\ M_{z1} \end{Bmatrix} = \frac{EI_z}{L\left(\dfrac{L^2}{12} + \eta\right)} \begin{bmatrix} 1 & -\dfrac{L}{2} & -1 & -\dfrac{L}{2} \\ & \left(\dfrac{L^2}{3} + \eta\right) & \dfrac{L}{2} & \left(\dfrac{L^2}{6} - \eta\right) \\ & \text{Sym.} & 1 & \dfrac{L}{2} \\ & & & \left(\dfrac{L^2}{3} + \eta\right) \end{bmatrix} \begin{Bmatrix} v_2 \\ \theta_{z2} \\ v_1 \\ \theta_{z1} \end{Bmatrix} \tag{7.48}$$

Shearing deformation may be included in general frame analysis by replacing terms in Equation 4.34 by the corresponding terms of Equation 7.48 and their counterparts in the x-z plane. For most framework members of practical length the influence of transverse shear deformation may be disregarded as negligible. It can, however, be significant in members of small span-to-depth ratio or—a similar and often related criterion—a region in which the shear-to-moment ratio is high. For these, use of approximate values of the equivalent shear area is usually satisfactory.

In Reference 7.9 it is shown that A_s for the rectangular section (Figure 7.12*a*) may be calculated as either $2bd/3$ or $5bd/6$, depending upon how resistance to cross-sectional warping due to transverse shear is accounted for in the assumed behavioral model. The difference is small. In the same reference it is suggested that, for the wide flange section of Figure 7.12*b* bent about its horizontal axis, the equivalent shear area may be taken as

$$A_s = \frac{8b_1I_z}{bd^2 - d_1^2(b - b_1)} \tag{7.49}$$

Reference 7.10 contains several similar suggestions, including one that satisfactory approximate results for the wide flange section may be obtained by taking for A_s the area of the web.

Application of these recommendations is illustrated in Examples 7.10 and 7.11. In the first of these it is shown that, for an end-loaded cantilever beam of rectangular cross section, the effect of shearing deformation is small, even for a relatively deep beam. The second example illustrates that the ratio of equivalent shear area to full area may be relatively small in wide flange sections, indicating that shearing deformation may require more frequent consideration in members of this type. In each of the sections considered, the two recommendations for equivalent area yielded approximately the same results, indicating that either should be satisfactory in design.

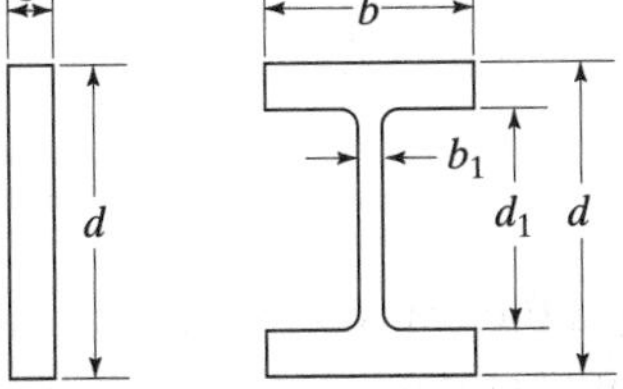

(*a*) (*b*) **Figure 7.12** Cross section dimensions.

EXAMPLE 7.10

The beam shown has a rectangular cross section. Illustrate the effect of depth-to-span ratio on the relative magnitude of flexural and shearing deformation.

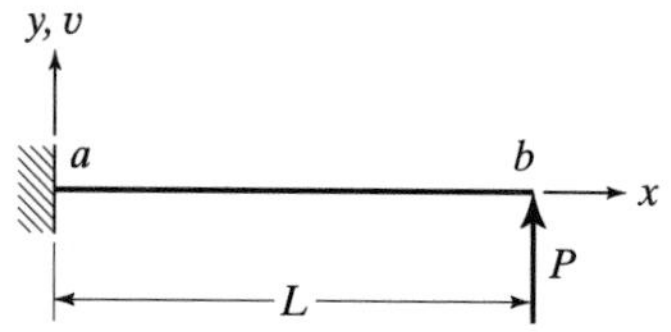

From Equation 7.47

$$v_b = \frac{PL^3}{3EI}\left(1 + \frac{3EI}{GL^2A_s}\right)$$

For

$$I = \frac{bd^3}{12}, \qquad A_s = \frac{2}{3}bd, \qquad \text{and } \nu = 0.3$$

$$v_b = \frac{PL^3}{3EI}\left(1 + 0.975\left(\frac{d}{L}\right)^2\right) = \gamma\frac{PL^3}{3EI}$$

where,

$$\text{for } \frac{d}{L} = \frac{1}{5}, \qquad \gamma = 1.039$$

$$\text{for } \frac{d}{L} = \frac{1}{10}, \qquad \gamma = 1.010$$

EXAMPLE 7.11

Calculate A_s for W27 × 102 and W10 × 100 sections using Eq. 7.49. Compare with the web area.

Sections:

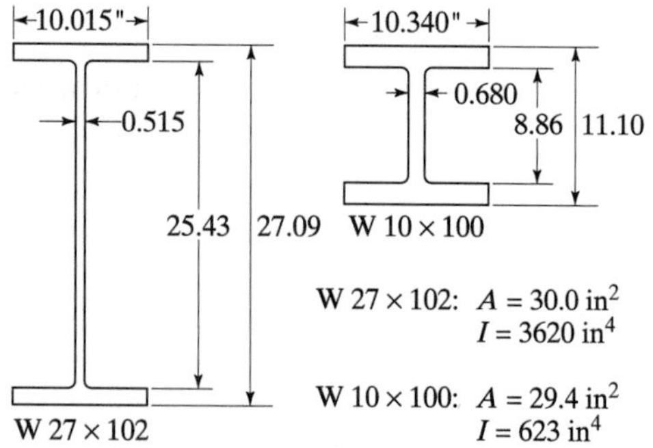

W27 × 102:

Equation 7.49 $$A_s = \frac{8(0.515)(3620)}{[(10.015)(27.09)^2 - (25.43)^2(10.015 - 0.515)]} = 12.37 \text{ in.}^2$$

Web area $$A_s = (0.515)(25.43) = 13.10 \text{ in.}^2$$

W10 × 100:

Equation 7.49 $$A_s = \frac{8(0.680)(623)}{[(10.34)(11.10)^2 - (8.86)^2(10.34 - 0.680)]} = 6.57 \text{ in.}^2$$

Web area $$A_s = (0.680)(8.86) = 6.02 \text{ in.}^2$$

7.6.2.2 The Circular Ring Beam To illustrate the application of the principle of virtual forces to a more general line element we will formulate a flexibility matrix and define the relationship needed to construct the corresponding stiffness matrix for an arc of the circular ring beam of Figure 7.13. Elementary flexure theory will be used; axial behavior, transverse shear deformation, and curved beam theory will be disregarded.

A stable, statically determinate support condition is defined by fixing point 2. The applicable expression for the internal virtual work is Equation 6.34:

$$\int_0^{\psi} \delta M_z \left(\frac{M_z}{EI_z}\right) ds$$

where $ds = R\,d\phi$. From statics, the relationship for the internal bending moment, M_z is

$$M_z = -F_{x1}\,R\sin\phi + F_{y1}R(1 - \cos\phi) - M_{z1}$$

or, in matrix form,

$$M_z = \lfloor -\sin\phi \quad (1-\cos\phi) \quad -1 \rfloor \begin{Bmatrix} F_{x1}\cdot R \\ F_{y1}\cdot R \\ M_{z1} \end{Bmatrix} = \lfloor \mathbf{Q} \rfloor \{\mathbf{F}_f\}$$

The virtual forces are chosen of the same form:

$$\delta M_z = \lfloor -\sin\phi \quad (1-\cos\phi) \quad -1 \rfloor \begin{Bmatrix} \delta F_{x1}\cdot R \\ \delta F_{y1}\cdot R \\ \delta M_{z1} \end{Bmatrix} = \lfloor \mathbf{Q} \rfloor \{\delta\mathbf{F}_f\}$$

We can now write

$$\delta W^*_{\text{int}} = \int_0^{\psi} \delta M_z (M_z/EI_z) R\,d\phi$$

$$= \frac{R}{EI_z} \lfloor \delta F_{x1}R \quad \delta F_{y1}R \quad \delta M_{z1} \rfloor \left[\int_0^{\psi} \begin{Bmatrix} -\sin\phi \\ (1-\cos\phi) \\ -1 \end{Bmatrix} \lfloor -\sin\phi \quad (1-\cos\phi) \quad -1 \rfloor \,d\phi \right] \begin{Bmatrix} F_{x1}R \\ F_{y1}R \\ M_{z1} \end{Bmatrix}$$

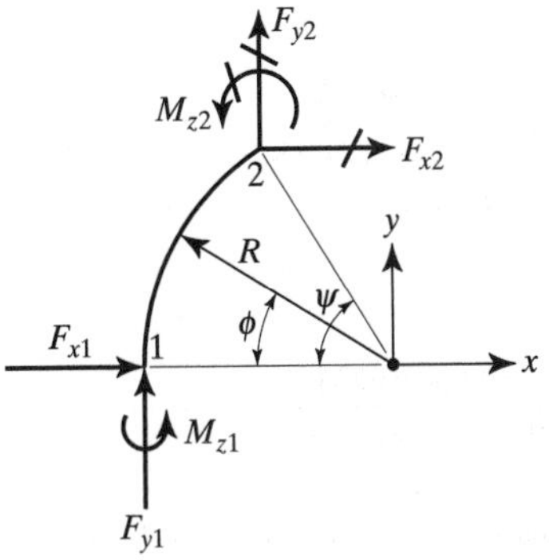

Figure 7.13 Ring beam.

which, upon integration, gives $\lfloor \delta \mathbf{F}_f \rfloor [\mathbf{d}] \{\mathbf{F}_f\}$ where

$$[\mathbf{d}] = \frac{R}{EI_z}\begin{bmatrix} \frac{\psi}{2} - \frac{1}{4}\sin 2\psi & \cos\psi - \frac{3}{4} - \frac{1}{4}\cos 2\psi & 1 - \cos\psi \\ & \frac{3}{2}\psi - 2\sin\psi + \frac{1}{4}\sin 2\psi & \sin\psi - \psi \\ \text{Sym} & & \psi \end{bmatrix} \tag{7.50}$$

and $[\mathbf{d}]$ is the desired flexibility matrix. For convenience we have chosen to write the joint direct forces as $F_{x1}R$ and $F_{y1}R$ rather than F_{x1} and F_{y1}. The corresponding displacement vector is, therefore, $\lfloor u_1/R \quad v_1/R \quad \theta_{z1} \rfloor$.

The establishment of the element stiffness matrix requires, in addition to the inverse of $[\mathbf{d}]$, the static equilibrium relationship, or

$$\begin{Bmatrix} F_{x2}R \\ F_{y2}R \\ M_{z2} \end{Bmatrix} = \begin{bmatrix} -1 & 0 & 0 \\ 0 & -1 & 0 \\ -\sin\psi & 1-\cos\psi & -1 \end{bmatrix} \begin{Bmatrix} F_{x1}R \\ F_{y1}R \\ M_{z1} \end{Bmatrix} \tag{7.51}$$

Application of these equations is illustrated in Example 7.12.

EXAMPLE 7.12

For the circular arch shown, $I = 150 \times 10^6$ mm^4, $E = 200{,}000$ MPa

1. Determine an element stiffness matrix
2. Analyze the structure

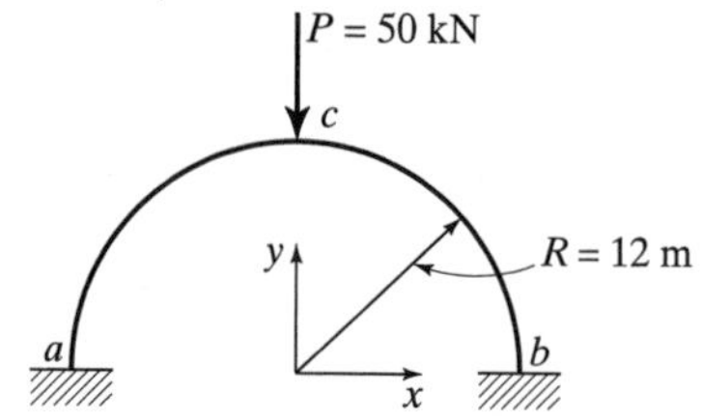

1. From Equations 7.50 and 7.51, for $\psi = 90°$

$$[\mathbf{d}] = \frac{R}{4EI}\begin{bmatrix} \pi & -2 & 4 \\ -2 & (3\pi - 8) & (4 - 2\pi) \\ 4 & (4 - 2\pi) & 2\pi \end{bmatrix} \qquad [\boldsymbol{\Phi}] = \begin{bmatrix} -1 & 0 & 0 \\ 0 & -1 & 0 \\ -1 & 1 & -1 \end{bmatrix}$$

From Equation 4.25

$$[\mathbf{k}] = 10^7\begin{bmatrix} 10.719 & 9.843 & -3.247 & -10.719 & -9.843 & 2.371 \\ & 10.719 & -2.371 & -9.843 & -10.719 & 3.247 \\ & & 1.365 & 3.247 & 2.371 & -0.429 \\ \text{Sym} & & & 10.719 & 9.843 & -2.371 \\ & & & & 10.719 & -3.247 \\ & & & & & 1.365 \end{bmatrix}$$

2. From symmetry, the only nonzero displacement is v_c. Therefore, analyzing one half the structure:

Displacements: $\{\boldsymbol{\Delta}_f\} = [\mathbf{k}_{ff}]^{-1}\{\mathbf{P}_f\}$

$$\frac{v_c}{R} = \frac{1}{10.719 \times 10^7}(-25R) \therefore v_c = -33.59 \text{ mm}$$

Reactions: $\{\mathbf{P}_s\} = [\mathbf{k}_{sf}]\{\boldsymbol{\Delta}_f\}$

$$\therefore R_{xa} = 22.96 \text{ kN},\ R_{ya} = 25 \text{ kN},\ \text{M}_{za} = -66.4 \text{ kN} \cdot \text{m}$$

Moment diagram:

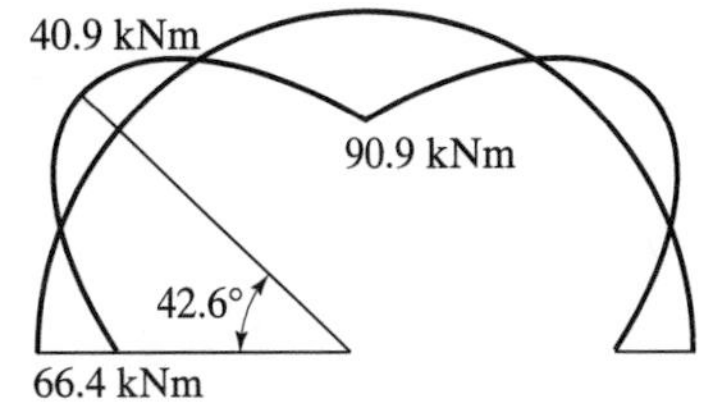

7.7 PROBLEMS

7.1 Using the quartic polynomial, $v = a_1 + a_2x + a_3x^2 + a_4x^3 + a_5x^4$, construct the shape function for the illustrated flexural element.

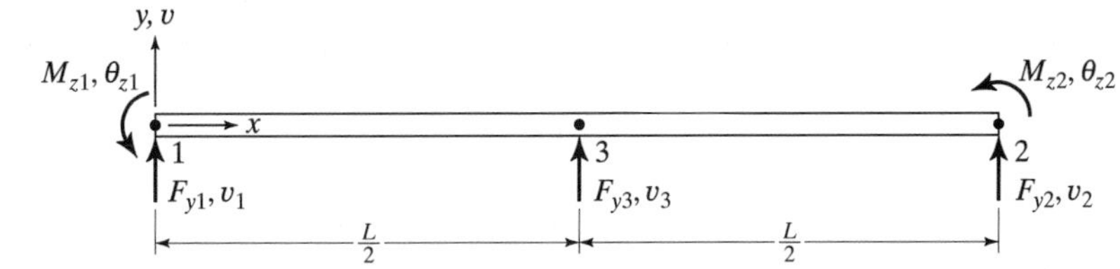

Problem 7.1

7.2 Formulate the shape function for the four-jointed axial member using $u = a_1 + a_2x + a_3x^2 + a_4x^3$.

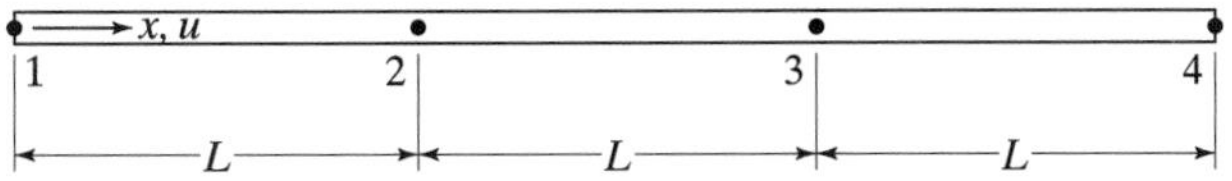

Problem 7.2

7.3 Using the quintic polynomial $v = a_1 + a_2x \cdots a_6x^5$, construct the shape function for a beam segment of length L. The node point degrees of freedom are

$$v_i \qquad \left.\frac{dv}{dx}\right|_i \qquad \left.\frac{d^2v}{dx^2}\right|_i \qquad (i = 1, 2)$$

7.4 Complete Example 7.2 by establishing the explicit form of the shape functions for v in terms of v_1, v_2, θ_{z1}, θ_{z2}.

7.5 Formulate the shape functions for the flexural element of Figure 7.3 on the basis of the following:

$$v = a_1 + a_2x + a_3 \sin\frac{\pi x}{2L} + a_4 \sin\frac{3\pi x}{2L}$$

7.6 Formulate the stiffness matrix for the element of Problem 7.1.

7.7 Formulate the stiffness matrix for a four-noded torsional element, using for the rotational displacement θ_x the following shape function.

$$\theta_x = N_1\theta_{x1} + N_2\theta_{x2} + N_3\theta_{x3} + N_4\theta_{x4}$$

where

$$N_1 = \frac{(L - x)(2L - x)(3L - x)}{6L^3} \qquad N_3 = \frac{x(L - x)(x - 3L)}{2L^3}$$

$$N_2 = \frac{x(2L - x)(3L - x)}{2L^3} \qquad N_4 = \frac{x(L - x)(2L - x)}{6L^3}$$

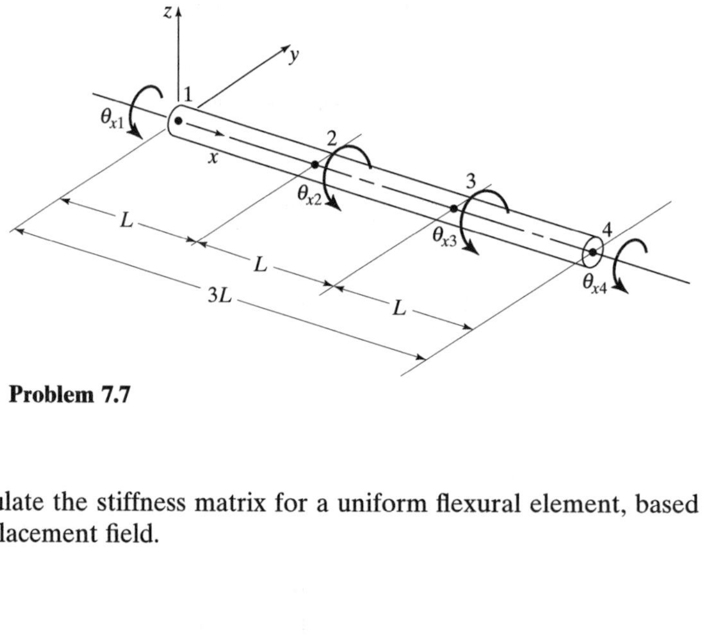

Problem 7.7

7.8 Formulate the stiffness matrix for a uniform flexural element, based on the following displacement field.

$$\begin{aligned} v = &\left[1 - \frac{3}{4}\left(\sin\frac{\pi x}{2L} - \frac{\pi x}{2L}\right) + \frac{1}{4}\left(\sin\frac{3\pi x}{2L} - \frac{3\pi x}{2L}\right)\right]v_1 \\ &+ \left[\frac{3}{4}\left(\sin\frac{\pi x}{2L} - \frac{\pi x}{2L}\right) - \frac{1}{4}\left(\sin\frac{3\pi x}{2L} - \frac{3\pi x}{2L}\right)\right]v_2 \\ &+ \left[x + \frac{L}{2\pi}\left(\sin\frac{\pi x}{2L} - \frac{\pi x}{2L}\right) + \frac{L}{2\pi}\left(\sin\frac{3\pi x}{2L} - \frac{3\pi x}{2L}\right)\right]\theta_{z1} \\ &- \left[\frac{L}{2\pi}\left(1 + \frac{3\pi}{2}\right)\left(\sin\frac{\pi x}{2L} - \frac{\pi x}{2L}\right) + \frac{L}{2\pi}\left(1 - \frac{\pi}{2}\right)\left(\sin\frac{3\pi x}{2L} - \frac{3\pi x}{2L}\right)\right]\theta_{z2} \end{aligned}$$

7.9 The governing differential equation for beam bending, representing the condition of vertical force equilibrium, is $(d^2/dx^2)EI_z(d^2v/dx^2) = 0$. Test the shape function and moment of inertia variation used in Example 7.3 for satisfaction of this condition.

7.10 Establish the exact displacement expression and shape functions for the given tapered axial member. Formulate the relevant stiffness matrix.

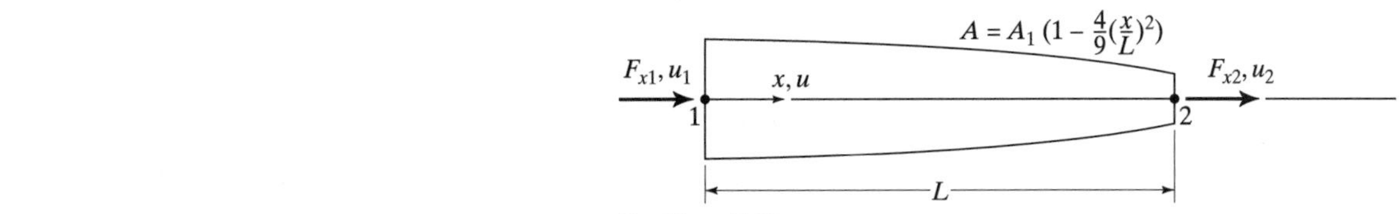

Problem 7.10

7.11 Formulate the stiffness matrix for the tapered axial member of Problem 7.10 using the shape function of Equation 7.6.

7.12 Form the stiffness equations for F_{x2} and F_{x3} for the illustrated three-noded axial member, using for the total displacement the expression:

$$u = \left(1 - \frac{3x}{L} + \frac{2x^2}{L^2}\right)u_1 + 4\left(\frac{x}{L} - \frac{x^2}{L^2}\right)u_2 + \left(\frac{2x^2}{L^2} - \frac{x}{L}\right)u_3$$

and for the virtual displacement (δu) the following expression.

$$\delta u = \frac{1}{\sqrt{2}}\left[\sin\frac{\pi x}{2L} + \sin\frac{3\pi x}{2L}\right]\delta u_2 + \frac{1}{2}\left[\sin\frac{\pi x}{2L} - \sin\frac{3\pi x}{2L}\right]\delta u_3$$

Discuss the implications of the result.

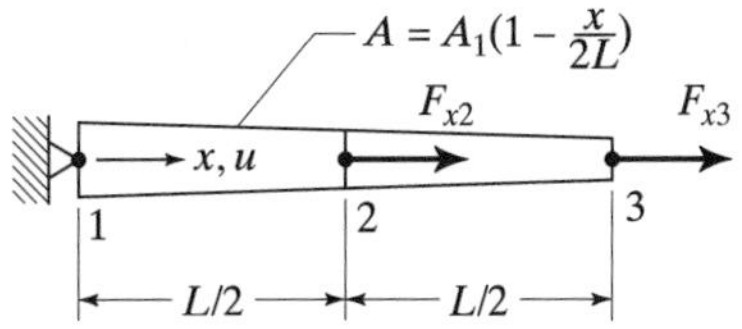

Problem 7.12

7.13 Compare the torsional response of a W10 × 45 and a W21 × 44 section for warping free and warping fixed end conditions. $E = 29{,}000$ ksi and $\nu = 0.3$.

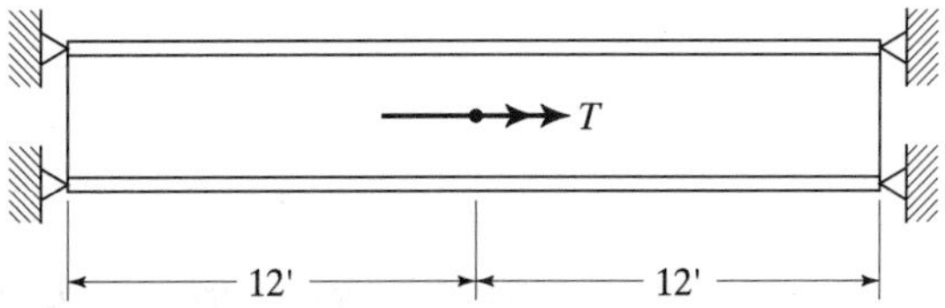

Problem 7.13

7.14 Repeat Problem 7.13 for warping fixed conditions at one end and warping free at the other.

7.15 An axial member is subjected to distributed loads as illustrated. The shape functions for Case a of the figures are the linear expressions $(1 - x/L)$ and x/L. For Case b they are given by u in Problem 7.12. Discuss the relationship of the results obtained with those established by simple prorating of forces to the joints.

$$q(x) = q_1\left(1 + \frac{x}{2L} + \left(\frac{x}{2L}\right)^2\right) \quad \text{and} \quad q(x) = q_2\left(\sin\frac{2\pi x}{L} + \sin\frac{4\pi x}{L}\right)$$

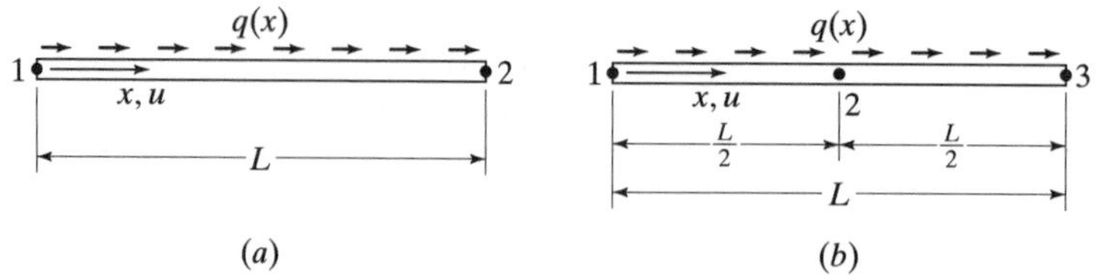

Problem 7.15

7.16 Formulate, by use of the work-equivalent load approach, the effective nodal forces for the beam flexure in selected cases of Table 5.1.

7.17 Formulate, by use of the work-equivalent load approach, the effective nodal forces for the beam flexure case for the following loads in the interval 1-2.

(a) $q = q_{nm}\left[1 - \left(\frac{x}{L}\right)^n\right]\left(\frac{x}{L}\right)^m$

(b) $q = q_n \sin^2 \frac{n\pi x}{L}$

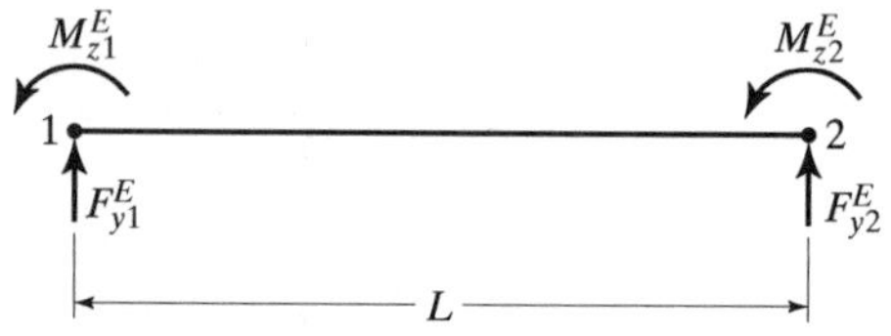

Problem 7.17

7.18 A space-frame element is subjected to the illustrated distributed loads. Calculate the relevant effective nodal forces.

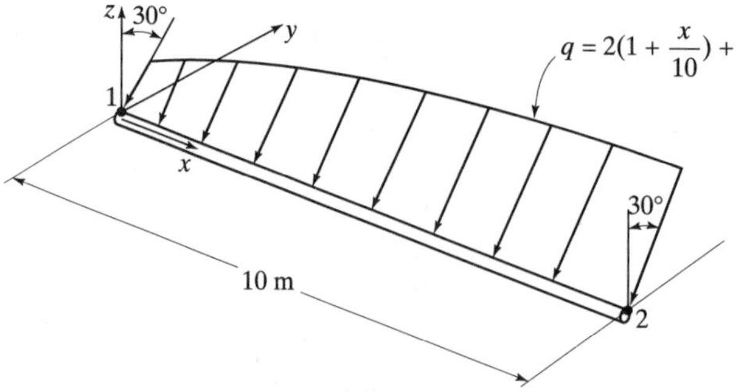

Problem 7.18

7.19 The axial element of Example 7.1 is subjected to a temperature change given by $T = (1 - x/2L)(T_1) + x/2L(T_3)$, where (T_1) and (T_3) are the temperature changes at points 1 and 3. Calculate the relevant initial force vector.

7.20 Calculate, for the tapered flexural element of Example 7.2, the vector of initial forces for an initial transverse displacement of the form:

$$v^i = v_A^i \sin \frac{\pi x}{L} + v_B^i \sin \frac{3\pi x}{L}$$

where v_A^i and v_B^i are measured amplitudes of initial displacement.

7.21 The temperature distribution across the depth of practical flexural elements is usually nonlinear. Also, such members are not symmetric about the z axis. Using the symbolism shown, formulate the expression for initial curvature, κ^i, to be used in the calculation of initial forces by means of Equation 7.38.

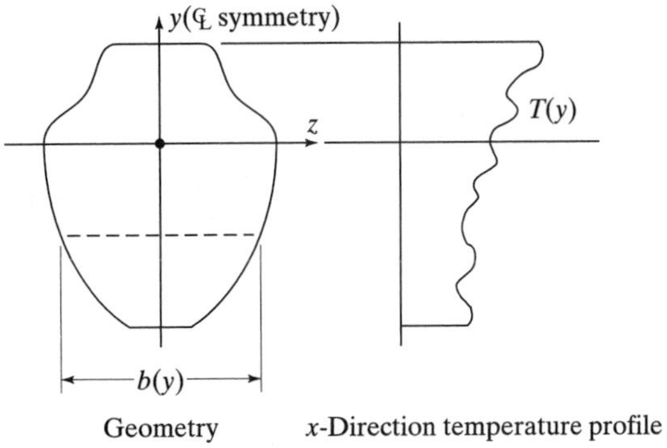

Problem 7.21

7.22 Develop a flexibility matrix for the beam shown.

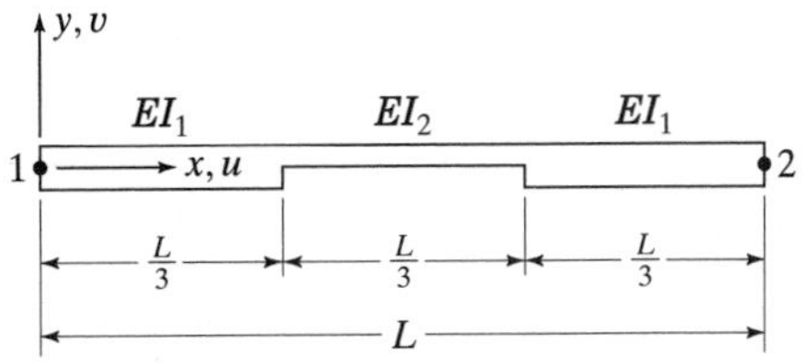

Problem 7.22

7.23 Formulate the flexibility matrix for the ring element of Section 7.6.2.2, accounting for both simple flexure and axial deformation. Designate the cross-sectional area of the member as A.

7.24 Construct the final algebraic form of the stiffness relationships for the ring beam element of Example 7.12.

7.25 Formulate the flexibility matrix for the element shown below. Consider only simple flexural behavior. The moment of inertia I and area A are constant.

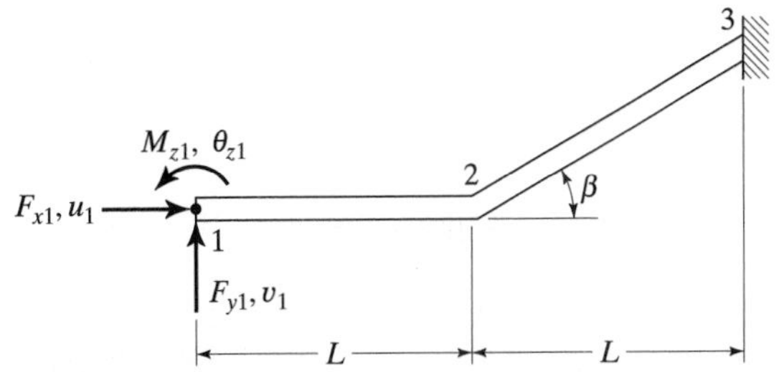

Problem 7.25

7.26 Formulate the flexibility matrix for the element shown below. Consider only flexural and torsional behavior. All section properties are constant.

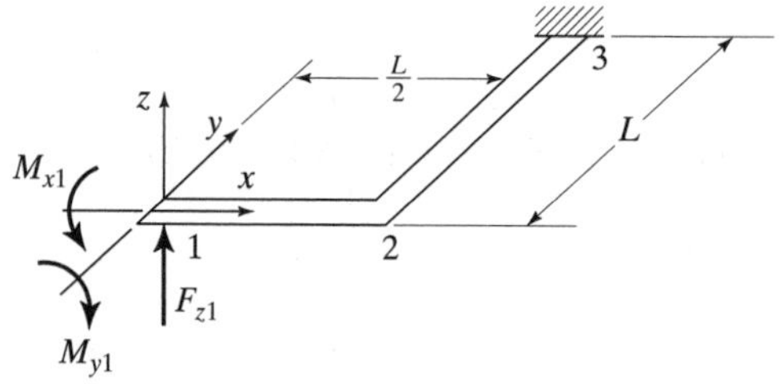

Problem 7.26

7.27 The wide flange beam shown is loaded by a concentrated force at the center. Calculate the length ratio, L/h, for which the central deflection caused by shear becomes 20% of that due to bending.

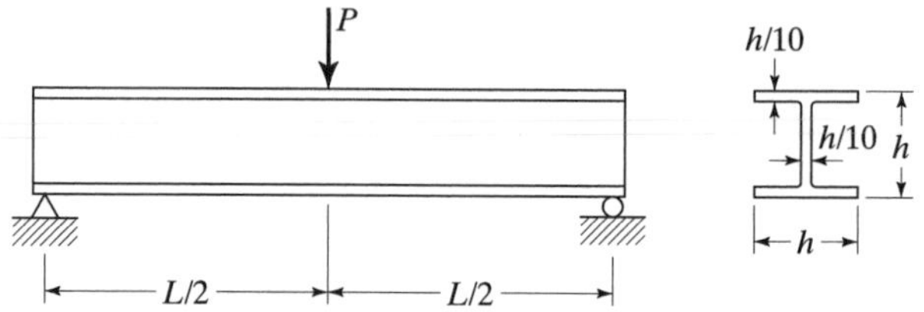

Problem 7.27

7.28 Repeat Problem 7.27 for a uniformly distributed load, q, instead of the concentrated load.

7.29 Calculate the displacement under the load at point A of the continuous circular arch shown. Consider only flexural behavior.

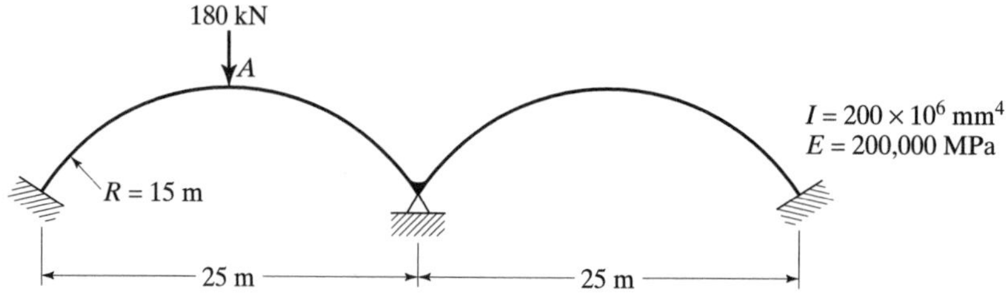

Problem 7.29

REFERENCES

7.1 R. H. Gallagher, *Finite Element Analysis Fundamentals* Prentice-Hall, Englewood Cliffs, N.J., 1975.

7.2 V. Z. Vlasov, *Thin Walled Elastic Beams*, 2nd Edition, Israel Program for Scientific Translations, Jerusalem, 1961.

7.3 S. P. Timoshenko and J. M. Gere, *Theory of Elastic Stability*, 2nd Edition, McGraw-Hill, New York, 1961.

7.4 P. A. Seaburg and C. J. Carter, *Torsional Analysis of Structural Steel Members*, Steel Design Guide Series 9, American Institute of Steel Construction, Chicago, 1996.

7.5 R. S. Barsoum and R. H. Gallagher, "Finite Element Analysis of Torsional and Torsional-Flexural Stability Problems," *Intl Jl for Num Meth in Engr*, Vol. 2, 335–352, 1970.

7.6 Y.-B. Yang and W. McGuire, "A Procedure for Analyzing Space Frames with Partial Warping Restraint," *Intl Jl for Num Meth in Engr*, Vol. 20, 1377–1398, 1984.

7.7 D. A. Nethercot and K. C. Rockey, "A Unified Approach to the Elastic Lateral Buckling of Beams" *Engr Jl, Amer. Inst of Steel Const.*, Vol. 9, No. 3, 1972.

7.8 R. S. Barsoum, "A Finite Element Formulation for the General Stability Analysis of Thin Walled Members," Ph.D. Thesis, Cornell University, Ithaca, N.Y., 1970.

7.9 S. Timoshenko, *Strength of Materials, Part I*, 3rd Edition, Van Nostrand, Princeton, N.J., 1955.

7.10 R. J. Roark and W. C. Young, *Formulas for Stress and Strain*, 5th Edition, McGraw-Hill, New York, 1975.

Chapter 8

Nonlinear Analysis of Frames—An Introduction

In this chapter we outline some of the ways in which the direct stiffness method of matrix analysis may be used in the nonlinear analysis of framed structures. The chapter starts with a brief discussion of nonlinear behavior, the aims of nonlinear analysis, and its place in structural engineering. Sources of nonlinearity that are of common concern in frame design are listed, and the levels of analysis to be considered are defined. To provide perspective and a basis for comparison with the matrix methods, established classical solutions of some elementary problems are illustrated. Following this a symbolic matrix stiffness method equation for each of the levels of analysis is presented. Strategies for the numerical solution of nonlinear equations are outlined, but details of equation solution are left to Chapter 12. The chapter ends with a brief discussion of problems associated with formulating the equations of equilibrium on the deformed structure, but a more thorough exploration of equation formulation is deferred to Chapters 9 and 10.

8.1 NONLINEAR BEHAVIOR, ANALYSIS, AND DESIGN

Most civil engineering structures behave in a linear elastic fashion under service loads. Exceptions are slender structures such as some suspension systems, arches and tall buildings, and structures subject to early localized yielding or cracking. But prior to reaching their limit of resistance, almost all structures would exhibit significant nonlinear response. Therefore, if linear elastic analysis is the highest level available, the design engineer must find another way to account for the effects that the analysis is incapable of simulating. The answer may lie in any of the following: a) individual judgment; b) code formulas that accept the results of a linear elastic or simpler analysis and make allowance for nonlinearity in some empirical or semi-empirical way; or c) supplementary theoretical or experimental studies.

In nonlinear analysis an attempt is made to improve the analytical simulation of the behavior of a structure in some respect. The fundamental aim is to improve the quality of design by providing the engineer with a more reliable prediction of the performance of a system that is under design or investigation. In making this closer link between structural analysis and actual behavior, the traditional distinction between the terms "analysis"—the determination of forces and displacements under given loads—and "design"—the proportioning of members and connections to resist the determined effects—becomes blurred (see Ref. 8.11). We will emphasize the analytical side of the problem but the determination of some aspect of behavior is the primary objective in the structures we consider.

Reflecting on the contents of the previous chapters we can see that, in linear elastic analysis, there is art in the reduction of the actual frame to a line diagram and in the assignment of boundary conditions and member properties. But when that has been done, the result is an analytical problem that has a unique solution, the one that satisfies the requirements of compatibility and equilibrium on the undeformed structure. That is all that is meant by the term *exact solution* in reference to a linear analysis. In the two preceding chapters we have seen that there may be mathematical difficulties in the formulation of the linear equations and that we may have to settle for an approximation of that solution. Nevertheless, although there are many different approaches to linear elastic analysis, and although they may differ in mode and amount of computational effort, all methods that are capable of producing "an exact solution" rest on the same few principles of mechanics. But the premise of linear elastic behavior forecloses the possibility of revealing any manifestation of nonlinearity: A problem has been solved, but the solution may not tell us all we would like to know regarding the structure. Indeed, crucial information may be missing.

In using nonlinear analysis the uncertainty regarding actual behavior may be reduced. In the process, however, the element of art in modeling the structure and in handling the equations of analysis is increased. In modeling, the analyst must decide what sources of nonlinearity are apt to be significant and how to represent them. In dealing with the resulting nonlinear equations, decisions must be made regarding how to reduce them to a system suitable for practical computation and, then, the method for solving the reduced system. In this and the following two chapters, procedures for the treatment of common situations will be presented. But the conditions encountered in practice are so varied, and the devices for dealing with them so many, that the coverage cannot be comprehensive.

8.1.1 Sources of Nonlinearity

In linear elastic analysis the material is assumed to be unyielding and its properties invariable, and the equations of equilibrium are formulated on the geometry of the unloaded structure or, in the case of self-strained structures, on an initial reference configuration (see Section 5.3). Subsequent deformations are assumed to be so small as to be insignificant in their effect on the equilibrium and mode of response of the system. One consequence of this was our ability to treat axial force, bending moments, and torques as uncoupled actions in developing the stiffness equations for elements of bisymmetrical section in Chapters 4 and 7.

Nonlinear analysis offers several options for addressing problems resulting from the above assumptions. We may consider only the *geometric nonlinearity*. That is, we may continue to treat the structural material as elastic but include the effects of deformations and finite displacements in formulating the equations of equilibrium. It is also possible to consider only the *material nonlinearity*, that is, the effect of changes in member material properties under load. And, as a third general option, we may include effects of both *geometric* and *material nonlinearity* in the analysis. In each case, the possibility of coupling of internal actions must be considered; it may be a dominant feature of the analysis. Among the many sources of each class of nonlinearity are the following.

Geometrical effects:

1. Initial imperfections such as member camber and out-of-plumb erection of a frame.

2. The P-Δ effect, a destabilizing moment equal to a gravity load times the horizontal displacement it undergoes as a result of the lateral displacement of the supporting structure.
3. The P-δ effect, the influence of axial force on the flexural stiffness of an individual member.[1]

Material effects:

1. Plastic deformation of steel structures.
2. Cracking or creep of reinforced concrete structures.
3. Inelastic interaction of axial force, bending, shear, and torsion.

Combined effects:

1. Plastic deformation plus P-Δ and/or P-δ effects.
2. Connection deformations.
3. Panel zone deformations.
4. Contributions of infilling and secondary systems to strength and stiffness.

8.1.2 Levels of Analysis

Rarely, if ever, is it possible to model all sources of nonlinearity and portray the actual behavior of a practical structure in all of its detail. Normally, the problem is one of selecting a method that falls short of the ideal in one way or another but that does provide adequate analytical simulation of the case at hand. The most common levels of analysis are represented in Figure 8.1 by schematic response curves for a statically loaded frame. The degree to which they can model true behavior differs, but each can yield information of value to the engineer.

By definition, *first-order (linear) elastic analysis* excludes nonlinearity, but it generally represents conditions at service loads very well.

The *elastic critical load* is the load at which both the original and an alternative loading path become mathematically valid[2] and it can be shown that the path taken from that point will be the alternative one. The load is commonly determined from an *eigenvalue analysis* of an idealized elastic model of the structure[3]. The analysis also yields the *eigenvector*, that is, the shape that the system assumes in the post-critical state, but it doesn't determine its amplitude. The *inelastic critical load* is similarly defined and calculated, but it is one in which the possibility of precritical inelastic material behavior is considered in the analysis. A critical load analysis may not include the nonlinear phenomena that must be taken into account to determine accurately the magnitude of the load that would cause failure, but frequently, it gives adequate—and valuable—pictures of possible types of failure.

In *second-order elastic analysis*[4] the effects of finite deformations and displacements

---

[1]The P-Δ and P-δ effects described here may be encountered and treated as distinct phenomena so we define them as such. But they often occur in combination, justifying a blanket "P-delta" definition such as "the secondary effect of column axial loads and lateral deflection on the moments in members" (from the Glossary of Reference 8.1).

[2]This state is generally referred to as *bifurcation* of the loading path.

[3]In this and other comparable methods of analysis it may be found that a number of values of the applied load satisfy the bifurcation condition (e.g., see Example 8.3). Each may be called a "critical load." Here, we are reserving the term for the smallest physically meaningful value.

[4]Following what has become common in civil engineering structural analysis, we use the term *second-order* not in a precise mathematical sense but merely as shorthand to designate any method in which the effects referred to are accounted for.

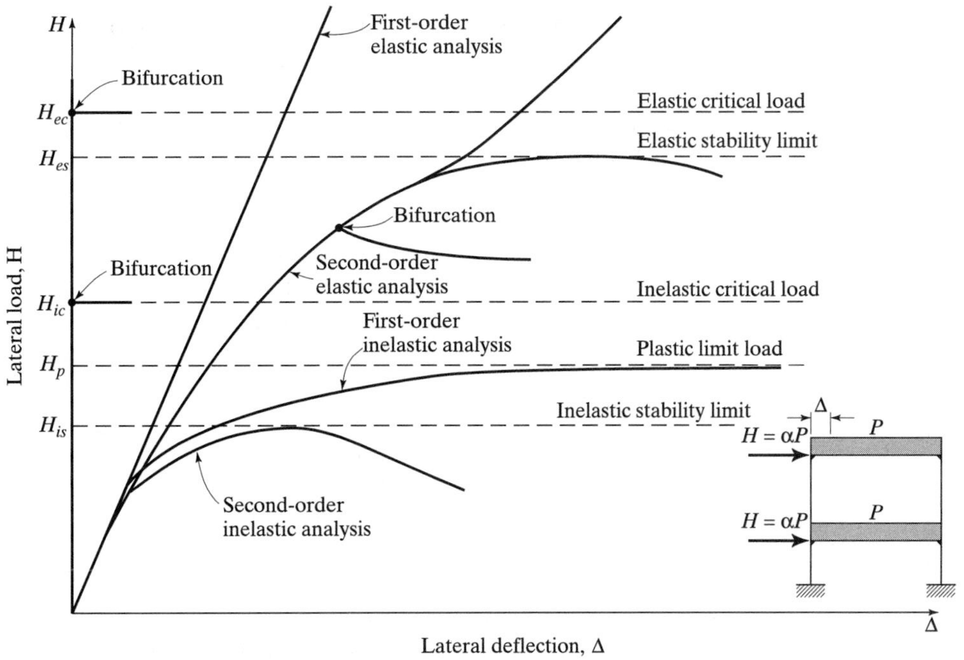

Figure 8.1 Levels of analysis.

of the system are accounted for in formulating the equations of equilibrium. A second-order elastic analysis can produce an excellent representation of destabilizing influences such as the P-Δ effect, but it has no provisions for detecting material nonlinearity. Several of the possible modes of nonlinear elastic behavior are indicated in Figure 8.1: (a) bifurcation (branching) of the loading path with the system following an alternative path in the post-critical state; (b) gradually increasing nonlinearity culminating in elastic instability at a limit point; and (c) increasing stiffness either from the onset of loading or, as shown, following a period of gradual softening.

In *first-order inelastic analysis* the equations of equilibrium are written in terms of the geometry of the undeformed structure. Inelastic regions can develop gradually or, if the plastic hinge concept is adopted, as abrupt changes in the structure's response. When the destabilizing effects of finite displacements are relatively insignificant, first-order inelastic analysis can produce an excellent representation of simple elastic-plastic behavior and failure through mechanism formation, that is, the simple *plastic limit load* of Figure 8.1. But it has no provisions for detecting geometric nonlinear effects and, of paramount concern, their influence on the stability of the system.

In *second-order inelastic analysis* the equations of equilibrium are written in terms of the geometry of the deformed system. It has the potential for accommodating all of the geometric, elastic, and material factors that influence the response of a structure. Thus, in principle—and in a deterministic sense—it enables the preparation of analytical models capable of faithfully simulating actual behavior and calculating the inelastic stability limit, that is, the point at which a system's capacity for resistance to additional load is exhausted and continued deformation results in a decrease in load-resisting capacity.

The path of increasingly nonlinear response (elastic or inelastic) culminating in instability is probably the most common mode of failure in civil engineering structures. For this reason, methods for detecting limit points will be a major concern of this text.

8.1.3 Examples from Established Theory

Underlying contemporary nonlinear matrix analysis are long established methods of structural mechanics and applied plasticity. A few of these are applied to elementary systems in the following examples. The purpose is to illustrate types of physical behavior and some of the concepts and mathematical techniques used in dealing with them. The examples will also be used in appraising the matrix methods to be developed in subsequent chapters.

In some of the examples large changes in geometry are studied for the purpose of defining or illustrating a particular phenomenon, but it is important to stress that even at the ultimate, practical civil engineering structures are rarely called upon to resist more than *moderately large displacements* and *small strains*. Both of these terms are qualitative; they depend upon the individual situation and cannot be defined precisely. In the interpretation of the examples to follow—and in any nonlinear analysis—it may be helpful to think of a moderately large displacement as one in the range of a hundredth to a tenth of the span or height of the structure, and a small strain as anything from the infinitesimal to one of the order of ten times the proportional limit of the material.

In Example 8.1 an ideally straight and rigid member supported by an elastic spring at the top is subjected to two types of loading. In the first, in which the applied horizontal force is zero, a critical load is determined by postulating a lateral displacement and studying the equilibrium of the displaced system to determine the conditions under which that displacement could occur. The system is stable for P less than kL, but when the two become equal, the loading path can bifurcate. Since in the deflected position the overturning moment, $P\Delta$, will be greater than the restraining moment, $k\Delta L \cos\theta$, the member will deflect. Therefore $P = kL$ is the critical load.

EXAMPLE 8.1

Member ab is a rigid bar and member bc has a constant axial stiffness coefficient k. Analyze the behavior of the system for:

1. $\alpha = 0$
2. $\alpha \neq 0$

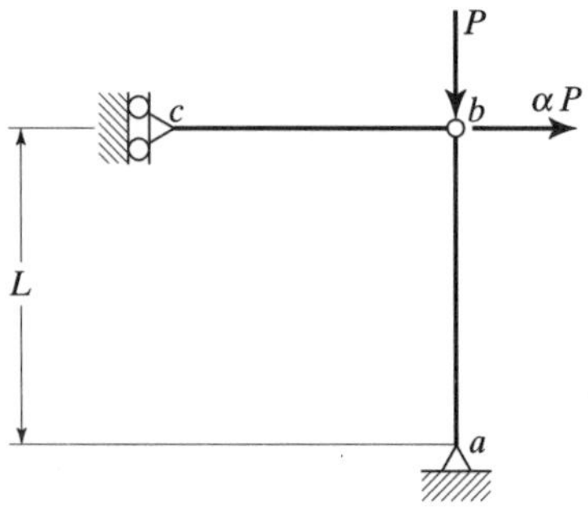

1. $\alpha = 0$. Assume the possibility of lateral displacement of b and consider the equilibrium of ab'.

ΣM_a: $P\Delta - k\Delta L \cos\theta = 0$, or $\Delta(P - kL\cos\theta) = 0$

For $\theta = 0$, two equilibrium states exist:

(1) $\Delta = 0$, the undeflected configuration

(2) $P = P_{cr} = kL$, the critical condition

The critical condition is unstable since, for $\Delta \neq 0$

$$P = kL\cos\theta < P_{cr} \qquad \text{(a)}$$

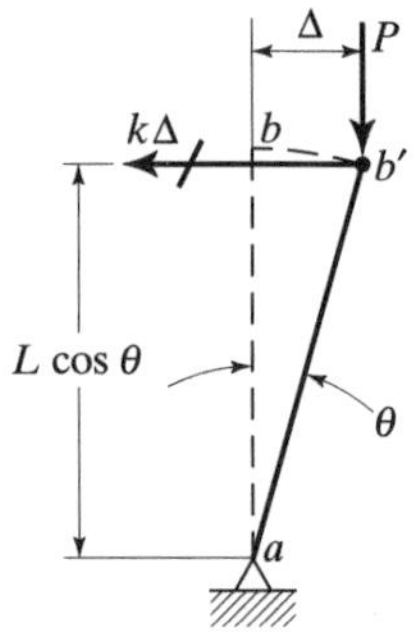

2. For $\alpha \neq 0$ there is lateral displacement from the outset

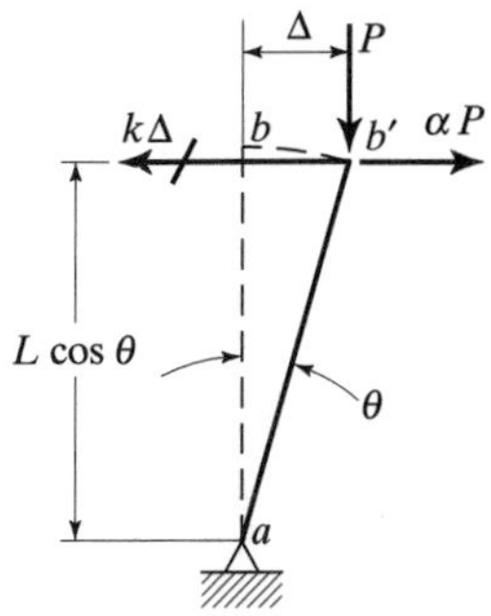

ΣM_a: $P\Delta + \alpha PL \cos\theta - k\Delta L \cos\theta = 0$ or, letting $\Delta = L \sin\theta$

$$P = \frac{kL \cos\theta}{1 + \alpha \cot\theta} \qquad \text{(b)}$$

Solving for limits: $\dfrac{d(P/kL)}{d\theta} = 0$ when

$$\theta = \tan^{-1}\sqrt[3]{\alpha} \qquad \text{(c)}$$

From Equation b, the equilibrium paths for several values of α are:

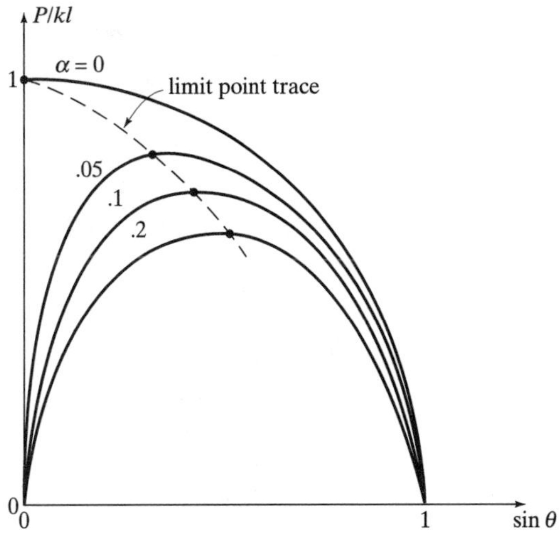

In the second case of Example 8.1, the presence of an active horizontal load results in lateral deflection from the outset. The destabilizing P-Δ effect is nonlinear and it is seen that stability limits that depend on the relative values of the horizontal and vertical loads are obtained.

The system of Example 8.2 is a shallow elastic arch consisting of two axial force members. Analysis of the geometry and the equilibrium of the displaced structure, without restricting it to small displacements, leads to the nonlinear elastic response illustrated by the solid curve. From the unloaded state, the rate of resistance to downward load decreases until, at stage 1, an elastic stability limit is reached. Deflection under decreasing load can then take place until, at stage 2, the structure is flat and unloaded but self-compressed. From this stage until stage 4, further deflection would require an upward load—with a maximum at stage 3—to equilibrate the decreasing compression in the members. At stage 4 the structure would again be unloaded and unstressed. It could then deflect further under increasing downward load. Point 1 is an obvious elastic stability limit, and the instability of response from points 1 to 3 could be demonstrated by further analysis (Example 8.8). But even without such an analysis, it is clear that attainment of equilibrium beyond point 1 would require either an external agent to control the unloading or the displacement subsequent to that stage or a constant load and a system capable of undergoing the displacements required to "snap through" to stage 5, at which point it would function as a stable suspension system under further loading.

EXAMPLE 8.2

The system is a shallow three-hinged arch. Members ab and bc have a constant axial stiffness coefficient k. Analyze the behavior of the system.

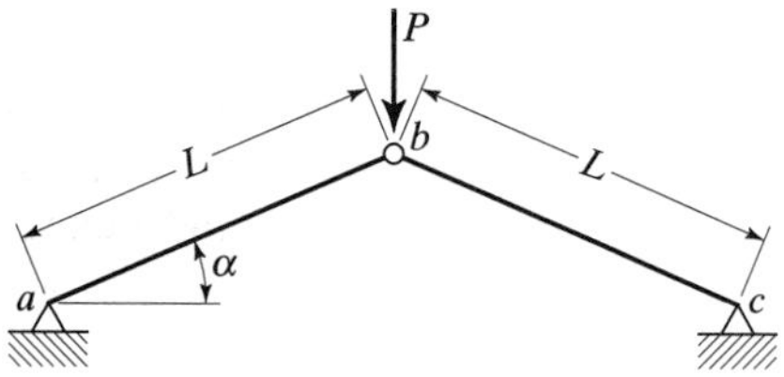

Consider the equilibrium of the displaced structure:

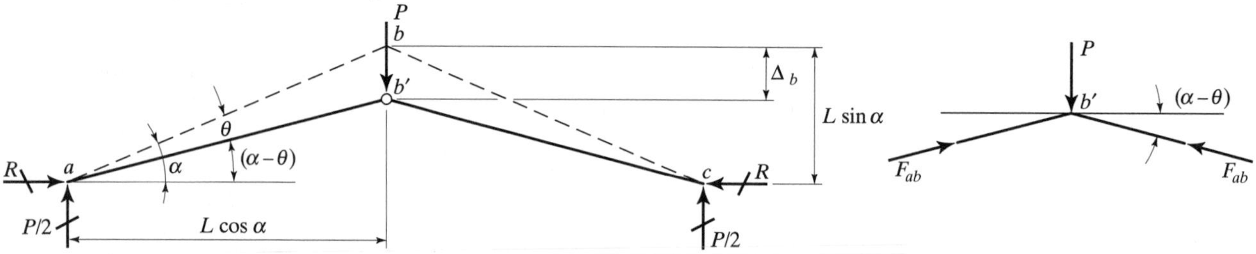

From the geometry of the displaced structure

$$\Delta_b = L[\sin\theta \sec(\alpha - \theta)] \qquad \text{(a)}$$

$$\Delta L_{ab} = L[1 - \cos\alpha \sec(\alpha - \theta)] \qquad \text{(b)}$$

From the equilibrium of joint b'

$$P = 2F_{ab}\sin(\alpha - \theta)$$

Letting $F_{ab} = k\,\Delta L_{ab}$

$$P = 2kL[\sin(\alpha - \theta) - \cos\alpha \tan(\alpha - \theta)] \qquad \text{(c)}$$

Solving for limits, $dP/d(\alpha - \theta) = 0$ when

$$\cos^3(\alpha - \theta) = \cos\alpha \qquad \text{(d)}$$

The equilibrium path for given values of α, L, and k is, from Equations a and c:

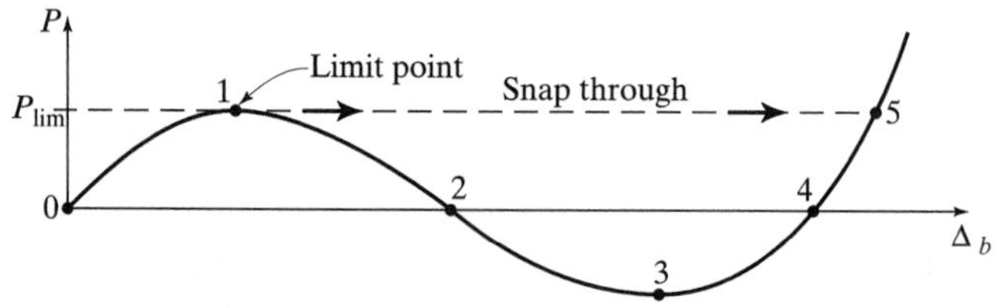

Displaced shape at key points:

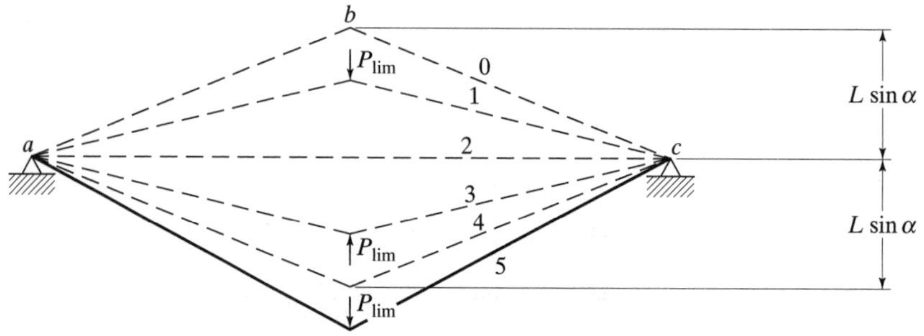

In Example 8.3 stability is provided by the flexural resistance of the member itself, and we limit consideration to the classical small displacement theory. The technique used to solve the first case is the same as that used in the first case of Example 8.1: The possibility of a small lateral displacement of an ideally straight, axially loaded member is *assumed*. Then, by studying the requirements of equilibrium for the straight and flexed states, the mathematical condition under which they can coexist is determined and, either by further mathematical analysis of the displaced condition or by logical deduction, a conclusion is reached regarding the relative stability of the two states. This technique was developed by Euler 250 years ago. Equations a–d summarize his theory and Equation c, which represents the governing state—neutral equilibrium in the deflected position—is usually called the *Euler load* in his honor.

Except for Euler's use of a physically undefined elastic constant, C, rather than the flexural rigidity, EI, Equation c comes directly from Reference 8.4. The same theory has been used as the basis for the determination of an inelastic critical load, P_{cri}, a value generally obtained by employing the *tangent modulus*, E_t, rather than E in the Euler equation. Physically, this is a means of accounting for pre-critical inelastic behavior of an ideal member (see Ref. 8.5 and Section 10.3). Computationally, an iterative solution is generally required since E_t is a function of the axial stress in the member.

EXAMPLE 8.3

Member ab is an ideally straight elastic bar. Summarize the small displacement theory of its flexural behavior for:

1. $M_0 = 0$
2. $M_0 = \alpha PL$

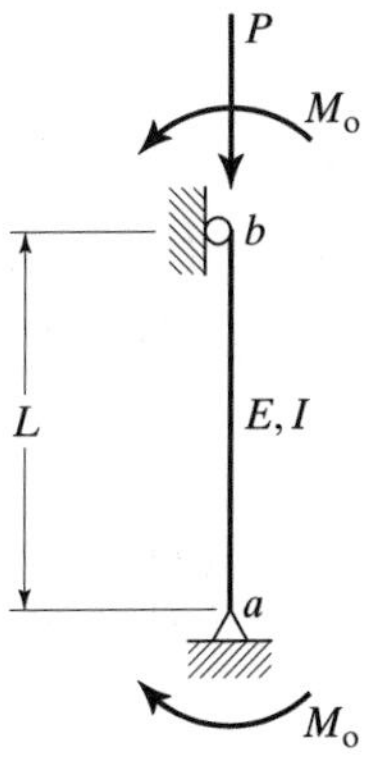

1. For $\alpha = 0$ assume the possibility of lateral displacement. The bending moment at x is $M = -Pv$ and, from small displacement theory,

$$\frac{d^2v}{dx^2} + \frac{Pv}{EI} = 0 \tag{a}$$

In the general solution

$$v = C_1 \sin\sqrt{\frac{P}{EI}}\,x + C_2 \cos\sqrt{\frac{P}{EI}}\,x \tag{b}$$

The boundary conditions $x = 0$, $v = 0$ and $x = L$, $v = 0$ yield

$$C_2 = 0 \qquad \text{and} \qquad C_1 \sin\sqrt{\frac{P}{EI}}\,L = 0$$

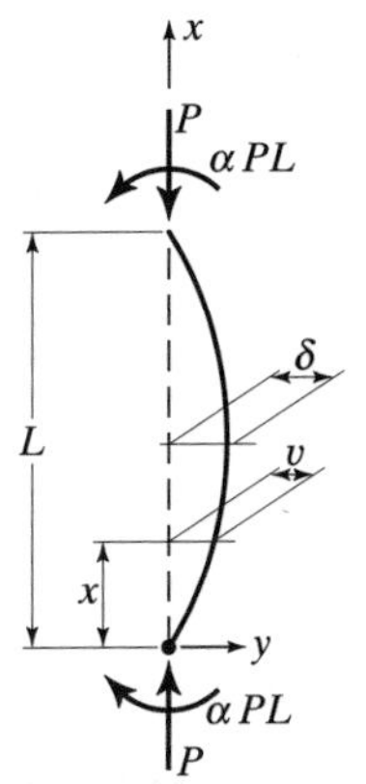

The smallest load that will admit deflection, i.e., $C_1 \neq 0$ is the Euler load

$$P_{\text{cre}} = \frac{\pi^2 EI}{L^2} \tag{c}$$

From Eq. b, the equation of the deflected curve is

$$v = C_1 \sin \frac{\pi x}{L} \tag{d}$$

2. For $\alpha \neq 0$ there is displacement from the outset. $M = -Pv - \alpha PL$, or

$$\frac{d^2v}{dx^2} + \frac{Pv}{EI} = -\frac{\alpha PL}{EI} \tag{e}$$

with the general solution

$$v = C_1 \sin\sqrt{\frac{P}{EI}}\,x + C_2 \cos\sqrt{\frac{P}{EI}}\,x - \alpha L$$

Utilizing the boundary conditions and recognizing that, by symmetry, $v_{\text{max}} = v|_{x=L/2} = \delta$

$$\delta = \alpha L\left[\sec\sqrt{\frac{P}{EI}}\frac{L}{2} - 1\right] \tag{f}$$

Graphically:

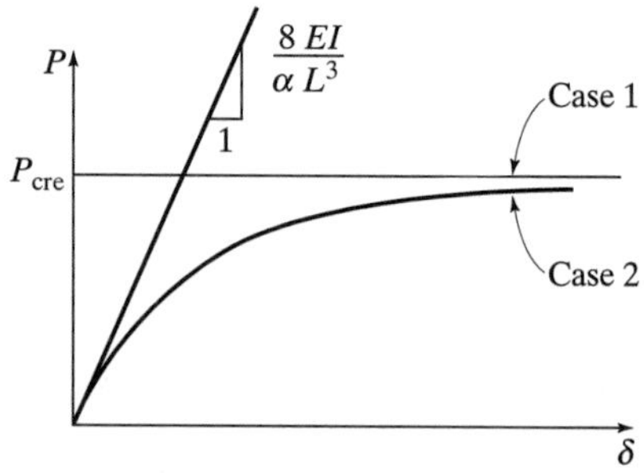

The second case of Example 8.3 also parallels that of Example 8.1 except that, whereas in Example 8.1 large displacements were accounted for, in this example small displacement theory is used. As a consequence, the calculated response becomes asymptotic to the critical load rather than exhibiting limit point characteristics. As a practical matter there can be no objection to this since any nonlinear elastic behavior encountered in realistic civil engineering structures will be represented adequately by small displacement theory.

The first two cases of Example 8.4 are identical in principle to the corresponding parts of Example 8.3. The mathematical difference is in the boundary conditions and the practical difference is that, in the second case, for members of equal length and flexural rigidity the relative influence of the axial force on lateral bending is greater in this example than in the previous one. There is a nonlinear effect in both members, but as mentioned in Section 8.1.1, it is often called the P-Δ effect when lateral translation of the applied load is possible, as in this example, and the P-δ effect when it is prevented, as in Example 8.3[5].

[5]In members with translating joints, the term P-δ is also used to describe the nonlinear deflection of the member with respect to the chord joining its two ends.

EXAMPLE 8.4

Member ab is an ideally straight elastic bar. Summarize the small displacement theory of its flexural behavior for:

1. $\alpha = 0$
2. $\alpha \neq 0$
3. $\alpha \neq 0$ and P directed upward

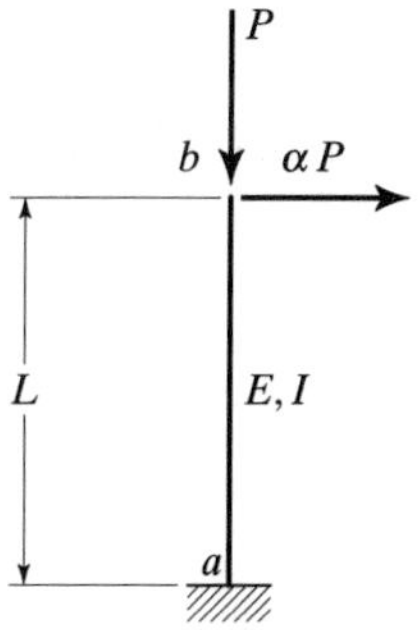

1. For $\alpha = 0$. Proceeding as in Example 8.3, the same differential equation is obtained. The boundary conditions are: $x = 0$, $v = 0$ and $x = L$, $dv/dx = 0$. Solving and satisfying the boundary conditions results in

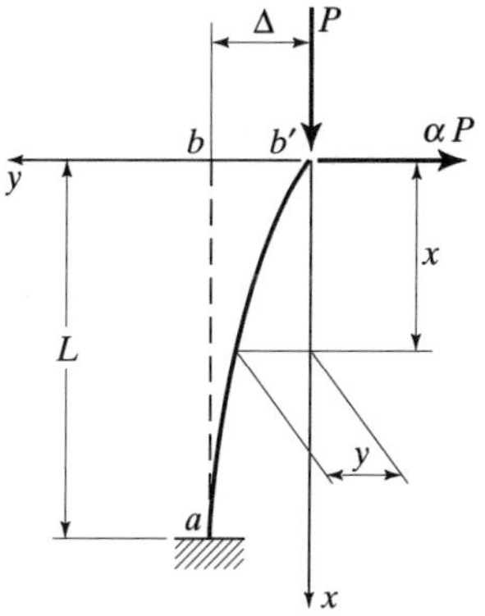

$$P_{\text{cre}} = \frac{\pi^2 EI}{4L^2} \tag{a}$$

and the equation of the deflected curve

$$v = C_1 \sin \frac{\pi x}{2L} \tag{b}$$

2. For $\alpha \neq 0$, proceeding as in Example 8.3 leads to the differential equation

$$\frac{d^2v}{dx^2} + \frac{Pv}{EI} = -\frac{\alpha P x}{EI} \tag{c}$$

and the general solution

$$v = C_1 \sin \sqrt{\frac{P}{EI}}\, x + C_2 \cos \sqrt{\frac{P}{EI}}\, x - \alpha x \tag{d}$$

The boundary conditions yield $C_2 = 0$ and $C_1 = \dfrac{\alpha}{\sqrt{\dfrac{P}{EI}} \cos \sqrt{\dfrac{P}{EI}}\, L}$

or, from Eq. (d), the deflection

$$\Delta = v|_{x=L} = \alpha L \left[\frac{\tan \sqrt{\dfrac{P}{EI}}\, L}{\sqrt{\dfrac{P}{EI}}\, L} - 1 \right] \tag{e}$$

3. For $\alpha \neq 0$ and P upward, $M = Pv - \alpha Px$ or, proceeding as above

$$v = C_1 \sinh\sqrt{\frac{P}{EI}}x + C_2 \cosh\sqrt{\frac{P}{EI}}x + \alpha x \quad \text{(f)}$$

and

$$\Delta = v|_{x=L} = \alpha L\left[1 - \frac{\tanh\sqrt{\frac{P}{EI}}L}{\sqrt{\frac{P}{EI}}L}\right] \quad \text{(g)}$$

Graphically:

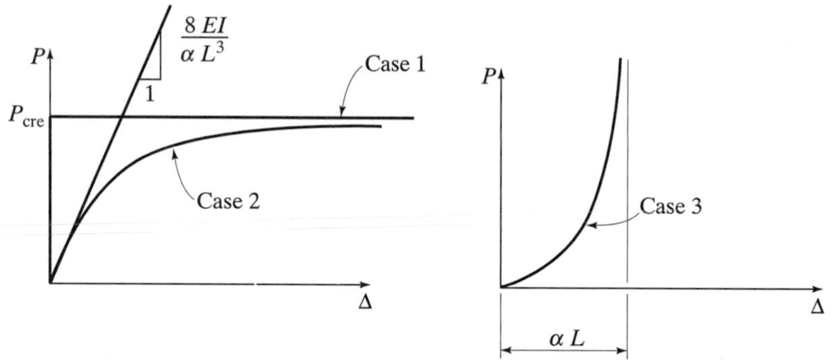

The third case of Example 8.4 is similar to the second but for the fact that the vertical component of load is a stabilizing one. As a consequence of starting from the small deflection equation of flexure, in this case the calculated response becomes asymptotic to the displacement that would obtain in a pure (unflexed) axial force member. All three cases of this example will be referred to in later applications of numerical analysis, but since they are well documented in standard texts (e.g., see Ref. 8.3), only those steps considered essential to an appreciation of the theory and its implications have been included here.

Example 8.5 is a study of the behavior of a rectangular section steel beam in which the material is assumed to have elastic–perfectly plastic characteristics. The nonlinear effect of gradually spreading plasticity is traced through the use of elementary equations that imply satisfaction of equilibrium on the undeformed structure. The theoretical limit of resistance is reached when the bending moment at the left end becomes the *plastic moment*, M_p, which is equal to the yield stress, σ_y, times the *plastic section modulus* ($bd^2/4$ in this case). At this point the cross section is fully plastified, and a *"plastic hinge"* forms. The theoretical limit is approached asymptotically, but in the numerical illustration resistance to further loading becomes insignificant by the time the end deflection becomes equal to a tenth of the span.

EXAMPLE 8.5

1. Determine the load-deflection relationship for the beam shown. The material is elastic–perfectly plastic. Consider partial plastification.
2. Calculate the result for $L = 8$ ft., $b = 2$ in., $d = 8.25$ in., $\sigma_y = 36$ ksi, $E = 29{,}000$ ksi

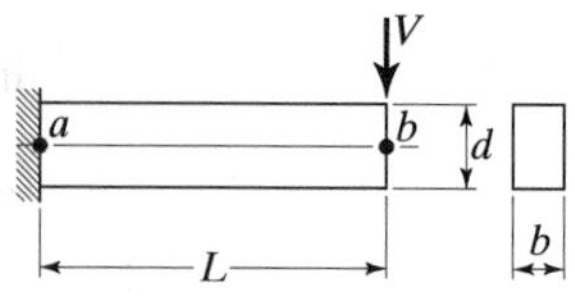

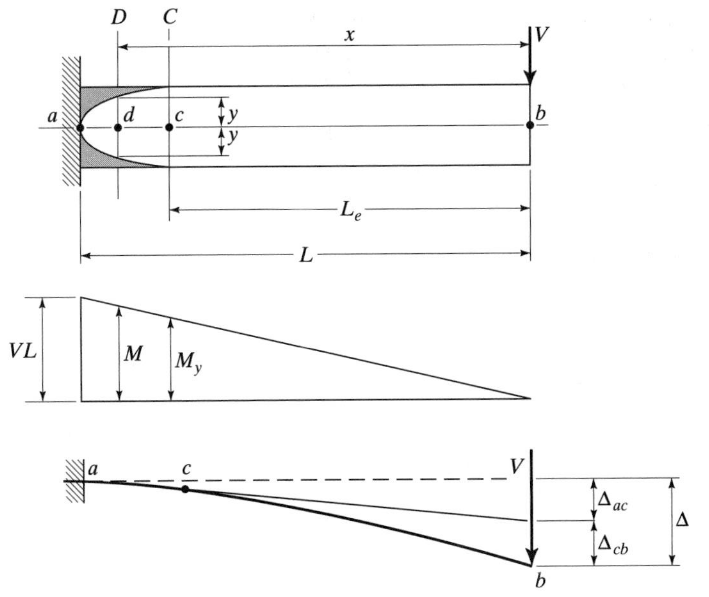

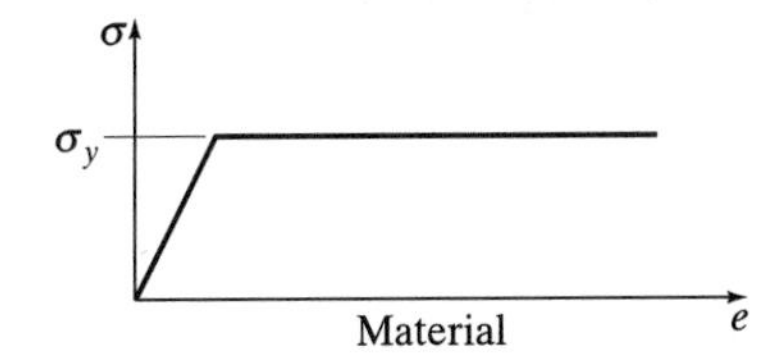

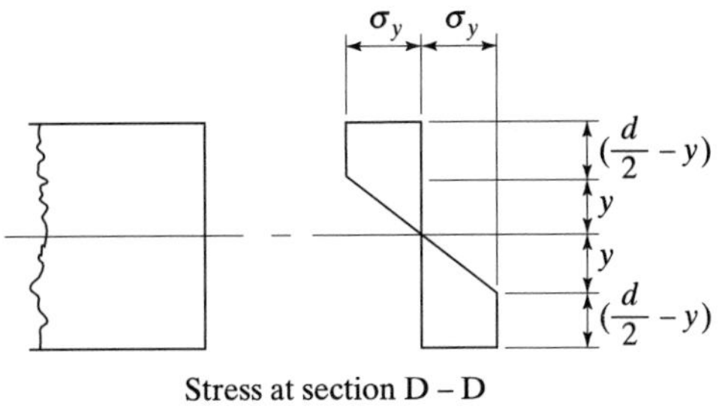

1. By moment-area principles:

$$\Delta = \Delta_{ac} + \Delta_{cb} = \int_{L_e}^{L} \frac{Mx\,dx}{EI_{eff}} + \frac{VL_e^3}{3EI} \tag{a}$$

where

$$V = \frac{M_y}{L_e} = \frac{bd^2\sigma_y}{6L_e} \tag{b}$$

At Section D-D:

$$M = Vx = \left[\frac{d^2}{4} - \frac{y^2}{3}\right]b\sigma_y$$

Using Eq. (b)

$$y = \frac{\sqrt{3}d}{2}\left[1 - \frac{2x}{3L_e}\right]^{1/2} \tag{c}$$

Also,

$$I_{eff} = \frac{2by^3}{3} \tag{d}$$

From Eqs. (a)–(d)

$$\Delta = \frac{2\sigma_y}{3E\,dL_e}\left[\frac{1}{\sqrt{3}}\int_{L_e}^{L} \frac{x^2\,dx}{\left[1 - \frac{2x}{3L_e}\right]^{3/2}} + L_e^3\right] \tag{e}$$

Integrating and evaluating limits

$$\Delta = \frac{2L^2\sigma_y}{3E\,d}\left[\frac{18(L_e/L)^2 - 6(L_e/L) - 1}{\sqrt{3\left(1 - \frac{2L}{3L_e}\right)}} - 10\left(\frac{L_e}{L}\right)^2\right] \tag{f}$$

2. For the given dimensions

$$S = 22.69 \text{ in}^3,\ Z = 34.03 \text{ in}^3$$

For $L_e = 96$ in:

$$M_y = 816.8 \text{ in. kips},\ V_y = 8.51 \text{ kips}$$
$$\Delta = \Delta_y = 0.925 \text{ in}$$

For $L_e = 64$ in

$$M_p = 1225.3 \text{ in. kips},\ V_p = 12.76 \text{ kips}$$
$$\Delta \to \infty$$

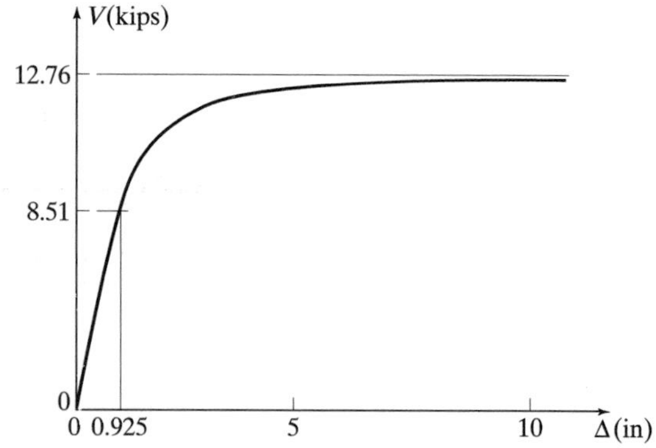

Example 8.6 is an exercise in simple plastic analysis, a method in which the development of zones of plasticity is ignored and cross sections are assumed to be either fully elastic or fully plastic. In the beam-column studied, mechanism motion can occur when plastic hinges form at the fixed end and under the transverse load. The interaction equation used to define full plastification of the cross section (Eq. 10.5 from Section 10.1.2), is one commonly used for steel wide flange sections subjected to axial force and major axis bending (see Ref. 8.6). The plastic limit load is obtained by assuming the hinge locations to be fully plastified and satisfying equations of equilibrium written for the member in its undeformed position.

EXAMPLE 8.6

For the member shown, determine the plastic limit load by conventional plastic analysis. Use the interaction curve of Equation 10.5.

Member properties: $A = 19.1$ in.^{2}, $I = 533$ in.^{4}, $Z = 96.8$ in.^{3}, $E = 29{,}000$ ksi, $\sigma_y = 50$ ksi

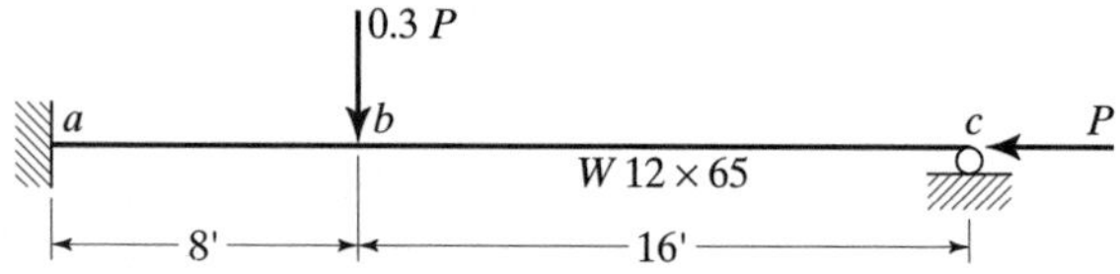

(a) The first hinge will form at a

$$P_y = 19.1 \times 50 = 955 \text{ kips}$$
$$M_p = 96.8 \times 50 = 4840 \text{ in. kips}$$

From Eq. 10.5:

$$\frac{P}{955} + \frac{0.85M_a}{4840} = 1$$

$\therefore P = 259.3$ kips, $M_a = 4148$ in. kips, $M_b = 2212$ in. kips

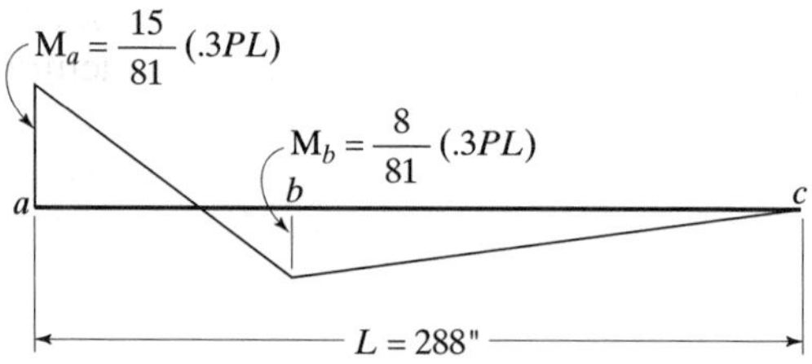

Elastic moment-diagram

(b) The second hinge will form at b, resulting in a mechanism.

Satisfy the equilibrium requirements:

$$\Sigma M_{br}\text{: } R_c = M_b/192$$
$$\Sigma M_{bl}\text{: } (M_a + M_b) = (.3P - R_c)96$$

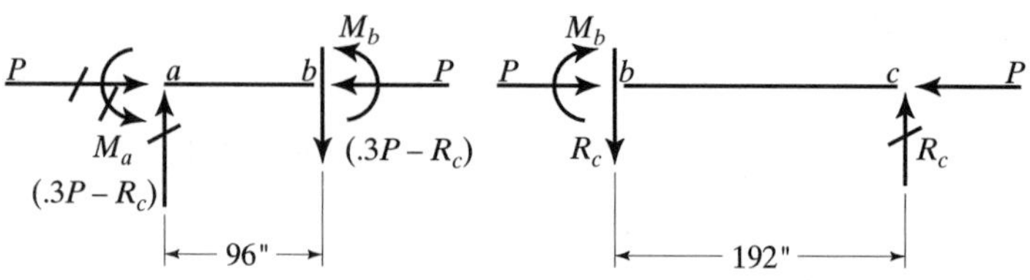

For hinges at a and b, $M_b = M_a$, or

$$M_a = 11.52P$$

Thus, from Equation 10.5

$$P = 325.7 \text{ kips},\ M_a = M_b = 3752 \text{ in. kips}$$

(c) Results:

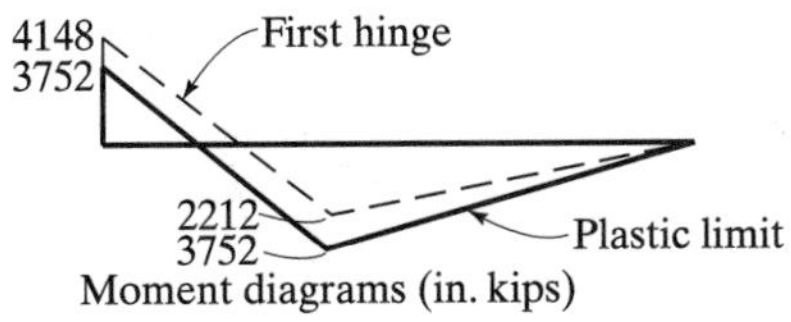

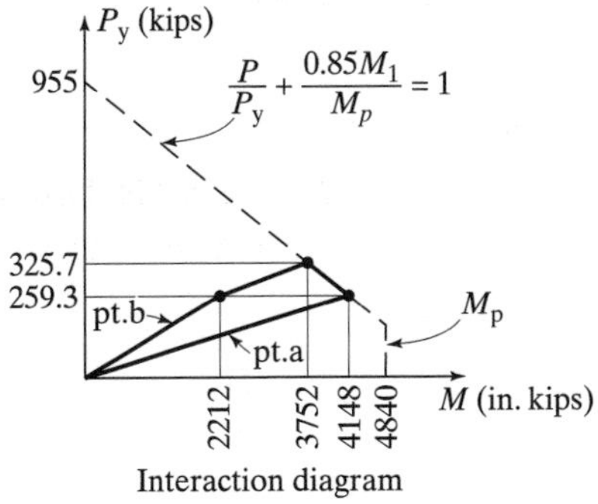

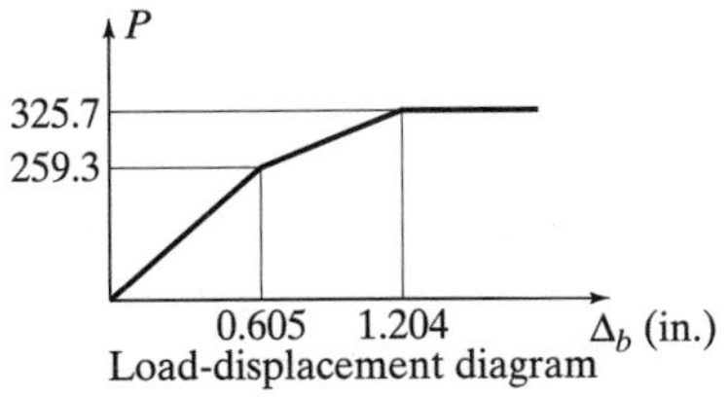

Before the limit is reached, formation of the first hinge results in redistribution of internal resistance. This is illustrated in two sketches in the example: (1) a comparison of elastic and final moment diagrams; and (2) an interaction diagram containing traces of the loading paths (the *force points*) of the end and interior cross sections. Also shown is a response curve, which indicates that since equilibrium equations are written on the undeformed structure and the member is assumed to behave elastically between plastic hinges, the response between levels of hinge formation is linear.

Example 8.7 is an analysis of the combined effect of the geometric nonlinear behavior of an elastic beam-column and the material nonlinearity of a semirigid connection of prescribed moment-rotation characteristics. Equilibrium is formulated on the deflected structure and classical small deflection elastic theory is used for the beam column. The nonlinear response equation, Equation c, is seen to have a limit. In the numerical illustration the system becomes unstable at an applied vertical load equal to 63% of the elastic critical load of the same member in a fixed base condition.

EXAMPLE 8.7

Member *ab* is an elastic bar connected to its base by a nonlinear rotational spring having the moment-rotation characteristic:

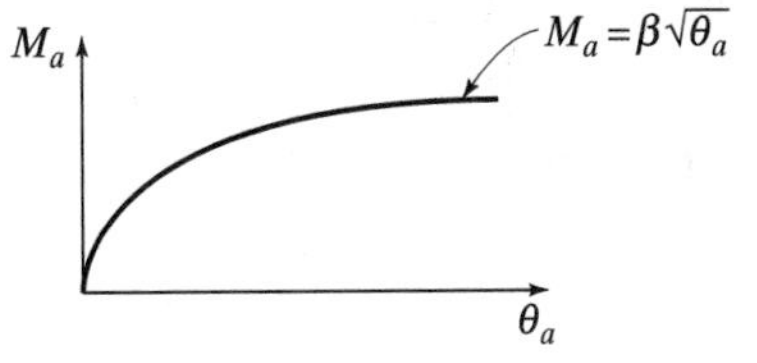

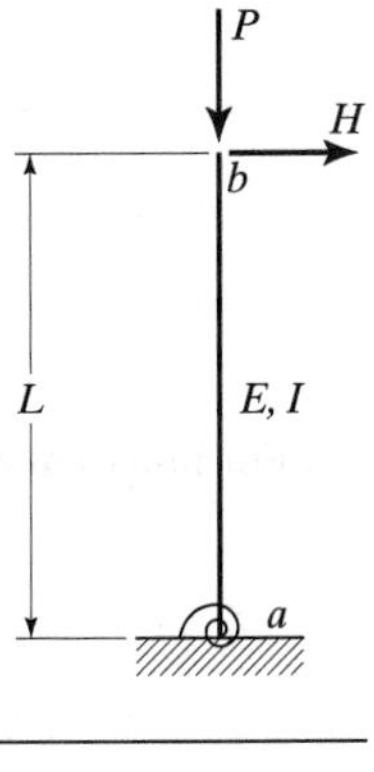

Analyze the bar under the action of a fixed force H and a varying force P

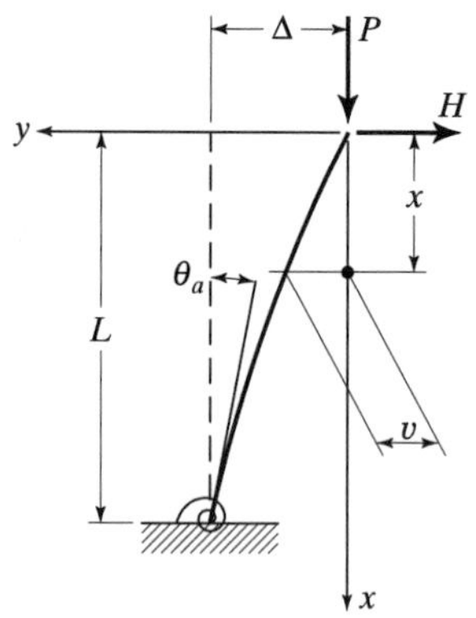

From Equation d of Example 8.4, letting $\alpha = H/P$

$$v = C_1 \sin\sqrt{\frac{P}{EI}}\,x + C_2 \cos\sqrt{\frac{P}{EI}}\,x - \frac{Hx}{P}$$

For $x = 0$, $v = 0$ and $C_2 = 0$. For $x = L$, $\dfrac{dv}{dx} = \theta_a$, or

$$C_1 = \frac{H/P + \theta_a}{\sqrt{\dfrac{P}{EI}}\cos\sqrt{\dfrac{P}{EI}}\,L}$$

Thus

$$v = \frac{(H/P + \theta_a)\sin\sqrt{\dfrac{P}{EI}}\,x}{\sqrt{\dfrac{P}{EI}}\cos\sqrt{\dfrac{P}{EI}}\,L} - \frac{Hx}{P}$$

But, for $x = L$, $v = \Delta$, or

$$\Delta = \frac{(H/P)\left(\tan\sqrt{\dfrac{P}{EI}}\,L - \sqrt{\dfrac{P}{EI}}\,L\right) + \theta_a \tan\sqrt{\dfrac{P}{EI}}\,L}{\sqrt{P/EI}} \qquad \text{(a)}$$

Also, $M_a = HL + P\Delta$, or

$$\Delta = \frac{\beta\sqrt{\theta_a} - HL}{P} \qquad \text{(b)}$$

Solving for Δ from Equations a and b:

$$\Delta = \frac{H}{EI\left(\dfrac{P}{EI}\right)^{3/2}}\left[\left(\tan\sqrt{\frac{P}{EI}}\,L - \sqrt{\frac{P}{EI}}\,L\right) + \frac{\beta^2\left(1 \pm \sqrt{1 - 4(HEI\tan^2\sqrt{P/EI}L)/\beta^2}\right)^2}{4HEI\tan\sqrt{\dfrac{P}{EI}}\,L}\right] \qquad \text{(c)}$$

Plot Equation c for: $L = 12$ ft, $I = 88.6$ in.^{4}, $E = 29{,}000$ ksi, $\beta = 10 \times 10^4$ in. kips/rad, and $H = 10$ kips

(Note that for a fixed base $P_{cre} = \dfrac{\pi^2(29{,}000)88.6}{4(144)^2} = 305.7$ kips)

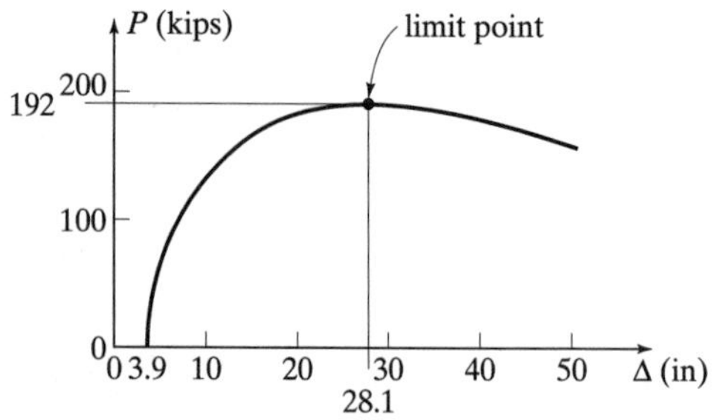

8.1.4 A Commentary on Stability

A major theme of this section has been structural instability, ways in which it may occur, and analytical methods for predicting—and forestalling—the possibility of its occurrence. We have used a case study approach in which individual situations have been considered and states of stable and unstable equilibrium deduced from the physical results of analyses of the case at hand. More mathematical criteria for stability exist but none is all encompassing, and in engineering applications it is rarely necessary to resort to them since stable equilibrium is usually evident from the physical facts. But to establish a relationship between the findings of our individual studies and to illustrate their underlying coherence, we present the following summary of a basic concept of stability and a criterion based on the total potential energy of a system. Works on stability such as References 8.2, 8.7, 8.8, and 8.9 contain more extensive treatment of this and other mathematical criteria.

Stability of equilibrium may be defined by reference to three states of a ball at rest on a surface (Fig. 8.2). If the ball is on a concave spherical surface (Fig. 8.2*a*), any displacement from the position shown will require a certain amount of work; the potential energy of the system will be raised. On this basis it is reasoned that it is in *stable equilibrium* in the defined position. If it is on a horizontal plane (Fig. 8.2*b*), it may be displaced without the expenditure of work; there will be no change in the potential energy of the system. In this case it is said to be in *indifferent* or *neutral equilibrium*. In the third case (Fig. 8.2*c*), the ball is on a convex spherical surface. Any displacement will lower it; the potential energy of the system will be decreased. It is in *unstable equilibrium*.

This definition of the stability of equilibrium of a rigid body has been extended to deformable ones by introducing the concept of *total potential energy* from theoretical mechanics:

$$\Pi = U + V$$

in which Π is the potential energy of the system, U is its strain energy, and V is the potential of the applied loads.

We may relate this expression to virtual work principles by recalling the virtual displacement statement of equilibrium

$$\delta W = \delta W_{\text{int}} - \delta W_{\text{ext}} = 0 \tag{6.13}$$

in which we have changed the signs of both the internal and external virtual work terms of the original Equation 6.13 to facilitate comparison. We first recognize that the internal virtual work of Equation 6.13 is the change in the strain energy of a structure as it passes from a configuration in a state of equilibrium to one defined by infinitesimally small virtual displacements and that the external virtual work is the negative of the corresponding change in potential of the applied loads. Comparing δW and Π it follows that, at equilibrium, the total potential energy assumes a stationary value, that is, in variational terms, $\delta\Pi = 0$.

Although this is a mathematical condition for static equilibrium, it is not sufficient to determine whether the equilibrium state is stable, neutral, or unstable in the sense defined above. But reasoning in the same way, it is seen that this depends upon whether, at equilibrium, the total potential energy is, respectively, a minimum, neutral, or a maximum and, further, that the distinction can be made by investigating the second or higher variations of Π. The process will be illustrated by application to two of the earlier examples of this chapter. Its extension and use in the analysis of the stability of multiple-degree-of-freedom systems may be found in the references cited.

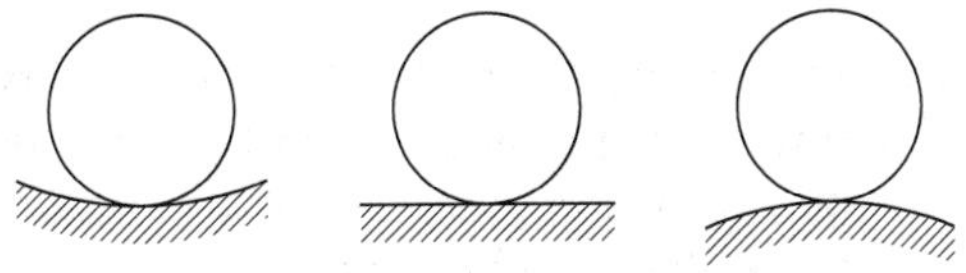

(*a*) (*b*) (*c*)

Figure 8.2 Equilibrium states.

In Example 8.8 the physical conclusions regarding the stability of the system of Example 8.2 are reexamined using the total potential energy criterion. The members are constant stiffness springs whose strain energy is a function of their shortening and the stiffness coefficient. The datum for measurement of the potential energy of the load is the unloaded configuration. Thus the total potential energy can be expressed as a function of the single variable θ and variations of the potential as derivatives with respect to θ. When equated to zero, the first derivative confirms the equilibrium path determined in Example 8.2. The second derivative produces the condition for stability of equilibrium. For any point on the equilibrium path at which the second derivative is positive the total potential energy is a minimum, therefore the equilibrium is stable. When the second derivative is zero, equilibrium is neutral, which in this case identifies the limit points 1 and 3 of Example 8.2. Similarly, negative values of the second derivative between points 1 and 3 signify maxima of the total energy in that region, that is, an unstable state.

EXAMPLE 8.8

For the structure of Example 8.2 use total potential energy criteria to:

1. Determine the equilibrium equation
2. Analyze the stability of the equilibrium path

---

1. Let θ of Example 8.2 be the basic displacement variable. Using Eq. 4.4a for U and letting $V = 0$ in the unloaded state

$$\Pi = U + V = 2\left(\frac{1}{2} k \, \Delta L_{ab}^2\right) - P\Delta_b$$

$$= kL^2[1 - \cos\alpha \sec(\alpha - \theta)]^2 - PL[\sin\theta \sec(\alpha - \theta)] \tag{a}$$

$$\delta\Pi = \frac{d\Pi}{d\theta} = 2kL^2\left\{\frac{[\cos(\alpha - \theta) - \cos\alpha][\cos\alpha \tan(\alpha - \theta)]}{\cos^2(\alpha - \theta)}\right\} - \frac{PL\cos\alpha}{\cos^2(\alpha - \theta)} \tag{b}$$

Letting $\delta\Pi = 0$ and solving for P

$$P = 2kL[\sin(\alpha - \theta) - \cos\alpha \tan(\alpha - \theta)] \tag{c}$$

as in Example 8.2

2. The second variation of Π is, from Eq. (b)

$$\frac{d^2\Pi}{d\theta^2} = \frac{\cos\alpha}{\cos^4(\alpha - \theta)}\{2kL^2[\cos\alpha + 2\cos\alpha \sin^2(\alpha - \theta) - \cos^3(\alpha - \theta) - 2\cos(\alpha - \theta)\sin^2(\alpha - \theta)] + 2PL[\cos(\alpha - \theta)\sin(\alpha - \theta)]\} \tag{d}$$

Substituting Eq. (c) for P, $\dfrac{d^2\Pi}{d\theta^2}$ on the equilibrium path is

$$\frac{d^2\Pi}{d\theta} = \frac{2kL^2[\cos\alpha - \cos^3(\alpha - \theta)]}{\cos^4(\alpha - \theta)} \tag{e}$$

Thus stability is governed by the sign of the numerator within the physical range $|\alpha - \theta| < \pi/2$:

$$\frac{d^2\Pi}{d\theta^2} > 0 \text{ for } \cos^3(\alpha - \theta) < \cos\alpha \qquad \text{the system is stable}$$

$$\frac{d^2\Pi}{d\theta^2} = 0 \text{ for } \cos^3(\alpha - \theta) = \cos\alpha \qquad \text{equilibrium is neutral}$$

$$\frac{d^2\Pi}{d\theta^2} < 0 \text{ for } \cos^3(\alpha - \theta) > \cos\alpha \qquad \text{the system is unstable}$$

which confirms the physical evidence of Example 8.2.

A similar reexamination of the first part of Example 8.4 is conducted in Example 8.9. We draw on an energy approach to the determination of the critical load in Reference 8.7 for the definition of U and V. Again, the total potential energy can be expressed as a function of a single variable, in this case the tip deflection, Δ. The bifurcation condition of Example 8.4 is confirmed, as is the neutrality of equilibrium in the deflected state. In both examples, however, there remains the question of the physical significance of "neutral equilibrium." In Example 8.4 we addressed it by analyzing the member under the presence of a small lateral force and demonstrating the destabilizing effect of a compressive load. In Example 8.9 our reliance upon small displacement theory limits the application of the total potential energy criterion. Closer examination of the static equilibrium of an ideal elastic bar in the deflected position is needed, and this requires consideration of larger displacements in an approach commonly known as the *theory of the elastica* (Ref. 8.7). By that theory it may be shown that the bifurcation load of an ideal elastic bar is the same as in the small displacement theory and that the deflected configuration is stable—but only weakly so. A very small increase in the load would cause a large deflection. In a structure of realistic dimensions, either the elastic displacement would become untenable or the P-Δ effect would result in destabilizing inelastic behavior. In either event, the critical load can be viewed as a practical limit on the stability of a framework element.

EXAMPLE 8.9

Use total potential energy criteria to analyze the stability of equilibrium of the column of Example 8.4

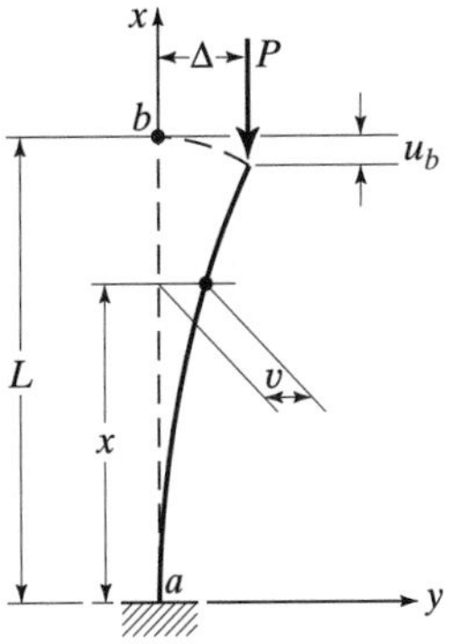

For present purposes use the coordinate system shown. From Example 8.4, the buckled configuration is

$$v = \Delta\left(1 - \cos\frac{\pi x}{2L}\right)$$

The bending moment at x is

$$M = P(\Delta - v) = P\Delta\cos\frac{\pi x}{2L}$$

In Reference 8.7 it is shown that

$$U = \int_0^L \frac{M^2\,dx}{2EI} = \frac{P^2\,\Delta^2 L}{4EI}$$

and that, for small displacements

$$u_b = \frac{1}{2}\int_0^L \left(\frac{dv}{dx}\right)^2 = \frac{\Delta^2\pi^2}{16L}$$

Thus, letting $V = 0$ prior to buckling

$$V = -Pu_b = -\frac{P\,\Delta^2\pi^2}{16L}$$

Therefore:

$$\Pi = \frac{P^2\,\Delta^2 L}{4EI} - \frac{P\,\Delta^2\pi^2}{16L} \tag{a}$$

$$\frac{d\Pi}{d\Delta} = \frac{P^2 L\Delta}{2EI} - \frac{P\pi^2\Delta}{8L} \tag{b}$$

$$\frac{d^2\Pi}{d\Delta^2} = \frac{P^2 L}{2EI} - \frac{P\pi^2}{8L} \tag{c}$$

Letting $\frac{d\Pi}{d\Delta} = 0$ in Equation b gives, for $\Delta \neq 0$, $P_{cre} = \frac{\pi^2 EI}{4L^2}$, which confirms the Euler solution.

Substituting P_{cre} in Equation c gives, for $\Delta \neq 0$, $\frac{d^2\Pi}{d\Delta^2} = 0$, which indicates a state of neutral equilibrium at the critical load.

8.2 A MATRIX APPROACH

In the next two chapters a number of the concepts and techniques of nonlinear analysis demonstrated in the above examples will be developed and applied in the context of the matrix stiffness method. For coherence and versatility, the approach taken is keyed to the levels of analysis illustrated in Figure 8.1. Each level can be defined by a symbolic matrix equation, as follows.

Linear elastic analysis will continue to be required. To emphasize its nature, we shall designate the linear elastic stiffness matrix as $[\mathbf{K_e}]$ or, in global analysis terms

$$[\mathbf{K_e}]\{\mathbf{\Delta}\} = \{\mathbf{P}\} \tag{8.1}$$

If we assume that, in the higher-level methods, the underlying nonlinear equations of equilibrium have been reduced to a set for which we can adapt techniques used for solving simultaneous linear algebraic equations, then behavior can be traced incrementally and each method can be stated symbolically as a variant of the global stiffness equation

$$[\mathbf{K_t}]\{\mathbf{d\Delta}\} = \{\mathbf{dP}\} \tag{8.2}$$

In Equation 8.2 $[\mathbf{K_t}]$ is a *tangent stiffness matrix*, $\{\mathbf{d\Delta}\}$ is a vector of incremental nodal point displacements, and $\{\mathbf{dP}\}$ is a vector of incremental nodal point loads and reactions. Typically, $[\mathbf{K_t}]$ has a linear elastic component and one or more additional components that are functions of the loads and/or the displacements in existence at the start of the increment. Thus the computational problem can be addressed as the conventional one of solving for unknown displacements and back substitution of the results in element stiffness equations to determine element forces, but doing so in a stepwise fashion in which the total response is determined through summation of increments.

The various levels differ in the types of nonlinearity included in $[\mathbf{K_t}]$, the way in which the equilibrium equations are formulated, the details of equation solution, and the way in which members may have to be subdivided into discrete elements to achieve satisfactory results.

In second-order elastic analysis the effects of finite deformations and displacements are accounted for in formulating the equations of equilibrium and Equation 8.2 becomes

$$[\mathbf{K_e} + \mathbf{K_g}]\{\mathbf{d\Delta}\} = \{\mathbf{dP}\} \tag{8.3}$$

in which $[\mathbf{K_g}]$, the *geometric stiffness matrix*, represents the change in stiffness that results from these effects. The matrix $[\mathbf{K_g}]$ can be developed in a number of ways, and various procedures can be used to solve Equation 8.3. Generally, members must be subdivided into several elements to produce satisfactory results.

In first-order inelastic analysis the equations of equilibrium are written in terms of the geometry of the undeformed structure, and Equation 8.2 becomes

$$[\mathbf{K_e} + \mathbf{K_m}]\{\mathbf{d\Delta}\} = \{\mathbf{dP}\} \tag{8.4}$$

where $[\mathbf{K_m}]$, which we shall call the *plastic reduction matrix*, represents the change in stiffness that results from inelastic behavior of the system.

In second-order inelastic analysis both geometric and material nonlinearity are accounted for. The equations of equilibrium are written in terms of the geometry of the deformed system and Equation 8.2 becomes

$$[\mathbf{K_e} + \mathbf{K_g} + \mathbf{K_m}]\{\mathbf{d\Delta}\} = \{\mathbf{dP}\} \tag{8.5}$$

For the calculation of elastic critical loads the global stiffness equation is cast in the form of a generalized *eigenvalue problem* in which the equation of equilibrium at the critical state is

$$[\mathbf{K_e} + \lambda\hat{\mathbf{K}}_\mathbf{g}]\{\mathbf{\Delta}\} = \{\mathbf{0}\} \tag{8.6}$$

where $[\hat{\mathbf{K}}_\mathbf{g}]$ is the geometric stiffness matrix computed for a reference load $\{\mathbf{P}_{ref}\}$[6], λ (an *eigenvalue*) is a load factor with respect to $\{\mathbf{P}_{ref}\}$, and $\{\mathbf{\Delta}\}$ (an *eigenvector*) is the buckled shape. The lowest value of λ that satisfies the equation for $\{\mathbf{\Delta}\} \neq \mathbf{0}$ yields the elastic critical load vector $\lambda\{\mathbf{P}_{ref}\}$, and the corresponding $\{\mathbf{\Delta}\}$ defines the buckled configuration. By modifying the material constants in $[\mathbf{K_e}]$, Equation 8.6 can be used to calculate inelastic critical loads in agreement with an accepted concept such as the tangent modulus theory. Methods commonly used for the calculation of both eigenvalues and eigenvectors are based on the following reduced and manipulated form of Equation 8.6.

$$\frac{1}{\lambda}\{\mathbf{\Delta}_f\} = -[\mathbf{K}_{ef}]^{-1}[\hat{\mathbf{K}}_{gf}]\{\mathbf{\Delta}_f\} \tag{8.7}$$

in which the subscript f symbolizes that the respective matrices and vectors relate to the free degrees of freedom only, and it is recognized that $[\mathbf{K}_{ef}]$ must thus be nonsingular. In general, however, the product $[\mathbf{K}_{ef}]^{-1}[\hat{\mathbf{K}}_{gf}]$ is not symmetric. Therefore, to cast Equation 8.7 in a form amenable to convenient solution it must be transformed in a way that avoids the complications of dealing with unsymmetric matrices. Processes by which this may be accomplished and the resulting equations solved are presented in Section 12.8. In the examples of the following chapters we will continue to refer to Equation 8.7 as the source of calculations of critical loads and buckled configurations. But except for simple cases in which the answers are self-evident, Equation 8.7 is being referred to in a symbolic sense; the actual solution requires some intermediate steps and an iterative (or comparable) procedure.

8.3 THE EQUATIONS OF ANALYSIS AND THEIR SOLUTION

In developing methods of linear elastic analysis in earlier chapters we dealt with assemblages of one-dimensional framing members for which exact or approximate solutions of the ordinary differential equations for each member are well known. The solutions were cast in the form of linear algebraic relationships between forces and displacements at the end of each element. As we have seen in the illustrative examples, in nonlinear analysis the response of the elements is a continuously or intermittently changing function of the applied loads. Unfortunately, it is often impossible to solve the underlying equations in any direct analytical way. And even if such solutions did exist, we would be confronted with the problem of solving a set of algebraically non-

[6]As defined here, the internal forces upon which $[\hat{\mathbf{K}}_\mathbf{g}]$ is based are calculated by a linear elastic analysis for the reference load $\{\mathbf{P}_{ref}\}$ and it is assumed that this distribution obtains at the critical load, that is, any internal redistribution of forces between the reference and critical load is disregarded. For this reason, this approach is often called *linear stability analysis* even though it is one device for the analysis of a physically nonlinear phenomenon.

linear or transcendental equations, another forbidding task. Normally, therefore, it is necessary to deal with them in some piecewise linear way, as in Equations 8.2–8.5.

The strategy we shall use is indicated by the way in which those equations are formulated. Initially, it requires obtaining symbolic analytical solutions of the differential equations of elements and reducing the degree of these inherently nonlinear solutions by eliminating higher-order terms deemed to be insignificant. Then, as in earlier work, assumed shape functions are used to cast the reduced solutions in the form of algebraic relationships between the forces and displacements at the element ends. In this case, however, the force-displacement relationships may, in themselves, be functions of the end forces. It also involves adaptation of features of plasticity theory such as the plastic hinge concept and the normality criterion for plastic flow. The immediate objective is the development of a load-dependent element tangent stiffness matrix that can be evaluated for given element forces and is a reliable predictor of a type of behavior in the neighborhood of the loads to which those forces correspond.[7] Conventional stiffness method procedures are then used to assemble a linear global stiffness equation. The process to this point encompasses the first three parts of the structural analysis procedure defined at the start of Chapter 1: basic mechanics, finite element mechanics, and equation formulation. The fourth part, equation solution, can use a variety of methods. Following is a summary of a few of them, with an emphasis on those we shall use later.

8.3.1 Equation Solution—The Options

The available choices for solving Equations 8.3 to 8.5, and some of the particular problems associated with each choice are illustrated in Figure 8.3.

The most elementary approach is an Euler or simple-step one (Fig. 8.3*b*), in which the tangent stiffness matrix is formulated for a given set of forces, increments of either load or displacement are specified, and the equation is solved for the corresponding unknowns. These are then used in the element equations to determine new values of internal forces, the tangent stiffness matrix is reformulated using these forces and the deformed geometry, and the process is repeated. The total loads, forces, and displacements at the end of each increment are determined by summing these entities over all increments. Equilibrium between external loads and internal forces is not checked, therefore the calculated response curve can drift from the true one. However, if the incremental steps are adequately small and for structures of moderate nonlinearity—which is generally the case for civil engineering systems—good accuracy may be obtained with a modest amount of computational effort. Single step methods are described further in Section 12.2.

A higher level of computational sophistication is shown schematically in Figure 8.3*c*. It requires use of one of the multitude of multistep or iterative schemes, of which some of the more prominent are described in Section 12.3. The basic objective of all of these schemes is the establishment of equilibrium at the end of the load increment. They do so by analyzing the imbalance between the applied loads at the end of a linear step and the internal (element-end) forces calculated from the results of that step. Each scheme is intended to correct, in an iterative fashion, for the imbalance between the

[7]Implicit in this discussion and throughout Chapters 9 and 10 is the use of an "updated" analysis in which at each step all quantities are referred to a loaded and deformed reference state. In incremental analysis an alternative approach is to refer all quantities to the initial configuration, which may be self-strained. In that case it is necessary to include an *initial strain matrix* in the tangent stiffness matrix (see Sections A.1, A.2, and Reference 8.10).

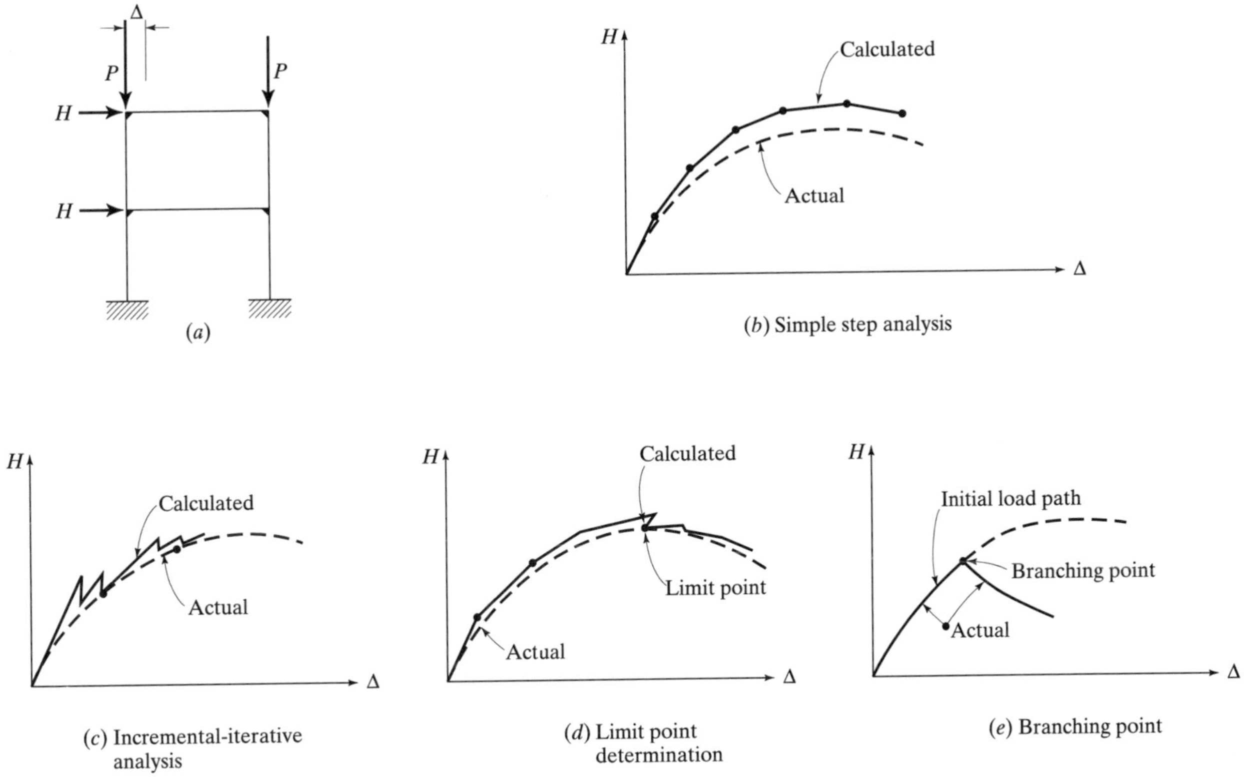

Figure 8.3 Equation solution methods and problems.

linear approximation and the actual nonlinear response or, more commonly, to reduce it to a tolerable level.

Common to all of these approaches are

1. The problems of numerical convergence just mentioned.
2. Detection of limit points and tracing post-limit point behavior (Figure 8.3*d*).
3. Detection of states of bifurcation, that is, points at which the response may branch off from its initial stable mode to a different one—either stable or unstable—with little or no numerical forewarning (Figure 8.3*e*).
4. Detection of changes in the constitutive relationships, such as the onset of material yielding.

The first three will be treated in Chapter 12 and the fourth will be dealt with in Chapter 10.

8.3.2 A Fundamental Problem

In nonlinear analysis the aim is to trace the history of all material points in a structure as it undergoes progressive loading. We have argued that, generally, the only practical way to do this is by a series of linear analyses that employs one or more of the equation solution schemes referred to above. But inherent to this approach is the fundamental problem illustrated by Figure 8.4, in which we examine the response of a typical member of the frame of Figure 8.4*a*, the rafter *ab*. Its initial, unloaded, undeformed state (Configuration 0) can be defined either in terms of the fixed global coordinate system,

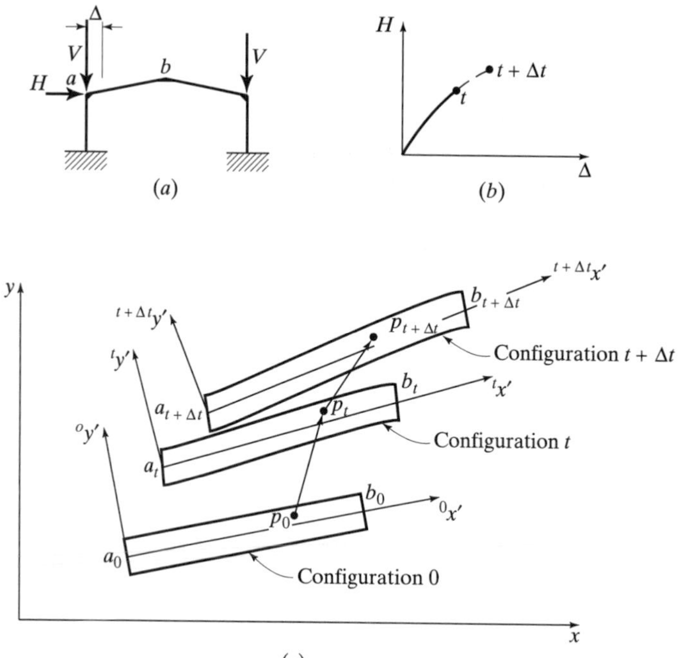

Figure 8.4 Element deformations.

x, y, or the local coordinate system ${}^0x'$, ${}^0y'$, in which ${}^0x'$ is directed through the member ends a_0, b_0 (Fig. 8.4c). Let us accept for the moment that, following the procedure of defining properties in local coordinates, transforming and assembling them for global analysis, and using one of the forms of incremental solution, we have been successful in tracing the exact behavior of the structure to Configuration t. The frame's correct load-displacement relationship is as indicated in Figure 8.4b, and now the properties of the rafter can be expressed either in terms of the global coordinate system or the new, updated local coordinate system ${}^tx'$, ${}^ty'$, with ${}^tx'$ directed through the ends of the displaced member. The member will have changed shape and size in the process, but the equations of equilibrium formulated on this shape are satisfied and the position of any material point, p, has been traced from its initial position, p_0, to its new position, p_t. But, in starting the next step (and in reality, of course, any previous step) we are faced with a dilemma. We would like to formulate the equations of equilibrium on Configuration $t + \Delta t$, but we can't since that configuration is still unknown.

To circumvent this problem we shall use the approach implied in the previous subsection. Configuration t is defined as a "reference configuration" and all stresses, loads, and displacements of the member in the next step are expressed in terms of this local ${}^tx'$, ${}^ty'$ coordinate system. The global tangent stiffness matrix is comprised of element matrices determined from their local reference configurations and the global equilibrium equation is therefore an approximate one that is linear in the incremental displacements. If a simple step solution procedure is adopted, the updated configuration determined from a single application of this equation becomes the new reference configuration and the process advances to consideration of the next load increment. But if one of the iterative schemes is followed, the first updated configuration becomes a trial "deformed configuration" that may then be used as the basis for a successive approximation scheme designed to converge on the correct Configuration $t + \Delta t$.

In the development of the requisite element matrices in the next two chapters we shall generally use physical arguments appropriate to the case at hand. They will satisfy our objective: the construction of equations that are reliable and useful in the design of civil engineering structures. But in a few instances, notably in accounting for finite rotations and the nonlinear interaction of bending moments and torques in Chapter 9, it will be necessary to turn to a more general approach, one that deals directly with the fundamental problem of formulating the equations of equilibrium on the deformed structure. We will use the results of that approach in Chapter 9, but we leave the details of its development to Appendix A.

8.4 PROBLEMS

8.1 The system shown has a small initial displacement Δ_0. Member *ab* is a rigid bar and member *bc* has a constant axial stiffness *k*. Determine the system's response as the load *P* is applied. Compare with the results of Example 8.1.

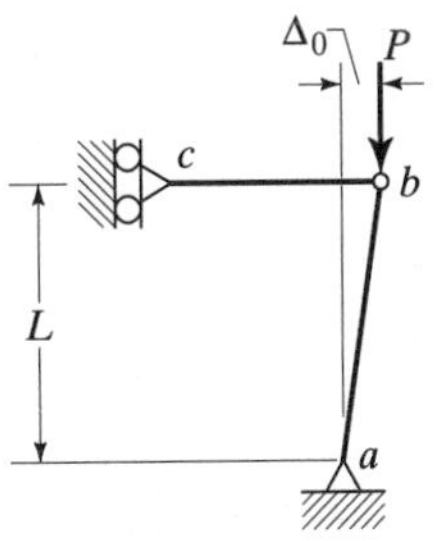

Problem 8.1

8.2 (a) Write the linear elastic global stiffness equation for the single-degree-of-freedom system shown. Compare with the comparable equation of Example 8.2.

(b) Nonlinear elastic behavior of this system will be analyzed in Example 9.2 using Equation 8.3. Explain how the same result could, in principle, be obtained solely from the equation written in answer to part (a). Comment on the computational problems that would be encountered and methods by which they might be solved in obtaining such a solution.

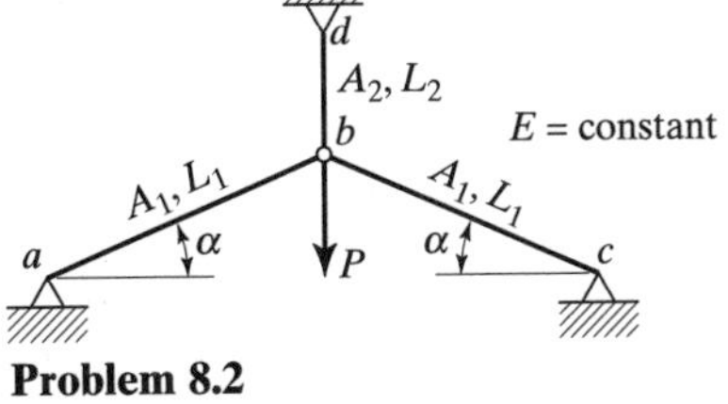

Problem 8.2

8.3 The conventional linear theory of flexure used in Examples 8.3 and 8.4 is based on the small displacement approximation of the radius of curvature (Eq. 4.28). Recall the nature of the approximation and review the significance of the neglected term by studying the more rigorous "theory of the elastica" as presented in texts on stability such as Reference 8.7. Comment on the practical differences. In particular, in what sense and to what degree does the neglected term affect the critical load and post-critical behavior of an elastic column?

8.4 An elastic member has the small initial curvature shown. Using small displacement theory, analyze its response as the load P is applied. Compare with the results of part 2 of Example 8.4.

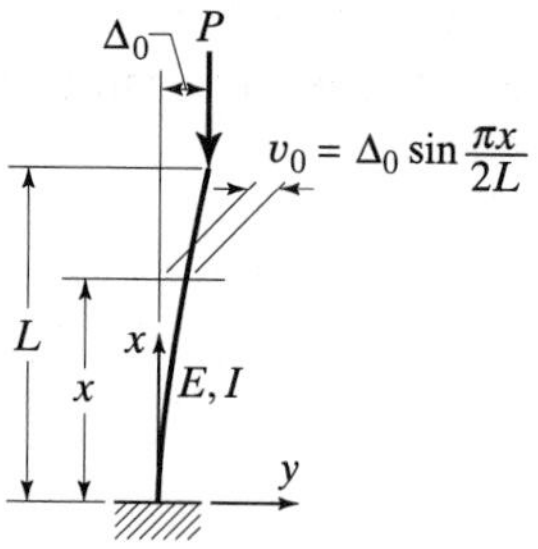

Problem 8.4

8.5 Using small displacement theory, analyze the response of the member shown. Compare the result with Example 8.3.

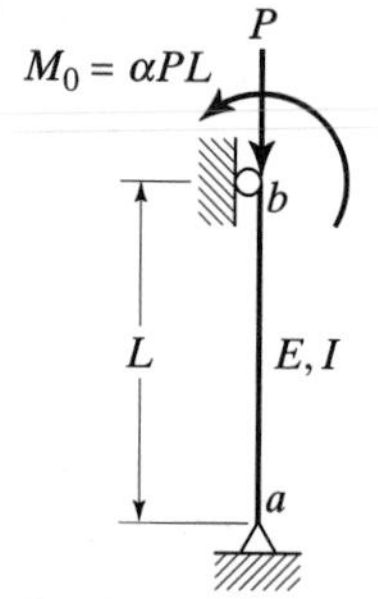

Problem 8.5

8.6 Member ab is elastic, but members bd and cd are rigid. Determine the critical load of the system. Compare the result with Example 8.4.

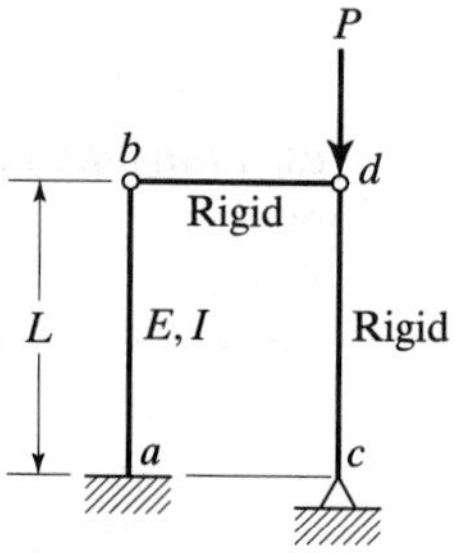

Problem 8.6

8.7 Member ab is a rigid bar connected to a fixed base by a linear rotational spring of stiffness $k = M/\theta$. (1) For $\alpha = 0$ determine the critical load. (2) For $\alpha \neq 0$ determine the load-displacement response (consider small displacements only; assume $\cos \theta = L$).

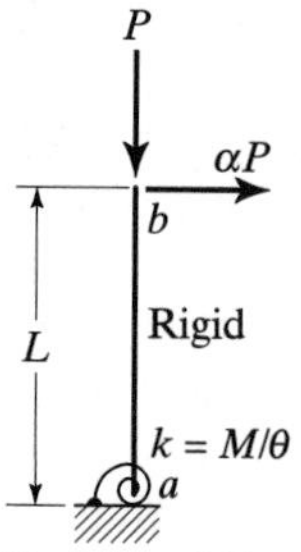

Problem 8.7

8.8 The beam shown is assumed to be made of elastic-perfectly plastic steel with $\sigma_y = 36$ ksi and $E = 29{,}000$ ksi. For cross section (a), a wide flange section, analyze it by the conventional elastic-plastic hinge method and plot the curve of P versus the deflection at point b. Sketch the corresponding curve for the rectangular cross section (b). Note the relative values of first yielding, second hinge formation, and plastic limit load.

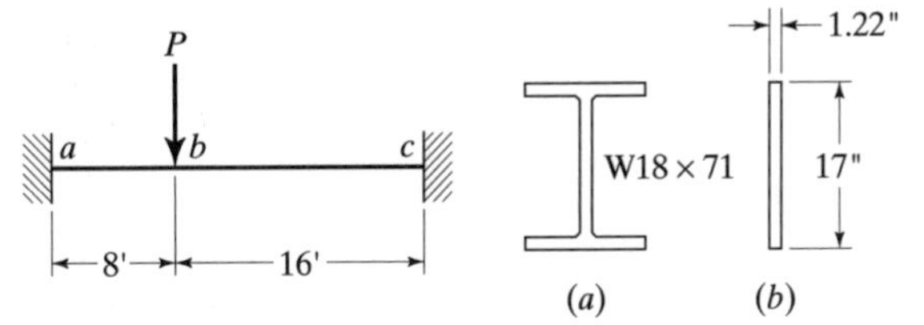

Problem 8.8

8.9 Repeat Example 8.6 for a uniformly distributed load, $q = .02P$ kips/ft instead of the concentrated transverse load.

8.10 Repeat Example 8.7 but with member *ab* assumed to be a rigid bar (consider large displacements).

REFERENCES

8.1 *Load and Resistance Factor Design Specification for Steel Buildings*, American Institute of Steel Construction, Chicago, 1993.

8.2 J. G. Croll and A. C. Walker, *Elements of Structural Stability*, John Wiley and Sons, New York, 1972.

8.3 S. P. Timoshenko, *Strength of Materials*, Parts I and II, 3rd edition, Van Nostrand, Princeton, 1956.

8.4 **A.** L. Euler, "Methodus Inveniendi Lineas Curvas...", Bousquet, Lausanne, 1744.

B. W. A. Oldfather, C. A. Ellis, and D. M. Brown, *Leonhard Euler's Elastic Curves, Translated and Annotated*, ISIS, Vol. XX, 1933, Saint Catherine Press, Bruges.

8.5 T. V. Galambos, editor, *Guide to Stability Design Criteria for Metal Structures*, 5th edition, John Wiley and Sons, New York, 1998.

8.6 *Plastic Design in Steel, A Guide and Commentary*, Manual of Practice No 41, American Society of Civil Engineers, New York, 1971.

8.7 S. P. Timoshenko and J. M. Gere, *Theory of Elastic Stability*, 3rd edition, McGraw-Hill, New York, 1961.

8.8 D. O. Brush and B. O. Almroth, *Buckling of Bars, Plates, and Shells*, McGraw-Hill, New York, 1975.

8.9 H. Ziegler, *Principles of Structural Stability*, Birkhauser, Basel, 1977.

8.10 K.-J. Bathe, *Finite Element Procedures*, Prentice Hall, Englewood Cliffs, NJ, 1996.

8.11 R. D. Ziemian, W. McGuire, and G. G. Deierlein, "Inelastic Limit States Design, Part I: Planar Frame Studies, Part II: Three-Dimensional Frame Study," Jl. of Struct. Engr., ASCE, Vol. 118, No. 9, 1992.

Chapter 9

Geometric Nonlinear and Elastic Critical Load Analysis

A stiffness method approach to the geometrical nonlinear analysis of planar and space structures and the calculation of elastic critical loads is developed in this chapter. Although the results of the two levels of analysis differ—a continuous response curve in the first case and one point in the response of an ideal model of a system in the second—they employ the same form of the elastic and geometric stiffness matrices. And both are useful in design.

The approach is based on Equations 8.3 and 8.6. The required elastic stiffness matrices will be drawn from earlier chapters, but geometric stiffness matrices will be derived in this chapter using virtual displacement principles and assumed shape functions for the purpose. In-plane behavior will be treated before out-of-plane behavior for two reasons: (1) whereas, currently, geometric nonlinear methods are used rather widely in the design of planar systems, their use in space structures is still limited; and (2) consideration of finite rotations and the interaction of torsion and flexure, which is essential to the nonlinear analysis of a space system, presents problems that require special attention. Accordingly, in Section 9.1 we'll develop and apply geometric stiffness matrices for the axial force element and the element subjected to axial force and bending about one axis. The interaction of axial force and St. Venant torsion is considered in Section 9.2. The geometric nonlinear analysis of three-dimensional frames is then taken up in Section 9.3. Some elements of the requisite 12-degree-of-freedom geometric stiffness matrix are drawn from the findings of the two preceding sections and some, specifically those relating to the interaction of torsion and bending, from Appendix A. Proper treatment of this interaction, which involves subtleties in the establishment of equilibrium on the deformed structure, requires a more mathematical, structural mechanics approach than that which has sufficed to this point. The approach referred to is important from the standpoint of satisfying our immediate need and, even more so, for its clarification of the fundamentals of nonlinear analysis. We present it in an appendix, however, because to develop it here would delay the application of the results, which is equally important to the appreciation of their significance. These results are illustrated in Section 9.4 through a series of examples of geometric nonlinear analysis and elastic buckling.

As in Chapters 4 and 7, all of the stiffness matrices developed in this chapter are in local coordinates.

9.1 GEOMETRIC STIFFNESS MATRICES FOR PLANAR ELEMENTS

We consider a prismatic element of bisymmetrical cross section subjected to an axial force and flexed about its z axis (Fig. 9.1*a*) and, in doing so, we'll disregard shear deformation. This element and loading furnish a simple illustration of the formulation of stiffness properties for geometric nonlinear analysis while giving insight into key aspects of analysis common to all structural forms. As load is applied to the structure of which the element is a part, its displacement from the reference configuration of Figure 9.1*a* is a function of how much it: (1) rotates as a rigid body in compliance with the relative displacement of its ends; (2) stretches or shortens; and (3) bends. All three occur simultaneously and they may influence each other. But to focus on the most significant effects, we will neglect or approximate some minor coupling and treat the total displacement as the resultant of two sequential actions: (1) stretching and rigid body rotation of the axis (Figure 9.1*b*); and (2) flexure of the element with respect to the chord joining the displaced ends (Figure 9.1*c*). And for the same purpose we shall develop two geometrical stiffness matrices: (1) for the case in which there is no bending and the problem reduces to that of a straight member undergoing the rotation and axial straining of Figure 9.1*b*; and (2) for a member subjected to combined axial force and bending, in which case the deformations of Figures 9.1*c* must also be accounted for.

9.1.1 Axial Force Member

In contrast to the infinitesimal strains that formed the basis of linear elastic analysis, here we start with a study of small finite strain of a differential length of the element's material. We are assuming the element to be straight and strain free in the reference

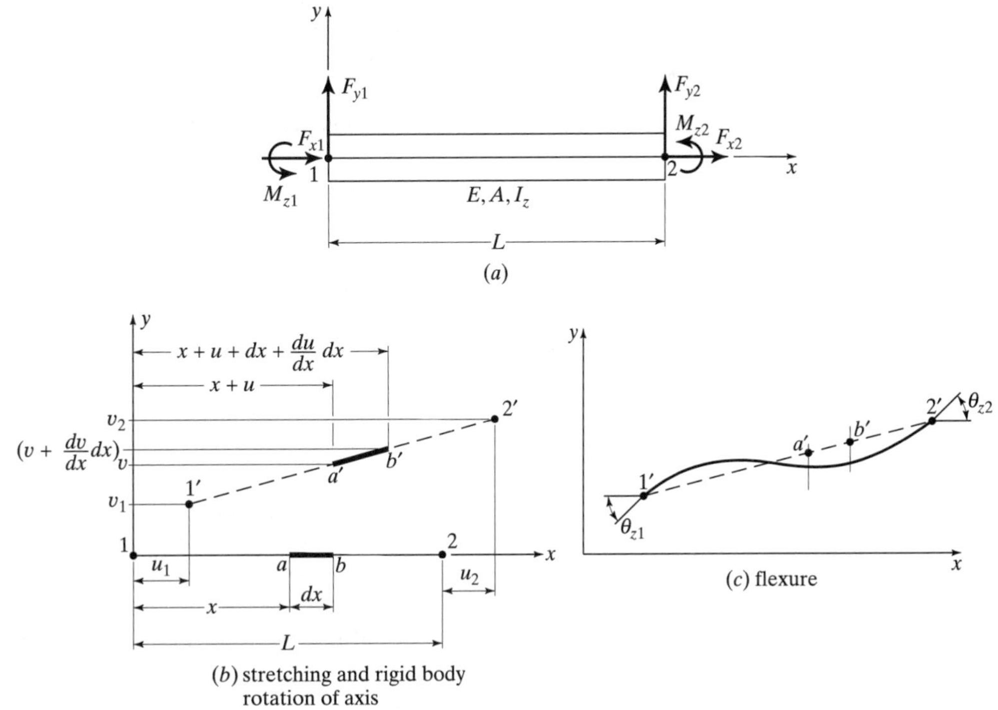

Figure 9.1 Finite strain of planar element.

configuration and, since we consider only axial strains, we can again treat the problem as mathematically one dimensional.

In Figure 9.1*b*, the length of the material segment, *dx*, in the reference configuration is designated *ab*. But after rigid body rotation and axial straining, its length is

$$a'b' = \left[1 + 2\frac{du}{dx} + \left(\frac{du}{dx}\right)^2 + \left(\frac{dv}{dx}\right)^2\right]^{1/2} dx$$

Designating the increment, $2\dfrac{du}{dx} + \left(\dfrac{du}{dx}\right)^2 + \left(\dfrac{dv}{dx}\right)^2$, as d_{ab}, we have

$$\frac{a'b'}{dx} = [1 + d_{ab}]^{1/2}$$

Or, expanding by the binomial theorem and neglecting higher-order terms

$$\frac{a'b'}{dx} = 1 + \frac{du}{dx} + \frac{1}{2}\left[\left(\frac{du}{dx}\right)^2 + \left(\frac{dv}{dx}\right)^2\right]$$

Now, defining the extension per unit length, $(a'b' - ab)/dx$, as the finite strain, e_{fin}, referred to the reference configuration,[1]

$$e_{\text{fin}} = \frac{du}{dx} + \frac{1}{2}\left[\left(\frac{du}{dx}\right)^2 + \left(\frac{dv}{dx}\right)^2\right] \tag{9.1}$$

Applying the principle of virtual displacements to the reference configuration (compare with Equation 6.15a)

$$\delta W_{\text{int}} = \int_{\text{vol}} \sigma_x \, \delta e_{\text{fin}} \, d(\text{vol}) \tag{9.2}$$

Or, using Equation 9.1 and integrating over the depth of the member, we have

$$\delta W_{\text{int}} = \int_0^L \sigma_x A\left(\frac{d\delta u}{dx}\right) dx + \frac{1}{2}\int_0^L \sigma_x A\left[\delta\left(\frac{du}{dx}\right)^2 + \delta\left(\frac{dv}{dx}\right)^2\right] dx \tag{9.3}$$

In the first integral of Equation 9.3, we have let $\delta(du/dx) = (d\delta u/dx)$, which is valid for infinitesimal displacement. Now, using the conventional elastic stress-strain relationship in the first integral and letting $\sigma_x A = F_{x2}$ in the second, the result is

$$\delta W_{\text{int}} = \int_0^L \left(\frac{du}{dx}\right) EA\left(\frac{d\delta u}{dx}\right) dx + \frac{1}{2} F_{x2}\int_0^L \left[\delta\left(\frac{du}{dx}\right)^2 + \delta\left(\frac{dv}{dx}\right)^2\right] dx \tag{9.4}$$

It may be seen that Equation 4.26a, the linear elastic stiffness matrix of the axial force element, follows directly from the first integral of Equation 9.4 (see also Equation 6.15c). Therefore, we need only concern ourselves with the second integral, which produces the geometric stiffness matrix $[\mathbf{k_g}]$. Noting that the virtual operator δ may be treated as a differential operator with respect to the variables du/dx and dv/dx (References 9.6 and 9.7), the internal virtual work for this term may be written as

$$\delta W_{\text{int } g} = F_{x2}\int_0^L \left[\left(\frac{d\delta u}{dx}\frac{du}{dx}\right) + \left(\frac{d\delta v}{dx}\frac{dv}{dx}\right)\right] dx \tag{9.5}$$

---

[1]The significance of neglecting higher-order terms in the binomial expansion is discussed in Section A.1 of Appendix A. There it is shown that, whereas the use of strains based on Equation 9.1 (and Equation 9.1a of Section 9.2) in virtual displacement equations written on the reference configuration is satisfactory for present purposes, the result is strictly valid only for small finite strains.

From the development of the virtual displacement formula for an element stiffness matrix through the use of shape functions in Section 7.2 it follows that

$$[\mathbf{k_g}] = F_{x2}\int_0^L [\{\mathbf{N}'_u\}\lfloor\mathbf{N}'_u\rfloor + \{\mathbf{N}'_v\}\lfloor\mathbf{N}'_v\rfloor]\,dx \tag{9.6}$$

where $\lfloor\mathbf{N}_u\rfloor$ and $\lfloor\mathbf{N}_v\rfloor$ are shape functions for the respective displacement coordinates, which in this case are those for a simple axially loaded member. Thus, from Equation 7.6 (with $\xi = x/L$)

$$u = (1-\xi)u_1 + \xi u_2 \quad \text{and} \quad v = (1-\xi)v_1 + \xi v_2 \tag{9.7}$$

Therefore

$$\lfloor\mathbf{N}'_u\rfloor = \left\lfloor -\frac{1}{L} \;\; \frac{1}{L}\right\rfloor \quad \text{and} \quad \lfloor\mathbf{N}'_v\rfloor = \left\lfloor -\frac{1}{L} \;\; \frac{1}{L}\right\rfloor \tag{9.8}$$

Use of Equations 9.7 and 9.8 in Equation 9.6 results in

$$[\mathbf{k_g}] = \frac{F_{x2}}{L}\begin{bmatrix} 1 & 0 & -1 & 0 \\ 0 & 1 & 0 & -1 \\ -1 & 0 & 1 & 0 \\ 0 & -1 & 0 & 1 \end{bmatrix} \quad (\text{columns: } u_1,\ v_1,\ u_2,\ v_2) \tag{9.9}$$

Thus the geometric stiffness matrix is a function of the total axial force acting on the element in the reference configuration. The components of nodal point displacement are incremental values referred to this configuration; in the equations of nonlinear analysis, they will be specifically designated as such. Application of Equation 9.9 is illustrated in Examples 9.1 and 9.2 of Section 9.1.3.

9.1.2 Combined Bending and Axial Force

To include the effects of bending on the geometric stiffness matrix, the strain due to flexure (Fig. 9.1c) must be added to Equation 9.1. In doing this we shall neglect the effects of the rotation and stretching of the element's axis on this quantity and use Equation 4.28, the elementary mechanics approximation for infinitesimal flexural strain. Thus the total finite strain becomes

$$e_{\text{fin}} = \frac{du}{dx} - y\left(\frac{d^2v}{dx^2}\right) + \frac{1}{2}\left[\left(\frac{du}{dx}\right)^2 + \left(\frac{dv}{dx}\right)^2\right] \tag{9.10}$$

Adding the new term to Equation 9.2, letting $\sigma_x = -yM_z/I_z$ in that term, noting that $\int_A y^2\,dA = I_z$, integrating over the depth of the member, and letting $\delta(d^2v/dx^2) = (d^2\delta v/dx^2)$, we have, in place of Equation 9.3

$$\begin{aligned}\delta W_{\text{int}} = &\int_0^L \sigma_x A\left(\frac{d\delta u}{dx}\right)dx + \int_0^L M_z\left(\frac{d^2\delta v}{dx^2}\right)dx \\ &+ \frac{1}{2}\int_0^L \sigma_x A\left[\delta\left(\frac{du}{dx}\right)^2 + \delta\left(\frac{dv}{dx}\right)^2\right]dx\end{aligned} \tag{9.11}$$

Using the elastic stress-strain relationship of Equation 4.31b in the new term and treating the others as before:

$$\begin{aligned}\delta W_{\text{int}} = &\int_0^L \left(\frac{du}{dx}\right)EA\left(\frac{d\delta u}{dx}\right)dx + \int_0^L \left(\frac{d^2v}{dx^2}\right)EI_z\left(\frac{d^2\delta v}{dx^2}\right)dx \\ &+ \frac{1}{2}F_{x2}\int_0^L \left[\delta\left(\frac{du}{dx}\right)^2 + \delta\left(\frac{dv}{dx}\right)^2\right]dx\end{aligned} \tag{9.12}$$

We have dealt with the first integral in the previous subsection, and we now note that Equation 4.32, a linear elastic stiffness matrix, follows from the second integral (see Eq. 6.21a). Thus reevaluation of the third integral using appropriate flexural shape functions is the only task remaining. For this purpose we adopt the functions used for the analysis of an element in simple flexure or, from Equation 7.9,

$$\begin{aligned} u &= (1-\xi)u_1 + \xi u_2 \\ v &= (1 - 3\xi^2 + 2\xi^3)v_1 + (1 - 2\xi + \xi^2)x\theta_{z1} + (3\xi^2 - 2\xi^3)v_2 - (\xi - \xi^2)x\theta_{z2} \end{aligned} \tag{9.13}$$

Use of the derivatives of these shape functions in Equation 9.6 results in

$$[\mathbf{k}_g] = \frac{F_{x2}}{L}\begin{array}{c} \begin{matrix} u_1 & v_1 & \theta_{z1} & u_2 & v_2 & \theta_{z2} \end{matrix} \\ \begin{bmatrix} 1 & 0 & 0 & -1 & 0 & 0 \\ & \frac{6}{5} & \frac{L}{10} & 0 & -\frac{6}{5} & \frac{L}{10} \\ & & \frac{2L^2}{15} & 0 & -\frac{L}{10} & -\frac{L^2}{30} \\ & & & 1 & 0 & 0 \\ & \text{Sym.} & & & \frac{6}{5} & -\frac{L}{10} \\ & & & & & \frac{2L^2}{15} \end{bmatrix} \end{array} \tag{9.14}$$

9.1.3 Examples of Plane Structure Analysis

Application of Equation 8.3 for geometric nonlinear analysis and Equation 8.7 for the determination of elastic critical loads is illustrated in the following elementary but representative examples. The appropriate equations of Chapter 4 are used for the formation of element elastic stiffness matrices and Equations 9.9 and 9.14 for the geometric ones. The computer results shown have been obtained from the program MASTAN2 and one of the equation solution schemes explained in Sections 12.2 and 12.3 or, for critical loads and buckled configurations, the algorithm for calculating eigenvalues and eigenvectors described in Section 12.8.

Whereas in the examples of linear numerical analysis in earlier chapters it was generally feasible to include extensive details of the formulation and solution of the problem, normally this is not the case in nonlinear analysis. Even in simple problems, a solution process composed of several increments and the use of subdivided members may make detailed explanation impractical. The aim will be to give a clear statement of the problem, key features of the approach used to solve it, and major results (e.g., a numerical answer or a graphical display of a particular behavioral phenomenon). The statement of the problem and approach taken should be sufficient to enable the reader to obtain the same results or, recognizing the possibility of reasonable differences in nonlinear analysis, at least comparable ones.

Example 9.1 is similar to Example 8.1 except for features essential for computer analysis: numerical specification of data—including a very large but not infinite area for member *bc*—and a pinned rather than a vertically sliding support at *a*. For the case of a vertical load only, Equation 8.7 becomes elementary if it is assumed that the effect of the vertical component of displacement on the critical load can be neglected. The critical load becomes identical to that of Example 8.1 and the horizontal component, u_b, remains indefinite. For the case of a combined vertical and horizontal load, a computer solution of Equation (e) results in the detection of a limit point at 85% of the critical load, which is approximately the same as the limit determined by Equations (b) and (c) of Example 8.1 for the comparable case of that example. The difference can be attributed to the difference in specified boundary conditions, and more important, the fact that, whereas the elastic stiffness coefficient of member *ab* was defined as a constant in Example 8.1, it was continually updated in the present example. For a small displacement problem, the change would be negligible, but in this large displacement one it is significant.

EXAMPLE 9.1

Both members are axial force members

$$A_{ab} = 2\ \text{mm}^2,\ A_{bc} = 5 \times 10^3\ \text{mm}^2,\ E = 200{,}000\ \text{MPa}$$

Analyze the response of the system for

1. $\alpha = 0$
2. $\alpha = 0.05$

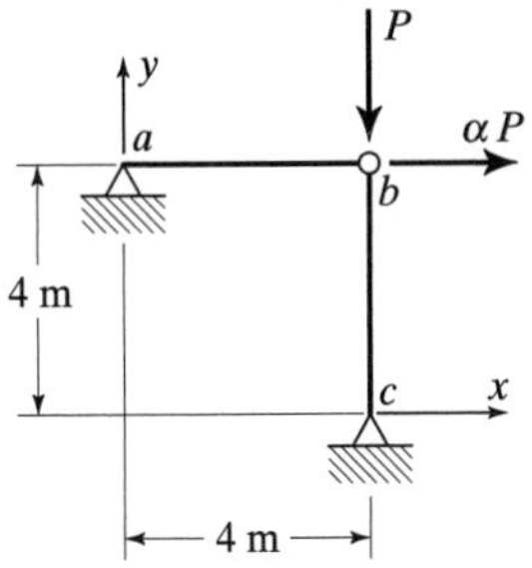

1. $\alpha = 0$ Calculate the critical load using Equation 8.7.
   Equation 4.26b:

$$[\mathbf{K}_{ef}] = 200\left[\begin{bmatrix} \frac{2}{4000} & 0 \\ 0 & 0 \end{bmatrix} + \begin{bmatrix} 0 & 0 \\ 0 & \frac{5 \times 10^3}{4000} \end{bmatrix}\right] = \frac{1}{10}\begin{bmatrix} 1 & 0 \\ 0 & 2.5 \times 10^3 \end{bmatrix} \begin{matrix} u_b \\ v_b \end{matrix} \tag{a}$$

Equation 9.9:

$$[\hat{\mathbf{K}}_{gf}] = -\frac{P_{\text{ref}}}{4000}\begin{bmatrix} 1 & 0 \\ 0 & 1 \end{bmatrix} \begin{matrix} u_b \\ v_b \end{matrix} \tag{b}$$

Equation 8.7

$$\frac{1}{\lambda}\begin{Bmatrix} u_b \\ v_b \end{Bmatrix} = \frac{P_{\text{ref}}}{400}\begin{bmatrix} 1 & 0 \\ 0 & 4 \times 10^{-4} \end{bmatrix}\begin{Bmatrix} u_b \\ v_b \end{Bmatrix} \tag{c}$$

Assuming $v_b = 0$ and $u_b \neq 0$

$$P_{\text{cre}} = \lambda P_{\text{ref}} = 400\ \text{kN} \tag{d}$$

From Example 8.1

$$P_{\text{cre}} = kL = 2 \times 200 = 400\ \text{kN}$$

2. $\alpha = 0.05$
   Equation 8.3:

$$[\mathbf{K}_{ef} + \mathbf{K}_{gf}]\{d\boldsymbol{\Delta}\} = \{d\mathbf{P}\} \tag{e}$$

$$\left[200\begin{bmatrix} \frac{2}{L_{ab}} & 0 \\ 0 & \frac{5 \times 10^3}{L_{bc}} \end{bmatrix} + \frac{F_{bc}}{L_{bc}}\begin{bmatrix} 1 & 0 \\ 0 & 1 \end{bmatrix} + \frac{F_{ab}}{L_{ab}}\begin{bmatrix} 1 & 0 \\ 0 & 1 \end{bmatrix}\right]\begin{Bmatrix} du_b \\ dv_b \end{Bmatrix} = dP\begin{Bmatrix} 0.05 \\ 1 \end{Bmatrix}$$

Solving iteratively:

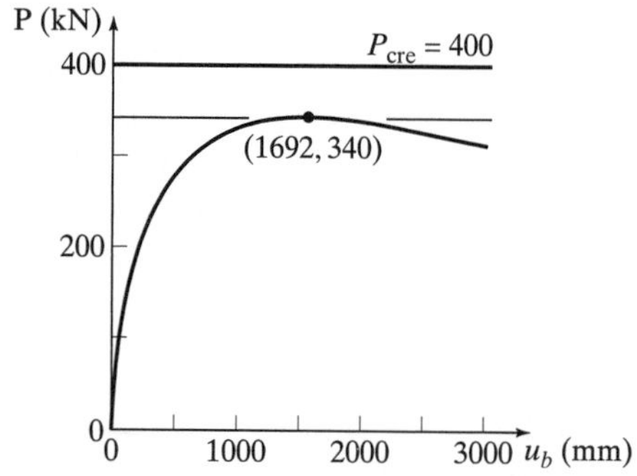

Example 9.2 is similarly related to Example 8.2. The member *bd* has been added, and dimensions and properties have been specified. The computer solutions illustrated contain provisions for continually updating stiffness coefficients and tracking physically unstable regions of the response curves (see Section 12.7). In recognition of its symmetry, the system is analyzed as a single-degree-of-freedom one. Increasing the stiffness of *bd* changes the response from a mode that is directly comparable to that of Example 8.2 ($A_{bd} = 0$) to one in which the region of instability is gradually decreased and finally eliminated. It is of interest to note that in the case of $A_{bd} = 0$ the numerically calculated values of the displacement at the stability limit point (212 mm) and the stability limit (2564 kN) are in agreement with the terms of the analytical solution (Examples 8.2 and 8.8).

EXAMPLE 9.2

All members are axial force members. $E = 200{,}000$ MPa. Analyze the large displacement response of the system for A_{ab} and $A_{bc} = 1000$ mm² and $A_{bd} = 0, 15, 30,$ and 45 mm².

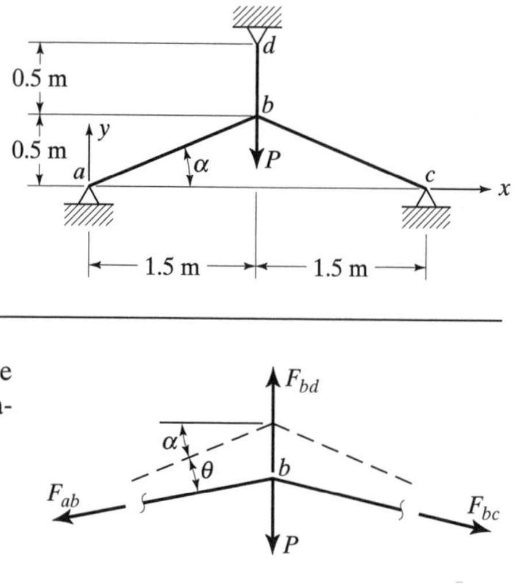

For equilibrium in a reference configuration—using symmetry and assuming the geometry, member lengths, and bar forces have been updated—the stiffness matrices are:

$$[\mathbf{K_e}] = \left[\frac{2A_{ab}E}{L_{ab}}\sin^2(\alpha - \theta) + \frac{A_{bd}E}{L_{bd}}\right]$$

$$[\mathbf{K_g}] = \left[\frac{2F_{ab}}{L_{ab}}\sin^2(\alpha - \theta) + \frac{F_{bd}}{L_{bd}}\right]$$

From Equation 8.3:

$$\left[2\left(\frac{A_{ab}E + F_{ab}}{L_{ab}}\right)\sin^2(\alpha - \theta) + \left(\frac{A_{bd}E + F_{bd}}{L_{bd}}\right)\right]dv_b = dP$$

Computed results:

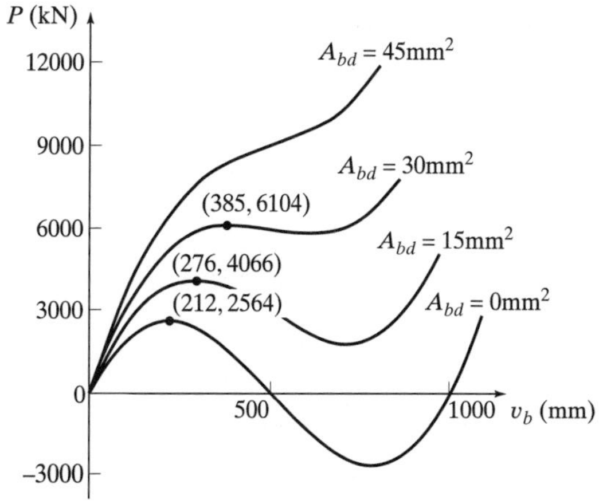

Example 9.3 is essentially a computerized version of Example 8.3. The member is subdivided into two elements of equal length and the global stiffness equations (Equation 8.7 for the first case and Equation 8.3 for the second) are assembled from the basic element equations (Equation 4.34 for the elastic stiffness matrix and Equation 9.14 for the geometric one). The two-element idealization proves to yield an excellent approximation of the critical load.

EXAMPLE 9.3

Member *ab* is a straight elastic bar

$$A = 1.27 \times 10^4 \text{ mm}^2,\ I = 3.66 \times 10^7 \text{ mm}^4,\ E = 200{,}000 \text{ MPa}$$

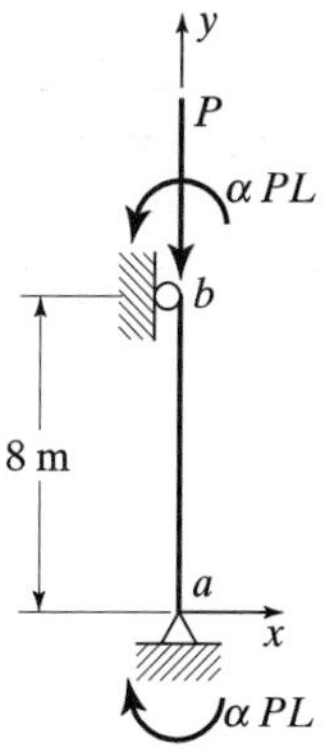

Analyze its behavior for

1. $\alpha = 0$
2. $\alpha = 1.25 \times 10^{-4}$

Model the member as two elements:

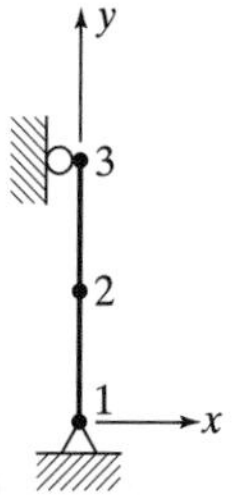

1. $\alpha = 0$
From Example 8.3

$$P_{\text{cre}} = \frac{\pi^2 EI}{L^2} = 1{,}129 \text{ kN}$$

From Equation 8.7

$$\lambda P_{\text{ref}} = 1{,}137 \text{ kN}$$

2. $\alpha = 1.125 \times 10^{-4}$
Computed results are:

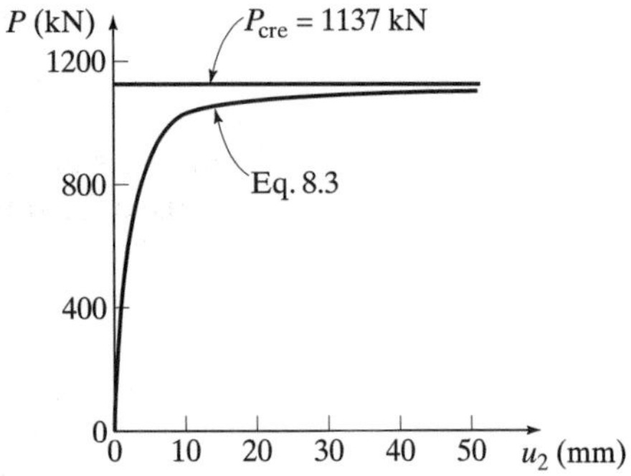

Example 9.4 is similarly related to Example 8.4. The nonlinear analyses are quantitative demonstrations of the destabilizing effect of a compressive force and the stiffening influence of a tensile one.

EXAMPLE 9.4

The member is a straight elastic bar

$$A = 1.27 \times 10^4 \text{ mm}^2, I = 3.66 \times 10^7 \text{ mm}^4, E = 200{,}000 \text{ MPa}$$

Analyze its behavior for:

1. $\alpha = 0$;
2. $\alpha = 0.005$;
3. $\alpha = 0.005$ and P directed upward.

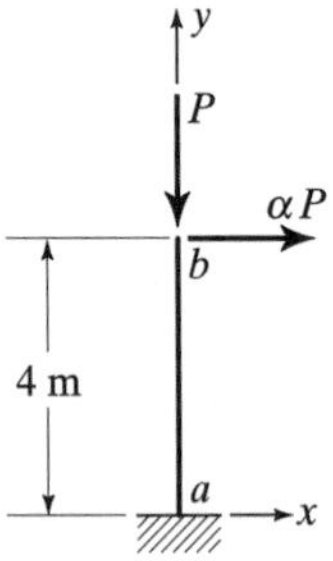

Model the member as two elements

1. $\alpha = 0$
From Example 8.4:

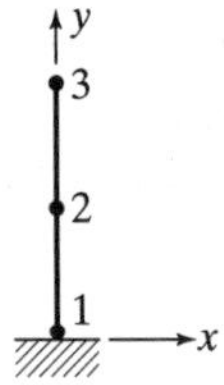

$$P_{cre} = \frac{\pi^2 EI}{4L^2} = 1{,}129 \text{ kN}$$

From Equation 8.7

$$\lambda P_{ref} = 1{,}129 \text{ kN}$$

2 and 3. For $\alpha = .005$ computed results are

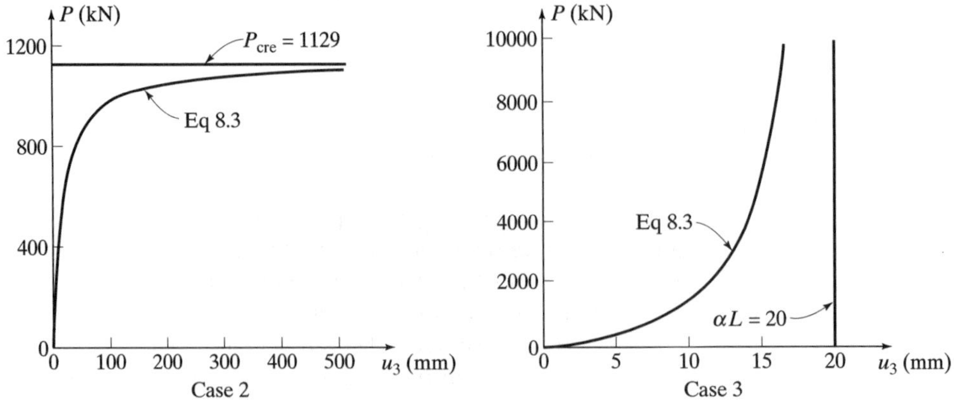

The behavior of a slender two-story frame is studied in Example 9.5, first by calculating the elastic critical load under vertical load alone, and then by using first- and second-order methods to analyze it under 60% of that load plus proportionately applied horizontal loads. A substantial P-Δ effect is observed; the second-order theory moments are significantly greater than the first-order ones. Analyses of this type may be used in satisfying the requirements for determining the second-order effects of lateral translation of building frame joints in standard design specifications (see Chapter C of Reference 9.2). The effect of column bowing has not been included in the display of results, and in this case it would not be significant. It could be approximated by using the shape function implicit in the geometric stiffness matrix of Equation 8.3 as an interpolation function. But where member bowing might be important, a preferred approach would be to subdivide the member. The refined model would give a more accurate representation of the member's stiffness and, as illustrated in Example 9.6, it yields the displacements of internal loads directly.

EXAMPLE 9.5

Members are straight elastic bars.

ab and bc: $A = 2.5 \times 10^4$ mm^2, $I = 6.36 \times 10^8$ mm^4

bd and ce: $A = 1.76 \times 10^4$ mm^2, $I = 8.61 \times 10^8$ mm^4

$E = 200{,}000$ MPa

1. $\alpha = 0$ Compute $P_{cre} = \lambda P_{ref}$
2. $\alpha = 0.01$ and $P = 0.6P_{cre}$
   Compare first- and second-order elastic analyses.

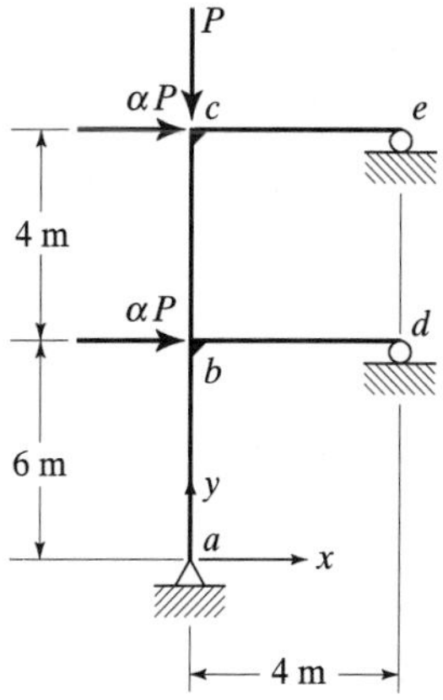

Model each member as a single element.

1. $\alpha = 0$ From Equation 8.7 $\lambda P_{ref} = 6630$ kN
2. $\alpha = 0.01$, $P = 3978$ kN
   Computed results are

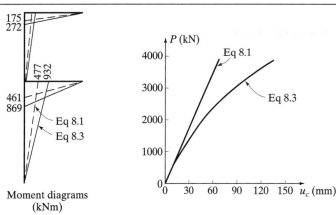

Moment diagrams (kNm)

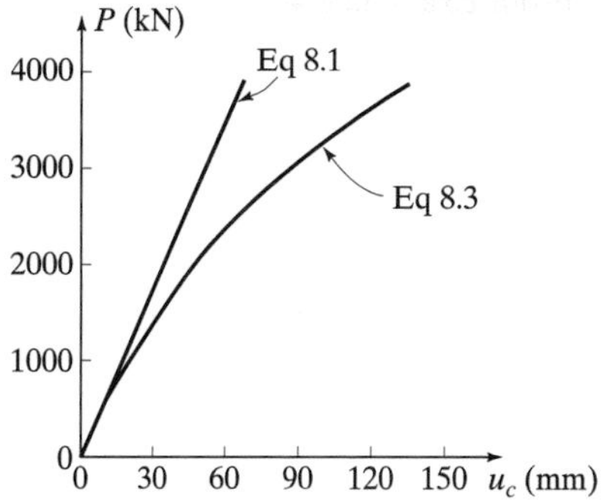

The single-story, single-bay portal frame analyzed by first- and second-order elastic methods in Example 9.6 will be analyzed by inelastic methods in Chapter 10 (Example 10.5). In all cases the columns are modeled by two elements and each segment of the girder by a single element. As in the preceding example, a significant second-order elastic effect is observed. The elastic redistribution of bending moments may also be noted: the moment at b changes sign, for example, and there is a shift in the location of the point of inflection in girder segment cd.

EXAMPLE 9.6

Members are steel wide flange sections

ab and de: $A = 13.3 \text{ in.}^2$, $I = 248 \text{ in.}^4$

bc and cd: $A = 24.8 \text{ in.}^2$, $I = 2850 \text{ in.}^4$

$E = 29{,}000 \text{ ksi}$

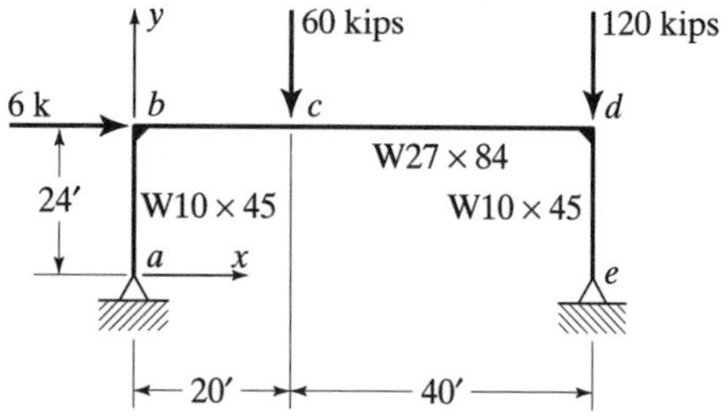

1. Compute $P_{cre} = \lambda P_{ref}$
2. Compare first- and second-order elastic analyses

Model columns as two elements and girder segments as one element.

1. From Equation 8.7 with P_{ref} equal to the vector of applied loads, $\lambda = 2.20$
2. Computed results are

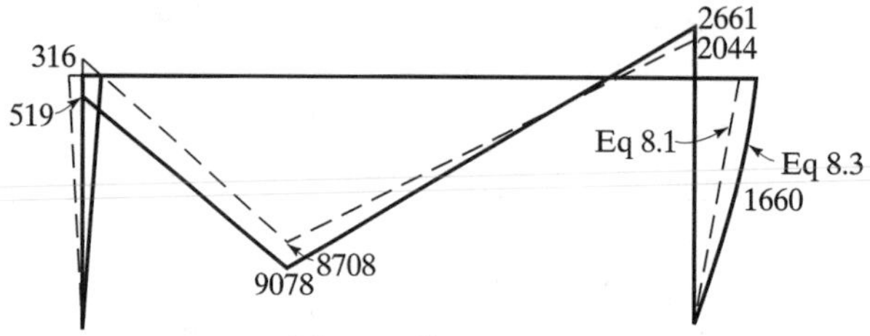

Moment diagrams
(in. kips)

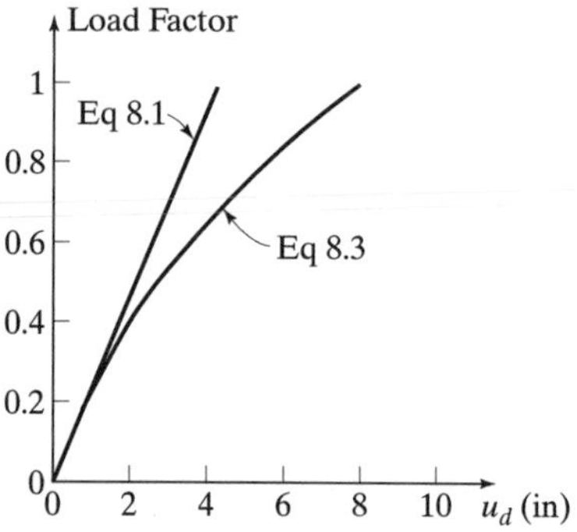

Example 9.7 presents the results of critical load and first- and second-order analyses of a planar trussed framework. It is one of seven slender frames studied analytically and in the laboratory by Hoff et al. (Ref. 9.3). They calculated critical loads by determining the limit of convergence of the Hardy Cross moment distribution process as applied to members subjected to axial force and bending, and they then compared their analytical and test results. For the specimen considered here the calculated and measured values were 203 pounds and 220 pounds, respectively. In the present computer analysis, each lower chord member has been modeled by two elements and all of the other members by one. The critical load obtained by Equation 8.7 is 210 pounds. As shown by a representative load-deflection curve, nonlinear response becomes perceptible at less than one-half the critical load. Also, this particular deflection component reverses direction under the secondary effect and increases markedly in magnitude as the critical as approached.

EXAMPLE 9.7

Members are elastic bars with

$A = 9.3482 \times 10^{-2} \text{ in.}^2$, $I = 6.9542 \times 10^{-4} \text{ in.}^4$,

$E = 29{,}000 \text{ ksi}$

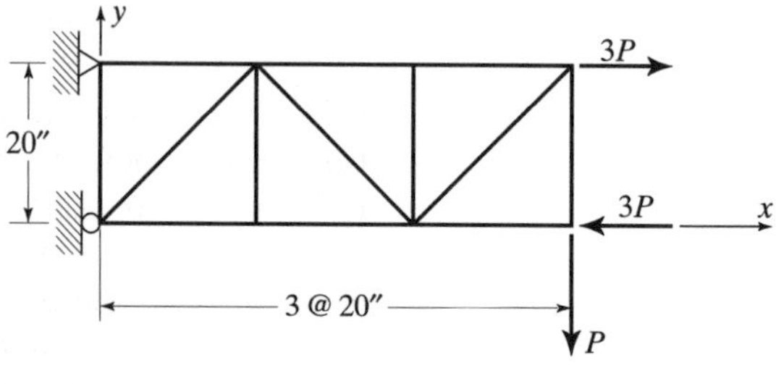

Compute $P_{cre} = \lambda P_{ref}$ and compare first- and second-order elastic analyses.

Model bottom chord members as two elements each and other members as single elements.

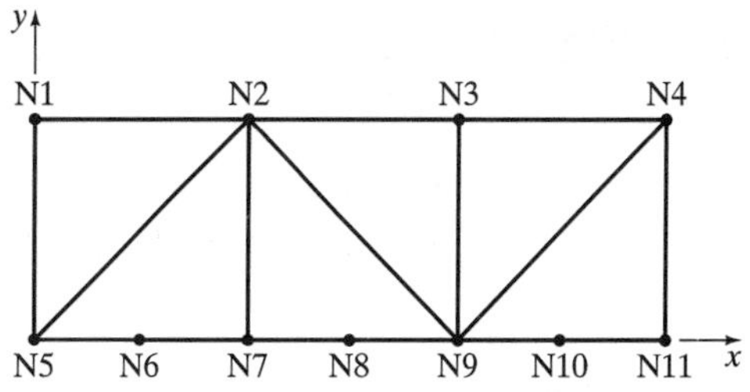

From Equation 8.7 $P_{cre} = 210$ lbs

The buckled configuration from Equation 8.7 and the deflected shape from Equation 8.3 at $P = 210$ lbs are:

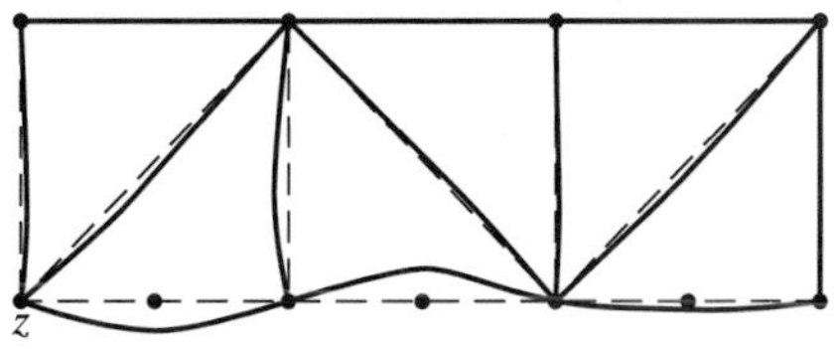

Buckled shape
Eq 8.7

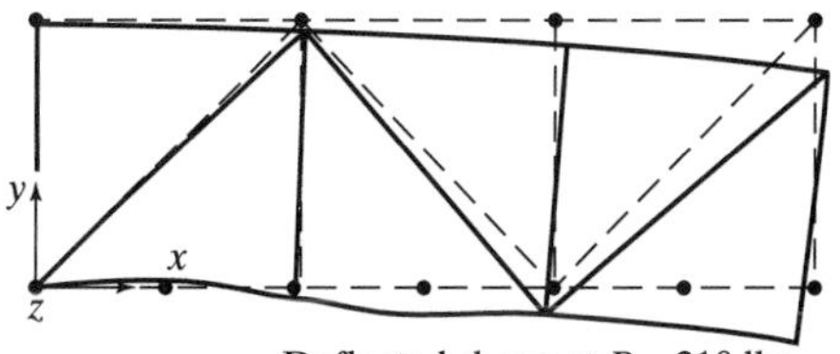

Deflected shape at $P = 210$ lbs
Eq 8.3

Vertical deflection at Node 6:

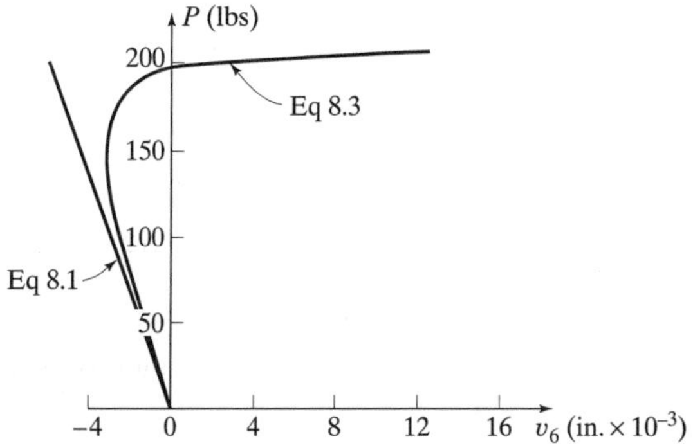

The effectiveness of a lateral brace of a compressed member is a function of its strength and stiffness. Example 9.8 illustrates methods for calculating each in the case of a column braced at midheight. It can be shown that the elastic critical load for an ideally straight column of this type is $\pi^2 EI/L^2$, provided that the stiffness of the brace in the lateral direction is equal to or greater than two times that load divided by L (see Ref. 9.4 or 9.5). For a smaller stiffness, the column will buckle in a single half wave at a lesser load, with a minimum of $\pi^2 EI/4L^2$ for no brace at all. This is demonstrated in the first part of the example. Since in the ideal case there is no force in the brace at incipient buckling, analysis of the strength demand requires the assumption of an imperfect system. This is done in the second part of the example, in which the midpoint of the column is assumed to be initially offset 1/500th of the total length (and nodes at the quarter points 3/4 as much). Second-order elastic analysis results in a brace force of approximately 2 kips at the critical load of 215 kips. It is seen that, in this particular example, both the stiffness and the strength required for the effective bracing of the elastic column are modest.

EXAMPLE 9.8

Member *ac* is an elastic bar braced at midheight by the elastic strut *bd*. $E = 29{,}000$ ksi.

1. Assume *ac* ideally straight. Determine the variation of the critical load with the axial stiffness of the brace.
2. Assume an initial offset of $2L/500$ at point *b* and a brace area, A_{br}, of 1.27 in.^{2}. Determine the variation of force in the brace with the applied load.

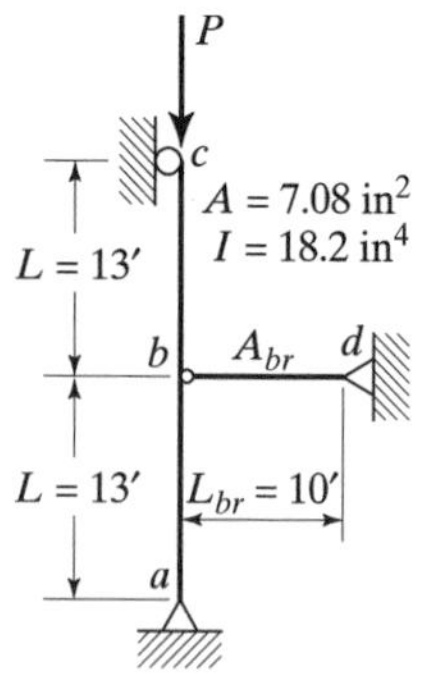

1. Buckling in two half waves will occur for $k_{br} \geq 2P_e/L$, in which $P_e = \pi^2EI/L^2$ and $k_{br} = A_{br}E/L_{br}$. The minimum condition for this is

$$k_{br} = \left(\frac{2}{156}\right)\frac{\pi^2(29{,}000)(18.2)}{(156)^2} = 2.744 \text{ kips/in.}$$

or

$$A_{br} = \frac{(2.744)(120)}{29{,}000} = 0.01136 \text{ in.}^2$$

Modeling *ac* by four elements and applying Equation 8.7 for a series of values of A_{br} produces the P_{cre} vs. A_{br} relationship of sketch (a).

2. For *b* offset $(2 \times 156)/500 = 0.624$ in. and $A_{br} = 1.27$ in.^{2}, application of Eq. 8.3 to the same model produces the *P* vs. F_{br} relationship of sketch (b).

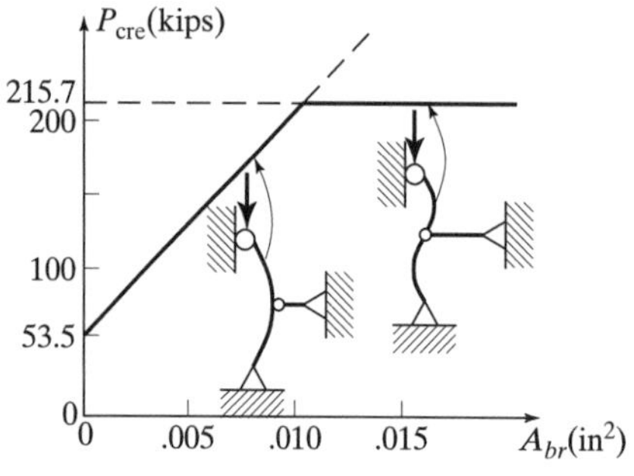

(*a*)

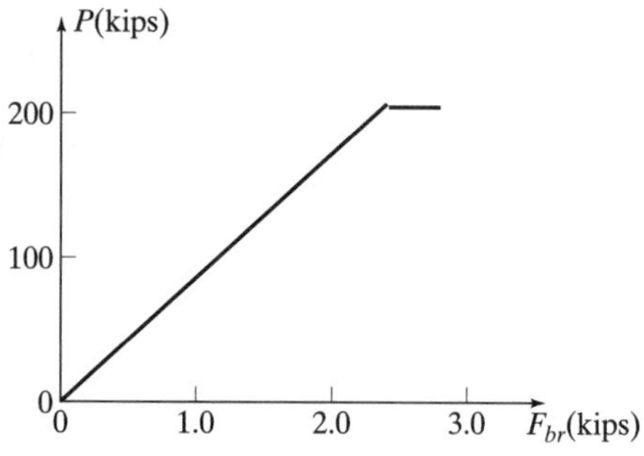

(*b*)

9.2 COMBINED TORSION AND AXIAL FORCE

In Section 9.1 we treated structures in which the response to an action was in a single plane. As seen in Sections 4.6 and 7.4, even in the simplest cases twisting about a straight axis can be accompanied by displacement of points in the body of the element in three coordinate directions. The interaction between torque and other components of the applied force can have a profound effect on the behavior of a system. The problem is a three-dimensional one that will be dealt with in Section 9.3, but one important factor, the influence of axial force on the twisting of a shaft in which the only mode of resistance is St. Venant torque, can be analyzed by a simple extension of the approach of Section 9.1.

Consider the rectangular tubular element of Figure 9.2 subjected to an axial force, F_{x2}, and torque, M_{x2}. Twisting and net longitudinal displacement of the left end are

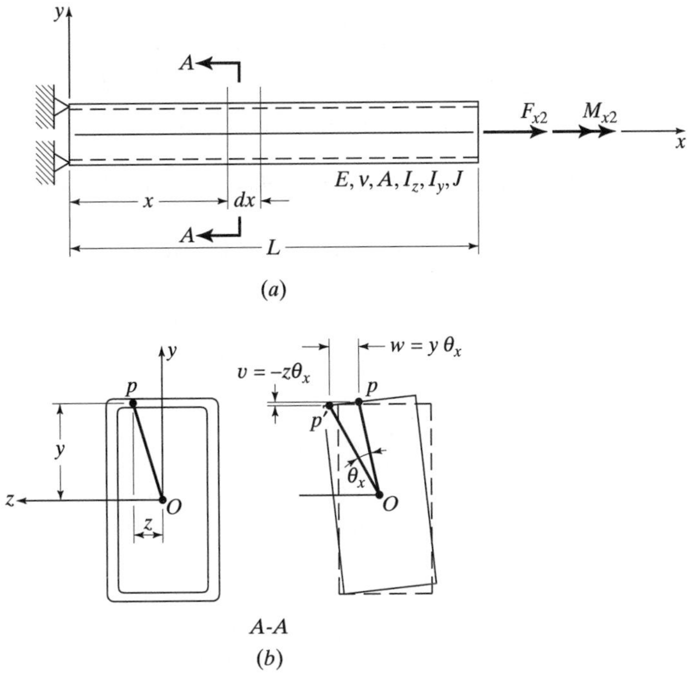

Figure 9.2 Combined Torsion and Axial Force.

prevented. A cross section at a distance x from that end will rotate about the centroidal axis through an angle θ_x and a point p in the tube's wall will translate as shown in Figure 9.2*b*. For a small rotation, the components of translation in the plane of the cross section are

$$v = -z\theta_x \qquad \text{and} \qquad w = y\theta_x$$

Thus

$$\frac{dv}{dx} = -z\,\frac{d\theta_x}{dx} \qquad \text{and} \qquad \frac{dw}{dx} = y\,\frac{d\theta_x}{dx} \tag{9.15}$$

By an obvious extension of Equation 9.1, the finite strain of a longitudinal fiber passing through point p is

$$e_{\text{fin}} = \frac{du}{dx} + \frac{1}{2}\left[\left(\frac{du}{dx}\right)^2 + \left(\frac{dv}{dx}\right)^2 + \left(\frac{dw}{dx}\right)^2\right] \tag{9.1a}$$

Limiting consideration to the geometric effect of interest here we have, from Equation 9.15

$$e_{\text{fin g}} = \frac{1}{2}\left[z^2\left(\frac{d\theta_x}{dx}\right)^2 + y^2\left(\frac{d\theta_x}{dx}\right)^2\right] = \frac{1}{2}\,[z^2 + y^2]\left(\frac{d\theta_x}{dx}\right)^2$$

This is a warping strain, which, in the presence of an axial stress, is a source of internal virtual work. Thus, following Equation 9.2, the relevant virtual work equation may be written as

$$\delta W_{\text{int g}} = \frac{1}{2}\int_{\text{vol}} \sigma_x\, \delta\left(\frac{d\theta_x}{dx}\right)^2 [z^2 + y^2]\, dA\, dx$$

or, letting $\sigma_x = F_{x2}/A$ and integrating over the cross section

$$\delta W_{\text{int g}} = \frac{1}{2}\frac{F_{x2}I_p}{A}\int_0^L \delta\left(\frac{d\theta}{dx}\right)^2 dx \tag{9.16}$$

in which $\int_A (z^2 + y^2)\, dA = I_p$, the polar moment of inertia of the cross section.

From Equations 9.4–9.8 it may be seen that, if a linear shape function is assumed for θ_x, Equation 9.16 leads to the stiffness matrix

$$[\mathbf{k_g}] = \frac{F_{x2}I_p}{AL}\begin{array}{c} \begin{array}{cc} \theta_{x1} & \theta_{x2} \end{array} \\ \begin{bmatrix} 1 & -1 \\ -1 & 1 \end{bmatrix} \end{array} \tag{9.17}$$

In Example 9.9, this equation is used in conjunction with Equation 4.27a to determine the torsional critical load of a column, a bifurcation condition in which twisting about the axis of an axially loaded member becomes possible. A cruciform section is used for illustration because, since the ratio of its torsional to flexural rigidity is relatively low, it is a shape prone to becoming unstable through twisting. For consistency with earlier numerical examples Equation 8.7 is used in the formulation of the problem, but in this case it is seen to lead immediately to the classical analytical solution (see Ref. 9.4). Also, although for most structural shapes simple torsional buckling is normally not the controlling mode, in the numerical example used for illustration, comparison with the flexural critical load shows that it would be in this particular case.

EXAMPLE 9.9

The member shown is a bisymmetrical steel cruciform section fixed against twisting at the top and bottom.

$$A = 48 \text{ in.}^2,\ I_y = I_z = 1152 \text{ in.}^4$$

$$I_p = 2304 \text{ in.}^4,\ J = 16 \text{ in.}^4,\ E = 29{,}000 \text{ ksi},\ \nu = 0.3$$

Determine the elastic critical load in pure torsion. Compare with the flexural critical load.

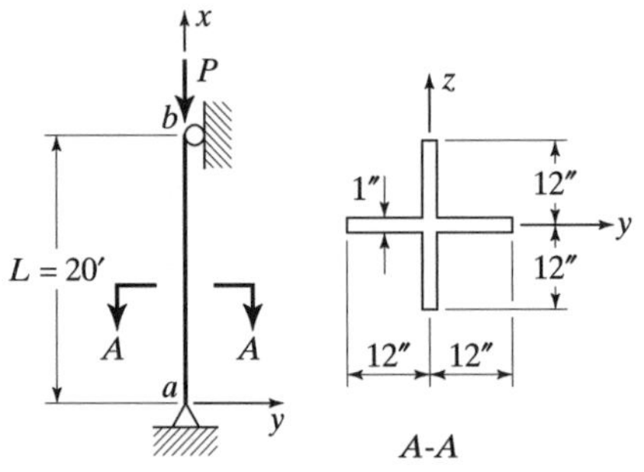

Model the member as two elements

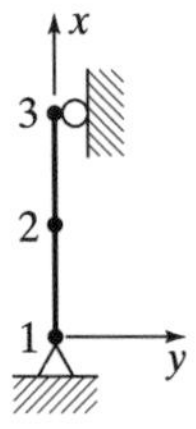

From Equation 4.27a

$$[\mathbf{K}_{ef}] = \left[\frac{GJ}{L}\right]$$

From Equation 9.17

$$[\hat{\mathbf{K}}_{gf}] = \left[-\frac{P_{\text{ref}}I_p}{AL}\right]$$

From Equation 8.7

$$\frac{1}{\lambda}\{\theta_{x2}\} = \frac{P_{\text{ref}}I_p}{GJA}\{\theta_{x2}\}$$

Thus

$$P_{\text{crt}} = \lambda P_{\text{ref}} = \frac{GJA}{I_p} = \frac{(29000)(16)(48)}{2(1.3)(2304)} = 3718 \text{ kips}$$

compared with

$$P_{\text{cry}} = \frac{\pi^2 EI_y}{L^2} = \frac{\pi^2(29000)(1152)}{(240)^2} = 5724 \text{ kips}$$

9.3 THREE DIMENSIONAL GEOMETRIC NONLINEAR ANALYSIS—AN OVERVIEW

As may be seen from Chapters 4 and 5, in linear elastic analysis the difference between planar system and three-dimensional system analysis is essentially quantitative: The computations are lengthier, but the basic equations are the same. In nonlinear analysis the quantitative differences remain, and are indeed magnified, but there are also qualitative differences. We already have the three-dimensional elastic stiffness matrix of Equation 4.34 and we may write a three-dimensional geometric stiffness matrix simply by adding to Equation 9.14 terms obtained by applying the same approach to a member bending in the x-z plane and the effect of St. Venant torsion–axial force interaction. With due attention to signs, the result is

$$[\mathbf{k_g}] = \frac{F_{x2}}{L}
\begin{array}{c}
\begin{array}{cccccccccccc}
u_1 & v_1 & w_1 & \theta_{x1} & \theta_{y1} & \theta_{z1} & u_2 & v_2 & w_2 & \theta_{x2} & \theta_{y2} & \theta_{z2}
\end{array}\\
\left[\begin{array}{cccccccccccc}
1 & 0 & 0 & 0 & 0 & 0 & -1 & 0 & 0 & 0 & 0 & 0\\
 & \frac{6}{5} & 0 & 0 & 0 & \frac{L}{10} & 0 & -\frac{6}{5} & 0 & 0 & 0 & \frac{L}{10}\\
 & & \frac{6}{5} & 0 & -\frac{L}{10} & 0 & 0 & 0 & -\frac{6}{5} & 0 & -\frac{L}{10} & 0\\
 & & & \frac{I_p}{A} & 0 & 0 & 0 & 0 & 0 & -\frac{I_p}{A} & 0 & 0\\
 & & & & \frac{2L^2}{15} & 0 & 0 & 0 & \frac{L}{10} & 0 & -\frac{L^2}{30} & 0\\
 & & & & & \frac{2L^2}{15} & 0 & -\frac{L}{10} & 0 & 0 & 0 & -\frac{L^2}{30}\\
 & & & & & & 1 & 0 & 0 & 0 & 0 & 0\\
 & & & & & & & \frac{6}{5} & 0 & 0 & 0 & -\frac{L}{10}\\
 & & & \text{Sym} & & & & & \frac{6}{5} & 0 & \frac{L}{10} & 0\\
 & & & & & & & & & \frac{I_p}{A} & 0 & 0\\
 & & & & & & & & & & \frac{2L^2}{15} & 0\\
 & & & & & & & & & & & \frac{2L^2}{15}
\end{array}\right]
\end{array} \tag{9.18}$$

Equation 4.34 remains valid and Equation 9.18 is satisfactory insofar as it goes. It may be noted that it contains a second order term corresponding to every first-order term of Equation 4.34. In addition to accounting for the interaction of axial force and

St. Venant torsion just mentioned, it has provisions for accounting for the geometric interaction of axial force and bending in the principal directions as the element displaces and, since shearing forces are treated as reactions to the variation in bending moments, for accounting for axial force—flexural shear interaction. The element stiffness matrices of Equations 4.34 and 9.18, when used in conjunction with the global equations of Chapter 8, are adequate for the analysis of many of the three-dimensional geometric nonlinear frame problems encountered in practice. Nevertheless, they are less than fully comprehensive.

Among the ways in which these equations fall short are some that are beyond the scope of this text, such as the treatment of nonsymmetrical cross sections. Others, however, relate to features that should be considered in any introductory work on nonlinear frame analysis. They include the need to consider the interaction of other components of force, such as bending moments and torque, the need to ensure the ability to satisfy equilibrium on the deflected structure, and the desirability of including provisions for the analysis of nonuniform torsion. To address these needs and desires means going beyond the condition-specific physical arguments used to this point and calling on more comprehensive fundamental principles of the mechanics of structures and finite displacements. Full understanding of the equations of analysis and the assumptions made in their development requires that the theory behind them be explained. But, for the reasons given in the introduction to this chapter, we are putting a review of the essential theory and the development of the additional terms in Appendix A. The terms relevant to the interaction of St. Venant torsion and flexure are summarized in Equations A.38 and A.48. Adding these to Equation 9.18, we have:

$$
[\mathbf{k_g}] = \begin{array}{c} \begin{array}{cccccccccccc} u_1 & v_1 & w_1 & \theta_{x1} & \theta_{y1} & \theta_{z1} & u_2 & v_2 & w_2 & \theta_{x2} & \theta_{y2} & \theta_{z2} \end{array} \\ \left[\begin{array}{cccccccccccc}
\frac{F_{x2}}{L} & 0 & 0 & 0 & 0 & 0 & -\frac{F_{x2}}{L} & 0 & 0 & 0 & 0 & 0 \\
 & \frac{6F_{x2}}{5L} & 0 & \frac{M_{y1}}{L} & \frac{M_{x2}}{L} & \frac{F_{x2}}{10} & 0 & -\frac{6F_{x2}}{5L} & 0 & \frac{M_{y2}}{L} & -\frac{M_{x2}}{L} & \frac{F_{x2}}{10} \\
 & & \frac{6F_{x2}}{5L} & \frac{M_{z1}}{L} & -\frac{F_{x2}}{10} & \frac{M_{x2}}{L} & 0 & 0 & -\frac{6F_{x2}}{5L} & \frac{M_{z2}}{L} & -\frac{F_{x2}}{10} & -\frac{M_{x2}}{L} \\
 & & & \frac{F_{x2}I_p}{AL} & -\frac{2M_{z1}-M_{z2}}{6} & \frac{2M_{y1}-M_{y2}}{6} & 0 & -\frac{M_{y1}}{L} & -\frac{M_{z1}}{L} & -\frac{F_{x2}I_p}{AL} & -\frac{M_{z1}+M_{z2}}{6} & \frac{M_{y1}+M_{y2}}{6} \\
 & & & & \frac{2F_{x2}L}{15} & 0 & 0 & -\frac{M_{x2}}{L} & \frac{F_{x2}}{10} & -\frac{M_{z1}+M_{z2}}{6} & -\frac{F_{x2}L}{30} & \frac{M_{x2}}{2} \\
 & & & & & \frac{2F_{x2}L}{15} & 0 & -\frac{F_{x2}}{10} & -\frac{M_{x2}}{L} & \frac{M_{y1}+M_{y2}}{6} & -\frac{M_{x2}}{2} & -\frac{F_{x2}L}{30} \\
 & & & & & & \frac{F_{x2}}{L} & 0 & 0 & 0 & 0 & 0 \\
 & & & \text{Sym.} & & & & \frac{6F_{x2}}{5L} & 0 & -\frac{M_{y2}}{L} & \frac{M_{x2}}{L} & -\frac{F_{x2}}{10} \\
 & & & & & & & & \frac{6F_{x2}}{5L} & -\frac{M_{z2}}{L} & \frac{F_{x2}}{10} & \frac{M_{x2}}{L} \\
 & & & & & & & & & \frac{F_{x2}I_p}{AL} & \frac{M_{z1}-2M_{z2}}{6} & -\frac{M_{y1}-2M_{y2}}{6} \\
 & & & & & & & & & & \frac{2F_{x2}L}{15} & 0 \\
 & & & & & & & & & & & \frac{2F_{x2}L}{15}
\end{array}\right] \end{array}
\tag{9.19}
$$

9.4 EXAMPLES OF THREE-DIMENSIONAL STRUCTURE ANALYSIS

Examples 9.10–9.12 are companion examples of the calculation of elastic critical loads in elementary structures: a simply supported beam subjected to uniform bending about the principal axis, an end-loaded cantilever beam, and a beam-column subjected to a combination of axial load and uniform bending about the principal axis. In each, buckling occurs through a combination of lateral bending and twisting. Results obtained by the application of Equation 8.7 are compared with classical solutions (see Ref. 9.4). Since provision for nonuniform torsion is not included in the geometric stiffness matrix used in the analysis (Equation 9.19) the neglect of warping resistance requires comment. In the first two examples the cross section is a narrow rectangle whose resistance to warping is essentially nonexistent. A wide flange member with significant warping resistance is used in the third example. However, as shall be shown in the analysis of the results, for the prescribed loading condition neglect of warping restraint has only a small effect on the calculated critical load. Nonuniform torsion in nonlinear analysis will be treated in Section A.5 of Appendix A. Included in Example A.3 of that appendix is an analysis of the member of Example 9.12 under loading conditions in which the warping restraint effect is substantial.

In Example 9.10 a three-element discretization of the member produces a calculated critical load within 5% of the theoretical value, which is often satisfactory for practical purposes. Also shown in the example are plots of the midspan lateral displacement and twist obtained from a second-order elastic analysis of the four-element model of the member in which a small proportionate lateral (out-of-plane) load was added at midspan. It is seen that the response becomes essentially asymptotic to the critical load, lending support to that value as the member's limit of elastic stability. It should be noted, however, that there are cases in which beams can undergo substantial in-plane deflection before buckling laterally and that this can have a significant influence on the critical load (see Appendix A, Section A.5, Example A.2). The prebuckling deflection effect was not included in the development of the classical solution referred to in Example 9.10, nor in the formulation of the stiffness equations, in which the undeflected member is taken as the reference configuration.

EXAMPLE 9.10

The member shown is a rectangular steel section fixed against twisting at both ends.

$$A = 20 \text{ in.}^2,\ I_z = 166.67 \text{ in.}^4,\ I_y = 6.67 \text{ in.}^4$$
$$J = 26.67 \text{ in.}^4,\ E = 29{,}000 \text{ ksi},\ \nu = 0.3$$

1. Determine the critical lateral buckling load.
2. Determine second-order response (add a perturbing load at midspan).

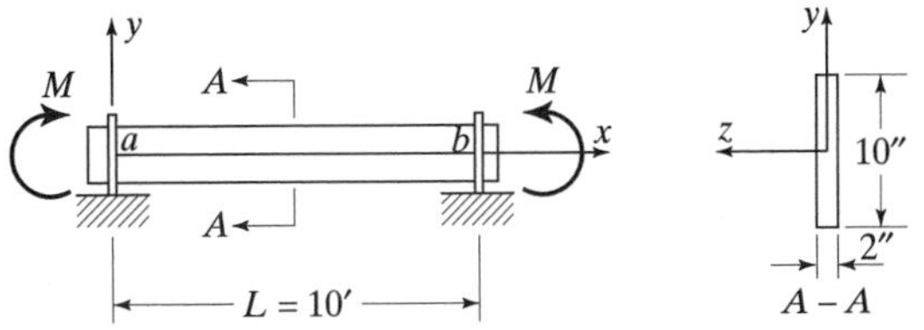

1. The theoretical critical moment is (Ref 9.4)

$$M_{cr} = \frac{\pi}{L}\sqrt{EI_yGJ} = \frac{29000\pi}{120}\sqrt{\frac{6.67 \times 26.67}{2.6}} = 6{,}280 \text{ in. kips}$$

The calculated critical moments using Equations 4.34, 9.19, and 8.7, and three levels of discretization are

No. elements	Critical moment (in. kips)	Percentage error (%)
3	6570	5
4	6440	3
8	6320	1

2. Midspan response
(Using 8-element model with a lateral load of $M/800\ L$ at midspan)

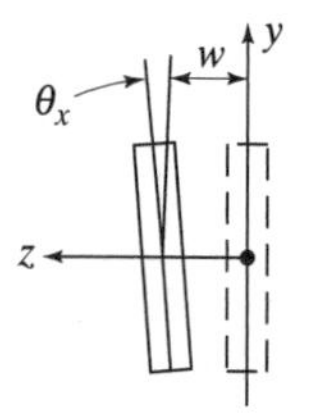

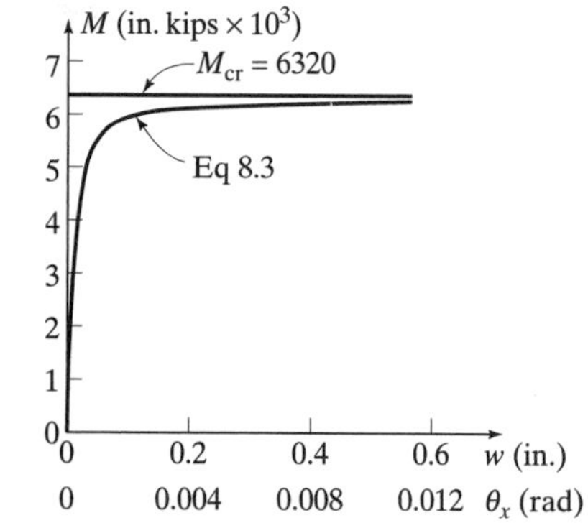

Example 9.11 is similar to Example 9.10 with perhaps the most significant difference being one that is not obvious from the summary of results. The solution of the differential equation of the end loaded cantilever is much less straightforward than that of the uniformly flexed, simply supported beam (see Ref. 9.4) whereas in the formulation and solution of the matrix equations there is no difference between the two.

EXAMPLE 9.11

The member shown is a cantilever beam of the same cross section as in Example 9.10. Determine the critical lateral buckling load.

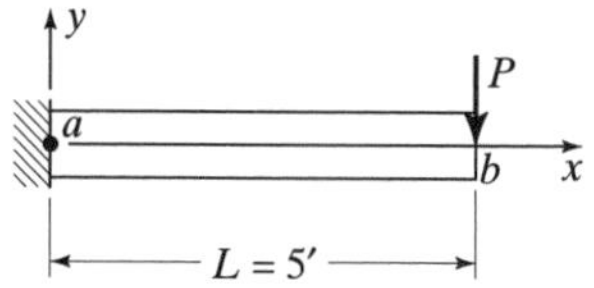

The theoretical elastic critical load is (Ref. 9.4)

$$P_{cr} = \frac{4.013}{L^2}\sqrt{EI_yGJ} = \frac{4.013 \times 29{,}000}{60^2}\sqrt{\frac{6.67 \times 26.67}{2.6}} = 267 \text{ kips}$$

The calculated critical load using Equations 4.34, 9.19, and 8.7 and three levels of discretization are

No elements	Critical load (kips)	Percentage error (%)
2	286	7
3	275	3
4	271	1

The buckled configuration of the member in Example 9.12 is again a combination of twisting and flexing. Equation A.58 of Appendix A is the classical critical load equation for the member and loading shown. Controlling parameters in that equation are P_{cf} and P_{ct}, the critical loads in minor axis bending and torsion, respectively. For given end moments, one of the two solutions to the resulting quadratic equation must be less than P_{cf} and P_{ct} and the other larger than both. Thus, the governing torsional-flexural critical load, P_{ctf}, is the lesser of the two solutions. In Example 9.12 the warping resistance term in P_{ct} has been neglected. Also, end moments proportional to the axial load have been prescribed, and a value of the proportionality factor that results in a critical load equation in which one root is infinitely large has been used.

The finite root is the simple function of P_{cf} and P_{ct} shown. This value is approximated to within 2% by the matrix analysis of a three-element discretization of the member.

EXAMPLE 9.12

Member *ab* is a wide flange section subjected to an axial force and bending about its major axis.

$$A = 22.4 \text{ in.}^2,\ I_z = 2100 \text{ in.}^4,\ I_y = 82.5 \text{ in.}^4,$$
$$J = 2.68 \text{ in.}^4,\ C_w = 11{,}100 \text{ in.}^6,\ E = 29{,}000,\ \nu = 0.3$$

Calculate the critical torsional-flexural critical load. Neglect warping resistance.

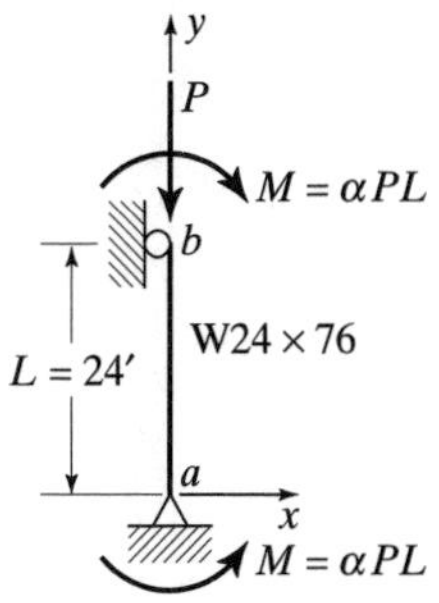

The critical load is the smallest load that satisfies the following equation (Ref. 9.4 and Equation A.58)

$$(P - P_{cf})(P - P_{ct}) = \frac{AM^2}{I_\rho} \tag{a}$$

in which

$$P_{cf} = \frac{\pi^2 EI_y}{L^2},\ \text{and } P_{ct} = \frac{GJA}{I_\rho} \text{ (neglecting warping resistance)}$$

Letting $M = \alpha PL$, Equation a becomes

$$\left(1 - \frac{\alpha^2 AL^2}{I_\rho}\right)P^2 - (P_{cf} + P_{ct})P + (P_{cf}P_{ct}) = 0 \tag{b}$$

For illustration, let $\frac{\alpha^2 AL^2}{I_\rho} = 1$. Thus

$$P_{ctf} = \frac{P_{cf}P_{ct}}{P_{cf} + P_{ct}} \tag{c}$$

in which

$$P_{cf} = \frac{\pi^2(29{,}000)(82.5)}{(24 \times 12)^2} = 284.7 \text{ kips and } P_{ct} = \frac{29{,}000}{2.6}\,\frac{(2.68 \times 22.4)}{(2100 + 82.5)} = 306.8 \text{ kips}$$

Thus

$$P_{ctf} = \frac{(284.7)(306.8)}{(284.7 + 306.8)} = 147.7 \text{ kips}$$

The calculated critical load using Equations 4.34, 9.19, and 8.7 and three levels of discretization is

No. elements	Critical load (kips)
2	155.3
3	151.1
4	149.6

It was noted that the validity of this example would be influenced by the fact that torsional warping resistance was neglected. Although true, the effect is not great in this case because the dominant contribution to displacement was minor axis bending. Application of Equation A.58 to the same problem, but with warping resistance included, shows that, whereas P_{ct} is more than twice the value used here, the critical torsional flexural buckling load is only about 7% higher.

Example 9.13 is similar to the first part of Example 9.8 in that it is a study of the effectiveness of lateral bracing, but in this case it involves the possibility of out-of-plane buckling of a planar system. The question is whether the tension diagonal has sufficient stiffness to prevent transverse displacement of the compression diagonal at its elastic critical load. It is found that if the tension diagonal is pin-connected at the junction of the two and the compression diagonal is continuous, it does not. But if the tension diagonal is continuous it does, regardless of the rigidity of the compression diagonal's connections. The reason for the small difference between the critical loads of cases 2 and 3 is that, as a continuous member, the compression diagonal derives some increased resistance from flexural restraint at the junction.

EXAMPLE 9.13

All members are straight elastic bars. Beam-to-column joints are rigid. Diagonals are pinned at each end and connected at *e*. $I_{diagonals}$ = 10.7 in.^{4}, E = 29,000 ksi.

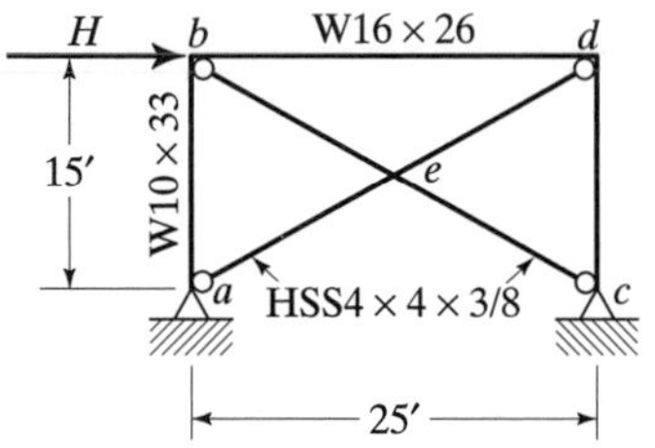

1. Calculate the elastic critical load, *H*, for
   (a) Tension diagonals pinned at *e*.
   (b) Compression diagonals pinned at *e*.
   (c) Both diagonals continuous at *e*.
2. Compare results with classical theory.

1. Model all members as two elements and assume the beam is braced out of plane. From Equation 8.7 the critical loads and buckled shapes are:

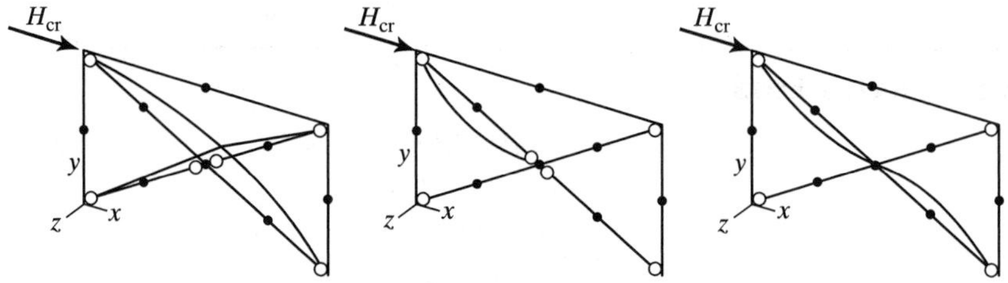

(*a*) H_{cr} = 86.0 kips (*b*) H_{cr} = 150.0 kips (*c*) H_{cr} = 164.0 kips

2. Compare axial force in *be* at H = 150 kips with its critical load as a pin-ended column:
   From linear analysis for H = 150 kips, F_{be} = 101.0 kips.
   For *be*, L = 174.9 in.:

$$P_{cr} = \frac{\pi^2(29{,}000)(10.7)}{(174.9)^2} = 100.1 \text{ kips}$$

Elastic force redistribution of internal forces in a space frame is illustrated by Example 9.14. As member *ae* buckles under load, the remainder of the system, particularly member *ab*, assumes an increasing portion of the resistance to further loading.

EXAMPLE 9.14

The member properties of the frame shown are

ab, ac, ad: $A = 2 \times 10^3$ mm^2, $I_y = I_z = 3 \times 10^6$ mm^4, $J = 6 \times 10^6$ mm^4

ae: $A = 2 \times 10^2$ mm^2, $I_y = I_z = 3 \times 10^5$ mm^4, $J = 3 \times 10^5$ mm^4

$E = 200{,}000$ MPa, $\nu = 0.3$.

1. Calculate the elastic critical load.
2. Determine the second-order response under the load P and a vertical load of $P \times 10^{-6}$ at the midspan of ae.

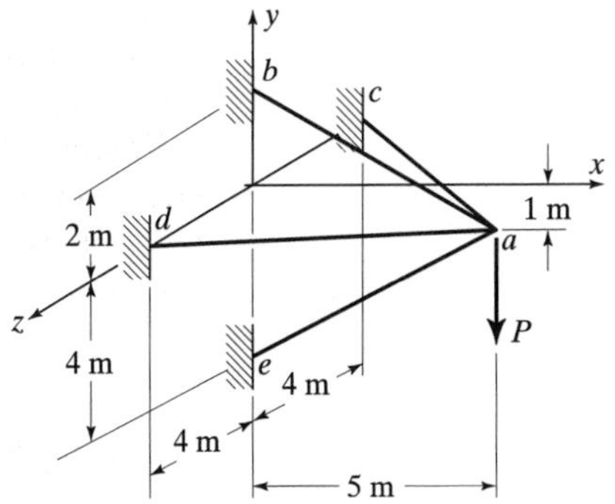

1. Model all members by two elements.
   From Equation 8.7, $P_{cre} = 116.1$ kN (sketch (a))
2. From Equation 8.3, force redistribution starts at approximately 110 kN (sketch (b))

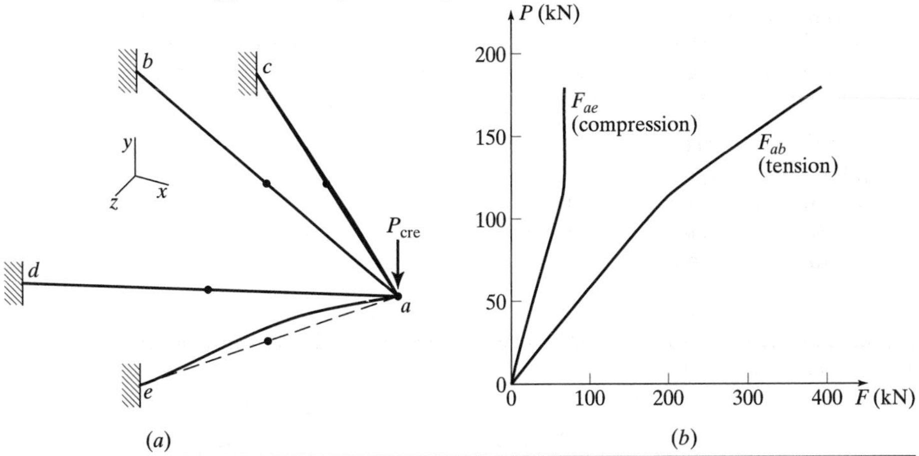

9.5 PROBLEMS

Note: In the following all references are to elastic methods of analysis and, unless otherwise mentioned, to in-plane behavior only. Members are aligned or oriented for major axis bending.

9.1 Compute the critical load and the nonlinear response for the truss shown (member ab is elastic and bc is essentially rigid). Use $\alpha = 0$ in the critical load analysis and $\alpha = 0.0025$ in the nonlinear analysis.

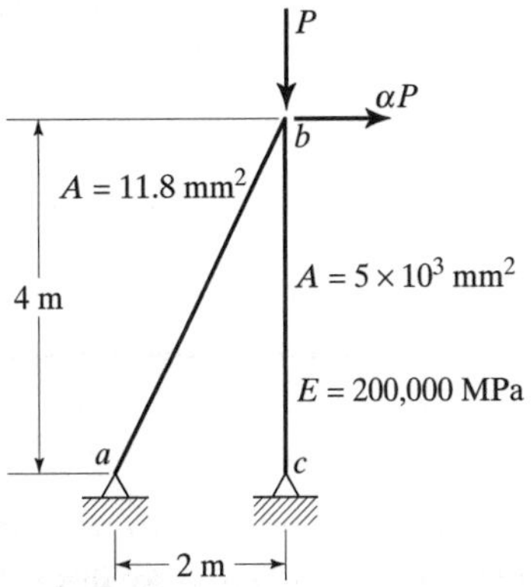

Problem 9.1

9.2 For the frame shown compute the critical load factor for the following reference loads:

	P_b	P_c
1	2000 kN	0
2	1333	667
3	1000	1000

For each case use several levels of element discretization and compare results.

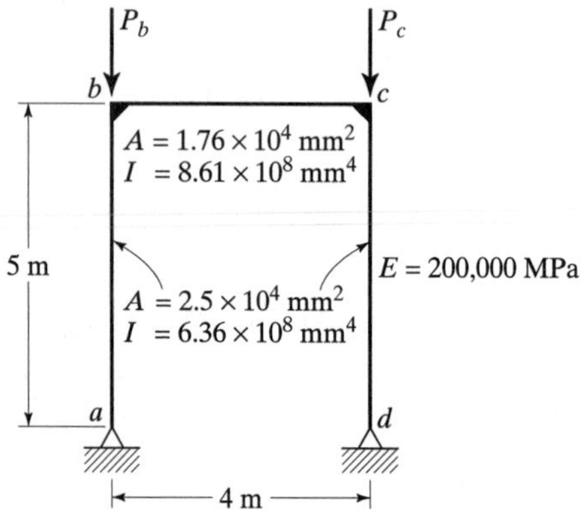

Problem 9.2

9.3 The same as Problem 9.2 except for the reduction of the moment of inertia of member *cd* to 3.18×10^8 mm^4 and use of the following reference loads:

	P_b	P_c
1	2000 kN	0
2	1000	1000
3	0	2000

9.4 (a) Compute the critical load and the buckled shape of the frame shown. (b) Compare the results of linear and nonlinear analyses for $P = 0.5P_{cre}$.

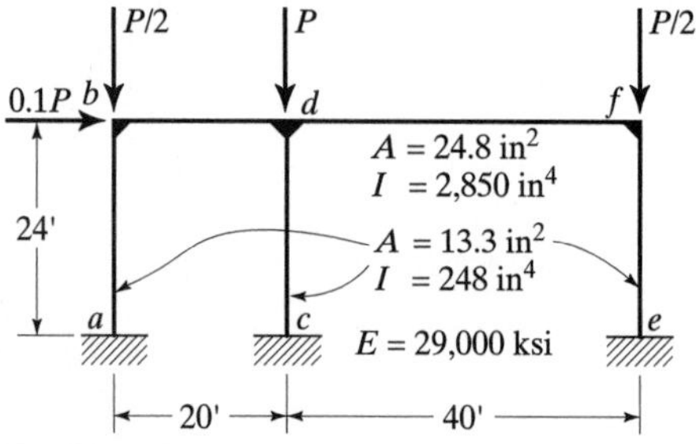

Problem 9.4

9.5 Compare the results of linear and nonlinear analyses for the frame and loading shown.

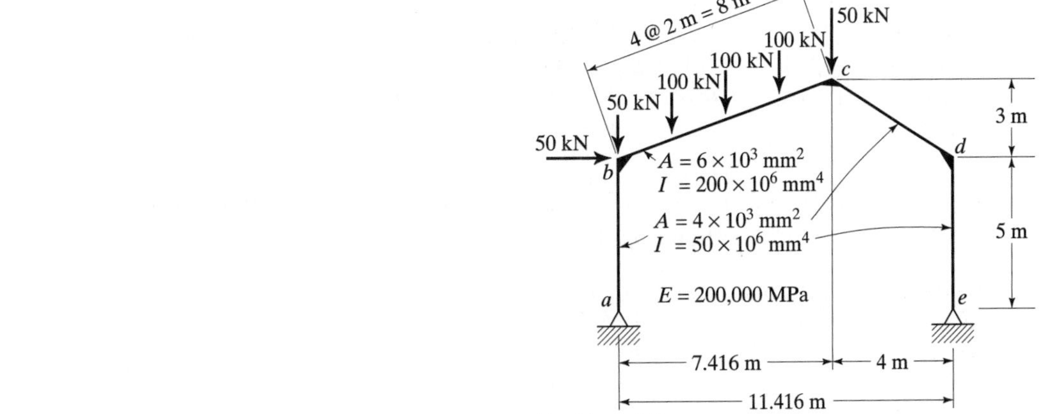

Problem 9.5

9.6 Use nonlinear analysis to calculate the ratio of the maximum moment in the span to M_b for the following cases: $M_a/M_b = -1.0, -0.5, 0, 0.5, 0.8$ and P/P_{cre} from 0 to 0.9, in which $P_{cre} = \pi^2 EI/L^2$. Compare results with classical solutions (see the Commentary of Reference 9.2 and Reference 9.4)

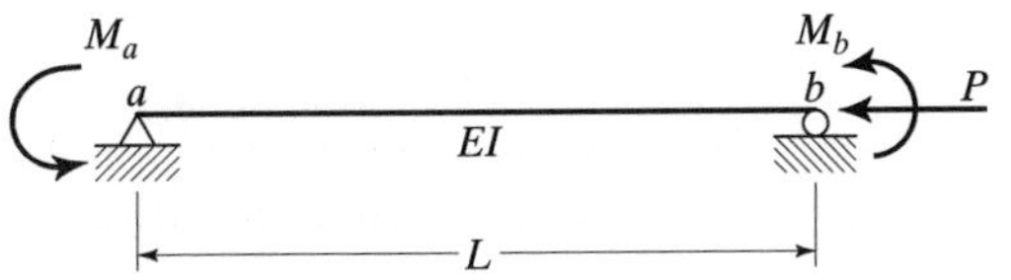

Problem 9.6

9.7 (Note: The frame shown is one of those analyzed and tested by Hoff, et al.; see Reference 9.3.) (a) Compute the critical load for the rigidly connected frame shown. (b) Compare linear and nonlinear analyses of the frame. (c) Compare the above results with those obtained by assuming the system to be a pin-connected truss.

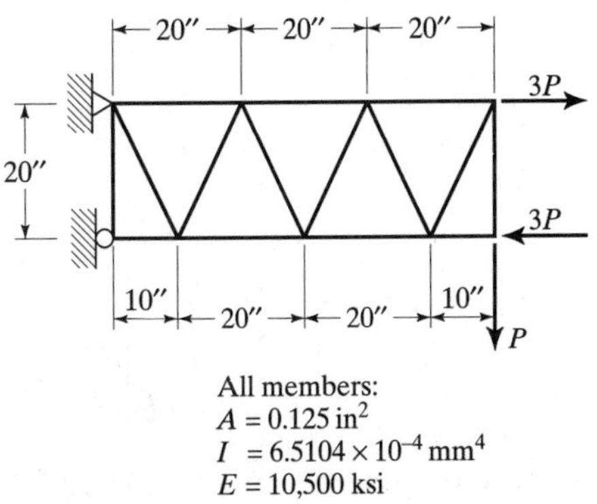

Problem 9.7

9.8 The member shown is restrained from twisting at each end. Compute its critical lateral buckling load. Compare the result with a theoretical solution in which it is shown that $P_{cre} = 16.94\sqrt{EI_yGJ}/L^2$ (Reference 9.4).

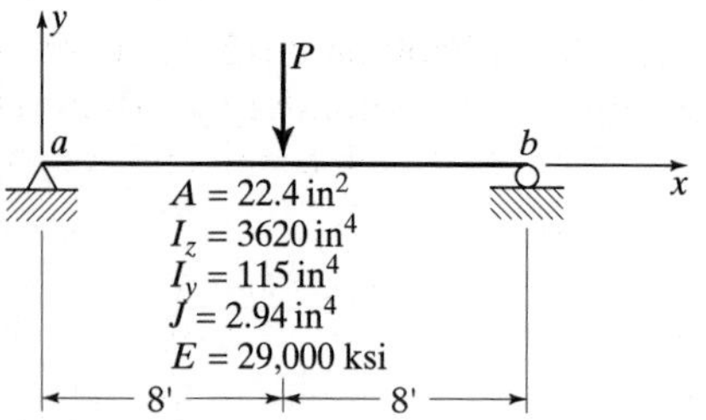

Problem 9.8

9.9 Compute the torsional-flexural buckling load of the frame shown. The frame is prevented from twisting at a and c and it is laterally supported at b.

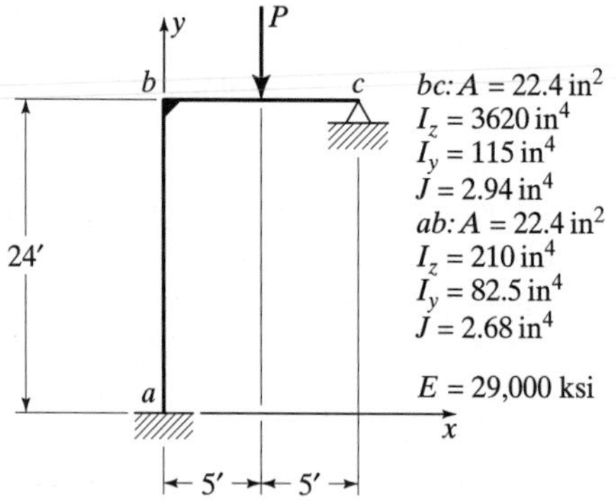

Problem 9.9

9.10 Compare linear and nonlinear analyses of the space frame shown.

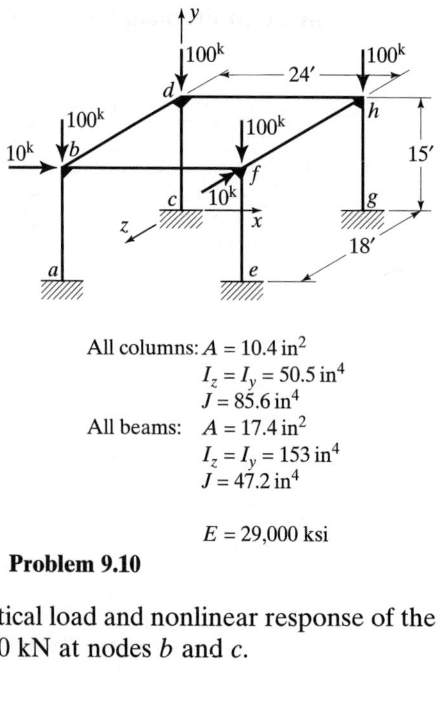

Problem 9.10

9.11 Compute the critical load and nonlinear response of the system of Problem 4.15 loaded by loads of 1500 kN at nodes b and c.

9.12 Study the response of the space frame of Problem 5.10(b). Vary the boundary conditions and the lateral support for the individual members. Compute the critical load and full nonlinear response for each case studied and comment on the results, for example, on the mode of behavior and the factors contributing to strength or weakness.

9.13 As in Problem 9.12, study the response of the space frame of Problem 5.10(d). Do not vary the boundary conditions in this study but do consider the effects of varying the relative axial, flexural, and torsional properties of the two members.

9.14 Member *ab* is a straight elastic column supported by a linear spring at *b*. $E =$ 200,000 MPa for both members. (a) Determine the minimum area of the strut required to develop the Euler load of the column (See Examples 8.1 and 8.3). (b) Analyze the resistance of the column for smaller values of the strut area. What is the resistance for larger values? (c) In all cases, what is the force at incipient buckling?

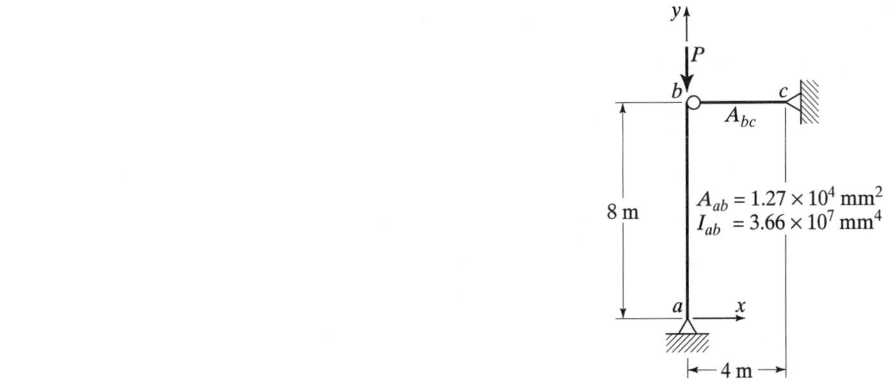

Problem 9.14

9.15 The system is the same as in Problem 9.14 except that *b* is initially offset 15 mm in the *x* direction. Analyze the resistance of the column for: (a) the minimum value of A_{bc} determined in Problem 9.14; (b) one-tenth of the minimum; and (c) 10 times the minimum.

9.16 Member *ab* is a straight, slender nonprismatic column. $E = 200{,}000$ MPa, $A_1 = 1.27 \times 10^4$ mm^2, $I_1 = 3.66 \times 10^7$ mm^4, $A_2 = 3.175 \times 10^4$ mm^2, $I_2 = 9.15 \times 10^7$ mm^4. Calculate the critical load. Compare the result with a theoretical solution (see Reference 9.4) in which it is shown that $P_{cre} = mEI_2/L^2$ and that the parameter $m = 6.68$ for the proportions of this problem ($a = 0.4$ and $I_1/I_2 = 0.4$).

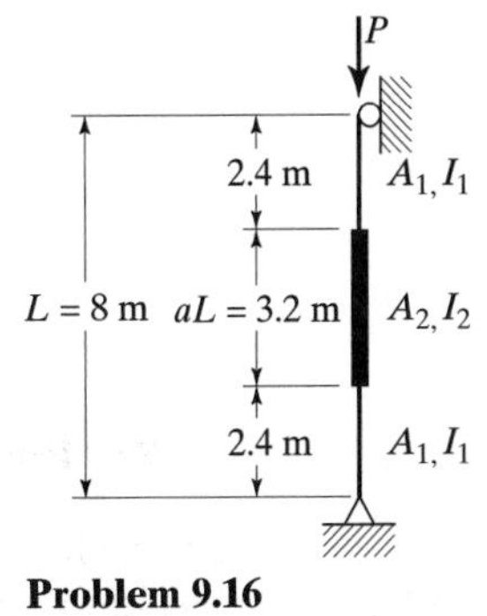

Problem 9.16

9.17 Compare the linear and nonlinear response of this simplified suspension system for (a) The load P applied at: (1) successive panel points; and (2) simultaneously at all panel points; and (b) I_g = 5,000, 10,000, and 20,000 in^4.

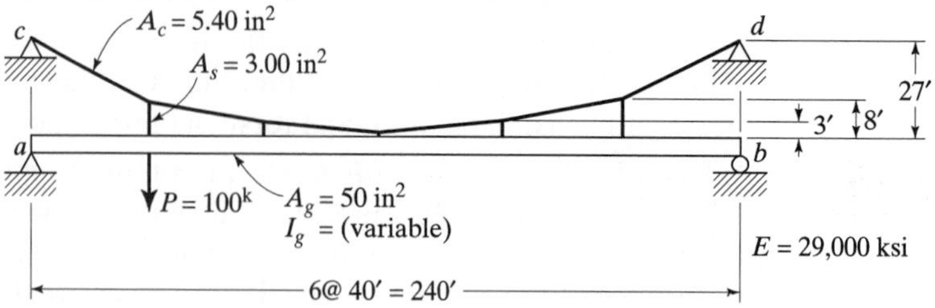

Problem 9.17

9.18 (a) Compare the linear and nonlinear response of this simplified cable-stayed system for the load P applied successively at 40-ft intervals along the span and simultaneously at all points. (b) Calculate the critical load for the system with P applied simultaneously at all points.

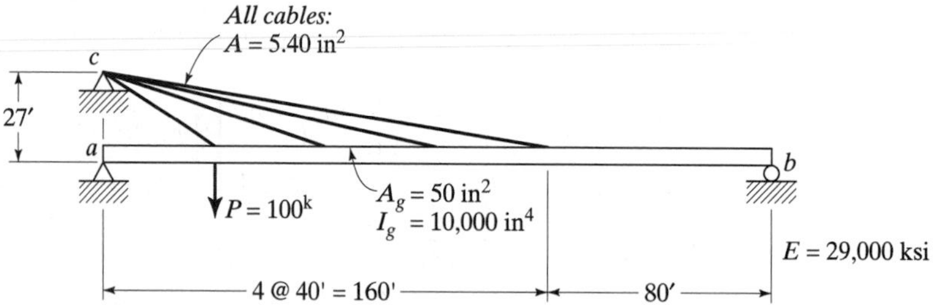

Problem 9.18

REFERENCES

9.1 Y. C. Fung, *A First Course in Continuum Mechanics*, 2nd edition, Prentice-Hall, Englewood Cliffs, N.J., 1977.

9.2 *Load and Resistance Factor Design Specification for Steel Buildings*, American Institute of Steel Construction, Chicago, 1993.

9.3 H. J. Hoff, B. A. Boley, S. V. Nardo, and S. Kaufman, "Buckling of Rigid Jointed Plane Trusses," *Trans., American Society of Civil Engineers*, Vol. 116, 1951.

9.4 S. P. Timoshenko and J. M. Gere, *The Theory of Elastic Stability*, 2nd edition, McGraw-Hill, New York, 1961.

9.5 W. McGuire, *Steel Structures*, Prentice-Hall, Englewood Cliffs, N.J., 1968.

9.6 J. T. Oden and E. A. Ripperger, *Mechanics of Elastic Structures*, Second Edition, Hemisphere Publishing Corp., New York, 1981.

9.7 K.-J. Bathe, *Finite Element Procedures*, Prentice-Hall, Englewood Cliffs, N.J., 1996.

Chapter 10

Material Nonlinear Analysis

A stiffness method approach to the inelastic analysis of planar frames is developed in this chapter. Equation 8.4 will be the base for considering the effects of material nonlinearity alone and Equation 8.5 for the treatment of combined material and geometric nonlinearity. The focus is on ductile systems: structures with members that, after a period of initial elastic response, can undergo substantial inelastic deformation without fracturing or becoming locally unstable. The material is assumed to be elastic–perfectly plastic and coverage is limited to plastic hinge methods; that is, methods in which cross sections are assumed to be either fully elastic or perfectly plastic. These methods are sufficient for the analysis of a great number of the problems encountered in the design of framed structures. Detailed analysis of strain hardening and distributed plasticity effects is generally accomplished through more complex nonlinear finite element programs, but procedures for bringing them within the purview of the concentrated plasticity approach have been advanced. Several will be summarized in Section 10.5 of this chapter.

Aspects of the theory of plasticity and ways in which they have been adapted for use in structural analysis are outlined in Section 10.1. Emphasis is on the background and justification for the plastic hinge theory. In Section 10.2 a particular adaptation of yield surface theory and the normality criterion is used in the development of a plastic reduction matrix suitable for incorporation in Equations 8.4 and 8.5. The tangent modulus theory approach to the determination of the inelastic critical loads of ideal structures and its use in material nonlinear analysis is described in Section 10.3. Section 10.4 contains examples illustrative of the methods presented in the previous sections.

10.1 NONLINEAR MATERIAL BEHAVIOR

Whereas many of the sources of geometric nonlinearity are sufficiently similar that the geometric stiffness matrix concept introduced in the previous chapter has widespread applicability in both elastic and inelastic nonlinear analysis, the sources of nonlinear material behavior differ in basic ways and thus require specialized treatment. For example, yielding of steel and cracking of concrete are fundamentally different phenomena, and further, each may be handled in a variety of ways. In steel the treatment of yielding may range from a simple rigid member–plastic hinge analysis to a three-dimensional finite element analysis based on rigorous continuum mechanics and plasticity theory. In reinforced concrete there are numerous schemes for either modeling individual cracks or for "smearing" a field of cracks to obtain their gross effect. We will concentrate on the development and application of one approach to the matrix analysis of frames of ductile material. Although limited in scope, the approach taken is of significant practical utility, and it provides insight into problems common to all types of material nonlinear analysis and guidance for their treatment.

In the engineering of framed structures we consider behavior at four levels: (1) at a point and the differential region around it; (2) on a cross section; (3) throughout the member's length; and (4) the system as a whole. In the chapters on linear elastic analysis consideration of stresses at a point was bypassed by starting with accepted engineering theories of their integrated effect on cross sections: axial force, bending moment, and torque. These stress resultants were used in the development of element equations, which then became the basis for system analysis. Since the subject was structural analysis alone and the sequence followed was a straightforward adaptation of classical displacement-based methods of elastic analysis, there was no need for explicit categorization. But there is in nonlinear analysis, in which, at each of the four levels, there are options for modeling some aspect of behavior. Proper interpretation of the final analysis depends on an understanding of the choices made and where they were made.

This may be illustrated and the role of various approaches to nonlinear analysis clarified by the following summary of some of the ideas used in modeling inelastic behavior and the schemes by which they are incorporated in systems analysis.

10.1.1 Plasticity Theory

The portion of the theory of plasticity of interest starts with consideration of a differential element at a point in a body of elastic–perfectly plastic material subjected to multidirectional stress. The material is presumed to have a known uniaxial stress-strain relationship (Fig. 10.1*a*) and, at the point considered, the principal stresses of Figure 10.1*b*. The theory involves the definition of a *yield function* and a corresponding *flow rule*. The most commonly accepted yield function is derived from the *von Mises criterion* and the flow rule from the *normality criterion* (e.g., Refs. 10.1 and 10.2).

The von Mises criterion postulates that the differential element will be elastic provided that the function

$$f = (\sigma_1 - \sigma_2)^2 + (\sigma_2 - \sigma_3)^2 + (\sigma_1 - \sigma_3)^2 \tag{10.1}$$

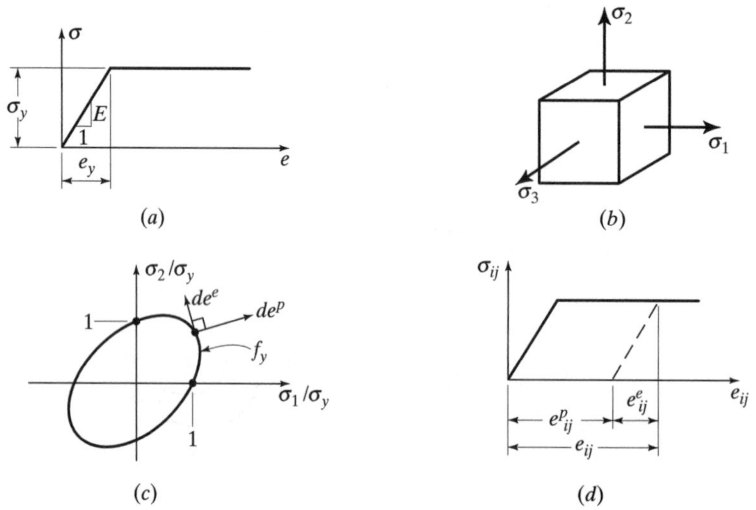

Figure 10.1 Elements of plasticity theory.

is less than $2\sigma_y^2$, that yielding can occur when it is equal to $2\sigma_y^2$ and that, for the ideally plastic material, values of f greater than $2\sigma_y^2$ are inadmissible. Depending on the direction of subsequent loading, an element in a yielded state will respond in one of three ways: 1) simple plastic straining; 2) a combination of plastic and elastic straining; or 3) simple elastic straining (a return to an elastic state).

To simplify visualization of this process, consider the case of plane stress in which σ_3 is equal to zero. For this, the yield function, f_y, can be written nondimensionally as

$$\frac{\sigma_1^2 - \sigma_1\sigma_2 + \sigma_2^2}{\sigma_y^2} = 1 \tag{10.2}$$

This is the ellipse shown in Figure 10.1*c*. It is generally called the *yield locus* or, interpreting it as a symbol of the full three-dimensional theory, the *yield surface*.

The flow rule, which relates plastic strain increments to current stresses and stress increments, is based on the idea that, in the plastic domain, the strain in a component direction is the sum of a recoverable elastic strain and an irrecoverable plastic strain, or, as in Figure 10.1*d*

$$e_{ij} = e_{ij}^e + e_{ij}^p \tag{10.3}$$

For stress on the von Mises yield surface, the normality criterion states that, if plastic strain occurs, its resultant must be normal to the surface at that point. Thus, for the planar case of Figure 10.1*c*, its incremental components are

$$de_{ij}^p = \lambda \frac{\partial f_y}{\partial \sigma_{ij}} \tag{10.4}$$

in which the factor λ is indefinite if plastic flow is not constrained and a function of the resistance offered by the surrounding body if it is constrained. It follows that the incremental elastic strain must be tangent to the surface or, if there is no plastic strain, directed inwards.

For a material that is not ideally plastic in the yielded state, for example, a material with the uniaxial bilinear elasto-plastic characteristics of Figure 10.2*a*, the same rules apply, but in addition, a *hardening rule* is required to specify how the yield surface is modified during plastic flow. There are a number of these, but the most elementary, and probably most commonly used, are *isotropic hardening*, in which the yield surface is assumed to expand uniformly, and *kinematic hardening*, in which it translates in space (Fig. 10.2*b*).

It is seen that these concepts—the yield surface, a flow rule, and a hardening rule—were developed from consideration of conditions at a point in a body of ductile

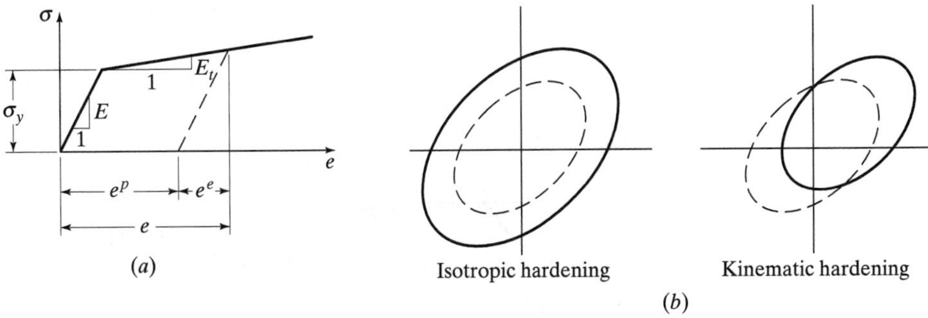

Figure 10.2 Hardening material.

material. Supported by experimental evidence, they and the underlying idea that strains can be treated as the sum of elastic and plastic parts are the basis for much of the theory of plasticity applicable to structural analysis. They may, for example, be incorporated in the constitutive relations of inelastic finite element programs, carried to the member level through a process of numerical integration and the system level by global analysis (Ref. 10.3).

10.1.2 Plastic Analysis

What is commonly called "plastic analysis" in structural engineering treats of an ideally elastic-plastic material (Figure 10.1*a*) and, in its simplest form, involves two concepts: the *plastic hinge* and *mechanism formation* (Ref. 10.4).

The plastic hinge concept postulates that the cross section of a member, say, the section of a steel beam subjected to axial force and bending as in Figure 10.3*a*, can have only two states: (1) completely elastic if the maximum stress as computed by common engineering theory is equal to or less than σ_y, or (2) fully plastic under a distribution of tensile and compressive stresses of magnitude σ_y that equilibrates the forces on the section. The latter case defines a plastic hinge, a section that can undergo indefinite plastic strain under these forces if it is not constrained by the resistance of the remainder of the system. The conditions under which such a hinge can form are cross-section dependent. A commonly used bilinear formula for the case of a wide flange beam bent about its major axis is (see Ref. 10.4):

$$M = 1.18\left(1 - \frac{P}{P_y}\right)M_p \leq M_p \tag{10.5}$$

Where P_y, the *squash load*, and M_p, the *plastic moment*, are, respectively, the section's area and plastic section modulus times σ_y. The graphical representation of Equation 10.5 in Figure 10.3*b* is an *interaction diagram* or, by extension of the terminology of plasticity theory, a *stress resultant yield surface.*

Mechanism formation is a member or system process. In its conceptually most elementary form, *rigid-plastic analysis*, elastic deformations are neglected and a search is made for the smallest value of the load required to form a mechanism, that is, a pattern of plastic hinges that will satisfy the conditions of equilibrium of the undeformed structure and, at the same time, admit departure from that configuration. Application of this process for finding the *plastic limit load* was illustrated in Example 8.6, which employed the yield surface defined by Equation 10.5. That example was a very simple one, however; the potential location of critical plastic hinges and the corresponding

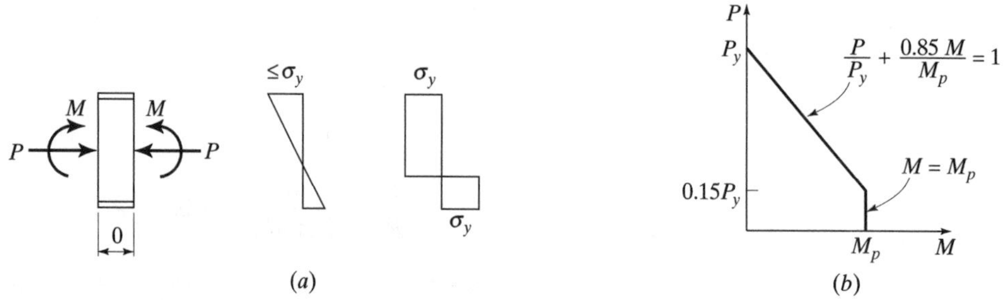

Figure 10.3 Plastic hinge formation.

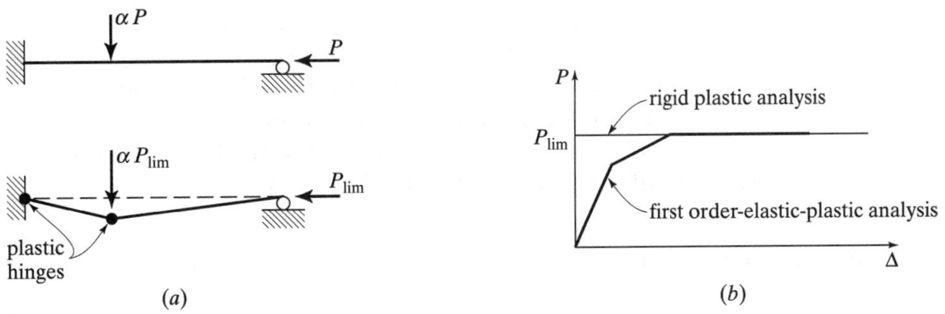

Figure 10.4 Mechanism formation and first-order elastic–plastic analysis.

requirements for satisfying equilibrium on the undeformed structure were obvious (Fig. 10.4*a*). That is rarely the case. Therefore, in routine plastic analysis it is generally more practical to use a first-order inelastic analysis that produces the same limit load and gives a more detailed picture of system behavior (Figure 10.4*b*). Essentially, it is a piecewise linear elastic analysis with the capability for detecting plastic hinges as they form and modifying them to deform plastically under further loading, with redistribution of the components of cross-sectional resistance following the dictates of the yield surface. Preferably, it should also be capable of detecting and accommodating the possibility of an elastic return from the yield surface.

10.1.3 Further Considerations

Plastic analysis as just described is practical, but it does not account for the gradual development of inelastic zones (distributed plasticity as in Example 8.5) nor for the second-order effects that contribute to instability (Example 8.7). There are finite element programs that include provisions for geometric nonlinearity and plasticity theory-based material nonlinearity that do not have these fundamental limitations. However, their use in general structural engineering practice may be limited by the need for specialized expertise in such tasks as the definition of realistic constitutive relationships and boundary conditions, selection and control of the equation solution process, and interpretation of results.

In conventional plastic design it has been customary to compensate for the inadequacies of simple plastic analysis in some hybrid fashion. For example, under the plastic design provisions of the 1989 AISC Specification for Structural Steel Buildings (Ref. 10.5), the maximum strength of one and two story steel frames "may be determined by a routine plastic analysis procedure and the frame instability effect may be ignored" and "members subjected to combined axial force and bending moment shall be proportioned to satisfy the following interaction formulas":

$$
\begin{aligned}
&\frac{P}{P_{cr}} + \frac{C_m M}{\left(1 - \dfrac{P}{P_e}\right) M_m} \leq 1.0 \\
&\frac{P}{P_y} + \frac{M}{1.18 M_p} \leq 1.0, \qquad M \leq M_p
\end{aligned}
\qquad (10.6)
$$

in which P and M are the calculated axial force and maximum moment in the member. The first of these equations is an empirical one that contains several coefficients and terms defined in the AISC specification and intended to account for member destabilizing effects that simple plastic analysis is incapable of simulating. The second, however, defines a plastic hinge (Eq. 10.5) and thus will be an integral part of the analysis.

10.2 A PLASTIC HINGE METHOD FOR DUCTILE FRAMES

To develop a plastic-hinge-based method of analysis, we again consider the element studied in Section 9.1, but in this case we define it to be of a ductile material. The assumptions made in that section apply to this one as well, and for the treatment of material nonlinear behavior, we add the following:

1. Plastic deformations are confined to zero-length plastic zones at the element ends (Fig. 10.5*a*).
2. The material is assumed to be linearly elastic–perfectly plastic with no strain hardening (Fig. 10.5*b*).
3. The effects of shearing stresses and direct stresses normal to the axis of the member on plastic deformation are disregarded.
4. As a consequence of the above, the end sections can undergo an abrupt transition from a fully elastic to a fully plastic state. Combined forces (axial force and bending) that initiate yielding on a cross section are assumed to produce full plastification of the section (Fig. 10.5*c*).
5. Plastic deformations are governed by the normality criterion.

10.2.1 The Yield Surface and a Plastic Reduction Matrix

An initial task is the development of a plastic reduction matrix for use in the global analysis equations. We will follow an approach devised by Porter and Powell (Ref.

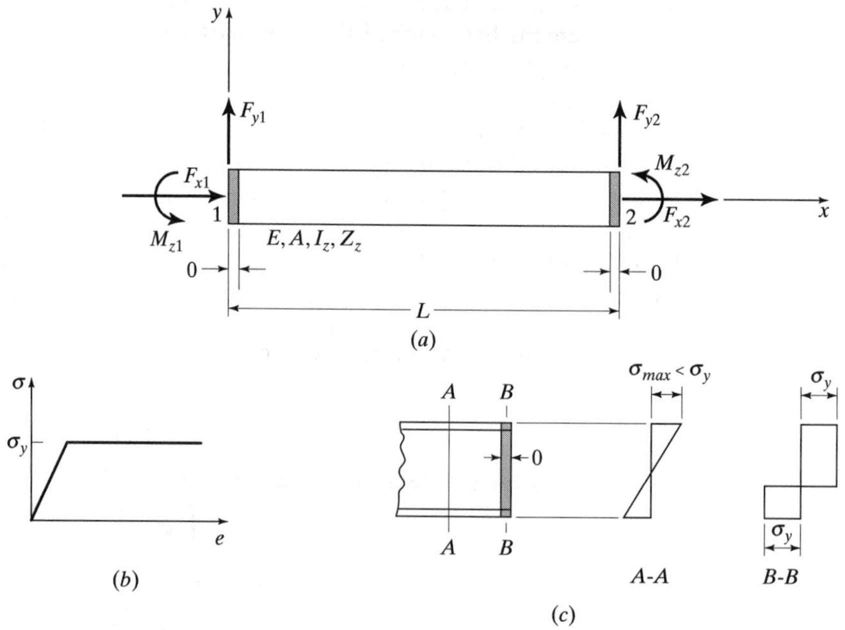

Figure 10.5 Concentrated plasticity (plastic hinge) element.

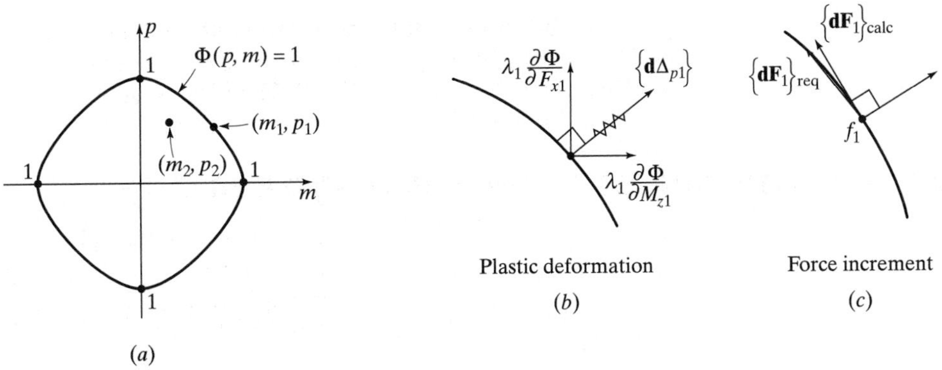

Figure 10.6 Yield surface, force increments, and plastic deformation.

10.7) which, in turn, uses features of plasticity theory and plastic analysis summarized in Section 10.1.

Included is the concept of a stress resultant yield surface, which, for the element considered, is assumed to be a continuous, convex function of the axial force and bending moment on a cross section (Fig. 10.6*a*). Symbolically, it can be represented as

$$\Phi(p, m) = 1 \tag{10.7}$$

in which $p = F_x/P_y$, the ratio of the axial force to the squash load $(A\sigma_y)$, and $m = M_z/M_{pz}$, the ratio of the bending moment to the plastic moment $(Z_z\sigma_y)$.

As noted earlier, for practical application the function Φ must be determined from studies of actual sections. Several possibilities will be considered later. But of interest at the moment is the basic concept that: (1) cross sections for which the *force point* (a point with the coordinates p, m) lies within the surface are elastic; (2) sections with force points on the surface are fully plastic; and (3) points outside the surface are not admissible because of the assumption of an elastic–perfectly plastic material.

To illustrate, let us assume that at a particular stage in a global analysis, the "1" end of the element has reached the yield surface while the "2" end is still elastic (Fig. 10.6*a*). Under the next increment of loading the force point at the 2 end may move in any direction, but there are only two possibilities for the 1 end. Depending on the response of the rest of the structure it may either remain plastic with a force point constrained to move along the yield surface or unload elastically with its force point returning to the interior. Both possibilities have to be considered, but we'll first deal with continued plastic action. It is assumed that the resultant incremental displacement at a plastified end can be treated as the sum of an elastic and a plastic contribution:

$$\{\mathbf{d\Delta}\} = \{\mathbf{d\Delta_e}\} + \{\mathbf{d\Delta_p}\} \tag{10.8}$$

It is also assumed that the normality criterion applies. Thus from Figure 10.6*b*, for end 1:

$$\{\mathbf{d\Delta}_{p1}\} = \lambda_1\{\mathbf{G}_1\} \tag{10.9}$$

where $\{\mathbf{G}_1\}$ is the gradient to the surface at that point

$$\{\mathbf{G}_1\} = \begin{Bmatrix} \dfrac{\partial \Phi}{\partial F_{x1}} \\ \dfrac{\partial \Phi}{\partial M_{z1}} \end{Bmatrix} \tag{10.10}$$

and λ_1 is the magnitude of the plastic deformation.[1] For an otherwise unrestrained location, such as the fixed end of a cantilever beam, it would be theoretically unlimited. In a stable system it will be constrained by contiguous elements and its value determined by Equation 10.14 below.[2] Admitting the possibility of plastification of both ends of the element, we have

$$\{\mathbf{d\Delta_p}\} = \begin{Bmatrix} \{\mathbf{d\Delta}_{p1}\} \\ \{\mathbf{d\Delta}_{p2}\} \end{Bmatrix} = \begin{bmatrix} \{\mathbf{G}_1\} & \mathbf{0} \\ \mathbf{0} & \{\mathbf{G}_2\} \end{bmatrix} \begin{Bmatrix} \lambda_1 \\ \lambda_2 \end{Bmatrix} = [\mathbf{G}]\{\boldsymbol{\lambda}\} \tag{10.11}$$

The matrix $[\mathbf{G}]$ is at the core of the development of the plastic reduction matrix. Its only nonzero elements are at plastified element ends. At these points it reduces the axial and rotational resistance and ensures that the other contribution to the total deformation, the elastic one, is tangent to the yield surface at the force point. For this reason, any calculated change in the force vector at such a point must follow the elastic relationship:

$$\{\mathbf{dF}\} = [\mathbf{k_e}]\{\mathbf{d\Delta_e}\} \tag{10.12}$$

where $[\mathbf{k_e}]$ is the element's elastic stiffness matrix. Thus, for a point on the yield surface the increment in force is tangent to it. As shown in Figure 10.6*c*, for a curved convex surface this violates the requirement that the force must move along the surface. Procedures for correcting for this discrepancy are discussed in Section 12.6.

From Equation 10.11 and the orthogonality of the plastic deformation and the calculated incremental force vectors, we have

$$\lfloor \mathbf{d\Delta_p} \rfloor \{\mathbf{dF}\} = \lfloor \boldsymbol{\lambda} \rfloor [\mathbf{G}]^{\mathrm{T}} \{\mathbf{dF}\} = 0$$

or, since $\lfloor \boldsymbol{\lambda} \rfloor$ is arbitrary,

$$[\mathbf{G}]^{\mathrm{T}}\{\mathbf{dF}\} = 0 \tag{10.13}$$

From the above relationships, equations that relate the magnitude of plastic deformation and the force increment to the total deformation increment may be developed by straightforward matrix manipulation. Thus, using Equations 10.8, 10.11, 10.12, and 10.13, and solving for $\{\boldsymbol{\lambda}\}$ yields

$$\{\boldsymbol{\lambda}\} = [[\mathbf{G}]^{\mathrm{T}}[\mathbf{k_e}][\mathbf{G}]]^{-1}[\mathbf{G}]^{\mathrm{T}}[\mathbf{k_e}]\{\mathbf{d\Delta}\} \tag{10.14}$$

Similarly, using Equations 10.8, 10.11, 10.12, and 10.14 and solving for $\{\mathbf{dF}\}$ results in

$$\{\mathbf{dF}\} = [[\mathbf{k_e}] + [\mathbf{k_m}]]\{\mathbf{d\Delta}\} \tag{10.15}$$

in which

$$[\mathbf{k_m}] = -[\mathbf{k_e}][\mathbf{G}][[\mathbf{G}]^{\mathrm{T}}[\mathbf{k_e}][\mathbf{G}]]^{-1}[\mathbf{G}]^{\mathrm{T}}[\mathbf{k_e}] \tag{10.16}$$

is the element plastic reduction matrix. It is left to the reader to verify the development of Equations 10.14 and 10.16 (Problem 10.1a).

If, after a plastic hinge has formed at a particular section, the force redistribution

---

[1] Justification for the use of the normality criterion in stress resultant space may be found in Reference 10.8. To illustrate its application symbolically as in Figures 10.6*b* and *c*, a dimensional rather than a nondimensional surface is employed.

[2] Although the "magnitude" of a deformation over "zero length" has no physical significance, the λ's are proportionality factors that indicate the magnitude of the theoretical "kink" in the deformed configuration at a plastic hinge. More important, their signs indicate whether the calculated plastic flow can take place or not (see Eq. 10.14 and its interpretation in the examples of Section 10.4).

that takes place under subsequent loading of the structure is such as to reduce the resultant action on that section, it may unload elastically. The appearance of a negative element in the $\{\boldsymbol{\lambda}\}$ vector computed by Equation 10.14 at the end of a load increment is a signal that this may be occurring and that the increment should be re-done and the analysis continued after returning the corresponding surface gradient to a null vector.

10.2.2 Definition of the Yield Surface

Application of Equation 10.16 requires the definition of the stress resultant yield surface. Its dimension is equal to the number of stress resultants considered in the determination of cross-section plastification and its form is a function of the section's shape. In principle, therefore, it could be a six-dimensional "hypersurface." But in a plastic-hinge analysis of ductile frames in which the effects of shearing strain due to torsion and transverse shear are not considered, it reduces to a three-dimensional one: axial force and bending about two axes. The function's form is established by application of the corresponding three equations of cross-sectional equilibrium to a given shape of given material, say an elastic–perfectly plastic material in which the yield points in tension and compression are assumed to be equal. A number of studies of this type are presented in Reference 10.9. For a bisymmetrical wide flange shape, they may be represented symbolically by one octant of a three-dimensional surface (Figure 10.7*a*). Traces of the surface for a W12 × 31 section on the major and minor axis bending planes are shown in Figure 10.7*b*.

Also plotted on Figure 10.7*b* are some equations useful in system analysis. The major axis bending equation, Equation 10.5, and its companion minor axis equation from Reference 10.4

$$M = 1.19\left[1 - \left(\frac{P}{P_y}\right)^2\right]M_p \leq M_p \tag{10.17}$$

are seen to be close approximations of the two traces. In the numerical examples of Section 10.4 the major axis bending terms of the following continuous function have been used.

$$\Phi(p, m_y, m_z) = p^2 + m_z^2 + m_y^4 + 3.5p^2m_z^2 + 3p^6m_y^2 + 4.5m_z^4m_y^2 = 1 \tag{10.18}$$

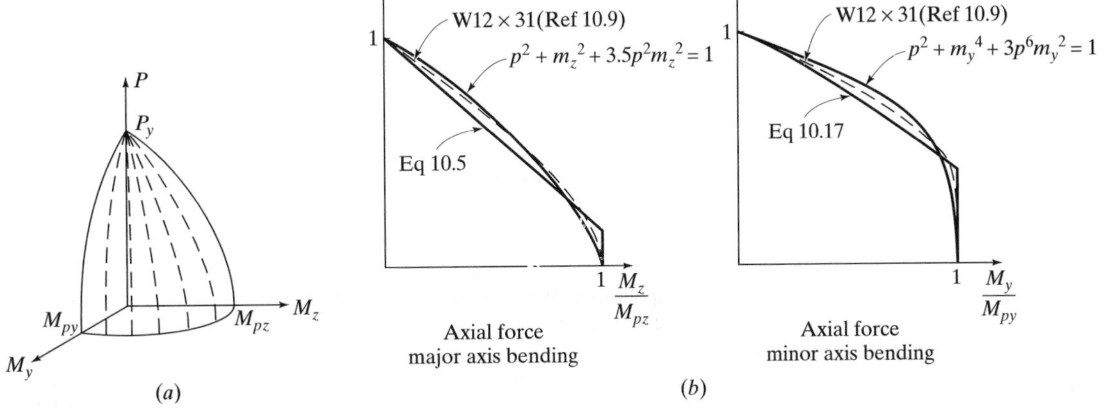

Figure 10.7 Wide flange section yield surface.

It is seen that this, too, is a good approximation of the traces. Equation 10.18 is a modified version of a yield surface described in Reference 10.10, in which application of the surface under conditions of biaxial as well as uniaxial bending is demonstrated.

10.3 INELASTIC CRITICAL LOAD THEORY

In Example 8.3 it was shown that the elastic critical load of the ideally straight column of Figure 10.8*a* is the Euler load, $P_{cre} = \pi^2 EI/L^2$. If the column were made of the elastic–perfectly plastic material of Figure 10.1, its theoretical limit of resistance would be the lesser of that load or the squash load, $P_y = A\sigma_y$. A long slender column would buckle, and a short stocky one would fail by starting to squash. It is known, however, that even short columns can buckle. The generally accepted explanation for this is the *tangent modulus theory* (Ref. 10.11). It is based on the observation that the load-shortening relationship for a section made of a material such as structural steel is affected by the residual stresses that result inevitably from the manufacturing process. Thus the effective stress-strain diagram of the material is as in Figure 10.8*b*. Below a proportional limit, σ_p, it is elastic. Above that point it is inelastic with a gradually decreasing resistance, measured by the tangent modulus, E_t. The theory postulates that, for an ideally straight column with an elastic critical stress greater than σ_p, bifurcation of equilibrium can occur and the column will start to bend at a load

$$P_{cri} = \frac{\pi^2 E_t I}{L^2} \tag{10.19}$$

in which E_t is the tangent modulus at the stress $\sigma_{cri} = P_{cri}/A$.

Except for the use of E_t rather than E, the development leading to Equation 10.19 is mathematically identical to that leading to the Euler load (Example 8.3). Physically there is a difference in that, whereas E is only a function of the type of material, E_t is also stress dependent. A frequently used relationship between the two moduli is

$$E_t = 4E\left[\frac{\sigma}{\sigma_y}\left(1 - \frac{\sigma}{\sigma_y}\right)\right] \tag{10.20}$$

This is an empirical expression designed to represent the performance of structural steel columns in the inelastic range (Reference 10.11). Implicit in it is the assumption that the proportional limit is half the yield point. In Examples 10.1–10.3 of the following section it will be used in the calculation of inelastic critical loads of ideal systems by Equation 8.6 and the iterative algorithm of Section 12.8. Equation 10.20 may also be invoked in second-order inelastic analysis to approximate the reduction in stiffness of regions that are within the yield surface but subjected to an average axial stress

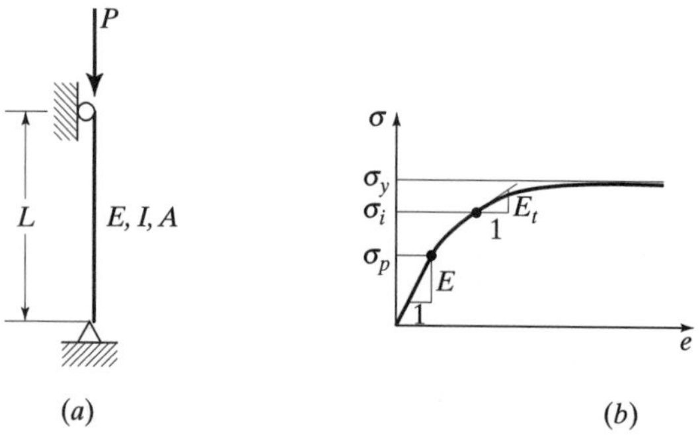

Figure 10.8 Tangent modulus concept.

greater than one half the yield point. This, too, is done in the examples of the next section.

10.4 EXAMPLES

Example 10.1 is a study in the behavior of a pinned end column. In the first part of the example, elastic and inelastic critical loads are calculated by Euler and tangent modulus theory for a length of 10 feet, which places the member in the intermediate slenderness range. It is seen that the tangent modulus load can be obtained directly from the simultaneous solution of Equations 10.19 and 10.20. A four-element model of the column of the same length, but with an assumed parabolic imperfection of amplitude equal to 1/1000 of its length, is then analyzed by second-order elastic and inelastic methods. In the inelastic solution, the effective modulus under axial stress greater than 25 ksi is reduced according to Equation 10.20. For the case analyzed, the calculated limit point is almost identical to the inelastic critical load.

EXAMPLE 10.1

Consider weak axis behavior of the column shown

$$A = 9.13\text{ in.}^2,\ I = 37.1\text{ in.}^4,\ Z = 14.1\text{ in.}^3,\ r = 2.02\text{ in.},\ E = 29{,}000\text{ ksi},\ \sigma_y = 50\text{ ksi}$$

1. For $L = 10'$
   (a) Calculate the elastic and inelastic critical loads.
   (b) Assume an initial parabolic bow of $L/1000$ amplitude. Analyze by second-order elastic and inelastic methods.

2. Use second-order inelastic analysis to determine the resistance of the bowed member for $0 \le L/r \le 200$. Compare with Equation 10.21.

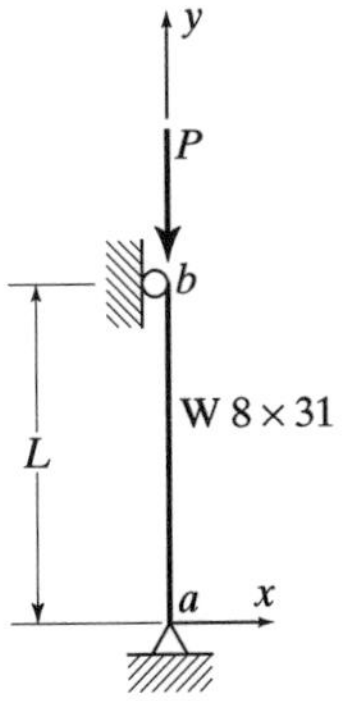

---

1. Use four element modeling

(a) $$P_{\text{cre}} = \frac{\pi^2 EI}{L^2} = \frac{\pi^2(29{,}000)(37.1)}{(120)^2} = 737.4\text{ kips}$$

From Equation 10.19 $$\sigma_{\text{cri}} = \frac{P_{\text{cri}}}{A} = \frac{\pi^2\, E_t(37.1)}{(120)^2(9.13)} \quad (1)$$

From Equation 10.20 $$E_t = 4(29{,}000)\left[\frac{\sigma_{\text{cri}}}{50}\left(1 - \frac{\sigma_{\text{cri}}}{50}\right)\right] \quad (2)$$

From (1) and (2):

$$E_t = 15{,}200\text{ ksi},\ \sigma_{\text{cri}} = 42.26\text{ ksi},\ P_{\text{cri}} = 385.8\text{ kips}$$

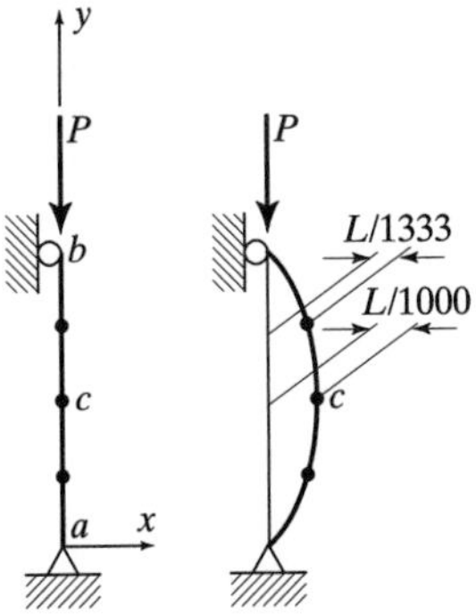

(b) Response curves

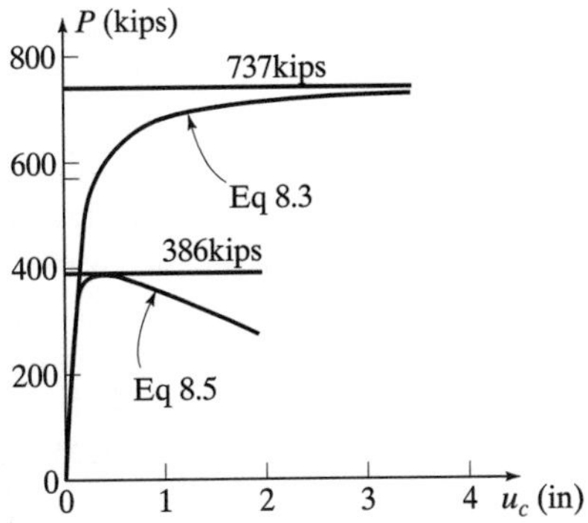

2. Resistance curves

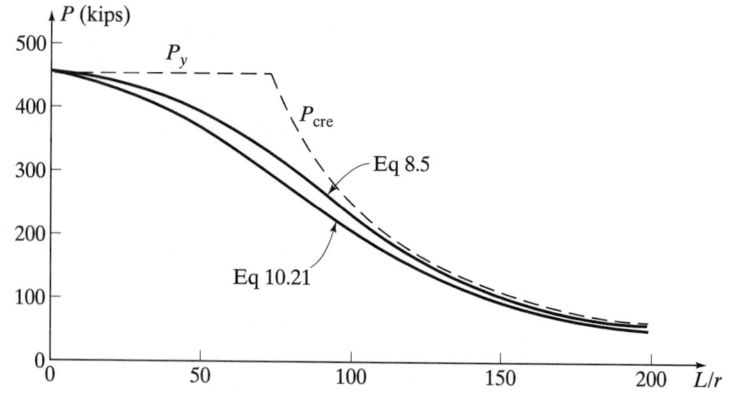

In the second part of Example 10.1, the inelastic analysis procedure is used to calculate the resistance of the imperfect member over the range of L/r, the slenderness ratio, from zero to two hundred. The result is compared with the AISC LRFD column formula (see Ref. 10.6)

$$\sigma_{cr} = (0.658^{\lambda^2})\sigma_y \qquad \text{for } \lambda \le 1.5$$
$$\sigma_{cr} = \left[\frac{0.877}{\lambda^2}\right]\sigma_y \qquad \text{for } \lambda > 1.5 \tag{10.21}$$

in which

$$\lambda = \frac{KL}{\pi r}\sqrt{\frac{\sigma_y}{E}}$$

For the effective length factor $K = 1$, and the assigned properties, the two parts of the column equation become

$$\sigma_{cr} = (0.658^{0.000175(L/r)^2})\sigma_y$$

and

$$\sigma_{cr} = \left[\frac{5020.2}{(L/r)^2}\right]\sigma_y$$

respectively.

In comparing results it should be recognized that the resistance of a column is a function of its cross-sectional shape, the nature of its supports, and its geometrical imperfections and residual stresses. Even among simply supported columns, there is a wide range of behavior (see Ref. 10.11). Behind the reduction of this complex behavior to the single design formula, Equation 10.21, there is an extensive body of analytical and experimental research. In contrast to this, the calculated curve is the result of a plastic hinge analysis of a single, arbitrarily selected, section. The agreement between the two curves is good: the difference in resistance is generally less than 10% and the maximum is 14%, which is within the range of variability—the "scatter band"—of the actual resistance of a column of given nominal dimensions and properties. This could be viewed as fortuitous, but a more favorable interpretation is that it is evidence that, in cases in which the columns are components of a system with members and boundary conditions that can be reasonably modeled, an inelastic analysis of the type illustrated can be a practically useful predictor of nominal strength.

In Example 10.2, the proportions of the ideally straight two-story columns are such that, at the critical load, the upper story is inelastic while the lower one remains elastic.

It demonstrates that a mixture of "Euler" and "tangent modulus" behavior can be accommodated by the algorithm of Section 12.8. The calculated critical load is between the values for the two spans treated as independent pin ended columns. The inference is that the upper, lower capacity, segment derives some restraint from the lower one.

EXAMPLE 10.2

For both spans of this continuous column, $E = 29{,}000$ ksi and $\sigma_y = 33$ ksi. For bc, $A = 9.13$ in.^{2}, $I = 37.1$ in.^{4} For ab, $A = 17.6$ in.^{2}, $I = 116.0$ in.^{4}

1. Determine the critical load of the system and the corresponding stress in each span.
2. Compare with the result obtained by treating each span as an isolated pin ended column.

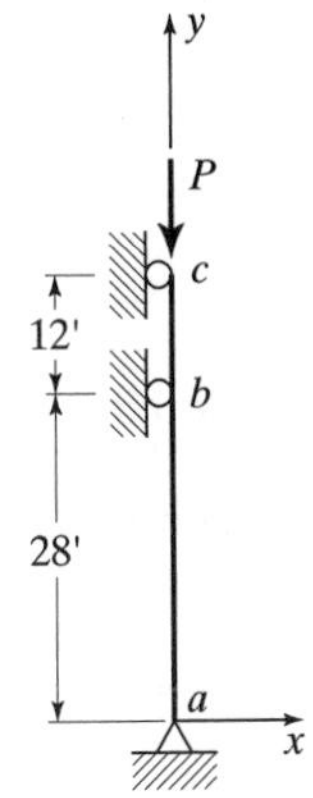

1. From Equation 8.7, using a four-element idealization for each span

$$P_{cr} = 265.5 \text{ kips}$$

$$bc: \quad \sigma_{cr} = 265.5/9.13 = 29.1 \text{ ksi}$$

$$ab: \quad \sigma_{cr} = 265.5/17.6 = 15.1 \text{ ksi}$$

span bc is inelastic and ab is elastic at the critical load.

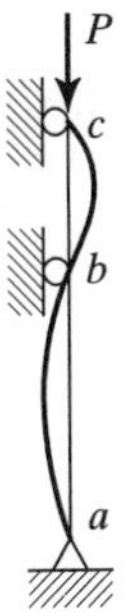

2. As pin ended columns the critical loads would be
bc: Following Example 10.1, $E_t = 14{,}500$ ksi, $\sigma_{cri} = 28.1$ ksi, and

$$P_{cri} = \frac{\pi^2(14{,}500)(37.1)}{(144)^2} = 256.1 \text{ kips} < P_{cr}$$

ab:

$$P_{cre} = \frac{\pi^2(29{,}000)(116)}{(336)^2} = 294.1 \text{ kips} > P_{cr}.$$

The "no sidesway" and "sidesway" elastic and inelastic critical loads of two frames are the subject of Example 10.3. In its sidesway-prevented state the columns of the first frame are effectively "short," and the material strength is the determining factor in its resistance. When sidesway is not prevented, the column effective length is much greater; the strength of the material has little effect on resistance. The second frame presents a different situation. Even without an external sidesway constraint, its proportions and loading are such that the exterior columns brace the heavily loaded interior one. The inelastic critical load at which the interior column is essentially fully supported is the governing condition.

EXAMPLE 10.3

Determine critical loads and buckling modes for the following frames. $E = 29{,}000$ ksi, $\sigma_y = 36$ ksi. Use Equation 8.7 and two-element idealization for all members.

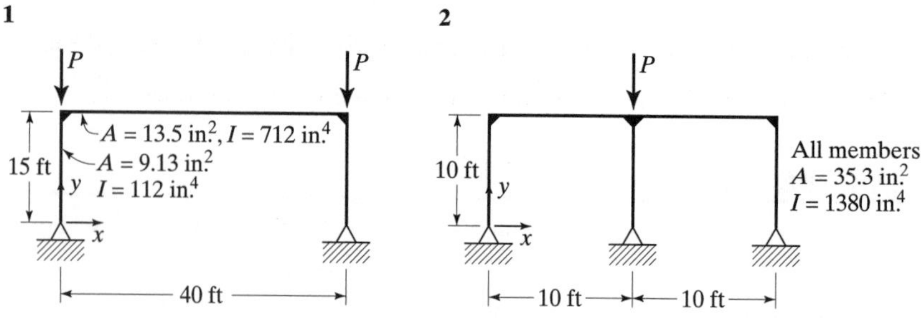

Critical loads and buckled configurations:

1. Consider sidesway prevented and sidesway permitted cases.

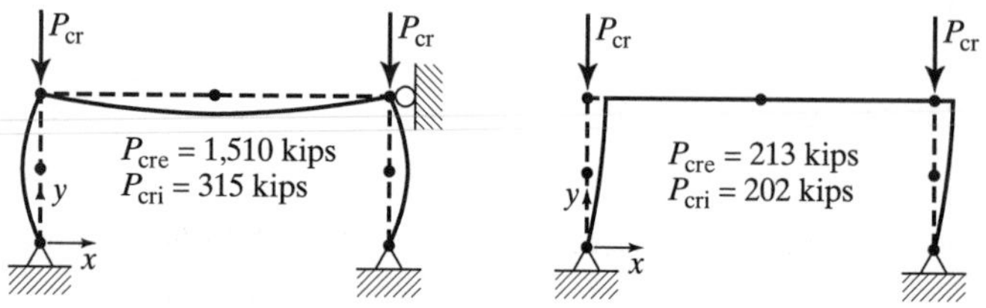

2. Consider sidesway permitted case.

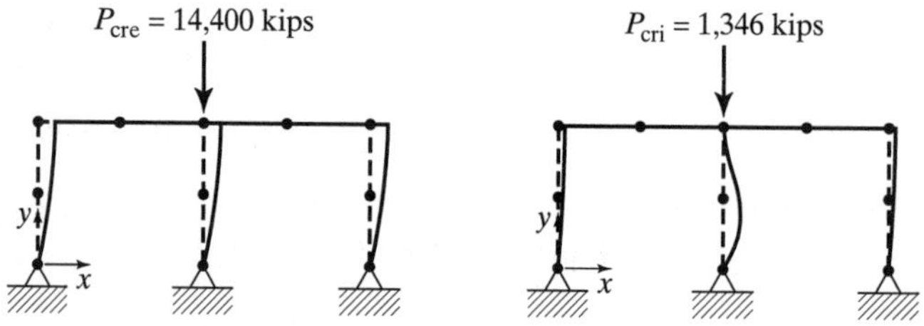

Application of Equation 8.4 for material nonlinear analysis and Equation 8.5 for geometric and material analysis is illustrated in Examples 10.4–10.8. The appropriate equations of Chapter 4 are used for the formation of element elastic stiffness matrices, Equation 9.18 for the geometric ones, and Equation 10.16 for plastic reduction matrices. The basic equation solution routines are the same as those of the examples of Chapter 9, but additional features associated with consideration of material nonlinearity include (1) detection of plastic hinge formation; (2) tracking the yield surface; and (3) detection of possible elastic unloading from the yield surface. Summaries of devices used for treating these conditions are presented in Chapter 12.

EXAMPLE 10.4

The member and loading are those of Example 8.6.

1. Compare first- and second-order inelastic analyses.
2. Determine the plastic deformation vector and the incremental force relationship at first hinge formation in the first-order analysis.

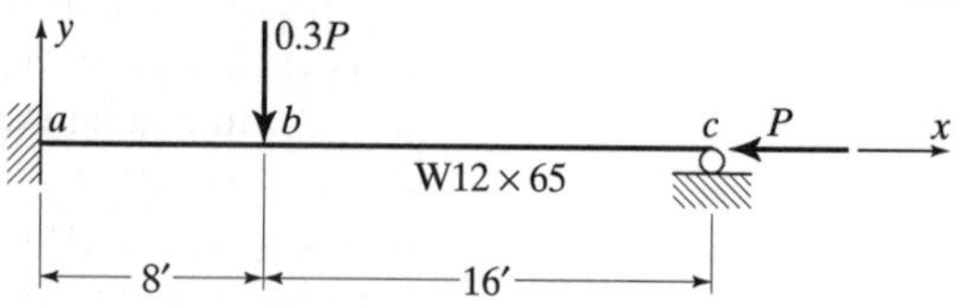

1. Idealize *ab* and *bc* as single elements.
From Example 8.6 $P_y = 955$ kips $M_p = 4840$ in. kips

From Equation 10.18, the yield surface is $\left(\frac{P}{955}\right)^2 + \left(\frac{M}{4840}\right)^2 + 3.5\left(\frac{PM}{4.62 \times 10^6}\right)^2 = 1 \qquad (a)$

From Equations 8.4 and 8.5, response curves and force point traces are:

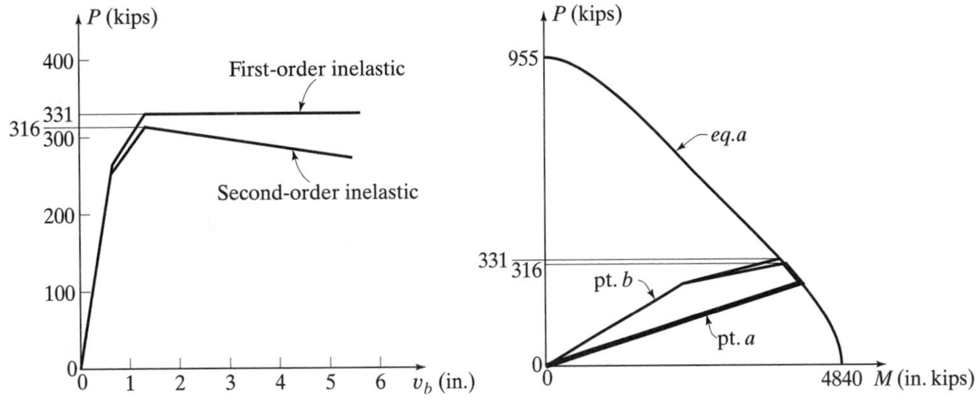

2. From first order analysis, first hinge forms at *a* when $P = 259.7$ kips, $M_a = 4151.2$ in. kips

$$p = 259.7/955 = .2719 \qquad m = 4151.2/4840 = 0.8507$$

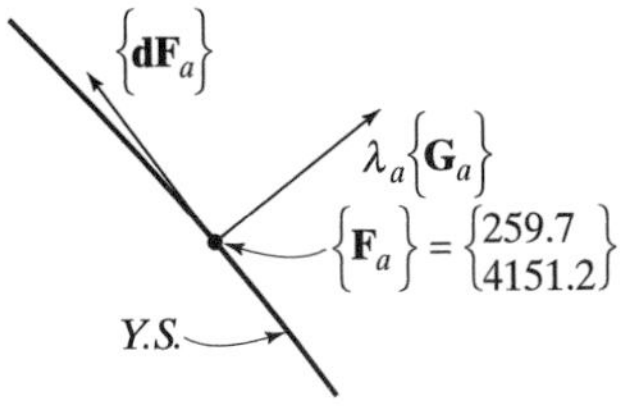

$$\frac{\partial \Phi}{\partial P} = \frac{2p}{P_y} + \frac{7pm^2}{P_y} = 0.002036$$

$$\frac{\partial \Phi}{\partial M} = \frac{2m}{M_p} + \frac{7p^2m}{M_p} = 0.0004461$$

$$\text{Equation 10.9: } \{\mathbf{d\Delta}_{pa}\} = \lambda_a\{\mathbf{G}_a\} = \lambda_a\begin{Bmatrix} 0.002036 \\ 0.000446 \end{Bmatrix}$$

$$\text{Equation 10.13: } \lfloor \mathbf{G}_a \rfloor \begin{Bmatrix} dP_a \\ dM_a \end{Bmatrix} = \lfloor .002036, .0004461 \rfloor \begin{Bmatrix} dP_a \\ dM_a \end{Bmatrix} = 0.$$

$$\therefore dM_a = -4.564\, dP_a$$

Example 10.4 is the same as Example 8.6 except for the use of a different yield surface and both first- and second-order matrix analyses. As expected from the close agreement of the yield surfaces, the difference in first-order analysis results is inconsequential. In the second-order matrix analysis the limit point is found to be approximately 5% less than the simple plastic limit load; the destabilizing P-δ effect is perceptible but not very large.

In this example the member is subdivided into the elements *ab* and *bc* in advance recognition of the potential location of plastic hinges. Initially the $[\mathbf{K_m}]$ matrix of Equation 8.4 or 8.5 is null and the analysis starts out as a conventional elastic one. The first hinge is detected with the aid of an algorithm based on the determination of the distance from a calculated state to the yield surface, plus an iterative procedure designed to converge on that surface to within an acceptable tolerance (see Section 12.6). The situation that exists once that has been accomplished is illustrated by the case of the first plastic hinge in the first-order inelastic analysis. The form of the plastic deformation vector and the incremental force relationship at that hinge are determined from

Equations 10.9 and 10.13, respectively. The magnitude of the components of plastic deformation and incremental force remain unknown until computed in the next step of the analysis, but $[\mathbf{K_m}]$ is no longer null. When calculated by Equation 10.16 and used in the global analysis through Equation 10.15, the force point at a is constrained to move tangent to the yield surface at that point. Steps that may be taken to correct for the spurious outward drift of a hinge point from a convex yield surface are described in Section 12.6.

The plastic reduction matrix of Equation 10.16 is seen to have its roots in elementary plastic analysis. For this simple case, the elementary approach of Example 8.5 is as expeditious as any, but its application to structures of any size verges from tedious to impossible. The matrix approach has its own numerical and behavioral problems: those illustrated by this example and also, for example, the detection and treatment of elastic unloading at a hinge and the avoidance of spurious "joint mechanisms" (a false signal of instability from the formation of plastic hinges on two sides of a node). But most of these are problems that can be dealt with effectively in the computer programs that are essential to its application.

The structure of Example 10.5 is the same as that of Example 9.6, but in this case it is subjected to two levels of inelastic analysis. In both first- and second-order inelastic analysis, failure is the result of plastic hinging at the same locations. The P-Δ effect is seen to reduce the load carrying capacity by 12%. It should be noted that the yield surface is represented nondimensionally because the force point traces pictured are for two different cross sections. The first plastic hinge forms at the top of a column under a combination of bending moment and a substantial axial force, the second hinge in the girder, in which the axial force is insignificant. Between the formation of the first and second hinges the moment at the column top decreases as the axial force increases, as dictated by the yield surface. In the plot of the response curves, results of the elastic analysis of Example 9.6 are added for comparison.

EXAMPLE 10.5

The frame and loading are the same as in Example 9.6. $\sigma_y = 36$ ksi.

Compare first and second-order inelastic analyses. Use the yield surface of Equation 10.18

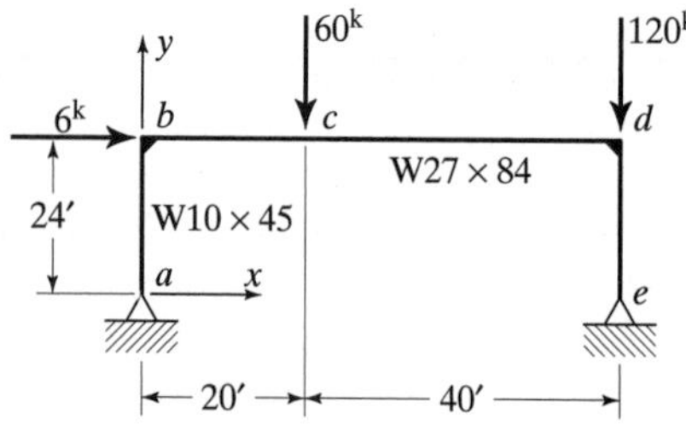

Model columns as two elements and girder segments as single elements

W10 × 45: $P_y = A\sigma_y = 13.3 \times 36 = 478.8$ kips, $M_p = Z\sigma_y = 54.9 \times 36 = 1976$ in. kips

W27 × 84: $P_y = 24.8 \times 36 = 892.8$ kips, $M_p = 244.0 \times 36 = 8784$ in. kips

From Equation 10.18: $\left(\frac{P}{P_y}\right)^2 + \left(\frac{M}{M_p}\right)^2 + 3.5\left(\frac{PM}{P_yM_p}\right)^2 = 1$

From Equations 8.4 and 8.5, results are:

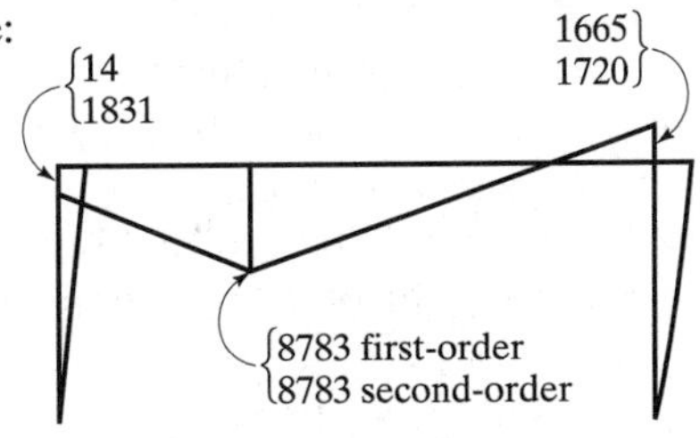

Limiting moment diagrams

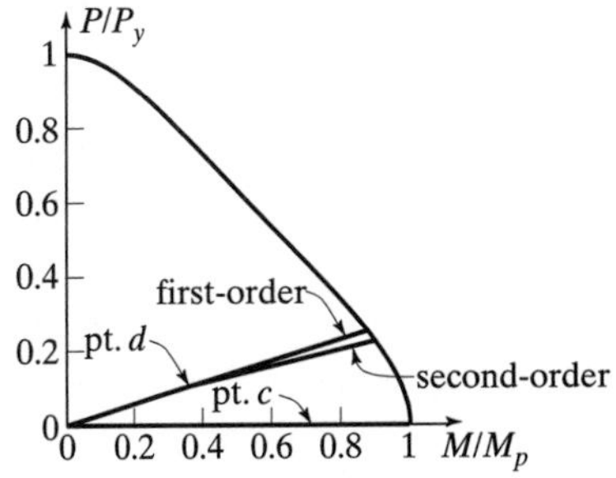

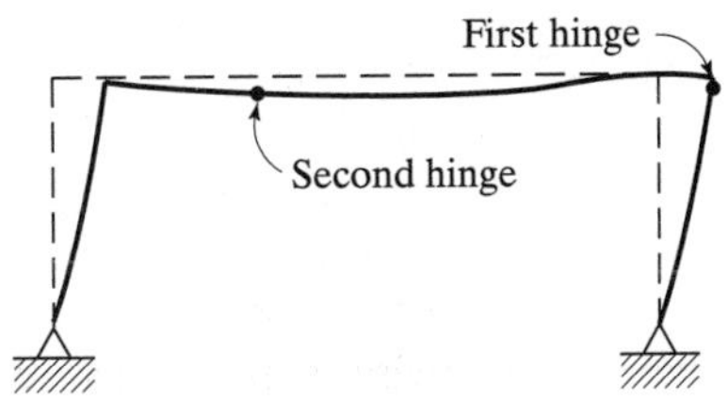

Mechanism

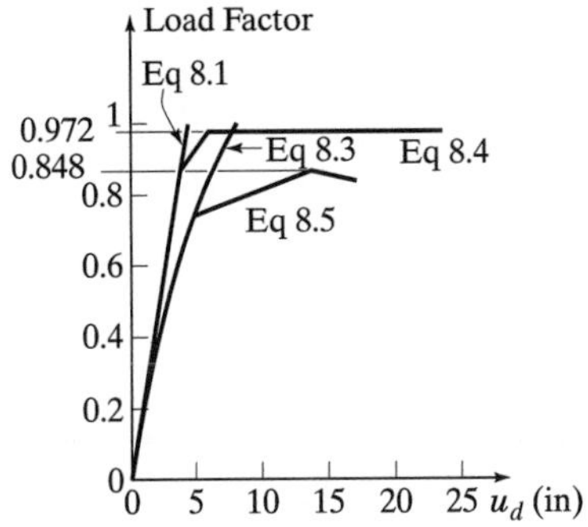

Example 10.6 is similar to Example 10.5 except for the addition of a column and specification of a loading pattern designed to illustrate further the P-Δ effect. The two levels of analysis—first- and second-order inelastic—yield two different failure modes. In the first case it is a simple story-wide panel mechanism. Under the combined destabilizing effects of geometric and material nonlinearity included in the second case, a stability limit is reached prior to mechanism formation.

EXAMPLE 10.6

Members *ab*, *cd*, and *ef* are W10 × 45 sections. Members *bd* and *df* are W27 × 84 sections (see Example 10.5 for section properties). $E = 29{,}000$ ksi, $\sigma_y = 36$ ksi

Compare first and second-order inelastic analyses. Use the yield surface of Equation 10.18.

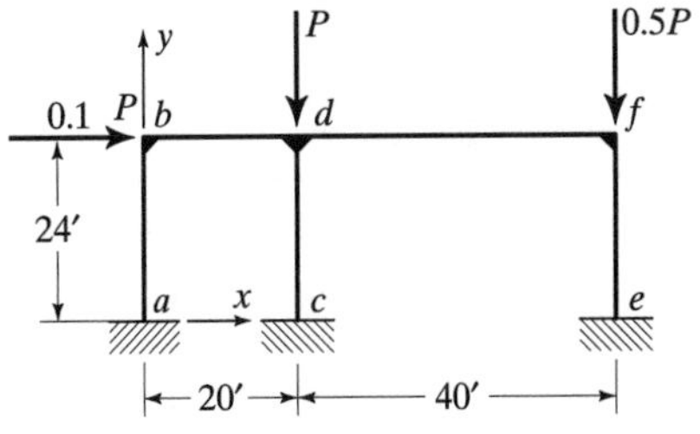

Model columns as two elements and girders as single elements. From Equations 8.4 and 8.5 results are:

Limiting moment diagrams (in. kips).

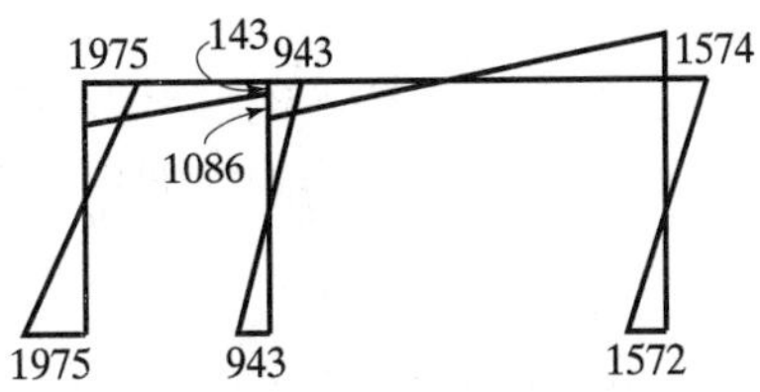

First-order inelastic

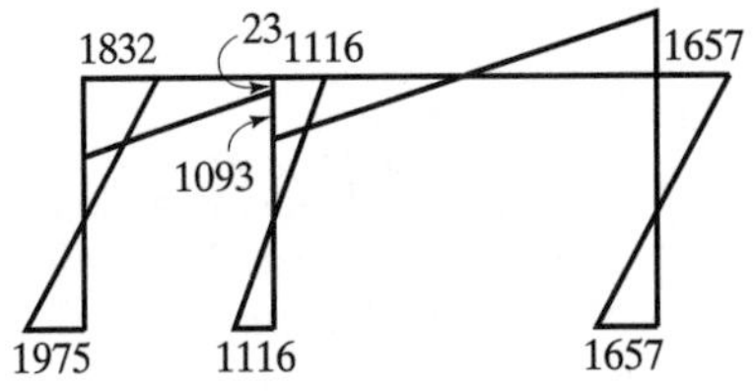

Second-order inelastic

Plastic hinge formation sequence:

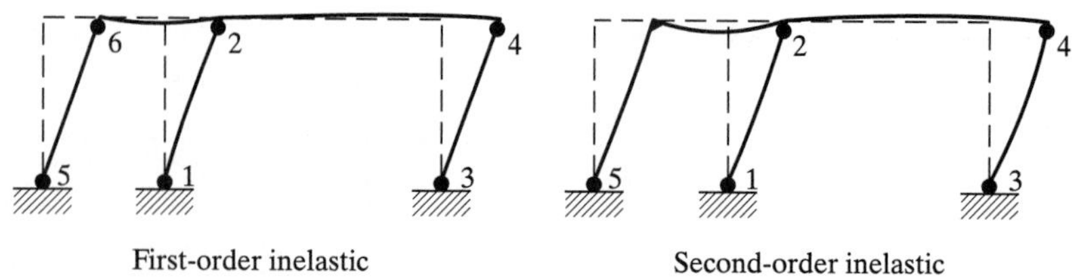

Response curves:

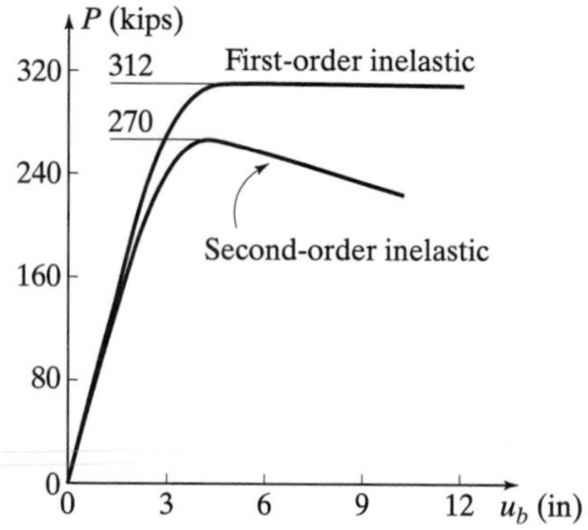

Example 10.7 presents the results of a second-order inelastic analysis of a frame subjected to non-proportional loading. It is one studied analytically and experimentally by Arnold et al. (Ref. 10.12). Beam loads of 20 kips each were applied first and held on the structure while 60 kip loads were placed on each column. With these loads in place, the frame was then tested to failure under horizontal loading. Shown in the example are a moment diagram, deflected configuration, and a response curve calculated by Equation 8.5. Member sizes and yield strengths reported in Reference 10.12 were used in the analysis. The agreement between analysis and test is good. The reported failure load was 7% higher than the calculated value, a difference attributable mainly to the actual finite size of the joints as compared to the point assumed in the analysis. The test frame was also somewhat more flexible than the analytical model. This may be attributed to small column base rotation observed in the test.

EXAMPLE 10.7

The properties of the frame shown are:

W5 × 18.5: $A = 5.45$ in.^{2}, $I = 25.4$ in.^{4},
$Z = 11.4$ in.^{3}, $\sigma_y = 56.2$ ksi

S10 × 25.4: $A = 7.38$ in.^{2}, $I = 122.1$ in.^{4},
$Z = 28.0$ in.^{3}, $\sigma_y = 38.6$ ksi
$E = 29{,}000$ ksi

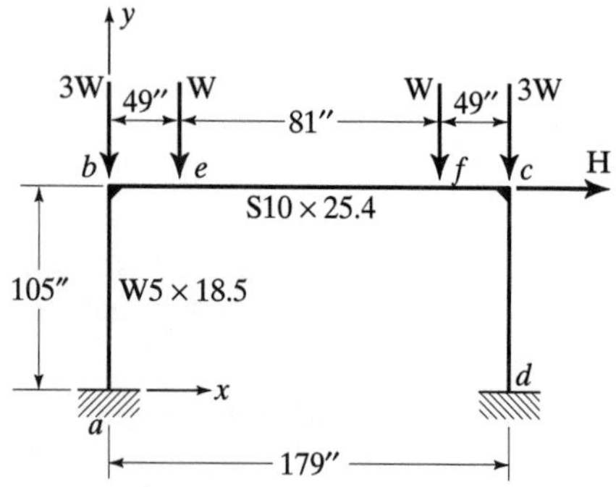

Loading sequence:

1. Apply beam loads, $W = 20$ kips.
2. Add column loads, $3W = 60$ kips.
3. Add H and load to collapse.

Calculate the response of the frame by second-order inelastic analysis.

Model columns as two elements; *be*, *ef*, and *fc* as one element. Each from Equation 8.5; using the yield surface of Equation 10.18:

Limiting moment diagram (in. kips)

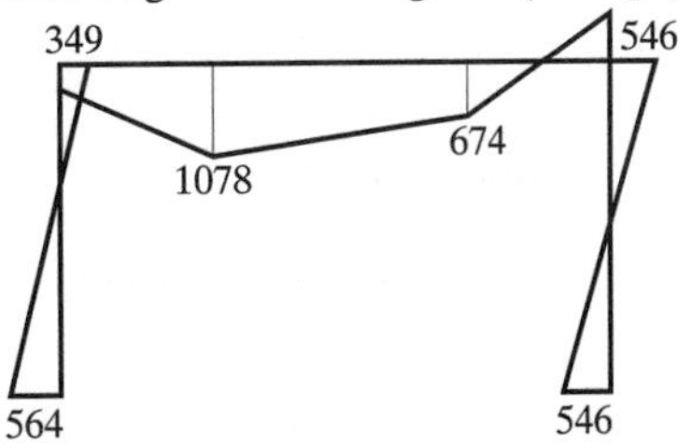

Plastic hinge formation sequence

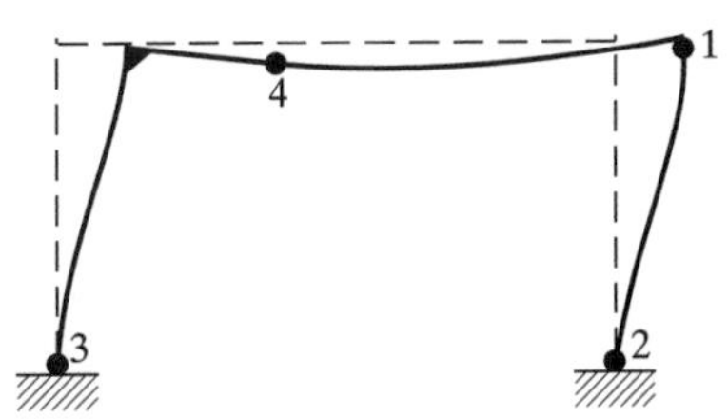

Response:

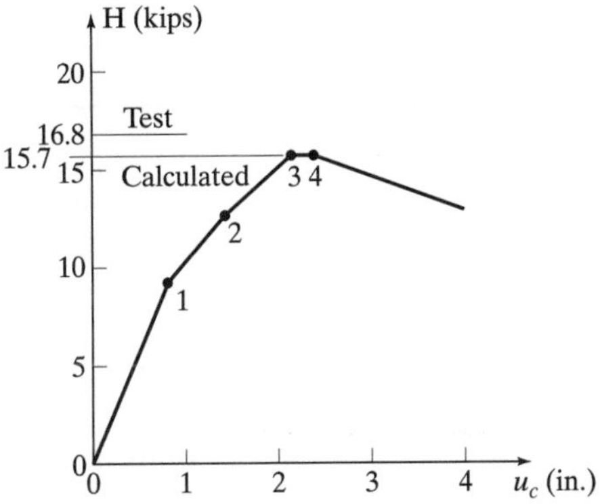

The use of distributed rather than concentrated loads in nonlinear analysis is demonstrated in Example 10.8. In this, as in all the systems analyzed in this section, plastic hinges can only form at defined nodal points. A burden is placed on the analyst to define nodes reasonably close to potential hinge locations. Few practical systems are so sensitive to hinge location that this is a problem, however.

EXAMPLE 10.8

All members are W30 × 108 sections

$A = 31.7$ in.^{2}, $I = 4470$ in.^{4}, $Z = 346$ in.^{3}, $\sigma_y = 50$ ksi, $E = 29{,}000$ ksi

Assume proportional loading.

Compare first and second-order inelastic analyses.

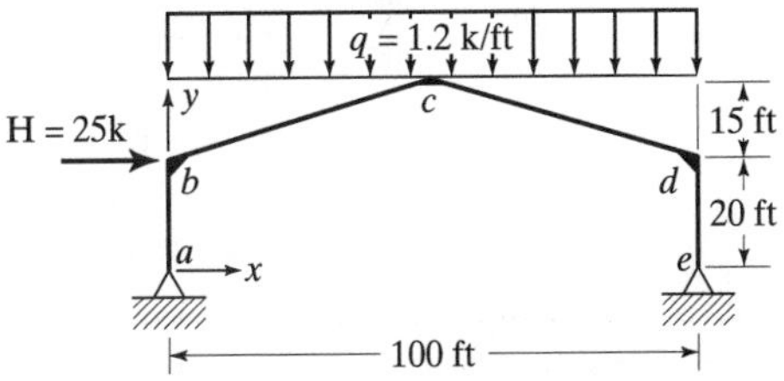

Model columns as two elements and girders as four elements. Use the yield surface of Equation 10.18.

Define uniform load in local coordinates:

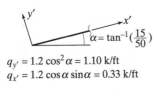

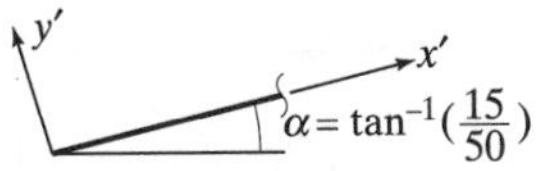

$q_{y'} = 1.2\cos^2\alpha = 1.10$ k/ft
$q_{x'} = 1.2\cos\alpha\sin\alpha = 0.33$ k/ft

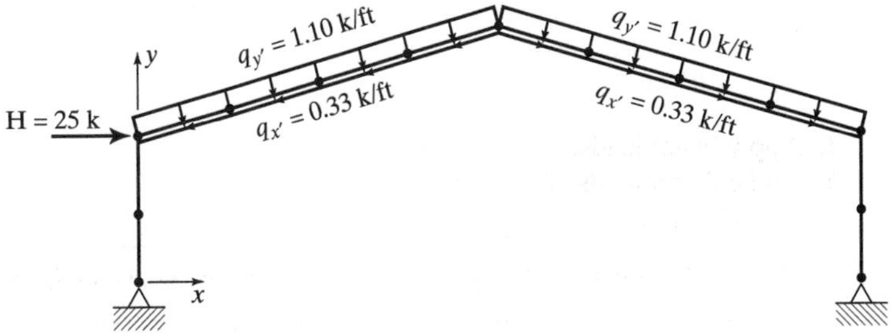

From Equations 8.4 and 8.5:

Plastic hinge formation sequence

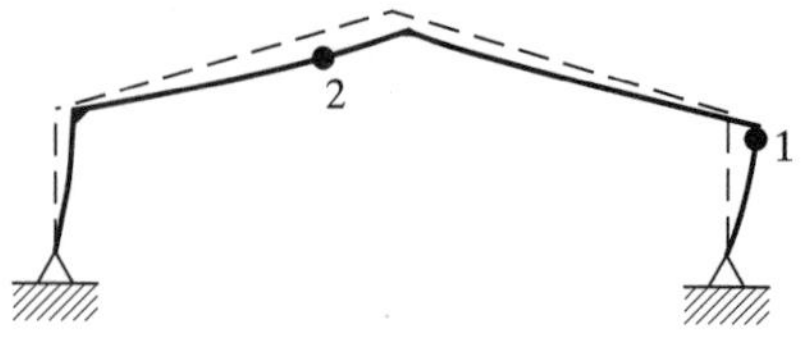

Response curves

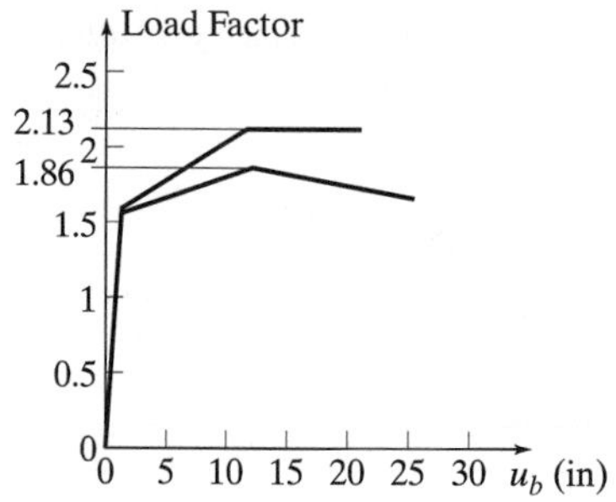

Example 10.9 is a study in *deflection stability*, a phenomenon related to the repeated application of a sequence of loads. If, as in the example, a two-span beam is subjected to the loading pattern shown, it will remain elastic provided the maximum moment under either loading is less than the plastic moment. At the other extreme, a plastic limit will be reached under the initial application of the first load if it is sufficiently large to cause a beam mechanism to form. For loads between the elastic and inelastic limits, plastification at either potential hinge point may occur and then be relieved as the sequence progresses. Because of the irrecoverable plastic rotation at an active hinge, the member will be permanently deformed and a residual moment will remain after the removal of the particular load. The situation will stabilize (colloquially "shakedown") if the magnitude of the load is equal to or less than a certain value, otherwise the permanent deformation will increase continuously, with the potential of becoming excessive. It can be shown that shakedown will occur if, at points of maximum moment, the absolute value of the sum of the residual moment, M_r, and the elastic moment, M_e, produced by the loads does not exceed the plastic moment, M_p. Thus, for a general point i

$$|(M_r)_i + (M_e)_i| \leq M_p \tag{10.22}$$

The illustrative example used is from Reference 10.4. The only possible residual moment diagram is the one shown. In the response curves an application or removal of a pattern of loads is designated as a step in the "loading sequence." It is seen that, at the shakedown load, stability is obtained after the first full cycle whereas for a load between it and the plastic limit, permanent deflection increases continually.[3]

EXAMPLE 10.9

The continuous beam is subjected to the repeated loading pattern shown

$$I = 533 \text{ in.}^4,\ Z = 96.8 \text{ in.}^3,\ \sigma_y = 50 \text{ ksi},\ E = 29{,}000 \text{ ksi}$$

1. Determine the elastic limit, the plastic limit, and the shakedown load.
2. Study the system using first-order inelastic analysis for
   (a) the shakedown load
   (b) $P = 95$ kips

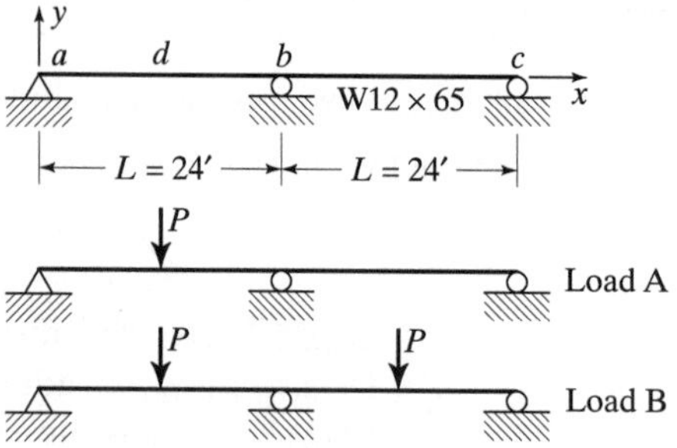

[3] Although, as shown, progressive plastification of this sort is possible, it is not often a factor that requires investigation in design. See Reference 10.4 for further discussion of this subject.

1. Moment diagrams (elastic ordinates shown for Loads A and B)

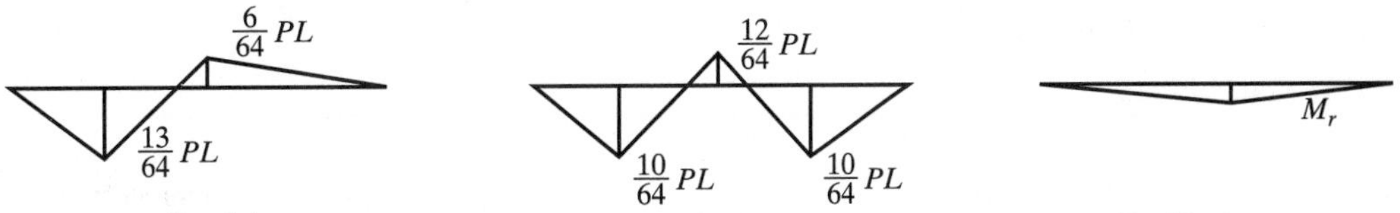

Load A   Load B   Residual moment

Limiting conditions for $M_p = 96.8 \times 50 = 4840$ in. kips:

Elastic limit:

$$P_e = \frac{64}{13}\frac{M_p}{L} = \frac{64(4840)}{13(288)} = 82.73 \text{ kips}$$

Plastic limit:

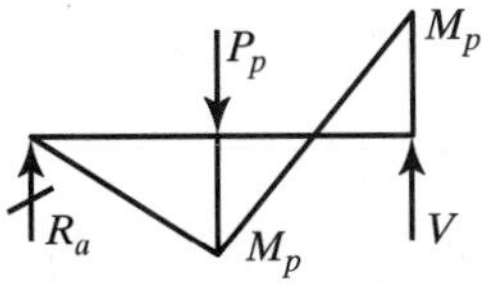

$$M_p + R_a L - \frac{P_p L}{2} = 0$$

and

$$M_p - \frac{R_a L}{2} = 0$$

$$\therefore P_p = 6M_p/L = 100.83 \text{ kips}$$

Shakedown load: From Eq. 10.22

$$\frac{13}{64} P_s L + \frac{1}{2} M_r = M_p$$

and

$$\frac{12}{64} P_s - M_r = M_p$$

$$\therefore M_r = -\frac{M_p}{19} = 254.8 \text{ in. kips and } P_s = \frac{96}{19}\frac{M_p}{L} = 84.91 \text{ kips}$$

2. Response
Loading sequence: (1) Load A, (2) unload, (3) Load B, (4) unload, etc.

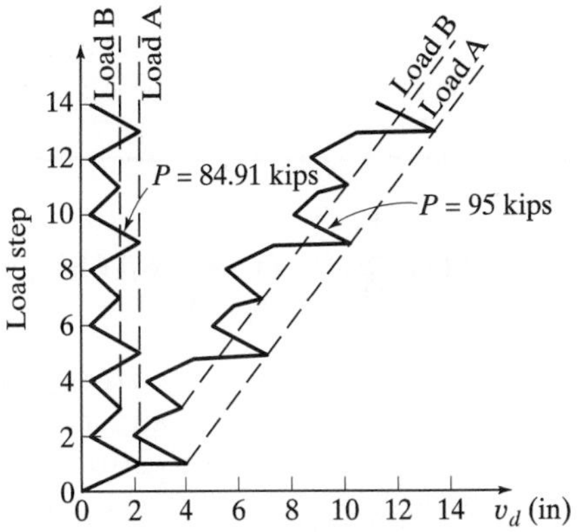

Example 10.10 is a summary of the results of a comparative study of a "fully braced" and an "eccentrically braced" frame. The two differ only in the insertion of a flexure and shear resistant link in the center of the second. Both are simplified representations of framing and loading in a lower story of a multistory diagonally braced system. Although the example is unrealistic to the extent that the effect of shear on yielding is not included in the second-order inelastic analyses to which the frames are subjected, there are lessons in the different modes of behavior that are of use in actual design. Frame A is stiffer and slightly stronger than Frame B, but its collapse is precipitated by inelastic buckling of member *cd* and it has little post limit buckling resistance. Frame B absorbs a greater amount of energy through progressive yielding of its beams; its failure is more gradual. Considerations of this sort are particularly important in

earthquake-resistant design in which a key factor is the amount of energy required to destabilize a system.

EXAMPLE 10.10

Compare the behavior of the following frames, $E = 29{,}000$ ksi. For beams, $\sigma_y = 33$ ksi. For columns and diagonals, $\sigma_y = 100$ ksi.

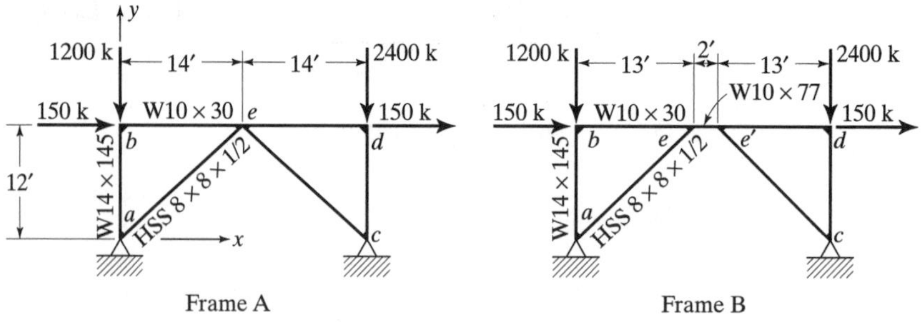

Model ee' by one element, all other members by two elements. Use Equation 8.5. Account for elastic shearing deformation (see Example 7.11).

Failure modes and plastic hinge formation sequence (by load factor):

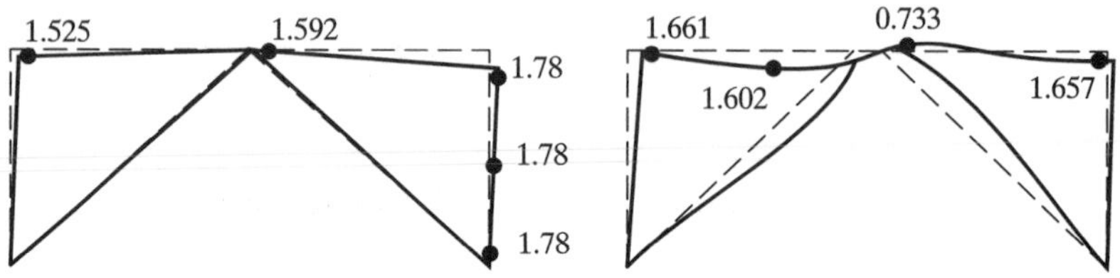

Response curves:

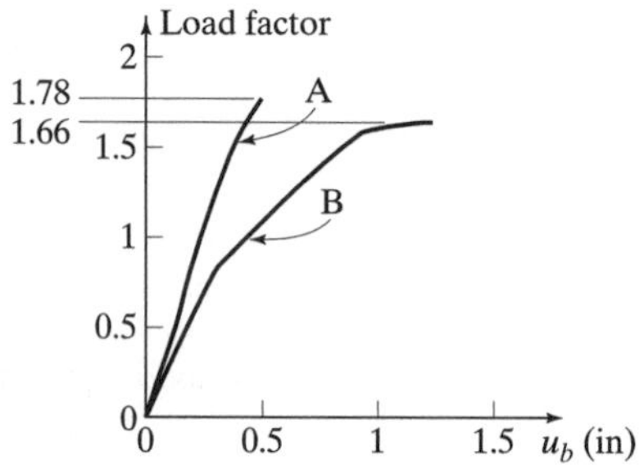

10.5 THE YIELD SURFACE CONCEPT—A BRIEF SURVEY OF FURTHER APPLICATIONS

The stress resultant yield surface concept has been adapted, in various ways, to the treatment of additional problems in material nonlinear analysis, including the spread of plasticity, strain hardening, and the analysis of reinforced concrete members. For several reasons—limited applicability, the need for extensive empirical support, or special computational requirements—application of these extensions will not be undertaken here. But the following summaries of a few of them illustrate the concept's potential.

10.5.1 Spread of Plasticity

A zone of gradually developing plasticity of the type studied in Example 8.5 can be reduced to a "quasi–plastic hinge" and analyzed by the matrix methods of this chapter following a procedure presented in Reference 10.13. The reduction is accomplished through a process in which nonlinear equations for the force-strain relationships of partially plastified cross sections are integrated numerically along the element to obtain inelastic flexibility coefficients, which are then used in generating an inelastic stiffness matrix. To illustrate, we consider the beam-column element of Figure 10.9a with

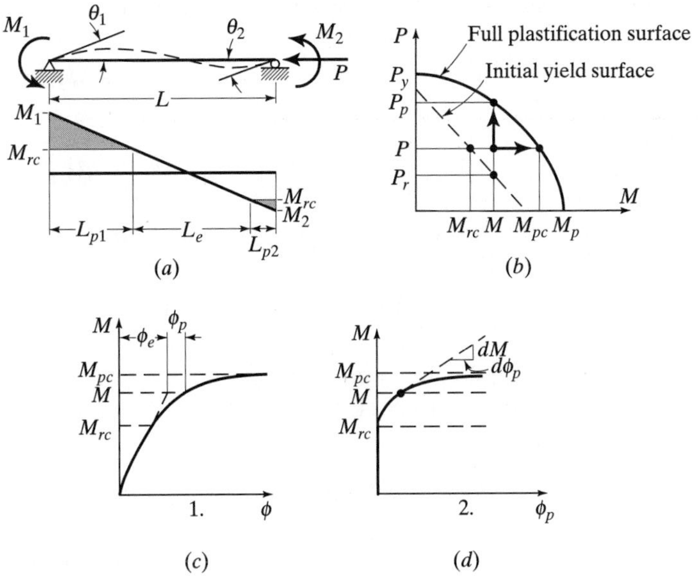

Figure 10.9 Quasi-plastic hinge model.

regions near each end in which the maximum stress is greater than the proportional limit. As shown in Figure 10.9*b*, the force point (M, P) of a typical cross section in one of these regions is between the *initial yield surface*, a surface bounding the region of elastic behavior, and the *full plastification surface*. The initial yield surface is determined from elementary engineering theory (taking into account the effect of residual stress) and the full plastification surface from the appropriate empirical relationship (e.g., Eq. 10.18). For a given axial force, the moment-curvature relationship of the material considered has the general shape illustrated in Figure 10.9*c* and, after subtracting the elastic curvature, the shape of Figure 10.9*d*. The axial force–strain relationships for a given bending moment are assumed to have similar shapes.

In Reference 10.13, the moment-plastic curvature and axial force–plastic strain relationships are modeled by the equations

$$\frac{d\phi_p}{dM} = \frac{c}{M_{pc} - M_{rc}} \left(\frac{M - M_{rc}}{M_{pc} - M} \right)^n \tag{10.23a}$$

and

$$\frac{de_p}{dP} = \frac{c_p}{P_p - P_r} \left(\frac{P - P_r}{P_p - P} \right)^{n_p} \tag{10.23b}$$

in which c and n are functions of P/P_p and c_p and n_p are functions of M/M_{pc}. These parameters are obtained through calibration with the results of analyses of the stress distribution on partially plastified cross sections conducted on an independent *fiber element program*, a program in which the cross section is subdivided into a number of small elements. Adding the inelastic relationships of Equation 10.23 to the conventional elastic ones in the partially plastified regions, integrating over the length of each region and satisfying the compatibility and boundary conditions for the regions, expressions for the elongation and the end rotations of the member can be obtained. For example:

$$\theta_1 = \frac{L}{6EI}(2M_1 - M_2) - \frac{(M_1 + M_2)}{L^2}[I_3(M_1) + I_3(M_2) - LI_2(M_1)] \tag{10.24}$$

in which

$$I_2(M) = \frac{L^2}{(M_1 + M_2)^2} \int_{M_{rc}}^{M} (M - m)F(m)\, dm$$

$$I_3(M) = \frac{L^3}{2(M_1 + M_2)^3} \int_{M_{rc}}^{M} (M - m)^2 F(m)\, dm$$

where m is an integration variable for moments and

$$F(m) = \frac{c}{(M_{pc} - M_{rc})} \left(\frac{m - M_{rc}}{M_{pc} - m} \right)^n$$

The incremental flexibility equation for the member of Figure 10.9*a*, which may be obtained by differentiation of Equation 10.24 and the comparable equations for θ_2 and u is, symbolically

$$\begin{Bmatrix} du \\ d\theta_1 \\ d\theta_2 \end{Bmatrix} = \begin{bmatrix} \partial u/\partial P & \partial u/\partial M_1 & \partial u/\partial M_2 \\ \partial \theta_1/\partial P & \partial \theta_1/\partial M_1 & \partial \theta_1/\partial M_2 \\ \partial \theta_2/\partial P & \partial \theta_2/\partial M_1 & \partial \theta_2/\partial M_2 \end{bmatrix} \begin{Bmatrix} dP \\ dM_1 \\ dM_2 \end{Bmatrix} \tag{10.25}$$

The four flexibility terms $\partial\theta_1/\partial P$, $\partial\theta_2/\partial P$, $\partial u/\partial M_1$, and $\partial u/\partial M_2$ have only plastic contributions whereas the remainder have both elastic and plastic parts. Studies in Reference 10.13 indicate that the purely plastic interaction terms have an insignificant effect on in-plane response. Neglecting them and using the remainder of the flexibility matrix in the flexibility to stiffness transformation process of Section 4.4.1 leads to the element stiffness equation

$$[\mathbf{k}_{\mathbf{em}}]\{\mathbf{d\Delta}\} = \{\mathbf{dP}\} \tag{10.26}$$

where $[\mathbf{k}_{\mathbf{em}}]$ is a symmetric 6×6 elastic-plastic matrix in which the zones of distributed plasticity have been converted into equivalent plastic hinges. Equation 10.26 is a formula for first-order inelastic analysis. For second-order inelastic analysis it may be modified to

$$[\mathbf{k}_{\mathbf{em}} + \mathbf{k}_{\mathbf{g}}]\{\mathbf{d\Delta}\} = \{\mathbf{dP}\} \tag{10.27}$$

in which $[\mathbf{k}_{\mathbf{g}}]$ is a conventional geometric stiffness matrix.

This approach requires a precalibration process to determine the cross section parameters of Equations 10.23, modification of Equation 10.24 and its companion equations for the case of single curvature, and numerical integration for the determination of the terms of $[\mathbf{k}_{\mathbf{em}}]$. Nevertheless, it can be computationally more efficient and possess an accuracy comparable to that of a fiber element program (see Ref. 10.13).

10.5.2 Multiple Yield Surfaces

There is a large body of research, both in plasticity theory and structural analysis, directed toward modeling the behavior of a material as its stress-strain relationships change under load. And much of it is based on theories or experimental evidence of how the yield surface changes. The representation of strain hardening by isotropically or kinematically hardening surfaces in Figure 10.2 is an illustration of two such theories. In the following we summarize a proposal for adapting one theory of plasticity, the *Mroz hardening theory* (Ref. 10.14), to inelastic structural analysis by combining the plastic hinge concept with the notion of multiple stress resultant yield surfaces.

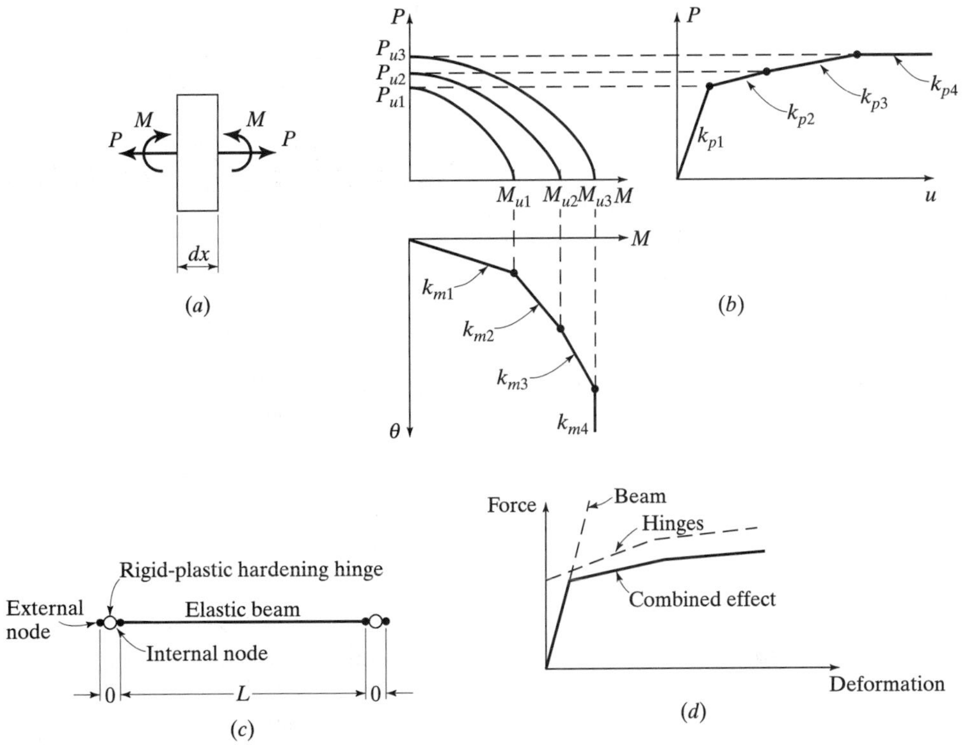

Figure 10.10 Generalized plastic hinge model.

The Mroz theory is more complex than the theories pictured in Figure 10.2 in that it postulates a series of yield surfaces with constant, but direction dependent, hardening between each pair. By extrapolation to a differential segment of a framework element subjected to axial force and bending it would imply the differential hardening behavior illustrated in Figure 10.10*b*. In Reference 10.15, Powell and Chen apply this idea to the analysis of a framework element and, in doing so, they assume that yielding and hardening, which may actually be present along the length of the member, can be treated as concentrated zero-length hinge subelements at the element ends (Fig. 10.10*c*). The resultant action of the element is thus a combination of the response of an elastic beam and inelastic hinges, as indicated in Figure 10.10*d*. The analysis involves inclusion of the plastic flexibility of the hinges and, ultimately, inversion of flexibility equations to obtain a stiffness relationship for the full element. Assumptions regarding hinge behavior or calibration against typical patterns of distributed plasticity are required for application of the method. Examples of both are presented in Reference 10.15.

10.5.3 Reinforced Concrete Members

Generally, the material ductility upon which the validity of the inelastic analysis method of Section 10.2 depends cannot be relied upon in conventional reinforced concrete structures. But the concept of a stress resultant *failure surface*, which is similar to a yield surface in appearance and in some of the ways it may be used in analysis, does have a place in the design of reinforced concrete structures. An approach commonly used in the development of a failure surface may be summarized by consideration of a doubly reinforced rectangular section subjected to axial force and bending about a principal axis (Figure 10.11*a*). Basic assumptions, which are implicit in Figure

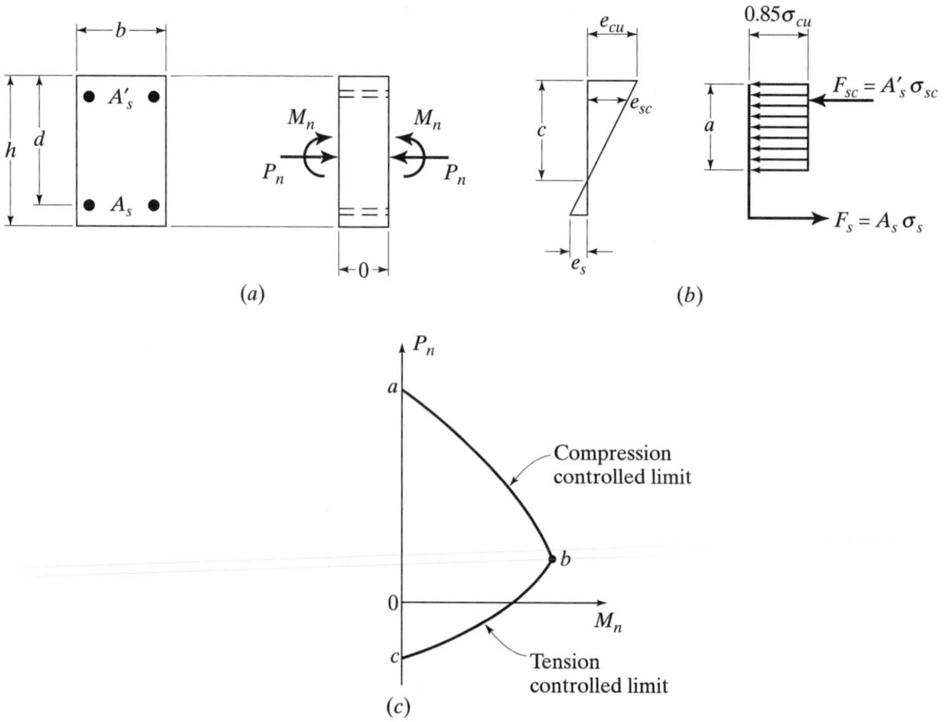

Figure 10.11 Reinforced concrete section failure surface.

10.11*b*, are: 1) plane sections remain plane in bending; 2) tensile resistance of the concrete is ignored; 3) the ultimate compressive resistance of the concrete can be represented by an *equivalent uniform stress block*; and 4) the limiting resistance of the reinforcement is its yield strength. These assumptions, plus the conditions of axial force and flexural equilibrium, are sufficient to calculate the resistance of a section of given dimensions and material properties; that is, to generate the surface of Figure 10.11*c*. In the region *ab* the limit of resistance is governed by compression in the concrete and in region *bc* by tension in the lower reinforcement. Point *b* is the *balanced failure point*, the condition under which crushing of the concrete and yielding of the lower row of steel would occur simultaneously.

The above approach and the physical consequences of the different modes of failure initiation are treated in detail in texts on reinforced concrete structures such as Reference 10.16. A three dimensional surface of the same type, but for the design of biaxially loaded steel-concrete composite columns, is developed in Reference 10.17. Included is a generalized empirical equation for the representation of the surface.

10.6 PROBLEMS

10.1 (a) Verify the development of Equations 10.14 and 10.16. (b) For element *ab* of Example 10.4, develop the $[\mathbf{k_m}]$ matrix for the case of a plastic hinge at end *a* using a symbolic mathematics program.

10.2 Equation 10.18 is a polynomial approximation of the yield surface of a W12 × 30 section of elastic–perfectly plastic steel subjected to axial force and bending. Using elementary beam theory, write a program to generate the surface and compare the results with Equation 10.18.

10.3 Determine the elastic and inelastic critical loads for the pin jointed plane frame shown.

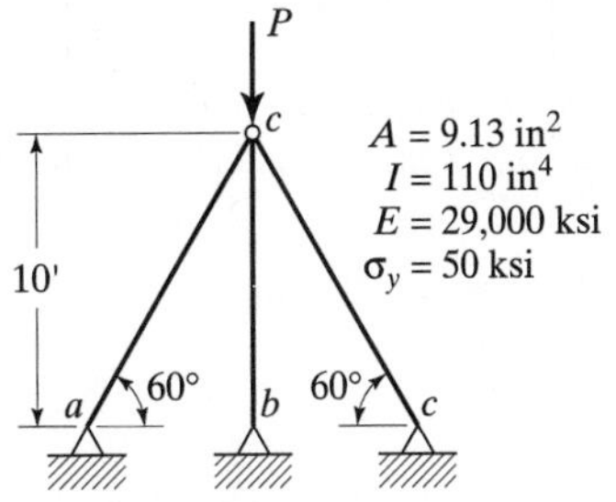

Problem 10.3

10.4 Determine the major axis elastic and inelastic critical loads for the columns shown. All loads are applied concentrically. σ_y = 50 ksi, E = 29,000 ksi.

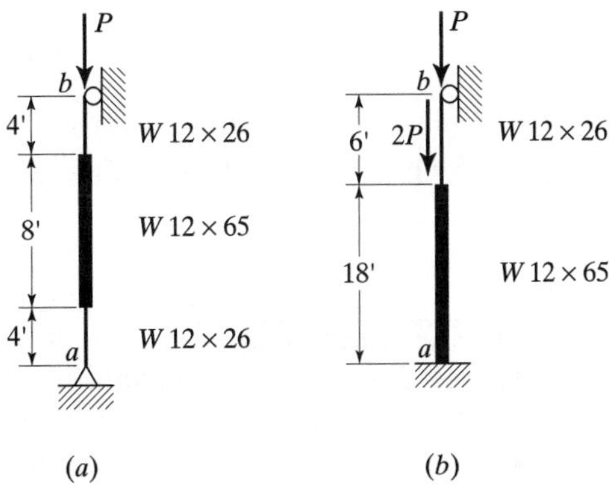

Problem 10.4

10.5 Determine the major axis elastic and inelastic critical loads for the continuous columns shown. All loads are applied concentrically. σ_y = 50 ksi, E = 29,000 ksi.

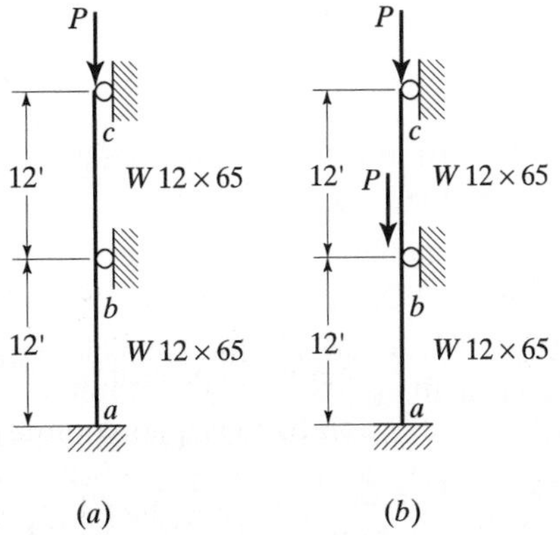

Problem 10.5

10.6 Calculate the first order inelastic response of the laterally supported beams shown. $\sigma_y = 50$ ksi, $E = 29{,}000$ ksi. For case (d) determine an approximate maximum strength-to-weight ratio by analyzing the system for several values of x.

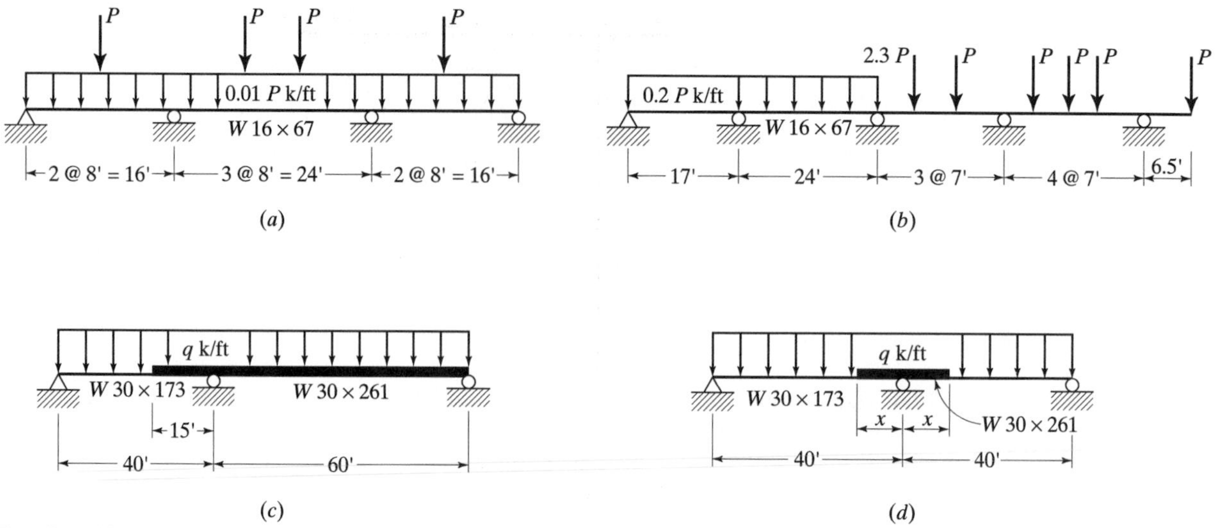

Problem 10.6

10.7 Calculate the second order inelastic response of the laterally supported beam-columns shown. $\sigma_y = 50$ ksi, $E = 29{,}000$, ksi.

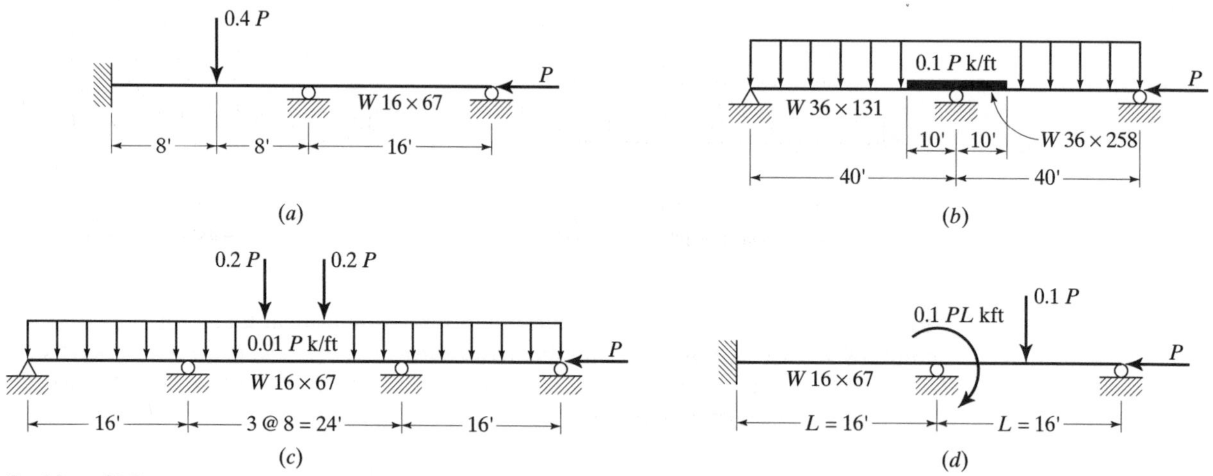

Problem 10.7

10.8 For the frame and loading conditions of Problem 9.2, calculate the inelastic critical loads. $\sigma_y = 250$ MPa.

10.9 Calculate the inelastic critical loads for the frame and loading conditions of Problem 9.3. $\sigma_y = 250$ MPa.

10.10 (a) Compute the inelastic critical load for the frame of Problem 9.4. $\sigma_y = 36$ ksi. (b) Calculate the response of the frame using first- and second-order inelastic analyses. Compare the results. Beams $Z = 244$ in^3, columns $Z = 54.9$ in^3, $\sigma_y = 36$ ksi.

10.11 Use first- and second-order inelastic methods to analyze the frame of Problem 9.5. Compare the results. Rafters $Z = 1.28 \times 10^6$ mm^3, columns $Z = 4.24 \times 10^5$ mm^3, $\sigma_y = 250$ MPa.

10.12 Analyze the frame shown by first- and second-order elastic and first- and second-order inelastic methods. $\sigma_y = 60$ ksi for columns and 36 ksi for all other members, $E = 29{,}000$ ksi. (Account for elastic shear deformation).

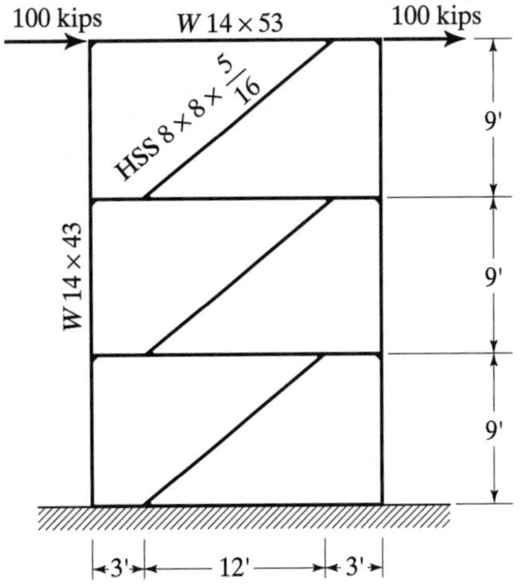

Problem 10.12

10.13 The rigid frames shown are laterally braced. $\sigma_y = 36$ ksi, $E = 29{,}000$. Analyze each frame by first and second order inelastic analysis methods and compare the results.

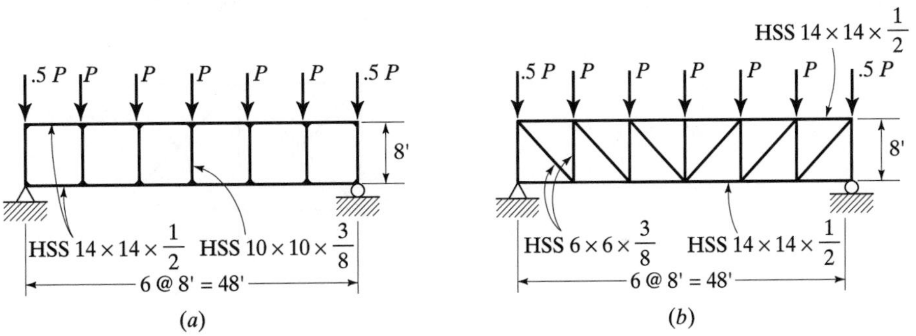

Problem 10.13

10.14 The beams, columns, and bracing members of the rigidly jointed frame shown have the sizes indicated. $\sigma_y = 36$ ksi, $E = 29{,}000$ ksi. Analyze the frame by first- and second-order inelastic methods.

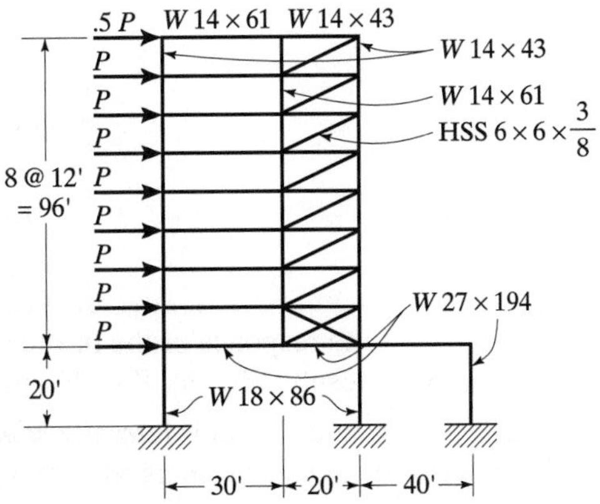

Problem 10.14

10.15 Study the deflection stability of the frame shown under the following loading pattern: 1) a vertical load (P = 30 kips) applied at c and retained on the structure; and 2) a horizontal load, H = 15 kips, applied repeatedly at b (H = 0, 15, 0, 15, etc). σ_y = 50 ksi., E = 29,000 ksi.

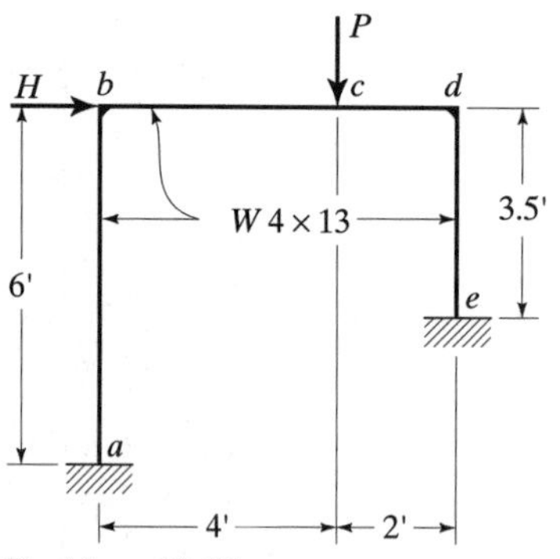

Problem 10.15

10.16 The rigid frame shown is laterally braced. σ_y = 36 ksi, E = 29,000 ksi. Analyze it by a second-order inelastic method. Attempt to improve the overall strength-to-weight ratio of the structure by considering and analyzing it for several different sizes of web members.

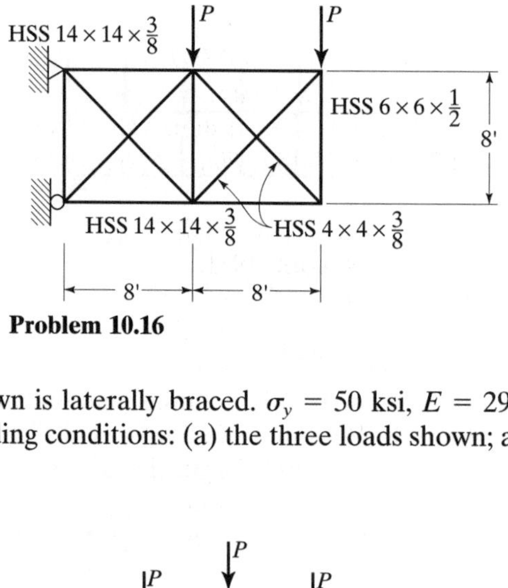

Problem 10.16

10.17 The arch shown is laterally braced. σ_y = 50 ksi, E = 29,000 ksi. Compare its resistance to two loading conditions: (a) the three loads shown; and (b) a load at point b only.

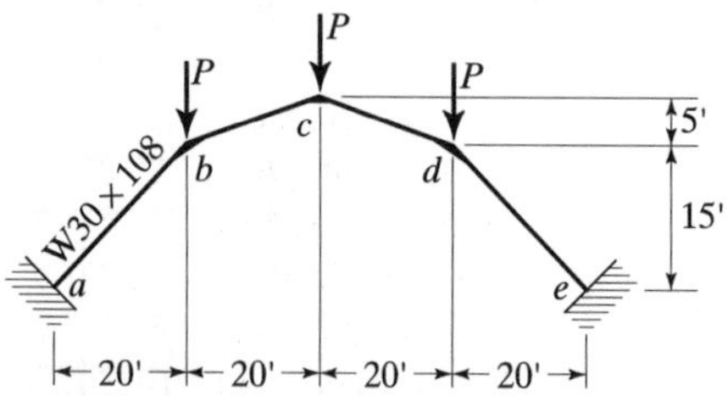

Problem 10.17

10.18 The frame shown is laterally braced, σ_y = 50 ksi, E = 29,000 ksi. Analyze it by a second-order inelastic method. Attempt to improve the overall strength-to-weight ratio by considering and analyzing it for several different sets of member sizes (vary member sizes individually if desired).

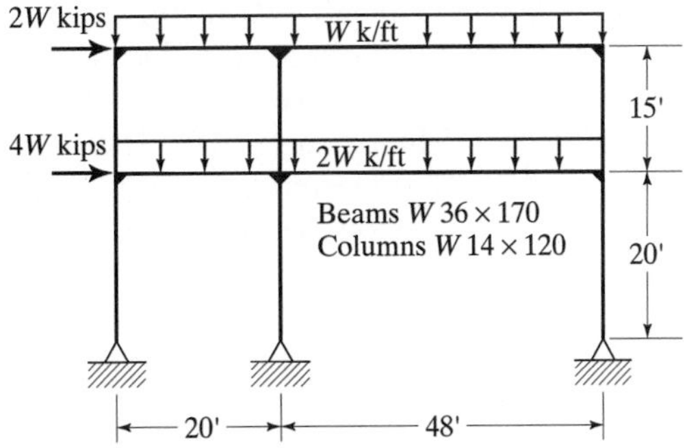

Problem 10.18

10.19 Analyze the system of Problem 9.17 by first- and second-order inelastic methods. Study the effect of varying girder size. Girder $\sigma_y = 50$ ksi, cables $\sigma_y = 150$ ksi.

10.20 Analyze the system of Problem 9.18 by first- and second-order inelastic methods. Study the effect of varying cable and girder sizes. Girder $\sigma_y = 50$ ksi, cables $\sigma_y = 150$ ksi.

REFERENCES

10.1 A. Mendelson, *Plasticity: Theory and Application*, Macmillan, New York, 1968.

10.2 M. Zyczkowski, *Combined Loadings in the Theory of Plasticity*, PWN—Polish Scientific Publishers (ARS Polona-Ruch distributors), Warsaw, 1981.

10.3 K.-J. Bathe, *Finite Element Procedures*, Prentice-Hall, Englewood Cliffs, N.J., 1996.

10.4 *Plastic Design in Steel, A Guide and Commentary*, ASCE Manual No. 41, ASCE, New York, 1971.

10.5 *Manual of Steel Construction, Allowable Stress Design*, 9th edition, American Institute of Steel Construction, Chicago, 1989.

10.6 *Manual of Steel Construction, Load and Resistance Factor Design*, 2nd edition, American Institute of Steel Construction, Chicago, 1994.

10.7 L. Porter and G. H. Powell, "Static and Dynamic Analysis of Inelastic Framed Structures," Report No. EERC 71-3, Earthquake Engineering Research Center, Univ. of California, Berkeley, 1971.

10.8 P. G. Hodge, *Plastic Analysis of Structures*, McGraw-Hill, New York, 1959.

10.9 W.-F. Chen and T. Atsuta, *Theory of Beam-Columns, Vol. 2: Space Behavior and Design*, McGraw-Hill, New York, 1977.

10.10 J. G. Orbison, W. McGuire, and J. F. Abel, "Yield Surface Applications in Nonlinear Steel Frame Analysis" *Computer Methods in Applied Mechanics and Engineering*, Vol. 33, Nos. 1–3, 1982.

10.11 T. V. Galambos, editor, John Wiley and Sons, *Guide to Stability Design Criteria for Metal Structures*, 5th edition, New York, 1998.

10.12 P. Arnold, P. F. Adams, and L.-W. Lu, "Strength and Behavior of an Inelastic Hybrid Frame," *Jl. of the Struct. Div.*, ASCE, Vol. 94, No. ST1, 1968.

10.13 M. R. Attalla, G. G. Deierlein, and W. McGuire, "Spread of Plasticity: A Quasi-Plastic-Hinge Approach," *Jl. of Struct. Engr.*, ASCE, Vol. 120, No. 8, 1994.

10.14 Z. Mroz, *Mathematical Models of Inelastic Material Behavior*, Solid Mechanics Division, University of Waterloo, Waterloo, Ontario, 1973.

10.15 G. H. Powell and P. F. S. Chen, "3D Beam Column Element with Generalized Plastic Hinges," *Jl. of Engr. Mech.*, ASCE, Vol. 112, No. 7, 1986.

10.16 J. G. MacGregor, *Reinforced Concrete, Mechanics and Design*, Prentice Hall, Saddle River, New Jersey, 1997.

10.17 P. R. Munoz and C-T T. Hsu, "Biaxially Loaded Concrete Encased Composite Columns: Design Equation," *Jl. of Struct. Eng.*, ASCE, Vol. 123, No. 12, Dec., 1997, Reston, Virginia.

Chapter 11

Solution of Linear Algebraic Equations

We have been concerned mainly with the physical aspects of the structural analysis problem—the formation of equations that satisfy the laws of equilibrium, compatibility and the constitutive relations of the material, and the interpretation of the solution of such equations. We now shift our attention to the mathematics of equation solving.

In Chapters 2–5, first-order elastic analysis included solving a system of linear simultaneous equations for the unknown displacement components. In Chapter 8, it was shown that more realistic modeling of structural systems results in nonlinear equilibrium equations. As a practical means for solving these algebraic and transcendental equations, the analysis is reduced to a series of linear increments in which the displacement components are determined by employing the same equation solution strategies as those used in first-order elastic analysis.

Although the subject of equation solving may be more in the realm of numerical methods or computer science than in structures, the algorithm employed can have a dramatic impact on many aspects of analysis, ranging from overall efficiency to accuracy and reliability of the results. For the larger problems common in practice, it has been estimated that 20% to 50% of the computer execution time may be devoted to solving systems of equations. Where a computer is limited to representing all values by a fixed number of significant figures, numerical results can be susceptible to accumulated error. Since primary responsibility for the entire analysis generally belongs to the structural engineer, more than a superficial acquaintance with the different possibilities for solving equations is required.

We begin this chapter with a review of the basic problem. Direct and iterative methods for solving large systems of equations are described and illustrated. We then present ways to improve solution efficiency by taking advantage of peculiar characteristics of most stiffness coefficient matrices such as symmetry, bandedness, and sparseness. Finally, we consider the questions of accuracy and error, and what may be done about them. We shall see that there are ways to detect systems of equations that may be particularly troublesome numerically, to estimate whether the solution errors may be intolerable and, if so, to avoid or at least ameliorate the situation.

11.1 THE BASIC CHOICE—DIRECT INVERSION VERSUS ELIMINATION OR ITERATION

The problem in numerical analysis that we have addressed many times can be stated as

$$[\mathbf{A}]\{\mathbf{x}\} = \{\mathbf{b}\} \tag{11.1}$$

where the elements of [**A**] and {**b**} are known and those of {**x**} unknown. We limit consideration to cases in which the coefficient matrix [**A**], while not necessarily symmetrical, is square and nonsingular. The object is to solve for {**x**}. The solution may be written symbolically as

$$\{\mathbf{x}\} = [\mathbf{A}]^{-1}\{\mathbf{b}\} \tag{11.2}$$

Interpreting $[\mathbf{A}]^{-1}$ literally as the inverse of [**A**], it can be evaluated through Cramer's rule as the general equation

$$[\mathbf{A}]^{-1} = \frac{[adj\ \mathbf{A}]}{|\mathbf{A}|} \tag{11.3}$$

In Equation 11.3, [*adj* **A**] is the *adjoint* matrix of [**A**]; that is, the matrix of transposed *cofactors*,[1] and |**A**| is the determinant of [**A**]. The derivation of Equation 11.3 can be found in most texts on matrix algebra and many on numerical analysis (see Ref. 11.1). Direct inversion of [**A**] in this fashion is straightforward. Unfortunately, for almost any matrix of practical interest, it is an inefficient way to solve Equation 11.1 since it requires the evaluation of a number of determinants of high order in calculating $[\mathbf{A}]^{-1}$. Many more efficient schemes for determining {**x**} without constructing the inverse of [**A**] explicitly have been devised. Generally, the most attractive are divided into two groups, direct elimination methods and iterative methods.

11.2 DIRECT ELIMINATION METHODS

Elimination methods employ the concept of factoring the matrix [**A**] into products of triangular and—sometimes—diagonal matrices. The *decomposition theorem* (Ref. 11.1) states that a general square matrix [**A**] is expressible in the form:

$$[\mathbf{A}] = [\mathbf{L}][\mathbf{D}][\mathbf{U}] \tag{11.4}$$

where [**L**] and [**U**] are unit lower and upper triangular matrices, and [**D**] is a diagonal matrix.

Variants of this basic decomposition equation can be obtained by associating [**D**] with either [**L**] or [**U**]. Further, limitless possibilities are opened up by considering [**D**] as the product of two diagonal matrices and associating one of this diagonal pair with [**L**] and the other with [**U**]. Thus [**A**] may be factored into lower and upper triangular matrices:

$$[\mathbf{A}] = [\mathbf{L}][\mathbf{U}] \tag{11.5}$$

where in this formulation, [**L**] and [**U**] are not necessarily the same as the [**L**] and [**U**] of Equation 11.4. More importantly, they are not unique. Instead, they depend on the

[1]The cofactor of a matrix element a_{ij} is the signed determinant defined as $A_{ij} = (-1)^{i+j}M_{ij}$, where M_{ij}, the *minor* of a_{ij}, is the determinant of the matrix formed by deleting the ith row and the jth column of [**A**].

way in which the [**D**] matrix of Equation 11.4 is distributed between the lower and upper triangular matrices of Equation 11.5. For a 3 × 3 matrix, the expanded form of Equation 11.5 is

$$\begin{bmatrix} a_{11} & a_{12} & a_{13} \\ a_{21} & a_{22} & a_{23} \\ a_{31} & a_{32} & a_{33} \end{bmatrix} = \begin{bmatrix} l_{11} & 0 & 0 \\ l_{21} & l_{22} & 0 \\ l_{31} & l_{32} & l_{33} \end{bmatrix} \begin{bmatrix} u_{11} & u_{12} & u_{13} \\ 0 & u_{22} & u_{23} \\ 0 & 0 & u_{33} \end{bmatrix} \tag{11.6}$$

In Equation 11.5 either [**L**], [**U**], or neither may be a unit triangular matrix. Usually the diagonals of both are not unity.

Differences among the many elimination methods found in the literature are simply differences in the way in which [**A**] is factored into [**L**] and [**U**]. Before showing how [**L**] and [**U**] are formed under two different procedures and describing the other features of these equation-solving techniques, we should point out the general advantage of matrix decomposition.

Assume that we have a square, nonsingular, possibly nonsymmetrical [**A**] matrix, and its corresponding [**L**] and [**U**] matrices have been generated by some proper decomposition scheme. Equation 11.1 may be rewritten:

$$[\mathbf{A}]\{\mathbf{x}\} = [\mathbf{L}][\mathbf{U}]\{\mathbf{x}\} = \{\mathbf{b}\} \tag{11.7}$$

For convenience, let us introduce an unknown auxiliary vector {**y**} which is defined by

$$[\mathbf{U}]\{\mathbf{x}\} = \{\mathbf{y}\} \tag{11.8}$$

From Equations 11.7 and 11.8, we have

$$[\mathbf{L}]\{\mathbf{y}\} = \{\mathbf{b}\} \tag{11.9}$$

Since [**L**] is a known lower triangular matrix and {**b**} is given, we may determine {**y**} element by element starting from the top equation in a simple process of forward substitution

$$y_i = \frac{b_i - \sum_{j=1}^{i-1} l_{ij} y_j}{l_{ii}} \qquad \text{for } i = 1, \ldots, n \tag{11.9a}$$

With [**U**] a known upper triangular matrix and {**y**} now determined, we return to Equation 11.8 and solve for {**x**} element by element starting from the bottom equation in a process of back substitution

$$x_i = \frac{y_i - \sum_{j=i+1}^{n} u_{ij} x_j}{u_{ii}} \qquad \text{for } i = n, \ldots, 1 \tag{11.8a}$$

We note that the whole process of solving for the unknown {**x**} has been completed without calculating the inverse of [**A**]. Instead, Equation 11.1 has been solved with two basic steps inherent to most elimination methods: (1) decomposition of the coefficient matrix, typically done by a process of forward or backward elimination; and (2) forward and/or backward substitution. These steps should be clear in the descriptions of the most popular Gauss and Cholesky methods that follow. As for the relative merits of decomposition and substitution versus explicit inversion, studies in which the required arithmetical operations have been counted and compared attest to the general superiority of the elimination method for solving systems of practical interest[2] (see Ref. 11.2).

11.2.1 Gauss Elimination

Basic Gauss elimination commences with a forward elimination process. Given a set of equations such as

$$\begin{aligned} a_{11}x_1 + a_{12}x_2 + a_{13}x_3 &= b_1 \\ a_{21}x_1 + a_{22}x_2 + a_{23}x_3 &= b_2 \\ a_{31}x_1 + a_{32}x_2 + a_{33}x_3 &= b_3 \end{aligned} \tag{11.10}$$

eliminate the first unknown from the second and third equations by multiplying the first equation by a_{21}/a_{11} and subtracting it from the second and then multiplying the original first equation by a_{31}/a_{11} and subtracting it from the third, yielding

$$\begin{aligned} \left(a_{22} - \frac{a_{21}a_{12}}{a_{11}}\right)x_2 + \left(a_{23} - \frac{a_{21}a_{13}}{a_{11}}\right)x_3 &= b_2 - \frac{a_{21}}{a_{11}}b_1 \\ \left(a_{32} - \frac{a_{31}a_{12}}{a_{11}}\right)x_2 + \left(a_{33} - \frac{a_{31}a_{13}}{a_{11}}\right)x_3 &= b_3 - \frac{a_{31}}{a_{11}}b_1 \end{aligned} \tag{11.11a}$$

Operating similarly on these two equations, x_2 can be eliminated, leaving one equation with the unknown x_3

$$\left[\left(a_{33} - \frac{a_{31}a_{13}}{a_{11}}\right) - a'\left(a_{23} - \frac{a_{21}a_{13}}{a_{11}}\right)\right]x_3 = \left(b_3 - \frac{a_{31}}{a_{11}}b_1\right) - a'\left(b_2 - \frac{a_{21}}{a_{11}}b_1\right) \tag{11.11b}$$

where

$$a' = \left(a_{32} - \frac{a_{31}a_{12}}{a_{11}}\right) \Big/ \left(a_{22} - \frac{a_{21}a_{12}}{a_{11}}\right)$$

The completed elimination—or reduction—process has resulted in converting the original system of equations to the form

$$\begin{aligned} u_{11}x_1 + u_{12}x_2 + u_{13}x_3 &= y_1 \\ u_{22}x_2 + u_{23}x_3 &= y_2 \\ u_{33}x_3 &= y_3 \end{aligned} \tag{11.12}$$

where $u_{11} = a_{11}$, $u_{22} = (a_{22} - a_{21}a_{12}/a_{11})$, $y_1 = b_1$, $y_2 = (b_2 - a_{21}b_1/a_{11})$, etc. Since we have performed only a series of algebraic manipulations, the solution to the system shown in Equation 11.12 is identical to that of Equation 11.10.

This solution is found by solving the third equation of the system shown in Equation 11.12—or equivalently, Equation 11.11b—for its only unknown x_3. The remaining unknowns x_2 and x_1 can now be obtained, in that order, by backward substitution

[2]The total number of floating point operations, *flops*, is often used to compare the computational efficiency of different solution algorithms. If n is the number of equations to be solved, an approximate number of flops required for the procedures presented in this section are as follows: Cramer's Rule, $n!$; Gauss elimination, $n^3/3$; Cholesky decomposition, $n^3/6$; and banded Cholesky decomposition, $n \cdot HBW^2$, where the half-bandwidth HBW is defined in Section 11.4.

working from the bottom up in Equation 11.12. The same procedure may be applied to systems of any size, provided the coefficient of the leading unknown at any stage of the elimination does not equal zero, a case we will consider later in this section.

Comparing Equations 11.8 and 11.12, we see that they are identical; the former is just a matrix statement of the latter. It is clear, therefore, that during the Gauss elimination process the coefficient matrix, [**A**], has been factored into upper and lower triangular matrices and, further, the lower triangular matrix has been inverted and postmultiplied by the right-hand side vector, {**b**}.

Gauss step-by-step decomposition of [**A**] into the product [**L**][**U**] can be represented algorithmically in several ways. Where [**A**] is typically the stiffness matrix in the structural analysis, one convenient scheme for performing the decomposition of this symmetric matrix is the following:

$$\begin{aligned} u_{ji} &= \frac{l_{ij}}{l_{jj}} && \text{for } j = 1, \ldots, i-1 \\ u_{ii} &= 1 && \\ l_{ji} &= a_{ij} - \sum_{k=1}^{i-1} u_{ki} l_{jk} && \text{for } j = i, \ldots, n \end{aligned} \tag{11.13}$$

This scheme is applied and verified in Example 11.1. Example 11.2 illustrates the matrix interpretation of Gauss elimination through Equations 11.7–11.9.

EXAMPLE 11.1

Given

$$[\mathbf{A}] = \begin{bmatrix} 100 & 30 & 20 & 10 \\ 30 & 178 & 201 & -36 \\ 20 & 201 & 485 & 21 \\ 10 & -36 & 21 & 350 \end{bmatrix}$$

factor into [**L**][**U**] using Equation 11.13 and verify results.

Column 1: $u_{11} = 1$

$$\begin{aligned} l_{11} &= a_{11} = 100 \\ l_{21} &= a_{12} = 30 \\ l_{31} &= a_{13} = 20 \\ l_{41} &= a_{14} = 10 \end{aligned}$$

Column 2: $u_{12} = l_{21}/l_{11} = 30/100 = 3/10$

$$\begin{aligned} u_{22} &= 1 \\ l_{22} &= a_{22} - u_{12} \times l_{21} = 178 - (\tfrac{3}{10}) \times 30 = 169 \\ l_{32} &= a_{23} - u_{12} \times l_{31} = 201 - (\tfrac{3}{10}) \times 20 = 195 \\ l_{42} &= a_{24} - u_{12} \times l_{41} = -36 - (\tfrac{3}{10}) \times 10 = -39 \end{aligned}$$

Column 3: $u_{13} = l_{31}/l_{11} = 20/100 = 1/5$

$$\begin{aligned} u_{23} &= l_{32}/l_{22} = 195/169 = 15/13 \\ u_{33} &= 1 \\ l_{33} &= a_{33} - u_{13} \times l_{31} - u_{23} \times l_{32} = 485 - (\tfrac{1}{5}) \times 20 - (\tfrac{15}{13}) \times 195 = 256 \\ l_{43} &= a_{34} - u_{13} \times l_{41} - u_{23} \times l_{42} = 21 - (\tfrac{1}{5}) \times 10 - (\tfrac{15}{13}) \times (-39) = 64 \end{aligned}$$

Column 4: $u_{14} = l_{41}/l_{11} = 10/100 = 1/10$

$$u_{24} = l_{42}/l_{22} = -39/169 = -3/13$$
$$u_{34} = l_{43}/l_{33} = 64/256 = 1/4$$
$$u_{44} = 1$$
$$l_{44} = a_{44} - u_{14} \times l_{41} - u_{24} \times l_{42} - u_{34} \times l_{43}$$
$$= 350 - (\tfrac{1}{10}) \times 10 - (-\tfrac{3}{13}) \times (-39) - (\tfrac{1}{4}) \times 64 = 324$$

Check:
$$\begin{bmatrix} 100 & 0 & 0 & 0 \\ 30 & 169 & 0 & 0 \\ 20 & 195 & 256 & 0 \\ 10 & -39 & 64 & 324 \end{bmatrix} \begin{bmatrix} 1 & 3/10 & 1/5 & 1/10 \\ 0 & 1 & 15/13 & -3/13 \\ 0 & 0 & 1 & 1/4 \\ 0 & 0 & 0 & 1 \end{bmatrix} = \begin{bmatrix} 100 & 30 & 20 & 10 \\ 30 & 178 & 201 & -36 \\ 20 & 201 & 485 & 21 \\ 10 & -36 & 21 & 350 \end{bmatrix}$$

$$[\mathbf{L}][\mathbf{U}] = [\mathbf{A}] \qquad \text{Q.E.D.}$$

EXAMPLE 11.2

Given

$$\begin{aligned} 100x_1 + 30x_2 + 20x_3 + 10x_4 &= 490 \\ 30x_1 + 178x_2 + 201x_3 - 36x_4 &= -347 \\ 20x_1 + 201x_2 + 485x_3 + 21x_4 &= -1112 \\ 10x_1 - 36x_2 + 21x_3 + 350x_4 &= 651 \end{aligned}$$

solve by Gauss elimination and verify results.

The coefficient matrix is identical to [**A**] of Example 11.1—use results of that example. From Equation 11.9, [**L**][**y**] = {**b**}, solve for {**y**} by forward substitution

$$\begin{bmatrix} 100 & 0 & 0 & 0 \\ 30 & 169 & 0 & 0 \\ 20 & 195 & 256 & 0 \\ 10 & -39 & 64 & 324 \end{bmatrix} \begin{Bmatrix} y_1 \\ y_2 \\ y_3 \\ y_4 \end{Bmatrix} = \begin{Bmatrix} 490 \\ -347 \\ -1112 \\ 651 \end{Bmatrix}$$

$$y_1 = 490/100 = 49/10$$
$$y_2 = [-347 - 30 \times (\tfrac{49}{10})]/169 = -38/13$$
$$y_3 = [-1112 - 20 \times (\tfrac{49}{10}) - 195 \times (-\tfrac{38}{13})]/256 = -5/2$$
$$y_4 = [651 - 10 \times (\tfrac{49}{10}) - (-39) \times (-\tfrac{38}{13}) - 64 \times (-\tfrac{5}{2})]/324 = 2$$

From Equation 11.8, [**U**]{**x**} = {**y**}, solve for {**x**} by backward substitution

$$\begin{bmatrix} 1 & 3/10 & 1/5 & 1/10 \\ 0 & 1 & 15/13 & -3/13 \\ 0 & 0 & 1 & 1/4 \\ 0 & 0 & 0 & 1 \end{bmatrix} \begin{Bmatrix} x_1 \\ x_2 \\ x_3 \\ x_4 \end{Bmatrix} = \begin{Bmatrix} 49/10 \\ -38/13 \\ -5/2 \\ 2 \end{Bmatrix}$$

$$x_4 = 2$$
$$x_3 = -\tfrac{5}{2} - (\tfrac{1}{4}) \times 2 = -3$$
$$x_2 = -\tfrac{38}{13} - (\tfrac{15}{13}) \times (-3) - (-\tfrac{3}{13}) \times 2 = 1$$
$$x_1 = \tfrac{49}{10} - (\tfrac{3}{10}) \times 1 - (\tfrac{1}{5}) \times (-3) - (\tfrac{1}{10}) \times 2 = 5$$

Check:

$$\begin{bmatrix} 100 & 30 & 20 & 10 \\ 30 & 178 & 201 & -36 \\ 20 & 201 & 485 & 21 \\ 10 & -36 & 21 & 350 \end{bmatrix} \begin{Bmatrix} 5 \\ 1 \\ -3 \\ 2 \end{Bmatrix} = \begin{Bmatrix} 490 \\ -347 \\ -1112 \\ 651 \end{Bmatrix}$$

$$[\mathbf{A}]\{\mathbf{x}\} = \{\mathbf{b}\} \qquad \text{Q.E.D.}$$

It was mentioned earlier that if at any stage in the elimination process the coefficient of the first equation for the leading unknown, often termed the *pivot* element, becomes zero, the basic process will not work. When checks are not made to avoid the resulting division by zero that will occur in Equations 11.11–11.13, the above process is called "naive" Gauss elimination. This difficulty can be avoided by employing *partial pivoting*, that is, by interchanging the offending equation with another having a nonzero coefficient for the same unknown. This will always be possible if the submatrix under consideration is nonsingular. Note that partial pivoting will destroy any existing symmetry of a coefficient matrix, thereby making it impossible to use algorithms based on symmetry, such as the one expressed by Equation 11.13. To preserve symmetry, this concept may be extended to *complete pivoting* where both rows and columns are interchanged. As discussed later in this chapter, pivoting strategies are also often employed to improve solution accuracy.

In Chapter 13 we will present static condensation and substructuring as one way of handling large systems of equations. The role of that technique in equation solving will not be discussed here other than to define matrix condensation as a process of contracting the size of a system of equations by eliminating certain degrees of freedom. The relationship between this definition and Gauss elimination should be fairly obvious. The tie between static condensation, substructuring, and Gauss elimination is developed in Reference 11.3 and numerous papers in the literature.

11.2.2 Cholesky Method

Although there exists Gauss decomposition schemes that can be applied to any square coefficient matrix, the Cholesky method applies only to cases where [**A**] is symmetrical and positive definite.[3] Its distinguishing feature is that [**A**] can be factored efficiently into upper and lower triangular matrices that are the transpose of each other; thus

$$[\mathbf{A}] = [\mathbf{L}][\mathbf{L}]^{\mathrm{T}} \tag{11.14}$$

Once this decomposition has been performed, equation solution proceeds by forward and backward substitution following Equations 11.7–11.9, with [**U**] in those equations becoming $[\mathbf{L}]^{\mathrm{T}}$.

---

[3]For all nonzero vectors {**z**}, [**A**] is a *positive definite* matrix as long as the inner product $\{\mathbf{z}\}^{\mathrm{T}}[\mathbf{A}]\{\mathbf{z}\}$ is always greater than zero. If this inner product is always greater than or equal to zero, [**A**] is *positive semi-definite* and if it can be less than zero, [**A**] is *indefinite*. Alternatively, the definiteness of a matrix can be defined in terms of its eigenvalues (see Section 12.8). If all the eigenvalues of [**A**] are positive, it is positive definite. [**A**] is positive semi-definite if its eigenvalues are greater than or equal to zero. And, [**A**] is indefinite when its eigenvalues are either positive, zero, or negative.

A convenient recurrence relation for Cholesky decomposition is as follows. Each successive column of [**L**] is calculated as

$$l_{ii} = \sqrt{a_{ii} - \sum_{k=1}^{i-1} l_{ik}^2}$$

$$l_{ji} = \frac{a_{ji} - \sum_{k=1}^{i-1} l_{jk} l_{ik}}{l_{ii}} \qquad \text{for } j = i+1, \ldots, n \tag{11.15}$$

It should be noted that if the above evaluation of l_{ii} involves taking the square root of a negative number, this implies that [**A**] is indefinite (footnote 3). With respect to structural analysis in which [**A**] = [$\mathbf{K}_{ff}$], this indicates that the system is unstable (see Section 8.1.4). An additional benefit of working with positive definite symmetric matrices is that a zero will never exist in the denominator of Equation 11.15; hence this algorithm can be employed without the use of pivoting.

Equation 11.15 may be modified to a slightly more efficient root-free version of Cholesky decomposition. In this scheme the coefficient matrix is factored into the form of Equation 11.4 with [**U**] being $[\mathbf{L}]^{\mathrm{T}}$; thus

$$[\mathbf{A}] = [\mathbf{L}][\mathbf{D}][\mathbf{L}]^{\mathrm{T}} \tag{11.16}$$

Starting with $i = 1$, each column of [**D**] and [**L**] can be calculated as

$$d_{ii} = a_{ii} - \sum_{k=1}^{i-1} d_{kk} l_{ik}^2$$

$$l_{ii} = 1 \tag{11.17}$$

$$l_{ji} = \frac{a_{ji} - \sum_{k=1}^{i-1} d_{kk} l_{jk} l_{ik}}{d_{ii}} \qquad \text{for } j = i+1, \ldots, n$$

With [**A**] factored into the above triple product, a solution can still be obtained following Equations 11.7–11.9 with [**U**] in those equations now becoming $[\mathbf{D}][\mathbf{L}]^{\mathrm{T}}$. If this method is employed with a pivoting strategy, it can also be used to solve symmetric indefinite systems. Beyond being a more general solution method, root-free Cholesky decomposition can provide insight for solving eigenproblems associated with critical load analysis (see Section 12.8).

Equations 11.15 and 11.17 are illustrated and verified in Example 11.3.

EXAMPLE 11.3

Given

$$[\mathbf{A}] = \begin{bmatrix} 100 & 30 & 20 & 10 \\ 30 & 178 & 201 & -36 \\ 20 & 201 & 485 & 21 \\ 10 & -36 & 21 & 350 \end{bmatrix}$$

1. Factor into $[\mathbf{L}][\mathbf{L}]^{\mathrm{T}}$ using Equation 11.15 and verify results.
2. Factor into $[\mathbf{L}][\mathbf{D}][\mathbf{L}]^{\mathrm{T}}$ using Equation 11.17 and verify results.

1. Column 1: $l_{11} = \sqrt{a_{11}} = \sqrt{100} = 10$

$l_{21} = a_{21}/l_{11} = 30/10 = 3$

$l_{31} = a_{31}/l_{11} = 20/10 = 2$

$l_{41} = a_{41}/l_{11} = 10/10 = 1$

Column 2: $l_{22} = \sqrt{a_{22} - l_{21}^2} = \sqrt{178 - 3^2} = 13$

$l_{32} = [a_{32} - (l_{31} \times l_{21})]/l_{22} = [201 - (2 \times 3)]/13 = 15$

$l_{42} = [a_{42} - (l_{41} \times l_{21})]/l_{22} = [-36 - (1 \times 3)]/13 = -3$

Column 3: $l_{33} = \sqrt{a_{33} - l_{31}^2 - l_{32}^2} = \sqrt{485 - 2^2 - 15^2} = 16$

$l_{43} = [a_{43} - (l_{41} \times l_{31}) - (l_{42} \times l_{32})]/l_{33} = [21 - (1 \times 2) - (-3 \times 15)]/16 = 4$

Column 4: $l_{44} = \sqrt{a_{44} - l_{41}^2 - l_{42}^2 - l_{43}^2} = \sqrt{350 - 1^2 - (-3)^2 - 4^2} = 18$

Check:

$$\begin{bmatrix} 10 & 0 & 0 & 0 \\ 3 & 13 & 0 & 0 \\ 2 & 15 & 16 & 0 \\ 1 & -3 & 4 & 18 \end{bmatrix} \begin{bmatrix} 10 & 3 & 2 & 1 \\ 0 & 13 & 15 & -3 \\ 0 & 0 & 16 & 4 \\ 0 & 0 & 0 & 18 \end{bmatrix} = \begin{bmatrix} 100 & 30 & 20 & 10 \\ 30 & 178 & 201 & -36 \\ 20 & 201 & 485 & 21 \\ 10 & -36 & 21 & 350 \end{bmatrix}$$

$[\mathbf{L}][\mathbf{L}]^T = [\mathbf{A}]$ Q.E.D.

2. Column 1: $d_{11} = a_{11} = 100$

$l_{11} = 1$

$l_{21} = a_{21}/d_{11} = 30/100 = 3/10$

$l_{31} = a_{31}/d_{11} = 20/100 = 1/5$

$l_{41} = a_{41}/d_{11} = 10/100 = 1/10$

Column 2: $d_{22} = a_{22} - d_{11} \times l_{21}^2 = 178 - 100 \times (\tfrac{3}{10})^2 = 169$

$l_{22} = 1$

$l_{32} = [a_{32} - d_{11} \times l_{31} \times l_{21}]/d_{22} = [201 - 100 \times (\tfrac{1}{5}) \times (\tfrac{3}{10})]/169 = 15/13$

$l_{42} = [a_{42} - d_{11} \times l_{41} \times l_{21}]/d_{22} = [-36 - 100 \times (\tfrac{1}{10}) \times (\tfrac{3}{10})]/169 = -3/13$

Column 3: $d_{33} = a_{33} - d_{11} \times l_{31}^2 - d_{22} \times l_{32}^2 = 485 - 100 \times (\tfrac{1}{5})^2 - 169 \times (\tfrac{15}{13})^2 = 256$

$l_{33} = 1$

$l_{43} = [a_{43} - d_{11} \times l_{41} \times l_{31} - d_{22} \times l_{42} \times l_{32}]/d_{33}$

$= [21 - 100 \times (\tfrac{1}{10}) \times (\tfrac{1}{5}) - 169 \times (-\tfrac{3}{13}) \times (\tfrac{15}{13})]/256 = 1/4$

Column 4: $d_{44} = a_{44} - d_{11} \times l_{41}^2 - d_{22} \times l_{42}^2 - d_{33} \times l_{43}^2$

$= 350 - 100 \times (\tfrac{1}{10})^2 - 169 \times (-\tfrac{3}{13})^2 - 256 \times (\tfrac{1}{4})^2 = 324$

$l_{44} = 1$

Check:

$$\begin{bmatrix} 1 & 0 & 0 & 0 \\ \frac{3}{10} & 1 & 0 & 0 \\ \frac{1}{5} & \frac{15}{13} & 1 & 0 \\ \frac{1}{10} & -\frac{3}{13} & \frac{1}{4} & 1 \end{bmatrix} \begin{bmatrix} 100 & 0 & 0 & 0 \\ 0 & 169 & 0 & 0 \\ 0 & 0 & 256 & 0 \\ 0 & 0 & 0 & 324 \end{bmatrix} \begin{bmatrix} 1 & \frac{3}{10} & \frac{1}{5} & \frac{1}{10} \\ 0 & 1 & \frac{15}{13} & -\frac{3}{13} \\ 0 & 0 & 1 & \frac{1}{4} \\ 0 & 0 & 0 & 1 \end{bmatrix} = \begin{bmatrix} 100 & 30 & 20 & 10 \\ 30 & 178 & 201 & -36 \\ 20 & 201 & 485 & 21 \\ 10 & -36 & 21 & 350 \end{bmatrix}$$

$[\mathbf{L}][\mathbf{D}][\mathbf{L}]^T = [\mathbf{A}]$ Q.E.D.

11.3 ITERATIVE METHODS

For larger systems, typically in excess of 1000 equations, the elimination methods can sometimes be inadequate for reasons that include available computer memory, execution time, and accumulated round-off error. To circumvent these problems, iterative methods can often be used as an alternative solution scheme. The primary reason these methods are effective is that a solution can be obtained to within some prespecified error tolerance.

11.3.1 Gauss-Seidel and Jacobi Iterations

One of the most common iterative schemes is the Gauss-Seidel method. The method begins with an initial estimate or guess at the solution vector, {**x**}. Then each of the system's equations is repeatedly used to solve for or update one of the unknowns in terms of previous values of the others. For the system provided in Equation 11.10, this yields

$$\begin{aligned} x_1^m &= (b_1 - a_{12}x_2^{m-1} - a_{13}x_3^{m-1})/a_{11} \\ x_2^m &= (b_2 - a_{21}x_1^m - a_{23}x_3^{m-1})/a_{22} \\ x_3^m &= (b_3 - a_{31}x_1^m - a_{32}x_2^m)/a_{33} \end{aligned} \tag{11.18}$$

where m and $m - 1$ are the present and previous iteration numbers and, for example, x_2^{m-1} and x_2^m are the previous and new estimates of the second element of the solution vector. Note that the original equations must be ordered so that the diagonal of [**A**] contains only nonzero values, a situation that will always be possible to attain in the analysis of stable structures.

In general, the Gauss-Seidel iterative form for a system of n-equations can be stated as

$$x_i^m = \frac{1}{a_{ii}}\left(b_i - \sum_{k=1}^{i-1} a_{ik}x_k^m - \sum_{k=i+1}^{n} a_{ik}x_k^{m-1}\right) \tag{11.19}$$

For convenience, a vector of zeros is often used as an initial guess at the solution. It should be noted, however, that in a nonlinear incremental analysis the solution vector from the previous load step can often provide an excellent initial guess for the current system of equations. Iterations are repeated until the solution converges "close enough" to the true solution. For an acceptable percent error tolerance ζ, a recommended convergence criterion is

$$\varepsilon_a < \zeta \tag{11.20}$$

with

$$\varepsilon_a = \max_{i=1,n}\left|\frac{x_i^m - x_i^{m-1}}{x_i^m}\right| 100\% \tag{11.20a}$$

and

$$\zeta = 0.5 \times 10^{2-q}\ \% \tag{11.20b}$$

where q is the desired number of accurate significant figures (Ref. 11.4).

Obviously the number of iterations required and the possibility of convergence can depend on several factors, including the initial guess and the required convergence tolerance. For systems where [**A**] is positive definite, however, the iteration process will always converge, barring any significant truncation error. To improve the rate of convergence, relaxation schemes can be employed. In this approach, a newly calculated value is modified by taking the weighted average of its current and previous values:

$$x_i^m = \beta x_i^m + (1 - \beta)x_i^{m-1} \tag{11.21}$$

where β is the relaxation or weighting factor. When β is defined between 1 and 2, it is assumed that the current value x_i^m is slowly approaching the exact solution and hence if more weight is placed on it the solution will converge faster. This modification is called *overrelaxation*. Setting β to a value between 0 and 1 is called *underrelaxation*, an adjustment that is often employed to damp out oscillations and attempt to make a nonconvergent system converge. Although the proper choice of β is problem specific, overrelaxation schemes are generally applied in solving structural analysis problems.

Instead of immediately using the most recently calculated elements of the solution vector, *Jacobi iteration* calculates all of the new values on the basis of all of previous values

$$x_i^m = \frac{1}{a_{ii}}\left(b_i - \sum_{\substack{k=1 \\ k \neq i}}^{n} a_{ik}x_k^{m-1}\right) \tag{11.22}$$

Although the Gauss-Seidel method typically converges faster, Jacobi iteration is worth mentioning since it is ideally suited for computer implementations that use parallel processing.

The Gauss-Seidel method with and without overrelaxation is illustrated in Example 11.4.

EXAMPLE 11.4

Using an acceptable error tolerance of $\zeta = 0.1\%$, solve the system of equations given in Example 11.2 by the Gauss-Seidel method for relaxation factors of:

1. $\beta = 1.0$
2. $\beta = 1.2$

In each case, combining equations 11.19 and 11.21 yields

$$x_1^m = \beta[b_1 - (a_{12}x_2^{m-1} + a_{13}x_3^{m-1} + a_{14}x_4^{m-1})]/a_{11} + (1 - \beta)x_1^{m-1}$$
$$x_2^m = \beta[b_2 - (a_{21}x_1^m + a_{23}x_3^{m-1} + a_{24}x_4^{m-1})]/a_{22} + (1 - \beta)x_2^{m-1}$$
$$x_3^m = \beta[b_3 - (a_{31}x_1^m + a_{32}x_2^m + a_{34}x_4^{m-1})]/a_{33} + (1 - \beta)x_3^{m-1}$$
$$x_4^m = \beta[b_4 - (a_{41}x_1^m + a_{42}x_2^m + a_{43}x_3^m)]/a_{44} + (1 - \beta)x_4^{m-1}$$

1. Using an initial guess of $\{\mathbf{x}^0\} = \begin{Bmatrix} 0 \\ 0 \\ 0 \\ 0 \end{Bmatrix}$ and $\beta = 1.0$

   The first iteration is

$$x_1^1 = 1.0[490 - (30 \times 0 + 20 \times 0 + 10 \times 0)]/100 + (1 - 1) \times 0 = 4.900$$
$$x_2^1 = 1.0[-347 - (30 \times 4.9 + 201 \times 0 + (-36) \times 0)]/178 + (1 - 1) \times 0 = -2.775$$
$$x_3^1 = 1.0[-1112 - (20 \times 4.9 + 201 \times (-2.775) + 21 \times 0]/485 + (1 - 1) \times 0 = -1.345$$
$$x_4^1 = 1.0[651 - (10 \times 4.9 + (-36) \times (-2.775) + 21 \times (-1.345))]/350 + (1 - 1) \times 0 = 1.515$$

A second iteration yields

$$\{\mathbf{x}^2\} = \begin{Bmatrix} 5.850 \\ -1.111 \\ -2.139 \\ 1.707 \end{Bmatrix}$$

with a maximum percent relative error of $\epsilon_a = \left|\dfrac{-1.111 - (-2.775)}{-1.111}\right| 100\% = 149.8\%$

After 14 iterations the error tolerance $\zeta = 0.1\%$ is satisfied with

$$\{\mathbf{x}^{14}\} = \begin{Bmatrix} 5.0004 \\ 0.9993 \\ -2.9996 \\ 1.9999 \end{Bmatrix} \text{ compared to the true solution } \{\mathbf{x}_T\} = \begin{Bmatrix} 5 \\ 1 \\ -3 \\ 2 \end{Bmatrix}$$

2. Using the same initial guess and $\beta = 1.2$
The first iteration yields

$$\{\mathbf{x}^1\} = \begin{Bmatrix} 5.880 \\ -3.529 \\ -1.287 \\ 1.688 \end{Bmatrix}$$

The second iteration would proceed with

$$\begin{aligned} x_1^2 &= 1.2[490 - (30 \times (-3.529) + 20 \times (-1.287) + 10 \times 1.688)]/100 + (1 - 1.2)5.880 \\ &= 6.081 \end{aligned}$$

Calculating the remaining elements of $\{\mathbf{x}^2\}$ gives

$$\{\mathbf{x}^2\} = \begin{Bmatrix} 6.081 \\ -0.709 \\ -2.530 \\ 1.781 \end{Bmatrix}$$

with a maximum percent relative error of $\epsilon_a = \left|\dfrac{-0.709 - (-3.529)}{-0.709}\right| 100\% = 397.7\%$

The benefits of over-relaxation are realized by obtaining a solution in only 8 iterations that satisfies the given error tolerance

$$\{\mathbf{x}^8\} = \begin{Bmatrix} 4.9999 \\ 1.0001 \\ -3.0000 \\ 2.0000 \end{Bmatrix}$$

11.3.2 Conjugate Gradient Method

The conjugate gradient method is based on the mathematical property that any vector $\{\mathbf{x}\}$ with n elements can be expressed in an n-dimensional space by a scaled combination of n linearly independent or basis vectors $\{\mathbf{s}_i\}$

$$\{\mathbf{x}\} = \sum_{i=1}^{n} \alpha_i \{\mathbf{s}_i\} \tag{11.23}$$

Hence, the true solution to the system of n equations posed in Equation 11.1 can be expressed by Equation 11.23 or by

$$\{\mathbf{x}\} = \{\mathbf{x}^0\} + \sum_{i=1}^{n} \alpha_i\{\mathbf{s}_i\} \tag{11.24}$$

where $\{\mathbf{x}^0\}$ is an initial guess at the solution vector and the α_i scale factors of Equations 11.23 and 11.24 are not necessarily the same.

With this in mind, Equation 11.24 can be estimated in a first iteration as

$$\{\mathbf{x}^1\} = \{\mathbf{x}^0\} + \alpha_1\{\mathbf{s}_1\} \tag{11.25}$$

Closer approximations could be achieved in successive iterations by calculating and including additional scaled basis vectors according to

$$\{\mathbf{x}^i\} = \{\mathbf{x}^{i-1}\} + \alpha_i\{\mathbf{s}_i\} \tag{11.26}$$

This process could then be repeated until a solution is obtained that is once again considered "close enough." It is important to note that there exists a limit to the number of iterations required; the exact solution will be obtained in a maximum of n iterations since all n basis vectors would be included.

Clearly, the crux of the matter is to provide a scheme for calculating each additional basis vector $\{\mathbf{s}_i\}$ and its corresponding scale factor α_i. To find these entities, the conjugate gradient method employs an optimization strategy. Assuming that $[\mathbf{A}]$ is a positive definite symmetric matrix, the solution to Equation 11.1 will occur when the quadratic

$$F(\mathbf{x}) = \frac{1}{2}\{\mathbf{x}\}^{\mathrm{T}}[\mathbf{A}]\{\mathbf{x}\} - \{\mathbf{b}\}^{\mathrm{T}}\{\mathbf{x}\} \tag{11.27}$$

is at a minimum. Where the gradient vector or slope of this function is defined by

$$\{\mathbf{G}\} = [\mathbf{A}]\{\mathbf{x}\} - \{\mathbf{b}\} \tag{11.28}$$

we are simply restating Equation 11.1 by indicating that Equation 11.27 will be at a minimum when $\{\mathbf{G}\}$ is the null vector. It is important to note that in structural analysis terms, Equation 11.27 represents the total potential of the system and the minimum given by setting Equation 11.28 to the null vector results in stiffness equations of equilibrium (see Section 8.1.4).

Instead of calculating the inverse of $[\mathbf{A}]$ to locate this minimum, it can be shown that for a given positive definite quadratic, such as the one in Equation 11.27, the unique minimum point may be defined by (Ref. 11.5)

$$\{\mathbf{x}\} = \{\mathbf{x}^0\} + \sum_{i=1}^{n} \frac{-\{\mathbf{s}_i\}^{\mathrm{T}}\{\mathbf{G}^{i-1}\}}{\{\mathbf{s}_i\}^{\mathrm{T}}[\mathbf{A}]\{\mathbf{s}_i\}}\{\mathbf{s}_i\} \tag{11.29}$$

where $\{\mathbf{s}_i\}$ are the n linearly independent **A**-conjugate directions.[4] Comparing Equations 11.24 and 11.29, the scale factor α_i has now been defined in terms of its corresponding conjugate direction $\{\mathbf{s}_i\}$

$$\alpha_i = \frac{-\{\mathbf{s}_i\}^{\mathrm{T}}\{\mathbf{G}^{i-1}\}}{\{\mathbf{s}_i\}^{\mathrm{T}}[\mathbf{A}]\{\mathbf{s}_i\}} \tag{11.30}$$

---

[4]The *conjugate direction* vectors for an $n \times n$ coefficient matrix $[\mathbf{A}]$ are all orthogonal in the inner product of $[\mathbf{A}]$ and are defined by

$$\{\mathbf{s}_i\}^{\mathrm{T}}[\mathbf{A}]\{\mathbf{s}_j\} = 0 \qquad \text{for } i, j = 1 \text{ to } n \text{ with } i \neq j$$

To determine the successive conjugate directions that will be used in each iteration, we introduce a residual vector

$$\{\mathbf{r}^{i-1}\} = \{\mathbf{b}\} - [\mathbf{A}]\{\mathbf{x}^{i-1}\} \tag{11.31}$$

Note that this residual vector is a measure of how well the solution obtained in the previous iteration satisfies Equation 11.1. This vector will either be zero, which means we have found the true solution, or nonzero and hence linearly independent of all of the previously calculated conjugate directions. Where the residual vector is the negative of Equation 11.28, Equation 11.30 may be written more conveniently as

$$\alpha_i = \frac{\{\mathbf{s}_i\}^{\mathrm{T}}\{\mathbf{r}^{i-1}\}}{\{\mathbf{s}_i\}^{\mathrm{T}}[\mathbf{A}]\{\mathbf{s}_i\}} \tag{11.32}$$

By further employing this residual vector, it can be shown that the conjugate direction for the ith iteration may be calculated using the recurrence equation

$$\{\mathbf{s}_i\} = \{\mathbf{r}^{i-1}\} - \frac{\{\mathbf{s}_{i-1}\}^{\mathrm{T}}[\mathbf{A}]\{\mathbf{r}^{i-1}\}}{\{\mathbf{s}_{i-1}\}^{\mathrm{T}}[\mathbf{A}]\{\mathbf{s}_{i-1}\}}\{\mathbf{s}_{i-1}\} \tag{11.33}$$

For the first iteration $\{\mathbf{s}_1\}$ would be assumed equal to $\{\mathbf{r}^0\}$. A derivation of Equation 11.33 is provided in Reference 11.2.

To minimize the number of matrix-to-vector multiplications, two simplifications are often employed. First, an auxiliary row vector $\lfloor \mathbf{y}_i \rfloor = \{\mathbf{s}_i\}^{\mathrm{T}}[\mathbf{A}]$ is retained during each iteration and used in the denominators of Equations 11.32 and 11.33. If the coefficient matrix $[\mathbf{A}]$ is symmetric, the transpose of this row vector $\lfloor \mathbf{y}_i \rfloor^{\mathrm{T}} = [\mathbf{A}]\{\mathbf{s}_i\}$ may be used to determine the residual vector for the next iteration by a simple vector addition according to

$$\{\mathbf{r}^i\} = \{\mathbf{b}\} - [\mathbf{A}]\{\mathbf{x}^i\} = \{\mathbf{b}\} - [\mathbf{A}](\{\mathbf{x}^{i-1}\} + \alpha_i\{\mathbf{s}_i\}) = \{\mathbf{r}^{i-1}\} - \alpha_i[\mathbf{A}]\{\mathbf{s}_i\} \tag{11.34a}$$

or

$$\{\mathbf{r}^i\} = \{\mathbf{r}^{i-1}\} - \alpha_i\lfloor \mathbf{y}_i \rfloor^{\mathrm{T}} \tag{11.34b}$$

Iterations in the conjugate gradient method are often repeated until the magnitude of the residual vector drops below a prescribed error tolerance according to

$$\varepsilon < \zeta \tag{11.35}$$

with

$$\varepsilon = \sqrt{\{\mathbf{r}^i\}^{\mathrm{T}}\{\mathbf{r}^i\}} \tag{11.35a}$$

and ζ being a quantity rather than a percentage as defined in Equation 11.20b. Similar to the Gauss-Seidel method, the number of iterations required to converge on an acceptable solution is a function of several factors, including the initial guess and the prescribed error tolerance. In the absence of a good estimate of the solution, the null vector is often used as a starting point. It is important to note that the condition[5] of the coefficient matrix can also have a major impact on the rate of convergence. To this end, several *preconditioning* schemes have been developed to improve systems of equations that are poorly conditioned (see Reference 11.2). One such method is presented in Section 11.6.3 on error analysis. Use of the conjugate gradient method as presented in Equations 11.26 and 11.31–11.35 is illustrated in Example 11.5.

[5]Briefly stated, the condition of a system of equations can be defined as the significance of small changes in the coefficients that often occur during the solution process. This topic will be explored in more detail in Section 11.6.

EXAMPLE 11.5

Solve the system of equations given in Example 11.2 by the conjugate gradient method using $\zeta = 0.001$.

With $[\mathbf{A}] = \begin{bmatrix} 100 & 30 & 20 & 10 \\ 30 & 178 & 201 & -36 \\ 20 & 201 & 485 & 21 \\ 10 & -36 & 21 & 350 \end{bmatrix}$, $\{\mathbf{b}\} = \begin{Bmatrix} 490 \\ -347 \\ -1112 \\ 651 \end{Bmatrix}$, and an initial guess $\{\mathbf{x}^0\} = \begin{Bmatrix} 0 \\ 0 \\ 0 \\ 0 \end{Bmatrix}$

First iteration:

Equation 11.31, $\{\mathbf{r}^0\} = \{\mathbf{b}\} - [\mathbf{A}]\{\mathbf{x}^0\} = \{\mathbf{b}\}$

Assume $\{\mathbf{s}_1\} = \{\mathbf{r}^0\} = \begin{Bmatrix} 490 \\ -347 \\ -1112 \\ 651 \end{Bmatrix}$

Calculate auxiliary row vector $\lfloor \mathbf{y}_1 \rfloor = \{\mathbf{s}_1\}^T[\mathbf{A}]$

Equation 11.32, $\alpha_1 = \dfrac{\{\mathbf{s}_1\}^T\{\mathbf{r}^0\}}{\lfloor \mathbf{y}_1 \rfloor\{\mathbf{s}_1\}} = 2.2235 \times 10^{-3}$

Equation 11.26, $\{\mathbf{x}^1\} = \{\mathbf{x}^0\} + \alpha_1\{\mathbf{s}_1\} = \begin{Bmatrix} 1.090 \\ -0.772 \\ -2.473 \\ 1.448 \end{Bmatrix}$

Equation 11.34b, $\{\mathbf{r}^1\} = \{\mathbf{r}^0\} - \alpha_1\lfloor \mathbf{y}_1 \rfloor^T$

Equation 11.35, $\epsilon = \sqrt{\{\mathbf{r}^1\}^T\{\mathbf{r}^1\}} = 589.86 \nless \zeta = 0.001$

Second iteration:

Equation 11.33, $\{\mathbf{s}_2\} = \{\mathbf{r}^1\} - \dfrac{\lfloor \mathbf{y}_1 \rfloor\{\mathbf{r}^1\}}{\lfloor \mathbf{y}_1 \rfloor\{\mathbf{s}_1\}}\{\mathbf{s}_1\} = \begin{Bmatrix} 523.536 \\ 246.999 \\ -1.378 \\ 269.710 \end{Bmatrix}$

$\alpha_2 = 5.0182 \times 10^{-3}$

$\{\mathbf{x}^2\} = \{\mathbf{x}^1\} + \alpha_2\{\mathbf{s}_2\} = \begin{Bmatrix} 3.716 \\ 0.468 \\ -2.479 \\ 2.801 \end{Bmatrix}$ with $\varepsilon = 355.50 \nless \zeta = 0.001$

Third iteration:

$\{\mathbf{s}_3\} = \begin{Bmatrix} 316.029 \\ 147.127 \\ -137.174 \\ -199.628 \end{Bmatrix}$, $\alpha_3 = 3.9667 \times 10^{-3}$, $\{\mathbf{x}^3\} = \begin{Bmatrix} 4.970 \\ 1.052 \\ -3.024 \\ 2.009 \end{Bmatrix}$, $\varepsilon = 4.009 \nless \zeta = 0.001$

Fourth iteration:

$\{\mathbf{s}_4\} = \begin{Bmatrix} 1.8400 \\ -3.1986 \\ 1.4636 \\ -0.5650 \end{Bmatrix}$, $\alpha_4 = 1.6118$, $\{\mathbf{x}^4\} = \begin{Bmatrix} 5 \\ 1 \\ -3 \\ 2 \end{Bmatrix}$, $\varepsilon = 0 < \zeta = 0.001$

As expected, the true solution is obtained after $n = 4$ iterations.

11.4 SPARSENESS AND BANDEDNESS

As indicated in Section 3.3, the global stiffness matrix is rather sparsely or weakly populated in most structural analysis problems. Again, this is because each row or equilibrium equation for a particular degree of freedom is only influenced by degrees of freedom associated with the often small number of elements connecting to that degree of freedom. All other degrees of freedom for the remaining unattached elements have no effect on this equilibrium equation and hence have zero coefficients in that row.

Sparseness often results in the stiffness matrix having a banded form. This is illustrated in Figure 11.1, in which crosses in the figure show the location of potentially nonzero stiffness coefficients. The bandedness of the stiffness matrix is indicated by the clustering of these coefficients about the main diagonal. As shown in Figure 11.1*b*, we will denote *HBW* as the half-bandwidth of a symmetric matrix.

The opportunity to gain efficiency in solving linear algebraic equations can be realized if we note that all coefficients outside the bandwidth will always retain a zero value during any of the solution procedures described in the previous sections. Hence, the performance of these schemes can be improved by modifications that avoid the storage and manipulation of these useless zeros. For example, Equation 11.15 may be modified as follows:

$$l_{ii} = \sqrt{a_{ii} - \sum_{k=\max(1,i-HBW)}^{i-1} l_{ik}^2}$$

$$l_{ji} = \frac{a_{ji} - \sum_{k=\max(1,j-HBW)}^{i-1} l_{jk} l_{ik}}{l_{ii}} \quad \text{for } j = i+1, \ldots, \min(i+HBW, n) \tag{11.36a}$$

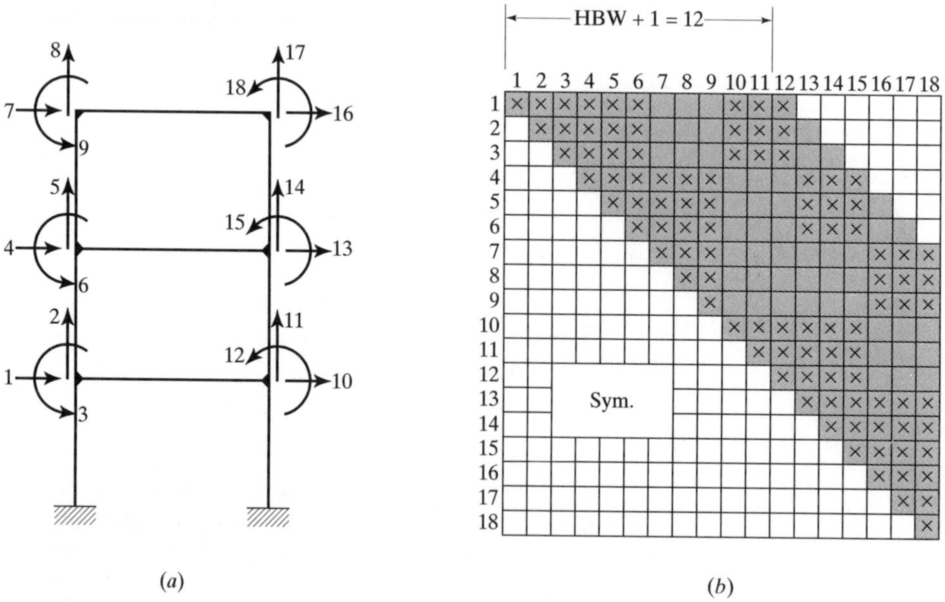

Figure 11.1 Banded $[\mathbf{K}_{ff}]$ stiffness matrix ($HBW = 11$).

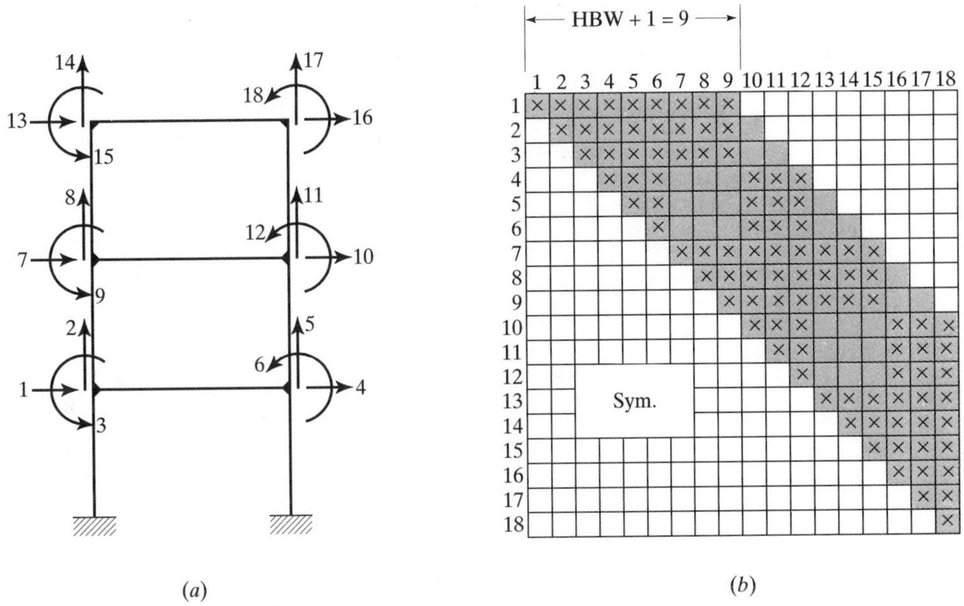

Figure 11.2 Improved bandedness ($HBW = 8$).

Equations 11.13 and 11.17 can be adjusted in a similar manner. In all cases, the half-bandwidth may be calculated as

$$HBW = \max_{1 \le el \le m} (maxdof_{el} - mindof_{el}) \tag{11.36b}$$

where m is the total number of structural elements and $maxdof_{el}$ and $mindof_{el}$ are the maximum and minimum global degree of freedom numbers associated with element *el*.

The shaded area of the matrix in Figure 11.1 covers all the stiffness coefficients of $[\mathbf{K}_{ff}]$ that need to be stored and manipulated. Therefore, it is obvious that it would be advantageous to renumber the degrees of freedom in such a way that the bandwidth of the system is minimized. For most structures, near optimum numbering is easily attainable through visual inspection of the system. This can be done by noting that the half-bandwidth of the global stiffness matrix parallels the maximum difference in global degree-of-freedom numbers pertaining to any element in the structure. Figure 11.2 illustrates the improvement in bandedness that can result from small changes to the degree-of-freedom numbering shown in Figure 11.1. Automatic numbering schemes that attempt to minimize the bandwidth are employed in most commercially available computer programs for structural analysis. For reviews of solution procedures that exploit bandedness and employ minimum storage schemes including skyline or envelope storage and banded storage, the reader is referred to References 11.2 and 11.6–11.8.

11.5 FRONTAL SOLVERS

In performing direct elimination procedures, such as Gauss elimination or others, it is important to note that we do not need to work with, and hence computationally store, all of the stiffness equations at one time. That is, we can assemble the stiffness equations on an element-by-element basis while at the same time successively eliminating degrees of freedom.

Starting with an element having one or more degrees of freedom for which all stiffness contributions have been assembled, we eliminate these degrees of freedom by expressing them as functions of the degrees of freedom common to this and another element. We then add, in proper fashion, the remaining, reduced stiffnesses of the first element to those of the second element. We may then eliminate any degrees of freedom for which the stiffnesses are fully assembled in the combined system. We proceed in this fashion, adding one element at a time—and nodal forces as they are encountered—until the entire structure has been covered. We will then have developed a set of equations that can be solved for one unknown at a time by back substitution. The process is called a *wave front solution*, since it may be depicted as a wave spreading gradually over the structure.

It is apparent that, in a frontal solution, the entire assembled global structure stiffness matrix is not needed. Although frontal solvers may be used effectively with banded systems, banding is irrelevant. Instead of node numbering being important, as in band solvers, the order in which elements are introduced becomes critical.

Before discussing features of frontal solvers, typical characteristics of a wave front solution will be illustrated through Example 11.6. To demonstrate the irrelevance of banding in a frontal solution, the degrees of freedom are labeled in an unusual way. Although it is not needed in the subsequent solution and, normally, would not be assembled, the global stiffness matrix is presented to show how a clumsy numbering system can obscure the natural bandedness of a structure.

EXAMPLE 11.6

Write the global stiffness matrix. Using a wave front approach, develop equations for calculating the displacements, $L = 5$ m, $E = 200{,}000$ MPa, $I = 100 \times 10^6$ mm^4.

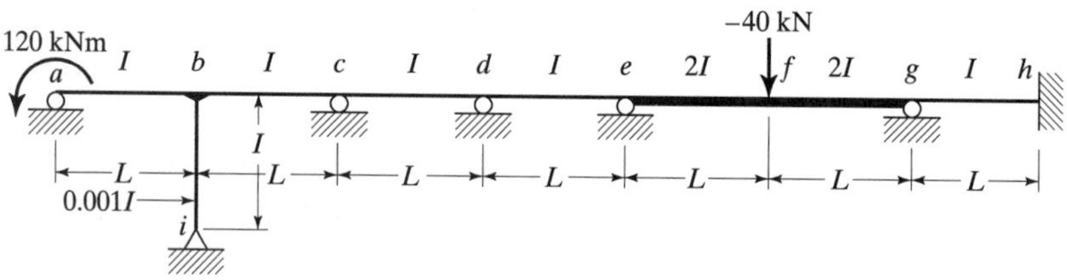

Degrees of freedom

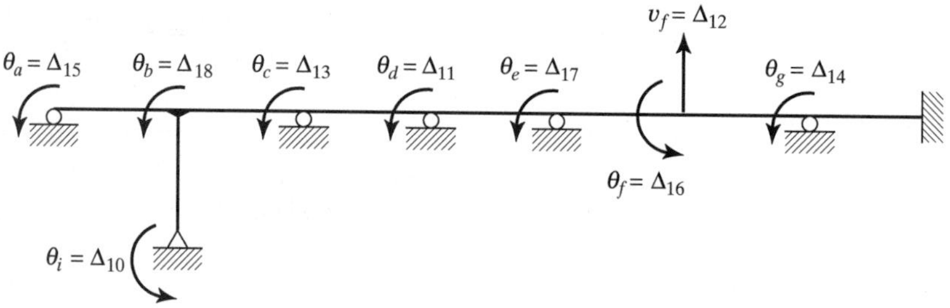

Global stiffness matrix

$$
[\mathbf{K}] = \frac{EI}{L}
\begin{array}{c}
\begin{array}{ccccccccc}
\Delta_{10} & \Delta_{11} & \Delta_{12} & \Delta_{13} & \Delta_{14} & \Delta_{15} & \Delta_{16} & \Delta_{17} & \Delta_{18}
\end{array} \\
\left[\begin{array}{ccccccccc}
0.004 & 0 & 0 & 0 & 0 & 0 & 0 & 0 & 0.002 \\
 & 8 & 0 & 2 & 0 & 0 & 0 & 2 & 0 \\
 & & 19.2 \times 10^{-7} & 0 & 2.4 \times 10^{-3} & 0 & 0 & -2.4 \times 10^{-3} & 0 \\
 & & & 8 & 0 & 0 & 0 & 0 & 2 \\
 & & & & 12 & 0 & 4 & 0 & 0 \\
 & & \text{Sym.} & & & 4 & 0 & 0 & 2 \\
 & & & & & & 16 & 4 & 0 \\
 & & & & & & & 12 & 0 \\
 & & & & & & & & 8.004
\end{array}\right]
\end{array}
$$

Wave front solution. Use primes to indicate degrees of freedom where stiffness coefficients are incomplete at stage considered. (See page 321 for a step-by-step description of the stages.)

1. Member *ib*:

$$\frac{EI}{L}\begin{bmatrix} 0.004 & 0.002 \\ 0.002 & 0.004 \end{bmatrix}\begin{Bmatrix} \Delta_{10} \\ \Delta'_{18} \end{Bmatrix} = \begin{Bmatrix} 0 \\ 0 \end{Bmatrix}$$

which reduces to

$$\begin{bmatrix} 1 & 0 \\ 0.5 & 1 \end{bmatrix}\begin{bmatrix} 0.004 & 0.002 \\ 0 & 0.003 \end{bmatrix}\begin{Bmatrix} \Delta_{10} \\ \Delta'_{18} \end{Bmatrix} = \begin{Bmatrix} 0 \\ 0 \end{Bmatrix} \quad \text{or} \quad \begin{bmatrix} 0.004 & 0.002 \\ 0 & 0.003 \end{bmatrix}\begin{Bmatrix} \Delta_{10} \\ \Delta'_{18} \end{Bmatrix} = \begin{Bmatrix} 0 \\ 0 \end{Bmatrix}$$

2. Add member *ab*:

$$\frac{EI}{L}\begin{bmatrix} 4 & 2 \\ 2 & 4.003 \end{bmatrix}\begin{Bmatrix} \Delta_{15} \\ \Delta'_{18} \end{Bmatrix} = \begin{Bmatrix} 120 \times 10^3 \\ 0 \end{Bmatrix}$$

reduces to

$$\begin{bmatrix} 4 & 2 \\ 0 & 3.003 \end{bmatrix}\begin{Bmatrix} \Delta_{15} \\ \Delta'_{18} \end{Bmatrix} = \frac{L}{EI}\begin{Bmatrix} 120 \times 10^3 \\ -60 \times 10^3 \end{Bmatrix}$$

3. Add member *bc*:

$$\frac{EI}{L}\begin{bmatrix} 7.003 & 2 \\ 2 & 4 \end{bmatrix}\begin{Bmatrix} \Delta_{18} \\ \Delta'_{13} \end{Bmatrix} = \begin{Bmatrix} -60 \times 10^3 \\ 0 \end{Bmatrix}$$

reduces to

$$\begin{bmatrix} 7.003 & 2 \\ 0 & 3.4288 \end{bmatrix}\begin{Bmatrix} \Delta_{18} \\ \Delta'_{13} \end{Bmatrix} = \frac{L}{EI}\begin{Bmatrix} -60 \times 10^3 \\ 17.136 \times 10^3 \end{Bmatrix}$$

4. Add member *cd*:

$$\frac{EI}{L}\begin{bmatrix} 7.4288 & 2 \\ 2 & 4 \end{bmatrix}\begin{Bmatrix} \Delta_{13} \\ \Delta'_{11} \end{Bmatrix} = \begin{Bmatrix} 17.136 \times 10^3 \\ 0 \end{Bmatrix}$$

reduces to

$$\begin{bmatrix} 7.4288 & 2 \\ 0 & 3.4616 \end{bmatrix}\begin{Bmatrix} \Delta_{13} \\ \Delta'_{11} \end{Bmatrix} = \frac{L}{EI}\begin{Bmatrix} 17.136 \times 10^3 \\ -4.6134 \times 10^3 \end{Bmatrix}$$

5. Add member *de*:

$$\frac{EI}{L}\begin{bmatrix} 7.4616 & 2 \\ 2 & 4 \end{bmatrix}\begin{Bmatrix} \Delta_{11} \\ \Delta'_{17} \end{Bmatrix} = \begin{Bmatrix} -4.6134 \times 10^3 \\ 0 \end{Bmatrix}$$

reduces to

$$\begin{bmatrix} 7.4616 & 2 \\ 0 & 3.4639 \end{bmatrix}\begin{Bmatrix} \Delta_{11} \\ \Delta'_{17} \end{Bmatrix} = \frac{L}{EI}\begin{Bmatrix} -4.6134 \times 10^3 \\ 1.2366 \times 10^3 \end{Bmatrix}$$

6. Add element *ef*:

$$\frac{EI}{L}\begin{bmatrix} 11.4639 & -2.4 \times 10^{-3} & 4 \\ -2.4 \times 10^{-3} & 9.6 \times 10^{-7} & -2.4 \times 10^{-3} \\ 4 & -2.4 \times 10^{-3} & 8 \end{bmatrix}\begin{Bmatrix} \Delta_{17} \\ \Delta'_{12} \\ \Delta'_{16} \end{Bmatrix} = \begin{Bmatrix} 1.2366 \times 10^3 \\ 0 \\ 0 \end{Bmatrix}$$

Eliminating Δ_{17}, this reduces to

$$\begin{bmatrix} 11.4639 & -2.4 \times 10^{-3} & 4 \\ 0 & 4.5755 \times 10^{-7} & -1.5626 \times 10^{-3} \\ 0 & -1.5626 \times 10^{-3} & 6.6043 \end{bmatrix} \begin{Bmatrix} \Delta_{17} \\ \Delta'_{12} \\ \Delta'_{16} \end{Bmatrix} = \frac{L}{EI} \begin{Bmatrix} 1.2366 \times 10^{3} \\ 0.259 \\ -4.3148 \times 10^{2} \end{Bmatrix}$$

7. Add element fg:

$$\frac{EI}{L} \begin{bmatrix} 1.4176 \times 10^{-6} & 8.3740 \times 10^{-4} & 2.4 \times 10^{-3} \\ 8.3740 \times 10^{-4} & 14.6043 & 4 \\ 2.4 \times 10^{-3} & 4 & 8 \end{bmatrix} \begin{Bmatrix} \Delta_{12} \\ \Delta_{16} \\ \Delta'_{14} \end{Bmatrix} = \begin{Bmatrix} -39.741 \\ -4.3148 \times 10^{2} \\ 0 \end{Bmatrix}$$

Eliminating Δ_{12} and Δ_{16}, this reduces to

$$\begin{bmatrix} 1.4176 \times 10^{-6} & 8.3740 \times 10^{-4} & 2.4 \times 10^{-3} \\ 0 & 14.110 & 2.5823 \\ 0 & 0 & 3.4642 \end{bmatrix} \begin{Bmatrix} \Delta_{12} \\ \Delta_{16} \\ \Delta'_{14} \end{Bmatrix} = \frac{L}{EI} \begin{Bmatrix} -39.741 \\ 23.044 \times 10^{3} \\ 63.064 \times 10^{3} \end{Bmatrix}$$

8. Add member gh:

$$\frac{EI}{L}[7.4642]\{\Delta_{14}\} = \{63.064 \times 10^{3}\}$$

reduces to

$$[7.4642]\{\Delta_{14}\} = \frac{L}{EI}\{63.064 \times 10^{3}\}$$

9. Collect complete set of equations:

$$\begin{aligned}
4\Delta_{10} + 2\Delta_{18} &= 0 \\
4\Delta_{15} + 2\Delta_{18} &= 120 \times 10^{3} \times L/EI \\
7.003\Delta_{18} + 2\Delta_{13} &= -60 \times 10^{3} \times L/EI \\
7.4288\Delta_{13} + 2\Delta_{11} &= 17.136 \times 10^{3} \times L/EI \\
7.4616\Delta_{11} + 2\Delta_{17} &= -4.6134 \times 10^{3} \times L/EI \\
11.4639\Delta_{17} - 2.4 \times 10^{-3}\Delta_{12} + 4\Delta_{16} &= 1.2366 \times 10^{3} \times L/EI \\
1.4176 \times 10^{-6}\Delta_{12} + 8.374 \times 10^{-4}\Delta_{16} + 2.4 \times 10^{-3}\Delta_{14} &= -39.741 \times L/EI \\
14.110\Delta_{16} + 2.5823\Delta_{14} &= 23.044 \times 10^{3} \times L/EI \\
7.4642\Delta_{14} &= 63.064 \times 10^{3} \times L/EI
\end{aligned}$$

10. Solve by back substitution.

$$\{\boldsymbol{\Delta}\} = \begin{Bmatrix} \Delta_{10} \\ \Delta_{11} \\ \Delta_{12} \\ \Delta_{13} \\ \Delta_{14} \\ \Delta_{15} \\ \Delta_{16} \\ \Delta_{17} \\ \Delta_{18} \end{Bmatrix} = \begin{bmatrix} 1.1366 \times 10^{-3} \text{ rad} \\ 0.4348 \times 10^{-3} \text{ rad} \\ -10.597 \text{ mm} \\ 0.4596 \times 10^{-3} \text{ rad} \\ 2.112 \times 10^{-3} \text{ rad} \\ 8.6366 \times 10^{-3} \text{ rad} \\ 0.02173 \times 10^{-3} \text{ rad} \\ -2.199 \times 10^{-3} \text{ rad} \\ -2.2732 \times 10^{-3} \text{ rad} \end{bmatrix}$$

In the example, the elements are considered in the order *ib*, *ab*, *bc*, *cd*, *de*, *ef*, *fg*, and *gh*. The element stiffness equations are written, added to equations from previous stages that remain incompletely reduced, and the combination is then further reduced to the extent permissible at the current state. Gauss elimination is employed, but other reduction algorithms could have been used. The extent to which the current set of equations can be reduced depends upon whether all stiffness coefficients relating to a given degree of freedom have been accumulated, and whether we have duly entered all loads acting at the nodal points for which we are compiling—in piecemeal fashion—equilibrium equations. For instance, in considering the initial element, *ib*, the first equation is complete. Δ_{10} may be expressed as a function of Δ_{18}. Once this has been done, Δ_{10} may be considered passive, and the equation containing it may be stored for later use in back substitution. In element *ib*, the second equation is not complete since Δ_{18} is at an interface with the deformable elements *ab* and *bc*. Its stiffness coefficients have components yet to be accumulated from these elements. Thus, at this stage, this equation can be only partially reduced by considering the effect of Δ_{10} (the rotation at *i*) upon Δ_{18}. Primes are used as temporary indicators of degrees of freedom appearing in equations in which the stiffness coefficients or the load vector term is incomplete. When the second element *ab* is brought into consideration, Δ_{15} appears and may be eliminated immediately since it is internal to the two-element system and the load vector term is the proper one for node *a*. The elimination of Δ_{18} must await the addition of the third element, *bc*, and the completion of the reduced load term at node *b*. Later, when we encounter element *ef*, only one of the three variables under consideration, that is, Δ_{17}, may be eliminated. The other two, Δ_{12} and Δ_{16}, must remain active until all of the equations in which they appear are complete through the addition of element *fg* and the load at node *f*.

Working through the structure in the above way, we arrive at nine equations that can be solved, one at a time, by back substitution.

The frontal solution concept is pictured in Figure 11.3. The wave front is the region between the two dashed lines in each sketch. The degrees of freedom in this region are the *active variables*. We start with the wave covering element *ib*. As soon as Δ_{10} has been eliminated (Figure 11.3a), Δ_{18} is the only active variable until the wave spreads over *ab* to include Δ_{15} in the front (Figure 11.3b). Progress at selected later stages is indicated in the remaining sketches. It may be noted that, each time an additional element is added, new variables become active. Some of these may be eliminated forthwith, for example, Δ_{15}; others, such as Δ_{18}, may remain active, that is, within the wave front, through the addition of a number of elements.

There is, of course, much more to practical frontal solvers than can be illustrated by a simple example. We have focused on the basics of the approach to equation solving that may be depicted by the wave front concept. Reference 11.9 presents a sophisticated procedure for solving large systems of equations.

Frontal solvers generally provide for the passage of the decomposition array through high-speed core memory three times–once when it is developed, once during forward elimination, and once during backward substitution. Among the advantages of frontal solvers are those already observed: (1) only a very small portion of the entire stiffness matrix need be considered at any one time; the entire matrix is never assembled; (2) efficiency does not depend upon banding; and (3) variables may be eliminated in a different order from that in which they are activated. In using frontal solvers, one is apt to encounter a rather formidable amount of bookkeeping, that is, programming related to data handling. Solvers that exploit bandedness are generally simpler in this respect.

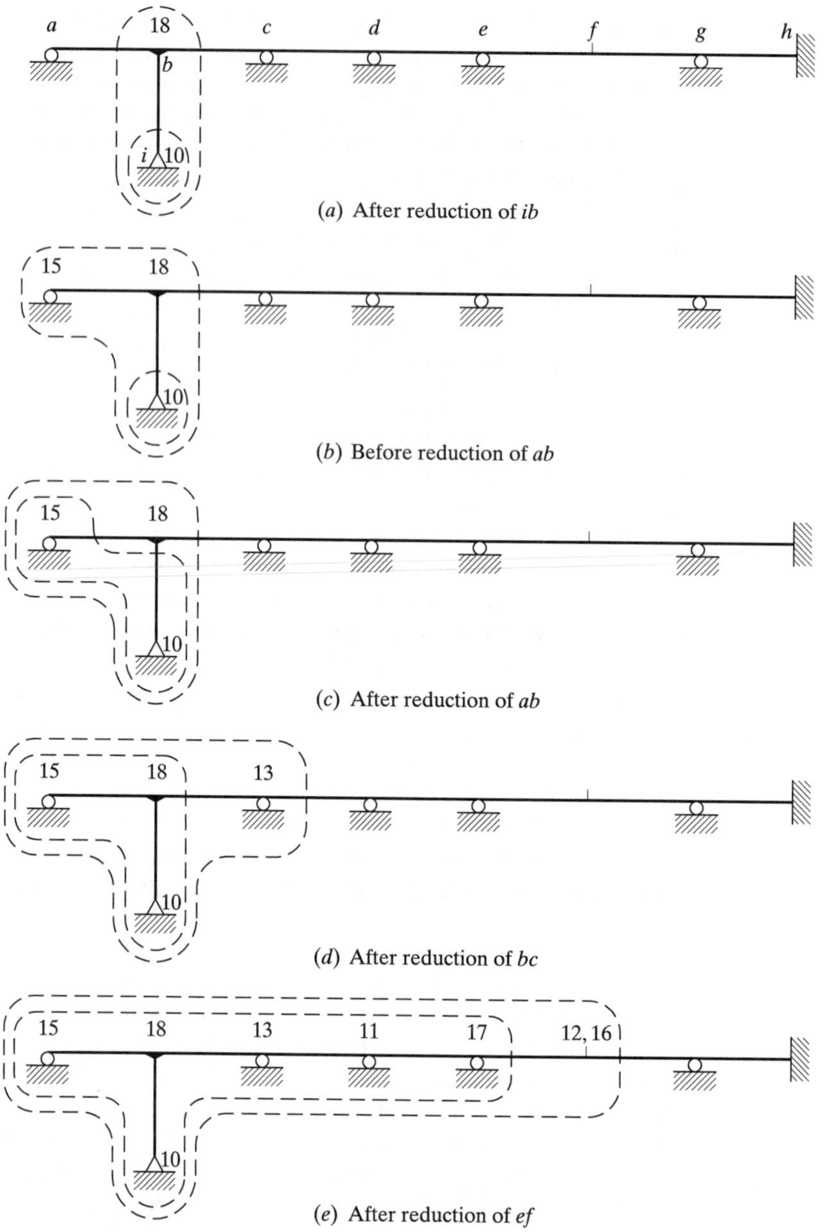

Figure 11.3 Wave front concept.

11.6 ERRORS

In structural analysis, as in modeling all physical problems, results are never exact. Recognizing that perfection is unattainable, the analysis must, however, provide answers sufficiently accurate to make a reliable estimate of structural behavior and, in turn, permit safe and economical design. Major sources of error in framework analysis may be grouped as follows:

Idealization. The process whereby an actual structure is converted to a computational model. Accuracy is affected by all simplifications and personal judgments in representing the loading, geometric and material properties, and support conditions.

Input/Output. These include errors that range from mistakes made while entering the idealized data into a computer analysis program to interpreting the results provided by such a program. Note that the two are intrinsically related; input errors almost always lead to incorrect output.

Procedural. Errors that arise due to approximations employed in the solution algorithms. These can be both theoretically and numerically based.

Manipulation. Those errors that occur during computational processes when irrational numbers, and in some cases rational numbers, are represented by an approximate value that has a fixed number of significant figures. They derive fundamentally from truncation and round-off operations.

It should be clear that this brief catalog is not comprehensive, nor is it completely definitive. There is in reality no sharp interface between categories. Although all of the above sources of error are important, we will now focus on errors of the manipulation type. The relevance of procedural errors will be illustrated in the next chapter.

The significance of manipulation errors in solving systems of equations is directly related to the condition of the problem (for a treatise on the effect of manipulation errors in algebraic processes see Ref. 11.10). If small errors in the coefficients of the equations, or in the solving process, have little effect on the solution, the problem is *well-conditioned*. On the other hand, if the effects are large, the problem is *ill-conditioned*. As shown in Figure 11.4, we may picture these ideas graphically by considering the system

$$a_{11}x_1 + a_{12}x_2 = b_1$$
$$a_{21}x_1 + a_{22}x_2 = b_2$$

If the equations—or rows of the coefficient matrix—are linearly dependent, no solution exists, the coefficient matrix is *singular*. This is illustrated in Figure 11.4*a* by two

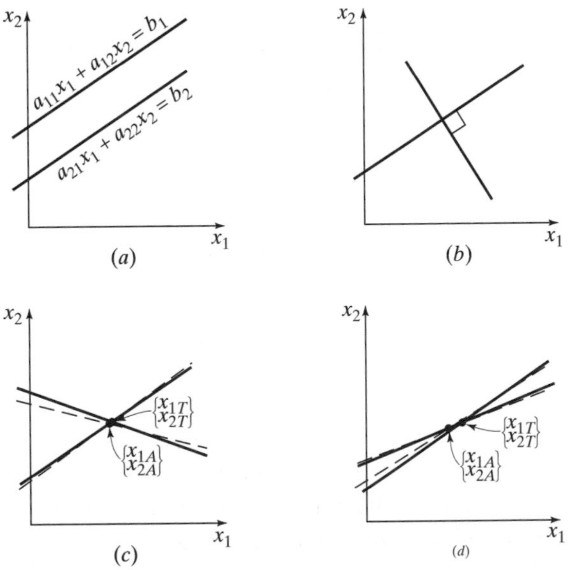

Figure 11.4 Problem solution. (a) Singularity. (b) Orthogonality. (c) Well-conditioned. (d) Ill-conditioned.

parallel lines. This should only occur in a framework analysis when the structure is unstable. Optimum conditioning is obtained if the two lines are orthogonal, a rare condition that is illustrated in Figure 11.4*b*.

The normal well-conditioned problem is represented by the intermediate case shown in Figure 11.4*c*. Although there will be inevitable solution errors and the problem actually solved will not be the mathematical one presented by the solid lines, but rather the imperfect physical one indicated by dashed lines, the error should be tolerable. That is, the calculated solution vector $\{\mathbf{x}_A\}$ will be sufficiently close to the true solution $\{\mathbf{x}_T\}$. When the angle between the lines becomes small, as in Figure 11.4*d*, the problem becomes ill-conditioned. The solution is then sensitive to small errors of representation or computation and it may be quite inaccurate. In other words, a wide range of answers can closely satisfy an ill-conditioned system of equations (see Example 11.12).

These concepts apply with equal validity to problems of large, practical size. In framed structures, conditioning problems may be encountered in systems where: (1) adjacent elements have widely varying stiffness; (2) it is possible for large rigid body rotations to occur without causing significant strain; or (3) there are many degrees of freedom (perhaps thousands). In Example 11.7, the behavior of a number of combinations of two axial force members of different stiffness is compared. The range of stiffness ratios is extreme and designed only to bring out different numerical characteristics.

EXAMPLE 11.7

For the structure shown, vary the relative stiffness of the two elements and determine the relative errors of 10-digit and 5-digit Gauss elimination solutions. Use a 10-digit closed form solution as a standard of comparison. In each case, input and round-off to the defined number of digits after every addition, subtraction, multiplication, and division. Let $E_1A_1/L_1 = k_1$ and $E_2A_2/L_2 = k_2$.

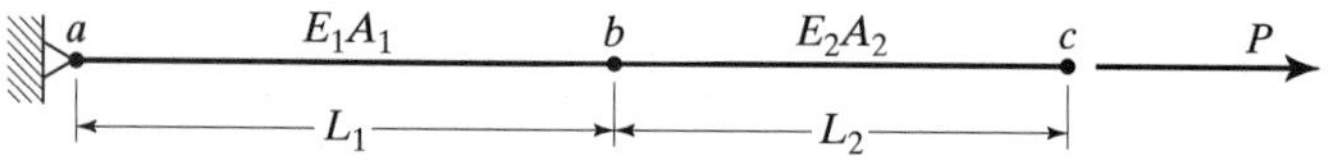

Closed form solution:

$$\begin{bmatrix} (k_1 + k_2) & -k_2 \\ -k_2 & k_2 \end{bmatrix} \begin{Bmatrix} \Delta_b \\ \Delta_c \end{Bmatrix} = \begin{Bmatrix} 0 \\ P \end{Bmatrix}$$

Thus

$$\begin{Bmatrix} \Delta_b \\ \Delta_c \end{Bmatrix} = \begin{bmatrix} \dfrac{1}{k_1} & \dfrac{1}{k_1} \\ \dfrac{1}{k_1} & \left(\dfrac{1}{k_1} + \dfrac{1}{k_2}\right) \end{bmatrix} \begin{Bmatrix} 0 \\ P \end{Bmatrix} = \begin{Bmatrix} \dfrac{P}{k_1} \\ P\left(\dfrac{1}{k_1} + \dfrac{1}{k_2}\right) \end{Bmatrix} \tag{a}$$

Gauss elimination solution:

$$\begin{aligned} (k_1 + k_2)\Delta_b - k_2\Delta_c &= 0 \\ -k_2\Delta_b + k_2\Delta_c &= P \end{aligned}$$

$$\Delta_c = \frac{P}{\left[k_2 - \dfrac{(-k_2)(-k_2)}{(k_1 + k_2)}\right]} \tag{b}$$

$$\Delta_b = \frac{k_2\Delta_c}{(k_1 + k_2)}$$

Comparison:

1. Let $k_1 = \sqrt{3}$. Vary k_2 from $\sqrt{2} \times 10^{-6}$ to $\sqrt{2} \times 10^{10}$.
2. *Closed form solution.* Use Equation a and 10-digit accuracy.
3. *10-digit Gauss elimination.* Use Equation b and 10-digit accuracy.
4. *5-digit Gauss elimination.* Use Equation b and 5-digit accuracy.

Results:

Case	k_1/k_2	Δ	Closed Form Δ for $P = 1$	10-Digit Gauss Δ for $P = 1$	10-Digit Gauss % Relative Error*	5-Digit Gauss Δ for $P = 1$	5-Digit Gauss % Relative Error
1	$\sqrt{3}/\sqrt{2} \times 10^{-6}$	Δ_b	0.5773502690	0.5773502692	-3.46×10^{-8}	0.57733	3.51×10^{-3}
	$= 1.225 \times 10^{6}$	Δ_c	707107.3588	707107.3589	-1.41×10^{-8}	707110.	-3.74×10^{-4}
2	$\sqrt{3}/\sqrt{2} \times 10^{-4}$	Δ_b	0.5773502690	0.5773502694	-6.93×10^{-8}	0.57734	1.78×10^{-3}
	$= 1.225 \times 10^{4}$	Δ_c	7071.645164	7071.645166	-2.83×10^{-8}	7071.6	6.39×10^{-4}
3	$\sqrt{3}/\sqrt{2} \times 10^{-2}$	Δ_b	0.5773502690	0.5773502688	3.46×10^{-8}	0.57736	-1.69×10^{-3}
	$= 1.225 \times 10^{2}$	Δ_c	71.28802841	71.28802840	1.40×10^{-8}	71.291	-4.17×10^{-3}
4	$\sqrt{3}/\sqrt{2}$	Δ_b	0.5773502690	0.5773502694	-6.93×10^{-8}	0.57736	-1.69×10^{-3}
	$= 1.225$	Δ_c	1.284457050	1.284457051	-7.79×10^{-8}	1.2845	-3.34×10^{-3}
5	$\sqrt{3}/\sqrt{2} \times 10^{2}$	Δ_b	0.5773502690	0.5773502585	1.82×10^{-6}	0.57773	-6.58×10^{-2}
	$= 1.225 \times 10^{-2}$	Δ_c	0.5844213368	0.5844213262	1.81×10^{-6}	0.58480	-6.48×10^{-2}
6	$\sqrt{3}/\sqrt{2} \times 10^{4}$	Δ_b	0.5773502690	0.5773498363	7.49×10^{-5}	0.49993	13.41
	$= 1.225 \times 10^{-4}$	Δ_c	0.5774209797	0.5774205469	7.50×10^{-5}	0.50000	13.41
7	$\sqrt{3}/\sqrt{2} \times 10^{6}$	Δ_b	0.5773502690	0.5773664984	-2.81×10^{-3}	∞	$-\infty$
	$= 1.225 \times 10^{-6}$	Δ_c	0.5773509761	0.5773672055	-2.81×10^{-3}	∞	$-\infty$
8	$\sqrt{3}/\sqrt{2} \times 10^{8}$	Δ_b	0.5773502690	0.5882352870	-1.89	∞	$-\infty$
	$= 1.225 \times 10^{-8}$	Δ_c	0.5773502761	0.5882352941	-1.89	∞	$-\infty$
9	$\sqrt{3}/\sqrt{2} \times 10^{10}$	Δ_b	0.5773502690	∞	$-\infty$	∞	$-\infty$
	$= 1.225 \times 10^{-10}$	Δ_c	0.5773502691	∞	$-\infty$	∞	$-\infty$

% Relative Error = 100 (true − approximate)/true

The reasons that problems are encountered in the Gauss solution process but not in the closed form solution can be explained. First, it is emphasized that both solutions are algebraically equivalent; the closed form solution (Equation a) is a simplification of the expressions obtained by Gauss elimination (Equation b). Hence, manipulation errors are the only source of any problems.

In the closed form solution the displacement Δ_c depends on the sum of the reciprocals of the two stiffnesses. The only manipulation errors are those that result from truncating the smaller reciprocal term when it is negligible and rounding of the calculated sum and its product with the axial force P. The truncation error directly corresponds to the physics of the problem; if the two stiffnesses vary significantly, the member with smaller stiffness will control the size of Δ_c.

In using a numerical equation-solving scheme such as Gauss elimination, the solution first requires stiffnesses k_1 and k_2 to be added. If one value is much smaller than the other, significant information is lost irretrievably through the necessity of truncating this sum to the number of figures retained by the computer. If the mismatch is so great that the significance of k_1 is lost in this addition, the displacement Δ_c may become extremely inaccurate or even unattainable. To make matters worse, any error in Δ_c is propagated into the calculation of displacement Δ_b as a result of the back-substitution phase of the Gauss solution process.

Perhaps it is necessary to indicate that Example 11.7 is not intended to imply that element stiffnesses are actually known to 10 digits. The example is intended only to illustrate cases of well-, poorly-, and ill-conditioned problems. Rarely can we be confident of more than two- or three-digit knowledge of stiffness quantities. The point is that, for a given level of confidence in these quantities, how can we detect and avoid or compensate for attrition produced by numerical error and thus assure a solution with at least the same level of reliability?

11.6.1 The Condition Number

We now turn to searching for general methods for detecting ill-conditioning in large-order systems. At the same time we will be estimating the effect of poor conditioning on the accuracy of results obtained in computations that retain only a fixed number of significant figures. Truncation errors will be emphasized because it has been found that initial truncation errors appear to be the main source of manipulation errors in problems of structural analysis (Ref. 11.11). To adequately describe and assess this error we first have to take a brief excursion into the subject of norms.

A *norm* is a function that provides a measure of the size or length of a vector or matrix. For present purposes a norm must exhibit properties of

(i) Scaling: $\|\alpha\{\mathbf{x}\}\| = |\alpha|\ \|\mathbf{x}\|$ (11.37a)

(ii) Positiveness: $\|\mathbf{x}\| > 0 \quad \text{for any } \{\mathbf{x}\} \neq \mathbf{0}$ (11.37b)

(iii) Triangle Inequality: $\|\{\mathbf{x}\} + \{\mathbf{y}\}\| \leq \|\mathbf{x}\| + \|\mathbf{y}\|$ (11.37c)

As examples, the Euclidean norm for a vector is expressed as

$$\|\mathbf{x}\|_e = \sqrt{\{\mathbf{x}\}^{\mathrm{T}}\{\mathbf{x}\}} = \sqrt{\sum_{i=1}^{n} x_i^2} \tag{11.38}$$

and the Frobenius norm for an $n \times n$ matrix $[\mathbf{A}]$ is

$$\|\mathbf{A}\|_F = \sqrt{\sum_{i=1}^{n}\sum_{j=1}^{n} a_{ij}^2} \tag{11.39}$$

Note that the convergence criterion given earlier in Equation 11.35a can be expressed as simply the Euclidean norm of the residual vector $\|\mathbf{r}^i\|_e$. Another popular and simpler norm to apply is the uniform or infinity norm. For a vector this norm is expressed as the maximum magnitude

$$\|\mathbf{x}\|_\infty = \max_{1 \leq i \leq n} |x_i| \tag{11.40}$$

and for an $n \times n$ matrix $[\mathbf{A}]$ as the maximum row-sum

$$\|\mathbf{A}\|_\infty = \max_{1 \leq i \leq n} \sum_{j=1}^{n} |a_{ij}| \tag{11.41}$$

Finally, if a vector norm and a matrix norm are consistent then it can be shown that (Ref. 11.2)

$$\|[\mathbf{A}]\{\mathbf{x}\}\| \leq \|\mathbf{A}\|\ \|\mathbf{x}\| \tag{11.42}$$

With this understanding of norms, an error analysis can commence by assuming that Equation 11.1 could be solved exactly for $\{\mathbf{x}\}$. First let us suppose that the coefficients in {**b**} are also known exactly but those in [**A**] are subject to some small uncertainty [δ**A**]. Equation 11.1 is modified to

$$[\mathbf{A} + \delta\mathbf{A}]\{\bar{\mathbf{x}}\} = \{\mathbf{b}\} \tag{11.43}$$

The difference between the solutions to Equations 11.1 and 11.43, {**x**} and {$\bar{\mathbf{x}}$} respectively, would in general reflect how sensitive Equation 11.1 is to small changes in the coefficient matrix [**A**]. If we denote this difference by

$$\{\delta\mathbf{x}\} = \{\bar{\mathbf{x}}\} - \{\mathbf{x}\} \tag{11.44}$$

and substitute this in Equation 11.43, we obtain

$$[\mathbf{A} + \delta\mathbf{A}]\{\mathbf{x} + \delta\mathbf{x}\} = \{\mathbf{b}\} \tag{11.45}$$

Solving this for the error {δ**x**} gives

$$\{\delta\mathbf{x}\} = -[\mathbf{A}]^{-1}[\delta\mathbf{A}]\{\mathbf{x} + \delta\mathbf{x}\} \tag{11.46}$$

Taking the norm of both sides of Equation 11.46 and employing Equations 11.37 and 11.42 yields

$$\|\delta\mathbf{x}\| \le \|\mathbf{A}^{-1}\|\,\|\delta\mathbf{A}\|\,\|\mathbf{x} + \delta\mathbf{x}\| \tag{11.47}$$

This can be simplified to a comparison of relative errors given by

$$\frac{\|\delta\mathbf{x}\|}{\|\mathbf{x} + \delta\mathbf{x}\|} \le \kappa \frac{\|\delta\mathbf{A}\|}{\|\mathbf{A}\|} \tag{11.48}$$

or by further employing Equation 11.37c

$$\frac{\|\delta\mathbf{x}\|}{\|\mathbf{x}\|} \le \frac{\kappa}{1 - \kappa \dfrac{\|\delta\mathbf{A}\|}{\|\mathbf{A}\|}} \frac{\|\delta\mathbf{A}\|}{\|\mathbf{A}\|} \tag{11.48a}$$

where κ, the *condition number* of the matrix [**A**] with respect to a given norm, is given by

$$\kappa = \|\mathbf{A}\|\,\|\mathbf{A}^{-1}\| \tag{11.49}$$

By further subjecting the coefficients in both [**A**] and {**b**} to uncertainties [δ**A**] and {δ**b**}, it can be shown that

$$\frac{\|\delta\mathbf{x}\|}{\|\mathbf{x}\|} \le \frac{\kappa}{1 - \kappa \dfrac{\|\delta\mathbf{A}\|}{\|\mathbf{A}\|}} \left(\frac{\|\delta\mathbf{A}\|}{\|\mathbf{A}\|} + \frac{\|\delta\mathbf{b}\|}{\|\mathbf{b}\|}\right) \tag{11.50}$$

It is important to note that in both Equations 11.48a and 11.50 the "error amplification" term $\kappa/(1 - \kappa\|\delta\mathbf{A}\|/\|\mathbf{A}\|)$ can be approximated by the condition number κ when $\|\delta\mathbf{A}\|/\|\mathbf{A}\|$ is sufficiently small. Hence, it should be clear that a bound can be placed on the relative error of the solution and more importantly it is directly proportional to the condition number κ of the coefficient matrix [**A**]. Obviously, the smaller the condition number (that is, the closer it is to unity) the more well-conditioned the system.

11.6.2 Estimate of Condition Number

To compute the condition number using Equation 11.49 would be a formidable task because it requires determining $[\mathbf{A}]^{-1}$. This is further complicated by the fact that if [**A**] is ill-conditioned, the inverse $[\mathbf{A}]^{-1}$ would be most likely computed with large errors and hence the accuracy of the condition number would be at best questionable. Thus we customarily rely on estimates or bounds on the condition number.

One reasonable estimate can be obtained from a lower bound estimate of the norm of $[\mathbf{A}]^{-1}$ (Ref. 11.5)

$$\|\mathbf{A}^{-1}\| \geq \frac{1}{\|\mathbf{A} - \mathbf{B}\|} \tag{11.51}$$

where $[\mathbf{B}]$ is a singular matrix which is "near" or close to the nonsingular matrix $[\mathbf{A}]$. Combining Equations 11.49 and 11.51 provides a bound on the condition number

$$\kappa \geq \frac{\|\mathbf{A}\|}{\|\mathbf{A} - \mathbf{B}\|} \tag{11.52}$$

The singular matrix $[\mathbf{B}]$ may be formed by first setting it equal to $[\mathbf{A}]$ and then making small changes so that it becomes singular. For example, this can usually be done by choosing $[\mathbf{B}]$ to have the same elements as $[\mathbf{A}]$ except that the last row is replaced by a linear combination of all of its other rows such that it matches the last row of $[\mathbf{A}]$ in all but one position. Example 11.8 illustrates the use of Equations 11.41 and 11.52.

EXAMPLE 11.8

Use Equations 11.41 and 11.52 to estimate the condition number for the following:

$$\begin{aligned} 10x_1 + 8x_2 + 7x_3 + 7x_4 &= 32 \\ 8x_1 + 10x_2 + 9x_3 + 6x_4 &= 33 \\ 7x_1 + 9x_2 + 10x_3 + 5x_4 &= 31 \\ 7x_1 + 6x_2 + 5x_3 + 5x_4 &= 23 \end{aligned}$$

$$[\mathbf{A}] = \begin{bmatrix} 10 & 8 & 7 & 7 \\ 8 & 10 & 9 & 6 \\ 7 & 9 & 10 & 5 \\ 7 & 6 & 5 & 5 \end{bmatrix}$$

$$\|\mathbf{A}\|_\infty = \max\begin{pmatrix} |10| + |8| + |7| + |7| \\ |8| + |10| + |9| + |6| \\ |7| + |9| + |10| + |5| \\ |7| + |6| + |5| + |5| \end{pmatrix} = 33$$

To find a singular matrix $[\mathbf{B}]$ close to matrix $[\mathbf{A}]$:

Partition $[\mathbf{A}]$ according to $\begin{bmatrix} [\mathbf{S}]^T & [\mathbf{R}] \\ [\mathbf{T}]^T & 5 \end{bmatrix}$ with $[\mathbf{S}] = \begin{bmatrix} 10 & 8 & 7 \\ 8 & 10 & 9 \\ 7 & 9 & 10 \end{bmatrix}$, $[\mathbf{T}] = \begin{bmatrix} 7 \\ 6 \\ 5 \end{bmatrix}$, and $[\mathbf{R}] = \begin{bmatrix} 7 \\ 6 \\ 5 \end{bmatrix}$

Solve $[\mathbf{S}]\{\mathbf{y}\} = [\mathbf{T}]$, $\{\mathbf{y}\} = \begin{Bmatrix} 0.6029 \\ 0.2500 \\ -0.1471 \end{Bmatrix}$

Let [**B**] be identical to [**A**] except $B_{n,n} = \{\mathbf{y}\}^T[\mathbf{R}]$

$$B_{4,4} = \begin{Bmatrix} 0.6029 \\ 0.2500 \\ -0.1471 \end{Bmatrix}^T \begin{bmatrix} 7 \\ 6 \\ 5 \end{bmatrix} = 4.9853$$

$$\text{Hence, } [\mathbf{B}] = \begin{bmatrix} 10 & 8 & 7 & 7 \\ 8 & 10 & 9 & 6 \\ 7 & 9 & 10 & 5 \\ 7 & 6 & 5 & 4.9853 \end{bmatrix} \text{ which is close to } [\mathbf{A}] \text{ and singular.}$$

$$\|\mathbf{A} - \mathbf{B}\|_\infty = |A_{n,n} - B_{n,n}| = |5 - 4.9853| = 0.0147$$

With

$$\frac{\|\mathbf{A}\|}{\|\mathbf{A} - \mathbf{B}\|} = \frac{33}{0.0147} = 2.24 \times 10^3$$

$\kappa \geq 2.24 \times 10^3$ which indicates the system is ill-conditioned.

Other more general methods for estimating a condition number are presented in the literature (see Ref. 11.2). An in-depth study of solution error (in Ref. 11.5) shows that the lower bound on the condition number is

$$\kappa \geq \frac{\max\limits_{1 \leq i \leq n} |\lambda_i|}{\min\limits_{1 \leq i \leq n} |\lambda_i|} \tag{11.53}$$

where λ_i are the eigenvalues of the coefficient matrix [**A**]. Eigenvalues may be obtained by expanding the determinant

$$|[\mathbf{A}] - \lambda[\mathbf{I}]| = 0 \tag{11.54}$$

where [**I**] is the identity matrix, and then solving for the roots of the resulting characteristic equation. Since eigenvalues are of fundamental importance in critical load analysis, an introduction to them and to eigenproblems in general will be presented in Chapter 12.

It should be noted, however, that as the number of equations n increases, calculating the eigenvalues using Equation 11.54 can become a task as formidable as calculating $[\mathbf{A}]^{-1}$. Fortunately, there are simple ways of estimating these values. For example, an upper bound on the largest eigenvalue can be obtained from the infinity norm

$$\max_{1 \leq i \leq n} |\lambda_i| \leq \|\mathbf{A}\|_\infty \tag{11.55}$$

An estimate of the smallest eigenvalue can be calculated using the relatively simple inverse iteration procedure presented in Section 12.8.4.

11.6.3 Error Estimates and Preconditioning

An approximate condition number is often all that is needed in estimating potential error. Suppose that the coefficient matrix [**A**] is represented in the computation by p decimal places, that is $\|\delta\mathbf{A}\|/\|\mathbf{A}\| = 10^{-p}$, and similarly, s is the resulting number of correct decimal places in the solution, $\|\delta\mathbf{x}\|/\|\mathbf{x}\| = 10^{-s}$. From Equation 11.48, the relationship between p and s—the effect of the initial truncation error—may then be estimated as

$$s \geq p - \log_{10}\kappa \tag{11.56}$$

Since the formula contains $\log_{10}\kappa$ and, at most, s and p are on the order of ten, it is clear that great precision in calculating κ is not required. The use of Equations 11.53 and 11.56 is illustrated in Example 11.9.

EXAMPLE 11.9

Using the stiffness matrix of Example 11.7, calculate the eigenvalues, condition numbers, and estimated number of correct decimal places in Cases 1, 4, and 7 of that example. Compare with results of Example 11.7.

Eigenvalue calculation:

$$[[\mathbf{K}] - \lambda[\mathbf{I}]]\{\mathbf{x}\} = 0$$

$$\left[\begin{bmatrix} (k_1 + k_2) & -k_2 \\ -k_2 & k_2 \end{bmatrix} - \lambda[\mathbf{I}]\right]\{\mathbf{x}\} = 0$$

$$\begin{bmatrix} (k_1 + k_2 - \lambda) & -k_2 \\ -k_2 & (k_2 - \lambda) \end{bmatrix}\{\mathbf{x}\} = 0$$

Characteristic equation:

$$(k_1 + k_2 - \lambda)(k_2 - \lambda) - (-k_2)^2 = 0$$

$$\lambda^2 - (k_1 + 2k_2)\lambda + k_1k_2 = 0$$

Solving for λ,

$$\lambda_1 = \left(\frac{k_1}{2} + k_2\right) + \sqrt{\left(\frac{k_1}{2}\right)^2 + k_2^2}$$

$$\lambda_2 = \left(\frac{k_1}{2} + k_2\right) - \sqrt{\left(\frac{k_1}{2}\right)^2 + k_2^2}$$

Results:

Case	$\frac{k_1}{k_2}$	λ_1	λ_2	$\kappa = \lambda_1/\lambda_2$	$\log_{10}\kappa$	$p = 10$		$p = 5$	
						s $(p - \log_{10}\kappa)$	s (Ex. 11.7)	s $(p - \log_{10}\kappa)$	s (Ex. 11.7)
1	$\frac{\sqrt{3}}{\sqrt{2} \times 10^{-6}}$	1.73	1.41×10^{-6}	1.22×10^6	6.1	4	9	0	4
4	$\frac{\sqrt{3}}{\sqrt{2}}$	3.94	0.62	6.33	0.8	9	9	4	4
7	$\frac{\sqrt{3}}{\sqrt{2} \times 10^6}$	2.83×10^6	0.87	3.27×10^6	6.5	4	4	0	0

Example 11.9 indicates that Equation 11.56 is accurate most of the time. However, in some cases, such as Case 1, it does forecast ill-conditioning even though the results of Example 11.7 indicate this problem is well-conditioned. This is a situation of *artificial ill-conditioning*, or *skewness*. This misleading behavior may be detected and rectified by a procedure known as *preconditioning* or *equilibrating*. It is important to note that

preconditioning does not change the nature of the problem; a truly ill-conditioned system will remain so even after preconditioning. Hence, when the condition number of a properly equilibrated matrix is used in Equation 11.56, it should provide a reliable estimate of the initial truncation error.

Preconditioning can be explained as the following series of matrix manipulations. First, we premultiply both sides of Equation 11.1 by $[\mathbf{D}_1]$

$$[\mathbf{D}_1][\mathbf{A}]\{\mathbf{x}\} = [\mathbf{D}_1]\{\mathbf{b}\} \tag{11.57}$$

and then substitute the solution vector $\{\mathbf{x}\}$ according to

$$\{\mathbf{x}\} = [\mathbf{D}_2]\{\overline{\mathbf{x}}\} \tag{11.58}$$

where $[\mathbf{D}_1]$ and $[\mathbf{D}_2]$ are square matrices to be prescribed. The resulting preconditioned system of equations, one with a more reliable and often improved condition number, is

$$[\overline{\mathbf{A}}]\{\overline{\mathbf{x}}\} = \{\overline{\mathbf{b}}\} \tag{11.59}$$

where $[\overline{\mathbf{A}}] = [\mathbf{D}_1][\mathbf{A}][\mathbf{D}_2]$ and $\{\overline{\mathbf{b}}\} = [\mathbf{D}_1]\{\mathbf{b}\}$.

Although it goes beyond our present needs, it may be observed that, for any $[\mathbf{D}_1]$ and $[\mathbf{D}_2]$, Equation 11.59 may be solved for $\{\overline{\mathbf{x}}\}$ and the result substituted in Equation 11.58 to find the solution $\{\mathbf{x}\}$ to the original system of equations. As suggested in Section 11.3.2, employing such a preconditioning strategy within the conjugate gradient method can make this iterative solution scheme more effective by significantly improving its rate of convergence.

Several schemes have been proposed for the selection of $[\mathbf{D}_1]$ and $[\mathbf{D}_2]$. In Reference 11.12 it is suggested that good results can be obtained by letting $[\mathbf{D}_1] = [\mathbf{D}_2] = [\mathbf{D}]$ where $[\mathbf{D}]$ is a diagonal matrix whose elements are defined by

$$d_{ii} = 1/\sqrt{a_{ii}} \qquad \text{for } i = 1, \ldots, n \tag{11.60}$$

In Example 11.10 this preconditioning scheme is applied to the cases presented in Example 11.9. A dramatic change in κ for Case 1 is observed and the system is now correctly predicted as well-conditioned. The ill-conditioning of Case 7 properly remains.

Example 11.11 is an illustration of the above statement that preconditioning does not change the nature of the problem. Several cases treated in Example 11.7 are solved again after equilibration by the scheme just described. There is seen to be no essential difference in the results obtained in the two examples.

EXAMPLE 11.10

Repeat Example 11.9 using a stiffness matrix equilibrated by Equation 11.60.

Preconditioning:

$$[\overline{\mathbf{K}}] = [\mathbf{D}][\mathbf{K}][\mathbf{D}] = \begin{bmatrix} \dfrac{1}{\sqrt{k_1 + k_2}} & 0 \\ 0 & \dfrac{1}{\sqrt{k_2}} \end{bmatrix} \begin{bmatrix} (k_1 + k_2) & -k_2 \\ -k_2 & k_2 \end{bmatrix} \begin{bmatrix} \dfrac{1}{\sqrt{k_1 + k_2}} & 0 \\ 0 & \dfrac{1}{\sqrt{k_2}} \end{bmatrix}$$

$$= \begin{bmatrix} 1 & -\dfrac{\sqrt{k_2}}{\sqrt{k_1 + k_2}} \\ -\dfrac{\sqrt{k_2}}{\sqrt{k_1 + k_2}} & 1 \end{bmatrix}$$

Determine eigenvalues of equilibrated matrix:

$$[[\overline{\mathbf{K}}] - \lambda[\mathbf{I}]]\{\mathbf{x}\} = \begin{bmatrix} (1-\lambda) & -\dfrac{\sqrt{k_2}}{\sqrt{k_1+k_2}} \\ -\dfrac{\sqrt{k_2}}{\sqrt{k_1+k_2}} & (1-\lambda) \end{bmatrix}\{\mathbf{x}\} = 0$$

Characteristic equation:

$$(1-\lambda)^2 - \frac{k_2}{k_1+k_2} = 0$$

$$\lambda^2 - 2\lambda + \frac{k_1}{k_1+k_2} = 0$$

Solving for λ,

$$\lambda_1 = 1 + \sqrt{\frac{k_2}{k_1+k_2}} \quad \text{and} \quad \lambda_2 = 1 - \sqrt{\frac{k_2}{k_1+k_2}}$$

Results:

Case	$\dfrac{k_1}{k_2}$	λ_1	λ_2	$\kappa = \lambda_1/\lambda_2$	$\log_{10}\kappa$	$p = 10$: s $(p - \log_{10}\kappa)$	$p = 10$: s (Ex. 11.7)	$p = 5$: s $(p - \log_{10}\kappa)$	$p = 5$: s (Ex. 11.7)
1	$\dfrac{\sqrt{3}}{\sqrt{2}\times 10^{-6}}$	1.001	0.999	1.002	≈ 0	10	9	5	4
4	$\dfrac{\sqrt{3}}{\sqrt{2}}$	1.67	0.33	5.07	0.7	9	9	4	4
7	$\dfrac{\sqrt{3}}{\sqrt{2}\times 10^{6}}$	2.00	6.12×10^{-7}	3.27×10^{6}	6.5	4	4	0	0

EXAMPLE 11.11

For Cases 1, 4, and 7 of Example 11.7, investigate the accuracy of results obtained by 10- and 5-digit Gauss elimination following preconditioning by Equations 11.57 to 11.60.

Start with equations that have been preconditioned algebraically:

$$\begin{bmatrix} 1 & -\dfrac{\sqrt{k_2}}{\sqrt{k_1+k_2}} \\ -\dfrac{\sqrt{k_2}}{\sqrt{k_1+k_2}} & 1 \end{bmatrix}\begin{Bmatrix} \overline{\Delta}_b \\ \overline{\Delta}_c \end{Bmatrix} = \begin{Bmatrix} 0 \\ \dfrac{P}{\sqrt{k_2}} \end{Bmatrix} \quad \text{and} \quad \begin{Bmatrix} \Delta_b \\ \Delta_c \end{Bmatrix} = \begin{Bmatrix} \dfrac{\overline{\Delta}_b}{\sqrt{k_1+k_2}} \\ \dfrac{\overline{\Delta}_c}{\sqrt{k_2}} \end{Bmatrix}$$

Inserting numerical values and solving for $\{\overline{\boldsymbol{\Delta}}\}$ and then $\{\boldsymbol{\Delta}\}$, the results are

Case	$\dfrac{k_1}{k_2}$	Δ	Closed Form Δ for $P = 1$	10-Digit Gauss Preconditioned Equations	5-Digit Gauss Preconditioned Equations
1	$\dfrac{\sqrt{3}}{\sqrt{2}\times 10^{-6}}$	Δ_b	0.5773502690	0.5773502695	0.57733
		Δ_c	707107.3588	707107.3586	707110.
4	$\dfrac{\sqrt{3}}{\sqrt{2}}$	Δ_b	0.5773502690	0.5773502698	0.57735
		Δ_c	1.284457050	1.284457051	1.2845
7	$\dfrac{\sqrt{3}}{\sqrt{2}\times 10^{6}}$	Δ_b	0.5773502690	0.5775121819	∞
		Δ_c	0.5773509761	0.5775128889	∞

11.6.4 Detecting, Controlling, and Correcting Error

We close this section on error analysis by shifting our attention from predicting possible error to detecting and, if possible, controlling and even correcting for numerical errors actually committed.

Perhaps the most obvious and commonly used procedure for error detection is substituting the solution vector obtained, say, $\{\mathbf{x}_A\}$, back in the original system of equations and calculating the corresponding right-hand side vector $\{\mathbf{b}_A\}$. If the elements of $\{\mathbf{b}_A\}$ differ substantially from those of the original right-hand side vector $\{\mathbf{b}\}$, it is clear that significant errors have occurred in the solution process and we have not found the true solution $\{\mathbf{x}_T\}$. Unfortunately, the converse is not necessarily true; small differences between $\{\mathbf{b}_A\}$ and $\{\mathbf{b}\}$ do not prove that $\{\mathbf{x}_A\}$ is accurate. By the mere definition of ill-conditioning given earlier, several solutions which may vary widely can provide a $\{\mathbf{b}_A\}$ that is acceptably close to $\{\mathbf{b}\}$. In this same regard, according to

$$\{\mathbf{x}_T\} - \{\mathbf{x}_A\} = [\mathbf{A}]^{-1}(\{\mathbf{b}\} - \{\mathbf{b}_A\}) = [\mathbf{A}]^{-1}\{\delta\mathbf{b}\} \tag{11.61}$$

the elements of $[\mathbf{A}]^{-1}$ may be so large—another sign of possible ill-conditioning—that a small discrepancy in the right-hand sides $\{\delta\mathbf{b}\}$ may be severely amplified, producing a wide range in solution vectors.

In a structural analysis, back substituting the solution vector is the equivalent of performing an equilibrium check. Although this check may be inconclusive, extending it to a check for displacement compatibility will confirm that a solution is the correct one for a given idealization. Where a complete equilibrium and compatibility check can be a time consuming process, a distributed sampling of checks can often confirm the merit of a solution.

When a poorly conditioned problem is encountered, several measures can be used to correct or at least control the amount of numerical error. One scheme is to use *scaled*[6] equations as a criterion for employing partial pivoting (see Section 11.2.1). Solution accuracy may then be improved by interchanging rows of original coefficient values so that the divisor is always the most significant coefficient in the submatrix column of the unknown being eliminated. This concept may be extended to *complete pivoting*, where columns are also interchanged so that the dividing coefficient is the largest in the entire submatrix of coefficients to be operated on. In complete pivoting, symmetry is preserved.

Another method for increasing solution accuracy is to employ a technique known as *iterative improvement* (Ref. 11.12). After factoring $[\mathbf{A}]$ according to algorithms such as Equations 11.13 or 11.15, the solution $\{\mathbf{x}^1\}$ to Equation 11.1 is first computed. The residual of this solution $\{\mathbf{r}^1\}$ is then computed according to Equation 11.31. Using $\{\mathbf{r}^1\}$ and the original decomposition of $[\mathbf{A}]$, a solution $\{\mathbf{z}^1\}$ is obtained for

$$[\mathbf{A}]\{\mathbf{z}^1\} = \{\mathbf{r}^1\} \tag{11.62}$$

A more accurate solution to the original system of equations can then be furnished by

$$\{\mathbf{x}^2\} = \{\mathbf{x}^1\} + \{\mathbf{z}^1\} \tag{11.63}$$

The process would then be repeated until the change in the solution $\{\mathbf{z}^i\}$ becomes negligible. In a structural analysis, this approach is equivalent to computing residual

---

[6]Scaling refers to normalizing the equations so that the maximum coefficient in a row equals unity. Because scaling may introduce rounding error, it is recommended that the scaled coefficients be used only as a criterion for pivoting and the original coefficient values be retained within the elimination and substitution computations.

loads from an equilibrium analysis and then applying these loads to the structure as forces to obtain displacement corrections.

Considering that a majority of numerical errors in framework analyses are the result of truncating valuable information, it is important to note that there are no schemes available for retrieving or appropriately compensating for this lost information. In this regard, perhaps the most effective approach in controlling numerical error is to use the greatest computational precision available. In doing so, one must be sure to formulate all load vectors, element properties, and stiffness coefficients with the same high-level precision used during the solution process.

Example 11.12 illustrates the above point that small differences in $\{\mathbf{b}_A\}$ and $\{\mathbf{b}\}$ do not prove that $\{\mathbf{x}_A\}$ is necessarily accurate. The use of iterative improvement is illustrated in Example 11.13.

EXAMPLE 11.12

Show that the following two solutions closely satisfy the system of equations given in Example 11.8 and that neither is close to the true solution $\{\mathbf{x}_T\} = \lfloor 1 \quad 1 \quad 1 \quad 1 \rfloor^T$

1. $\{\mathbf{x}_A\} = \lfloor 6 \quad 2.9 \quad -0.1 \quad -7.2 \rfloor^T$
2. $\{\mathbf{x}_A\} = \lfloor 1.50 \quad 1.19 \quad 0.89 \quad 0.18 \rfloor^T$

$$[\mathbf{A}]\{\mathbf{x}\} = \{\mathbf{b}\} \text{ with } [\mathbf{A}] = \begin{bmatrix} 10 & 8 & 7 & 7 \\ 8 & 10 & 9 & 6 \\ 7 & 9 & 10 & 5 \\ 7 & 6 & 5 & 5 \end{bmatrix} \text{ and } \{\mathbf{b}\} = \begin{Bmatrix} 32 \\ 33 \\ 31 \\ 23 \end{Bmatrix}$$

1. with $\{\mathbf{x}_A\} = \lfloor 6 \quad 2.9 \quad -0.1 \quad -7.2 \rfloor^T$

$$\{\mathbf{b}_A\} = [\mathbf{A}]\{\mathbf{x}_A\} = \begin{Bmatrix} 32.1 \\ 32.9 \\ 31.1 \\ 22.9 \end{Bmatrix}$$

$$\{\delta\mathbf{b}\} = \{\mathbf{b}\} - \{\mathbf{b}_A\} = \begin{Bmatrix} -0.1 \\ 0.1 \\ -0.1 \\ 0.1 \end{Bmatrix} \text{ when } \{\delta\mathbf{x}\} = \{\mathbf{x}_T\} - \{\mathbf{x}_A\} = \begin{Bmatrix} -5 \\ -1.9 \\ 1.1 \\ 8.2 \end{Bmatrix}$$

2. with $\{\mathbf{x}_A\} = \lfloor 1.50 \quad 1.19 \quad 0.89 \quad 0.18 \rfloor^T$

$$\{\mathbf{b}_A\} = [\mathbf{A}]\{\mathbf{x}_A\} = \begin{Bmatrix} 32.01 \\ 32.99 \\ 31.01 \\ 22.99 \end{Bmatrix}$$

$$\{\delta\mathbf{b}\} = \{\mathbf{b}\} - \{\mathbf{b}_A\} = \begin{Bmatrix} -0.01 \\ 0.01 \\ -0.01 \\ 0.01 \end{Bmatrix} \text{ when } \{\delta\mathbf{x}\} = \{\mathbf{x}_T - \mathbf{x}_A\} = \begin{Bmatrix} -0.5 \\ -0.19 \\ 0.11 \\ 0.82 \end{Bmatrix}$$

EXAMPLE 11.13

Use iterative improvement on the second solution in Example 11.12 to converge on the true solution.

With $\{\mathbf{x}_1\} = \lfloor 1.50 \quad 1.19 \quad 0.89 \quad 0.18 \rfloor^T$

$$\{\mathbf{r}_1\} = \{\mathbf{b}\} - [\mathbf{A}]\{\mathbf{x}_1\} = \begin{Bmatrix} -0.01 \\ 0.01 \\ -0.01 \\ 0.01 \end{Bmatrix}$$

$$\{\mathbf{z}_1\} = [\mathbf{A}]^{-1}\{\mathbf{r}_1\} = \begin{Bmatrix} -0.50 \\ -0.19 \\ 0.11 \\ 0.82 \end{Bmatrix}$$

$$\{\mathbf{x}_2\} = \{\mathbf{x}_1\} + \{\mathbf{z}_1\} = \begin{Bmatrix} 1 \\ 1 \\ 1 \\ 1 \end{Bmatrix}$$

$$\{\mathbf{r}_2\} = \{\mathbf{b}\} - [\mathbf{A}]\{\mathbf{x}_2\} = \begin{Bmatrix} 0 \\ 0 \\ 0 \\ 0 \end{Bmatrix}$$ indicating that $\{\mathbf{x}_2]$ is the true solution.

11.7 PROBLEMS

11.1 Solve the following systems of equations by (1) Gauss elimination; (2) the Cholesky method; and (3) the root-free Cholesky method.

(a)
$$\begin{aligned} 64x_1 + 32x_2 - 16x_3 &= 140 \\ 32x_1 + 416x_2 + 72x_3 &= 170 \\ -16x_1 + 72x_2 + 120x_3 &= 5 \end{aligned}$$

(b)
$$\begin{aligned} 25x_1 + 5x_2 - 2.5x_3 &= 60 \\ 5x_1 + 17x_2 + 11.5x_3 &= -8 \\ -2.5x_1 + 11.5x_2 + 13.25x_3 &= 11 \end{aligned}$$

11.2 Attempt to solve the following system of equations by the Cholesky method. What can be concluded?

$$\begin{aligned} 20x_1 + 10x_2 - 40x_3 &= 100 \\ 10x_1 + 15x_2 + 30x_3 &= -110 \\ -40x_1 + 30x_2 - 20x_3 &= -300 \end{aligned}$$

11.3 Solve the equations of Problem 11.2 by the root-free Cholesky method.

11.4 Solve the equations of Problem 11.1 by (1) Gauss-Seidel iteration; and (2) Jacobi iteration. Obtain a solution which is accurate to within three significant figures.

11.5 Solve the equations of Problem 11.1b by (1) Gauss-Seidel iteration; and (2) Gauss-Seidel iteration with overrelaxation $\beta = 1.4$. Use $\zeta = 5\%$.

11.6 For the equations of Problem 11.2, perform five iterations of the Gauss-Seidel method. Based on this solution and the conclusion made in Problem 11.2, comment on the apparent effectiveness of using this scheme.

11.7 Solve the equations of Problem 11.1 by (1) the conjugate gradient method; and (2) the conjugate gradient method with preconditioning. Use $\zeta = 0.1$. Discuss the effect that preconditioning has on the error calculated at the end of each iteration.

11.8 Solve the equations of Problem 11.2 by the conjugate gradient method. Use $\zeta = 0.1$.

11.9 Solve the following systems of equations by the Cholesky method using (1) Equation 11.15 and (2) Equation 11.36a, which accounts for bandedness.

(a)
$$\begin{aligned} 25x_1 + 100x_2 \qquad\qquad\qquad &= -100 \\ 100x_1 + 409x_2 - 9x_3 \qquad\quad &= -445 \\ -9x_2 + 10x_3 + 3x_4 &= 63 \\ 3x_3 + 13x_4 &= 74 \end{aligned}$$

(b)
$$\begin{aligned} 16x_1 - 48x_2 \qquad\qquad\qquad\qquad &= 16 \\ -48x_1 + 180x_2 + 72x_3 \qquad\qquad &= -192 \\ 72x_2 + 153x_3 + 36x_4 \qquad &= -171 \\ 36x_3 + 148x_4 + 4x_5 &= 492 \\ 4x_4 + 29x_5 &= 74 \end{aligned}$$

11.10 For the frames shown, label the degrees of freedom so that the bandwidth of $[\mathbf{K}_{ff}]$ is at a minimum. Use Equation 11.36b to calculate the half-bandwidth.

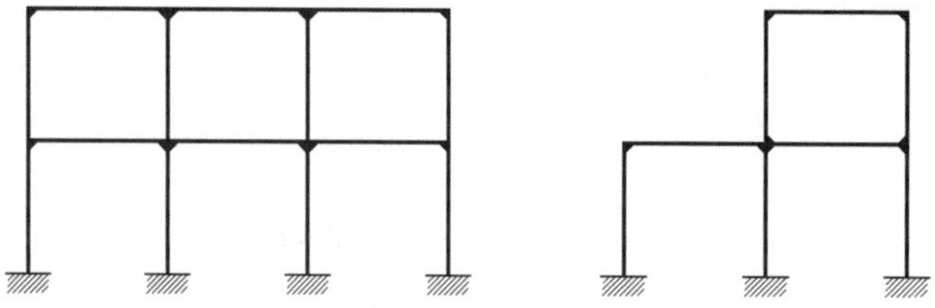

(*a*) (*b*)

Problem 11.10

11.11 Solve the structure shown using a wave front approach. $E = 200{,}000$ MPa

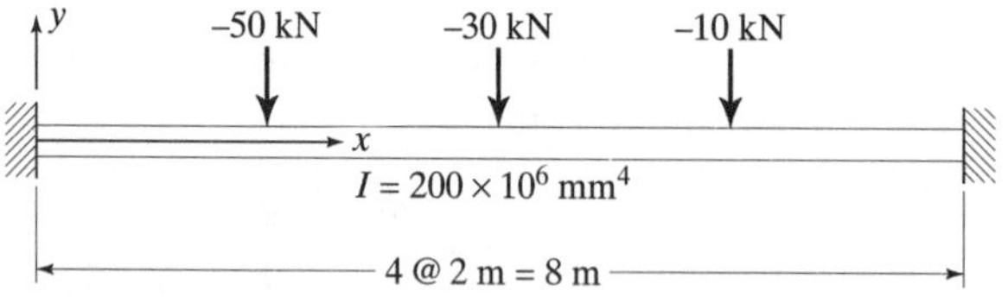

Problem 11.11

11.12 Use Equations 11.41 and 11.52 to comment on the condition of the equations of (1) Problem 11.1; and (2) Problem 11.2.

11.13 Comment on the condition of the following systems of equations using (1) Equation 11.52 with norms calculated by Equation 11.41; and (2) Equation 11.53.

(a)
$$\begin{aligned} 1.2x_1 + 0.4x_2 &= 2 \\ 23.9x_1 + 7.95x_2 &= 39.8 \end{aligned}$$

(b)
$$\begin{aligned} x_1 + 0.8x_2 &= 3.1 \\ 99.9x_1 + 1000x_2 &= -2.5 \end{aligned}$$

11.14 Repeat Problem 11.13 after applying preconditioning.

11.15 Solve the following system of equations and discuss the condition of the system:

$$
\begin{aligned}
1.00x_1 + 0.50x_2 + 0.33x_3 &= 6 \\
0.50x_1 + 0.33x_2 + 0.25x_3 &= 4 \\
0.33x_1 + 0.25x_2 + 0.20x_3 &= 2
\end{aligned}
$$

11.16 The following system of equations is similar to that in Problem 11.15. Solve it exactly by retaining the fractional notation. Compare your solution with that obtained in Problem 11.15. Approximately how many digits would have to be used in a Gauss elimination solution using decimal notation in order to obtain four-digit accuracy?

$$
\begin{aligned}
x_1 + \tfrac{1}{2}x_2 + \tfrac{1}{3}x_3 &= 6 \\
\tfrac{1}{2}x_1 + \tfrac{1}{3}x_2 + \tfrac{1}{4}x_3 &= 4 \\
\tfrac{1}{3}x_1 + \tfrac{1}{4}x_2 + \tfrac{1}{5}x_3 &= 2
\end{aligned}
$$

11.17 Repeat Example 11.9 using Equations 11.41 and 11.52 to calculate the condition numbers.

11.18 The structure shown consists of three axial force members in series. The forces are $P_b = P_d = 1$ and $P_c = 0$. The stiffnesses are $k_{ab} = k_{cd} = 1$ and $k_{bc} = 1 \times 10^4$. (1) Calculate the displacement vector using Gauss elimination and four-digit arithmetic (truncate to four digits after every operation—be sure that truncation is done and not merely on the calculator's display—this may require reentry after every operation). (2) Repeat using eight-digit arithmetic. (3) Based on these solutions, discuss the condition of the system.

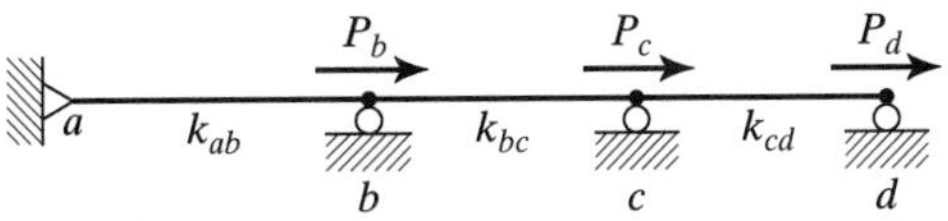

Problem 11.18

11.19 Use Equations 11.52 or 11.53 and Equation 11.56 to estimate the truncation error for Problem 11.18. Compare with results obtained in Problem 11.18.

11.20 Precondition the coefficient matrix of Problem 11.18, and repeat Problems 11.18 and 11.19.

11.21 Repeat Problems 11.18 to 11.20 with $k_{ab} = k_{cd} = 1 \times 10^4$ and $k_{bc} = 1$.

REFERENCES

11.1 L. Fox, *An Introduction to Numerical Linear Algebra*, Oxford University Press, New York, 1965.

11.2 G. H. Golub and C. F. Van Loan, *Matrix Computations*, 3rd edition, The Johns Hopkins University Press, Baltimore, Md., 1996.

11.3 K. J. Bathe, *Finite Element Procedures*, Prentice-Hall, Englewood Cliffs, N.J., 1996.

11.4 J. B. Scarborough, *Numerical Mathematical Analysis*, 6th edition, The Johns Hopkins University Press, Baltimore, Md., 1966.

11.5 J. L. Buchanan and P. R. Turner, *Numerical Methods and Analysis*, McGraw-Hill, New York, 1992.

11.6 A. George and J. W. H. Liu, *Computer Solution of Large Sparse Positive Definite Systems*, Prentice-Hall, Englewood Cliffs, N.J., 1981.

11.7 K. J. Bathe and E. L. Wilson, *Numerical Methods in Finite Element Analysis*, Prentice-Hall, Englewood Cliffs, N.J., 1976.

11.8 G. H. Paulino, I. F. M. Menezes, M. Gattass, and S. Mukherjee, "Node and Element Resequencing Using the Laplacian of a Finite Element Graph," *Intl. Jl. Num. Meth. in Engr.*, Vol. 37, No. 9, 1994, pp. 1511–1555.

11.9 B. M. Irons, "A Frontal Solution Program for Finite Element Analysis," *Intl Jl for Num. Methods in Engineering*, 2, 5–32, 1970.

11.10 J. H. Wilkinson, *Rounding Errors in Algebraic Processes*, Prentice-Hall, Englewood Cliffs, N.J., 1963.

11.11 R. A. Rosanoff, J. F. Gloudeman, and S. Levy, "Numerical Conditioning of Stiffness Matrix Formulations for Framed Structures," Technical Report AFFDL-TR-68-150, USAF Flight Dynamics Laboratory, Wright-Patterson Air Force Base, Ohio, 1968.

11.12 G. Forsythe and C. B. Moler, *Computer Solution of Linear Algebraic Systems*, Prentice-Hall, Englewood Cliffs, N.J., 1967.

Chapter 12

Solution of Nonlinear Equilibrium Equations

In Chapter 8 we outlined strategies for the numerical solution of the equilibrium equations associated with a nonlinear analysis. The main purpose of the current chapter is to provide details of these schemes. Two distinctly different mathematical approaches will be explored.

We begin with the incremental method and present the single-step and iterative (multi-step) procedures that are most commonly used to calculate nonlinear structural behavior. Particulars considered crucial to the successful implementation of these methods are examined. They include selection of appropriate load increment size, updating the deformed geometry, force recovery algorithms, and strategies for detecting limit points and tracing post-limit point behavior.

We then proceed to the second approach, in which a predicted limit point or critical load is determined from an eigenvalue analysis. After providing an introduction to basic concepts, the nonlinear equilibrium equations are cast in different forms of eigenproblems and direct and iterative schemes for solving them are presented.

This chapter is not intended to be comprehensive or to replace textbooks dedicated to these and related nonlinear solution topics. It should, however, provide an engineer with information sufficient to perform successful nonlinear analyses of frameworks.

12.1 INCREMENTAL ANALYSIS

As shown in the preceding chapters, formulation of equilibrium on the deformed and perhaps yielded geometry of a structure will result in a system of nonlinear stiffness equations. One method for solving these equations is to approximate their nonlinearity with a piecewise segmental fit (see Figure 12.1 in which the approach is illustrated by a representative load component-displacement diagram). In doing so, the total load on a structure, $\{\mathbf{P}\}$, or its equivalent, the product of a total load ratio λ and a given reference load $\{\mathbf{P}_{\text{ref}}\}$, is applied through a series of load increments $\{\mathbf{dP}_i\}$. Mathematically, this is written as

$$\{\mathbf{P}\} = \lambda\{\mathbf{P}_{\text{ref}}\} = \sum_{i=1}^{n} \{\mathbf{dP}_i\} \tag{12.1}$$

where n is the total number of load increments employed. The structure's corresponding displacement response $\{\boldsymbol{\Delta}\}$ then follows as

$$\{\boldsymbol{\Delta}\} = \sum_{i=1}^{n} \{\mathbf{d\Delta}_i\} \tag{12.2}$$

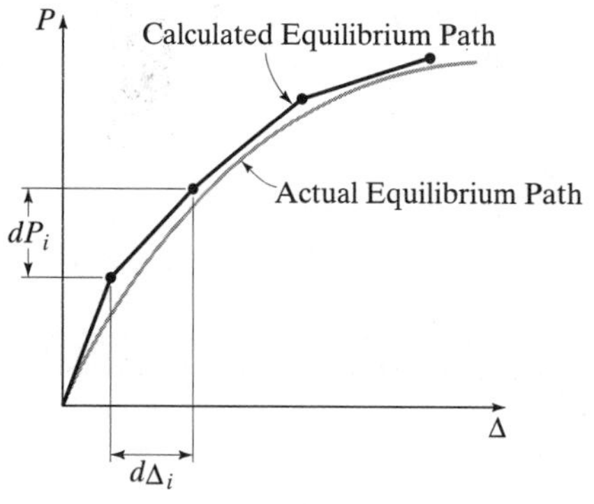

Figure 12.1 Piecewise segmental fit.

where $\{\mathbf{d\Delta}_i\}$ is the vector of displacements resulting from applying the ith load increment $\{\mathbf{dP}_i\}$.

The degree to which a piecewise fit approximates the actual equilibrium solution is clearly a function of how accurately the nonlinear relationship between $\{\mathbf{dP}_i\}$ and $\{\mathbf{d\Delta}_i\}$ is represented during each load increment. To adapt the relatively simple techniques presented in Chapter 11 for solving linear algebraic equations, the nonlinear behavior that occurs during each increment will be approximated by performing one or more linear analyses. The number of analyses performed and how their results are used to evaluate the displacement $\{\mathbf{d\Delta}_i\}$ produced by a given load increment $\{\mathbf{dP}_i\}$ is what distinguishes various nonlinear solution techniques. In this regard, general categories of single-step methods and iterative methods will be presented.

12.2 INCREMENTAL SINGLE-STEP METHODS

The single-step methods all employ a strategy that is analogous to solving systems of linear or nonlinear differential equations by the Runge-Kutta methods. In general they can be cast in the form

$$\{\mathbf{\Delta}_i\} = \{\mathbf{\Delta}_{i-1}\} + \{\mathbf{d\Delta}_i\} \tag{12.3}$$

where $\{\mathbf{\Delta}_{i-1}\}$ and $\{\mathbf{\Delta}_i\}$ are the total displacements at the end of the previous and current load increments, respectively. In Equation 12.3, the increment of unknown displacements $\{\mathbf{d\Delta}_i\}$ is found in a single step by solving the linear system of equations

$$[\overline{\mathbf{K}}_i]\{\mathbf{d\Delta}_i\} = \{\mathbf{dP}_i\} \tag{12.4}$$

where the load increment $\{\mathbf{dP}_i\}$ is given by

$$\{\mathbf{dP}_i\} = d\lambda_i\{\mathbf{P}_{\text{ref}}\} \tag{12.4a}$$

The size of the load ratio for the first increment, $d\lambda_1$, which must be prescribed by the analyst, should normally be about 10%–20% of the anticipated maximum applied load. The same value may be used for $d\lambda_i$ in subsequent increments of the analysis, but as shown in Section 12.4 there are also automated procedures for calculating this quantity that may be used to advantage.

In Equation 12.4, $[\overline{\mathbf{K}}_i]$ can be interpreted as the representative stiffness for the entire load increment. Taken as a weighted average, $[\overline{\mathbf{K}}_i]$ is written as

$$[\overline{\mathbf{K}}_i] = \sum_{j=1}^{m} \alpha_j[\mathbf{K}_j] \tag{12.5}$$

where α_j is the weighting coefficient corresponding to a stiffness $[\mathbf{K}_j]$ that was calculated at one of m sampling points within the increment. Each $[\mathbf{K}_j]$ can be defined by typical Runge-Kutta type recurrence relationships. That is, the sampling point used to calculate $[\mathbf{K}_j]$ is obtained by using some or all of the stiffness matrices that correspond to the previous $j - 1$ sampling points. The number of sampling points, m, employed in each step defines the order of the method. In general, the accuracy of the nonlinear relationship between $\{\mathbf{dP}_i\}$ and $\{\mathbf{d\Delta}_i\}$ will improve with an increase in this order.

By varying the weighting factors and the number and location of stiffness sampling points, various types of single-step strategies can be devised from Equation 12.5. Two popular schemes are the Euler and midpoint Runge-Kutta methods.

12.2.1 Euler Method

The most elementary of the single-step strategies is the Euler, or simple step, method. Using Equation 12.5 with $m = 1$ and $\alpha_1 = 1$, the Euler method can be cast as a first-order Runge-Kutta approach with

$$[\overline{\mathbf{K}}_i] = 1 \cdot [\mathbf{K}_1] \tag{12.6}$$

In this case, $[\mathbf{K}_1]$, the tangent stiffness matrix, is calculated using the deformed geometry and corresponding element forces that exist at the start of the increment. Hence, the displacement vector $\{\mathbf{d\Delta}_i\}$ appearing in Equation 12.3 is determined in each load increment by performing the single linear analysis represented by Equation 12.4. This method is illustrated in Figure 12.2.[1]

12.2.2 Second-Order Runge-Kutta Methods

The second-order form ($m = 2$) of Equation 12.5 is given by

$$[\overline{\mathbf{K}}_i] = \alpha_1[\mathbf{K}_1] + \alpha_2[\mathbf{K}_2] \tag{12.7}$$

where $[\mathbf{K}_1]$ is the tangent stiffness at the start of the increment and $[\mathbf{K}_2]$ is the stiffness calculated using the deformed geometry and corresponding element forces at some point within the increment. Noting that the displacements at this second sampling point are not initially known, an additional global frame analysis must be performed before

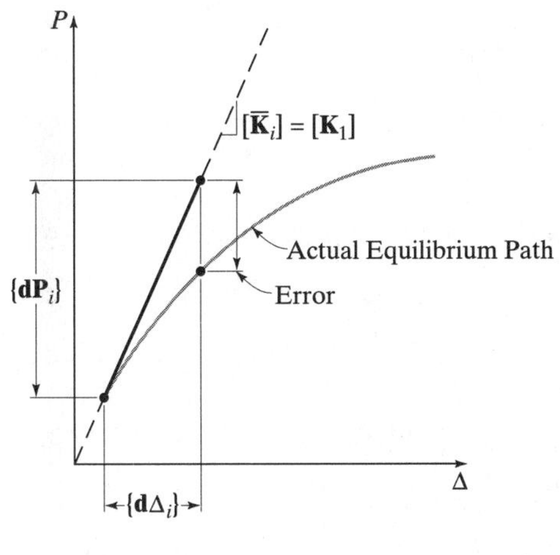

Figure 12.2 Euler method.

[1]Figures 12.2–12.6 are schematic load-displacement diagrams in which multi-dimensional response has been represented as uni-dimensional for illustrative purposes. Vector notation is employed to coordinate the forces, displacements, and stiffnesses shown with the related equations of the text.

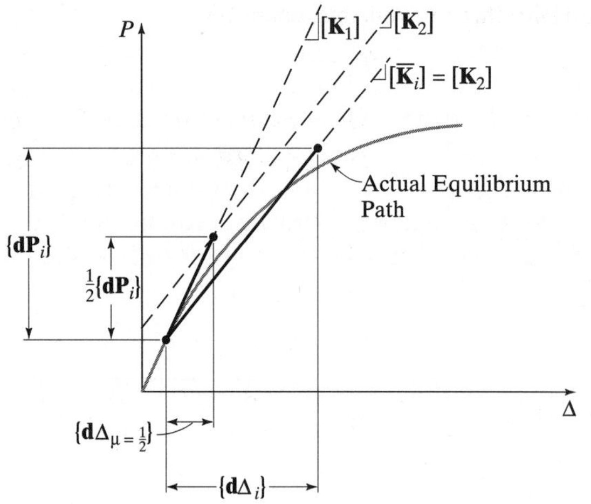

Figure 12.3 Midpoint Runge-Kutta method.

Equations 12.7 and 12.4 can be applied. This is achieved by first solving the following equation for $\{\mathbf{d\Delta}_\mu\}$:

$$[\mathbf{K}_1]\{\mathbf{d\Delta}_\mu\} = \mu\{\mathbf{dP}_i\} \qquad \text{with } 0 < \mu \leq 1 \tag{12.8}$$

where the prescribed load ratio μ defines the relative location of the sampling point within the load increment. The displacements at the second sampling point $\{\mathbf{\Delta}_2\}$, which are needed to formulate $[\mathbf{K}_2]$, are obtained by adding $\{\mathbf{d\Delta}_\mu\}$ to the displacements at the end of the previous increment $\{\mathbf{\Delta}_{i-1}\}$:

$$\{\mathbf{\Delta}_2\} = \{\mathbf{\Delta}_{i-1}\} + \{\mathbf{d\Delta}_\mu\} \tag{12.9}$$

With $[\mathbf{K}_1]$ and $[\mathbf{K}_2]$ evaluated and given a set of prescribed weighting factors α_1 and α_2, Equation 12.7 can be used to calculate the representative stiffness $[\overline{\mathbf{K}}_i]$. The incremental displacements $\{\mathbf{d\Delta}_i\}$ appearing in Equation 12.3 are then obtained by solving Equation 12.4.

One of the most common versions of the second-order schemes is the midpoint Runge-Kutta method. As the name implies, the second sampling point is taken at the midpoint of the load increment ($\mu = 1/2$). By further using weighting factors of $\alpha_1 = 0$ and $\alpha_2 = 1$, this method relies fully on using the midincrement stiffness as the representative stiffness for the entire load increment. Also known as the predictor-corrector or improved polygon method, this procedure is illustrated in Figure 12.3.

Other popular second-order schemes include Heun's method with a single corrector ($\mu = 1$, $\alpha_1 = \alpha_2 = 1/2$) and Ralston's method ($\mu = 0.75$, $\alpha_1 = 1/3$, and $\alpha_2 = 2/3$). Using a single-degree-of-freedom system, the Euler and midpoint Runge-Kutta methods are illustrated in Example 12.1.

EXAMPLE 12.1

A nonlinear spring has a stiffness that is related to the force in the spring by $k = 1/2\,(P + 1)$. Using three increments of the following methods, determine the extension of the spring Δ when subjected to a force of $P = 3$.

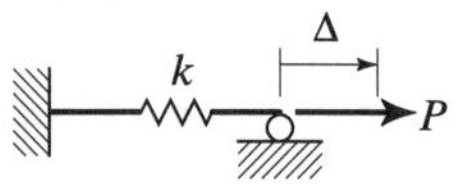

1. Euler method
2. Mid-point Runge-Kutta method
3. Compare both solutions to the exact solution $\Delta = (P + 1)^2 - 1$

1. Euler method with three increments, $d\lambda_i = 1/3$

First increment:

$$P_0 = 0 \qquad \bar{k}_1 = \frac{1}{2(P_0 + 1)} = 0.5$$

$$dP_1 = d\lambda_1 P = 1.0 \qquad d\Delta_1 = \bar{k}_1^{-1} dP_1 = 2.0$$

$$P_1 = P_0 + dP_1 = 1.0 \qquad \Delta_1 = \Delta_0 + d\Delta_1 = 2.0$$

Second increment:

$$P_1 = 1.0 \qquad \bar{k}_2 = \frac{1}{2(P_1 + 1)} = 0.25$$

$$dP_2 = d\lambda_2 P = 1.0 \qquad d\Delta_2 = \bar{k}_2^{-1} dP_2 = 4.0$$

$$P_2 = P_1 + dP_2 = 2.0 \qquad \Delta_2 = \Delta_1 + d\Delta_2 = 6.0$$

Third increment:

$$P_2 = 2.0 \qquad \bar{k}_3 = \frac{1}{2(P_2 + 1)} = 0.1667$$

$$dP_3 = d\lambda_3 P = 1.0 \qquad d\Delta_3 = \bar{k}_3^{-1} dP_3 = 6.0$$

$$P_3 = P_2 + dP_3 = 3.0 \qquad \Delta_3 = \Delta_2 + d\Delta_3 = 12.0$$

When $P = 3.0$, the deflection is $\Delta = 12.0$.

2. Mid-point Runge-Kutta method with three increments, $d\lambda_i = 1/3$ and $\mu = 1/2$

First Increment:

Predictor step

$$P_0 = 0 \qquad k_s = \frac{1}{2(P_0 + 1)} = 0.5$$

$$\mu dP_1 = \mu d\lambda_1 P = 0.5 \qquad d\Delta_\mu = k_s^{-1} \mu dP_1 = 1.0$$

$$P_m = P_0 + \mu dP_1 = 0.5 \qquad \Delta_m = \Delta_0 + d\Delta_\mu = 1.0$$

Corrector step

$$k_m = \frac{1}{2(P_m + 1)} = 0.3333 \qquad \bar{k}_1 = 0.0k_s + 1.0k_m = 0.3333$$

$$dP_1 = d\lambda_1 P = 1.0 \qquad d\Delta_1 = \bar{k}_1^{-1} dP_1 = 3.0$$

$$P_1 = P_0 + dP_1 = 1.0 \qquad \Delta_1 = \Delta_0 + d\Delta_1 = 3.0$$

Second increment:

Predictor step

$$P_1 = 1.0 \qquad k_s = 0.25$$

$$\mu dP_2 = 0.5 \qquad d\Delta_\mu = k_s^{-1} \mu dP_2 = 2.0$$

$$P_m = P_1 + \mu dP_2 = 1.5 \qquad \Delta_m = \Delta_1 + d\Delta_\mu = 5.0$$

Corrector step

$$k_m = 0.2 \qquad \bar{k}_2 = 0.0k_s + 1.0k_m = 0.2$$

$$dP_2 = 1.0 \qquad d\Delta_2 = \bar{k}_2^{-1} dP_2 = 5.0$$

$$P_2 = P_1 + dP_2 = 2.0 \qquad \Delta_2 = \Delta_1 + d\Delta_2 = 8.0$$

Third increment:

Predictor step

$$P_2 = 2.0 \qquad k_s = 0.1667$$

$$\mu dP_3 = 0.5 \qquad d\Delta_\mu = k_s^{-1} \mu dP_3 = 3.0$$

$$P_m = P_2 + \mu dP_3 = 2.5 \qquad \Delta_m = \Delta_2 + d\Delta_\mu = 11.0$$

Corrector step

$$k_m = 0.1429 \qquad \bar{k}_3 = 0.0k_s + 1.0k_m = 0.1429$$

$$dP_3 = 1.0 \qquad d\Delta_3 = \bar{k}_3^{-1} dP_3 = 7.0$$

$$P_3 = P_2 + dP_3 = 3.0 \qquad \Delta_3 = \Delta_2 + d\Delta_3 = 15.0$$

When $P = 3$, the deflection is $\Delta = 15.0$.

3. Comparison with exact solution

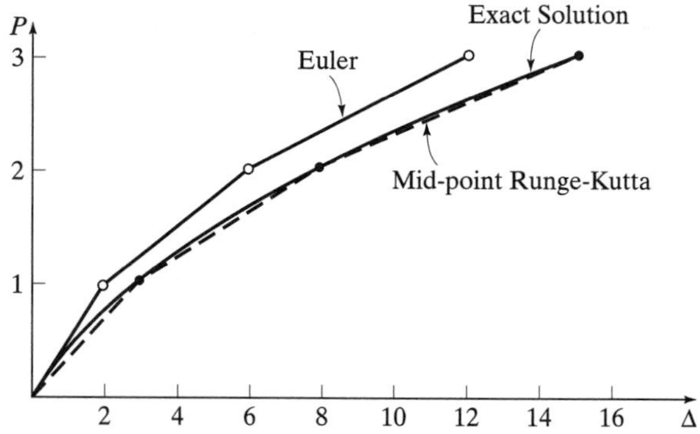

Major advantages of the single-step methods are their simplicity and efficiency. In most cases, only one or two analyses are performed in each increment. In this regard, these schemes, especially the midpoint Runge-Kutta method, are particularly attractive for the analysis of structures exhibiting moderate nonlinearity. A drawback of these techniques is that the error (see Fig. 12.2) resulting from the use of a single representative stiffness in each load increment can accumulate. Hence, the total internal element forces, which are a by-product of the potentially inaccurate displacements, are not necessarily in equilibrium with the externally applied forces. Although this so-called *drift-off* error can be reduced by using a smaller load ratio $d\lambda_i$, the additional number of increments required for analyzing highly nonlinear systems may become unreasonable. In these cases the use of an iterative scheme may be more appropriate. Examples 12.2 and 12.3 illustrate these concepts by comparing results obtained by the Euler and midpoint Runge-Kutta methods with results from an iterative solution method presented in the next section.

EXAMPLE 12.2

Perform second-order elastic analyses of the steel frame shown using

1. Euler method with $d\lambda = 0.5$
2. Euler method with $d\lambda = 0.25$
3. Euler method with $d\lambda = 0.1$
4. Work Control method

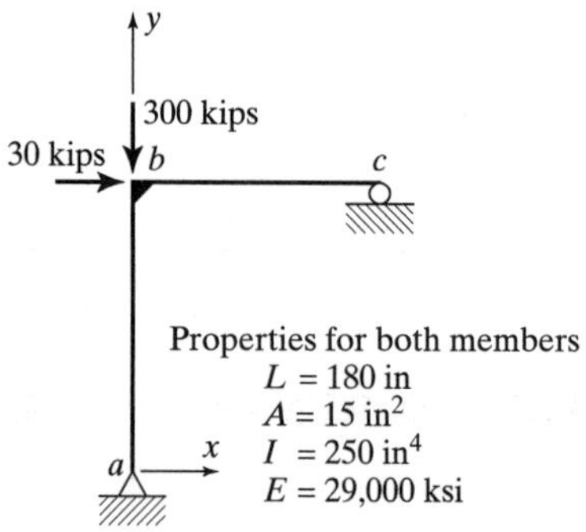

Modeling column as two elements and beam as a single element

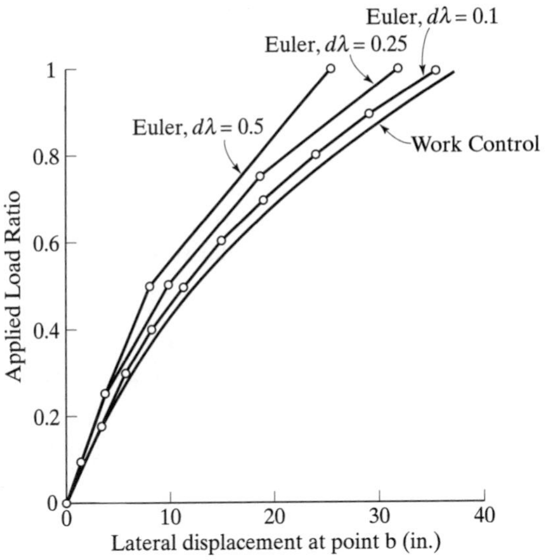

EXAMPLE 12.3

Using the frame shown in Example 12.2, perform second-order elastic analyses using

1. Euler method with $d\lambda = 0.25$
2. Midpoint Runge-Kutta method with $d\lambda = 0.5$
3. Work Control method

Modeling column as two elements and beams as a single element.

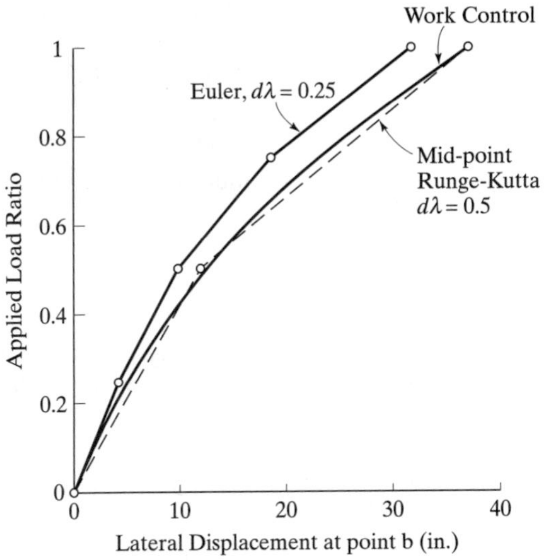

12.3 INCREMENTAL-ITERATIVE METHODS

In contrast to the single-step schemes, the iterative methods need not use a single representative stiffness in each load increment. Instead, increments are subdivided into a number of steps, each of which is a cycle in an iterative process aimed at satisfying the requirements of equilibrium to within a specified tolerance. Equation 12.3 is thus modified to

$$\{\mathbf{\Delta}_i\} = \{\mathbf{\Delta}_{i-1}\} + \sum_{j=1}^{m_i} \{\mathbf{d\Delta}_i^j\} \tag{12.10}$$

where m_i is the number of iterative steps required in the ith load increment.[2] In each step j, the unknown displacements $\{\mathbf{d\Delta}_i^j\}$ are found by solving the linear system of equations

$$[\mathbf{K}_i^{j-1}]\{\mathbf{d\Delta}_i^j\} = \{\mathbf{dP}_i^j\} + \{\mathbf{R}_i^{j-1}\} \tag{12.11}$$

where $[\mathbf{K}_i^{j-1}]$ is the stiffness evaluated using the deformed geometry and corresponding element forces up to and including the previous iteration, and $\{\mathbf{R}_i^{j-1}\}$ represents the imbalance between the existing external and internal forces. This unbalanced load vector can be calculated according to

$$\{\mathbf{R}_i^{j-1}\} = \{\mathbf{P}_i^{j-1}\} - \{\mathbf{F}_i^{j-1}\} \tag{12.12}$$

where $\{\mathbf{P}_i^{j-1}\}$ is the total external force applied and $\{\mathbf{F}_i^{j-1}\}$ is a vector of net internal forces produced by summing the existing element end forces at each global degree of freedom. Similar to Equation 12.4a, the load applied in each step is determined by

$$\{\mathbf{dP}_i^j\} = d\lambda_i^j\{\mathbf{P}_{\text{ref}}\} \tag{12.13}$$

where $d\lambda_i^j$ is now taken as the load ratio for the current iteration.

The methods presented in this section represent a sample of the different types of multistep or iterative methods available; for a comprehensive review, the reader is referred to Reference 12.1. All of these procedures will follow Equations 12.10–12.13. In discussing these methods *we will assume that the load ratio for the first step of the ith increment, $d\lambda_i^{j=1}$, has been prescribed according to either experience or an automatic strategy* (see Section 12.4). The difference between the procedures can then be defined simply by the algorithm or constraint equation used to determine the iterative load ratios $d\lambda_i^{j\geq 2}$ for the remaining steps in an increment.

To provide a summary of the iterative methods, it is convenient to replace Equation 12.11 with the following two equations:

$$[\mathbf{K}_i^{j-1}]\{\overline{\mathbf{d\Delta}_i^j}\} = \{\mathbf{P}_{\text{ref}}\} \tag{12.14a}$$

$$[\mathbf{K}_i^{j-1}]\{\overline{\overline{\mathbf{d\Delta}_i^j}}\} = \{\mathbf{R}_i^{j-1}\} \tag{12.14b}$$

Summing the solutions obtained in Equations 12.14a and 12.14b yields the displacement vector for each iteration

$$\{\mathbf{d\Delta}_i^j\} = d\lambda_i^j\{\overline{\mathbf{d\Delta}_i^j}\} + \{\overline{\overline{\mathbf{d\Delta}_i^j}}\} \tag{12.15}$$

Application of Equations 12.12–12.15 is illustrated schematically in Figure 12.4. The ith increment starts by assuming the unbalanced load $\{\mathbf{R}_i^0\} = 0$ and calculating the displacements for the first iteration $\{\mathbf{d\Delta}_i^1\}$ that correspond to an applied load $d\lambda_i^1\{\mathbf{P}_{\text{ref}}\}$ and a tangent stiffness $[\mathbf{K}_i^0]$, which is based on the results of the previous increment. After updating the deformed geometry and element forces (see Section 12.5), the second iteration begins by determining the unbalanced load $\{\mathbf{R}_i^1\}$. This load is then applied to the structure and the resulting displacements $\{\overline{\overline{\mathbf{d\Delta}_i^2}}\}$ are determined using an updated tangent stiffness $[\mathbf{K}_i^1]$. This stiffness $[\mathbf{K}_i^1]$ is then used again to calculate the displacements $\{\overline{\mathbf{d\Delta}_i^2}\}$ that correspond to an applied load $\{\mathbf{P}_{\text{ref}}\}$. The iterative load ratio

[2]A subscript will be used to indicate a particular increment and a superscript will represent an iterative step.

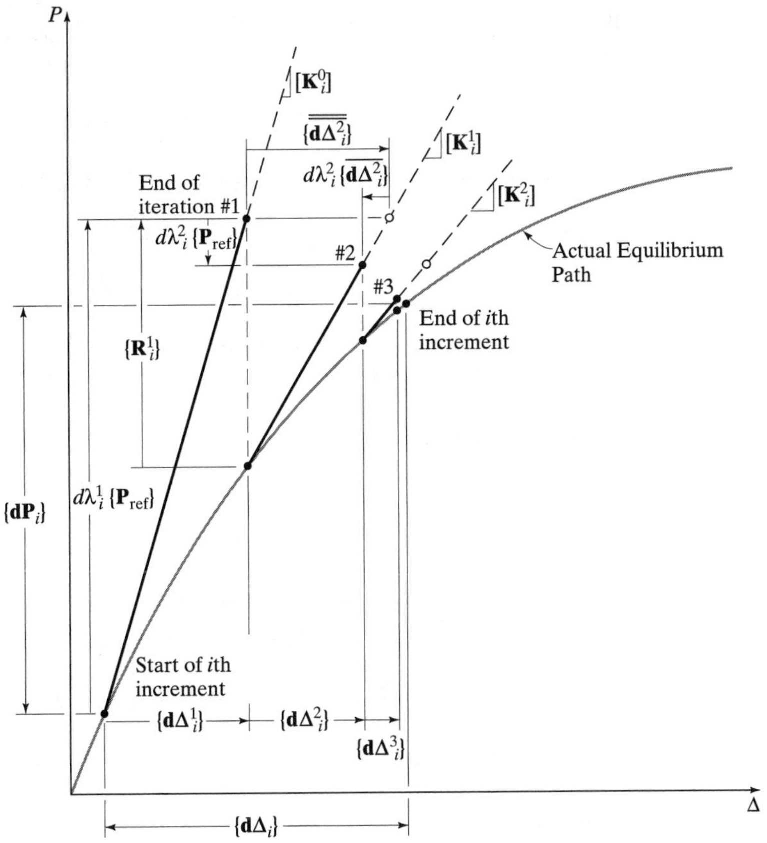

Figure 12.4 Schematic of iterations.

$d\lambda_i^2$ is computed using one of the methods of the following subsections (Equations 12.16, 12.17, 12.19 or 12.21). The net displacements for the second iteration $\{\mathbf{d\Delta}_i^2\}$ is the sum of $\{\overline{\overline{\mathbf{d\Delta}_i^2}}\}$ and $d\lambda_i^2\{\overline{\mathbf{d\Delta}_i^2}\}$. The second cycle of iteration concludes by checking convergence criteria (see Section 12.3.6). Necessary additional iterations are performed following the same procedure. The total load applied in the ith increment is $\{\mathbf{dP}_i\} = \left(\sum_{j=1}^{m_i} d\lambda_i^j\right)\{\mathbf{P}_{\text{ref}}\}$ and the resulting displacement is $\{\mathbf{d\Delta}_i\} = \sum_{j=1}^{m_i} \{\mathbf{d\Delta}_i^j\}$.

12.3.1 Load Control Method

In the load control or Newton-Raphson method, a fixed amount of load is employed in each increment. All of this load is applied in the first step $j = 1$ and additional iterations $j \geq 2$ are only performed in an attempt to satisfy equilibrium requirements. In this regard, the iterative load ratio in Equation 12.15 is given by

$$d\lambda_i^j = 0 \qquad \text{for } j \geq 2 \tag{12.16}$$

This procedure is illustrated in Figure 12.5*a*.

A significant shortcoming of this method becomes apparent when attempting to solve problems with limit points. Once a fixed load is defined in the first step, there is no way to modify the load vector should a limit point occur within the increment.

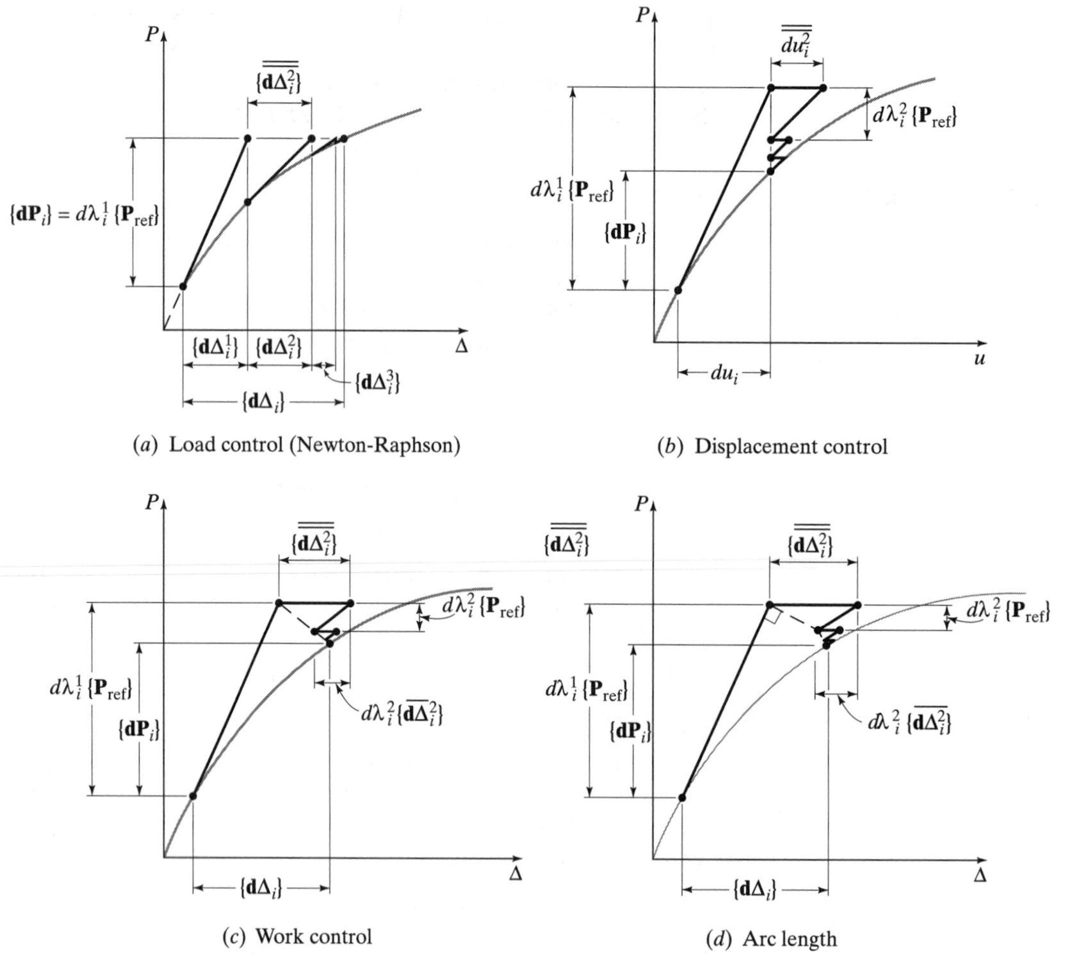

Figure 12.5 Iterative methods.

Although repeating the entire increment with a reduced initial load ratio can enable one to approach the limit point slowly, the resulting near singular nature of the stiffness matrix makes it difficult to trace the post-limit state response of a structure.

12.3.2 Displacement Control Methods

In a traditional displacement control method, the load ratio in the first step of an increment is defined so that a particular "key" displacement component will change only by a prescribed amount. The load ratios for the remaining iterations are then constrained so that this key displacement component will not change. By requiring Equation 12.15 to be zero for a particular free degree of freedom, say, du, the iterative load ratio is given by

$$d\lambda_i^j = -\overline{\overline{du_i^j}}/\overline{du_i^j} \qquad \text{for } j \geq 2 \tag{12.17}$$

where $\overline{du_i^j}$ and $\overline{\overline{du_i^j}}$ are single elements in the solution vectors of Equation 12.14. The method is illustrated in Figure 12.5*b*.

Clearly, the disadvantage of this method is having to select the key displacement component; in some cases it may be obvious but in others not. One simple technique is to use whichever degree of freedom incurred the absolute largest change during the first iteration of the current increment. Other more sophisticated techniques for overcoming this and other problems associated with the displacement control methods are available in the literature (see Refs. 12.2 and 12.3).

12.3.3 Work Control Method

As its name implies, the work control method uses both loads and displacements. In this case, the constraint condition requires a zero increment in external work for each equilibrium iteration. This is represented by

$$dW_i^j = \{\mathbf{dP}_i^j\}^{\mathrm{T}}\{\mathbf{d\Delta}_i^j\} = 0 \qquad \text{for } j \geq 2 \tag{12.18}$$

By substituting Equations 12.13 and 12.15 into Equation 12.18, the following iterative load ratio is obtained:

$$d\lambda_i^j = \frac{-\{\mathbf{P}_{\mathrm{ref}}\}^{\mathrm{T}}\{\overline{\overline{\mathbf{d\Delta}_i^j}}\}}{\{\mathbf{P}_{\mathrm{ref}}\}^{\mathrm{T}}\{\overline{\mathbf{d\Delta}_i^j}\}} \qquad \text{for } j \geq 2 \tag{12.19}$$

The work control method is illustrated in Figure 12.5*c*. Both this method and the displacement control schemes are suitable for calculating post-limit state response (see Ref. 12.3).

12.3.4 Constant Arc Length Methods

As with the work control procedure, the arc length methods iterate neither at fixed load nor at fixed displacement. One of the many constant arc length methods currently available is based on defining and further constraining an arbitrary arc length ds in each equilibrium iteration according to

$$ds^2 = \{\mathbf{d\Delta}_i^1\}^{\mathrm{T}}\{\mathbf{d\Delta}_i^j\} + d\lambda_i^1\, d\lambda_i^j = 0 \qquad \text{for } j \geq 2 \tag{12.20}$$

Substitution of Equation 12.15 into this constraint equation results in an orthogonality equation that can be used to calculate the following iterative load ratio:

$$d\lambda_i^j = \frac{-\{\mathbf{d\Delta}_i^1\}^{\mathrm{T}}\{\overline{\overline{\mathbf{d\Delta}_i^j}}\}}{\{\mathbf{d\Delta}_i^1\}^{\mathrm{T}}\{\overline{\mathbf{d\Delta}_i^j}\} + d\lambda_i^1} \qquad \text{for } j \geq 2 \tag{12.21}$$

A schematic of the constant arc length method is presented in Figure 12.5*d*. In many cases, arc length procedures can solve problems that not only exhibit limit point behavior, but also snapthrough and snapback response (see Ref. 12.4).

12.3.5 Modified Iterative Technique

In many cases the efficiency of the iterative methods can be enhanced by replacing the stiffness matrix $[\mathbf{K}_i^{j-1}]$ used in each iteration (see Eqs. 12.11 and 12.14) with the tangent stiffness $[\mathbf{K}_i^0]$ from the initial step of the increment. Although this modified iterative technique[3] will typically require more steps per load increment (see Figure 12.6), the additional effort is often offset by the substantial savings realized as a result of not having to assemble and decompose a new global stiffness matrix during each iteration. It should be noted, however, that the use of this technique may be inadequate

---

[3]This method is sometimes called a modified Newton-Raphson approach. We choose not to use this description since the technique can be applied to solutions schemes other than the Newton-Raphson method.

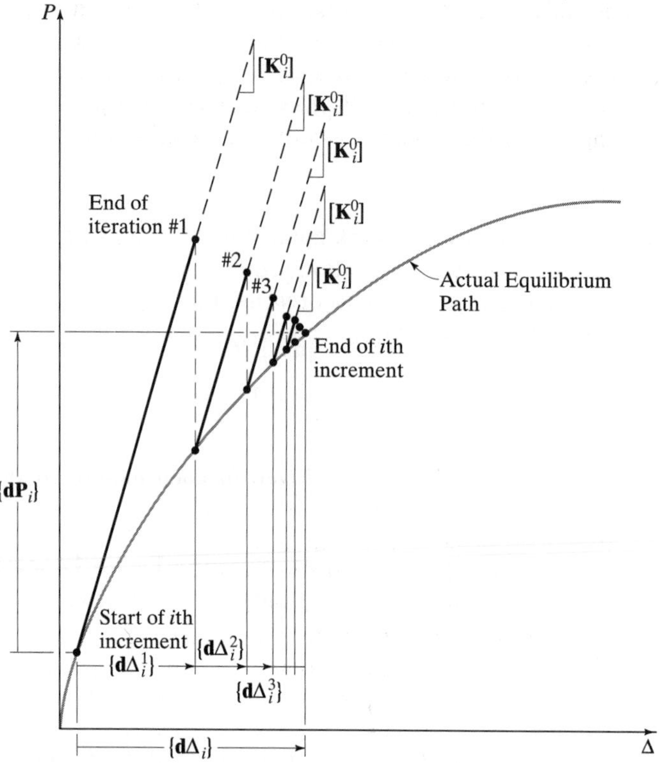

Figure 12.6 Modified iterative approach.

for the analysis of structures that exhibit extreme deformations or significant stiffening behavior as a result of large internal tensile forces.

12.3.6 Convergence Criteria

We now turn our attention to the determination of the appropriate number of iterations needed to satisfy the requirements of equilibrium. Given that the deformations in each iteration are at the heart of any equilibrium requirement, it seems reasonable to base convergence criteria on these calculated values. In Reference 12.5, three alternative norms are suggested for testing convergence in this way:

Modified absolute norm:

$$\|\varepsilon\| = \frac{1}{N}\sum_{k=1}^{N}\left|\frac{d\Delta_k}{\Delta_{\text{ref}}}\right| \tag{12.22}$$

Modified Euclidean norm:

$$\|\varepsilon\| = \sqrt{\frac{1}{N}\sum_{k=1}^{N}\left(\frac{d\Delta_k}{\Delta_{\text{ref}}}\right)^2} \tag{12.23}$$

Maximum norm:

$$\|\varepsilon\| = \max_{1\le k\le N}\left|\frac{d\Delta_k}{\Delta_{\text{ref}}}\right| \tag{12.24}$$

In these equations, N is the total number of unknown displacement components, $d\Delta_k$ is the kth element of the incremental displacement vector $\{\mathbf{d\Delta}_i^j\}$, and depending on the units of $d\Delta_k$, the value of Δ_{ref} is taken as either the largest component of translation or rotation within the total displacement vector $\{\mathbf{\Delta}_i\}$.

Using any of these norms, a convergence criterion of

$$\|\varepsilon\| \leq \zeta \tag{12.25}$$

can be used where the acceptable tolerance ζ is usually on the order of 10^{-2} to 10^{-6}, depending on the desired accuracy. As an alternative to comparing displacements, reasonable equilibrium criteria may also be based on assessing the unbalanced load vector (Ref. 12.6) or by studying the increment of internal work (Ref. 12.4).

12.4 AUTOMATIC LOAD INCREMENTATION

The size of the load ratio employed in each increment of the analysis can have a dramatic effect on the solution. In the single-step methods, proper selection of $d\lambda$ is the only means for controlling drift-off error. In the iterative schemes, a poor definition of $d\lambda_i^1$ could result in the solution not converging within a practical number of iterations. In both methods, an excessively small load ratio may result in significant computational effort with a negligible increase in accuracy. To provide assistance in determining $d\lambda_i^1$, there are several types of automated procedures that may be employed. Two of these schemes are presented. The first may be applied to both the single-step and iterative methods. The second is intended only for iterative procedures. In both schemes, it is assumed that the size of the load ratio for the first increment $d\lambda_1$ or the first iteration of the first increment $d\lambda_1^1$ has been prescribed by the analyst. As stated earlier, a value of 10%–20% of the anticipated maximum applied load is generally satisfactory. Section 12.6 also provides load ratio size constraints that are required specifically for a material nonlinear analysis based on the plastic hinge method.

12.4.1 Change in Stiffness

The load ratio $d\lambda$ at any point in the analysis should reflect the corresponding state of nonlinearity. A simple scalar measure of the degree of nonlinearity can be obtained from a *current stiffness parameter* given by (Ref. 12.7)

$$S_i = \frac{\{\mathbf{d\Delta}_1^1\}^{\mathrm{T}}\{\mathbf{P}_{\mathrm{ref}}\}}{\{\mathbf{d\Delta}_i^1\}^{\mathrm{T}}\{\mathbf{P}_{\mathrm{ref}}\}} \tag{12.26}$$

Since the parameter S_i will always have an initial value of unity, stiffening or softening of the structural system will be indicated by values greater than or less than one, respectively. With the exception of bifurcation points, S_i will become zero at a limit point. For a more sophisticated and perhaps more rigorous measure of nonlinearity, Reference 12.3 provides details on a *generalized stiffness parameter*.

Using either Equation 12.26 or the stiffness parameter of Reference 12.3, load ratios for a single-step method or the first cycle of an iterative method can be obtained from

$$d\lambda_i^1 = \pm d\lambda_1^1\,|S_i|^{\gamma} \tag{12.27}$$

where $d\lambda_1^1$ is the load ratio prescribed at the start of the analysis, and the exponent γ typically equals 0.5 to 1 (Ref. 12.1). Strategies for selecting the appropriate sign in Equation 12.27 will be discussed in Section 12.7.

12.4.2 Number of Iterations

In the multistep methods the number of iterations in each increment needed to satisfy equilibrium requirements is usually proportional to the size of the initial load ratio $d\lambda_i^1$. In this regard, the following parameter can be employed:

$$\psi_{i-1} = \frac{N_d}{N_{i-1}} \tag{12.28}$$

where N_d is the number of iterations the analyst desires for convergence and N_{i-1} is the number of iterations required for convergence in the previous or $i-1$ increment. Using this parameter, the initial load ratio is obtained by

$$d\lambda_i^1 = \pm d\lambda_{i-1}^1 \, |\psi_{i-1}|^\gamma \tag{12.29}$$

where the exponent γ typically ranges from 0.5 to 1.

Because the values provided in Equations 12.26 and 12.28 may result in initial load ratios significantly larger than 1, it is always desirable to prescribe an absolute maximum limit on the initial load step. It should be noted that other similar but more sophisticated schemes based on fixed increments of displacement, external work, or arc-length are available in the literature (see Ref. 12.1).

12.5 ELEMENT RESULT CALCULATIONS

12.5.1 Updating Deformed Geometry

To employ the stiffness relations of Chapters 8 and 9 in conjunction with the incremental analysis procedures presented in Sections 12.1–12.3 requires that the structural geometry include all accumulated deformations. For moderate displacement problems, this can be achieved by completing each step of the single-step or iterative method with a process of coordinate updating. That is, the coordinates of each node are modified or updated to include the translational displacement components that occurred during the step.

In updating the coordinates of nodes or element ends, we are representing the deformed geometry of the structure by changing the position and hence the orientation of each element with respect to the global coordinate system. This repositioning should also account for any changes in the element's web orientation angle (see Example 5.5). The web orientation may change in each step and for elements subjected to twisting it will vary along the element length. For moderate displacement problems this effect may be approximated satisfactorily by the average of the end rotations taken about the element's local x-axis, $(d\theta_{x1} + d\theta_{x2})/2$. In all cases the element is assumed to be straight at the start of the next step or iterative cycle.

12.5.2 Force Recovery

In linear analysis, element forces are conventionally calculated by substitution of displacements obtained from the global analysis in the element stiffness equation, Equation 3.11, which for present purposes may be written

$$\{\mathbf{F}\} = [\mathbf{k}]\{\mathbf{\Delta}\} \tag{12.30}$$

The components of $\{\mathbf{\Delta}\}$ may be in either the global axes directions or transformed to the element's local axes. In the first case $[\mathbf{k}]$ must be expressed in global coordinates, $\{\mathbf{F}\}$ will be likewise, and a further transformation will be required to convert $\{\mathbf{F}\}$ to the vector of axial forces, end shears, and moments that is usually needed for design. In the second case, [**k**] must be in local coordinates and it follows that {**F**} will be also.

We may illustrate the extension of this conventional procedure to nonlinear analysis by considering the element of Figure 12.7. The first sketch, Figure 12.7*a*, illustrates the element's orientation and the forces acting on it at the start of a linear step of a single-step or iterative procedure. Figure 12.7*b* shows its orientation at the end of the step with the corresponding forces referred to the initial local axes. Figure 12.7*c* is of the

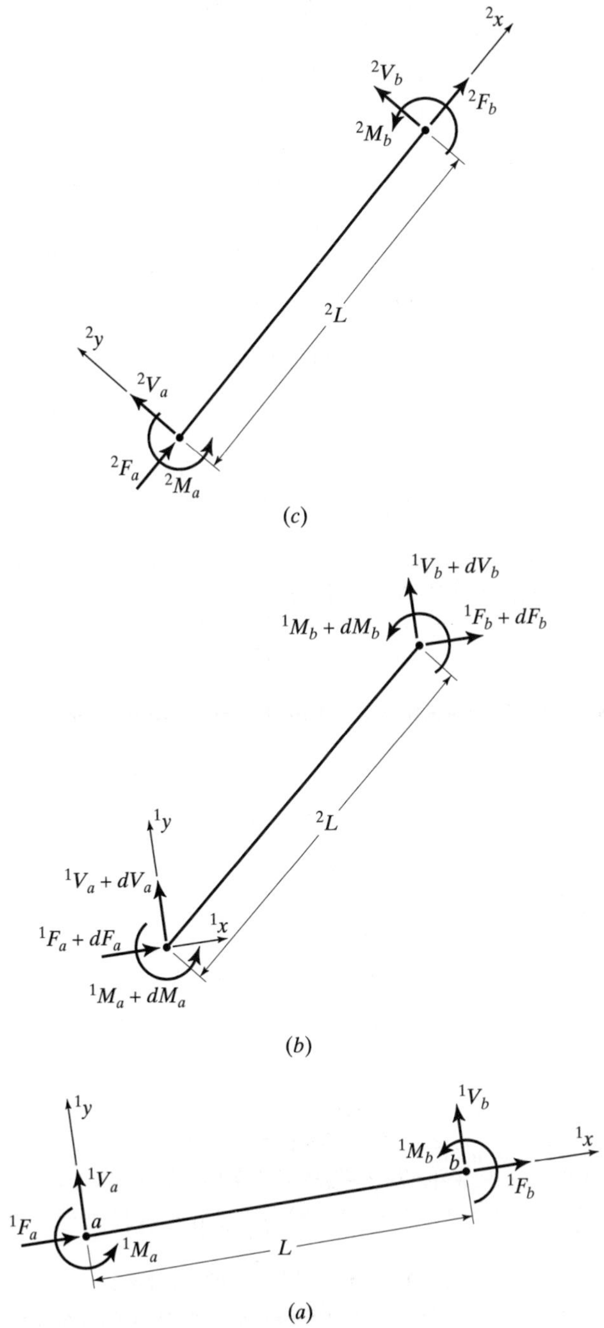

Figure 12.7 Element end forces.

same state with the forces transformed to the new local axes. This is the form needed for use in design as well as for updating the tangent stiffness matrix. To simplify the notation we have omitted all subscripts and superscripts except those needed to identify the ends of the element and its orientation at the beginning and end of a linear step. In discussing this figure the following vectors will be used:

1. The forces at the start of the step:

$$\{{}^1\mathbf{F}\} = \lfloor {}^1F_a \quad {}^1V_a \quad {}^1M_a \quad {}^1F_b \quad {}^1V_b \quad {}^1M_b \rfloor^{\mathrm{T}} \tag{12.31a}$$

2. The increment of forces:

$$\{\mathbf{dF}\} = \lfloor dF_a \quad dV_a \quad dM_a \quad dF_b \quad dV_b \quad dM_b \rfloor^{\mathrm{T}} \tag{12.31b}$$

3. The forces at the end of the step referred to the initial configuration:

$$\{{}^2_1\mathbf{F}\} = \{{}^1\mathbf{F}\} + \{\mathbf{dF}\} \tag{12.31c}$$

4. The forces at the end of the step referred to the final configuration:

$$\{{}^2\mathbf{F}\} = \lfloor {}^2F_a \quad {}^2V_a \quad {}^2M_a \quad {}^2F_b \quad {}^2V_b \quad {}^2M_b \rfloor^{\mathrm{T}} \tag{12.31d}$$

5. The displacement increments transformed to the initial configuration:

$$\{\mathbf{d\Delta}\} = \lfloor du_a \quad dv_a \quad d\theta_a \quad du_b \quad dv_b \quad d\theta_b \rfloor^{\mathrm{T}} \tag{12.31e}$$

Following Equation 8.2, we have $\{\mathbf{dF}\} = [\mathbf{k_t}]\{\mathbf{d\Delta}\}$ in which $[\mathbf{k_t}]$ is the element tangent stiffness matrix in the local coordinates at the start of the step. Thus for nonlinear elastic analysis Equation 12.31c becomes $\{{}^2_1\mathbf{F}\} = \{{}^1\mathbf{F}\} + [\mathbf{k_e} + \mathbf{k_g}]\{\mathbf{d\Delta}\}$ where the tangent stiffness matrix has been separated into its elastic and geometric components, both calculated at the start of the step. $\{{}^2\mathbf{F}\}$ may be obtained from $\{{}^2_1\mathbf{F}\}$ by the transformation

$$\{{}^2\mathbf{F}\} = [{}^2_1\mathbf{\Gamma}]\{{}^2_1\mathbf{F}\} \tag{12.32}$$

in which $[{}^2_1\mathbf{\Gamma}]$ is the element transformation matrix[4] (Ref. 12.8). Use of Equation 12.32 is illustrated in Example 12.4.

EXAMPLE 12.4

Given the geometry and force information shown, use Equation 12.32 to calculate the element's forces at the end of the current step.
Forces at start of step:

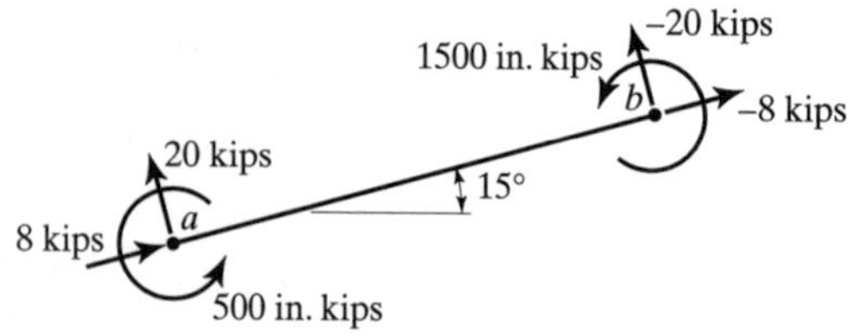

[4]Rather than constructing $[{}^2_1\mathbf{\Gamma}]$ directly, it is generally more convenient to calculate it as the product of a transformation from the initial local coordinates to the global system and one from the global system to the final local coordinates. Recognizing the orthogonality of the transformation matrices, it can be demonstrated readily that the result is

$$\{{}^2\mathbf{F}\} = [{}^2_{\mathbf{x}}\mathbf{\Gamma}][{}^1_{\mathbf{x}}\mathbf{\Gamma}]^{\mathrm{T}}\{{}^2_1\mathbf{F}\} \tag{12.33}$$

In these transformation matrices the subscript **x** refers to the global system.

Increment of element forces: $\{\mathbf{dF}_i\} = \lfloor 2 \quad 5 \quad 100 \quad -2 \quad -5 \quad 400 \rfloor^{\mathrm{T}}$
Forces at end of step referred to final configuration:

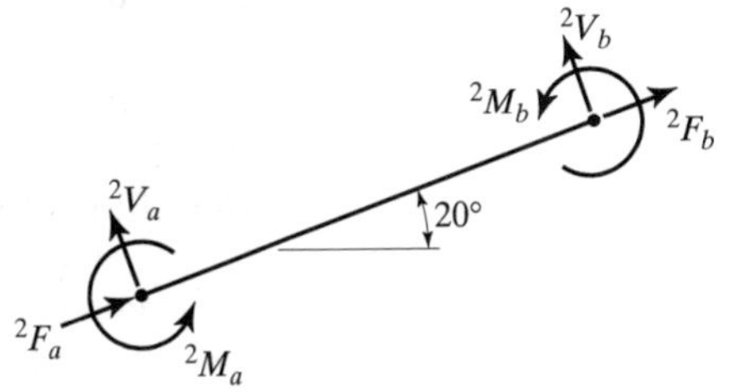

$$\{{}^1\mathbf{F}\} = \lfloor 8 \quad 20 \quad 500 \quad -8 \quad -20 \quad 1500 \rfloor^{\mathrm{T}}$$

$$\{{}^2_1\mathbf{F}\} = \{{}^1\mathbf{F}\} + \{\mathbf{dF}\} = \lfloor 10 \quad 25 \quad 600 \quad -10 \quad -25 \quad 1900 \rfloor^{\mathrm{T}}$$

$$\text{with } {}^1\phi = 15°,\ [{}^1\boldsymbol{\gamma}] = \begin{bmatrix} \cos 15 & \sin 15 & 0 \\ -\sin 15 & \cos 15 & 0 \\ 0 & 0 & 1 \end{bmatrix} = \begin{bmatrix} 0.966 & 0.259 & 0 \\ -0.259 & 0.966 & 0 \\ 0 & 0 & 1 \end{bmatrix}$$

$$[{}^1\boldsymbol{\Gamma}] = \begin{bmatrix} [{}^1\boldsymbol{\gamma}] & \mathbf{0} \\ \mathbf{0} & [{}^1\boldsymbol{\gamma}] \end{bmatrix}$$

$$\text{with } {}^2\phi = 20°,\ [{}^2\boldsymbol{\gamma}] = \begin{bmatrix} 0.940 & 0.342 & 0 \\ -0.342 & 0.940 & 0 \\ 0 & 0 & 1 \end{bmatrix} \text{ and } [{}^2\boldsymbol{\Gamma}] = \begin{bmatrix} [{}^2\boldsymbol{\gamma}] & \mathbf{0} \\ \mathbf{0} & [{}^2\boldsymbol{\gamma}] \end{bmatrix}$$

From Equations 12.32 and 12.33

$$\{{}^2\mathbf{F}\} = [{}^2_1\boldsymbol{\Gamma}]\{{}^2_1\mathbf{F}\} = [{}^2\boldsymbol{\Gamma}][{}^1\boldsymbol{\Gamma}]^{\mathrm{T}} \begin{Bmatrix} 10 \\ 25 \\ 600 \\ -10 \\ -25 \\ 1900 \end{Bmatrix} = \begin{Bmatrix} 12.14 \\ 24.03 \\ 600 \\ -12.14 \\ -24.03 \\ 1900 \end{Bmatrix}$$

This method of force recovery is straightforward in application and conventional in the sense that it is a direct extension of the standard linear elastic procedure. It will be recognized, however, that there is nothing in it to distinguish between displacements resulting from rigid body motion and those due to deformation: It is an approximate approach to force recovery. Methods that do distinguish between the two sources of displacement are discussed in Appendix B and the limitations of this conventional one are explored. It is shown that, for small strain, moderate displacement analysis, the rigid body motion effects are normally small. For this reason, and because it has been found adequate in many examples of both elastic and inelastic nonlinear analysis, conventional force recovery in conjunction with a single step method is considered suitable for general framework analysis. In cases of highly nonlinear structures, an iterative analysis method and the force recovery procedure defined in Appendix B are recommended.

12.6 PLASTIC HINGE CONSTRAINTS

For the material analysis model developed in Section 10.2, the load ratio is typically reduced to prevent plastic hinge formation from occurring within a load increment, and thereby avoid any accompanying abrupt changes in stiffness. To calculate a load ratio that results in plastic hinges forming only at the end of an increment requires a comparison of the total element forces at the end of each step with the plastic hinge criterion employed (e.g., the yield surface of Equation 10.18). If no hinges form, the solution scheme continues. In some cases, as illustrated in Figure 12.8, one or more sets of element end forces will breach the yield surface and signify the formation of plastic hinges at a fraction τ of the current load ratio. If it is the first step of the increment, the load ratio should be reset to the product of the minimum of these fractions $\tau_{\min}$ and the current load ratio. In situations where hinges are detected in analyses that follow the first step, it is suggested that the entire increment be repeated using an appropriately reduced load ratio.

Calculation of the fraction τ for a given set of element end forces reduces to a relatively simple rootfinding problem. For example, Equation 10.7 can be rewritten as

$$\Phi(p + \tau\, dp, m + \tau\, dm) - 1 = 0 \tag{12.34}$$

where p, dp, m, and dm are the known quantities of nondimensionalized forces and moments shown in Figure 12.8, and τ is the unknown root. To ensure proper convergence, a rootfinding method that always brackets the root should be used. The method of false position, or *regula falsi*, is recommended. Starting with initial guesses of $\tau_l = 0$ and $\tau_u = 1$, this technique provides an improved estimate of the root τ_r according to

$$\tau_r = \tau_u - \frac{[\Phi(\tau_u) - 1](\tau_l - \tau_u)}{\Phi(\tau_l) - \Phi(\tau_u)} \tag{12.35}$$

where, for example, $\Phi(\tau_u) = \Phi(p + \tau_u dp, m + \tau_u dm)$. Using Equation 12.35, successive iterations would continue with τ_r replacing whichever of the two guesses, τ_l or τ_u, provides a function value $\Phi(\tau) - 1$ with the same sign as $\Phi(\tau_r) - 1$. The use of Equations 12.34 and 12.35 is illustrated in Example 12.5.

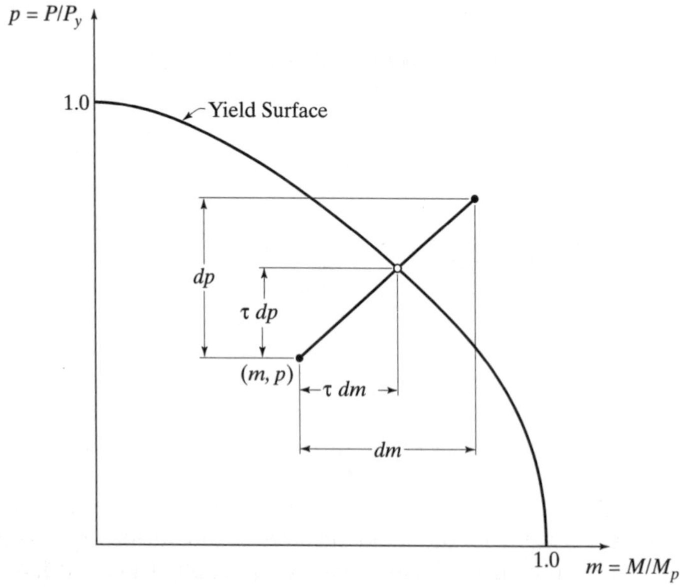

Figure 12.8 Plastic hinge formation at fraction τ of attempted load increment.

EXAMPLE 12.5

Given the non-dimensionalized quantities $p = 0.3$, $dp = 0.05$, $m = 0.8$, and $dm = 0.1$, use Equation 12.35 to calculate the fraction τ so that Equation 12.34 is satisfied to four decimal places. Assume $\Phi(p, m) = p^2 + m^2 + 3.5p^2m^2$.

First iteration:

$$\tau_l = 0 \quad \Phi(\tau_l) = \Phi(p + \tau_l dp, m + \tau_l dm) = 0.9316$$

$$\tau_u = 1 \quad \Phi(\tau_u) = \Phi(p + \tau_u dp, m + \tau_u dm) = 1.2798$$

From Equation 12.35, $\tau_r = 1 - \dfrac{(1.2798 - 1)(0 - 1)}{(0.9316 - 1.2798)} = 0.1964$

$$\Phi(\tau_r) = \Phi(p + \tau_r dp, m + \tau_r dm) = 0.9935$$

Second iteration:

$$\Phi(\tau_l) - 1 = -0.0684 \qquad \Phi(\tau_u) - 1 = 0.2798$$

$$\Phi(\tau_r) - 1 = -0.0065$$

since $\Phi(\tau_r) - 1$ and $\Phi(\tau_l) - 1$ are both negative, let $\tau_l = \tau_r$

$$\tau_l = 0.1964, \ \Phi(\tau_l) = 0.9935$$

$$\tau_u = 1.000, \ \Phi(\tau_u) = 1.2798$$

$$\tau_r = 1 - \frac{(1.2798 - 1)(0.1964 - 1)}{(0.9935 - 1.2798)} = 0.2146$$

$$\Phi(\tau_r) = 0.9994$$

Third iteration:

$$\Phi(\tau_l) - 1 = -0.0065 \qquad \Phi(\tau_u) - 1 = 0.2798$$

$$\Phi(\tau_r) - 1 = -0.0006$$

since $\Phi(\tau_r) - 1$ and $\Phi(\tau_l) - 1$ are both negative, let $\tau_l = \tau_r$

$$\tau_r = 1 - \frac{(1.2798 - 1)(0.2146 - 1)}{(0.9994 - 1.2798)} = 0.2163$$

$\Phi(\tau_r) = 0.9999$ After 3 iterations, $\tau = 0.2163$ satisfies Equation 12.34.

Another situation in a material nonlinear analysis that may require a reduction in the size of the load ratio occurs when plastic hinges exist at the start of an increment. The use of a plastic reduction matrix will constrain the element forces at a plastic hinge to move tangent to the yield surface. If the surface is curved, these forces will not remain on the surface with a change in applied load and, as a result, may exceed a tolerable amount of yield surface drift (see Figure 12.9, forces move from point A to B). To avoid this it may be necessary to once again reduce the current load ratio by a

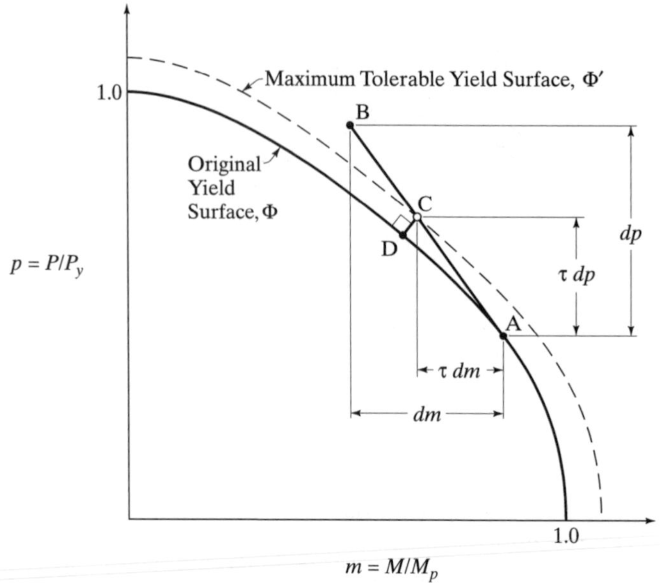

Figure 12.9 Controlling yield surface drift.

factor of τ, so that the increment in forces move from point A to C. This can be achieved by first isotropically expanding the original theoretical yield surface Φ by a prescribed amount to obtain a maximum tolerable surface Φ'. The above root-finding strategy can then be employed on this expanded surface Φ' to find the permissible fraction τ. Note that a yield surface can be isotropically expanded by artificially increasing the yield strength of the material.

Obviously, for the analysis to continue the forces residing on the tolerable yield surface must be returned to the original yield surface. For practical structures, these small deviations in force can probably be neglected without a significant impact on overall results. In any case, the drifted force point should be returned to point D on the yield surface via a path that is approximately normal to the surface (see Fig. 12.9). More sophisticated one-step and iterative procedures for circumventing problems of yield surface drift are found in References 12.9 and 12.10.

In addition to reaching the yield surface or moving tangent to it, the distribution of element forces may change under subsequent applied loading to the extent that one or more sections with plastic hinges may unload elastically (see Section 10.2.1). In this event it is suggested that the entire load increment be repeated using elastic properties for these sections.

12.7 LIMIT POINT AND POST-LIMIT ANALYSIS

In Section 8.1.2 a physical description of a stability limit point was given as *the point at which a system's capacity for resistance to additional load is exhausted and continued deformation results in a decrease in load-resisting capacity* (see Figure 12.10). It may be characterized numerically as a point in the analysis at which the global stiffness

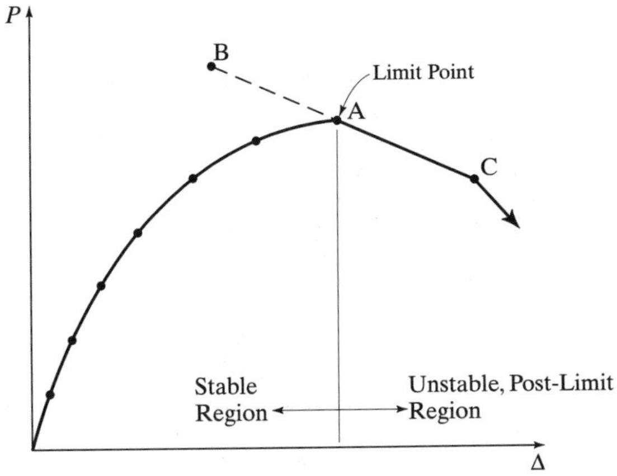

Figure 12.10 Limit point and tracing post-limit response.

matrix ceases to be positive definite.[5] Limit points can be detected by the presence of one or more nonpositive coefficients on the main diagonal of the stiffness matrix following a Gauss or a root-free Cholesky decomposition (see Section 11.2).

Once a limit point is reached, several methods are available for continuing the analysis into the post-limit region. A simple approach, given in Reference 12.11, is to continue using a positive initial load ratio $d\lambda_i^1$ and obtain a solution to the nonpositive definite system by performing the standard decomposition and substitution steps provided in Section 11.2. As shown in Figure 12.10, applying this action in the post-limit region results in moving from point A to B. Reversing the sign of the load increment and corresponding displacements correctly shifts the results from point A to C. This process is repeated until either the analysis is complete or the stiffness matrix returns to being positive definite. If the latter occurs, this stiffening behavior is represented by reversing the sign of the initial load ratio back to positive.

Note that the global stiffness matrix will be singular at the theoretical limit point, thereby making it impossible to continue directly into tracing post-limit state behavior. Although a zero-stiffness limit point is rarely encountered in the incremental methods commonly employed, the approach to it indicates that the numerical accuracy of the solution may diminish in the neighborhood of a limit point because of ill-conditioning. Techniques for detecting, controlling, and correcting for this situation were presented in Section 11.5.

As indicated in Figure 12.11, a bifurcation state can occur prior to reaching a smooth limit point. In these situations the response of the structure will branch off from its initial stable mode to a different stable or unstable mode with little to no numerical forewarning. Detection and treatment of bifurcation points is discussed in Reference 12.12. However, beyond recognizing this possibility, it is stressed that few structural systems display ideal bifurcation. The effect of inevitable imperfections in geometry, material, and loading greatly reduce the probability of its occurrence. As illustrated in several of the examples of Chapters 9 and 10, reasonable simulation of actual behavior can generally be obtained by a nonlinear computational analysis that incorporates small imperfections within the model.

[5] The global stiffness matrix may also become singular when all elements that are connected at a common node have formed a plastic hinge at this node. To prevent this local joint instability from controlling the overall limit point of the structure, an elastic element with a relatively small stiffness is often used in parallel to any elements with yielded ends.

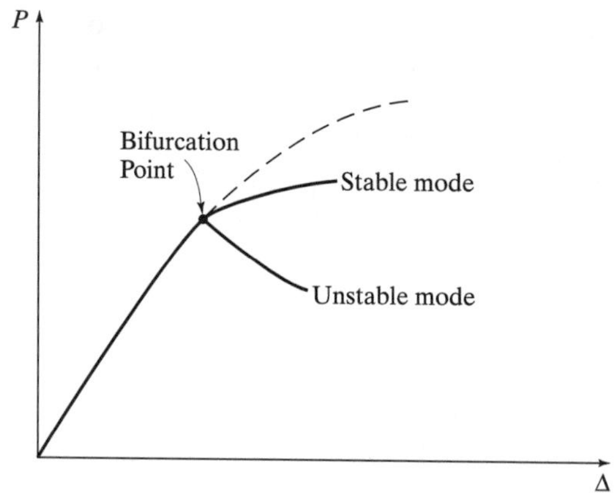

Figure 12.11 Bifurcation state.

12.8 CRITICAL LOAD ANALYSIS—AN EIGENPROBLEM

In Chapter 8 general concepts regarding the stability of structural systems were introduced. As shown in Chapters 9 and 10 and Section 12.7, the limit point or point of neutral equilibrium that usually identifies the transition from a stable to an unstable structure can be determined from an incremental analysis. Mathematically, this point may be characterized by Equation 11.50 when the condition number of the tangent stiffness matrix becomes infinite and small changes to the load vector will result in unbounded changes to the displacement vector. That is, the current system of equilibrium equations is singular and displacement vectors other than the null vector satisfy

$$[\mathbf{K}_{\mathbf{t},ff}]\{\boldsymbol{\Delta}_f\} = \{\mathbf{0}\} \tag{12.36}$$

in which $[\mathbf{K}_{\mathbf{t},ff}]$ is the tangent stiffness matrix and the subscript ff indicates that only the free degrees of freedom are included.

As an alternative to performing a complete incremental analysis, insight into the stability of many structural systems can be obtained by neglecting material nonlinear behavior (a qualification that will be modified below) and assuming that the relative distribution of internal force is the same at all ratios of applied load. Combining these simplifications with the fact that all of the element geometric stiffness matrices are linear functions of their end forces (see Eqs. 9.17–9.19) permits Equation 12.36 to be rewritten as

$$[\mathbf{K}_{\mathbf{e},ff} + \lambda\hat{\mathbf{K}}_{\mathbf{g},ff}]\{\boldsymbol{\Delta}_f\} = \{\mathbf{0}\} \tag{12.37}$$

or

$$[\mathbf{K}_{\mathbf{e},ff}]\{\boldsymbol{\Delta}_f\} = \lambda[-\hat{\mathbf{K}}_{\mathbf{g},ff}]\{\boldsymbol{\Delta}_f\} \tag{12.38}$$

where $[-\hat{\mathbf{K}}_{\mathbf{g},ff}]$ is calculated from element forces that were obtained from a linear elastic analysis for a known reference load $\{\mathbf{P}_{\text{ref}}\}$, and λ represents the ratio of the elastic critical load to this reference load.

Equation 12.38 is the general form of an eigenvalue problem. In most cases it is easier to solve this type of problem than to perform an incremental analysis. Three basic approaches for solving the problem as it relates to a stability or critical load analysis are provided in the following sections. The first two are based on conversion of the general eigenproblem of Equation 12.38 to one of the standard form $[\mathbf{H}]\{\mathbf{Y}\} = \omega\{\mathbf{Y}\}$. The third method, which is deduced from concepts presented in the first two, provides a solution scheme that does not require conversion.

Before proceeding, however, we note that the above process can be modified to include some degree of material nonlinear behavior (see Section 10.3). This will affect Equation 12.38 in three ways; (1) $[\mathbf{K}_{\mathbf{e},ff}]$ is replaced by $[\mathbf{K}_{\mathbf{t},ff}]$ reflecting any reduction in material stiffness at the critical load, (2) a nonlinear incremental analysis using $[\mathbf{K}_{\mathbf{t},ff}]$ is required to determine internal force distribution, and (3) the internal element forces and hence $[-\hat{\mathbf{K}}_{\mathbf{g},ff}]$ are no longer linear functions of the reference load $\{\mathbf{P}_{\text{ref}}\}$. With these points in mind, the objective of an inelastic critical load analysis is to determine the minimum load ratio $\bar{\lambda}$ that will satisfy the following equation with $\lambda = 1$:

$$[\mathbf{K}(\bar{\lambda}\mathbf{P}_{\text{ref}})_{\mathbf{t},ff} + \lambda\mathbf{K}(\bar{\lambda}\mathbf{P}_{\text{ref}})_{\mathbf{g},ff}]\{\mathbf{\Delta}_f\} = \{\mathbf{0}\} \tag{12.39}$$

This can be done systematically by prescribing different values of $\bar{\lambda}$ and performing the associated eigenvalue analyses until the load ratio λ in Equation 12.39 takes on a value of unity. The minimum load ratio $\bar{\lambda}$ meeting the requirements of Equation 12.39 represents the ratio of the inelastic critical load to the reference load.

12.8.1 Reduction to Standard Form

The standard form of a matrix eigenvalue problem is

$$[\mathbf{H}]\{\mathbf{Y}\} = \omega\{\mathbf{Y}\} \tag{12.40}$$

where $[\mathbf{H}]$ is a known $n \times n$ matrix, ω is an unknown constant, and $\{\mathbf{Y}\}$ is a vector with n unknowns. The objective in solving an eigenvalue problem is to go beyond the trivial solution $\{\mathbf{Y}\} = \{\mathbf{0}\}$ and determine those pairs of eigenvalues ω_i and eigenvectors $\{\mathbf{Y}_i\}$ that also satisfy Equation 12.40. There will be in general n different eigenpairs that satisfy this equation. In all cases, eigenvectors can only be determined to within a scalar multiple since both $\{\mathbf{Y}_i\}$ and $c\{\mathbf{Y}_i\}$ satisfy Equation 12.40, where c is a scalar quantity. It is important to note that a symmetric matrix $[\mathbf{H}]$ will always yield real eigenvalues and real orthogonal eigenvectors. This fact considerably simplifies the computational details associated with the more practical methods for solving eigenvalue problems.

With this introduction, these concepts will be applied specifically to critical load analysis. We begin by converting Equation 12.38 to standard form. One way of doing this is to premultiply both of its sides by the inverse of $[\mathbf{K}_{\mathbf{e},ff}]$ yielding

$$[\mathbf{K}_{\mathbf{e},ff}]^{-1}[-\hat{\mathbf{K}}_{\mathbf{g},ff}]\{\mathbf{\Delta}_f\} = \frac{1}{\lambda}\{\mathbf{\Delta}_f\} \tag{12.41}$$

Thus, this equation takes on the form of Equation 12.40 with $[\mathbf{H}] = [\mathbf{K}_{\mathbf{e},ff}]^{-1}[-\hat{\mathbf{K}}_{\mathbf{g},ff}]$ and $\omega = 1/\lambda$. Unfortunately, the product $[\mathbf{K}_{\mathbf{e},ff}]^{-1}[-\hat{\mathbf{K}}_{\mathbf{g},ff}]$ will not in general be symmetric. Since from a computational point of view a symmetric coefficient matrix $[\mathbf{H}]$ is desirable, an alternative conversion is presented.

Considering that $[\mathbf{K}_{\mathbf{e},ff}]$ is always positive definite, it can be factored according to the Cholesky method (see Section 11.2.2) as

$$[\mathbf{K}_{\mathbf{e},ff}] = [\mathbf{L}][\mathbf{L}]^{\mathrm{T}} \tag{12.42}$$

Noting also that $[\mathbf{L}^{-1}]^{\mathrm{T}} = [\mathbf{L}^{\mathrm{T}}]^{-1}$ and $[\mathbf{L}]^{-1}[\mathbf{L}] = [\mathbf{L}^{\mathrm{T}}]^{-1}[\mathbf{L}]^{\mathrm{T}} = [\mathbf{I}]$ where $[\mathbf{I}]$ is the identity matrix, Equation 12.38 is rewritten as

$$[\mathbf{L}]^{-1}[-\hat{\mathbf{K}}_{\mathbf{g},ff}][\mathbf{L}^{\mathrm{T}}]^{-1}[\mathbf{L}]^{\mathrm{T}}\{\mathbf{\Delta}_f\} = \omega[\mathbf{L}]^{\mathrm{T}}\{\mathbf{\Delta}_f\} \tag{12.43}$$

where $\omega = 1/\lambda$. Equation 12.43 can then be expressed in the standard form of Equation 12.40

$$[\mathbf{H}]\{\mathbf{Y}\} = \omega\{\mathbf{Y}\} \tag{12.40}$$

where

$$[\mathbf{H}] = [\mathbf{L}]^{-1}[-\hat{\mathbf{K}}_{\mathbf{g},ff}][\mathbf{L}^{\mathrm{T}}]^{-1} \tag{12.44a}$$

$$\{\mathbf{Y}\} = [\mathbf{L}]^{\mathrm{T}}\{\boldsymbol{\Delta}_f\} \tag{12.44b}$$

Since $[\hat{\mathbf{K}}_{\mathbf{g},ff}]$ is symmetric, it can be shown that $[\mathbf{H}]$ will also be symmetric, and we have met an essential objective of converting Equation 12.38 to standard form. Example 12.6 illustrates this process.

EXAMPLE 12.6

For the structural system shown, prepare the eigenproblem in standard form.

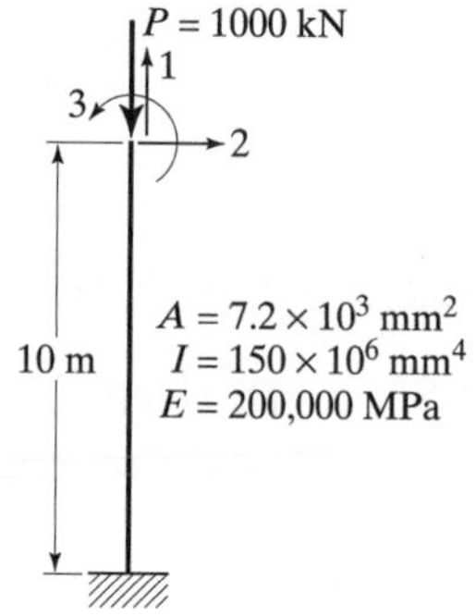

Using the degree-of-freedom labels provided

$$[\mathbf{K}_{\mathbf{e},ff}] = \begin{bmatrix} 144 & 0 & 0 \\ 0 & 0.36 & 1800 \\ 0 & 1800 & 1.2 \times 10^7 \end{bmatrix}$$

From Equation 11.15: $[\mathbf{L}] = \begin{bmatrix} 12 & 0 & 0 \\ 0 & 0.6 & 0 \\ 0 & 3000 & 1732.05 \end{bmatrix}$

From Equation 12.46: $[\mathbf{L}]^{-1} = \begin{bmatrix} 0.0833 & 0 & 0 \\ 0 & 1.67 & 0 \\ 0 & -2.89 & 5.77 \times 10^{-4} \end{bmatrix}$

With $P_{\text{ref}} = 1000$ kN, $F_{x2} = -1000$ kN

From Equation 9.14: $[\hat{\mathbf{K}}_{\mathbf{g},ff}] = \begin{bmatrix} -0.1 & 0 & 0 \\ 0 & -0.12 & -100 \\ 0 & -100 & -1.33 \times 10^6 \end{bmatrix}$

From Equation 12.44a: $[\mathbf{H}] = [\mathbf{L}]^{-1}[-\hat{\mathbf{K}}_{\mathbf{g},ff}][\mathbf{L}^{-1}]^{\mathrm{T}} = \begin{bmatrix} 6.944 \times 10^{-4} & 0 & 0 \\ 0 & 0.333 & -0.481 \\ 0 & -0.481 & 1.111 \end{bmatrix}$

Standard form of eigenproblem: $[\mathbf{H}]\{\mathbf{Y}\} = \omega\{\mathbf{Y}\}$

The eigenvalues ω calculated from Equations 12.43 and 12.40 are identical. The reciprocals of these values $1/\omega$ represent the load ratios λ that satisfy equation 12.38. Therefore, the reciprocal of the largest eigenvalue calculated from Equation 12.40 is equal to the smallest λ, that is, to the critical load ratio for the structural system. The corresponding eigenvectors $\{\mathbf{Y}\}$ obtained by solving Equation 12.44 are related to the buckled configurations or modes $\{\boldsymbol{\Delta}_f\}$ of the structure according to

$$\{\boldsymbol{\Delta}_f\} = [\mathbf{L}^{-1}]^{\mathrm{T}}\{\mathbf{Y}\} \tag{12.45}$$

Note that the Cholesky decomposition given in Equation 12.42 is one of the most time-consuming parts of this conversion process. This step can be avoided if we initially perform the same decomposition as part of the linear elastic analysis that is required to obtain forces for the calculation of $[-\hat{\mathbf{K}}_{g,ff}]$. After employing Equation 11.15 to obtain $[\mathbf{L}]$, each successive ith row of $[\mathbf{L}]^{-1}$ can be calculated from

$$l_{ii}^{-1} = 1/l_{ii}$$
$$l_{ij}^{-1} = -\frac{\sum_{k=j+1}^{i} l_{ik}^{-1}\, l_{kj}}{l_{jj}} \qquad \text{for } j = i-1, i-2, \ldots, 1 \tag{12.46}$$

We now proceed by presenting popular methods for solving standard and general forms of eigenvalue problems.

12.8.2 Polynomial Expansion

A formal solution for the eigenpairs can be obtained by rewriting Equation 12.40 in the form

$$[\mathbf{H} - \omega\mathbf{I}]\{\mathbf{Y}\} = \{\mathbf{0}\} \tag{12.47}$$

where $[\mathbf{I}]$ is an $n \times n$ identity matrix. A nontrivial solution can exist if and only if the matrix $[\mathbf{H} - \omega\mathbf{I}]$ is singular; that is, the determinant of the coefficient matrix is zero:

$$|\mathbf{H} - \omega\mathbf{I}| = 0 \tag{12.48}$$

Expansion of this determinant yields a polynomial of degree n in the unknown scalar ω, where n is the order of the coefficient matrix. Solution of this characteristic equation then produces n roots which are the eigenvalues. For each eigenvalue ω_i, an associated eigenvector $\{\mathbf{Y}_i\}$ can be obtained by assuming a value of unity for one of its elements, usually the nth element, and then using Equation 12.47 to solve for its remaining elements. Example 12.7 illustrates the use of polynomial expansion.

EXAMPLE 12.7

For the column and results provided in Example 12.6, use polynomial expansion to determine the critical buckling load and corresponding buckled shape.

$$[\mathbf{H} - \omega\mathbf{I}] = \begin{bmatrix} 6.944 \times 10^{-4} - \omega & 0 & 0 \\ 0 & 0.333 - \omega & -0.481 \\ 0 & -0.481 & 1.111 - \omega \end{bmatrix}$$

$$|\mathbf{H} - \omega\mathbf{I}| = (6.944 \times 10^{-4} - \omega)[(0.333 - \omega) \times (1.111 - \omega) - (-0.481)^2] = 0$$
$$= (6.944 \times 10^{-4} - \omega)[\omega^2 - 1.444\omega + 0.1386] = 0$$

The roots of this characteristic equation are

$$\omega_1 = 6.944 \times 10^{-4}$$
$$\omega_2 = \frac{1.444 + \sqrt{1.444^2 - (4 \times 0.1386)}}{2} = 1.341 \qquad \text{(maximum)}$$
$$\omega_3 = \frac{1.444 - \sqrt{1.444^2 - (4 \times 0.1386)}}{2} = 0.1036$$

Maximum ω corresponds to minimum critical load ratio

$$\lambda_{min} = 1/\omega_{max} = 0.746$$

$P_{cr} = \lambda P_{ref} = 746$ kN which is within 1.5% of the theoretical value $P_{cr} = \dfrac{\pi^2 EI}{4L^2} = 740$ kN

The corresponding buckled shape is obtained from Equation 12.47.

$$[\mathbf{H} - \omega\mathbf{I}]\{\mathbf{Y}\} = \{\mathbf{0}\}$$

$$\begin{bmatrix} -1.3399 & 0 & 0 \\ 0 & -1.0076 & -0.4810 \\ 0 & -0.4810 & -0.2296 \end{bmatrix} \begin{Bmatrix} Y_1 \\ Y_2 \\ Y_3 \end{Bmatrix} = \begin{Bmatrix} 0 \\ 0 \\ 0 \end{Bmatrix}$$

After performing Gauss Elimination (see Section 11.2.1) this system of equations reduces to

$$\begin{bmatrix} -1.3399 & 0 & 0 \\ 0 & -1.0076 & -0.4810 \\ 0 & 0 & 0 \end{bmatrix} \begin{Bmatrix} Y_1 \\ Y_2 \\ Y_3 \end{Bmatrix} = \begin{Bmatrix} 0 \\ 0 \\ 0 \end{Bmatrix}$$

As expected, this system of equations is singular, $|\mathbf{H} - \omega\mathbf{I}| = 0$, and we will need to assume a value for one element of $\{\mathbf{Y}\}$. With $Y_3 = 1$, back substitution yields $\{\mathbf{Y}\} = \begin{Bmatrix} 0 \\ -0.4774 \\ 1 \end{Bmatrix}$.

Buckled shape:

From Equation 12.45, with $[\mathbf{L}]^{-1}$ computed in Example 12.6

$$\{\mathbf{\Delta}_f\} = \begin{bmatrix} 0.0833 & 0 & 0 \\ 0 & 1.67 & 0 \\ 0 & -2.89 & 5.77 \times 10^{-4} \end{bmatrix}^T \begin{Bmatrix} 0 \\ -0.4774 \\ 1 \end{Bmatrix} = \begin{Bmatrix} 0 \\ -3.687 \\ 5.77 \times 10^{-4} \end{Bmatrix}$$

which may be expressed to within a scalar multiple as

$$\{\mathbf{\Delta}_f\} = \lfloor \Delta_1 \quad \Delta_2 \quad \Delta_3 \rfloor^{\mathrm{T}} = c\lfloor 0 \quad 1.0 \quad -1.56 \times 10^{-4} \rfloor^{\mathrm{T}}$$

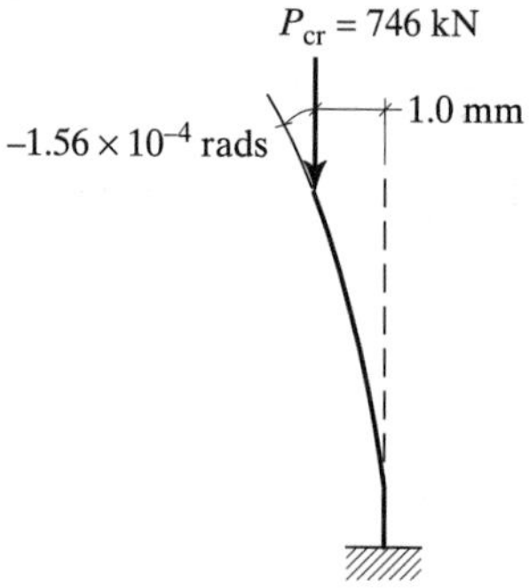

12.8.3 Power Method

Polynomial expansion is only recommended for situations where the order of $[\mathbf{H}]$ is very low ($n < 5$). Typically, this is not the case in critical load analyses of practical structures. Also, it is generally true that only one or a few of the largest eigenvalues and associated eigenvectors of Equation 12.40 are sought. For these reasons more efficient numerical schemes are desired.

The power method is often the simplest numerical algorithm to use when solving eigenproblems of standard form. It further suits our needs since it always provides the dominant or absolute maximum eigenvalue and its companion eigenvector. This iterative scheme starts with an initial guess $\{\mathbf{Y}^0\}$ as the dominant eigenvector. This estimate, typically a vector of ones, is then substituted into the left side of Equation 12.40 yielding

$$[\mathbf{H}]\{\mathbf{Y}^0\} = \{\hat{\mathbf{Y}}^1\} \tag{12.49}$$

A better estimate $\{\mathbf{Y}^1\}$ of the dominant eigenvector is obtained by *normalizing*, or dividing all elements of $\{\hat{\mathbf{Y}}^1\}$ by $\|\hat{\mathbf{Y}}^1\|_e$ (see Eq. 11.38). A first approximation ω^1 of the dominant eigenvalue is given by

$$\omega^1 = \{\mathbf{Y}^1\}^{\mathrm{T}}[\mathbf{H}]\{\mathbf{Y}^1\} \tag{12.50}$$

Using $\{\mathbf{Y}^1\}$, the procedure is then repeated in the next iteration to define the dominant eigenpair more precisely.

Iterations continue until a solution of acceptable accuracy is obtained. In this regard, the following convergence criterion based on successive approximations of the desired eigenvalue is recommended:

$$\varepsilon_a < \zeta \tag{12.51}$$

where

$$\varepsilon_a = \left| \frac{\omega^i - \omega^{i-1}}{\omega_i} \right| 100\% \tag{12.52}$$

and ζ is a prescribed percent error tolerance (see Eq. 11.20b). Example 12.8 illustrates the power method and the use of Equations 12.49 and 12.52.

EXAMPLE 12.8

Using the results of Example 12.6, use the power method to calculate the column's critical load ratio and buckled configuration. Use $\zeta = 0.1\%$.

Using an initial estimate of $\{\mathbf{Y}\} = \begin{Bmatrix} 1 \\ 1 \\ 1 \end{Bmatrix}$, the results of the first iteration are

$$\{\hat{\mathbf{Y}}^1\} = [\mathbf{H}]\{\mathbf{Y}^0\} = \begin{Bmatrix} -6.944 \times 10^{-4} \\ -0.1478 \\ -0.6300 \end{Bmatrix}$$

$$\{\mathbf{Y}^1\} = \{\hat{\mathbf{Y}}^1\}/\|\hat{\mathbf{Y}}^1\|_e = \{\hat{\mathbf{Y}}^1\}/0.6471 = \begin{Bmatrix} 0.0011 \\ -0.2284 \\ 0.9736 \end{Bmatrix}$$

and

$$\omega^1 = \{\mathbf{Y}^1\}^T[\mathbf{H}]\{\mathbf{Y}^1\} = 1.2845$$

Results of the second iteration:

$$\{\hat{\mathbf{Y}}^2\} = \lfloor 7.453 \times 10^{-7} \quad -0.5445 \quad 1.1916 \rfloor^T$$

$$\|\hat{\mathbf{Y}}^2\|_e = 1.3102$$

$$\{\mathbf{Y}^2\} = \lfloor 5.6884 \times 10^{-7} \quad -0.4156 \quad 0.9095 \rfloor^T$$

$$\omega^2 = 1.3405$$

and

$$\varepsilon_a = \left| \frac{1.3405 - 1.2845}{1.3405} \right| 100\% = 4.18\% \not< \zeta = 0.1\%$$

After a third iteration:

$$\{\mathbf{Y}^3\} = \lfloor 2.9465 \times 10^{-10} \quad -0.4297 \quad 0.9030 \rfloor^T$$

$$\omega^3 = 1.3409$$

$$\varepsilon_a = 0.03\% < \zeta = 0.1\%$$

Critical load ratio:

$$\lambda_{min} = 1/\omega^3 = 0.7458 \qquad P_{cr} = \lambda_{min} P_{ref} = 746 \text{ kN}$$

Buckled configuration:
From Equation 12.45

$$\{\boldsymbol{\Delta}_f\} = [\mathbf{L}^{-1}]^{\mathrm{T}}\{\mathbf{Y}^3\} = \begin{Bmatrix} 2.455 \times 10^{-11} \\ -3.3228 \\ 5.2132 \times 10^{-4} \end{Bmatrix}$$

which may be expressed to within a scalar multiple as

$$\{\boldsymbol{\Delta}_f\} = c\lfloor 0 \quad 1.0 \quad -1.57 \times 10^{-4} \rfloor^{\mathrm{T}}$$

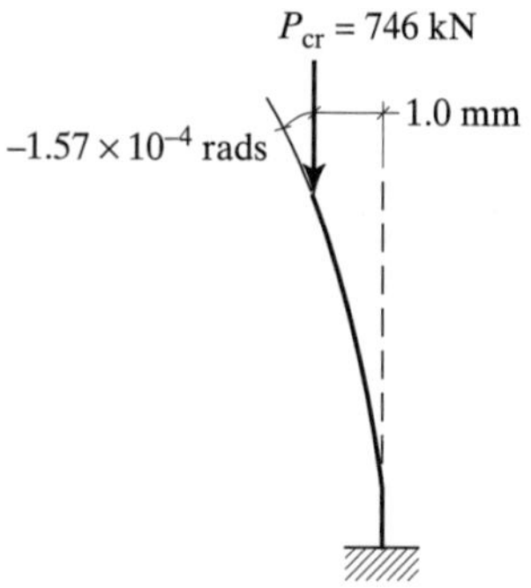

In some cases, the power method is slow to converge. One approach for accelerating this process is to use one of the relaxation schemes presented in Section 11.3.1. A better estimate of the current normalized eigenvector can be obtained by

$$\{\mathbf{Y}^i\} = \beta\{\mathbf{Y}^i\} + (1 - \beta)\{\mathbf{Y}^{i-1}\} \tag{12.53}$$

where the relaxation factor β has a value between 0 and 2. If successive iterations are found to result in oscillating estimates of an eigenvalue, underrelaxation ($0 < \beta < 1$) should improve the convergence rate. Overrelaxation ($1 < \beta \leq 2$) will provide similar benefits when successive estimates of this eigenvalue tend to change in a monotonic fashion.

For some structural systems, the first few critical load ratios and their corresponding buckled shapes are often desired. After calculating the dominant eigenpair ω_1 and $\{\mathbf{Y}_1\}$, subsequent eigenpairs can be found by employing a *deflation* scheme. In this approach $[\mathbf{H}]$ is transformed into a different coefficient matrix, say $[\mathbf{H}_2]$, which has the same eigenpairs as the original coefficient matrix $[\mathbf{H}]$ except that the current dominant eigenvalue becomes zero. Thus, the second largest eigenvalue ω_2 and the associated eigenvector $\{\mathbf{Y}_2\}$ of $[\mathbf{H}]$ can be obtained by using the power method to find the dominant eigenpair of $[\mathbf{H}_2]$. The process is then repeated until the desired number of eigenpairs of $[\mathbf{H}]$ is found. An example of a deflation scheme is Hotelling's method given by

$$[\mathbf{H}_{i+1}] = [\mathbf{H}_i] - \frac{\omega_i\{\mathbf{Y}_i\}\{\mathbf{Y}_i\}^{\mathrm{T}}}{\{\mathbf{Y}_i\}^{\mathrm{T}}\{\mathbf{Y}_i\}} \tag{12.54}$$

It should be noted that in all cases the buckled configurations $\{\boldsymbol{\Delta}_f\}$ corresponding to the calculated eigenvectors $\{\mathbf{Y}\}$ are obtained by employing Equation 12.45.

The Sturm sequence property provides a useful and convenient means for confirming the calculation of multiple eigenvalues. The property is based on performing an $[\mathbf{L}][\mathbf{D}][\mathbf{L}]^{\mathrm{T}}$ factorization (see Eq. 11.16) of $[\mathbf{H} - \omega\mathbf{I}]$ for a given value of ω. The number of negative elements in $[\mathbf{D}]$ will equal the number of eigenvalues less than ω. Examples 12.9 and 12.10 illustrate Hotelling's deflation method and the Sturm sequence property, respectively.

EXAMPLE 12.9

Using the results of Examples 12.6 and 12.8, employ Hotelling's deflation and the power method to determine the column's next sub-dominant critical load ratio and buckled configuration. Use $\zeta = 0.1\%$

From Equation 12.54 and using ω_1^3 and $\{\mathbf{Y}_1^3\}$

$$[\mathbf{H}_2] = [\mathbf{H}] - \frac{\omega_1^3\{\mathbf{Y}_1^3\}\{\mathbf{Y}_1^3\}^{\mathrm{T}}}{\{\mathbf{Y}_1^3\}^{\mathrm{T}}\{\mathbf{Y}_1^3\}} = 10^{-3} \times \begin{bmatrix} 0.6944 & 0 & 0 \\ 0 & 85.71 & 39.18 \\ 0 & 39.18 & 17.88 \end{bmatrix}$$

Using $[\mathbf{H}_2]$ in place of $[\mathbf{H}]$ in Equations 12.49 to 12.52, the power method converges after two iterations

$$\omega_2^2 = 0.1036$$
$$\{\mathbf{Y}_2^2\} = \lfloor 3.3892 \times 10^{-5} \quad 0.9095 \quad 0.4156 \rfloor^{\mathrm{T}}$$

ω_2^2 produces a critical load ratio of $\lambda_2 = \dfrac{1}{\omega_2^2} = 9.6511$ or

$$P_{\mathrm{cr},2} = \lambda_2 P_{\mathrm{ref}} = 9651 \text{ kN}$$

and a buckled configuration of $\{\mathbf{\Delta}_{f,2}\} = [\mathbf{L}^{-1}]\{\mathbf{Y}_2^2\} = \begin{Bmatrix} 2.8244 \times 10^{-6} \\ 0.3160 \\ 2.3997 \times 10^{-4} \end{Bmatrix}$

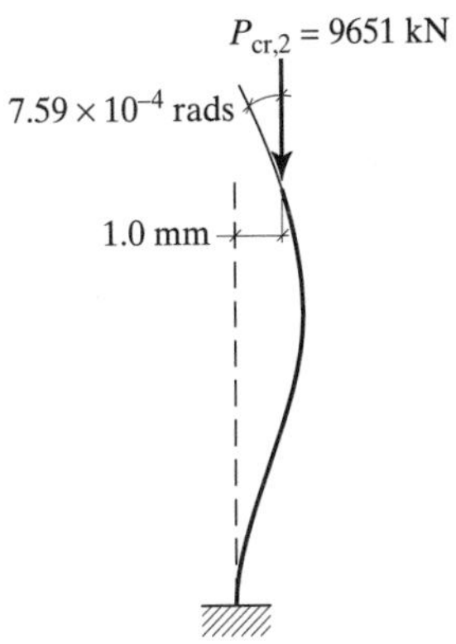

which may be expressed to within a scalar multiple as

$$\{\mathbf{\Delta}_{f,2}\} = c\lfloor 0 \quad 1.0 \quad 7.59 \times 10^{-4} \rfloor^{\mathrm{T}}$$

Note that $P_{\mathrm{cr},2} = 9651$ kN is approximately one and half times the theoretical solution of $P_{\mathrm{cr},2} = \dfrac{9\pi^2 EI}{4L^2} = 6662$ kN. By subdividing the column into more elements, a more accurate solution can be obtained. For example, using two elements provides $P_{\mathrm{cr},2} = 6884$ kN which is within 3.3% of the theoretical solution.

EXAMPLE 12.10

Using the results of Example 12.6, apply the Sturm sequence property to show that $[\mathbf{H}]$ has two eigenvalues that are smaller than $\omega = 0.5$

$$\text{With } [\mathbf{H}] = \begin{bmatrix} 6.944 \times 10^{-4} & 0 & 0 \\ 0 & 0.333 & -0.481 \\ 0 & -0.481 & 1.111 \end{bmatrix} \text{ and } \omega = 0.5,$$

$$[\mathbf{H} - \omega\mathbf{I}] = \begin{bmatrix} -0.4993 & 0 & 0 \\ 0 & -0.1667 & -0.4811 \\ 0 & -0.4811 & 0.6100 \end{bmatrix}$$

From Equation 11.17, $[\mathbf{H} - \omega\mathbf{I}] = [\mathbf{L}][\mathbf{D}][\mathbf{L}]^{\mathrm{T}}$

$$\text{with } [\mathbf{L}] = \begin{bmatrix} 1 & 0 & 0 \\ 0 & 1 & 0 \\ 0 & 2.8868 & 1 \end{bmatrix} \text{ and } [\mathbf{D}] = \begin{bmatrix} -0.4993 & 0 & 0 \\ 0 & -0.1667 & 0 \\ 0 & 0 & 1.999 \end{bmatrix}$$

The number of negative elements in $[\mathbf{D}]$ is two, which correctly indicates that $[\mathbf{H}]$ has two eigenvalues that are less than $\omega = 0.5$. These two eigenvalues were obtained in Example 12.7 as $\omega_1 = 6.944 \times 10^{-4}$ and $\omega_3 = 0.1036$.

12.8.4 Inverse Iteration

The simplicity and efficiency of the power method can be offset by the necessity of having to convert the eigenproblem to standard form. This, coupled with the fact that the bandedness of $[\mathbf{K}_{e,ff}]$ and $[-\mathbf{K}_{g,ff}]$ is not carried through in $[\mathbf{H}]$, suggests the need for an iterative algorithm that can be used to solve the eigenproblem in generalized form.

One such method is inverse iteration. It is also well suited for the eigenproblem defined by Equation 12.38 because it converges on the eigenvector that corresponds to the minimum eigenvalue or critical load ratio λ. The method begins by assuming an initial estimate $\{\mathbf{\Delta}_f^0\}$ of the controlling eigenvector, typically a vector of ones, and then calculating an auxiliary vector $\{\mathbf{y}^1\}$ according to

$$\{\mathbf{y}^1\} = [-\mathbf{K}_{g,ff}]\{\mathbf{\Delta}_f^0\} \tag{12.55}$$

An improved estimate $\{\mathbf{\Delta}_f^1\}$ of the eigenvector is obtained by first solving the system of equations

$$[\mathbf{K}_{e,ff}]\{\hat{\mathbf{\Delta}}_f^1\} = \{\mathbf{y}^1\} \tag{12.56}$$

and then normalizing the solution $\{\hat{\mathbf{\Delta}}_f^1\}$ according to

$$\{\mathbf{\Delta}_f^1\} = \frac{\{\hat{\mathbf{\Delta}}_f^1\}}{\sqrt{\{\hat{\mathbf{\Delta}}_f^1\}^{\mathrm{T}}[-\mathbf{K}_{g,ff}]\{\hat{\mathbf{\Delta}}_f^1\}}} \tag{12.57}$$

An estimate of the corresponding eigenvalue can be obtained from

$$\lambda^1 = \{\mathbf{\Delta}_f^1\}^{\mathrm{T}}[\mathbf{K}_{e,ff}]\{\mathbf{\Delta}_f^1\} \tag{12.58}$$

Similar to the power method, the procedure is repeated using $\{\mathbf{\Delta}_f^{i-1}\}$ as the initial estimate for the ith iteration. The scheme continues until the convergence criterion defined in Equations 12.51 and 12.52 is satisfied. Example 12.11 illustrates the use of the inverse iteration method to solve the problem defined in Example 12.6.

EXAMPLE 12.11

Employ inverse iteration to compute the critical load ratio and buckled configuration for the column defined in Example 12.6. Use $\zeta = 1\%$.

Using an initial estimate of $\{\mathbf{\Delta}_f^0\} = \begin{Bmatrix} 1 \\ 1 \\ 1 \end{Bmatrix}$, the results of the first iteration are

$$\{\mathbf{y}^1\} = [-\mathbf{K}_{g,ff}]\{\mathbf{\Delta}_f^0\} = \begin{bmatrix} 0.1 & 0 & 0 \\ 0 & 0.12 & 100 \\ 0 & 100 & 1.33 \times 10^6 \end{bmatrix} \begin{Bmatrix} 1 \\ 1 \\ 1 \end{Bmatrix} = \begin{Bmatrix} 0.10 \\ 100.12 \\ 1.334 \times 10^6 \end{Bmatrix}$$

Solve $[\mathbf{K}_{e,ff}]\{\hat{\mathbf{\Delta}}_f^1\} = \{\mathbf{y}^1\}$

$$\begin{bmatrix} 144 & 0 & 0 \\ 0 & 0.36 & 1800 \\ 0 & 1800 & 1.2 \times 10^7 \end{bmatrix} \{\hat{\mathbf{\Delta}}_f^1\} = \begin{Bmatrix} 0.10 \\ 100.12 \\ 1.334 \times 10^6 \end{Bmatrix}, \qquad \{\hat{\mathbf{\Delta}}_f^1\} = \begin{Bmatrix} 6.944 \times 10^{-4} \\ -1.110 \times 10^3 \\ 0.2776 \end{Bmatrix}$$

Normalizing according to Equation 12.57:

$$\{\mathbf{\Delta}_f^1\} = \frac{\{\hat{\mathbf{\Delta}}_f^1\}}{\sqrt{\{\hat{\mathbf{\Delta}}_f^1\}^{\mathrm{T}}[-\mathbf{K}_{g,ff}]\{\hat{\mathbf{\Delta}}_f^1\}}} = \begin{Bmatrix} 1.598 \times 10^{-6} \\ -2.553 \\ 6.386 \times 10^{-4} \end{Bmatrix}$$

A first estimate of the minimum eigenvalue is

$$\lambda^1 = \{\mathbf{\Delta}_f^1\}^T[\mathbf{K}_{e,ff}]\{\mathbf{\Delta}_f^1\} = 1.371$$

Results of the second iteration:

$$\{\mathbf{y}^2\} = \lfloor 1.598 \times 10^{-6} \quad -0.2425 \quad 596.161 \rfloor^T$$
$$\{\hat{\mathbf{\Delta}}_f^2\} = \lfloor 1.109 \times 10^{-9} \quad -3.689 \quad 6.030 \times 10^{-4} \rfloor^T$$
$$\{\mathbf{\Delta}_f^2\} = \lfloor 8.578 \times 10^{-10} \quad -2.852 \quad 4.662 \times 10^{-4} \rfloor^T$$
$$\lambda^2 = 0.750$$

and according to Equation 12.51

$$\varepsilon_a = \left|\frac{0.750 - 1.371}{0.750}\right| 100\% = 82.8\% \nless \zeta = 1\%$$

Results of third iteration:

$$\{\mathbf{y}^3\} = \lfloor 8.578 \times 10^{-11} \quad -0.2956 \quad 336.43 \rfloor^T$$
$$\{\hat{\mathbf{\Delta}}_f^3\} = \lfloor 5.957 \times 10^{-13} \quad -3.846 \quad 6.049 \times 10^{-4} \rfloor^T$$
$$\{\mathbf{\Delta}_f^3\} = \lfloor 4.444 \times 10^{-3} \quad -2.869 \quad 4.512 \times 10^{-4} \rfloor^T$$
$$\lambda^3 = 0.746$$

and

$$\varepsilon_a = \left|\frac{0.746 - 0.750}{0.746}\right| 100\% = 0.54\% < \zeta = 1\%$$

With a critical load ratio of $\lambda = 0.746$, $P_{cr} = \lambda P_{ref} = 746$ kN and the buckled configuration $\{\mathbf{\Delta}_f^3\}$ expressed to within a scalar multiple is

$$\{\mathbf{\Delta}_f^3\} = c\lfloor 0 \quad 1.0 \quad -1.57 \times 10^{-4} \rfloor^T$$

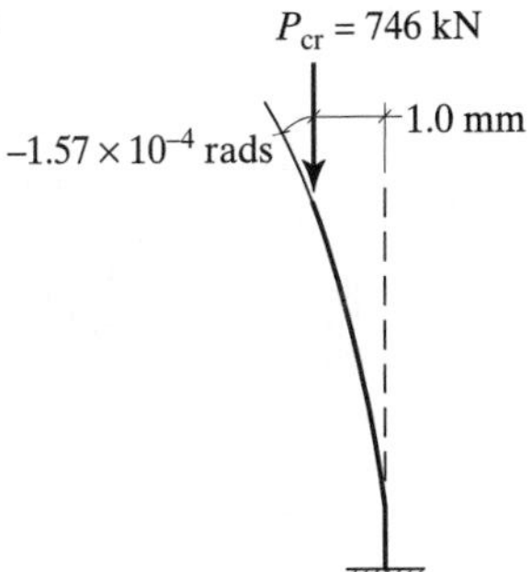

Finally, it should be noted that there are other more complex but possibly more efficient solution techniques available for solving eigenvalue problems. In general, methods for solving eigenproblems can be classified in one of four groups: vector iteration approaches such as the power and inverse iteration methods, transformation strategies, polynomial iteration techniques, and Sturm sequence–based methods. Since there are no direct methods for determining the roots of a polynomial of degree 5 or higher, all schemes are iterative. Several of them employ variations of the iterative methods as well as the acceleration and deflation schemes presented in this section. However, there is no single algorithm that is best suited for all eigenvalue problems. Reference 12.13 provides a review and criteria for selecting appropriate numerical algorithms for solving eigenvalue problems related to structural engineering.

12.9 PROBLEMS

12.1 For the nonlinear spring stiffness provided, use three increments of the following methods to determine the extension of the spring Δ when subjected to a force of $P = 3$

(a) Euler method
(b) Midpoint Runge-Kutta method
(c) Compare both solutions to the exact solution, $\Delta = (P + 2)^3 - 8$

$$k = \frac{1}{3(P+2)^2}$$

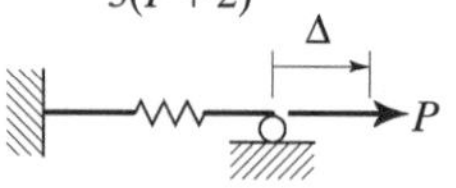

Problem 12.1

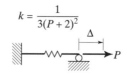

12.2 Repeat Problem 12.1 for $k = 1/e^P$ with an exact solution of $\Delta = e^P - 1$.

12.3 Repeat Problem 12.1 for

$$k = \left[\frac{3}{2}\cosh\left(\frac{3P}{2}\right)\right]^{-1}$$

with an exact solution of $\Delta = \sinh(3P/2)$.

12.4 Member ab is a rigid bar connected to its base by a nonlinear rotational spring with stiffness

$$k = \frac{(10^6 - M^2)^2}{10^6 + M^2}\ \frac{\text{in. kips}}{\text{rad}}$$

Using three increments of the following methods, determine the vertical displacement at b for $P = 8$ kips and $L_{ab} = 10$ ft.

(a) Euler method
(b) Midpoint Runge-Kutta method
(c) Compare both solutions to the exact solution,

$$\theta = \frac{M}{10^6 - M^2}$$

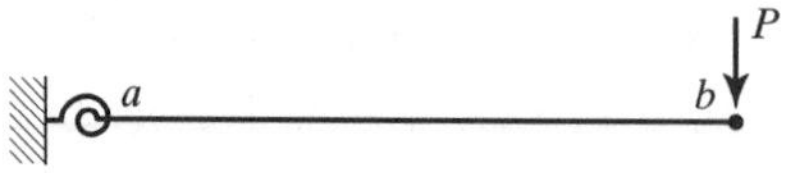

Problem 12.4

12.5 Use the Euler method to calculate the second-order elastic response of the cantilevered beam shown (adapted from Ref. 12.14). Start with an increment size of $d\lambda = 0.5$ and repeat analyses with the initial increment size reduced by one-half until there is a negligible change in the load-displacement response. Confirm your solution with the work control method.

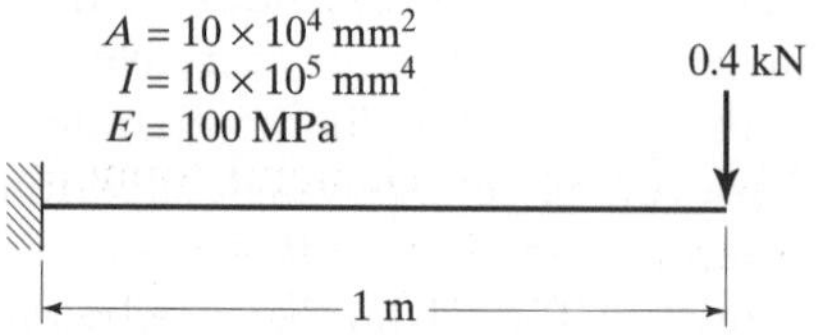

Problem 12.5

12.6 Repeat the convergence study of Problem 12.5 using the midpoint Runge-Kutta method. Compare the results.

12.7 Repeat Problems 12.5 and 12.6 for the frame shown (adapted from Ref. 12.15).

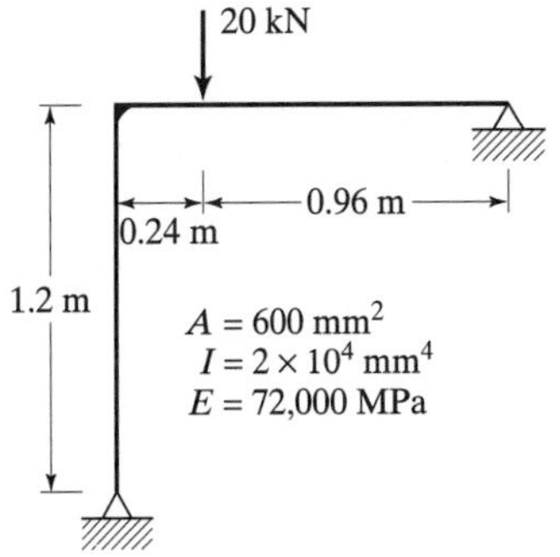

Problem 12.7

12.8 Use the Euler method to calculate the second-order inelastic response of the frame shown. Start with an increment size of $d\lambda = 0.5$ and repeat analyses with the initial increment size reduced by one-half until there is a negligible change the load-displacement response.

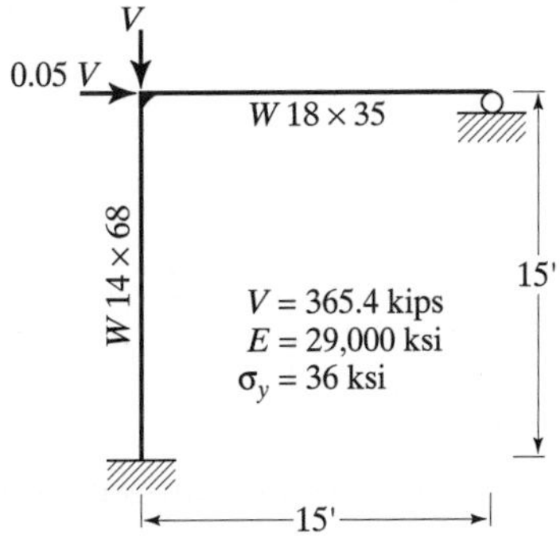

Problem 12.8

12.9 Repeat the convergence study of Problem 12.8 using the midpoint Runge-Kutta method. Compare the results.

12.10 Repeat Example 12.5 using $p = 0.6$, $dp = 0.1$, $m = -0.5$, and $dm = -0.08$.

12.11 Given that the nondimensionalized quantities $p = 0.378$ and $m = 0.756$ satisfy a yield surface defined by $\Phi(p, m) = p^2 + m^2 + 3.5p^2m^2 = 1$, determine the fraction τ that must be multiplied by increments $dp = 0.1$ and $dm = 0.2$ in order to satisfy a yield surface drift tolerance of 0.01. In this case, the maximum tolerable yield surface is

$$\Phi'(p, m) = \frac{1}{1.01^2}\left(p^2 + m^2 + \frac{3.5}{1.01^2}p^2m^2\right) = 1$$

12.12 Using two elements to model each of the columns shown, prepare the eigenproblem in standard form and solve for the elastic critical load ratio and corresponding buckled shape by (1) polynomial expansion and (2) the power method with $\zeta = 0.1\%$. Assume a compressive axial force of $P_{\text{ref}} = 500$ kips and let $E = 29{,}000$ ksi,

$I = 1200$ in.^{4}, and $L = 29$ ft. Neglect axial deformation and compare with exact solution provided.

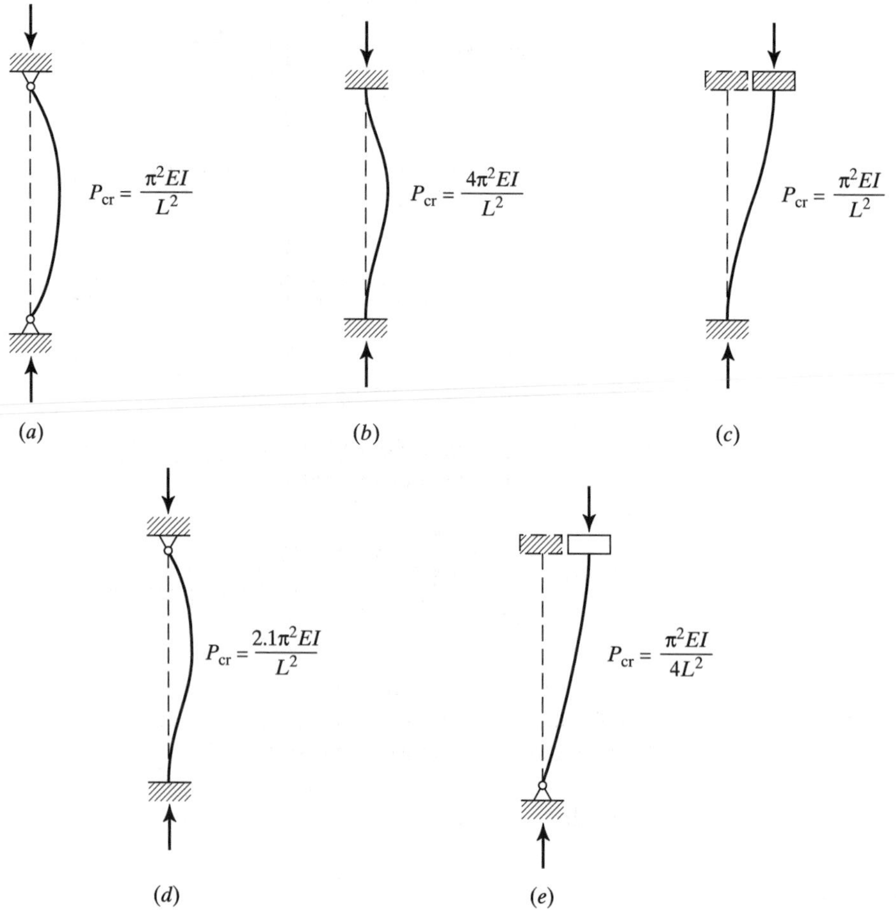

Problem 12.12

12.13 Using the results of part (2) of Problem 12.12, employ Hotelling's deflation and the power method to determine the next subdominant critical load ratio and buckled configuration. Use $\zeta = 0.1\%$.

12.14 Show that the dominant and subdominant eigenvectors obtained for each column in Problems 12.12 and 12.13, respectively, are orthogonal, $\{\mathbf{Y}_1\}^T\{\mathbf{Y}_2\} = 0$.

12.15 Apply the Sturm sequence property to the results of Problem 12.13 to confirm that the two eigenvalues calculated for each column are both less than $\omega = 1.05/\max(\lambda_1, \lambda_2)$.

12.16 For the columns shown in Problem 12.12, use inverse iteration to compute the critical load ratios and buckled configurations. Use $\zeta = 0.1\%$.

12.17 Repeat Problem 12.16 using an estimate for the initial eigenvector that is more reasonable than a vector of ones.

12.18 For the two-bar truss shown, calculate the critical load ratio and corresponding buckled configuration by employing the following methods with $\zeta = 0.1\%$:

(a) Reduction to standard form and the power method
(b) Inverse iteration
(c) Structural analysis program

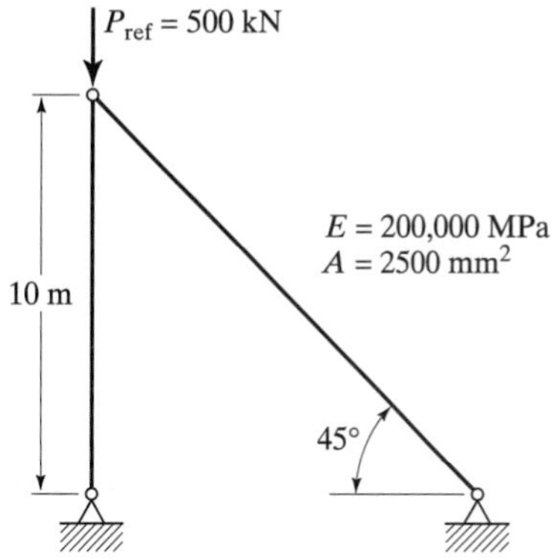

Problem 12.18

12.19 Repeat Example 9.10 by employing two elements and the following methods to calculate the beam's critical load ratio and corresponding buckled configuration. Use $\zeta = 0.1\%$:

(a) Reduction to standard form and the power method
(b) Inverse iteration

12.20 Repeat Problem 12.19 using one element to model the beam defined in Example 9.11.

12.21 Repeat Problem 9.16 by employing three elements and the following methods to calculate the column's critical load ratio and corresponding buckled configuration. Neglect axial deformation and use $\zeta = 0.1\%$.

(a) Reduction to standard form and the power method
(b) Inverse iteration

REFERENCES

12.1 M. J. Clarke and G. J. Hancock, "A Study of Incremental-Iterative Strategies for Nonlinear Analyses," *Intl. Jl. Num. Meth. in Engr.*, Vol. 29, 1990, pp. 1365–1391.

12.2 O. C. Zienkiewicz, "Incremental Displacement in Nonlinear Analysis," *Intl. Jl. Num. Meth. in Engr.*, Vol. 3, 1971, pp. 587–592.

12.3 Y-B. Yang and S-R. Kuo, *Theory and Analysis of Nonlinear Framed Structures*, Prentice-Hall, Englewood Cliffs, N.J., 1994.

12.4 M. A. Crisfield, *Non-linear Finite Element Analysis of Solids and Structures, Volume 1: Essentials*, John Wiley and Sons, New York, 1991.

12.5 P. G. Bergan and R. W. Clough, "Convergence Criteria for Iterative Processes," *AIAA Jl*, Vol. 10, 8, August, 1972, pp. 1107–1108.

12.6 K. J. Bathe, *Finite Element Procedures*, Prentice-Hall, Englewood Cliffs, N.J., 1996.

12.7 P. G. Bergan, G. Horrigmoe, B. Krakeland, and T. H. Søreide, "Solution Techniques for Nonlinear Finite Element Problems," *Intl. Jl. Num. Meth. in Engr.*, Vol. 12, 1978, pp. 1677–1696.

12.8 C. I. Pesquera, *Integrated Analysis and Design of Steel Frames with Interactive Computer Graphics*, Ph.D. Dissertation, Cornell University, Ithaca, N.Y., 1984.

12.9 J. G. Orbison, W. McGuire, and J. F. Abel, "Yield Surface Applications in Nonlinear Steel Frame Analysis," *Computer Meth. in Appl. Mech. and Engr.*, Vol. 33, 1982, pp. 557–573.

12.10 R. D. Krieg and D. B. Krieg, "Accuracies of Numerical Solution Methods for the Elastic-Perfectly Plastic Model," *Jl. of Pressure Vessel Tech., Trans. ASME*, November, 1977, pp. 510–515.

12.11 P. G. Bergan, "Automated Incremental-Iterative Solution Schemes," *Numerical Methods for Nonlinear Problems*, Proceedings of the International Conference, Swansea, September, 1980.

12.12 M. A. Crisfield, *Non-linear Finite Element Analysis of Solids and Structures, Volume 2: Advanced Topics*, John Wiley and Sons, New York, 1997.

12.13 K.-J. Bathe and E. L. Wilson, *Numerical Methods in Finite Element Analysis*, Prentice-Hall, Englewood Cliffs, N.J., 1976.

12.14 K. Mattiasson, "Numerical Results from Large Deflection Beam and Frame Problems Analysed by Means of Elliptic Integrals," *Intl. Jl. Num. Meth. in Engr.*, Vol. 17, No. 1, 1981.

12.15 S. L. Lee, F. S. Manuel, and E. C. Rossow, "Large Deflection and Stability of Elastic Frames" *Jl. Engr. Mech. Div., ASCE*, Vol. 94, No. EM2, April, 1968.

Chapter 13

Special Analysis Procedures

In earlier chapters many of the basic principles of structural mechanics were reviewed and the fundamentals of the stiffness method were developed. Now we wish to look at some of the techniques used in coping with complications encountered in the practical application of this method. The complications can usually be associated with one or more of the following factors: problem size, solution efficiency, and solution accuracy. Most real structures have a large number of degrees of freedom. A structure may be so large that an analysis can be achieved most effectively if it is broken down into parts, with each part analyzed separately and the results combined in a way that satisfies all of the equilibrium, compatibility, and boundary conditions. More often, the analyst may adopt this approach because the structure lends itself to it naturally. The second factor, solution efficiency, refers to the relative cost, in terms of programming and computer expenditures, of the many different ways of ordering the numerical operations leading from the input of data to the output of the solution. Solution accuracy refers to the relative precision of these different methods and to ways in which errors may be analyzed and corrected.

This chapter deals with some of the methods for treating problems of large size. It also covers a few logically or mathematically related techniques that are useful in the treatment of some special but frequently encountered problems. Most of the methods discussed are strictly valid for linear elastic analysis only, but some are unrestricted. Additional ways for treating large problems and the subjects of efficient and accurate solution of equations were treated in Chapters 11 and 12.

One way to handle large structures is to use a coarse structural idealization, that is, to disregard, suppress, or approximate the effect of degrees of freedom that, in the opinion of the analyst, have only a minor bearing on the result. The many different ways in which this may be done are so dependent upon the individual structure that they cannot be discussed usefully in a general text. Here we present some schemes for reducing the order of the systems of equations that have to be solved at any one time—once the structure has been idealized. This means that generally we will be discussing methods for reducing the order of the stiffness matrices to be inverted. We shall first discuss *matrix condensation* as a general technique. Then we shall show how it is employed as part of a routine—*the substructuring method*—for breaking the entire structure into parts, analyzing each part in turn, and putting the results together in the proper way.

After this, we shall look at some of the practically important features of support conditions and coordinate types, connection stiffness and joint size, and the application of the principles of symmetry and antisymmetry. It is a mixed group of topics. Some are related mathematically, but all have the merit of either contributing to the efficiency of an analysis or to the realism of the structural model. We shall also consider the problem of revising structural properties in the light of the results of preliminary analysis, an essential task in the normal iterative analysis and design process.

13.1 CONDENSATION

The term *condensation* refers to the contraction in size of a system of equations by elimination of certain degrees of freedom. The condensed equations are expressed in terms of preselected degrees of freedom $\{\mathbf{\Delta}_c\}$ that, together with the eliminated quantities $\{\mathbf{\Delta}_b\}$, comprise the total original set of degrees of freedom, that is, $\lfloor \mathbf{\Delta} \rfloor = \lfloor \mathbf{\Delta}_b \mid \mathbf{\Delta}_c \rfloor$. Mathematically, the process takes the original set of equations, $[\mathbf{K}]\{\mathbf{\Delta}\} = \{\mathbf{P}\}$, which may be expressed in partitioned form as

$$\begin{bmatrix} \mathbf{K}_{bb} & \mathbf{K}_{bc} \\ \mathbf{K}_{cb} & \mathbf{K}_{cc} \end{bmatrix} \begin{Bmatrix} \mathbf{\Delta}_b \\ \mathbf{\Delta}_c \end{Bmatrix} = \begin{Bmatrix} \mathbf{P}_b \\ \mathbf{P}_c \end{Bmatrix} \tag{13.1}$$

and condenses them to the form:

$$[\hat{\mathbf{K}}_{cc}]\{\mathbf{\Delta}_c\} = \{\hat{\mathbf{P}}_c\} \tag{13.2}$$

To reduce Equation 13.1 to Equation 13.2, first expand the upper partition and solve for $\{\mathbf{\Delta}_b\}$. Thus

$$\{\mathbf{\Delta}_b\} = -[\mathbf{K}_{bb}]^{-1}[\mathbf{K}_{bc}]\{\mathbf{\Delta}_c\} + [\mathbf{K}_{bb}]^{-1}\{\mathbf{P}_b\} \tag{13.3}$$

Substituting this value of $\{\mathbf{\Delta}_b\}$ in the expanded lower partition of Equation 13.1 yields

$$-[\mathbf{K}_{cb}][\mathbf{K}_{bb}]^{-1}[\mathbf{K}_{bc}]\{\mathbf{\Delta}_c\} + [\mathbf{K}_{cc}]\{\mathbf{\Delta}_c\} = \{\mathbf{P}_c\} - [\mathbf{K}_{cb}][\mathbf{K}_{bb}]^{-1}\{\mathbf{P}_b\} \tag{13.4}$$

Letting

$$[\mathbf{K}_{cc}] - [\mathbf{K}_{cb}][\mathbf{K}_{bb}]^{-1}[\mathbf{K}_{bc}] = [\hat{\mathbf{K}}_{cc}] \tag{13.5}$$

and

$$\{\mathbf{P}_c\} - [\mathbf{K}_{cb}][\mathbf{K}_{bb}]^{-1}\{\mathbf{P}_b\} = \{\hat{\mathbf{P}}_c\} \tag{13.6}$$

Equation 13.4 becomes identical to Equation 13.2.

One way of using the condensation procedure is immediately obvious. If Equations 13.1 are stiffness equations in which $\lfloor \mathbf{P} \rfloor = \lfloor \mathbf{P}_b \mid \mathbf{P}_c \rfloor$ is a known load vector, one may solve for the unknown displacements in two steps. First, from Equation 13.2,

$$\{\mathbf{\Delta}_c\} = [\hat{\mathbf{K}}_{cc}]^{-1}\{\hat{\mathbf{P}}_c\} \tag{13.7}$$

Second, by substituting the result in Equation 13.3,

$$\{\mathbf{\Delta}_b\} = [\mathbf{K}_{bb}]^{-1}\{\mathbf{P}_b\} - [\mathbf{K}_{bb}]^{-1}[\mathbf{K}_{bc}][\hat{\mathbf{K}}_{cc}]^{-1}\{\hat{\mathbf{P}}_c\} \tag{13.8}$$

Although it can be shown that the two-stage solution always requires a greater number of arithmetical operations than does the direct solution of the unpartitioned form of Equation 13.1, it has the virtue of enabling the solution to be effected through the inversion of matrices $[\mathbf{K}_{bb}]$ and $[\hat{\mathbf{K}}_{cc}]$, which are each of lower order than the original unpartitioned matrix $[\mathbf{K}]$. This can be important in treating problems of very large size or in using limited computer-core storage.

The concepts of condensation have already been used although they were not identified by that term. For instance, Examples 4.9 and 4.14 contain applications of condensation techniques. Example 13.1 is a further illustration of condensation.

EXAMPLE 13.1

From the results of Example 4.3, remove the degree of freedom θ_{z2} by condensation.

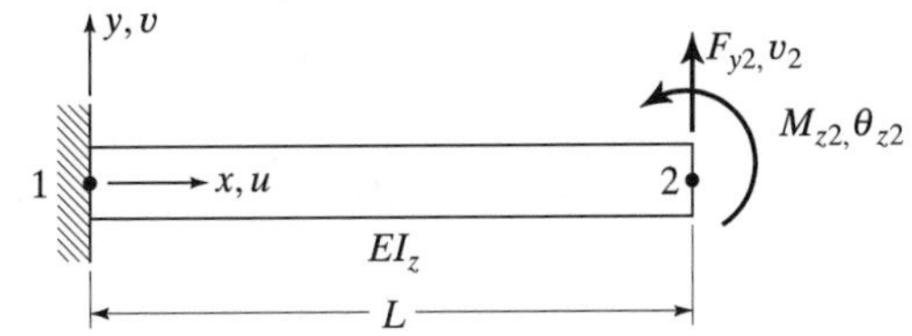

Solving for $\lfloor M_{z2} \quad F_{y2} \rfloor$ from Example 4.3,

$$\begin{Bmatrix} M_{z2} \\ F_{y2} \end{Bmatrix} = \frac{2EI_z}{L}\begin{bmatrix} 2 & -3/L \\ -3/L & 6/L^2 \end{bmatrix}\begin{Bmatrix} \theta_{z2} \\ v_2 \end{Bmatrix}$$

From Equation 13.5,

$$[\hat{\mathbf{K}}_{cc}] = \frac{2EI_z}{L}\left[\frac{6}{L^2} - \left(-\frac{3}{L}\right)\left(\frac{1}{2}\right)\left(-\frac{3}{L}\right)\right] = \frac{3EI_z}{L^3}$$

From Equation 13.6,

$$\{\hat{\mathbf{F}}_c\} = F_{y2} - \left(-\frac{3}{L}\right)\left(\frac{1}{2}\right)M_{z2} = F_{y2} + \frac{3}{2}\frac{M_{z2}}{L}$$

From Equation 13.2,

$$\left[\frac{3EI_z}{L^3}\right]\{v_2\} = \left\{F_{y2} + \frac{3}{2}\frac{M_{z2}}{L}\right\} \quad \text{or} \quad v_2 = \frac{F_{y2}L^3}{3EI_z} + \frac{M_{z2}L^2}{2EI_z}$$

It is of interest to observe that condensation of a stiffness matrix means satisfaction of the equilibrium equations corresponding to the elements eliminated. Also, it should be clear that the "eliminated" degrees of freedom are not discarded. They are expressed as functions of the corresponding forces, the remaining degrees of freedom, and the coefficients of the equations, with the result substituted in the original equations (see Equations 13.3 and 13.4). This process is the matrix analogue of traditional Gauss elimination for the solution of simultaneous linear algebraic equations (Section 11.2). Nothing is "lost" or "approximated" in the process of condensation. The effect of the forces that correspond to the eliminated degrees of freedom (the forces $\{\mathbf{P}_b\}$) on the remainder of the system is incorporated in the force vector $\{\hat{\mathbf{P}}_c\}$ of Equation 13.6. In Equations 13.7 and 13.8 all of the elements contributing to the stiffness of the system are properly represented. Finally, selection of the degrees of freedom to be eliminated (the displacement components $\{\mathbf{\Delta}_b\}$) is at the analyst's discretion. Usually, however, the selection is not arbitrary; there are generally logical reasons for choosing certain degrees of freedom for elimination.

13.2 SUBSTRUCTURING

When, for one of the reasons given in the introduction to this chapter, it is either impossible or undesirable to analyze a structure in its entirety in one stage, it may be analyzed by the *substructuring method*. In this method, major components of the structure, called *substructures*, are first analyzed separately and the results are then combined. The mathematics of substructuring follow directly from the equations of condensation.

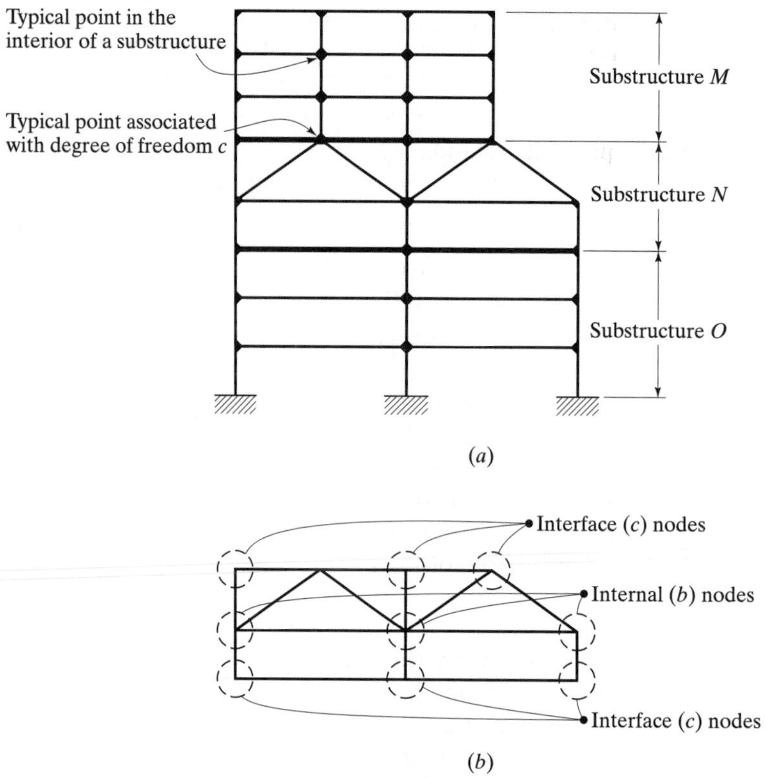

Figure 13.1 Substructure analysis. (*a*) Substructure analysis conditions (applied load not shown). (*b*) Typical interior and interface nodes—substructure *N*.

Figure 13.1*a* shows a structure partitioned into three major substructures M, N, and O. There is no physical significance to the fact that the interface members are shown heavier. They are just drawn that way to emphasize the divisions we have made for purposes of analysis. Consider first the stiffness properties of the Nth substructure. The following subscripts are employed:

b—free degrees of freedom internal to a substructure, that is, not associated with any other substructure.
c—free degrees of freedom located on an interface between two substructures.

Figure 13.1*b* illustrates the use of these subscripts in substructure N. The stiffness equations for this substructure can be written in the partitioned form:

$$\begin{Bmatrix} \mathbf{P}_b^N \\ \hline \mathbf{P}_c^N \end{Bmatrix} = \left[\begin{array}{c|c} \mathbf{K}_{bb}^N & \mathbf{K}_{bc}^N \\ \hline \mathbf{K}_{cb}^N & \mathbf{K}_{cc}^N \end{array}\right] \begin{Bmatrix} \mathbf{\Delta}_b^N \\ \hline \mathbf{\Delta}_c^N \end{Bmatrix} \tag{13.9}$$

Observing that, except for the addition of superscripts, Equation 13.9 is identical to Equation 13.1, we may use the results of Section 13.1 to write the equations that effect the elimination of the interior nodes of substructure N. From Equation 13.2,

$$[\hat{\mathbf{K}}_{cc}^N]\{\mathbf{\Delta}_c^N\} = \{\hat{\mathbf{P}}_c^N\} \tag{13.10}$$

where, from Equation 13.5,

$$[\hat{\mathbf{K}}_{cc}^N] = [\mathbf{K}_{cc}^N] - [\mathbf{K}_{cb}^N][\mathbf{K}_{bb}^N]^{-1}[\mathbf{K}_{bc}^N] \tag{13.11}$$

It is useful to list the contributions to $\{\hat{\mathbf{P}}_c^N\}$. Using Equation 13.6 to do this, Equation 13.10 becomes

$$[\hat{\mathbf{K}}_{cc}^N]\{\mathbf{\Delta}_c^N\} = \{\mathbf{P}_c^N\} - [\mathbf{K}_{cb}^N][\mathbf{K}_{bb}^N]^{-1}\{\mathbf{P}_b^N\}$$

To simplify the notation let

$$[\mathbf{K}_{cb}^N][\mathbf{K}_{bb}^N]^{-1}\{\mathbf{P}_b^N\} = \{\mathbf{R}_c^N\} \tag{13.12}$$

We then have

$$[\hat{\mathbf{K}}_{cc}^N]\{\mathbf{\Delta}_c^N\} = \{\mathbf{P}_c^N\} - \{\mathbf{R}_c^N\} \tag{13.13}$$

We see that the condensed stiffness matrix $[\hat{\mathbf{K}}_{cc}^N]$ relates the interface degrees of freedom $\{\mathbf{\Delta}_c^N\}$ to the forces acting at the interface nodes $\{\mathbf{P}_c^N\}$ plus fictitious nodal loads $-\{\mathbf{R}_c^N\}$, which are equivalent to the real forces acting at the interior nodes (compare with Equation 5.21). It should be clear from the development of Equation 13.13 that we may treat the entire substructure N as a single element having the *substructure stiffness matrix* $[\hat{\mathbf{K}}_{cc}^N]$ and that we may combine all such substructure stiffness matrices, interface nodal forces, and equivalent nodal loads to form the stiffness equations for the juncture degrees of freedom for the complete structure:

$$[\hat{\mathbf{K}}_{cc}]\{\mathbf{\Delta}_c\} = \{\mathbf{P}_c\} - \{\mathbf{R}_c\} \tag{13.14}$$

The rules of combination and manipulation of substructure equations are the same as those developed in Chapters 3–5 for single members. In forming $[\hat{\mathbf{K}}_{cc}]$, rigid body motion of the structure is prevented by taking into account the actual boundary conditions as, for example, the zero displacements at the bottoms of the columns of substructure O in Figure 13.1. The matrix $[\hat{\mathbf{K}}_{cc}]$ includes only the free degrees of freedom and is therefore nonsingular. Care must be taken to avoid inadvertent duplication of loads acting at interface nodes or stiffnesses of elements situated on the boundaries between substructures.

Solution of Equation 13.14 yields all of the displacements, $\{\mathbf{\Delta}_c\}$, of the interface nodes. Thus

$$\{\mathbf{\Delta}_c\} = [\hat{\mathbf{K}}_{cc}]^{-1}\{\mathbf{P}_c - \mathbf{R}_c\} \tag{13.15}$$

Extracting from $\{\mathbf{\Delta}_c\}$ the interface displacements for each of the substructures in turn, we may then obtain the displacements of interior nodes from Equation 13.3. Hence, for substructure N of Figure 13.1,

$$\{\mathbf{\Delta}_b^N\} = -[\mathbf{K}_{bb}^N]^{-1}[\mathbf{K}_{bc}^N]\{\mathbf{\Delta}_c^N\} + [\mathbf{K}_{bb}^N]^{-1}\{\mathbf{P}_b^N\} \tag{13.16}$$

Having all of the nodal-point displacements, member forces and reactions may be found in the usual way.

In Example 13.2, substructuring is used in the analysis of a fixed-end beam subjected to concentrated loads. The original system, which has six free nodal-point degrees of freedom, is arbitrarily divided into two substructures, each of which has four free degrees of freedom. By eliminating the interior degrees of freedom in each substructure, stiffness equations having as unknowns the two common degrees of freedom are formed and solved. Therefore, instead of solving the problem through the inversion of one six-by-six matrix, we arrive at the same result after inverting three two-by-two matrices. It may be observed that in the course of computing the equivalent nodal load contribution of each substructure, we calculate $[\mathbf{K}_{bb}^N]^{-1}\{\mathbf{P}_b^N\}$. This is a displacement vector that, when added to displacements resulting from movement of the interface nodes, $-[\mathbf{K}_{bb}^N]^{-1}[\mathbf{K}_{bc}^N]\{\mathbf{\Delta}_c^N\}$, yields the resultant displacement of the interior nodes of each substructure (see Eq. 13.16).

EXAMPLE 13.2

Analyze the fixed end beam shown using the substructure method. Calculate the deflection and slope at each load point. $E = 200{,}000$ MPa.

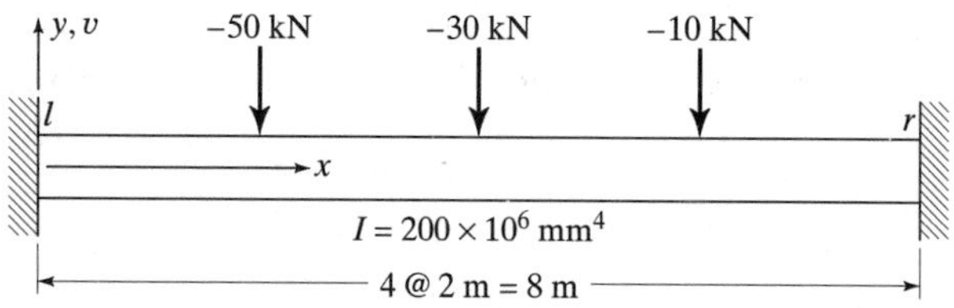

The degrees of freedom, substructures, and element stiffness matrix are as follows.

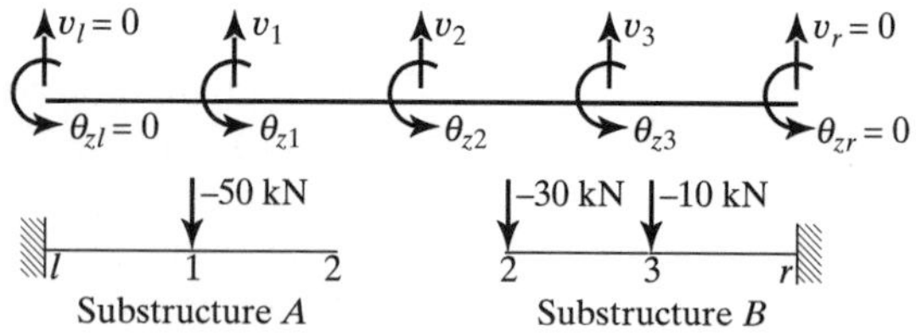

From Equation 4.34, the segment stiffness matrix is

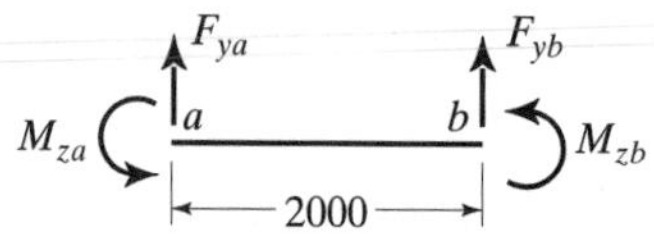

$$\begin{Bmatrix}F_{ya}\\M_{za}\\F_{yb}\\M_{zb}\end{Bmatrix} = \frac{200\times 2\times 10^8}{2000}\begin{bmatrix} \frac{12}{(2000)^2} & \frac{6}{(2000)} & -\frac{12}{(2000)^2} & \frac{6}{(2000)}\\ & 4 & -\frac{6}{(2000)} & 2\\ \text{Sym.} & & \frac{12}{(2000)^2} & -\frac{6}{(2000)}\\ & & & 4\end{bmatrix}$$

$$= 4\times 10^7\begin{bmatrix} 1.5\times10^{-6} & 1.5\times10^{-3} & -1.5\times10^{-6} & 1.5\times10^{-3}\\ & 2 & -1.5\times10^{-3} & 1\\ \text{Sym.} & & 1.5\times10^{-6} & -1.5\times10^{-3}\\ & & & 2\end{bmatrix}$$

Analysis of Substructure A. From Equation 13.9:

$$\begin{Bmatrix}\mathbf{P}_b^A\\ \hline \mathbf{P}_c^A\end{Bmatrix} = \left[\begin{array}{c|c}\mathbf{K}_{bb}^A & \mathbf{K}_{bc}^A\\ \hline \mathbf{K}_{cb}^A & \mathbf{K}_{cc}^A\end{array}\right]\begin{Bmatrix}\mathbf{\Delta}_b^A\\ \hline \mathbf{\Delta}_c^A\end{Bmatrix}$$

$$\begin{Bmatrix}-50\\0\\ \hline 0\\0\end{Bmatrix} = 4\times10^7\left[\begin{array}{cc|cc} 3.0\times10^{-6} & 0 & -1.5\times10^{-6} & 1.5\times10^{-3}\\ & 4 & -1.5\times10^{-3} & 1\\ \hline & & 1.5\times10^{-6} & -1.5\times10^{-3}\\ \text{Sym.} & & & 2\end{array}\right]\begin{Bmatrix}v_1\\ \theta_{z1}\\ \hline v_2\\ \theta_{z2}\end{Bmatrix}$$

From Equation 13.11, $[\hat{\mathbf{K}}_{cc}^A] = [\mathbf{K}_{cc}^A] - [\mathbf{K}_{cb}^A][\mathbf{K}_{bb}^A]^{-1}[\mathbf{K}_{bc}^A]$

$$[\hat{\mathbf{K}}_{cc}^A] = 4\times10^7\begin{bmatrix} 0.1875\times10^{-6} & -0.375\times10^{-3}\\ -0.375\times10^{-3} & 1\end{bmatrix}\quad (\text{columns: } v_2,\ \theta_{z2})$$

From Equation 13.12, $\{\mathbf{R}_c^A\} = [\mathbf{K}_{cb}^A][\mathbf{K}_{bb}^A]^{-1}\{\mathbf{P}_b^A\}$

$$\{\mathbf{R}_c^A\} = 4 \times 10^7 \begin{bmatrix} -1.5 \times 10^{-6} & -1.5 \times 10^{-3} \\ 1.5 \times 10^{-3} & 1 \end{bmatrix} \begin{Bmatrix} -\frac{5}{12} \\ 0 \end{Bmatrix} = \begin{Bmatrix} 25.0 \\ -2.5 \times 10^4 \end{Bmatrix}$$

Analysis of Substructure B. From Equation 13.9:

$$\begin{Bmatrix} \mathbf{P}_b^B \\ \hline \mathbf{P}_c^B \end{Bmatrix} = \left[\begin{array}{c|c} \mathbf{K}_{bb}^B & \mathbf{K}_{bc}^B \\ \hline \mathbf{K}_{cb}^B & \mathbf{K}_{cc}^B \end{array}\right] \begin{Bmatrix} \mathbf{\Delta}_b^B \\ \hline \mathbf{\Delta}_c^B \end{Bmatrix}$$

$$\begin{Bmatrix} -10 \\ 0 \\ \hline -30 \\ 0 \end{Bmatrix} = 4 \times 10^7 \left[\begin{array}{cc|cc} 3.0 \times 10^{-6} & 0 & -1.5 \times 10^{-6} & -1.5 \times 10^{-3} \\ & 4 & 1.5 \times 10^{-3} & 1 \\ \hline & & 1.5 \times 10^{-6} & 1.5 \times 10^{-3} \\ & \text{Sym.} & & 2 \end{array}\right] \begin{Bmatrix} v_3 \\ \theta_{z3} \\ \hline v_2 \\ \theta_{z2} \end{Bmatrix}$$

From Equation 13.11, $[\hat{\mathbf{K}}_{cc}^B] = [\mathbf{K}_{cc}^B] - [\mathbf{K}_{cb}^B][\mathbf{K}_{bb}^B]^{-1}[\mathbf{K}_{bc}^B]$

$$[\hat{\mathbf{K}}_{cc}^B] = 4 \times 10^7 \begin{bmatrix} 0.1875 \times 10^{-6} & 0.375 \times 10^{-3} \\ 0.375 \times 10^{-3} & 1 \end{bmatrix} \quad (\text{columns: } v_2,\ \theta_{z2})$$

From Equation 13.12, $\{\mathbf{R}_c^B\} = [\mathbf{K}_{cb}^B][\mathbf{K}_{bb}^B]^{-1}\{\mathbf{P}_b^B\}$

$$\{\mathbf{R}_c^B\} = 4 \times 10^7 \begin{bmatrix} -1.5 \times 10^{-6} & 1.5 \times 10^{-3} \\ -1.5 \times 10^{-3} & 1 \end{bmatrix} \begin{Bmatrix} -\frac{1}{12} \\ 0 \end{Bmatrix} = \begin{Bmatrix} 5.0 \\ 0.5 \times 10^4 \end{Bmatrix}$$

Analysis of Assembled Substructures. From Equation 13.14 $[\hat{\mathbf{K}}_{cc}]\{\mathbf{\Delta}_c\} = \{\mathbf{P}_c\} - \{\mathbf{R}_c\}$. Summing contributions of substructures A and B,

$$4 \times 10^7 \begin{bmatrix} 0.375 \times 10^{-6} & 0 \\ 0 & 2 \end{bmatrix} \begin{Bmatrix} v_2 \\ \theta_{z2} \end{Bmatrix} = \begin{Bmatrix} -30 \\ 0 \end{Bmatrix} - \begin{Bmatrix} 30.0 \\ -2 \times 10^4 \end{Bmatrix}$$

Solving,

$$\begin{Bmatrix} v_2 \\ \theta_{z2} \end{Bmatrix} = \frac{10^{-7}}{4} \begin{bmatrix} \frac{8}{3} \times 10^6 & 0 \\ 0 & \frac{1}{2} \end{bmatrix} \begin{Bmatrix} -60 \\ 2 \times 10^4 \end{Bmatrix} = \begin{Bmatrix} -4.0 \text{ mm} \\ 0.25 \times 10^{-3} \text{ rad} \end{Bmatrix}$$

From Equation 13.16, $\{\mathbf{\Delta}_b^A\} = -[\mathbf{K}_{bb}^A]^{-1}[\mathbf{K}_{bc}^A]\{\mathbf{\Delta}_c^A\} + [\mathbf{K}_{bb}^A]^{-1}\{\mathbf{P}_b^A\}$

$$\begin{Bmatrix} v_1 \\ \theta_{z1} \end{Bmatrix} = -\begin{bmatrix} \frac{1}{3} \times 10^6 & 0 \\ 0 & 0.250 \end{bmatrix} \begin{bmatrix} -1.5 \times 10^{-6} & 1.5 \times 10^{-3} \\ -1.5 \times 10^{-3} & 1 \end{bmatrix} \begin{Bmatrix} -4 \\ 0.25 \times 10^{-3} \end{Bmatrix} + \begin{Bmatrix} -\frac{5}{12} \\ 0 \end{Bmatrix}$$

$$= \begin{Bmatrix} -2.54 \text{ mm} \\ -1.56 \times 10^{-3} \text{ rad} \end{Bmatrix}$$

and

$$\{\mathbf{\Delta}_b^B\} = -[\mathbf{K}_{bb}^B]^{-1}[\mathbf{K}_{bc}^B]\{\mathbf{\Delta}_c^B\} + [\mathbf{K}_{bb}^B]^{-1}\{\mathbf{P}_b^B\} = \begin{Bmatrix} v_3 \\ \theta_{z3} \end{Bmatrix} = \begin{Bmatrix} -1.96 \text{ mm} \\ 1.48 \times 10^{-3} \text{ rad} \end{Bmatrix}$$

Deflected Structure.

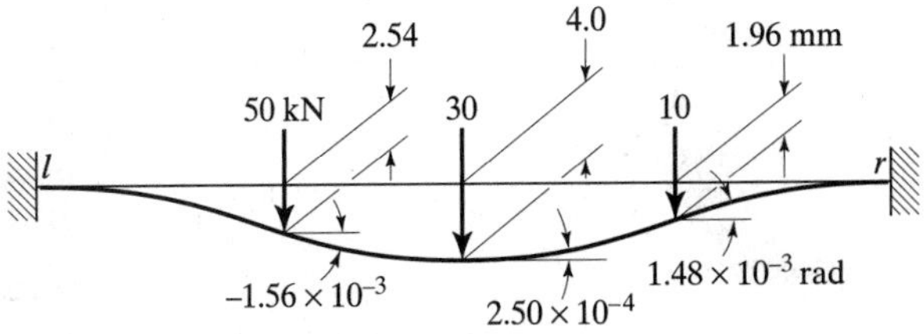

In Example 13.3, the same method is applied to the solution of a three-story frame. Neglecting axial deformations, there remain nine free degrees of freedom in this structure. Substructuring is again arbitrary. Two substructures, each having six free degrees of freedom, are identified. After condensation, stiffness equations in the three common degrees of freedom are formed and solved. The similarity between this solution and that of Example 13.2 should be apparent. So, too, should the similarity between the fictitious nodal forces $\{\mathbf{R}_c\}$ and the fixed end forces $\{\mathbf{P}^F\}$ discussed in Sections 5.2 and 7.5. A substructure carrying loads on interior nodes is treated in the same way as a single element loaded between nodal points. A single member is, in fact, a substructure of a larger system. Multimember substructures may be combined with single-element "substructures." Another way of saying the same thing is that, in condensation, we have a scheme for reducing multi-element systems to equivalent single members.

In Example 13.3 we draw numerical values of stiffness coefficients from previous examples, particularly Examples 5.9, 4.13, and 4.8. Figure 13.2 illustrates the relationships between the forces, reactions, and displacements we are dealing with. It also

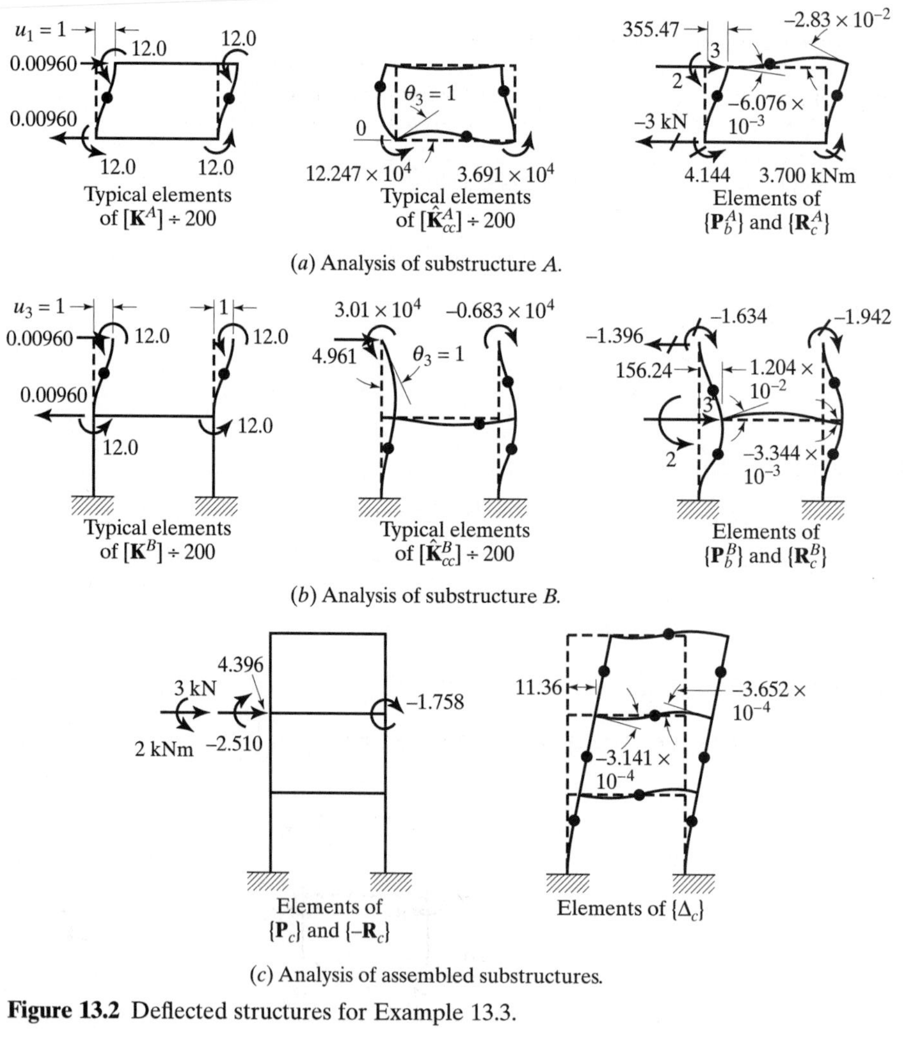

Figure 13.2 Deflected structures for Example 13.3.

pictures the physical response of the various subdivisions as well as of the total assemblage. Some of the elements of the **[K]** and **[K̂]** matrices of substructure A are shown in Figure 13.2a. The effect of forces acting on the interior nodes—assuming that the boundary nodes are restrained—is also shown. Substructure B is similarly analyzed in Figure 13.2b. Figure 13.2c shows the effect of the equivalent nodal forces on the combined system. The resultant displacement components of each node and the moment diagram, which is calculated from the member stiffness equations, are shown in the example itself. It is suggested that the example be studied in conjunction with Figure 13.2.

Examples 13.2 and 13.3 are, of course, merely illustrations of a technique. In practice, such simple systems would be solved without substructuring. The method comes into its own only in the analysis of large systems.

The relationships for a given substructure (Eqs. 13.1–13.6) are also useful in preliminary design studies. Such applications will be described in Section 13.7.

EXAMPLE 13.3

Analyze the rigid frame shown using the substructure method. Neglect axial deformations. I of each beam is 200×10^6 mm^4 and I of each column is 50×10^6 mm^4. $E = 200{,}000$ MPa. Refer to Examples 5.9, 4.13, and 4.8 for numerical values of element and substructure stiffness influence coefficients. See also Figure 13.2 for physical interpretation of selected influence coefficients.

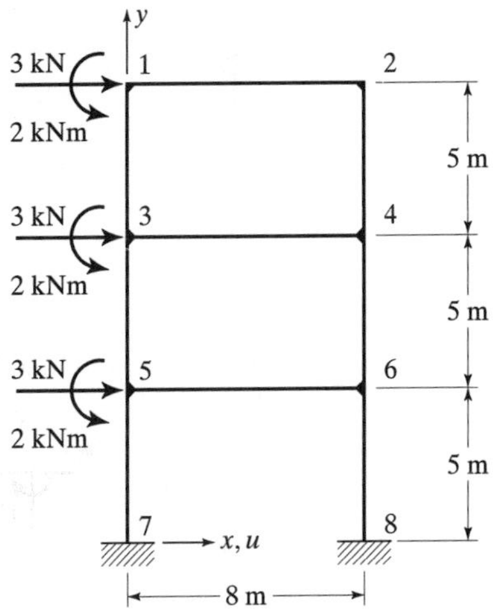

Degrees of freedom and substructures.

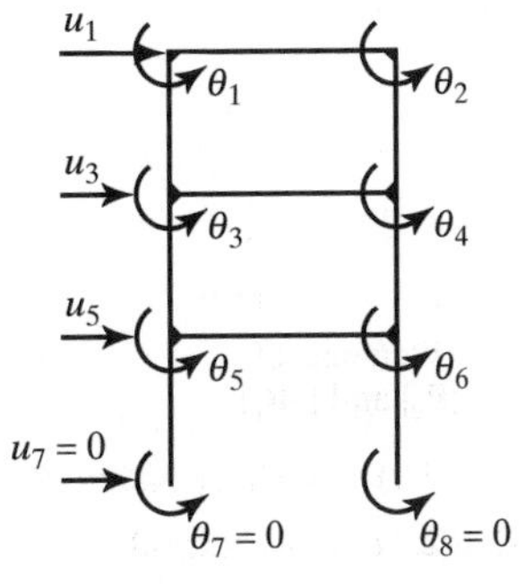

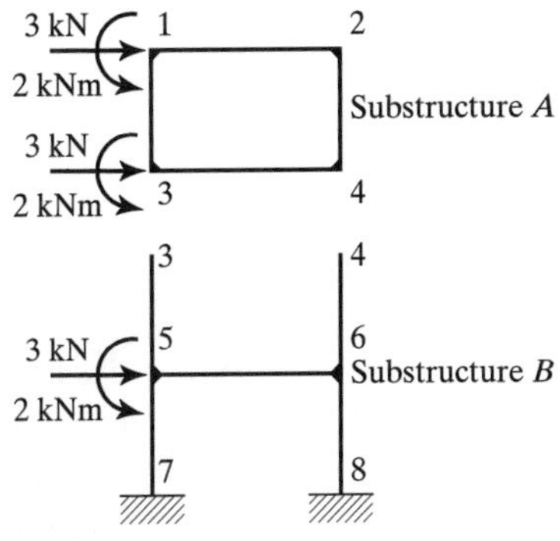

Analysis of Substructure A.

$$\left\{\begin{array}{c}\mathbf{P}_b^A\\ \hline \mathbf{P}_c^A\end{array}\right\} = \left[\begin{array}{c|c}\mathbf{K}_{bb}^A & \mathbf{K}_{bc}^A\\ \hline \mathbf{K}_{cb}^A & \mathbf{K}_{cc}^A\end{array}\right]\left\{\begin{array}{c}\boldsymbol{\Delta}_b^A\\ \hline \boldsymbol{\Delta}_c^A\end{array}\right\}$$

$$\left\{\begin{array}{c}3\\ 2\times10^3\\ 0\\ \hline 3\\ 2\times10^3\\ 0\end{array}\right\} = 200\left[\begin{array}{ccc|ccc}0.00960 & 12.0 & 12.0 & -0.00960 & 12.0 & 12.0\\ & 1.4\times10^5 & 0.5\times10^5 & -12.0 & 0.2\times10^5 & 0\\ & & 1.4\times10^5 & -12.0 & 0 & 0.2\times10^5\\ \hline & & & 0.00960 & -12.0 & -12.0\\ & & & & 1.4\times10^5 & 0.5\times10^5\\ \text{Sym.} & & & & & 1.4\times10^5\end{array}\right]\left\{\begin{array}{c}u_1\\ \theta_1\\ \theta_2\\ \hline u_3\\ \theta_3\\ \theta_4\end{array}\right\}$$

$$[\hat{\mathbf{K}}_{cc}^A] = [\mathbf{K}_{cc}^A] - [\mathbf{K}_{cb}^A][\mathbf{K}_{bb}^A]^{-1}[\mathbf{K}_{bc}^A] = 200\begin{array}{c}\begin{array}{ccc}u_3 & \theta_3 & \theta_4\end{array}\\ \begin{bmatrix}0 & 0 & 0\\ & 12.247\times10^4 & 3.691\times10^4\\ \text{Sym.} & & 12.247\times10^4\end{bmatrix}\end{array}$$

$$\{\mathbf{R}_c^A\} = [\mathbf{K}_{cb}^A][\mathbf{K}_{bb}^A]^{-1}\{\mathbf{P}_b^A\}$$

$$\begin{bmatrix}-0.0096 & -12.0 & -12.0\\ 12.0 & 0.2\times10^5 & 0\\ 12.0 & 0 & 0.2\times10^5\end{bmatrix}\begin{Bmatrix}355.47\\ -6.076\times10^{-3}\\ -2.83\times10^{-2}\end{Bmatrix} = \begin{Bmatrix}-3.00\\ 4.144\times10^3\\ 3.700\times10^3\end{Bmatrix}$$

Analysis of Substructure B.

$$\left\{\begin{array}{c}\mathbf{P}_b^B\\ \hline \mathbf{P}_c^B\end{array}\right\} = \left[\begin{array}{c|c}\mathbf{K}_{bb}^B & \mathbf{K}_{bc}^B\\ \hline \mathbf{K}_{cb}^B & \mathbf{K}_{cc}^B\end{array}\right]\left\{\begin{array}{c}\boldsymbol{\Delta}_b^B\\ \hline \boldsymbol{\Delta}_c^A\end{array}\right\}$$

$$\left\{\begin{array}{c}3\\ 2\times10^3\\ 0\\ \hline 0\\ 0\\ 0\end{array}\right\} = 200\left[\begin{array}{ccc|ccc}0.0192 & 0 & 0 & -0.00960 & -12.0 & -12.0\\ & 1.8\times10^5 & 0.5\times10^5 & 12.0 & 0.2\times10^5 & 0\\ & & 1.8\times10^5 & 12.0 & 0 & 0.2\times10^5\\ \hline & & & 0.00960 & 12.0 & 12.0\\ \text{Sym.} & & & & 0.4\times10^5 & 0\\ & & & & & 0.4\times10^5\end{array}\right]\left\{\begin{array}{c}u_5\\ \theta_5\\ \theta_6\\ \hline u_3\\ \theta_3\\ \theta_4\end{array}\right\}$$

$$[\hat{\mathbf{K}}_{cc}^B] = [\mathbf{K}_{cc}^B] - [\mathbf{K}_{cb}^B][\mathbf{K}_{bb}^B]^{-1}[\mathbf{K}_{bc}^B] = 200\begin{array}{c}\begin{array}{ccc}u_3 & \theta_3 & \theta_4\end{array}\\ \begin{bmatrix}0.00355 & 4.961 & 4.961\\ & 3.010\times10^4 & -0.683\times10^4\\ \text{Sym.} & & 3.010\times10^4\end{bmatrix}\end{array}$$

$$\{\mathbf{R}_c^B\} = [\mathbf{K}_{cb}^B][\mathbf{K}_{bb}^B]^{-1}\{\mathbf{P}_b^B\}$$

$$= \begin{bmatrix}-0.0096 & 12.0 & 12.0\\ -12.0 & 0.2\times10^5 & 0\\ -12.0 & 0 & 0.2\times10^5\end{bmatrix}\begin{Bmatrix}156.24\\ 1.204\times10^{-2}\\ -3.344\times10^{-3}\end{Bmatrix} = \begin{Bmatrix}-1.396\\ -1.634\times10^3\\ -1.942\times10^3\end{Bmatrix}$$

Analysis of Assembled Substructures: $[\hat{\mathbf{K}}_{cc}]\{\mathbf{\Delta}_c\} = \{\mathbf{P}_c\} - \{\mathbf{R}_c\}$

$$200\begin{bmatrix} 0.00355 & 4.961 & 4.961 \\ & 15.257 \times 10^4 & 3.008 \times 10^4 \\ \text{Sym.} & & 15.257 \times 10^4 \end{bmatrix}\begin{Bmatrix} u_3 \\ \theta_3 \\ \theta_4 \end{Bmatrix} = \begin{Bmatrix} 3 \\ 2 \times 10^3 \\ 0 \end{Bmatrix} - \begin{Bmatrix} -4.396 \\ 2.510 \times 10^3 \\ 1.758 \times 10^3 \end{Bmatrix}$$

$$\{\mathbf{\Delta}_c\} = \begin{Bmatrix} u_3 \\ \theta_3 \\ \theta_4 \end{Bmatrix} = \begin{Bmatrix} 11.36 \text{ mm} \\ -3.141 \times 10^{-4} \text{ rad} \\ -3.652 \times 10^{-4} \text{ rad} \end{Bmatrix}$$

$$\{\mathbf{\Delta}_b^A\} = \begin{Bmatrix} u_1 \\ \theta_1 \\ \theta_2 \end{Bmatrix} = -[\mathbf{K}_{bb}^A]^{-1}[\mathbf{K}_{bc}^A]\{\mathbf{\Delta}_c^A\} + [\mathbf{K}_{bb}^A]^{-1}\{\mathbf{P}_b^A\} = \begin{Bmatrix} 14.05 \text{ mm} \\ -0.573 \times 10^{-4} \text{ rad} \\ -1.570 \times 10^{-4} \text{ rad} \end{Bmatrix}$$

$$\{\mathbf{\Delta}_b^B\} = \begin{Bmatrix} u_5 \\ \theta_5 \\ \theta_6 \end{Bmatrix} = -[\mathbf{K}_{bb}^B]^{-1}[\mathbf{K}_{bc}^B]\{\mathbf{\Delta}_c^B\} + [\mathbf{K}_{bb}^B]^{-1}\{\mathbf{P}_b^B\} = \begin{Bmatrix} 6.04 \text{ mm} \\ -5.070 \times 10^{-4} \text{ rad} \\ -5.759 \times 10^{-4} \text{ rad} \end{Bmatrix}$$

Deflected structure and moment diagram.

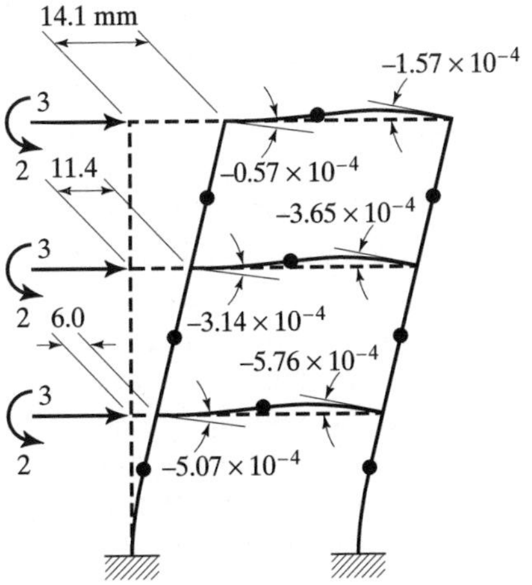

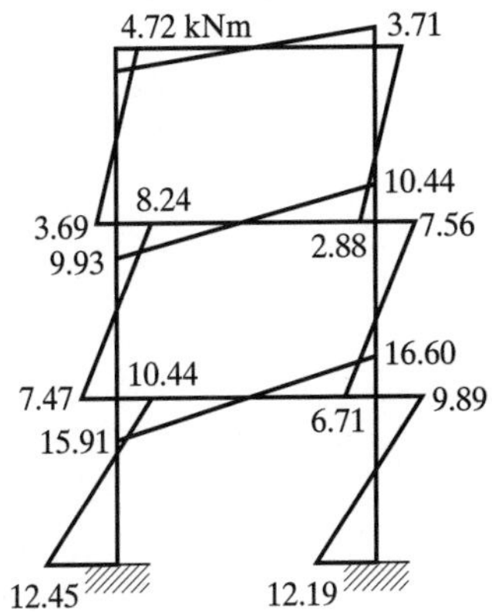

Moments obtained from Equation 4.34.
Beams:

$$M_a = \frac{200^2 \times 10^3}{8}(4\theta_a + 2\theta_b)$$

Columns:

$$M_a = \frac{50 \times 200 \times 10^3}{5}\left[4\theta_a + 2\theta_b - \frac{6}{5000}(u_b - u_a)\right]$$

13.3 CONSTRAINTS

We have used the term *constraint* to designate the suppression of a particular component of displacement. Now, we view the definition of constraint in the broad sense as a set of relationships between degrees of freedom that are supplemental to the basic stiffness relationships. Algebraically, a typical constraint equation—the ith such equation in a given problem—will be written in the form:

$$G_{i1}\Delta_1 + G_{i2}\Delta_2 + \cdots + G_{ij}\Delta_j \cdots + G_{in}\Delta_n = H_i \tag{13.17}$$

where G_{i1}, G_{i2}, G_{ij}, G_{in}, H_i are specified coefficients and Δ_1, Δ_2, Δ_j, Δ_n are the degrees of freedom of the structure. It should be noted that Equation 13.17 is a linear algebraic equation and is thus applicable only to the treatment of linear problems.

In the simplest type of constraint, the familiar one of a single constrained degree of freedom at a support point, all $G_{ij} = 0$ except the coefficient that multiplies that degree of freedom. The latter takes on the value 1.0. Also, $H_i = 0$. For example, if degree of freedom 2 is suppressed, we have $G_{i2} = 1$, and the constraint equation becomes

$$\Delta_2 = 0$$

Practical circumstances under which the general definition of a constraint given by Equation 13.17 plays a role are illustrated in Figure 13.3. In Figure 13.3*a*, for example, the column bases are assumed to settle in such a manner that their vertical displacements are related to one another as indicated. In Figure 13.3*b* we see that, because of the relatively great axial stiffness of the horizontal members of the frame, it is acceptable to set all horizontal displacements at a given level equal to one another.

Not only does the general constraint condition involve more than one degree of freedom, but a given structural analysis situation may feature more than one constraint condition. Thus, in developing the analytical approach to the treatment of constraints, we consider the case where r such conditions are present in a structure described by n degrees of freedom. The r linear constraint conditions are, for this case,

$$[\mathbf{G}]\{\boldsymbol{\Delta}\} = \{\mathbf{H}\} \tag{13.18}$$

where $[\mathbf{G}]$ is an $r \times n$ matrix of the constraint equation coefficients, $\{\boldsymbol{\Delta}\}$ is the list of joint degrees of freedom of the structure, and $\{\mathbf{H}\}$ is a vector of known constants. For simplicity, we deal only with the case $\{\mathbf{H}\} = \mathbf{0}$. Development of the more general case, $\{\mathbf{H}\} \neq \mathbf{0}$, is left as an exercise for the reader (Problem 13.11).

Now, each constraint equation can be viewed as a means to eliminate one of the degrees of freedom in favor of the remainder, simply by solving the constraint equation for the former and using the result in the stiffness equations. Hence, with r constraint conditions, it is possible to eliminate r degrees of freedom from the stiffness equations. To accomplish this elimination we partition Equation 13.18 as follows:

$$[\mathbf{G}_e \mid \mathbf{G}_c]\begin{Bmatrix} \boldsymbol{\Delta}_e \\ \hline \boldsymbol{\Delta}_c \end{Bmatrix} = \{\mathbf{0}\} \tag{13.19}$$

where $[\mathbf{G}_e]$ is a solvable (nonsingular) $r \times r$ matrix and $[\mathbf{G}_c]$ is the $r \times (n - r)$ remainder of the original matrix $[\mathbf{G}]$. $\{\boldsymbol{\Delta}_e\}$ and $\{\boldsymbol{\Delta}_c\}$ are the corresponding degrees of freedom. Solving Equation 13.19 for $\{\boldsymbol{\Delta}_e\}$,

$$\{\boldsymbol{\Delta}_e\} = -[\mathbf{G}_e]^{-1}[\mathbf{G}_c]\{\boldsymbol{\Delta}_c\}$$

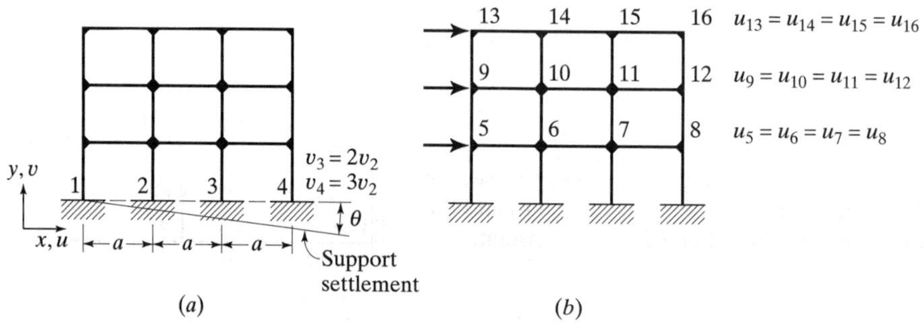

Figure 13.3 Representative constraint conditions.

or

$$\{\boldsymbol{\Delta}_e\} = [\boldsymbol{\Gamma}_{ec}]\{\boldsymbol{\Delta}_c\} \tag{13.20}$$

with

$$[\boldsymbol{\Gamma}_{ec}] = -[\mathbf{G}_e]^{-1}[\mathbf{G}_c] \tag{13.21}$$

To account for the constraint condition in the stiffness analysis, the stiffness equations must be partitioned to separate the degrees of freedom $\{\boldsymbol{\Delta}_e\}$ and $\{\boldsymbol{\Delta}_c\}$

$$\begin{Bmatrix} \mathbf{P}_e \\ \hline \mathbf{P}_c \end{Bmatrix} = \left[\begin{array}{c|c} \mathbf{K}_{ee} & \mathbf{K}_{ec} \\ \hline \mathbf{K}_{ce} & \mathbf{K}_{cc} \end{array}\right] \begin{Bmatrix} \boldsymbol{\Delta}_e \\ \hline \boldsymbol{\Delta}_c \end{Bmatrix} \tag{13.22}$$

and, by substitution of Equation 13.20,

$$\{\mathbf{P}_e\} = [[\mathbf{K}_{ee}][\boldsymbol{\Gamma}_{ec}] + [\mathbf{K}_{ec}]]\{\boldsymbol{\Delta}_c\} \tag{13.22a}$$

$$\{\mathbf{P}_c\} = [[\mathbf{K}_{ce}][\boldsymbol{\Gamma}_{ec}] + [\mathbf{K}_{cc}]]\{\boldsymbol{\Delta}_c\} \tag{13.22b}$$

The degrees of freedom to be eliminated, $\{\boldsymbol{\Delta}_e\}$, have been expressed in terms of the degrees of freedom to be retained, $\{\boldsymbol{\Delta}_c\}$, by the transformation relationship, Equation 13.20. Equation 13.22b could be solved, as given, for $\{\boldsymbol{\Delta}_c\}$, and this solution could then be inserted in Equation 13.20 to yield $\{\boldsymbol{\Delta}_e\}$. The matrix on the right side of Equation 13.22b is unsymmetric, however, so that the solution of this equation is relatively expensive. To produce a symmetric set of equation coefficients and to establish relationships between $\{\boldsymbol{\Delta}_c\}$ and the counterpart forces, there must be a corresponding, or contragredient, transformation of joint forces. This transformation is given by the transpose of the matrix that transformed the displacements; that is (see Section 5.1),

$$\{\hat{\mathbf{P}}_c\} = [\boldsymbol{\Gamma}_{ec}]^{\mathrm{T}}\{\mathbf{P}_e\} \tag{13.23}$$

We have placed the "hat" (^) over $\{\hat{\mathbf{P}}_c\}$ to emphasize that it represents the transference of forces $\{\mathbf{P}_e\}$ to the location of the degrees of freedom *c*. Such forces supplement the loads $\{\mathbf{P}_c\}$ initially at these degrees of freedom. Hence, the total effective loads at degree of freedom *c*, $\{\overline{\mathbf{P}}_c\}$ are

$$\{\overline{\mathbf{P}}_c\} = \{\mathbf{P}_c\} + \{\hat{\mathbf{P}}_c\} \tag{13.24}$$

and, by use of Equations 13.22,

$$\begin{aligned} \{\overline{\mathbf{P}}_c\} &= \{\mathbf{P}_c\} + [\boldsymbol{\Gamma}_{ec}]^{\mathrm{T}}\{\mathbf{P}_e\} \\ &= [[\mathbf{K}_{cc}] + [\mathbf{K}_{ce}][\boldsymbol{\Gamma}_{ec}] + [\boldsymbol{\Gamma}_{ec}]^{\mathrm{T}}[\mathbf{K}_{ec}] + [\boldsymbol{\Gamma}_{ec}]^{\mathrm{T}}[\mathbf{K}_{ee}][\boldsymbol{\Gamma}_{ec}]]\{\boldsymbol{\Delta}_c\} \end{aligned} \tag{13.25}$$

Equation 13.25 discloses that if the transformation of degrees of freedom, $[\boldsymbol{\Gamma}_{ec}]$ can be established from a set of constraint conditions, then its use in this equation will result in a reduction of the degrees of freedom to be solved in the subsequent stiffness analysis. This reduction is equal to the number of constraint conditions. The eliminated displacements can subsequently be determined from Equation 13.20. Example 13.4 illustrates this.

EXAMPLE 13.4

The beam shown is constrained to displace in such a manner that $v_2 = v_3$ and $\theta_2 = -\theta_3$. Calculate v_3 and the bending moments at points 2 and 3. EI = constant.

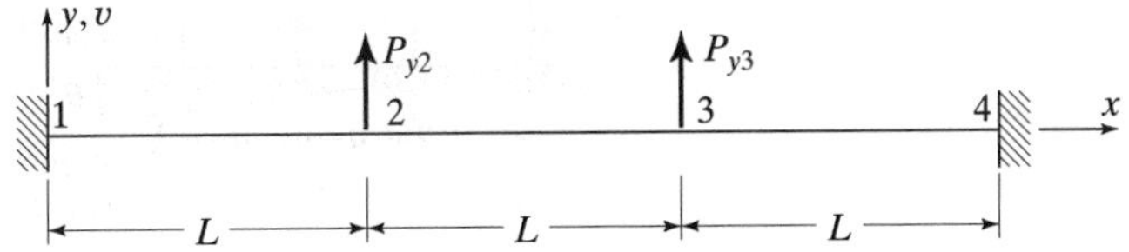

The constraint conditions will be used to eliminate the v_2 and θ_2 degrees of freedom. The global stiffness equation of the structure, arranged so as to facilitate the elimination, is

$$\begin{Bmatrix} P_{y2} \\ P_{m2} \\ \hline P_{y3} \\ P_{m3} \end{Bmatrix} = \frac{2EI}{L^3} \left[\begin{array}{cc|cc} 12 & 0 & -6 & 3L \\ 0 & 4L^2 & -3L & L^2 \\ \hline -6 & -3L & 12 & 0 \\ 3L & L^2 & 0 & 4L^2 \end{array}\right] \begin{Bmatrix} v_2 \\ \theta_2 \\ v_3 \\ \theta_3 \end{Bmatrix} = \left[\begin{array}{c|c} \mathbf{K}_{ee} & \mathbf{K}_{ec} \\ \hline \mathbf{K}_{ce} & \mathbf{K}_{cc} \end{array}\right] \begin{Bmatrix} v_2 \\ \theta_2 \\ \hline v_3 \\ \theta_3 \end{Bmatrix}$$

The stipulated constraints can be written as

$$\left[\begin{array}{cc|cc} 1 & 0 & -1 & 0 \\ 0 & 1 & 0 & 1 \end{array}\right] \begin{Bmatrix} v_2 \\ \theta_2 \\ \hline v_3 \\ \theta_3 \end{Bmatrix} = \begin{Bmatrix} 0 \\ 0 \end{Bmatrix}$$

Solving for v_2, θ_2,

$$\begin{Bmatrix} v_2 \\ \theta_2 \end{Bmatrix} = \begin{bmatrix} 1 & 0 \\ 0 & -1 \end{bmatrix} \begin{Bmatrix} v_3 \\ \theta_3 \end{Bmatrix} = [\boldsymbol{\Gamma}_{ec}] \begin{Bmatrix} v_3 \\ \theta_3 \end{Bmatrix}$$

Now, in accordance with Equation 13.25,

$$\{\mathbf{P}_c\} + [\boldsymbol{\Gamma}_{ec}]^{\mathrm{T}}\{\mathbf{P}_e\} = [[\mathbf{K}_{cc}] + [\mathbf{K}_{ce}][\boldsymbol{\Gamma}_{ec}] + [\boldsymbol{\Gamma}_{ec}]^{\mathrm{T}}[\mathbf{K}_{ec}] + [\boldsymbol{\Gamma}_{ec}]^{\mathrm{T}}[\mathbf{K}_{ee}][\boldsymbol{\Gamma}_{ec}]]\{\boldsymbol{\Delta}_c\}$$

$$\begin{Bmatrix} P_{y3} \\ P_{m3} \end{Bmatrix} + \begin{bmatrix} 1 & 0 \\ 0 & -1 \end{bmatrix} \begin{Bmatrix} P_{y2} \\ P_{m2} \end{Bmatrix} = \frac{2EI}{L^3} \left[\begin{bmatrix} 12 & 0 \\ 0 & 4L^2 \end{bmatrix} + \begin{bmatrix} -6 & -3L \\ 3L & L^2 \end{bmatrix} \begin{bmatrix} 1 & 0 \\ 0 & -1 \end{bmatrix} \right.$$
$$\left. + \begin{bmatrix} 1 & 0 \\ 0 & -1 \end{bmatrix} \begin{bmatrix} -6 & 3L \\ -3L & L^2 \end{bmatrix} + \begin{bmatrix} 1 & 0 \\ 0 & -1 \end{bmatrix} \begin{bmatrix} 12 & 0 \\ 0 & 4L^2 \end{bmatrix} \begin{bmatrix} 1 & 0 \\ 0 & -1 \end{bmatrix} \right] \begin{Bmatrix} v_3 \\ \theta_3 \end{Bmatrix}$$

And, after forming the products and collecting terms,

$$\begin{Bmatrix} P_{y3} + P_{y2} \\ P_{m3} - P_{m2} \end{Bmatrix} = \frac{2EI}{L^3} \begin{bmatrix} 12 & 6L \\ 6L & 6L^2 \end{bmatrix} \begin{Bmatrix} v_3 \\ \theta_3 \end{Bmatrix}$$

After inversion,

$$\begin{Bmatrix} v_3 \\ \theta_3 \end{Bmatrix} = \frac{L}{12EI} \begin{bmatrix} L^2 & -L \\ -L & 2 \end{bmatrix} \begin{Bmatrix} P_{y3} + P_{y2} \\ P_{m3} - P_{m2} \end{Bmatrix}$$

and, with $P_{m3}, P_{m2} = 0$,

$$v_3 = \frac{(P_{y3} + P_{y2})L^3}{12EI}$$

$$\theta_3 = -\frac{(P_{y3} + P_{y2})L^2}{12EI}$$

13.4 JOINT COORDINATES

There are many important practical circumstances under which the coordinate axes used to describe behavior at a joint in the global equations must be different from the global axes. Such *joint axes* were mentioned in Section 2.2. We now describe some of these circumstances and develop procedures to deal with them.

A typical situation is shown in Figure 13.4*a*. The behavior at joint i would usually be described in terms of the x- and y-direction displacements u_i and v_i. Because of the indicated support condition, however, it is necessary to impose a constraint in the y''

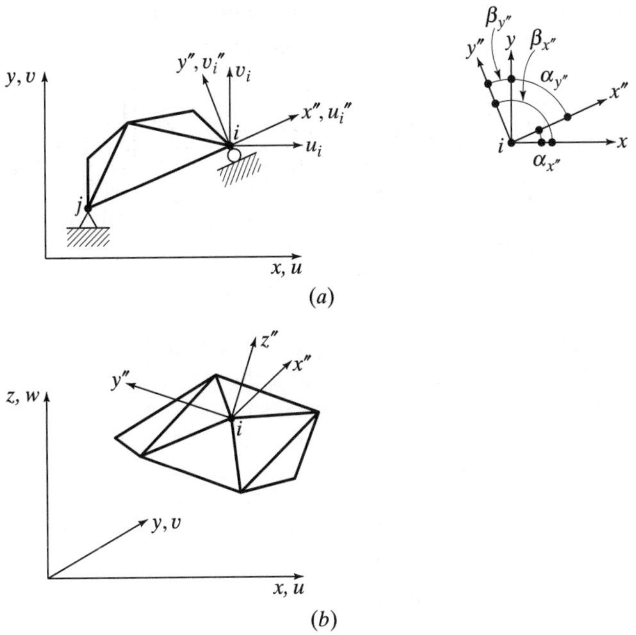

Figure 13.4 Joint coordinates.

direction, $v_i'' = 0$. This support constraint cannot be implemented merely by setting one global-direction degree of freedom to zero, thereby eliminating a single column of the global stiffness matrix. But it can be accomplished by using u_i'' and v_i'' as degrees of freedom in the global equations and then setting the single degree of freedom v_i'' equal to zero.

Another situation is described in Figure 13.4*b*. This portrays a segment of a planar truss that is a portion of a large space truss referred to global axes x-y-z. Local axes x''-y''-z'', differing from the global directions, are positioned at joint i of the planar truss with x'' and y'' in the plane of the truss. Clearly, the structure possesses no stiffness in the z'' direction. If the stiffness equations are written with respect to the global axes, this condition will not be apparent from inspection of such equations, since the x'' and y'' stiffnesses have components in all of the global directions. What is desired is to have a zero row and column in the global stiffness relationships, which can then be struck out before solution of the equations. This can be accomplished if the joint axes are used to describe, in the global equations, the behavior at joint i. The row and column corresponding to stiffness in the z'' direction will then be zero and can be eliminated directly.

The employment of joint coordinates requires a selective transformation of degrees of freedom and corresponding forces. These transformations must follow the algebraic procedures developed in Section 5.1.2 (Eqs. 5.12–5.16). Here we may view the process as one of transforming from a set of equations written entirely in global coordinates:

$$\{\mathbf{P}\} = [\mathbf{K}]\{\mathbf{\Delta}\} \tag{13.26}$$

to one that is written in a set of mixed coordinates, some global and some referred to particular joints. We designate this set as

$$\{\mathbf{P}''\} = [\mathbf{K}'']\{\mathbf{\Delta}''\} \tag{13.27}$$

A question of immediate significance is how [**K**] is transformed into [**K**″]. Following Equation 5.16, which is completely general, we may write

$$\{\mathbf{K}''\} = [\boldsymbol{\Gamma}]^{\mathrm{T}}[\mathbf{K}][\boldsymbol{\Gamma}] \tag{13.28}$$

where, now, the transformation matrix may be defined as a square matrix having: (1) unity on the diagonals and zeros elsewhere in all rows and columns corresponding to unaltered degrees of freedom; and (2) appropriate direction cosines in the positions corresponding to all pairs of altered degrees of freedom (one global and one joint degree of freedom).

The manner in which the global stiffness matrix is modified to accommodate joint coordinates may be detailed by using the problem of Figure 13.4*a* as an example. The direction angles linking the global axes to the joint coordinate axes at node *i* are defined in the figure. Using the notation of Section 5.1 and the coordinates of the points *i* and *j*, the direction cosines of the *x*-*y* axes with respect to the *x*″-*y*″ axes are $l_{x''} = m_{y''} = (x_i - x_j)/L$, $m_{x''} = -l_{y''} = -(y_i - y_j)/L$, where $L = \sqrt{(x_i - x_j)^2 + (y_i - y_j)^2}$. The transformation matrix

$$[\boldsymbol{\Gamma}] = \begin{array}{c} \\ \Delta_1 \\ u_i \\ v_i \\ \Delta_n \end{array} \begin{array}{c} \begin{array}{cccc} \Delta_1 & u_i'' & v_i'' & \Delta_n \end{array} \\ \begin{bmatrix} 1 & 0 & 0 & 0 \\ 0 & l_{x''} & m_{x''} & 0 \\ 0 & l_{y''} & m_{y''} & 0 \\ 0 & 0 & 0 & 1 \end{bmatrix} \end{array} \tag{13.29}$$

Substituting this value of [**Γ**] in Equation 13.28, the transformation may be made. The process is illustrated in Example 13.5.

EXAMPLE 13.5

The illustrated truss was analyzed in Examples 3.1 and 3.2 for support conditions of a hinge at point *c* and a roller on a horizontal plane at point *b*. Analyze the same structure for a hinge at point *c* and a roller on an inclined plane at point *b*.

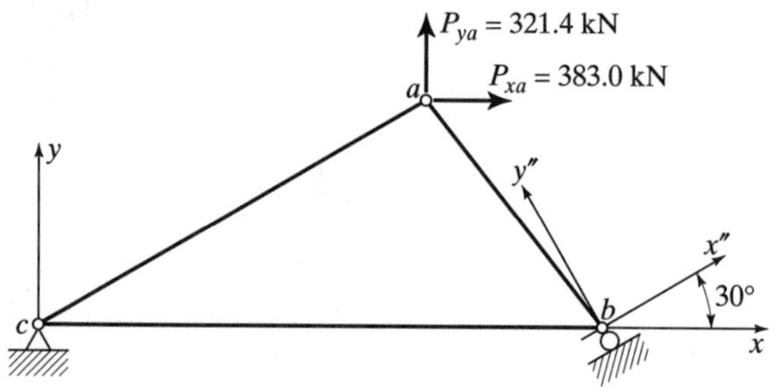

The global stiffness matrix for the unconstrained state of this structure was established in Example 3.1. For convenience, the boundary conditions at point *c*, $u_c = v_c = 0$, are applied to the matrix, since these conditions exist in both Example 3.2 and the present example and do not affect the operations to be performed. Thus, from Example 3.1,

$$\{\mathbf{P}\} = [\mathbf{K}]\{\boldsymbol{\Delta}\} \quad \begin{Bmatrix} P_{xa} \\ P_{ya} \\ P_{xb} \\ P_{yb} \end{Bmatrix} = 10^2 \begin{bmatrix} 6.348 & -1.912 & -3.536 & 3.536 \\ & 4.473 & 3.536 & -3.536 \\ & & 6.830 & -3.536 \\ \text{Sym.} & & & 3.536 \end{bmatrix} \begin{Bmatrix} u_a \\ v_a \\ u_b \\ v_b \end{Bmatrix}$$

Now, from the relationship between the above degrees of freedom and those pertaining to the inclined plane,

$$\{\boldsymbol{\Delta}\} = [\boldsymbol{\Gamma}]\{\boldsymbol{\Delta}''\} \quad \begin{Bmatrix} u_a \\ v_a \\ u_b \\ v_b \end{Bmatrix} = \begin{bmatrix} 1 & 0 & 0 & 0 \\ 0 & 1 & 0 & 0 \\ 0 & 0 & 0.866 & -0.500 \\ 0 & 0 & 0.500 & 0.866 \end{bmatrix} \begin{Bmatrix} u_a \\ v_a \\ u''_b \\ v''_b \end{Bmatrix}$$

Apply the transformation to the global stiffness matrix:

$$\{\mathbf{P}\} = [\mathbf{K}][\boldsymbol{\Gamma}]\{\boldsymbol{\Delta}''\} \quad \begin{Bmatrix} P_{xa} \\ P_{ya} \\ P_{xb} \\ P_{yb} \end{Bmatrix} = 10^2 \begin{bmatrix} 6.348 & -1.912 & -1.294 & 4.830 \\ -1.912 & 4.473 & 1.294 & -4.830 \\ -3.536 & 3.536 & 4.147 & -6.477 \\ 3.536 & -3.536 & -1.294 & 4.830 \end{bmatrix} \begin{Bmatrix} u_a \\ v_a \\ u''_b \\ v''_b \end{Bmatrix}$$

Now, let $\{\mathbf{P}''\} = [\boldsymbol{\Gamma}]^{\mathrm{T}}\{\mathbf{P}\}$, by Equation 5.14 (or Equation 13.28):

$$\{\mathbf{P}''\} = [\boldsymbol{\Gamma}]^{\mathrm{T}}[\mathbf{K}][\boldsymbol{\Gamma}]\{\boldsymbol{\Delta}''\} \quad \begin{Bmatrix} P_{xa} \\ P_{ya} \\ P''_{xb} \\ P''_{yb} \end{Bmatrix} = 10^2 \begin{bmatrix} 6.348 & -1.912 & -1.294 & 4.830 \\ -1.912 & 4.473 & 1.294 & -4.830 \\ -1.294 & 1.294 & 2.944 & -3.194 \\ 4.830 & -4.830 & -3.194 & 7.421 \end{bmatrix} \begin{Bmatrix} u_a \\ v_a \\ u''_b \\ v''_b \end{Bmatrix}$$

The remaining boundary condition, $v''_b = 0$, is now enforced. Thus

$$\begin{Bmatrix} 383.0 \\ 321.4 \\ 0 \end{Bmatrix} = 10^2 \begin{bmatrix} 6.348 & -1.912 & -1.294 \\ -1.912 & 4.473 & 1.294 \\ -1.294 & 1.294 & 2.944 \end{bmatrix} \begin{Bmatrix} u_a \\ v_a \\ u''_b \end{Bmatrix}$$

Solving:

$$\begin{Bmatrix} u_a \\ v_a \\ u''_b \end{Bmatrix} = \begin{Bmatrix} 0.928 \\ 1.143 \\ -0.094 \end{Bmatrix} \text{ mm}$$

It may be demonstrated readily that it is not necessary to carry out the matrix multiplication of Equation 13.28 in full, since only a limited number of rows and columns are affected. The effects on the rows and columns of the global stiffness matrix are as follows.

1. The column of the original global stiffness matrix multiplied by u_i is multiplied by $l_{x''}$, and added to the product of $l_{y''}$ and the column multiplied by v_i. The resulting column vector, which corresponds to u''_i, replaces the column vector corresponding to u_i. A similar operation replaces v_i with v''_i.
2. A new row in the P_{xi} location is constructed by multiplying the row corresponding to P_{xi} by $l_{x''}$ and the row corresponding to P_{yi} by $l_{y''}$, and adding the two. A new row in the P_{yi} location is also formed by performance of corresponding opera-

tions. In following the sequence described, the rows operated upon must contain u_i'' and v_i'' columns before products are formed.

These operations are illustrated schematically in Figure 13.5. It is suggested that the reader use them to verify Example 13.5.

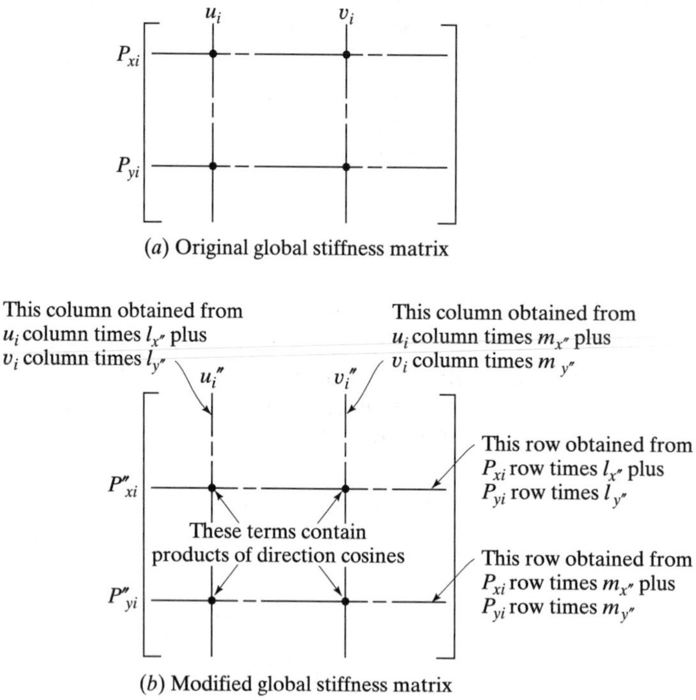

Figure 13.5 Replacement of global-coordinate displacements by local-coordinate displacements.

The element stiffness matrices could alternatively be formed directly in terms of the relevant local axes. This presents certain difficulties in input data assimilation because the support conditions are referred to the joints rather than to the elements, but the procedure is nevertheless concise and efficient. Formation of the global stiffness matrix and all other operations then proceed in the usual way. Still another approach is to treat Equations 13.26 and 13.27 as constraint conditions, in the manner described in the Section 13.3.

One scheme that avoids the algebraic complexity of the preceding approaches is to introduce a special boundary element, as shown in Figure 13.6. This element is then

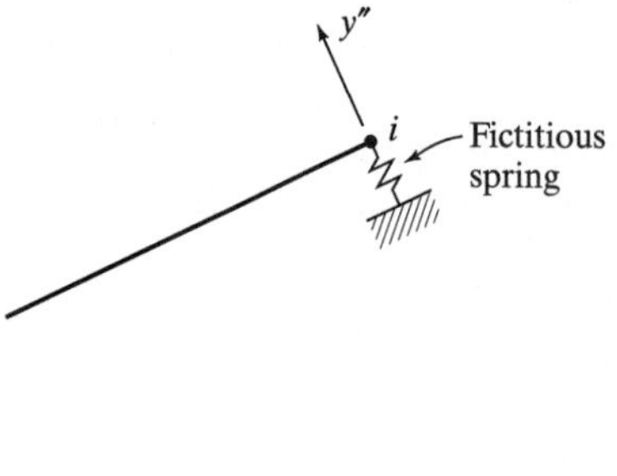

Figure 13.6 Artificial supports.

assigned a value that is three or four orders of magnitude greater than the other stiffness coefficients.

13.5 CONNECTIONS AND JOINTS

Throughout the text we have emphasized that the analytical model is an idealization that approximates the behavior of the actual structure. For example, members that have depth and width have in all cases been reduced to line elements. The objective of this section is to introduce basic concepts of modeling the behavior of connections. Emphasis is placed on elementary methods for incorporating connection stiffness and joint size within large-scale framework analyses.

13.5.1 Flexible Connections

With respect to stiffness, connections may be classified according to their moment-rotation relationships into one of three categories: ideally rigid, ideally pinned, and semi-rigid (see Figure 13.7). These curves express the moment M transmitted through the connection as a function of the relative rotation θ_r between two connected members. In a rigid or fully restrained connection, no relative rotation occurs for the moment resisted and the M-θ_r curve is a vertical line. For a pinned or unrestrained connection, no moment develops for the relative rotation imposed and hence the

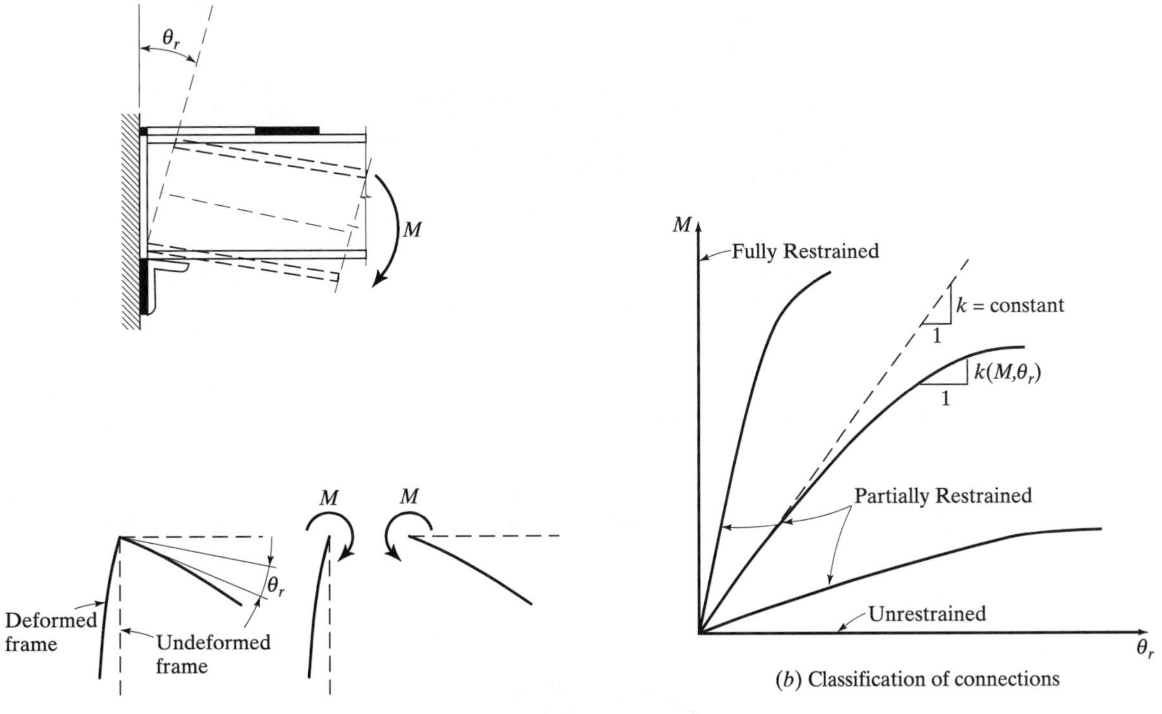

(*a*) Definition of M and θ_r

Figure 13.7 Connection characteristics.

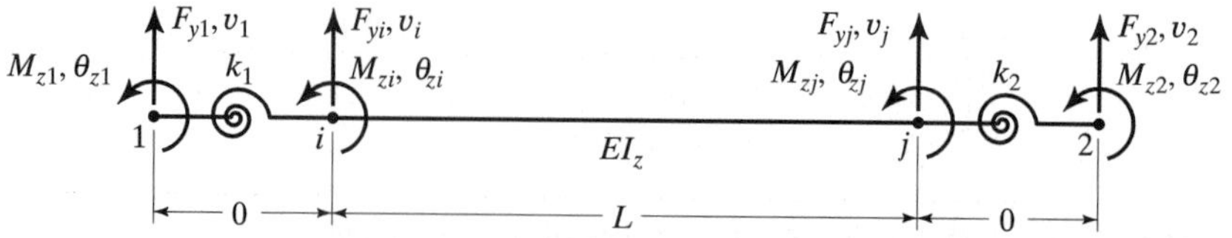

Figure 13.8 Beam element with rotational springs.

M-θ_r curve is a horizontal line. All intermediate conditions are, in principle, semi-rigid or partially retrained. In practice, however, moderate departures from the ideals are ignored, and most analyses are performed by assuming either rigid or pinned connections.

Other systems for classifying connections that also include strength and ductility attributes are available in the literature (see Refs. 13.1 and 13.2). The process of classifying a particular connection and, in turn, determining its corresponding M-θ_r curve is important and requires considerable judgment and knowledge of connection behavior. Since this process is not within the purview of this text, we will assume that the classification has been made and we'll focus on implementation of the results.

One method for approximating the effect of connection flexibility within an analytical model is to include a linear rotational spring at each end of a beam element (Figure 13.8). The stiffness k of these springs is the ratio of the transmitted moment to the rotation within the connection. That is,

$$M = k\theta_r \tag{13.30}$$

in which, at the left and right ends of the element shown in Figure 13.8, Equation 13.30 is $M_{z1} = k_1(\theta_{z1} - \theta_{zi})$ and $M_{z2} = k_2(\theta_{z2} - \theta_{zj})$, respectively. In this approximation, the length of the connection is ignored and the spring stiffness k is generally taken as an empirically determined initial stiffness of the connection (Figure 13.7b). By varying the stiffness of the springs from $k = 0$ to $k = \infty$, the above three general connection categories can be defined. This approach is usually limited to linear analysis under service loads.

By letting $k_1 = \alpha_1 EI_z/L$ and $k_2 = \alpha_2 EI_z/L$, the first-order elastic force-displacement relationship for a beam with flexible end-connections is

$$\begin{Bmatrix} F_{y1} \\ M_{z1} \\ F_{y2} \\ M_{z2} \end{Bmatrix} = \frac{\alpha EI_z}{L} \begin{bmatrix} \frac{12}{L^2}\left(1 + \frac{\alpha_1 + \alpha_2}{\alpha_1\alpha_2}\right) & \frac{6}{L}\left(1 + \frac{2}{\alpha_2}\right) & \frac{-12}{L^2}\left(1 + \frac{\alpha_1 + \alpha_2}{\alpha_1\alpha_2}\right) & \frac{6}{L}\left(1 + \frac{2}{\alpha_1}\right) \\ & 4\left(1 + \frac{3}{\alpha_2}\right) & \frac{-6}{L}\left(1 + \frac{2}{\alpha_2}\right) & 2 \\ & & \frac{12}{L^2}\left(1 + \frac{\alpha_1 + \alpha_2}{\alpha_1\alpha_2}\right) & \frac{-6}{L}\left(1 + \frac{2}{\alpha_1}\right) \\ \text{Sym.} & & & 4\left(1 + \frac{3}{\alpha_1}\right) \end{bmatrix} \begin{Bmatrix} v_1 \\ \theta_{z1} \\ v_2 \\ \theta_{z2} \end{Bmatrix} \tag{13.31}$$

where $\alpha = \alpha_1\alpha_2/(\alpha_1\alpha_2 + 4\alpha_1 + 4\alpha_2 + 12)$. The development of Equation 13.31 is provided in the following two examples. In Example 13.6, the flexibility-stiffness transformations presented in Section 4.4 are employed and in Example 13.7 the same result is obtained by the condensation method of Section 13.1.

EXAMPLE 13.6

Using Equation 4.25, develop the stiffness relationship for a beam with rotational springs at its ends ($M = k\theta$) and bent about its z axis.

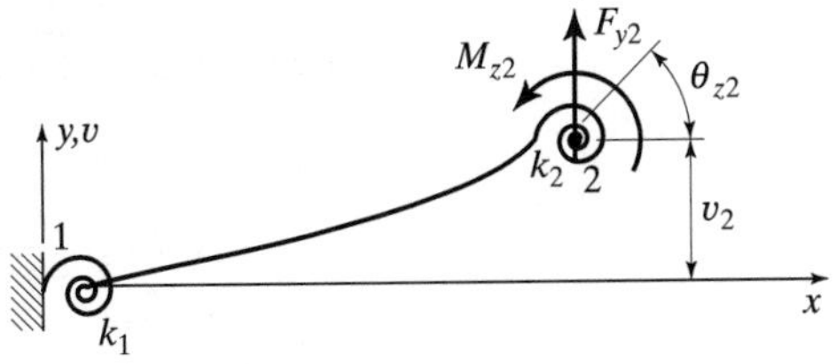

Modify the result of Example 4.3 to include the effects of the rotational springs, the flexibility influence coefficients are determined as:

$$v_2 = \frac{F_{y2}L^3}{3EI_z} + \frac{M_{z2}L^2}{2EI_z} + \frac{(F_{y2}L + M_{z2})}{k_1} L$$

$$\theta_{z2} = \frac{F_{y2}L^2}{2EI_z} + \frac{M_{z2}L}{EI_z} + \left(\frac{F_{y2}L + M_{z2}}{k_1}\right) + \frac{M_{z2}}{k_2}$$

With $k_1 = \alpha_1 \dfrac{EI_z}{L}$ and $k_2 = \alpha_2 \dfrac{EI_z}{L}$, assemble equations in matrix form $\{\boldsymbol{\Delta}\} = [\mathbf{d}]\{\mathbf{F}\}$:

$$\begin{Bmatrix} v_2 \\ \theta_{z2} \end{Bmatrix} = \frac{L}{EI_z} \begin{bmatrix} \dfrac{L^2}{3} + \dfrac{L^2}{\alpha_1} & \dfrac{L}{2} + \dfrac{L}{\alpha_1} \\ \dfrac{L}{2} + \dfrac{L}{\alpha_1} & 1 + \dfrac{1}{\alpha_1} + \dfrac{1}{\alpha_2} \end{bmatrix} \begin{Bmatrix} F_{y2} \\ M_{z2} \end{Bmatrix}$$

Modifying the results of Example 4.4, $[\mathbf{d}]^{-1}$ and $[\boldsymbol{\Phi}]$ are:

$$[\mathbf{d}]^{-1} = \frac{\alpha EI_z}{L} \begin{bmatrix} \dfrac{12}{L^2}\left(1 + \dfrac{\alpha_1 + \alpha_2}{\alpha_1\alpha_2}\right) & \dfrac{-6}{L}\left(1 + \dfrac{2}{\alpha_1}\right) \\ \dfrac{-6}{L}\left(1 + \dfrac{2}{\alpha_1}\right) & 4\left(1 + \dfrac{3}{\alpha_1}\right) \end{bmatrix}$$

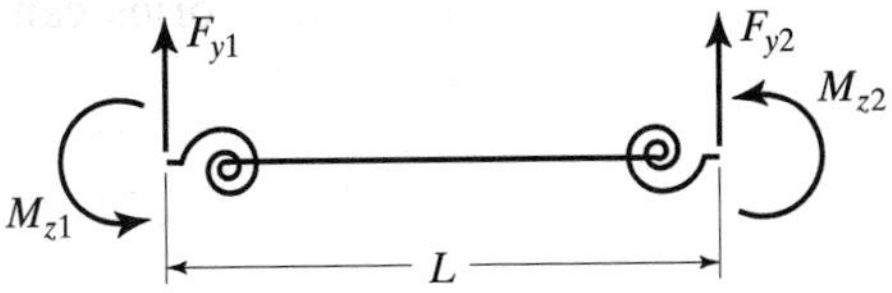

in which $\alpha = \alpha_1\alpha_2/(\alpha_1\alpha_2 + 4\alpha_1 + 4\alpha_2 + 12)$, and

$$\begin{Bmatrix} F_{y1} \\ M_{z1} \end{Bmatrix} = \begin{bmatrix} -1 & 0 \\ -L & -1 \end{bmatrix} \begin{Bmatrix} F_{y2} \\ M_{z2} \end{Bmatrix} \quad \text{with} \quad [\boldsymbol{\Phi}] = -\begin{bmatrix} 1 & 0 \\ L & 1 \end{bmatrix}$$

Apply Equation 4.25:
$$[\mathbf{k}] = \left[\begin{array}{c|c} [\mathbf{d}]^{-1} & [\mathbf{d}]^{-1}[\boldsymbol{\Phi}]^{\mathrm{T}} \\ \hline [\boldsymbol{\Phi}][\mathbf{d}]^{-1} & [\boldsymbol{\Phi}][\mathbf{d}]^{-1}[\boldsymbol{\Phi}]^{\mathrm{T}} \end{array}\right]$$

$$\begin{Bmatrix} F_{y2} \\ M_{z2} \\ F_{y1} \\ M_{z1} \end{Bmatrix} = \frac{\alpha EI_z}{L} \begin{bmatrix} \dfrac{12}{L^2}\left(1 + \dfrac{\alpha_1 + \alpha_2}{\alpha_1\alpha_2}\right) & \dfrac{-6}{L}\left(1 + \dfrac{2}{\alpha_1}\right) & \dfrac{-12}{L^2}\left(1 + \dfrac{\alpha_1 + \alpha_2}{\alpha_1\alpha_2}\right) & \dfrac{-6}{L}\left(1 + \dfrac{2}{\alpha_2}\right) \\ & 4\left(1 + \dfrac{3}{\alpha_1}\right) & \dfrac{6}{L}\left(1 + \dfrac{2}{\alpha_1}\right) & 2 \\ & & \dfrac{12}{L^2}\left(1 + \dfrac{\alpha_1 + \alpha_2}{\alpha_1\alpha_2}\right) & \dfrac{6}{L}\left(1 + \dfrac{2}{\alpha_2}\right) \\ \text{Sym.} & & & 4\left(1 + \dfrac{3}{\alpha_2}\right) \end{bmatrix} \begin{Bmatrix} v_2 \\ \theta_{z2} \\ v_1 \\ \theta_{z1} \end{Bmatrix}$$

EXAMPLE 13.7

Show that Equation 13.31 can also be obtained by removing the internal degrees of freedom (v_i, θ_{zi}, v_j, θ_{zj}) through the condensation method of Section 13.1.

With $k_1 = \alpha_1 \dfrac{EI_z}{L}$ and $k_2 = \alpha_2 \dfrac{EI_z}{L}$, formulate stiffness matrix for each component:

$$\begin{Bmatrix} M_{z1} \\ M_{zi} \end{Bmatrix} = \frac{\alpha_1 EI_z}{L}\begin{bmatrix} 1 & -1 \\ -1 & 1 \end{bmatrix}\begin{Bmatrix} \theta_{z1} \\ \theta_{zi} \end{Bmatrix} \quad \text{and} \quad \begin{Bmatrix} M_{zj} \\ M_{z2} \end{Bmatrix} = \frac{\alpha_2 EI_z}{L}\begin{bmatrix} 1 & -1 \\ -1 & 1 \end{bmatrix}\begin{Bmatrix} \theta_{zj} \\ \theta_{z2} \end{Bmatrix}$$

$$\begin{Bmatrix} F_{yi} \\ M_{zi} \\ F_{yj} \\ M_{zj} \end{Bmatrix} = \frac{EI_z}{L}\begin{bmatrix} \frac{12}{L^2} & \frac{6}{L} & \frac{-12}{L^2} & \frac{6}{L} \\ \frac{6}{L} & 4 & \frac{-6}{L} & 2 \\ \frac{-12}{L^2} & \frac{-6}{L} & \frac{12}{L^2} & \frac{-6}{L} \\ \frac{6}{L} & 2 & \frac{-6}{L} & 4 \end{bmatrix}\begin{Bmatrix} v_i \\ \theta_{zi} \\ v_j \\ \theta_{zj} \end{Bmatrix}$$

Using the equilibrium requirements $F_{y1} = F_{yi}$ and $F_{y2} = F_{yj}$ and the compatibility requirements $v_1 = v_i$ and $v_2 = v_j$, assemble stiffness relationships for the entire system:

$$\begin{Bmatrix} M_{zi} \\ M_{zj} \\ \hline F_{y1} \\ M_{z1} \\ F_{y2} \\ M_{z2} \end{Bmatrix} = \frac{EI_z}{L}\left[\begin{array}{cc|cccc} 4+\alpha_1 & 2 & \frac{6}{L} & -\alpha_1 & \frac{-6}{L} & 0 \\ 2 & 4+\alpha_2 & \frac{6}{L} & 0 & \frac{-6}{L} & -\alpha_2 \\ \hline \frac{6}{L} & \frac{6}{L} & \frac{12}{L^2} & 0 & \frac{-12}{L^2} & 0 \\ -\alpha_1 & 0 & 0 & \alpha_1 & 0 & 0 \\ \frac{-6}{L} & \frac{-6}{L} & \frac{-12}{L^2} & 0 & \frac{12}{L^2} & 0 \\ 0 & -\alpha_2 & 0 & 0 & 0 & \alpha_2 \end{array}\right]\begin{Bmatrix} \theta_{zi} \\ \theta_{zj} \\ \hline v_1 \\ \theta_{z1} \\ v_2 \\ \theta_{z2} \end{Bmatrix}$$

To remove θ_{zi} and θ_{zj}, use Equation 13.5 with

$$[\mathbf{K}_{bb}] = \frac{EI_z}{L}\begin{bmatrix} 4+\alpha_1 & 2 \\ 2 & 4+\alpha_2 \end{bmatrix} \quad \text{and} \quad [\mathbf{K}_{bb}]^{-1} = \frac{EI_z}{L(\alpha_1\alpha_2 + 4\alpha_1 + 4\alpha_2 + 12)}\begin{bmatrix} 4+\alpha_2 & -2 \\ -2 & 4+\alpha_1 \end{bmatrix}$$

$$[\mathbf{K}_{bc}] = [\mathbf{K}_{cb}]^{\mathrm{T}} = \frac{EI_z}{L}\begin{bmatrix} \frac{6}{L} & -\alpha_1 & \frac{-6}{L} & 0 \\ \frac{6}{L} & 0 & \frac{-6}{L} & -\alpha_2 \end{bmatrix}$$

and

$$[\mathbf{K}_{cc}] = \frac{EI_z}{L}\begin{bmatrix} \frac{12}{L^2} & 0 & \frac{-12}{L^2} & 0 \\ 0 & \alpha_1 & 0 & 0 \\ \frac{-12}{L^2} & 0 & \frac{12}{L^2} & 0 \\ 0 & 0 & 0 & \alpha_2 \end{bmatrix}$$

$$[\hat{\mathbf{K}}_{cc}] = [\mathbf{K}_{cc}] - [\mathbf{K}_{cb}][\mathbf{K}_{bb}]^{-1}[\mathbf{K}_{bc}]$$

$$[\hat{\mathbf{K}}_{cc}] = \frac{\alpha EI_z}{L}\begin{bmatrix} \frac{12}{L^2}\left(1+\frac{\alpha_1+\alpha_2}{\alpha_1\alpha_2}\right) & \frac{6}{L}\left(1+\frac{2}{\alpha_2}\right) & \frac{-12}{L^2}\left(1+\frac{\alpha_1+\alpha_2}{\alpha_1\alpha_2}\right) & \frac{6}{L}\left(1+\frac{2}{\alpha_1}\right) \\ & 4\left(1+\frac{3}{\alpha_2}\right) & \frac{-6}{L}\left(1+\frac{2}{\alpha_2}\right) & 2 \\ & & \frac{12}{L^2}\left(1+\frac{\alpha_1+\alpha_2}{\alpha_1\alpha_2}\right) & \frac{-6}{L}\left(1+\frac{2}{\alpha_1}\right) \\ \text{Sym.} & & & 4\left(1+\frac{3}{\alpha_1}\right) \end{bmatrix}$$

with $\alpha = \alpha_1\alpha_2/(\alpha_1\alpha_2 + 4\alpha_1 + 4\alpha_2 + 12)$ this is the stiffness matrix of Equation 13.31.

To predict the strength limit state of frames with semi-rigid connections, nonlinear M-θ_r relationships should be employed. One method for accomplishing this is to perform an incremental analysis (see Sections 12.1–12.3) with equation 13.30 modified to

$$dM = k\, d\theta_r \tag{13.30a}$$

where k is the rotational spring stiffness for a particular load increment (see Figure 13.7*b*). In this case, k would be a prescribed nonlinear function of either the total end moment M or relative rotation θ_r (see Example 8.7, $M_a = k\theta_a$ with $k = \beta/\sqrt{\theta_a}$). Note that if the coefficients α_1 and α_2 are defined as functions that can be calibrated to model the nonlinear connection stiffnesses k_1 and k_2, an incremental form of Equation 13.31 can be employed. It should also be noted that for a nonlinear analysis that includes elements with linear or nonlinear rotational springs at their ends, the condensation method of Section 13.1 may be employed. That is, the approach illustrated in Example 13.7 could be used with the element's first-order elastic stiffness replaced with its appropriate nonlinear stiffness matrix.

The special case of a pinned connection at one end of an element and a rigid connection at the other end can be derived from Equation 13.31. For example, substitution of $\alpha_1 = 0$ to represent a pin at end 1 and $\alpha_2 = \infty$ for a rigid connection at end 2 results in

$$\begin{Bmatrix} F_{y1} \\ M_{z1} \\ F_{y2} \\ M_{z2} \end{Bmatrix} = \frac{EI_z}{L}\begin{bmatrix} \frac{3}{L^2} & 0 & \frac{-3}{L^2} & \frac{3}{L} \\ & 0 & 0 & 0 \\ & & \frac{3}{L^2} & \frac{-3}{L} \\ \text{Sym.} & & & 3 \end{bmatrix}\begin{Bmatrix} v_1 \\ \theta_{z1} \\ v_2 \\ \theta_{z2} \end{Bmatrix} \tag{13.32}$$

Likewise, the stiffness relationship for a rigid connection at end 1 ($\alpha_1 = \infty$) and a pin at end 2 ($\alpha_2 = 0$) is

$$\begin{Bmatrix} F_{y1} \\ M_{z1} \\ F_{y2} \\ M_{z2} \end{Bmatrix} = \frac{EI_z}{L} \begin{bmatrix} \frac{3}{L^2} & \frac{3}{L} & \frac{-3}{L^2} & 0 \\ & 3 & \frac{-3}{L} & 0 \\ & & \frac{3}{L^2} & 0 \\ \text{Sym.} & & & 0 \end{bmatrix} \begin{Bmatrix} v_1 \\ \theta_{z1} \\ v_2 \\ \theta_{z2} \end{Bmatrix} \tag{13.33}$$

Equations 13.32 and 13.33 can also be obtained using the plastic hinge concepts presented in Section 10.2. The stiffness matrix for an element with the bending moment released at one or both ends may be obtained from

$$[\mathbf{k}] = [\mathbf{k}_{ij}] - [\mathbf{k}_{ij}][\mathbf{G}][[\mathbf{G}]^{\mathrm{T}}[\mathbf{k}_{ij}][\mathbf{G}]]^{-1}[\mathbf{G}]^{\mathrm{T}}[\mathbf{k}_{ij}] \tag{13.34}$$

where $[\mathbf{k}_{ij}]$ is the stiffness matrix for the unreleased flexural element and the matrix $[\mathbf{G}]$ only contains nonzero values for the rotational degrees of freedom being released. For example, the $[\mathbf{G}]$ matrix for a planar element subjected to axial force and flexure with a pinned connection or, equivalently, zero moment at end 1 is

$$[\mathbf{G}]^{\mathrm{T}} = [0 \quad 0 \quad 1 \quad 0 \quad 0 \quad 0] \tag{13.35}$$

Likewise, for a three-dimensional element with both y- and z-moments released at end 2, $[\mathbf{G}]$ is defined as

$$[\mathbf{G}]^{\mathrm{T}} = \begin{bmatrix} 0 & 0 & 0 & 0 & 0 & 0 & 0 & 0 & 0 & 0 & 1 & 0 \\ 0 & 0 & 0 & 0 & 0 & 0 & 0 & 0 & 0 & 0 & 0 & 1 \end{bmatrix} \tag{13.36}$$

A benefit of using Equation 13.34 to incorporate pinned connections is that the stiffness matrix $[\mathbf{k}]$ can be generalized for nonlinear analysis by accounting for geometric and/or material nonlinear effects directly within $[\mathbf{k}_{ij}]$ as described in Chapters 8–10. Equation 13.34 can also be used to incorporate pinned connections in a linear elastic analysis in which $[\mathbf{k}_{ij}]$ has been modified to account for shear deformations (see Section 7.6.2). It is left as an exercise for the reader to demonstrate that the use of Equations 13.34 and 13.35 result in the form of Equation 13.32 (Problem 13.16).

13.5.2 Finite Joint Size

The joints in an actual structure are not mathematical points or nodes as assumed in the typical analytical model. Three cases are illustrated in Figure 13.9*a*:

1. Structural joints are of a finite size that can be on the order of 5% to 10% of the lengths of the intersecting members. This region or panel zone may represent a structural component with a stiffness much greater than that of the connected members.
2. In floor systems, it is common for the tops of beams and not their centroidal axes to share a common elevation.
3. For reasons related to connection geometry, the centroidal axis of a diagonal member may not intersect the nodes representing its ends.

Other cases include situations where the shear center of the member does not pass through its centroidal axis or when applied loads are eccentric to the nodes or the centroids of members.

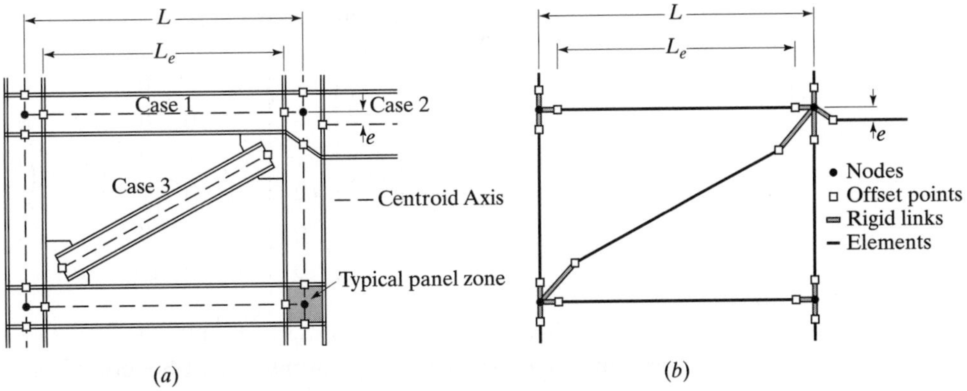

Figure 13.9 (*a*) Frame with moment-resisting connections. (*b*) Analysis model with rigid links used to model panel zones.

All of the above suggest the occasional need to offset the ends of elements from the arbitrary location of their defining nodes in the analytical model. As shown in Figure 13.9*b* and in detail in Figure 13.10*a*, one method for accomplishing this is to incorporate rigid links at the ends of an element. Equilibrium of the free-body diagrams in Figure 13.10*b* yields the relationship between forces at the ends of the element $\{\mathbf{F}_{ij}\}$ and the forces at the ends of the rigid links $\{\mathbf{F}_{12}\}$ as

$$\{\mathbf{F}_{12}\} = [\mathbf{T}]\{\mathbf{F}_{ij}\} \tag{13.37}$$

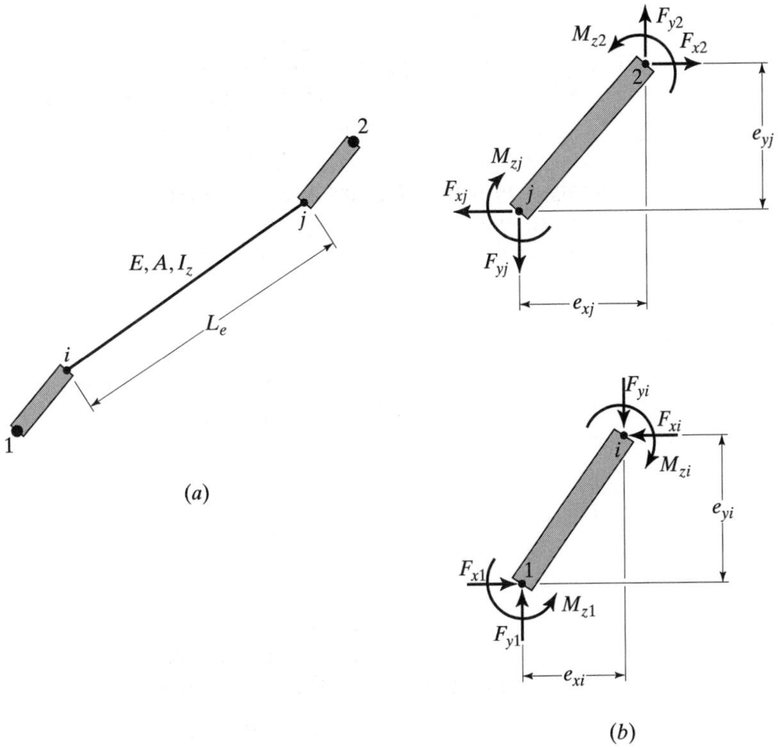

Figure 13.10 (*a*) Offset element with rigid links. (*b*) Free-body diagrams of rigid links.

where the force transformation matrix [T] is given by

$$[\mathbf{T}] = \begin{bmatrix} 1 & 0 & 0 & 0 & 0 & 0 \\ 0 & 1 & 0 & 0 & 0 & 0 \\ -e_{yi} & e_{xi} & 1 & 0 & 0 & 0 \\ 0 & 0 & 0 & 1 & 0 & 0 \\ 0 & 0 & 0 & 0 & 1 & 0 \\ 0 & 0 & 0 & e_{yj} & -e_{xj} & 1 \end{bmatrix} \tag{13.38}$$

Note that this matrix is not orthogonal, $[\mathbf{T}]^{-1} \neq [\mathbf{T}]^{\mathrm{T}}$. From the principle of virtual work, the contragredient transformation (see Section 5.1.3) for the displacements is

$$\{\boldsymbol{\Delta}_{ij}\} = [\mathbf{T}]^{\mathrm{T}}\{\boldsymbol{\Delta}_{12}\} \tag{13.39}$$

As illustrated in Example 13.8, these transformations can also be derived using the constraint method of Section 13.3.

EXAMPLE 13.8

By assuming rigid links at the ends of element ij, develop six linear constraint conditions in the form of Equation 13.20 with $\{\boldsymbol{\Delta}_e\} = \{\boldsymbol{\Delta}_{ij}\}$ and $\{\boldsymbol{\Delta}_c\} = \{\boldsymbol{\Delta}_{12}\}$. Show that these results may be used to develop a set of transformations equivalent to Equations 13.37 to 13.39.

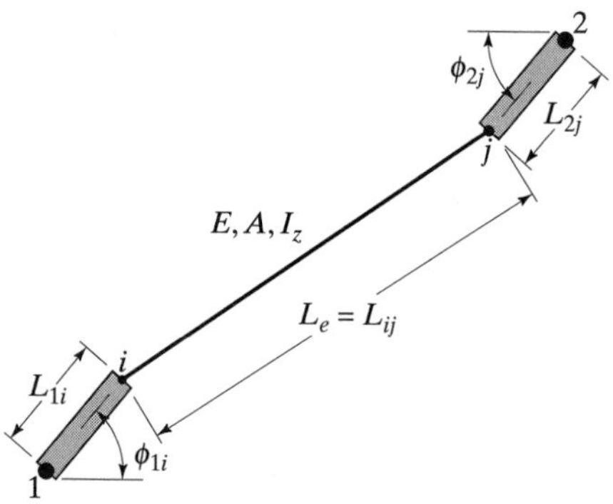

With $\{\boldsymbol{\Delta}_e\} = \{\boldsymbol{\Delta}_{ij}\} = \lfloor u_i \quad v_i \quad \theta_{zi} \quad u_j \quad v_j \quad \theta_{zj} \rfloor^{\mathrm{T}}$

$\{\boldsymbol{\Delta}_c\} = \{\boldsymbol{\Delta}_{12}\} = \lfloor u_1 \quad v_1 \quad \theta_{z1} \quad u_2 \quad v_2 \quad \theta_{z2} \rfloor^{\mathrm{T}}$

Write three constraint conditions for each rigid link.
Rigid link 1-i:

$$u_i = u_1 - [L_{1i}\cos\phi_{1i} - L_{1i}\cos(\phi_{1i} + \theta_{z1})]$$

with θ_{z1} small, $\cos\theta_{z1} \approx 1$ and $\sin\theta_{z1} \approx \theta_{z1}$

$$u_i = u_1 - (L_{1i}\sin\phi_{1i})\,\theta_{z1}$$

$$v_i = v_1 + [L_{1i}\sin(\phi_{1i} + \theta_{z1}) - L_{1i}\sin\phi_{1i}]$$

$$v_i = v_1 + (L_{1i}\cos\phi_{1i})\,\theta_{z1}$$

$$\theta_{zi} = \theta_{z1}$$

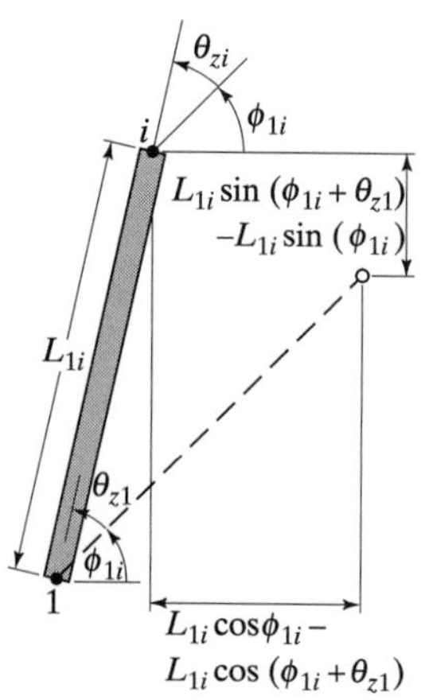

Rigid link 2-j:

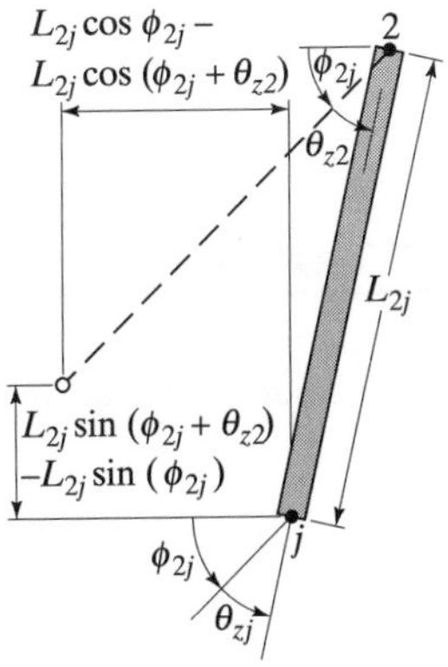

$$u_j = u_2 + [L_{2j}\cos\phi_{2j} - L_{2j}\cos(\phi_{2j} + \theta_{z2})]$$

with θ_{z2} small, $\cos\theta_{z2} \approx 1$ and $\sin\theta_{z2} = \theta_{z2}$

$$u_j = u_2 + (L_{2j}\sin\phi_{2j})\,\theta_{z2}$$

$$v_j = v_2 - [L_{2j}\sin(\phi_{2j} + \theta_{z2}) - L_{2j}\sin\phi_{2j}]$$

$$v_j = v_2 - (L_{2j}\cos\phi_{2j})\,\theta_{z2}$$

$$\theta_{zj} = \theta_{z2}$$

In the form of Equation 13.20

$$\begin{Bmatrix} u_i \\ v_i \\ \theta_{zi} \\ u_j \\ v_j \\ \theta_{zj} \end{Bmatrix} = \begin{bmatrix} 1 & 0 & -L_{1i}\sin\phi_{1i} & 0 & 0 & 0 \\ 0 & 1 & L_{1i}\cos\phi_{1i} & 0 & 0 & 0 \\ 0 & 0 & 1 & 0 & 0 & 0 \\ 0 & 0 & 0 & 1 & 0 & L_{2j}\sin\phi_{2j} \\ 0 & 0 & 0 & 0 & 1 & -L_{2j}\cos\phi_{2j} \\ 0 & 0 & 0 & 0 & 0 & 1 \end{bmatrix} \begin{Bmatrix} u_1 \\ v_1 \\ \theta_{z1} \\ u_2 \\ v_2 \\ \theta_{z2} \end{Bmatrix}$$

From virtual work, the contragredient transformation for the forces are

$$\begin{Bmatrix} F_{x1} \\ F_{y1} \\ M_{z1} \\ F_{x2} \\ F_{y2} \\ M_{z2} \end{Bmatrix} = \begin{bmatrix} 1 & 0 & 0 & 0 & 0 & 0 \\ 0 & 1 & 0 & 0 & 0 & 0 \\ -L_{1i}\sin\phi_{1i} & L_{1i}\cos\phi_{1i} & 1 & 0 & 0 & 0 \\ 0 & 0 & 0 & 1 & 0 & 0 \\ 0 & 0 & 0 & 0 & 1 & 0 \\ 0 & 0 & 0 & L_{2j}\sin\phi_{2j} & -L_{2j}\cos\phi_{2j} & 1 \end{bmatrix} \begin{Bmatrix} F_{xi} \\ F_{yi} \\ M_{zi} \\ F_{xj} \\ F_{yj} \\ M_{zj} \end{Bmatrix}$$

With reference to Figure 13.10(b) $e_{xi} = L_{1i}\cos\phi_{1i}$ $\quad e_{xj} = L_{2j}\cos\phi_{2j}$

$e_{yi} = L_{1i}\sin\phi_{1i}$ $\quad e_{yj} = L_{2j}\sin\phi_{2j}$

and hence Equations 13.37 to 13.39 are equivalent to the above transformations.

From Equations 13.37 and 13.39, the stiffness relationship in global coordinates for an element with rigid offsets can be expressed by

$$\{\mathbf{F}_{12}\} = [\mathbf{T}][\mathbf{\Gamma}_{ij}]^{\mathrm{T}}[\mathbf{k}'_{ij}][\mathbf{\Gamma}_{ij}][\mathbf{T}]^{\mathrm{T}}\{\mathbf{\Delta}_{12}\} \tag{13.40}$$

where the local stiffness matrix $[\mathbf{k}'_{ij}]$ and the rotational transformation matrix $[\mathbf{\Gamma}_{ij}]$ are both calculated using the length and orientation of the element as defined by offset points i and j (see Fig. 13.10a).

After nodal displacements have been calculated, use of Equation 13.39 yields the end forces associated and aligned with the offset element

$$\{\mathbf{F}'_{ij}\} = [\mathbf{k}'_{ij}]\{\mathbf{\Delta}'_{ij}\} = [\mathbf{k}'_{ij}][\mathbf{\Gamma}_{ij}]\{\mathbf{\Delta}_{ij}\} = [\mathbf{k}'_{ij}][\mathbf{\Gamma}_{ij}][\mathbf{T}]^{\mathrm{T}}\{\mathbf{\Delta}_{12}\} \tag{13.41}$$

Another method for incorporating the effects of finite joint sizes within a framework analysis is to represent the above rigid links by additional elements that are assigned stiffness coefficients three or four orders of magnitude greater than those of the offset elements. By employing this approach, Example 13.9 demonstrates that accounting for this factor typically increases the predicted stiffness and strength of the structure.

EXAMPLE 13.9

For the frame shown, compare results of analyses that do and do not include rigid links to model the panel zones.

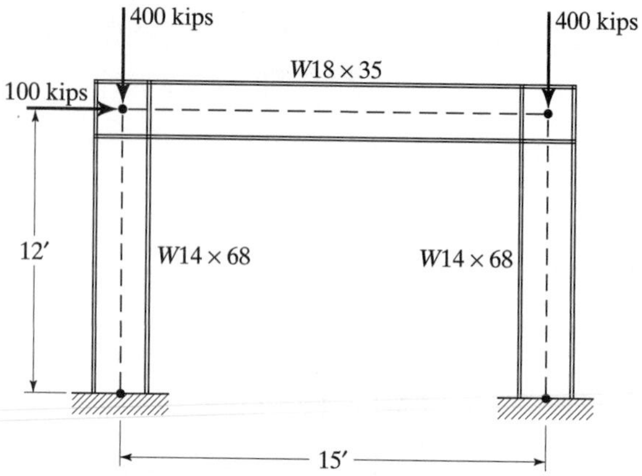

Analytical Models:

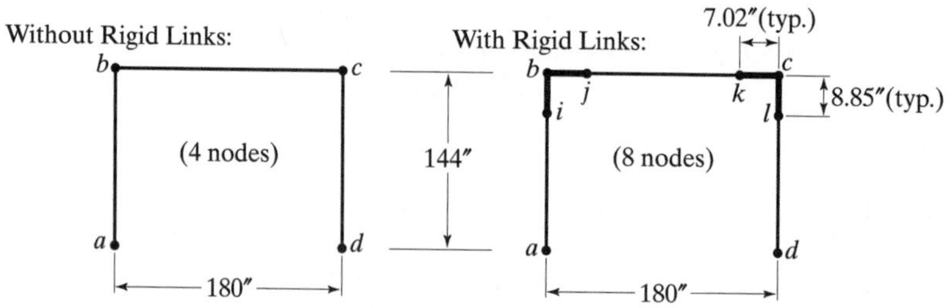

Material Properties:

All elements $E = 29{,}000$ ksi
$\sigma_y = 36$ ksi

Elements *ai*, *jk*, *dl* $E = 29 \times 10^3$ ksi
$\sigma_y = 36$ ksi
Elements *ib*, *bj*, *kc*, *lc* $E = 29 \times 10^6$ ksi
$\sigma_y = 36 \times 10^3$ ksi
(increased by several orders of magnitude to simulate rigid links)

Comparative Results:
(i) Elastic Analysis

Type	Quantity Measured	w/o Rigid Links	w/ Rigid Links	% Diff.
First-order	Max. Lateral Disp.	0.630 in.	0.543 in.	−16.0
	Max. Column Moment	3869 in. kips (pt. a)	3603 in. kips (pt. a)	−7.4
	Max. Beam Moment	2232 in. kips (pt. b)	2302 in. kips (pt. j)	+3.0
Second-order	Max. Lateral Disp.	0.659 in.	0.564 in.	−16.8
	Max. Column Moment	4020 in. kips (pt. a)	3723 in. kips (pt. a)	−8.0
	Max. Beam Moment	2338 in. kips (pt. b)	2394 in. kips (pt. k)	+2.3
Eigenvalue	Critical Load Ratio, λ	21.15	24.58	+14.0

(ii) Second-order Inelastic Analyses

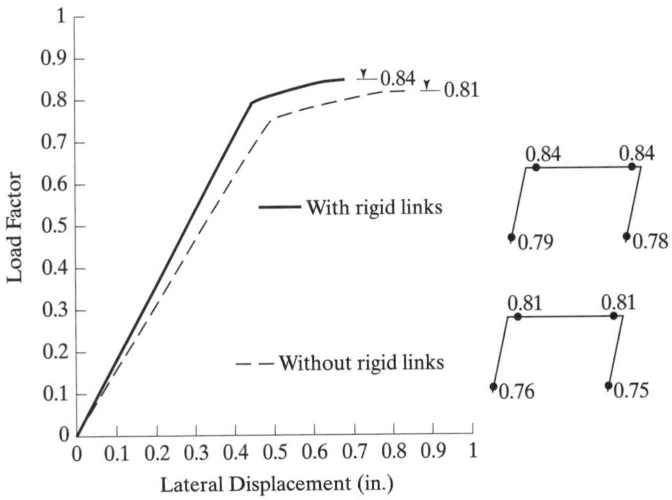

Finally, it should be noted that panel zones are not necessarily rigid components. Experimental and analytical studies have demonstrated that the high shear forces and corresponding deformations in these zones can have a pronounced effect on the ultimate strength of a frame. In this regard, nonlinear models that treat the panel zone as a subassemblage have been proposed. Reference 13.3 contains a comprehensive review of such models.

13.6 SYMMETRY AND ANTISYMMETRY

Many bridges, buildings, aircraft, ships, and structures such as cooling towers and radio telescopes possess some form of symmetry. Recognition of symmetry permits a complete structural analysis to be made by considering only a portion of the total structure: $\frac{1}{2}$, $\frac{1}{4}$, or even less, depending on the type and degree of symmetry. Example 5.9 was an elementary illustration of the application of the principles of symmetry. We now consider the subject in a somewhat more formal way.

There are three aspects to the problem: (1) recognition and definition of the type of symmetry; (2) manipulation of the loads and forces in a way that enables one to take advantage of symmetry; and (3) prescription of the proper boundary conditions on the segment of the total structure to be analyzed.

Structural symmetry involves some balanced arrangement of geometry, structural properties, and support conditions. In the dictionary sense it is the correspondence in size, form, and arrangement of parts on opposite sides of a plane, line, or point. Here we only consider the symmetry of structures that have a balanced arrangement about some plane and consist of members that have bisymmetrical cross-sections. Our examples are limited to plane structures. Reference 13.4 contains a useful further discussion of the principles of symmetry and a comprehensive list of publications on structural applications.

The plane structures of Figure 13.11 are obviously symmetrical. Their symmetry may be defined either in terms of rotation about the y axis or reflection of every material point through the y-z plane (the vertical plane that is normal to the paper and passes through point O). In Reference 13.4, Glockner notes that three common symmetry operations, reflection in a plane, rotation about an axis, and inversion through a center,

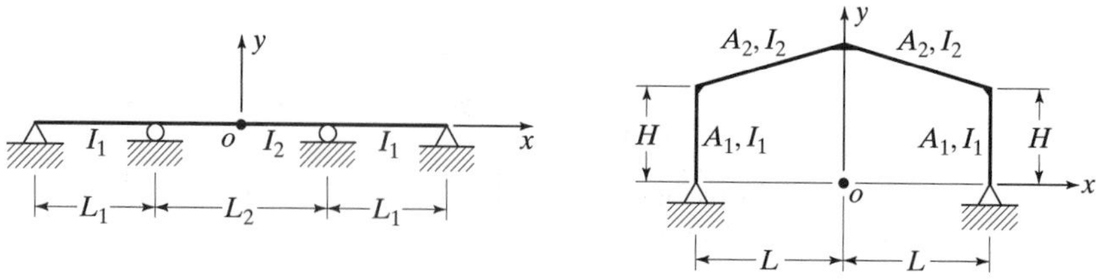

Figure 13.11 Symmetrical Structures.

are equivalent to one or more reflections. Thus reflection may be treated as the fundamental operation of symmetry. For simplicity, we use the concept of rotation in the following discussion, since it seems easier to visualize.

In working with symmetrical structures, it is useful to employ the concepts of symmetric and antisymmetric loads. The following definitions of the two conditions are similar to Glockner's:

1. A system of loads acting on a symmetric structure is *symmetric* if it is brought into an equivalent configuration as the symmetry conditions of the structure are applied to the system.
2. A system of loads acting on a symmetric structure is *antisymmetric* if it is brought into an equivalent configuration as the symmetry conditions of the structure are applied to the system and the signs of all load vectors are reversed.

These definitions are illustrated in Figure 13.12. For the symmetrical loads in Figure 13.12*a*, rotation about the *y* axis of structural symmetry produces an equivalent system of loads, except for the insignificant reversal of the identifying symbols. For the antisymmetrical loads in Figure 13.12*b*, rotation about the *y* axis produces the situation shown. Reversal of the direction of the rotated loads would produce a system equivalent to the original, unrotated one.

More often than not, the loads on a symmetric structure are neither symmetric nor antisymmetric. Nevertheless, it is still possible to take advantage of structural symmetry for all cases in which the principle of superposition applies. As shown in Figure 13.13, any load on a structure symmetrical with respect to a plane may be treated as the sum of a symmetrical loading (Figure 13.13*b*) and an antisymmetrical loading (Figure 13.13*c*).

Given symmetrical or antisymmetrical loads or having decomposed the natural loading into such components, there remains the problem of determining the boundary conditions that must be applied on the plane of symmetry to permit one-half of the structure to be analyzed properly. In terms of displacements these are:

1. For symmetric loading there can be
   (a) No translation normal to the plane of symmetry.
   (b) No rotations about two orthogonal axes in the plane of symmetry.

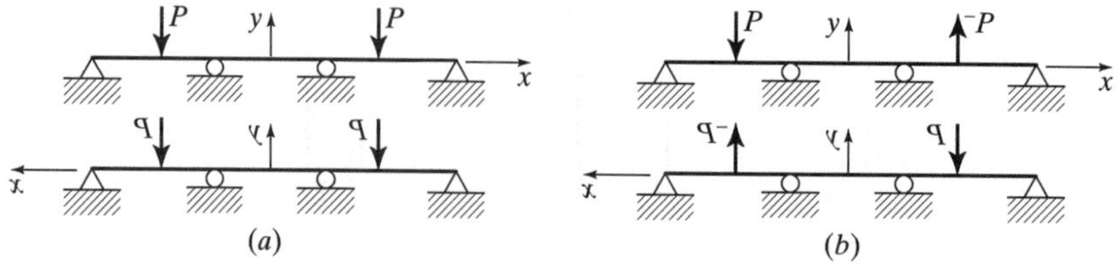

Figure 13.12 Load systems. (*a*) Symmetrical loads. (*b*) Antisymmetrical loads.

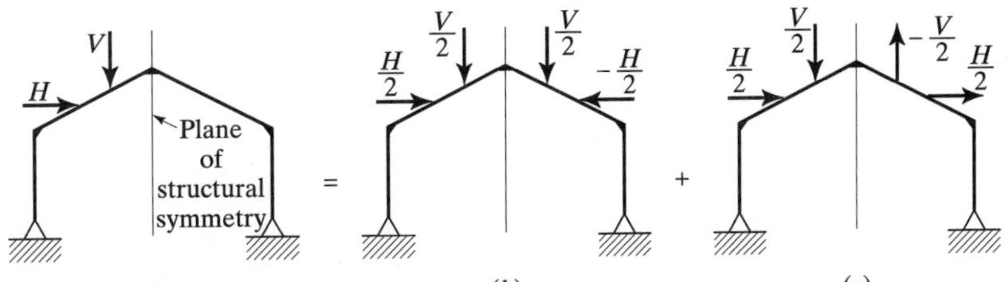

Figure 13.13 Decomposition of loading systems. (*a*) General loading. (*b*) Symmetrical part. (*c*) Antisymmetrical part.

2. For antisymmetric loading there can be
   (a) No translation in the plane of symmetry.
   (b) No rotation about an axis normal to the plane of symmetry.

Although offered without proof, most of these requirements should be clear on physical grounds. As an exercise in verifying them, consider the possible displacements of the node a of the planar symmetrical space frame of Figure 13.14 under the separate actions of the symmetrical load V and the antisymmetrical load H. Schemes for determining symmetry boundary conditions are discussed further in Reference 13.4.

To solve a problem once symmetry has been recognized and the loads have been suitably decomposed, the characteristic segment of the structure is analyzed twice: once for the symmetric loading condition and once for the antisymmetric condition. The results are superimposed using proper signs for all displacements and forces.

In Example 13.10, this procedure is applied to a symmetric continuous beam. In this case, nothing is gained through the application of symmetry since there are two unknown degrees of freedom in each analysis, which is the same as the number of unknowns (θ_b and θ_c) in the original structure. It is merely an illustration of the application of the principles of symmetry and a demonstration of the treatment of loads on an

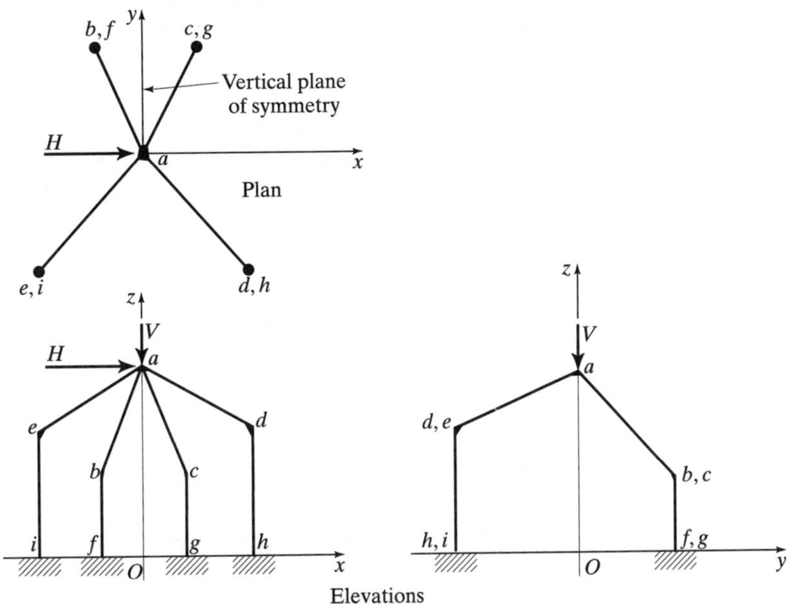

Figure 13.14 Planar symmetry conditions.

element bisected by a plane of symmetry. It should be clear, however, that had the number of spans been greater than three, the reduction in degrees of freedom would have been significant.

Example 13.11 is an application of symmetry principles to a rigid frame. The number of unknown degrees of freedom in the complete frame is nine, yet the complete analysis is accomplished by considering no more than five. Further reductions could have been effected by taking advantage of the usual frame assumption that axial deformations are negligible. This was not done since our interest at this point is simply in illustrating the principles of symmetry. Close attention should be paid to the signs used in combining displacement components. Note that in crossing the plane of symmetry, the signs of the u and θ symmetrical components, and the v antisymmetrical component, are reversed.

Finally, it will be observed that in most of the general-purpose computer programs in common use, it is possible to take advantage of symmetry by specifying appropriate boundary conditions.

EXAMPLE 13.10

Using symmetry principles, determine the slope at b and c for the beam shown. $I = 50 \times 10^6$ mm^4, $E = 200{,}000$ MPa for all members.

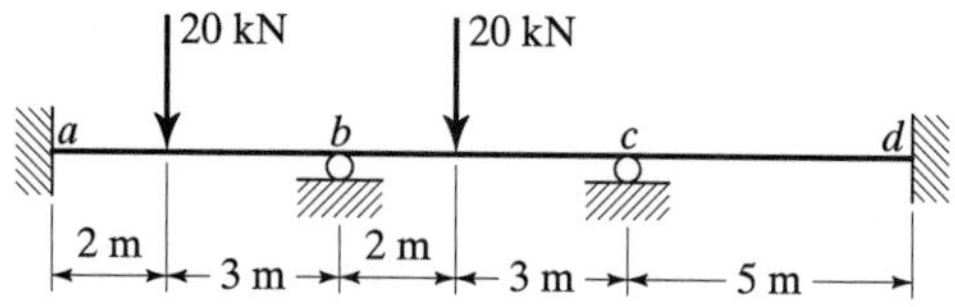

Symmetrical part:

By symmetry, $\theta_o = 0$.

Unknown degrees of freedom: θ_b, v_o.

Fixed-end forces:

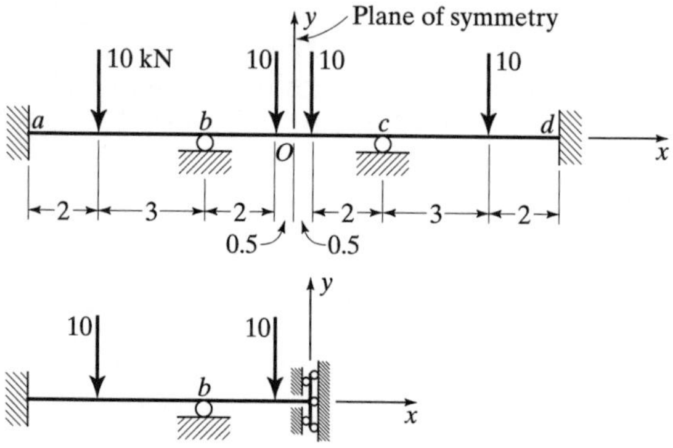

$$ab \qquad M_{zb}^F = -10 \times 2^2 \times 3/5^2 = -4.80 \text{ kNm}$$

$$bo \qquad M_{zb}^F = 10 \times 2 \times 0.5^2/2.5^2 = 0.80$$

$$M_{zo}^F = -10 \times 2^2 \times 0.5/(2.5)^2 = -3.20$$

$$F_{yb}^F = 1.04 \text{ kN} \qquad F_{yo}^F = 8.96 \text{ kN}$$

Element stiffness equations (from Equation 4.34):

$$ab \qquad M_{zb} = 200(4 \times 50 \times 10^6/5000)\theta_b = 200(4 \times 10^4)\theta_b$$

$$bo \qquad \begin{Bmatrix} M_{zb} \\ F_{yo} \end{Bmatrix} = 200\begin{bmatrix} \dfrac{4 \times 50 \times 10^6}{2500} & -\dfrac{6 \times 50 \times 10^6}{(2500)^2} \\ \text{Sym.} & \dfrac{12 \times 50 \times 10^6}{(2500)^3} \end{bmatrix}\begin{Bmatrix} \theta_b \\ v_o \end{Bmatrix}$$

$$= 200\begin{bmatrix} 8 \times 10^4 & -48.0 \\ -48.0 & 0.0384 \end{bmatrix}\begin{Bmatrix} \theta_b \\ v_o \end{Bmatrix}$$

Global stiffness equations:

$$\begin{Bmatrix} 0 \\ 0 \end{Bmatrix} = 200\begin{bmatrix} 12 \times 10^4 & -48.0 \\ -48.0 & 0.0384 \end{bmatrix}\begin{Bmatrix} \theta_b \\ v_o \end{Bmatrix} + \begin{Bmatrix} -4.00 \times 10^3 \\ 8.96 \end{Bmatrix}$$

Solving:

$$\begin{Bmatrix} \theta_b \\ v_o \end{Bmatrix} = \begin{Bmatrix} -0.600 \times 10^{-3} \text{ rad} \\ -1.917 \text{ mm} \end{Bmatrix}$$

Antisymmetrical part:

By antisymmetry, $v_o = 0$.

Unknown degrees of freedom: θ_b, θ_o.

Element stiffness equations (Equation 4.34):

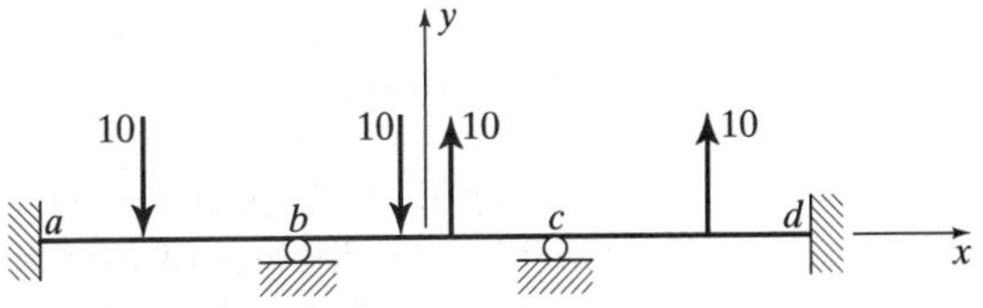

$$ab \qquad M_{zb} = 200(4 \times 10^4)\theta_b$$

$$bo \qquad \begin{Bmatrix} M_{zb} \\ M_{zo} \end{Bmatrix} = 200\begin{bmatrix} 8 \times 10^4 & 4 \times 10^4 \\ 4 \times 10^4 & 8 \times 10^4 \end{bmatrix}\begin{Bmatrix} \theta_b \\ \theta_o \end{Bmatrix}$$

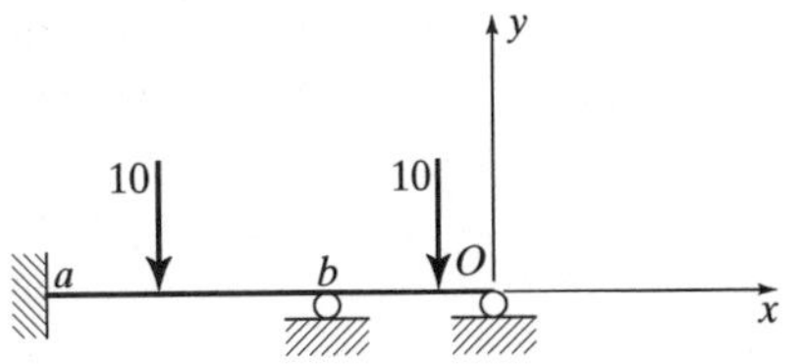

Global stiffness equations:

$$\begin{Bmatrix} 0 \\ 0 \end{Bmatrix} = 200\begin{bmatrix} 12 \times 10^4 & 4 \times 10^4 \\ 4 \times 10^4 & 8 \times 10^4 \end{bmatrix}\begin{Bmatrix} \theta_b \\ \theta_o \end{Bmatrix} + \begin{Bmatrix} -4.00 \times 10^3 \\ -3.20 \times 10^3 \end{Bmatrix}$$

Solving:

$$\begin{Bmatrix} \theta_b \\ \theta_o \end{Bmatrix} = \begin{Bmatrix} 0.120 \times 10^{-3} \\ 0.140 \times 10^{-3} \end{Bmatrix} \text{ rad}$$

Combined analysis:

$$\theta_b = (-0.600 + 0.120)10^{-3} = -0.480 \times 10^{-3} \text{ rad}$$

$$\theta_c = (+0.600 + 0.120)10^{-3} = +0.720 \times 10^{-3} \text{ rad}$$

$$v_o = -1.92 \text{ mm}$$

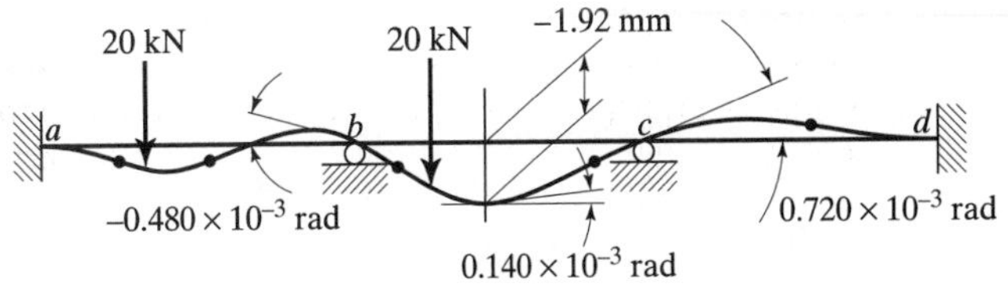

EXAMPLE 13.11

Using symmetry principles, determine the displacements at the joints of the rigid frame shown. For members *ab* and *de*, $A = 4 \times 10^3$ mm^2, $I = 50 \times 10^6$ mm^4. For members *bc* and *cd*, $A = 6 \times 10^3$ mm^2, $I = 200 \times 10^6$ mm^4. $E = 200{,}000$ MPa for all members. Use the results of Example 5.3.

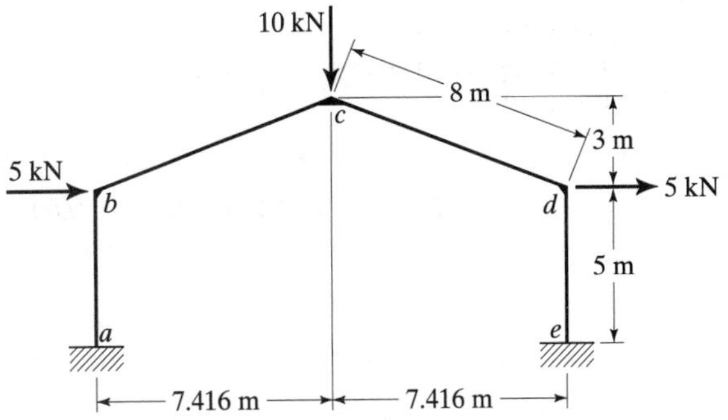

Symmetrical part:

By symmetry, $u_c = \theta_c = 0$.

Unknown degrees of freedom: u_b, v_b, θ_b, v_c.

From the element equations of Example 5.3, the global stiffness equations are

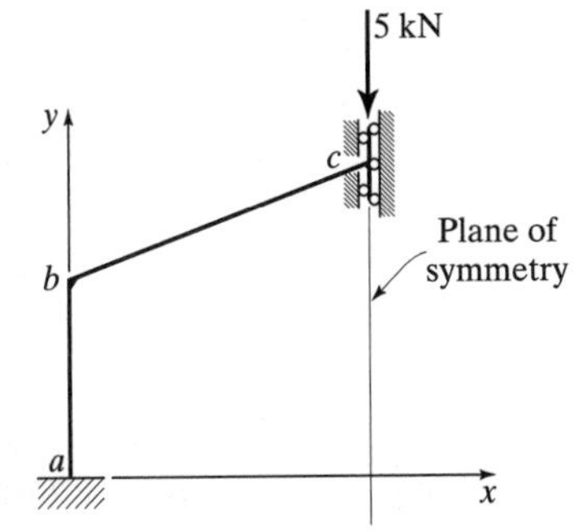

$$\begin{Bmatrix} 0 \\ 0 \\ 0 \\ -5 \end{Bmatrix} = 200\begin{bmatrix} 0.6500 & 0.2591 & 4.969 & -0.2591 \\ & 0.9095 & 17.381 & -0.1095 \\ \text{Sym.} & & 1.4 \times 10^5 & -17.381 \\ & & & 0.1095 \end{bmatrix}\begin{Bmatrix} u_b \\ v_b \\ \theta_b \\ v_c \end{Bmatrix}$$

Solving,

$$\lfloor u_b \quad v_b \quad \theta_b \quad v_c \rfloor = \lfloor -2.208 \text{ mm} \quad -0.0313 \text{ mm} \quad -0.611 \times 10^{-3} \text{ rad} \quad -5.583 \text{ mm} \rfloor$$

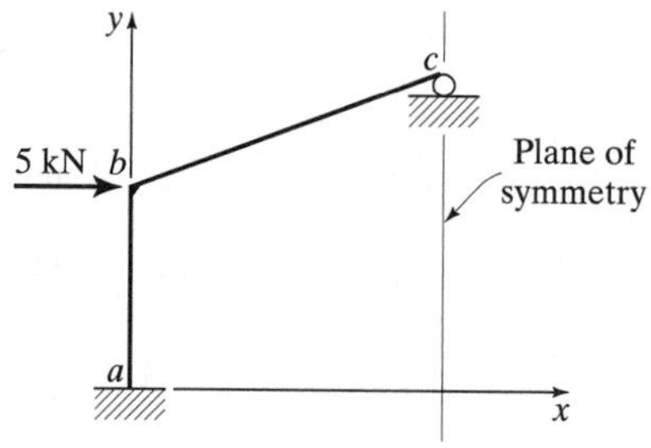

Antisymmetrical part:

By antisymmetry, $v_c = 0$.

Unknown degrees of freedom: $u_b, v_b, \theta_b, u_c, \theta_c$.

From the element equations of Example 5.3, the global stiffness equations are

$$\begin{Bmatrix} 5 \\ 0 \\ 0 \\ 0 \\ 0 \end{Bmatrix} = 200 \begin{bmatrix} 0.6500 & 0.2591 & 4.969 & -0.6452 & -7.0313 \\ & 0.9095 & 17.381 & -0.2591 & 17.381 \\ & & 1.4 \times 10^5 & 7.0313 & 0.5 \times 10^5 \\ & \text{Sym.} & & 0.6452 & 7.0313 \\ & & & & 1 \times 10^5 \end{bmatrix} \begin{Bmatrix} u_b \\ v_b \\ \theta_b \\ u_c \\ \theta_c \end{Bmatrix}$$

Solving,

$$\lfloor u_b \quad v_b \quad \theta_b \quad u_c \quad \theta_c \rfloor$$
$$= \lfloor 7.085 \text{ mm} \quad 0.0093 \text{ mm} \quad -0.7403 \times 10^{-3} \text{ rad} \quad 7.093 \text{ mm} \quad 0.3680 \times 10^{-3} \text{ rad} \rfloor$$

Combined analysis:

$u_b = -2.20 + 7.08 = 4.88$ mm

$v_b = -0.03 + 0.01 = -0.02$ mm

$\theta_b = (-0.611 - 0.740)10^{-3} = -1.351 \times 10^{-3}$ rad

$u_c = 0 + 7.09 = 7.09$ mm

$v_c = -5.58 + 0 = -5.58$ mm

$\theta_c = 0 + 0.368 \times 10^{-3} = 0.368 \times 10^{-3}$ rad

$u_d = +2.20 + 7.08 = 9.28$ mm

$v_d = -0.03 - 0.01 = -0.04$ mm

$\theta_d = (+0.611 - 0.740)10^{-3} = -0.129 \times 10^{-3}$ rad

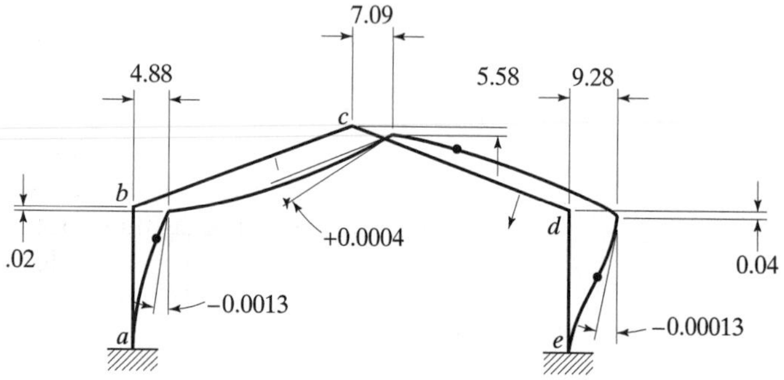

13.7 REANALYSIS TECHNIQUES

The term *reanalysis techniques* refers to procedures for assessment of the effect of modest structural design changes without incurring the computational costs of a complete reanalysis. Up until now, our work has been concerned with situations in which the arrangement and proportions of the structural members and the support conditions are known. In practice, however, the designer must first estimate the member sizes, analyze the structure, and compare the calculated and allowable forces and displacements. Then, if certain calculated values are greater than the allowables, or if the allowables are unacceptably greater than the calculated values, changes will be made in the structural analysis conditions. Indeed, it may prove necessary to repeat this cycle a number of times. It is useful, therefore, to consider procedures, alternative to complete conventional reanalysis, by which the results for the modified structures can be obtained.

In the case of support condition changes, the procedure already described for substructuring is applicable. Consider the structure shown in Figure 13.15*a*, where all column bases are assumed to be fixed. The degrees of freedom of the remaining joints, comprising all of the displacements to be determined in this analysis, are designated as $\{\mathbf{\Delta}_b\}$. The global stiffness equations, already modified to account for the support conditions, are then

$$\{\mathbf{P}_b\} = [\mathbf{K}_{bb}]\{\mathbf{\Delta}_b\} \tag{13.42}$$

and, after solution,

$$\{\mathbf{\Delta}_b\} = [\mathbf{K}_{bb}]^{-1}\{\mathbf{P}_b\} \tag{13.43}$$

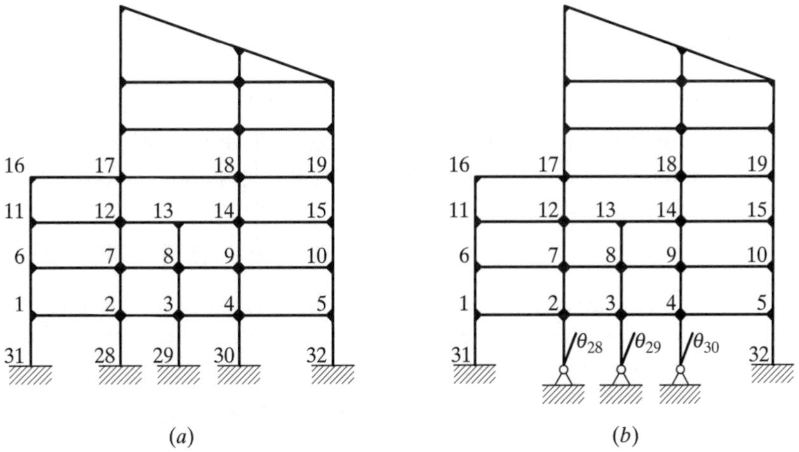

Figure 13.15 Modification of support conditions.

Now, if it is decided to examine the effect of changing the support conditions of the structure from fixed to hinged supports at points 28, 29, and 30 (see Fig. 13.15*b*), there is no change in the stiffness properties $[\mathbf{K}_{bb}]$ associated with the degrees of freedom $\{\mathbf{\Delta}_b\}$. One must supplement global stiffness equations, however, to account for the degrees of freedom $\{\mathbf{\Delta}_c\}$, where $\{\mathbf{\Delta}_c\} = \lfloor \theta_{28} \quad \theta_{29} \quad \theta_{30} \rfloor$. Thus the global stiffness equations become

$$\begin{Bmatrix} \mathbf{P}_b \\ \hline \mathbf{P}_c \end{Bmatrix} = \left[\begin{array}{c|c} \mathbf{K}_{bb} & \mathbf{K}_{bc} \\ \hline \mathbf{K}_{cb} & \mathbf{K}_{cc} \end{array}\right] \begin{Bmatrix} \mathbf{\Delta}_b \\ \hline \mathbf{\Delta}_c \end{Bmatrix} \tag{13.44}$$

These equations are exactly the same form as in substructure analysis, where the solution for $\{\mathbf{\Delta}_c\}$ is given by

$$\{\mathbf{\Delta}_c\} = [[\mathbf{K}_{cc}] - [\mathbf{K}_{cb}][\mathbf{K}_{bb}]^{-1}[\mathbf{K}_{bc}]]^{-1}[\{\mathbf{P}_c\} - [\mathbf{K}_{cb}][\mathbf{K}_{bb}]^{-1}\{\mathbf{P}_b\}] \tag{13.15a}$$

and

$$\{\mathbf{\Delta}_b\} = -[\mathbf{K}_{bb}]^{-1}[\mathbf{K}_{bc}]\{\mathbf{\Delta}_c\} + [\mathbf{K}_{bb}]^{-1}\{\mathbf{P}_b\} \tag{13.16a}$$

Note that $[\mathbf{K}_{bb}]^{-1}$ is already available from the solution of the original structure (Fig. 13.15*a*), and the solution for the modified structure (Fig. 13.15*b*) requires only the inverse of a matrix of the same order as the "relaxed" support degrees of freedom $\{\mathbf{\Delta}_c\}$.

A general procedure for investigating the effects of different support possibilities emerges from the above. All degrees of freedom that will not be among those considered for support conditions are first isolated and designated as $\{\mathbf{\Delta}_b\}$. The associated stiffness coefficients $[\mathbf{K}_{bb}]$ are evaluated and solved, $[\mathbf{K}_{bb}]^{-1}$. The degrees of freedom of any support condition to be examined are then designated as $\{\mathbf{\Delta}_c\}$, and the solution for the full set of joint displacements is obtained from Equations 13.15*a* and 13.16*a*.

We consider now reanalysis techniques intended to deal with situations in which the element stiffnesses are modified on account of changes in cross-sectional properties or material properties. Two classes of such techniques can be identified: exact and approximate procedures.

Exact reanalysis techniques have appeared often in the literature, described by various names and formulated through different strategies. Each depends upon exploitation of the zero terms of the matrix of changed element stiffnesses. One such approach is described as follows.

If the changes to the original stiffness matrix $[\mathbf{K}^\circ]$ are denoted as $[\delta\mathbf{K}^\circ]$, the new stiffness equations can be written in the form:

$$\{\mathbf{P}\} = [[\mathbf{K}^\circ] + [\delta\mathbf{K}^\circ]]\{\boldsymbol{\Delta}\} \tag{13.45}$$

Note that in this development the load vector $\{\mathbf{P}\}$ is assumed to be unaltered by the changes in dead load due to changes in member size represented as $[\delta\mathbf{K}^\circ]$. By factoring out $[\mathbf{K}^\circ]$ on the left side, this can be written as

$$\{\mathbf{P}\} = [\mathbf{K}^\circ][[\mathbf{I}] + [\mathbf{K}^\circ]^{-1}[\delta\mathbf{K}^\circ]]\{\boldsymbol{\Delta}\} \tag{13.45a}$$

and, by inversion (recalling that the inverse of a product of matrices is the product of their inverses in reversed order), we obtain the solution for the displacements $\{\boldsymbol{\Delta}\}$ of the modified structure:

$$\{\boldsymbol{\Delta}\} = [[\mathbf{I}] + [\mathbf{K}^\circ]^{-1}[\delta\mathbf{K}^\circ]]^{-1}[\mathbf{K}^\circ]^{-1}\{\mathbf{P}\} \tag{13.46}$$

Furthermore, since $[\mathbf{K}^\circ]^{-1}\{\mathbf{P}\} = \{\boldsymbol{\Delta}^\circ\}$, where $\{\boldsymbol{\Delta}^\circ\}$ symbolizes the solution for displacements of the original (unmodified structure),

$$\{\boldsymbol{\Delta}\} = [[\mathbf{I}] + [\mathbf{K}^\circ]^{-1}[\delta\mathbf{K}^\circ]]^{-1}\{\boldsymbol{\Delta}^\circ\} \tag{13.46a}$$

Since $\{\boldsymbol{\Delta}^\circ\}$ is already available from the solution of the original structure, it is only necessary to develop an expression for $[[\mathbf{I}] + [\mathbf{K}^\circ]^{-1}[\delta\mathbf{K}^\circ]]^{-1}$. In so doing, it is assumed that the changes in stiffness can all be collected in the upper left portion of $[\delta\mathbf{K}^\circ]$. In Figure 13.15*a*, for example, if member 7-12 is altered, the vector of displacements will be written as $\lfloor u_7 \quad v_7 \quad \theta_7 \quad u_{12} \quad v_{12} \quad \theta_{12} \cdots \rfloor$ with a corresponding arrangement of the vector of applied loads. If all degrees of freedom associated with stiffness changes are designated as $\{\boldsymbol{\Delta}_1\}$ and the remainder as $\{\boldsymbol{\Delta}_2\}$, we can rewrite Equation 13.45 in the partitioned form:

$$\begin{Bmatrix} \mathbf{P}_1 \\ \mathbf{P}_2 \end{Bmatrix} = \begin{bmatrix} \mathbf{K}_{11}^\circ & \mathbf{K}_{12}^\circ \\ \mathbf{K}_{21}^\circ & \mathbf{K}_{22}^\circ \end{bmatrix} \begin{Bmatrix} \boldsymbol{\Delta}_1 \\ \boldsymbol{\Delta}_2 \end{Bmatrix} + \begin{bmatrix} \delta\mathbf{K}_{11}^\circ & \mathbf{0} \\ \mathbf{0} & \mathbf{0} \end{bmatrix} \begin{Bmatrix} \boldsymbol{\Delta}_1 \\ \boldsymbol{\Delta}_2 \end{Bmatrix} \tag{13.45b}$$

where $[\delta\mathbf{K}_{11}^\circ]$ contains all of the modified element stiffness terms. Also, we write the already available inverse of $[\mathbf{K}^\circ]$ in the form:

$$[\mathbf{K}^\circ]^{-1} = \begin{bmatrix} \mathbf{D}_{11}^\circ & \mathbf{D}_{12}^\circ \\ \mathbf{D}_{21}^\circ & \mathbf{D}_{22}^\circ \end{bmatrix}$$

With the above expanded relationships, we can write $[[\mathbf{I}] + [\mathbf{K}^\circ]^{-1}[\delta\mathbf{K}^\circ]]$ as

$$\left[\begin{bmatrix} \mathbf{I} & \mathbf{0} \\ \mathbf{0} & \mathbf{I} \end{bmatrix} + \begin{bmatrix} \mathbf{D}_{11}^\circ & \mathbf{D}_{12}^\circ \\ \mathbf{D}_{21}^\circ & \mathbf{D}_{22}^\circ \end{bmatrix} \begin{bmatrix} \delta\mathbf{K}_{11}^\circ & \mathbf{0} \\ \mathbf{0} & \mathbf{0} \end{bmatrix}\right] = \begin{bmatrix} \mathbf{q} & \mathbf{0} \\ \mathbf{D}_{21}^\circ\,\delta\mathbf{K}_{11}^\circ & \mathbf{I} \end{bmatrix}$$

where $[\mathbf{q}] = [\mathbf{I} + \mathbf{D}_{11}^\circ\,\delta\mathbf{K}_{11}^\circ]$. Note that here the size of $[\mathbf{q}]$ and the identity matrix $[\mathbf{I}]$ is the same as the number of degrees of freedom associated with the modification. The inverse of the above is

$$\begin{bmatrix} \mathbf{q}^{-1} & \mathbf{0} \\ -\mathbf{D}_{21}^\circ\,\delta\mathbf{K}_{11}^\circ\mathbf{q}^{-1} & \mathbf{I} \end{bmatrix} = [[\mathbf{I}] + [\mathbf{K}^\circ]^{-1}[\delta\mathbf{K}^\circ]]^{-1} \tag{13.47}$$

Substitution of Equation 13.47 into 13.46a, with $\{\boldsymbol{\Delta}\}$ partitioned into $\lfloor \boldsymbol{\Delta}_1 \mid \boldsymbol{\Delta}_2 \rfloor^T$, gives

$$\{\boldsymbol{\Delta}\} = \begin{Bmatrix} \boldsymbol{\Delta}_1 \\ \boldsymbol{\Delta}_2 \end{Bmatrix} = \begin{Bmatrix} \mathbf{q}^{-1}\,\boldsymbol{\Delta}_1^\circ \\ \boldsymbol{\Delta}_2^\circ - \mathbf{D}_{21}^\circ\,\delta\mathbf{K}_{11}^\circ\mathbf{q}^{-1}\,\boldsymbol{\Delta}_1^\circ \end{Bmatrix} \tag{13.48}$$

Observe that the only matrix inverse appearing in this expression is that of $[\mathbf{q}]$. The necessary operations are otherwise matrix multiplications.

Example 13.12 illustrates the above procedure. In order to present the procedure in detail, the structure consists of only three members and the changes affect a large part of the total stiffness matrix. To be useful, however, the reanalysis situation should involve a relatively small number of members of the original structure.

EXAMPLE 13.12

The stiffness of member 1-2 is to be changed from $k = 1$ to $k = 2$. Using the "exact" procedure, calculate the displacements in the changed structure.

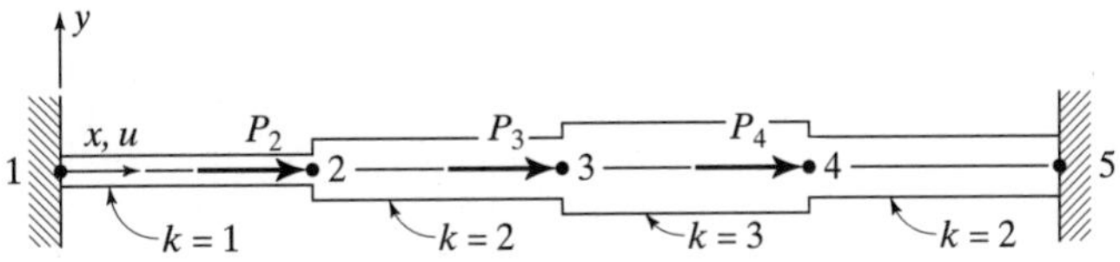

Original structure

Stiffness equations for the original structure are

$$\begin{Bmatrix} P_2 \\ P_3 \\ P_4 \end{Bmatrix} = \begin{bmatrix} 3 & -2 & 0 \\ -2 & 5 & -3 \\ 0 & -3 & 5 \end{bmatrix} \begin{Bmatrix} u_2 \\ u_3 \\ u_4 \end{Bmatrix}$$

Flexibility equations for the original structure, $\{\boldsymbol{\Delta}^\circ\} = [\mathbf{D}^\circ]\{\mathbf{P}\}$, are

$$\begin{Bmatrix} u_2 \\ \hline u_3 \\ u_4 \end{Bmatrix} = \frac{1}{28} \left[\begin{array}{c|cc} 16 & 10 & 6 \\ \hline 10 & 15 & 9 \\ 6 & 9 & 11 \end{array}\right] \begin{Bmatrix} P_2 \\ \hline P_3 \\ P_4 \end{Bmatrix} \tag{13.1}$$

$$[\mathbf{D}^\circ_{11}] = \frac{1}{28}(16) \qquad [\delta\mathbf{K}^\circ] = \begin{bmatrix} 1 & 0 & 0 \\ 0 & 0 & 0 \\ 0 & 0 & 0 \end{bmatrix} \qquad [\delta\mathbf{K}^\circ_{11}] = (1)$$

$$[\mathbf{q}] = [[\mathbf{I}] + [\mathbf{D}^\circ_{11}][\delta\mathbf{K}^\circ_{11}]] = \left((1) + \frac{1}{28}(16)(1)\right) = \frac{44}{28}$$

Thus, with $\{\boldsymbol{\Delta}^\circ_1\} = \frac{1}{28}\lfloor 16 \quad 10 \quad 6 \rfloor \begin{Bmatrix} P_2 \\ P_3 \\ P_4 \end{Bmatrix}$,

$$\{\boldsymbol{\Delta}_1\} = [\mathbf{q}]^{-1}[\boldsymbol{\Delta}^\circ_1] = \frac{28}{44} \cdot \frac{1}{28} \lfloor 16 \quad 10 \quad 6 \rfloor \begin{Bmatrix} P_2 \\ P_3 \\ P_4 \end{Bmatrix} = \left\lfloor \frac{16}{44} \quad \frac{10}{44} \quad \frac{6}{44} \right\rfloor \begin{Bmatrix} P_2 \\ P_3 \\ P_4 \end{Bmatrix}$$

Also, with $\{\boldsymbol{\Delta}^\circ_2\} = \frac{1}{28}\begin{bmatrix} 10 & 15 & 9 \\ 6 & 9 & 11 \end{bmatrix}\begin{Bmatrix} P_2 \\ P_3 \\ P_4 \end{Bmatrix}$, and $[\mathbf{D}^\circ_{21}] = \frac{1}{28}\begin{bmatrix} 10 \\ 6 \end{bmatrix}$,

$$\{\boldsymbol{\Delta}_2\} = \{\boldsymbol{\Delta}^\circ_2\} - [\mathbf{D}^\circ_{21}][\delta\mathbf{K}^\circ_{11}][\mathbf{q}]^{-1}\{\boldsymbol{\Delta}^\circ_1\}$$

$$= \frac{1}{28}\begin{bmatrix} 10 & 15 & 9 \\ 6 & 9 & 11 \end{bmatrix}\begin{Bmatrix} P_2 \\ P_3 \\ P_4 \end{Bmatrix} - \frac{1}{28}\begin{bmatrix} 10 \\ 6 \end{bmatrix}(1)\left\lfloor \frac{16}{44} \quad \frac{10}{44} \quad \frac{6}{44} \right\rfloor \begin{Bmatrix} P_2 \\ P_3 \\ P_4 \end{Bmatrix}$$

$$= \frac{1}{44}\begin{bmatrix} 10 & 20 & 12 \\ 6 & 12 & 16 \end{bmatrix}\begin{Bmatrix} P_2 \\ P_3 \\ P_4 \end{Bmatrix}$$

To summarize,

$$\begin{Bmatrix} \mathbf{\Delta}_1 \\ \hline \mathbf{\Delta}_2 \end{Bmatrix} = \begin{Bmatrix} u_2 \\ \hline u_3 \\ u_4 \end{Bmatrix} = \frac{1}{44}\left[\begin{array}{ccc} 16 & 10 & 6 \\ \hline 10 & 20 & 12 \\ 6 & 12 & 16 \end{array}\right]\begin{Bmatrix} P_2 \\ P_3 \\ P_4 \end{Bmatrix}$$

This result agrees with that obtained by direct inversion of the stiffness matrix of the changed structure.

An *approximate* reanalysis formula can be obtained as follows. In Equation 13.45 we first replace $\{\mathbf{\Delta}\}$, the total displacements of the modified structure, with $\{\mathbf{\Delta}^\circ\} + \{\delta\mathbf{\Delta}^\circ\}$, where $\{\mathbf{\Delta}^\circ\}$ designates the already-solved-for displacements of the original structure and $\{\delta\mathbf{\Delta}^\circ\}$ are the as-yet unknown changes in displacements due to the structural modifications. Equations 13.45 therefore is now written as

$$\{\mathbf{P}\} = [[\mathbf{K}^\circ] + [\delta\mathbf{K}^\circ]]\{\mathbf{\Delta}^\circ + \delta\mathbf{\Delta}^\circ\} \tag{13.45c}$$

Expansion of the right side gives

$$\{\mathbf{P}\} = [\mathbf{K}^\circ]\{\mathbf{\Delta}^\circ\} + [\delta\mathbf{K}^\circ]\{\mathbf{\Delta}^\circ\} + [\mathbf{K}^\circ]\{\delta\mathbf{\Delta}^\circ\} + [\delta\mathbf{K}^\circ]\{\delta\mathbf{\Delta}^\circ\} \tag{13.45d}$$

Now, $[\mathbf{K}^\circ]\{\mathbf{\Delta}^\circ\} = \{\mathbf{P}\}$. (The displacements $\{\mathbf{\Delta}^\circ\}$ are the solution to the problem for the original stiffness $[\mathbf{K}^\circ]$.) These terms therefore can be cancelled from the two sides of Equation 13.45d. Also, we discard the higher-order term $[\delta\mathbf{K}^\circ]\{\delta\mathbf{\Delta}^\circ\}$; this comprises the approximation being made. Equation 13.45d is then reduced to

$$[\delta\mathbf{K}^\circ]\{\mathbf{\Delta}^\circ\} + [\mathbf{K}^\circ]\{\delta\mathbf{\Delta}^\circ\} = \{\mathbf{0}\}$$

or

$$\{\delta\mathbf{\Delta}^\circ\} = -[\mathbf{K}^\circ]^{-1}[\delta\mathbf{K}^\circ]\{\mathbf{\Delta}^\circ\} \tag{13.49}$$

All terms on the right side of Equation 13.49, $[\mathbf{K}^\circ]^{-1}$, $[\delta\mathbf{K}^\circ]$, and $\{\mathbf{\Delta}^\circ\}$, are known, and the change in displacement $\{\delta\mathbf{\Delta}^\circ\}$ due to the modification $[\delta\mathbf{K}^\circ]$ is obtained simply from a matrix product. Equation 13.49 has been developed without any assumptions being made regarding the location of zero terms in $[\delta\mathbf{K}^\circ]$. In large-scale practical design analysis, however, the changes in stiffness may occupy a small portion of $[\delta\mathbf{K}^\circ]$, and in such cases it is possible to achieve significant efficiencies in formation of the product on the right side.

The calculated changes in displacement can be applied to the element stiffness matrices to determine the associated changes in the internal forces. In some cases, the analyst may wish to have the revised flexibility matrix. This can be obtained from

$$\begin{aligned} \{\mathbf{\Delta}^\circ\} + \{\delta\mathbf{\Delta}^\circ\} &= [\mathbf{K}^\circ]^{-1}\{\mathbf{P}\} - [\mathbf{K}^\circ]^{-1}[\delta\mathbf{K}^\circ]\{\mathbf{\Delta}^\circ\} \\ &= [\mathbf{K}^\circ]^{-1}\{\mathbf{P}\} - [\mathbf{K}^\circ]^{-1}[\delta\mathbf{K}^\circ][\mathbf{K}^\circ]^{-1}\{\mathbf{P}\} \\ &= [[\mathbf{K}^\circ]^{-1} - [\mathbf{K}^\circ][\delta\mathbf{K}^\circ][\mathbf{K}^\circ]^{-1}]\{\mathbf{P}\} \\ &= [\mathbf{K}^{\circ -1}][[\mathbf{I}] - [\delta\mathbf{K}^\circ][\mathbf{K}^\circ]^{-1}]\{\mathbf{P}\} \end{aligned} \tag{13.50}$$

Thus, the revised flexibility matrix is

$$[\mathbf{D}^\circ] + [\delta\mathbf{D}^\circ] = [\mathbf{K}^\circ]^{-1}[[\mathbf{I}] - [\delta\mathbf{K}^\circ][\mathbf{K}^\circ]^{-1}] \tag{13.51}$$

The use of the above approximate reanalysis technique is illustrated in Example 13.13. This problem, which was solved exactly in Example 13.12, shows that the approximate method is not always sufficiently accurate. The design analyst must exercise judgment in identifying such circumstances. To improve the accuracy of this method, Equation 13.49 can be employed iteratively, that is, $\{\boldsymbol{\Delta}^\circ\} + \{\delta\boldsymbol{\Delta}^\circ\}$ can be used in place of $\{\boldsymbol{\Delta}^\circ\}$ to obtain a better estimate of the displacement changes. This approach adds to the computational cost, however, which might then exceed the cost of a completely new analysis. Indeed, the cost of reanalysis techniques versus completely new analysis have been questioned in some studies (Refs. 13.5, 13.6). Note also that a wide variety of reanalysis techniques has been proposed (e.g., Refs. 13.5–13.10), of which the above is merely a simple example.

EXAMPLE 13.13

Using the data of Example 13.12 and the "approximate" method (Equation 13.50), calculate the displacements in the changed structure. Compare the results with the "exact" solution.

From Equation 13.50:

$$\{\boldsymbol{\Delta}^\circ\} + \{\delta\boldsymbol{\Delta}^\circ\} = [\mathbf{K}^\circ]^{-1}[[\mathbf{I}] - [\delta\mathbf{K}^\circ][\mathbf{K}^\circ]^{-1}]\{\mathbf{P}\}$$

$$\begin{Bmatrix} u_2 \\ u_3 \\ u_4 \end{Bmatrix}_{\text{approx.}} = \frac{1}{28}\begin{bmatrix} 16 & 10 & 6 \\ 10 & 15 & 9 \\ 6 & 9 & 11 \end{bmatrix}\left[\begin{bmatrix} 1 & 0 & 0 \\ 0 & 1 & 0 \\ 0 & 0 & 1 \end{bmatrix} - \frac{1}{28}\begin{bmatrix} 1 & 0 & 0 \\ 0 & 0 & 0 \\ 0 & 0 & 0 \end{bmatrix}\begin{bmatrix} 16 & 10 & 6 \\ 10 & 15 & 9 \\ 6 & 9 & 11 \end{bmatrix}\right]\begin{Bmatrix} P_2 \\ P_3 \\ P_4 \end{Bmatrix}$$

$$= \frac{1}{(28)^2}\begin{bmatrix} 192 & 120 & 72 \\ 120 & 320 & 192 \\ 72 & 192 & 272 \end{bmatrix}\begin{Bmatrix} P_2 \\ P_3 \\ P_4 \end{Bmatrix}$$

or, in decimal form,

$$\begin{Bmatrix} u_2 \\ u_3 \\ u_4 \end{Bmatrix}_{\text{approx.}} = \begin{bmatrix} 0.245 & 0.153 & 0.092 \\ 0.153 & 0.408 & 0.245 \\ 0.092 & 0.245 & 0.347 \end{bmatrix}\begin{Bmatrix} P_2 \\ P_3 \\ P_4 \end{Bmatrix}$$

The exact solution, in decimal form, is

$$\begin{Bmatrix} u_2 \\ u_3 \\ u_4 \end{Bmatrix} = \begin{bmatrix} 0.363 & 0.227 & 0.136 \\ 0.227 & 0.454 & 0.273 \\ 0.136 & 0.273 & 0.364 \end{bmatrix}\begin{Bmatrix} P_2 \\ P_3 \\ P_4 \end{Bmatrix}$$

13.8 PROBLEMS

13.1 Apply the condensation equations of Section 13.1 to the solution of Problem 3.4.

13.2 Condense the angular displacement θ_b from Example 5.7 to produce stiffness equations in terms of u_b and v_b; then solve for the latter.

13.3 Form the stiffness matrix for the three-jointed axial member of Example 7.1, and then eliminate joint 2 by use of the condensation procedure of Section 13.1. Compare the result with the axial member stiffness matrix.

13.4 Use substructuring to analyze the structures of Problem 4.9.

13.5 Use substructuring to analyze the structure of Problem 4.15.

13.6 Combine substructuring with, where appropriate, the methods of Section 5.2 to analyze the structures of Problem 5.8.

13.7 Use substructuring to analyze the structure shown.

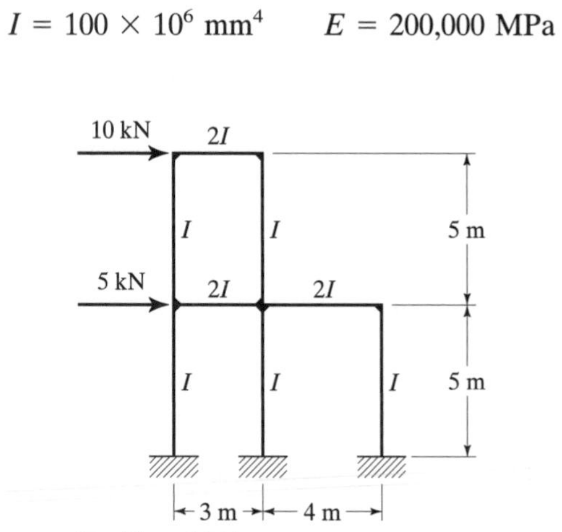

Problem 13.7

13.8 Use substructuring in conjunction with a computer program to analyze the trusses of Problem 3.15.

13.9 (a) Compute the displacements, reactions, and bar forces for the truss shown using substructuring. Cross-sectional areas (mm^2 $\times$ 10^3) are shown on each bar. E = 200,000 MPa. (Same as Problem 3.1).

(b) Solve the same problem but for a vertical load of 300 kN at d.

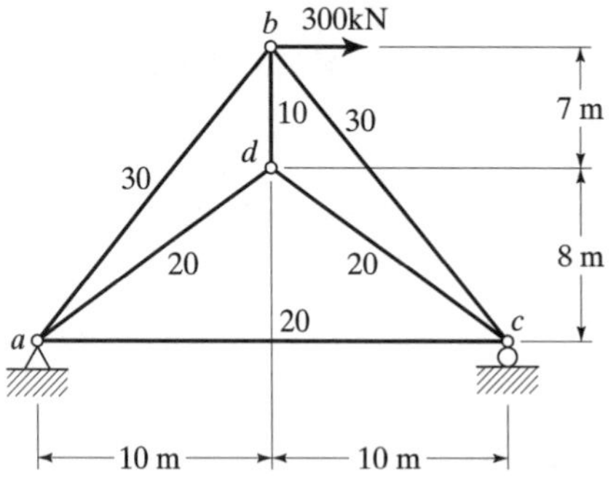

Problem 13.9

13.10 Remove the hinge from Example 4.14 by enforcement of the constraint condition $\theta_{zol} - \theta_{zor} = 0$ and demonstrate that the resulting stiffness equations are those of a member of length L.

13.11 Establish the matrix equations, similar to Equation 13.25, to accommodate constraint conditions of the type given by Equation 13.18 (i.e., $\{\mathbf{H}\} \neq \mathbf{0}$).

13.12 Assume that the joints a and b of Example 5.6 are so connected that $\theta_{za} = -6\theta_{zb}$. Enforce this constraint condition and solve for θ_{zb}.

13.13 A roller support is placed at joint b of the rigid frame of Example 4.13 as shown below. Using the "joint coordinate" approach of Section 13.4 and the data of Example 4.13, solve the problem to account for the presence of this support.

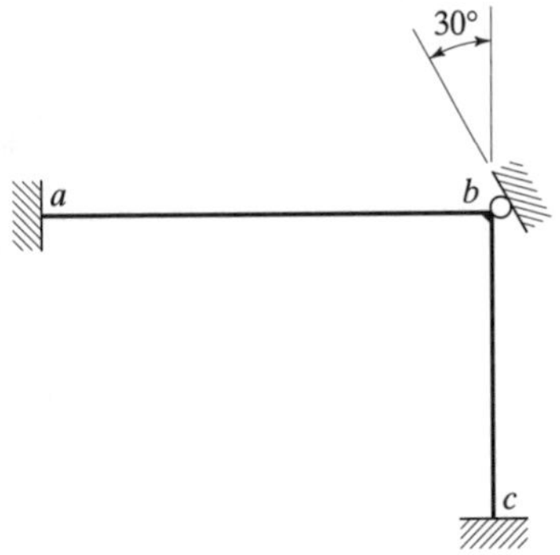

Problem 13.13

13.14 Re-solve Example 5.7 after the assembled stiffness matrix at joint b has been revised so that the coordinate axes are aligned in the direction of member ab and perpendicular to it.

13.15 Example 5.10 has been revised so that the hinge at point c is replaced by a roller that is oriented as shown below. Calculate the displacements for this revised support condition.

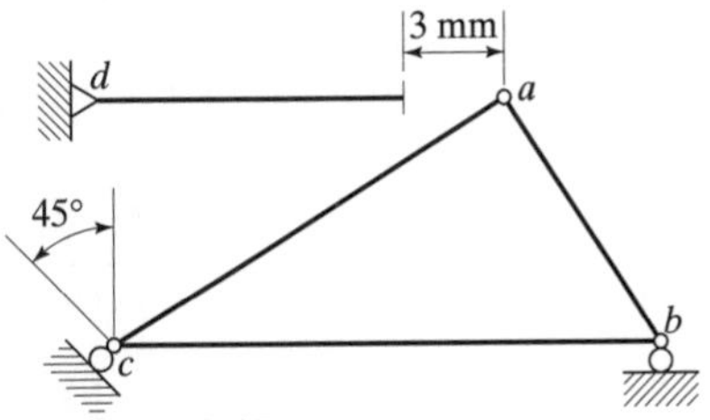

Problem 13.15

13.16 Demonstrate that the use of Equations 13.34 and 13.35 result in the form of Equation 13.32.

13.17 Compute the displacements, reactions, and internal forces for the beam shown. $E = 200{,}000$ MPa and $I = 10 \times 10^6$ mm^4.

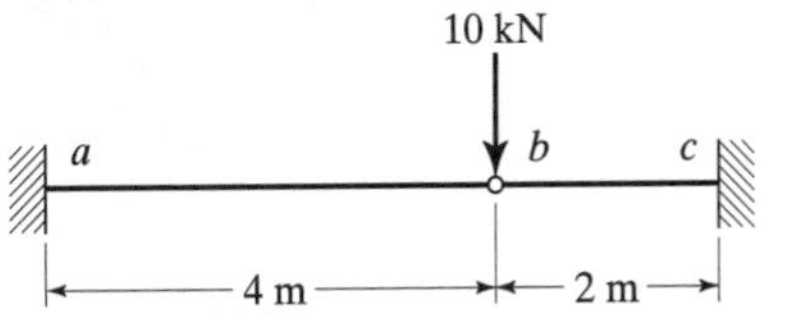

Problem 13.17

13.18 For the beam shown, plot the relationship between α_1 and the vertical displacement v_b. For what value of α_1 is the beam essentially fixed at a?

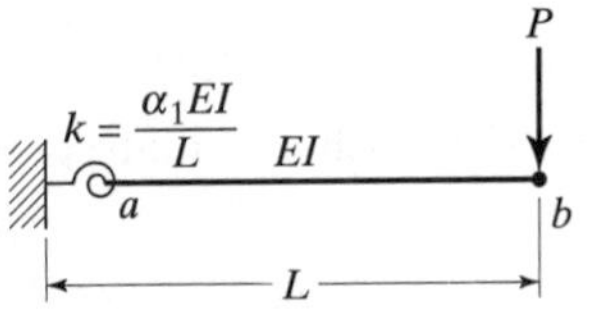

Problem 13.18

13.19 Compute the displacements, reactions, and internal forces for the frame shown. $E = 29{,}000$ ksi. Use the following connection stiffnesses:

(a) $k = 1200$ in. kips/rad
(b) $k = 12{,}000$ in. kips/rad
(c) $k = 120{,}000$ in. kips/rad

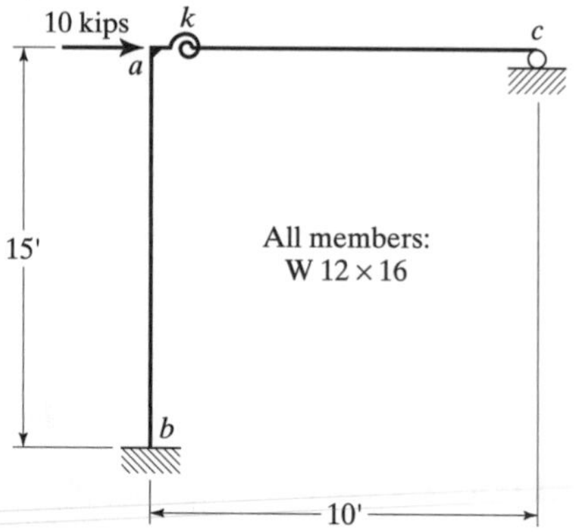

Problem 13.19

13.20 Compute the displacements, reactions, and internal forces for the frame shown. $E = 29{,}000$ ksi. Use the following connection stiffnesses:

(a) $k_1 = k_2 = 14.4 \times 10^4$ in. kips/rad
(b) $k_1 = 2k_2 = 14.4 \times 10^4$ in. kips/rad
(c) $k_1 = 0$ and $k_2 = \infty$
(d) $k_1 = 0$ and $k_2 = 14.4 \times 10^4$ in. kips/rad

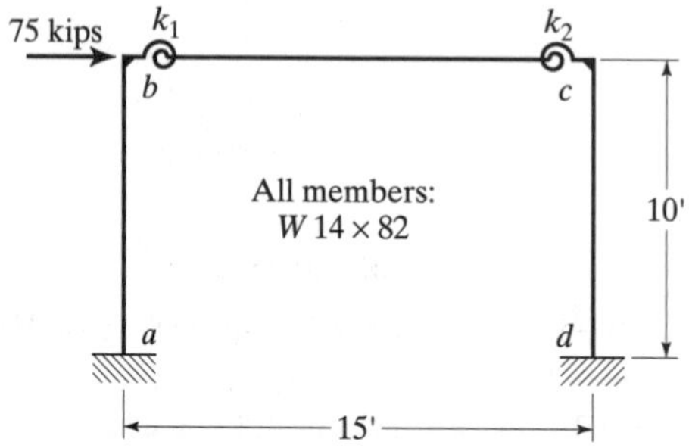

Problem 13.20

13.21 Use Equation 13.38 to analyze the beam-column shown. Compare results with an analysis that uses an equivalent axial force and moment acting at the centroid of the member. $E = 29{,}000$ ksi.

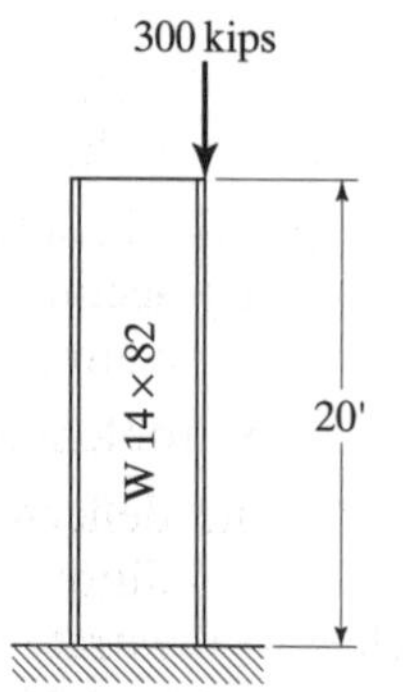

Problem 13.21

13.22 Compute the displacements, reactions, and internal forces for the beam-column shown. $E = 29{,}000$ ksi.

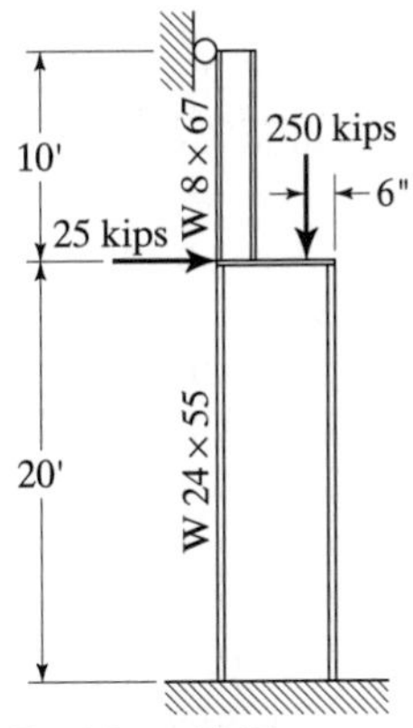

Problem 13.22

13.23 Repeat Example 13.9 for the frame shown. Assume all connections are rigid. $\sigma_y = 36$ ksi, $E = 29{,}000$ ksi.

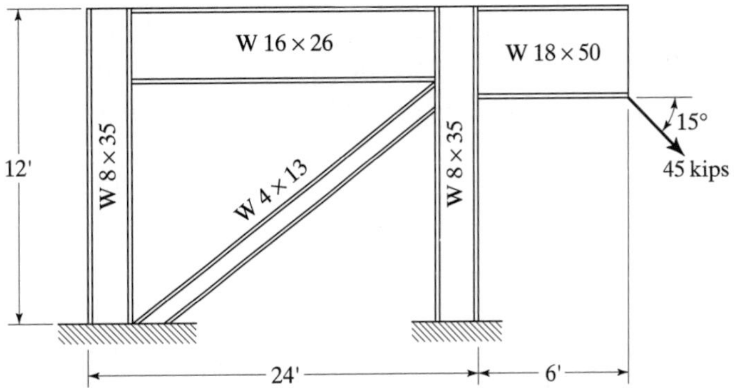

Problem 13.23

13.24 The area of member *ca* of Example 5.10 is changed from 15×10^3 mm^2 to 7.5×10^3 mm^2. Recalculate the displacements, using both the exact and approximate methods of Section 13.7.

13.25 Using the approximate method of Section 13.7, calculate the displacements and internal forces of the structure of Example 5.7, for a change in the size of member *ab* to $A = 800$ mm^2, $I = 300 \times 10^6$ mm^4.

13.26 Using the approximate method of Section 13.7, calculate the displacements and internal forces of the structure of Example 5.4 for a change in the size of member *ad* to 20×10^3 mm^2. Confirm the accuracy of the approximate solution by performing a computer analysis of the revised structure.

13.27 (a) Re-solve Problem 4.1*g* using symmetry and antisymmetry conditions. (b) Calculate the displacements at joint *b* of the second structure in Problem 4.9 using symmetry and antisymmetry conditions.

13.28 Re-solve Problem 5.8*c* using symmetry and antisymmetry conditions.

13.29 Sketch the deflected shapes of the structures in Problems 5.10*a*, 5.10*c*, and 5.10*d*. State the conditions of symmetry and antisymmetry that can be invoked in each. Use sketches to illustrate how they would be employed, that is, what the resulting idealized structures would look like.

REFERENCES

13.1 Y. Goto and S. Miyashita, "Classification System for Rigid and Semirigid Connections," ASCE, *Jl. of Struct. Engr.*, Vol. 124, No. 7, July, 1998.

13.2 R. Bjorhovde, A. Colson, and J. Brozzetti, "Classification System for Beam-to-Column Connections," ASCE, *Jl. of Struct. Engr.*, Vol. 116, No. 11, Nov., 1990.

13.3 W. F. Chen and E. Lui, *Stability Design of Steel Frames*, CRC Press, Boca Raton, Fla., 1991.

13.4 P. G. Glockner, "Symmetry in Structural Mechanics," ASCE, *Jl. of the Struct. Div.*, Vol. 99, No. ST1, Jan., 1973.

13.5 D. Kavlie and G. H. Powell, "Efficient Reanalysis of Modified Structures," ASCE, *Jl. of the Struct. Div.*, Vol. 97, No. ST1, Jan., 1971.

13.6 J. H. Argyris and J. R. Roy, "General Treatment of Structural Modifications," ASCE, *Jl. of the Struct. Div.*, Vol. 98, No. ST2, Feb., 1972.

13.7 U. Kirsch and M. F. Rubenstein, "Reanalysis for Limited Structural Design Modifications," ASCE, *Jl. of the Engr. Mech. Div.*, Vol. 98, No. EM1, Feb., 1972.

13.8 A. K. Noor and H. E. Lowder, "Approximate Techniques of Structural Reanalysis," *Computers and Structures*, Vol. 4, 1974.

13.9 J. Sobieszczanski, "Structural Modification by the Perturbation Method," ASCE, *Jl. of the Struct. Div.*, Vol. 94, No. ST12, Dec., 1968.

13.10 R. J. Melosh and R. Luik, "Multiple Configuration Analysis of Structures," ASCE, *Jl. of the Structural Div.*, Vol. 94, No. ST11, Nov., 1968.

Appendix A

Nonlinear Analysis—A Further Look

Geometric nonlinear response, and in particular the stability of framed structures, is a function of the way in which components of force interact with each other as the structure deforms and displaces under load. The effect of axial force on the uniaxial bending and simple twisting of a member is a major concern, but it is one that lent itself to the relatively simple physical analyses of Sections 9.1 and 9.2. Equally important are the phenomena whereby a beam subjected to primary bending about one principal cross sectional axis can become unstable through a combination of bending about the other one and twisting about the longitudinal axis, and the closely related mode in which the primary action is a combination of axial force and bending about one principal cross-sectional axis of a beam-column.[1] These, too, were treated in Chapter 9, but in the interest of obtaining tangible results we employed without proof various terms representing the spatial interaction of bending, torque, and other force components and we limited our analysis of torque to the pure St. Venant type. The requisite proofs require taking a close look at the way in which the stress resultants used in engineering analysis are derived from the principles of structural mechanics and the effect of displacements, particularly finite rotations, on their interaction.

Thus this appendix has a twofold purpose. The immediate one is the development of the terms of the geometric stiffness matrices of Chapter 9 that were beyond the reach of the physical arguments of that chapter and the analysis of nonuniform torsion. But in the course of meeting this objective, a fundamental approach to the development of the equations of elastic nonlinear analysis will be described and key problems encountered in applying it to framed structures will be treated. Therefore the broader purpose is to provide deeper insight into the process of nonlinear analysis.

The topics covered include

1. Use of virtual displacement principles in Lagrangian formulations.
2. Linearization of the virtual displacement equations.
3. Application of the linearized equations to the framework element.
4. The nature of finite rotations and the need to consider them in establishing equilibrium on the deformed structure.
5. Nonuniform torsion.

[1]Both are forms of *torsional-flexural buckling*, but that term is often reserved for the second, with the first commonly called *lateral buckling*.

A.1 VIRTUAL DISPLACEMENT PRINCIPLES IN LAGRANGIAN FORMULATIONS

In Section 8.3.2 we discussed the fundamental problem of formulating the equations of equilibrium on an unknown configuration and we described a scheme used to circumvent it in the analysis of the interaction of axial force with bending and twisting moments. In this appendix we review a more rigorous method: use of the principle of virtual displacements in a *Lagrangian formulation*.[2] This is an approach upon which broad formulations of the nonlinear analysis problem can be based. Here it will be used as the basis for the analysis of additional forms of interaction between components of force acting on the framework element. The review draws on References A.1, A.2, and A.3. Further details may be found in those works.

The rigorous use of virtual displacement theory to establish equilibrium on the deformed configuration would require that it be applied to that state, that is, to Configuration $t + \Delta t$ in Figure 8.4. The idea of applying a virtual displacement to an unknown configuration would appear to be compounding the problem. But it has been found possible to express these virtual displacements and the sought after real force and displacement increments in terms of the known conditions and coordinate system at configuration t. As shall be demonstrated, a particular measure of finite strain—the *Green-Lagrange* strain tensor—and stress—the second *Piola-Kirchhoff* stress tensor—are conjugate variables; that is, their scalar product (an operation defined below) is a measure of the real or virtual work done during a corresponding real or virtual displacement (see Section 5.3.1). And both tensors satisfy the condition of being expressible in terms of the known conditions at configuration t.

For simplicity of notation in the demonstration we will consider Configuration t to be the unknown deformed state and Configuration 0 the known unloaded, undeformed state, as in Figure A.1. At the end we will discuss the rather straightforward adaptation of the results to the case in which Configuration t is a known intermediate state and Configuration $t + \Delta t$ is sought. In Figure A.1 the components of all vectors and matrices

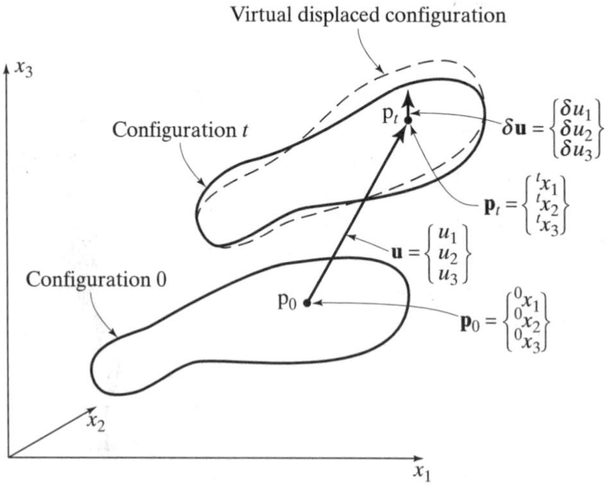

Figure A.1 Virtual displacements in nonlinear analysis.

---

[2]The term *Lagrangian formulation* refers to one that uses material coordinates defined in terms of the undeformed or a temporary reference configuration. The Lagrangian approach is in contrast to the *Eulerian formulation*, which employs spatial coordinates in the deformed configuration and is commonly used in fluid mechanics for analysis of the motion of material through a stationary control volume.

are stated in terms of a fixed Cartesian coordinate system. If we wish to formulate the equations of equilibrium on the deformed state and to use virtual displacement principles to do so, then the displacements should be applied to Configuration t, as indicated. Thus the internal virtual work is

$$\delta W_{\text{int}} = \int_{V_t} \mathbf{T}:\delta\mathbf{e}\, dV_t \tag{A.1}$$

in which $\mathbf{T}$ is the real (physical or "Cauchy") stress tensor and $\mathbf{e}$ is the infinitesimal strain tensor. $\mathbf{T}:\delta\mathbf{e}$ is defined as the scalar product of two tensors, that is, the sum of the nine products of components or, in the usual indicial notation, $T_{ij}\,\delta e_{ij}$, and integration is over the volume of the deformed configuration. In matrix form, the stress tensor may be written as

$$\mathbf{T} = \begin{bmatrix} \sigma_x & \tau_{xy} & \tau_{xz} \\ \tau_{yx} & \sigma_y & \tau_{yz} \\ \tau_{zx} & \tau_{zy} & \sigma_z \end{bmatrix} \text{ or } \begin{bmatrix} T_{11} & T_{12} & T_{13} \\ T_{21} & T_{22} & T_{23} \\ T_{31} & T_{32} & T_{33} \end{bmatrix} \tag{A.2}$$

Noting that ${}^t x_i = {}^0 x_i + u_i$, in which u_i is a Cartesian component of the increment of displacement between the undeformed and the deformed configuration, alternative forms of the infinitesimal strain matrix are[3]

$$\mathbf{e} = \begin{bmatrix} e_x & \frac{1}{2}\gamma_{xy} & \frac{1}{2}\gamma_{xz} \\ \frac{1}{2}\gamma_{yx} & e_y & \frac{1}{2}\gamma_{yz} \\ \frac{1}{2}\gamma_{zx} & \frac{1}{2}\gamma_{zy} & e_z \end{bmatrix} \text{ or } \begin{bmatrix} \frac{\partial u_1}{\partial {}^t x_1} & \frac{1}{2}\left(\frac{\partial u_1}{\partial {}^t x_2} + \frac{\partial u_2}{\partial {}^t x_1}\right) & \frac{1}{2}\left(\frac{\partial u_1}{\partial {}^t x_3} + \frac{\partial u_3}{\partial {}^t x_1}\right) \\ \frac{1}{2}\left(\frac{\partial u_2}{\partial {}^t x_1} + \frac{\partial u_1}{\partial {}^t x_2}\right) & \frac{\partial u_2}{\partial {}^t x_2} & \frac{1}{2}\left(\frac{\partial u_2}{\partial {}^t x_3} + \frac{\partial u_3}{\partial {}^t x_2}\right) \\ \frac{1}{2}\left(\frac{\partial u_3}{\partial {}^t x_1} + \frac{\partial u_1}{\partial {}^t x_3}\right) & \frac{1}{2}\left(\frac{\partial u_3}{\partial {}^t x_2} + \frac{\partial u_2}{\partial {}^t x_3}\right) & \frac{\partial u_3}{\partial {}^t x_3} \end{bmatrix} \tag{A.3}$$

or, in tensor component form:

$$e_{ij} = \frac{1}{2}\left(\frac{\partial u_i}{\partial {}^t x_j} + \frac{\partial u_j}{\partial {}^t x_i}\right) \tag{A.4}$$

Thus

$$\delta e_{ij} = \frac{1}{2}\left(\frac{\partial \delta u_i}{\partial {}^t x_j} + \frac{\partial \delta u_j}{\partial {}^t x_i}\right) \tag{A.5}$$

In the above equations the superscript t signifies that the tensors apply to the deformed configuration and that they are written as functions of the components of position vectors of points in that configuration. Also, integration is over the deformed volume. For these reasons, although Equation A.1 is a valid statement of the principle of virtual displacements, it can't be used directly to determine the physical stresses in configuration t since that configuration is still unknown. Therefore, we turn to an alternative expression of the internal virtual work

$$\delta W_{\text{int}} = \int_{V_0} {}^t_0\tilde{\mathbf{T}}:\delta^t_0\boldsymbol{\varepsilon}\, dV_0 \tag{A.6}$$

[3]To ensure that $\mathbf{e}$ transforms as a second-order tensor, the factor 1/2 is included in the definition of the shear strain terms (see Reference A.1 and compare with Figure 4.1*b* and Equation 4.1c).

in which $\tilde{\mathbf{T}}$ is the second Piola-Kirchhoff stress tensor and $\boldsymbol{\varepsilon}$ is the Green-Lagrange strain tensor. Here, the superscript t again signifies that the tensors apply to the deformed configuration, but the subscript 0 designates that they are referred to the undeformed configuration; that is, their components are written as functions of the known Cartesian coordinates of p_0 and integration is over the undeformed volume. Our immediate aim is to demonstrate that the internal virtual work symbolized by Equation A.6 is indeed identical to that of Equation A.1.[4] Throughout, we assume that the external virtual work is deformation independent (see Section A.3 for a comment on the implications of this restriction).

Consider strains first: The Green-Lagrange strain tensor is a measure of finite strain defined as

$$ {}^t_0\boldsymbol{\varepsilon} = \tfrac{1}{2}({}^t_0\mathbf{X}^{\mathrm{T}}\, {}^t_0\mathbf{X} - \mathbf{I}) \tag{A.7}$$

in which ${}^t_0\mathbf{X}$ is the *deformation gradient*

$$ {}^t_0\mathbf{X} = \begin{bmatrix} \dfrac{\partial^t x_1}{\partial^0 x_1} & \dfrac{\partial^t x_1}{\partial^0 x_2} & \dfrac{\partial^t x_1}{\partial^0 x_3} \\ \dfrac{\partial^t x_2}{\partial^0 x_1} & \dfrac{\partial^t x_2}{\partial^0 x_2} & \dfrac{\partial^t x_2}{\partial^0 x_3} \\ \dfrac{\partial^t x_3}{\partial^0 x_1} & \dfrac{\partial^t x_3}{\partial^0 x_2} & \dfrac{\partial^t x_3}{\partial^0 x_3} \end{bmatrix} \tag{A.8}$$

The deformation gradient describes the elongation and rotation that a material fiber undergoes between Configuration 0 and Configuration t. As illustrated in Figure A.2, a differential material element, $d^0\mathbf{x}$, at point p_0 would become $d^t\mathbf{x}$ at point p_t. Therefore, by the chain rule of differentiation

$$\{d^t\mathbf{x}\} = [{}^t_0\mathbf{X}]\{d^0\mathbf{x}\}$$

Also, noting that the squares of the lengths of the element in the two states are

$$(d^0x)^2 = \lfloor d^0\mathbf{x} \rfloor \{d^0\mathbf{x}\} \qquad \text{and} \qquad (d^t\mathbf{x})^2 = \lfloor d^t\mathbf{x} \rfloor \{d^t\mathbf{x}\}$$

it follows that

$$ (d^tx)^2 - (d^0x)^2 = \lfloor d^0\mathbf{x} \rfloor [{}^t_0\mathbf{X}^{\mathrm{T}}\, {}^t_0\mathbf{X} - \mathbf{I}]\{d^0\mathbf{x}\} \tag{A.9}$$

Through comparison of Equations A.7 and A.9 it may be seen that the Green-Lagrange strain tensor was defined to give the change in the squared length of the material element (see also Reference A.4). From Figure A.2 and the manner in which Equation A.9 was derived, it is clear that it contains information on the rotation of the

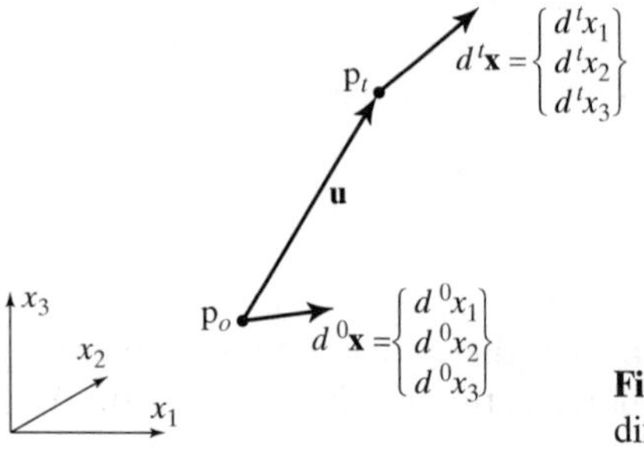

Figure A.2 Displacement, stretch, and rotation of a differential material element.

[4]It is suggested that the reader verify the key equations of this demonstration by carrying out the intermediate operations.

element as well as on its elongation. But it can be shown that the deformation gradient can be decomposed into the product of a symmetric stretch matrix, ${}^t_0\mathbf{S}$, and an orthogonal rotation matrix, ${}^t_0\mathbf{R}$, such that ${}^t_0\mathbf{X} = {}^t_0\mathbf{R}\,{}^t_0\mathbf{S}$. Therefore ${}^t_0\mathbf{X}^{\mathrm{T}}\,{}^t_0\mathbf{X} = {}^t_0\mathbf{S}\,{}^t_0\mathbf{R}^{\mathrm{T}}\,{}^t_0\mathbf{R}\,{}^t_0\mathbf{S} = {}^t_0\mathbf{S}\,{}^t_0\mathbf{S}$, which means that although the stretch tensor itself produces some rotation of all material vectors except those in the principal strain directions, the Green-Lagrange strain tensor is independent of their rigid body motions (see also Refs. A.1 and A.2).

Using Equation A.8 and letting ${}^tx_i = {}^0x_i + u_i$, the Green-Lagrange strain in tensor component form becomes

$$ {}^t_0\varepsilon_{ij} = \tfrac{1}{2}({}^t_0u_{i,j} + {}^t_0u_{j,i} + {}^t_0u_{k,i}\,{}^t_0u_{k,j}) \tag{A.10}$$

in which commas denote differentiation with respect to the coordinate following. For comparison with the infinitesimal strain tensor we have, for example:

$$ {}^t_0\varepsilon_{11} = \frac{\partial u_1}{\partial {}^0x_1} + \frac{1}{2}\left[\left(\frac{\partial u_1}{\partial {}^0x_1}\right)^2 + \left(\frac{\partial u_2}{\partial {}^0x_1}\right)^2 + \left(\frac{\partial u_3}{\partial {}^0x_1}\right)^2\right] \tag{A.11a}$$

$$ {}^t_0\varepsilon_{12} = \frac{1}{2}\left(\frac{\partial u_1}{\partial {}^0x_2} + \frac{\partial u_2}{\partial {}^0x_1}\right) + \frac{1}{2}\left[\frac{\partial u_1}{\partial {}^0x_1}\frac{\partial u_1}{\partial {}^0x_2} + \frac{\partial u_2}{\partial {}^0x_1}\frac{\partial u_2}{\partial {}^0x_2} + \frac{\partial u_3}{\partial {}^0x_1}\frac{\partial u_3}{\partial {}^0x_2}\right] \tag{A.11b}$$

Also from Equation A.7 we can write the virtual strain as

$$ \delta{}^t_0\boldsymbol{\varepsilon} = \tfrac{1}{2}[(\delta{}^t_0\mathbf{X}^{\mathrm{T}})({}^t_0\mathbf{X}) + ({}^t_0\mathbf{X}^{\mathrm{T}})(\delta{}^t_0\mathbf{X})] \tag{A.12}$$

Now, defining $\delta_t\mathbf{U}$ as

$$ \delta_t\mathbf{U} = \begin{bmatrix} \dfrac{\partial \delta u_1}{\partial {}^tx_1} & \dfrac{\partial \delta u_1}{\partial {}^tx_2} & \dfrac{\partial \delta u_1}{\partial {}^tx_3} \\ \dfrac{\partial \delta u_2}{\partial {}^tx_1} & \dfrac{\partial \delta u_2}{\partial {}^tx_2} & \dfrac{\partial \delta u_2}{\partial {}^tx_3} \\ \dfrac{\partial \delta u_3}{\partial {}^tx_1} & \dfrac{\partial \delta u_3}{\partial {}^tx_2} & \dfrac{\partial \delta u_3}{\partial {}^tx_3} \end{bmatrix} \tag{A.13}$$

from the chain rule for the partial differentiation of the elements of $\delta_t\mathbf{U}$ with respect to those of $\{\mathbf{p}_0\}$ it may be seen that

$$ \delta{}^t_0\mathbf{X} = \delta_t\mathbf{U}\,{}^t_0\mathbf{X} \tag{A.14}$$

Thus, from Equations A.12 and A.5

$$ \begin{aligned} \delta{}^t_0\boldsymbol{\varepsilon} &= \tfrac{1}{2}[{}^t_0\mathbf{X}^{\mathrm{T}}(\delta_t\mathbf{U})^{\mathrm{T}}\,{}^t_0\mathbf{X} + {}^t_0\mathbf{X}^{\mathrm{T}}(\delta_t\mathbf{U}){}^t_0\mathbf{X}] = {}^t_0\mathbf{X}^{\mathrm{T}}(\tfrac{1}{2}[(\delta_t\mathbf{U})^{\mathrm{T}} + (\delta_t\mathbf{U})]){}^t_0\mathbf{X} \\ &= {}^t_0\mathbf{X}^{\mathrm{T}}\,\delta\mathbf{e}\,{}^t_0\mathbf{X} \end{aligned} \tag{A.15}$$

which relates the virtual Green-Lagrange strains to the virtual infinitesimal strains.

Now consider the second Piola-Kirchhoff stress tensor. In the context of the two states considered it is defined as

$$ {}^t_0\tilde{\mathbf{T}} = \frac{{}^0\rho}{{}^t\rho}\,{}^0_t\mathbf{X}\,\mathbf{T}\,{}^0_t\mathbf{X}^{\mathrm{T}} \tag{A.16}$$

in which ${}^0\rho$ and ${}^t\rho$ are the material densities in the undeformed and deformed configurations respectively and ${}^0_t\mathbf{X}$ is the *spatial deformation gradient*, which may be verified to be the inverse of the deformation gradient (Equation A.8). In component form the stress tensor is

$$ {}^t_0\tilde{T}_{ij} = \frac{{}^0\rho}{{}^t\rho}\,{}^0_tx_{i,m}\,T_{mn}\,{}^0_tx_{j,n} \tag{A.17}$$

Substituting Equations A.15 and A.16 in Equation A.6, we have

$$\int_{V_0} {}_0^t\tilde{\mathbf{T}}:\delta_0^t\boldsymbol{\varepsilon}\,dV_0 = \int_{V_0} \frac{{}^0\rho}{{}^t\rho}\,[{}_t^0\mathbf{X}\,\mathbf{T}\,{}_t^0\mathbf{X}^{\mathrm{T}}]:[{}_0^t\mathbf{X}^{\mathrm{T}}\,\delta\mathbf{e}\,{}_0^t\mathbf{X}]\,dV_0 \tag{A.18}$$

Assuming no loss of mass in the transition, that is, that ${}^0\rho V_0 = {}^t\rho V_t$, and recognizing the orthogonality of the gradient matrices and the symmetry of the stress and strain tensors, we see that

$$\int_{V_0} {}_0^t\tilde{\mathbf{T}}:\delta_0^t\boldsymbol{\varepsilon}\,dV_0 = \int_{V_t} \mathbf{T}:\delta\mathbf{e}\,dV_t \tag{A.19}$$

which satisfies the objective of demonstrating that Equations A.6 and A.1 yield identical values of the internal virtual work.

As a commentary on the above it will be observed that Green-Lagrange strains are but one of several measures of finite strain, logarithmic (Hencky) strains being another. All have a physical base, but the Green-Lagrange variety is perhaps the most useful in the development of general methods of nonlinear analysis.[5]

As measures of stress, on the other hand, the Cauchy stresses are the ones that have physical meaning and thus they or their resultants are the ones we must ultimately calculate. As shown in References A.1 and A.2, second Piola-Kirchhoff stresses are ingenious constructions based on the transformation of stresses and the deformation of the element on which they act from one state to another, but they have little physical meaning. Their primary virtues are those just described: in combination with Green-Lagrange strains they are valid measurements of virtual work and thus provide a basis for the formulation of nonlinear equations of equilibrium on a known configuration.

The Lagrangian formulation is used in both Figures 8.4 and A.1 and each illustrates variations on the same basic idea of using material coordinates in the undeformed or a temporary reference configuration—as opposed to spatial coordinates in the deformed configuration—to trace the nonlinear behavior of a system. The process shown in Figure A.1 is generally termed a *total Lagrangian* one in that the reference configuration is the actual initial one, whereas that in Figure 8.4 is called an *updated Lagrangian* approach. All of the equations developed from Figure A.1 can be applied to the case of Figure 8.4 by redefining Configurations 0 and t of Figure A.1 as Configurations t and $t + \Delta t$, respectively.

---

[5]It may be noted that Equation 9.1a of Section 9.2 is apparently equivalent to Equation A.11a, the comparable Green-Lagrange strain component. There is a fundamental difference, however. The Green-Lagrange strain components involve only linear and quadratic terms of the displacement derivatives. No higher-order terms were neglected in the development of Equation A.11. The Green-Lagrange strain tensor is a complete finite strain tensor, and it is for this reason that there are no restrictions on its use in conjunction with second Piola-Kirchhoff stresses in the formulation of large displacement problems. The definition of strain as the unit extension of an element, as in Equations 9.1 and 9.1a, produced the same result, but those equations are approximations obtained by neglecting higher-order terms. When used in conjunction with Cauchy stresses in a virtual displacement equation formulated on the reference configuration, as we did in the development of Equations 9.4 and 9.16, it must be recognized that the result is only valid for small finite strains.

Equation 9.1 does furnish a simple geometric interpretation of the direct components of the Green-Lagrange strain. When strains are small, the measure of strain based on the change in the squared length of a material element is a close approximation of the unit extension. Similar analysis of the angle change between two material elements leads to the conclusion that, when the unit extensions and angle changes are small, Green-Lagrange components such as Equation A.11b are approximately one-half of the change in angle in the plane defined by the elements (see Refs. A.4 and A.5 for further geometrical interpretation of the strain components).

In Figure A.1 all variables have been stated in terms of components of a single Cartesian coordinate system, whereas in Figure 8.4 local element coordinate systems have been shown for each stage of loading, with the understanding that conventional transformation procedures will be used to transform the resulting matrices to a single, fixed system for the formation of the global equations of analysis. But the coordinate system employed is not a distinguishing feature of the total and updated Lagrangian approaches; the two schemes were used merely to illustrate the choices available.

Since, in practice, linearized approximations of the basic equations are normally used, both Lagrangian approaches will be applied incrementally. When used consistently, both yield the same results.[6]

A.2 AN UPDATED LAGRANGIAN FORMULATION AND ITS LINEARIZATION

An essential step between the statement of the principle of virtual displacements and its application to specific elements is the development of general incremental equations of equilibrium. An updated Lagrangian formulation of this type is outlined in this section.

As in Section 8.3, we start at an intermediate stage of the analysis, Configuration t, a known, loaded and deformed reference state; one in which the material's constitutive laws and the conditions of equilibrium and compatibility are presumed to have been satisfied by prior incremental analysis (see Figure 8.4). The objective is to write equations of equilibrium for the next increment, Configuration $t + \Delta t$, but with all static and kinematic variables expressed in terms of the reference state and, having done that, to reduce the equations to a linear form suitable for incremental and incremental-iterative analyses.

As noted above, the equations of Section A.1 may be applied to the present case symbolically by shifting 0 and t to t and $t + \Delta t$ respectively. Thus, for equilibrium on Configuration $t + \Delta t$, from Equation A.1 and the basic virtual work equation, $\delta W_{\text{int}} = \delta W_{\text{ext}}$ (Eq. 6.13), we have

$$\int_{V_{t+\Delta t}} \mathbf{T}:\delta\mathbf{e}\, dV_{t+\Delta t} = {}^{t+\Delta t}R$$

in which ${}^{t+\Delta t}R$ is the external virtual work. The expression for external virtual work will be retained in this form throughout the section. Implicit in this is the assumption that the external loading is deformation independent. The significance and limitations of this will be considered in Section A.3. From Equation A.19, the alternative expression for internal virtual work, we have

$$\int_{V_t} {}^{t+\Delta t}_{\ \ t}\tilde{\mathbf{T}}:\delta\, {}^{t+\Delta t}_{\ \ t}\boldsymbol{\varepsilon}\, dV_t = {}^{t+\Delta t}R \tag{A.20}$$

or, in indicial notation

$$\int_{V_t} {}^{t+\Delta t}_{\ \ t}\tilde{T}_{ij}\, \delta\, {}^{t+\Delta t}_{\ \ t}\varepsilon_{ij}\, dV_t = {}^{t+\Delta t}R \tag{A.20a}$$

in which component forms of ${}^{t+\Delta t}_{\ \ t}\tilde{T}_{ij}$ and $\delta\, {}^{t+\Delta t}_{\ \ t}\varepsilon_{ij}$ may be obtained from Equations A.17 and A.10 by the appropriate changes in sub- and superscripts.

[6] The effect of initial strains has not been included in this discussion. When present, they must be considered in the analysis of the undeformed state (see discussion of initial strain matrices in Ref. A.2).

The stresses may be decomposed as

$${}^{t+\Delta t}_{t}\tilde{T}_{ij} = {}^{t}_{t}\tilde{T}_{ij} + {}_{t}\tilde{T}_{ij} = {}^{t}T_{ij} + {}_{t}\tilde{T}_{ij}$$

in which ${}^{t}_{t}\tilde{T}_{ij}$, the second Piola-Kirchhoff stress at Configuration t, is identical to the Cauchy stress, ${}^{t}T_{ij}$, at that configuration, and ${}_{t}\tilde{T}_{ij}$ is the increment of Piola-Kirchhoff stress between Configurations t and $t + \Delta t$.

For Green-Lagrange strains and incremental displacements expressed in terms of the reference state, ${}^{t+\Delta t}_{t}\varepsilon_{ij} = {}_{t}\varepsilon_{ij}$. Thus, from Equation A.10

$${}_{t}\varepsilon_{ij} = \tfrac{1}{2}({}_{t}u_{i,j} + {}_{t}u_{j,i} + {}_{t}u_{k,i}\,{}_{t}u_{k,j}) \tag{A.21}$$

Letting ${}_{t}\varepsilon_{ij} = {}_{t}e_{ij} + {}_{t}\eta_{ij}$, Equation A.21 may be decomposed into linear and nonlinear components in which

$${}_{t}e_{ij} = \tfrac{1}{2}({}_{t}u_{i,j} + {}_{t}u_{j,i}) \quad \text{and} \quad {}_{t}\eta_{ij} = \tfrac{1}{2}({}_{t}u_{k,i}\,{}_{t}u_{k,j}) \tag{A.22}$$

The virtual strains may be decomposed in the same way.

Decomposing stresses and strains in Equation A.20a in this fashion we have

$$\int_{V_t} {}_{t}\tilde{T}_{ij}\,\delta_{t}\varepsilon_{ij}\,dV_t + \int_{V_t} {}^{t}T_{ij}\,\delta_{t}e_{ij}\,dV_t + \int_{V_t} {}^{t}T_{ij}\,\delta_{t}\eta_{ij}\,dV_t = {}^{t+\Delta t}R \tag{A.23}$$

Equation A.23 remains a consistent statement of the principle of virtual displacements in which all integrals are over the known volume of Configuration t. For a given displacement variation, the second integral is a known quantity. It is customary to transpose it to the right-hand side of the equation as an expression of the virtual work of the existing forces on Configuration t, but we shall leave it on the left-hand side for a reason to be explained in Section A.4.2. The third integral is linear in the incremental displacements, but the first one is not, so it must be approximated in some way for the purpose of practical computation. Approximations suggested in Reference A.2 are ${}_{t}\tilde{T}_{ij} = {}_{t}C_{ijrs}\,{}_{t}e_{rs}$ and $\delta_{t}\varepsilon_{ij} = \delta_{t}e_{ij}$, in which ${}_{t}C_{ijrs}$ is the incremental stress-strain tensor at Configuration t. The result is the linearized equation

$$\int_{V_t} {}_{t}C_{ijrs}\,{}_{t}e_{rs}\,\delta_{t}e_{ij}\,dV_t + \int_{V_t} {}^{t}T_{ij}\,\delta_{t}e_{ij}\,dV_t + \int_{V_t} {}^{t}T_{ij}\,\delta_{t}\eta_{ij}\,dV_t = {}^{t+\Delta t}R \tag{A.24}$$

The development of the revised integral and the approximations involved are explained further in Reference A.2. Equation A.24 admits incorporation of variable materials properties, but for linear elastic materials ${}_{t}C_{ijrs}$ is constant and the first integral leads to the conventional elastic stiffness matrix.

A.3 APPLICATION TO THE FRAMEWORK ELEMENT

Equation A.23 or a suitably linearized version of it, such as Equation A.24, can be the base for the development of finite element equations for the nonlinear analysis of many line, plate, and solid elements (see Reference A.2). Here we shall use Equation A.24 in a limited way: to fill in the gaps in the nonlinear framework element stiffness matrices that could not be developed through simple physical arguments.

Figure A.3, a copy of Figure 4.6. shows the bisymmetrical element and the nodal forces and degrees of freedom considered. All quantities shall be referred to the axes shown, thereby eliminating the need for the sub- and superscripts of Equation A.24. As before, deformations due to flexurally induced shear will be neglected and, in this section, we shall only consider simple St. Venant torsion. These limitations, plus the use of common engineering theory for converting stresses to cross-sectional stress resultants, admits considerable simplification of Equation A.24.

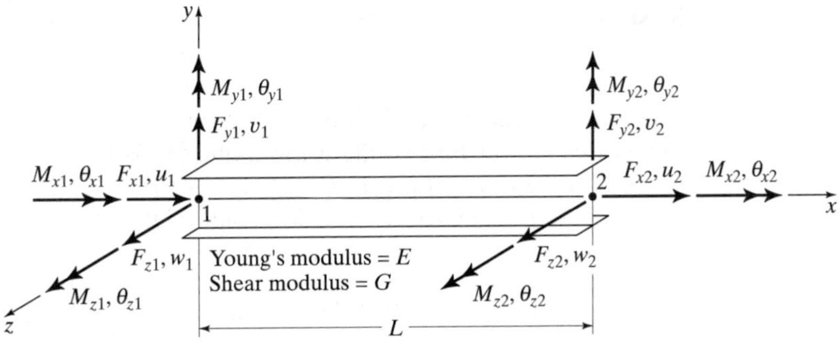

Figure A.3 Bisymmetrical framework element.

As shown in Figure A.4*a*, the stresses at a point on a cross section are σ_x, τ_{xy}, and τ_{xz}. Because σ_y, σ_z, and τ_{yz} are not considered, and since $\tau_{yx} = \tau_{xy}$ and $\tau_{zx} = \tau_{xz}$, these are the only independent stress components. They and the corresponding strain components can be written as simple vectors

$$\{\mathbf{T}\} = \lfloor \sigma_x \ \tau_{xy} \ \tau_{xz} \rfloor^{\mathrm{T}} \qquad \{\mathbf{e}\} = \lfloor e_x \ \gamma_{xy} \ \gamma_{xz} \rfloor^{\mathrm{T}} \qquad \{\eta\} = \lfloor \eta_{xx} \ \eta_{xy} \ \eta_{xz} \rfloor^{\mathrm{T}} \tag{A.25}$$

For elastic behavior, the stress-strain tensor becomes

$$\lceil \mathbf{C} \rfloor = \lceil E \ G \ G \rfloor \tag{A.26}$$

If applied forces are deformation independent, ${}^{t+\Delta t}R$, the external virtual work can be expressed as the scalar product $\lfloor {}^{t+\Delta t}\mathbf{F} \rfloor \{\delta \mathbf{u}\}$, with

$$\lfloor {}^{t+\Delta t}\mathbf{F} \rfloor = \lfloor F_{x1} \ F_{y1} \ F_{z1} \ M_{x1} \ M_{y1} \ M_{z1} \ F_{x2} \ F_{y2} \ F_{z2} \ M_{x2} \ M_{y2} \ M_{z2} \rfloor \tag{A.27}$$

in which $\lfloor {}^{t+\Delta t}\mathbf{F} \rfloor$ is the vector of forces at the *end* of the load step, and

$$\{\delta \mathbf{u}\} = \delta \lfloor u_1 \ v_1 \ w_1 \ \theta_{x1} \ \theta_{y1} \ \theta_{z1} \ u_2 \ v_2 \ w_2 \ \theta_{x2} \ \theta_{y2} \ \theta_{z2} \rfloor^{\mathrm{T}} \tag{A.28}$$

in which $\{\delta \mathbf{u}\}$ is the vector of virtual displacements from the *reference configuration*. The condition of deformation independence is satisfied by path independent conservative loads, which are the only loads to be considered here. Reference A.2 contains suggestions for treating forces that do not satisfy this restriction.

From the above it may be seen that the first integral of Equation A.24 produces the conventional elastic stiffness matrix and the second integral produces the forces acting on the element in the reference configuration. The components of these may be evaluated by the methods of the earlier chapters. Our immediate interest is therefore in

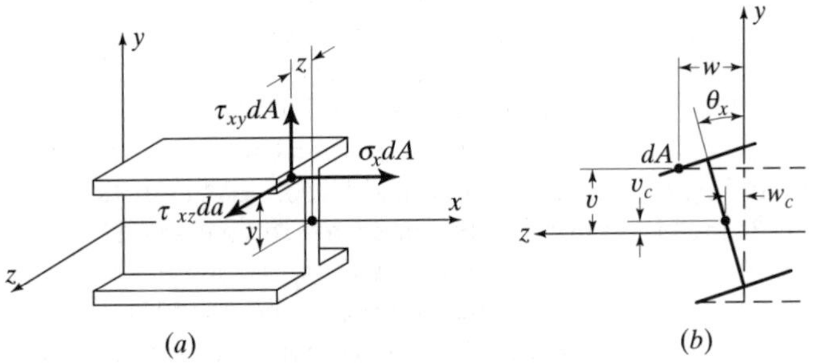

Figure A.4 Stresses and displacements.

the third integral alone. From the definition of strains and the use of Equation A.22 we have

$$\int_V T_{ij}\delta\eta_{ij}dV = \frac{1}{2}\int_V \sigma_x\delta\left[\left(\frac{\partial u}{\partial x}\right)^2 + \left(\frac{\partial v}{\partial x}\right)^2 + \left(\frac{\partial w}{\partial x}\right)^2\right]dV$$
$$+ \int_V \tau_{xy}\delta\left[\frac{\partial u}{\partial x}\frac{\partial u}{\partial y} + \frac{\partial w}{\partial x}\frac{\partial w}{\partial y}\right]dV + \int_V \tau_{xz}\delta\left[\frac{\partial u}{\partial x}\frac{\partial u}{\partial z} + \frac{\partial v}{\partial x}\frac{\partial v}{\partial z}\right]dV \quad \text{(A.29)}$$

In which the fact that $\tau_{yx} = \tau_{xy}$ and $\tau_{zx} = \tau_{xz}$ and the engineering assumption of no in-plane distortion that results in $\partial v/\partial y = \partial w/\partial z = 0$ have been accounted for.

In using Equation A.29, we shall need the incremental displacements at a point on the cross section and their derivatives. From Figure A.4, the displacements are

$$u = u_c - z\frac{\partial w_c}{\partial x} - y\frac{\partial v_c}{\partial x} \qquad v = v_c - z\theta_x \qquad w = w_c + y\theta_x \quad \text{(A.30a)}$$

with the derivatives

$$\frac{\partial u}{\partial x} = \frac{\partial u_c}{\partial x} - z\frac{\partial^2 w_c}{\partial x^2} - y\frac{\partial^2 v_c}{\partial x^2} \qquad \frac{\partial v}{\partial x} = \frac{\partial v_c}{\partial x} - z\frac{\partial \theta_x}{\partial x} \qquad \frac{\partial w}{\partial x} = \frac{\partial w_c}{\partial x} + y\frac{\partial \theta_x}{\partial x}$$
$$\frac{\partial u}{\partial y} = -\frac{\partial v_c}{\partial x} \qquad \frac{\partial v}{\partial y} = 0 \qquad \frac{\partial w}{\partial y} = \theta_x \quad \text{(A.30b)}$$
$$\frac{\partial u}{\partial z} = -\frac{\partial w_c}{\partial x} \qquad \frac{\partial v}{\partial z} = -\theta_x \qquad \frac{\partial w}{\partial z} = 0$$

We will also convert stresses to stress resultants by integration over the cross section, giving

$$\int_A \sigma_x dA = F_x \qquad \int_A \tau_{xy}dA = F_y = -\frac{M_{z1} + M_{z2}}{L} \qquad \int_A \tau_{xz}dA = F_z = \frac{M_{y1} + M_{y2}}{L}$$
$$\int_A \sigma_x z dA = M_y \qquad \int_A \sigma_x y dA = -M_z \qquad \int_A \tau_{xz}y dA = \alpha M_x \qquad -\int_A \tau_{xy}z dA = (1 - \alpha)M_x \quad \text{(A.31)}$$

in which the flexural shears are treated as reactions to bending moments, and we define by α and $(1 - \alpha)$ the portions of the total torque, M_x, resisted by the stresses τ_{xz} and τ_{xy}, respectively. For pure St. Venant torsion $\alpha = 1/2$, but for reasons to be explained later, we will retain it as an undetermined parameter for the present.

To summarize the development of the desired terms of the geometric stiffness matrix we will consider representative components of Equation A.29. For example, using Equation A.30 in the first integral of the right-hand side of that equation yields

$$\frac{1}{2}\int_V \sigma_x\delta\left[\left(\frac{\partial u}{\partial x}\right)^2 + \left(\frac{\partial v}{\partial x} - z\frac{\partial \theta_x}{\partial x}\right)^2 + \left(\frac{\partial w}{\partial x} + y\frac{\partial \theta_x}{\partial x}\right)^2\right]dV$$

in which the subscript c has been dropped to simplify the notation and the effect of the higher-order derivatives in $\partial u/\partial x$ is assumed to be negligible. Expanding the terms of this integral and using Equation A.31 to convert stresses to stress resultants, we have

$$\frac{1}{2}\int_L F_x\delta\left[\left(\frac{\partial u}{\partial x}\right)^2 + \left(\frac{\partial v}{\partial x}\right)^2 + \left(\frac{\partial w}{\partial x}\right)^2\right]dx + \frac{1}{2}\frac{F_x I_p}{A}\int_L \delta\left(\frac{\partial \theta_x}{\partial x}\right)^2 dx$$
$$- M_y\int_L \delta\left[\frac{\partial v}{\partial x}\frac{\partial \theta_x}{\partial x}\right]dx - M_z\int_L \delta\left[\frac{\partial w}{\partial x}\frac{\partial \theta_x}{\partial x}\right]dx \quad \text{(A.32)}$$

Comparing the first two integrals with Equations 9.4 and 9.16, it is seen that they are the source of the terms of Equation 9.18 that relate to axial straining and the interaction of axial force with flexure and torsion. The last two integrals, which remain to be evaluated, are sources of flexure-torsion interaction terms. To avoid repetition in the evaluation, we will consider only the last in detail. Noting that for constant shear $M_z = -M_{z1}\left(1 - \frac{x}{L}\right) + M_{z2}\left(\frac{x}{L}\right)$ gives

$$M_{z1}\int_L \delta\left[\frac{\partial w}{\partial x}\frac{\partial \theta_x}{\partial x}\right]dx - \left(\frac{M_{z1} + M_{z2}}{L}\right)\int_L x\,\delta\left[\frac{\partial w}{\partial x}\frac{\partial \theta_x}{\partial x}\right]dx \tag{A.33}$$

We'll return to the evaluation of the integrals in Equation A.33 after similarly analyzing the other integrals of Equation A.29. Using Equation A.30 in the second one on the right-hand side yields

$$\int_V \tau_{xy}\,\delta\left[\left(-\frac{\partial v}{\partial x}\right)\left(\frac{\partial u}{\partial x} - z\frac{\partial^2 w}{\partial x^2} - y\frac{\partial^2 v}{\partial x^2}\right) + \theta_x\left(\frac{\partial w}{\partial x} + y\frac{\partial \theta_x}{\partial x}\right)\right]dV$$

Neglecting products of the derivatives of the same variable as of higher order, noting that for a bisymmetrical cross section $\int_A (\tau_{xy}y)\,dA = 0$, and using Equation A.31 to convert stresses to stress resultants, we have

$$\int_L F_y\,\delta\left[\left(-\frac{\partial v}{\partial x}\right)\left(\frac{\partial u}{\partial x}\right) + \theta_x\left(\frac{\partial w}{\partial x}\right)\right]dx - \int_L (1-\alpha)M_x\,\delta\left[\frac{\partial v}{\partial x}\frac{\partial^2 w}{\partial x^2}\right]dx$$

Similarly for the third integral on the right-hand side of Equation A.29,

$$\int_L F_z\,\delta\left[\left(-\frac{\partial w}{\partial x}\right)\left(\frac{\partial u}{\partial x}\right) - \theta_x\left(\frac{\partial v}{\partial x}\right)\right]dx + \int_L \alpha M_x\,\delta\left[\frac{\partial w}{\partial x}\frac{\partial^2 v}{\partial x^2}\right]dx$$

In both of these, the first term in the first integral represents a degree of flexure-axial force interaction that is assumed to be insignificant and is therefore neglected. The remaining terms involve torsion-flexure interaction. Adding these terms, and replacing F_y and F_z with their equivalent end moments results in

$$\begin{aligned}&-\left(\frac{M_{z1} + M_{z2}}{L}\right)\int_L \delta\left[\theta_x\frac{\partial w}{\partial x}\right]dx - \left(\frac{M_{y1} + M_{y2}}{L}\right)\int_L \delta\left[\theta_x\frac{\partial v}{\partial x}\right]dx\\ &-\int_L (1-\alpha)M_x\,\delta\left[\frac{\partial v}{\partial x}\frac{\partial^2 w}{\partial x^2}\right]dx + \int_L \alpha M_x\,\delta\left[\frac{\partial w}{\partial x}\frac{\partial^2 v}{\partial x^2}\right]dx\end{aligned} \tag{A.34}$$

Application of the shape function approach to the determination of geometric stiffness matrix coefficients can be illustrated by selection of representative terms in Equations A.33 and A.34. For example, we may analyze the terms in which w is the real displacement component. Thus, first selecting the integrals in which w appears

$$\begin{aligned}M_{z1}\int_L \delta\left[\frac{\partial w}{\partial x}\frac{\partial \theta_x}{\partial x}\right]dx - \left(\frac{M_{z1} + M_{z2}}{L}\right)\left(\int_L x\,\delta\left[\frac{\partial w}{\partial x}\frac{\partial \theta_x}{\partial x}\right]dx + \int_L \delta\left[\theta_x\frac{\partial w}{\partial x}\right]dx\right)\\ -\int_L (1-\alpha)M_x\,\delta\left[\frac{\partial v}{\partial x}\frac{\partial^2 w}{\partial x^2}\right]dx + \int_L \alpha M_x\,\delta\left[\frac{\partial w}{\partial x}\frac{\partial^2 v}{\partial x^2}\right]dx\end{aligned}$$

and then retaining only the terms in which w is real, we have

$$
\begin{aligned}
M_{z1}\int_L \frac{\partial w}{\partial x}\,\delta\frac{\partial \theta_x}{\partial x}\,dx &- \left(\frac{M_{z1}+M_{z2}}{L}\right)\left(\int_L x\left(\frac{\partial w}{\partial x}\,\delta\frac{\partial \theta_x}{\partial x}\right)dx + \int_L \frac{\partial w}{\partial x}\,\delta\theta_x\,dx\right)\\
&- \int_L (1-\alpha)M_x\frac{\partial^2 w}{\partial x^2}\,\delta\frac{\partial v}{\partial x}\,dx + \int_L \alpha M_x\frac{\partial w}{\partial x}\,\delta\frac{\partial^2 v}{\partial x^2}\,dx
\end{aligned}
\tag{A.35}
$$

Later we will investigate the possibility that α might have a value other than 1/2 and the consequences of that event. But since in this section we are limiting consideration to pure St. Venant torsion, we will let M_x equal its equivalent, M_{x2} and α equal 1/2, and we'll assume a linear shape function for the twist. Cubic functions will be assumed for the transverse displacements (see Equation 7.9). Thus

$$
\begin{aligned}
\theta_x &= \lfloor \mathbf{N}_1 \rfloor\{\boldsymbol{\theta}_x\} = \lfloor (1-\xi)\ \ \xi \rfloor \lfloor \theta_{x1}\ \ \theta_{x2} \rfloor^{\mathrm{T}}\\
v &= \lfloor \overline{\mathbf{N}}_3 \rfloor\{\mathbf{v}\} = \lfloor (1-3\xi^2+2\xi^3)\ (1-2\xi+\xi^2)x\ (3\xi^2-2\xi^3)\ -(\xi-\xi^2)x \rfloor\\
&\quad \lfloor v_1\ \ \theta_{z1}\ \ v_2\ \ \theta_{z2} \rfloor^{\mathrm{T}}\\
w &= \lfloor \mathbf{N}_3 \rfloor\{\mathbf{w}\} = \lfloor (1-3\xi^2+2\xi^3)\\
&\quad -(1-2\xi+\xi^2)x\ (3\xi^2-2\xi^3)\ (\xi-\xi^2)x \rfloor \lfloor w_1\ \ \theta_{y1}\ \ w_2\ \ \theta_{y2} \rfloor^{\mathrm{T}}
\end{aligned}
\tag{A.36}
$$

in which $\xi = x/L$.

Following the procedures of Section 7.2, we have

$$
\begin{aligned}
&\lfloor \delta\boldsymbol{\theta}_x \rfloor\left[M_{z1}\int_L \{\mathbf{N}_1'\}\lfloor \mathbf{N}_3' \rfloor dx - \left(\frac{M_{z1}+M_{z2}}{L}\right)\left(\int_L x\{\mathbf{N}_1'\}\lfloor \mathbf{N}_3' \rfloor dx + \int_L \{\mathbf{N}_1\}\lfloor \mathbf{N}_3' \rfloor dx\right)\right]\{\mathbf{w}\}\\
&\lfloor \delta\mathbf{v} \rfloor\left[-\left(\frac{M_{x2}}{2}\right)\left(\int_L \{\overline{\mathbf{N}}_3'\}\lfloor \mathbf{N}_3'' \rfloor - \int_L \{\overline{\mathbf{N}}_3''\}\lfloor \mathbf{N}_3' \rfloor\right)\right]\{\mathbf{w}\}
\end{aligned}
\tag{A.37}
$$

Integrating and combining terms results in the geometric stiffness matrix

$$
[\mathbf{k_g}] =
\begin{matrix} w_1 & \theta_{y1} & w_2 & \theta_{y2} \end{matrix}
\begin{bmatrix}
0 & \dfrac{M_{x2}}{L} & 0 & -\dfrac{M_{x2}}{L}\\[2mm]
\dfrac{M_{z1}}{L} & \dfrac{M_{z1}+M_{z2}}{6} & -\dfrac{M_{z1}}{L} & -\dfrac{M_{z1}+M_{z2}}{6}\\[2mm]
\dfrac{M_{x2}}{L} & 0 & -\dfrac{M_{x2}}{L} & -\dfrac{M_{x2}}{2}\\[2mm]
0 & -\dfrac{M_{x2}}{L} & 0 & \dfrac{M_{x2}}{L}\\[2mm]
\dfrac{M_{z2}}{L} & -\dfrac{M_{z1}+M_{z2}}{6} & -\dfrac{M_{z2}}{L} & \dfrac{M_{z1}+M_{z2}}{6}\\[2mm]
-\dfrac{M_{x2}}{L} & \dfrac{M_{x2}}{2} & \dfrac{M_{x2}}{L} & 0
\end{bmatrix}
\tag{A.38}
$$

Analysis of the terms in which v and θ_x are the real displacement components and treating the remainder of the integrals of Equation A.29 in the same way will, with one exception, result in all of the flexure-torsion interaction terms considered to be significant. These are the contributions that have been added to Equation 9.18 in the production of Equation 9.19. The excepted terms, which have also been added to Equation 9.18, are developed below (Eq. A.48).

A.4 FINITE ROTATIONS AND EQUILIBRIUM IN THE DEFORMED CONFIGURATION

A problem not considered in the above is illustrated by Figure A.5. The structure shown is a plane frame subjected to a transverse load (Fig. A.5*a*). Assume that at a particular load step it has been established that joint *b* is in equilibrium. The bending moment, *M*, at the end of member *bc* is balanced by the torque, *T*, at the end of member *ab* (Fig. A.5*b*). If these quantities are determined by conventional engineering theories for beams and for shafts subjected to St. Venant torsion, each can be represented as in Figure A.5*c*, that is, by couples consisting of direct and shearing forces, *M* and *T*/2, acting on the ends of rigid links of unit length. If the joint is now subjected to a small rigid body rotation, θ_x, about the *x* axis, equilibrium will be destroyed because the incremental moment about the *y* axis generated by the bending moment is twice that generated by the torque (Fig. A.5*d*).

The problem is subtle because there are several suspects for the source of the difficulty:

1. Inconsistencies between the conventional engineering theories of flexure and torsion.
2. Flaws in the linearization of the basic continuum mechanics equation, for example in going from Equation A.23 to A.24.

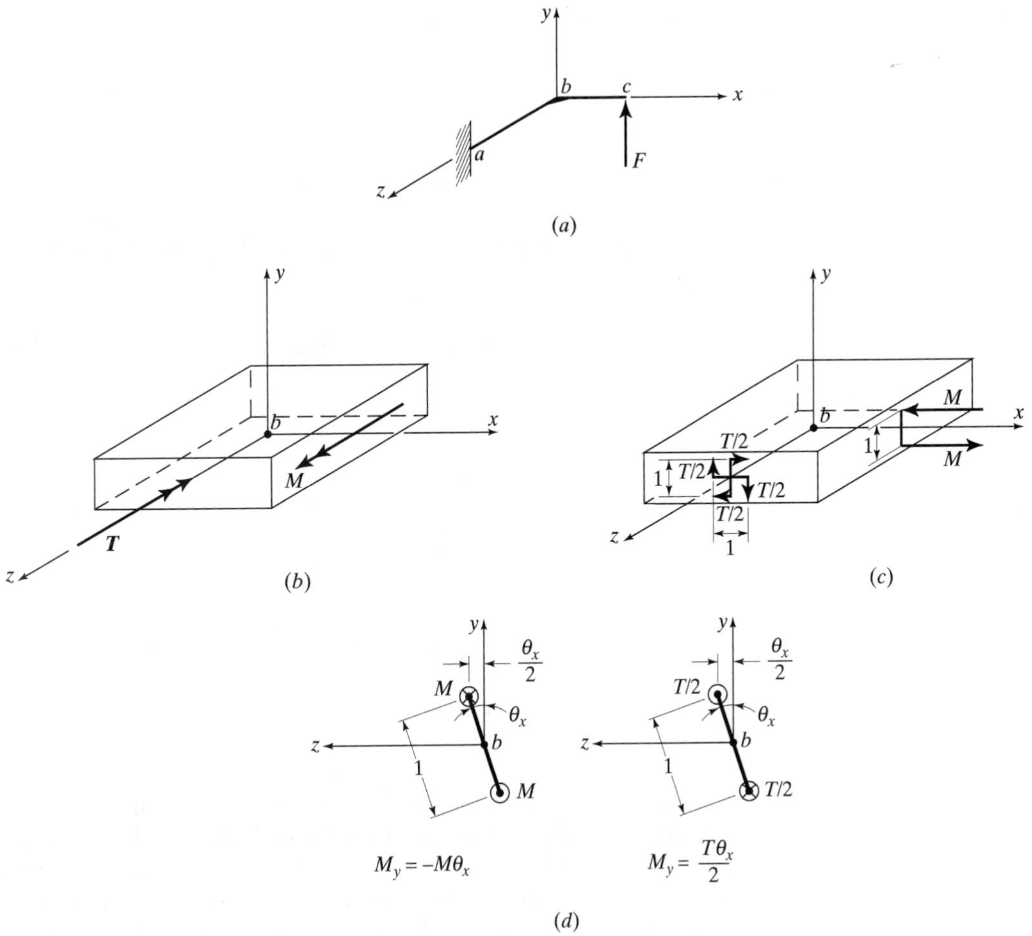

Figure A.5 Finite rotation of a joint.

3. The fact that finite rigid-body rotations do not obey the commutative law of addition and thus are not true vector quantities (see Ref. A.1).
4. The fact that in linear analysis in general, and in the use of the shape function approach in particular, components of rotation are approximated by displacement derivatives (see Eq. 2.1).

Several approaches to the treatment of the problem have been proposed. In matrix analysis most of them involve the formulation of the geometric stiffness matrix (Refs. A.6–A.9). The following, which is patterned after one in Reference A.9, considers both the nature of finite rotations and the distinct characteristics of the engineering theories of bending and torsion.

A.4.1 The Finite Rotation Equation

In 1775, using spherical trigonometry and a fixed coordinate system, Euler developed the following relationship between the locations of a vector as it is rotated through an angle φ about an arbitrary axis through the origin (Ref. A.10).

$$\begin{aligned}
\cos\zeta' &= \cos\zeta(\cos^2\alpha + \sin^2\alpha\cos\varphi) + \cos\eta(\cos\alpha\cos\beta(1-\cos\varphi) - \cos\gamma\sin\varphi)\\
&\quad + \cos\vartheta(\cos\alpha\cos\gamma(1-\cos\varphi) + \cos\beta\sin\varphi)\\
\cos\eta' &= \cos\eta(\cos^2\beta + \sin^2\beta\cos\varphi) + \cos\vartheta(\cos\beta\cos\gamma(1-\cos\varphi) - \cos\alpha\sin\varphi)\\
&\quad + \cos\zeta(\cos\alpha\cos\beta(1-\cos\varphi) + \cos\gamma\sin\varphi)\\
\cos\vartheta' &= \cos\vartheta(\cos^2\gamma + \sin^2\gamma\cos\varphi) + \cos\zeta(\cos\alpha\cos\gamma(1-\cos\varphi) - \cos\beta\sin\varphi)\\
&\quad + \cos\eta(\cos\beta\cos\gamma(1-\cos\varphi) + \cos\alpha\sin\varphi)
\end{aligned} \tag{A.39}$$

in which ζ, η, and ϑ, and ζ', η', and ϑ' are the direction angles of the vector before and after the rotation, α, β, and γ are the direction angles of the axis of rotation, and φ is the angle of rotation.

Euler's equations may be put in compact matrix form by referring them to Figure A.6 and making the following substitutions:

$$\begin{Bmatrix}\cos\alpha\\ \cos\beta\\ \cos\gamma\end{Bmatrix} = \begin{Bmatrix} l\\ m\\ n\end{Bmatrix} \qquad \begin{Bmatrix}\cos\zeta\\ \cos\eta\\ \cos\vartheta\end{Bmatrix} = \frac{1}{|\mathbf{a}|}\begin{Bmatrix}a_x\\ a_y\\ a_z\end{Bmatrix} \qquad \begin{Bmatrix}\cos\zeta'\\ \cos\eta'\\ \cos\vartheta'\end{Bmatrix} = \frac{1}{|\mathbf{b}|}\begin{Bmatrix}b_x\\ b_y\\ b_z\end{Bmatrix} \tag{A.40}$$

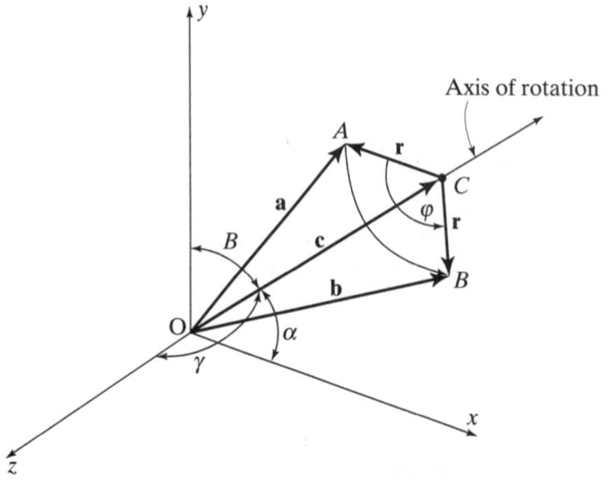

Figure A.6 Finite rotation in a global coordinate system.

Noting that $|\mathbf{a}| = |\mathbf{b}|$ and letting $(1 - \cos\varphi) = 2\sin^2\left(\frac{\varphi}{2}\right)$, Equation A.39 may be written

$$\{\mathbf{b}\} = \left[\mathbf{I} + \sin\varphi\mathbf{V} + 2\sin^2\frac{\varphi}{2}\mathbf{V}^2\right]\{\mathbf{a}\} \tag{A.41}$$

in which $\{\mathbf{a}\}$ and $\{\mathbf{b}\}$ are the position vectors of points A and B, and

$$\mathbf{V} = \begin{bmatrix} 0 & -n & m \\ n & 0 & -l \\ -m & l & 0 \end{bmatrix}$$

Further, by scaling the direction cosine vector of the rotation axis by the magnitude of the rotation, φ[7], that is, $\boldsymbol{\varphi} = \varphi\lfloor l\ m\ n \rfloor = \lfloor \varphi_x\ \varphi_y\ \varphi_z \rfloor$ we have

$$\{\mathbf{b}\} = \left[\mathbf{I} + \frac{\sin\varphi}{\varphi}\mathbf{W} + \frac{1}{2}\left(\frac{\sin(\varphi/2)}{\varphi/2}\right)^2\mathbf{W}^2\right]\{\mathbf{a}\} \tag{A.42}$$

in which

$$\mathbf{W} = \begin{bmatrix} 0 & -\varphi_z & \varphi_y \\ \varphi_z & 0 & -\varphi_x \\ -\varphi_y & \varphi_x & 0 \end{bmatrix} \tag{A.43}$$

Equations A.41 and A.42 are alternative ways of expressing Equation A.39, the original Euler equation. All apply to a rotation of any magnitude. For the small to moderate rotations of interest in normal structural engineering applications, the approximations $\sin\varphi = \varphi$ and $\sin\varphi/2 = \varphi/2$ are satisfactory. Thus, Equation A.41 reduces to $\{\mathbf{b}\} = [\mathbf{T}]\{\mathbf{a}\}$ in which

$$[\mathbf{T}] = [\mathbf{I} + \mathbf{W} + \tfrac{1}{2}\mathbf{W}^2] \tag{A.44}$$

It may be seen that it is the third matrix, $\mathbf{W}^2$, that makes the difference between viewing rotations as infinitely small and small but finite.

A.4.2 Application of the Finite Rotation Equation to the Framework Element

To investigate the role of finite rotations in the problem of interest, we start with the second integral on the left-hand side of Equation A.24

$$\int_V T_{ij}\,\delta e_{ij}\,dV$$

in which all terms refer to a known reference configuration. As mentioned in the discussion of Equations A.23 and A.24, for a given displacement variation it can be integrated to produce an expression of the virtual work of the forces acting on the element prior to the application of a load step. In doing this, the approximation of rotation components by displacement derivatives is used and the finite rotation effect is not considered. Nothing of consequence with respect to the interaction of axial force and bending moments or torque is lost in the approximation. But it has been shown in Reference A.9 that, in the determination of strains, inclusion of the finite rotation

[7]In view of the noncommutative nature of finite rotations, the vector $\boldsymbol{\varphi}$ has been designated variously as a "pseudo vector" and a "rotational pseudo vector" (see Ref. A6). We believe that in the context in which it is being used here, that is, in the analysis of the incremental response to a given set of loads, the term *rotation component vector* is more descriptive and not subject to misinterpretation.

effect as represented in the $\mathbf{W}^2$ matrix results in virtual work terms that are of the same order as those found in the analysis of $\int_V T_{ij}\delta\eta_{ij}dV$, the third integral of Equations A.23 and A.24.

Further, these additional terms can be used to explain and rectify the moment-torque equilibrium imbalance illustrated in Figure A.5 and to demonstrate that, to the order of terms considered in the analysis, the bending moment-torque interaction is not a function of the parameter α. A summary of these findings follows.

For the Cauchy stresses and infinitesimal strains of Equation A.25 and the expression of these strains in terms of displacements in Equation A.4, the above integral becomes

$$\int_V T_{ij}\delta e_{ij}dV = \int_V \sigma_x\delta\left(\frac{\partial u}{\partial x}\right)dV + \int_V \tau_{xy}\delta\left(\frac{\partial u}{\partial y}+\frac{\partial v}{\partial x}\right)dV + \int_V \tau_{xz}\delta\left(\frac{\partial u}{\partial z}+\frac{\partial w}{\partial x}\right)dV \tag{A.45}$$

To determine the relevant virtual strain terms, consider the finite rotation of a cross section of a bisymmetrical member (Figure A.7). The section is assumed to remain plane and the rotation is about the centroid. From Equation A.44

$$\{\mathbf{b}\} = [\mathbf{I} + \mathbf{W} + \tfrac{1}{2}\mathbf{W}^2]\{\mathbf{a}\}$$

in which, as above, $\{\mathbf{a}\}$ and $\{\mathbf{b}\}$ are the position vectors of a point on the section before and after rotation. The incremental displacement of the point is equal to $\{\mathbf{b}\} - \{\mathbf{a}\}$, or $\{\mathbf{u}\} = [\mathbf{W} + \mathbf{W}^2/2]\{\mathbf{a}\}$. Now assume that $[\mathbf{W}]$ has been accounted for in the analysis of infinitesimal strains and that the only displacements that are of interest here are the additional ones associated with the finite rotation effect. Thus let

$$\{\mathbf{u}\} = \tfrac{1}{2}[\mathbf{W}]^2\{\mathbf{a}\}$$

or in expanded form and in the notation of Figure A.7:

$$\begin{Bmatrix} u \\ v \\ w \end{Bmatrix} = \frac{1}{2}\begin{bmatrix} -(\theta_y^2+\theta_z^2) & (\theta_x\theta_y) & (\theta_x\theta_z) \\ (\theta_x\theta_y) & -(\theta_x^2+\theta_z^2) & (\theta_y\theta_z) \\ (\theta_x\theta_z) & (\theta_y\theta_z) & -(\theta_x^2+\theta_y^2) \end{bmatrix}\begin{Bmatrix} 0 \\ y \\ z \end{Bmatrix}$$

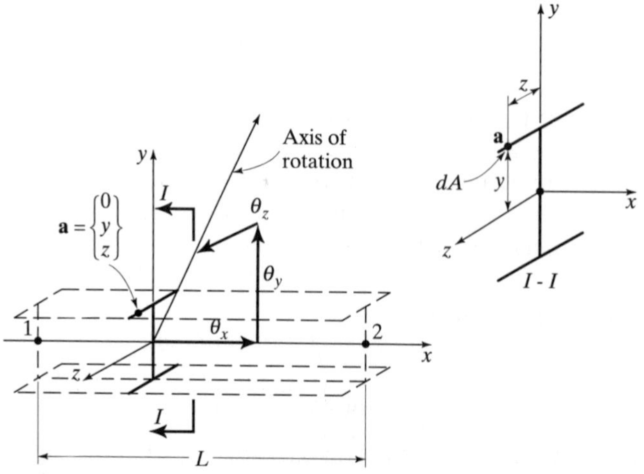

Figure A.7 Finite rotation of a cross-section.

The partial derivatives required in Equation A.45 are

$$\frac{\partial u}{\partial x} = \frac{1}{2}\left(\theta_x \frac{\partial \theta_y}{\partial x} + \theta_y \frac{\partial \theta_x}{\partial x}\right)y + \frac{1}{2}\left(\theta_x \frac{\partial \theta_z}{\partial x} + \theta_z \frac{\partial \theta_x}{\partial x}\right)z \qquad \frac{\partial u}{\partial y} = \frac{\theta_x\theta_y}{2} \qquad \frac{\partial u}{\partial z} = \frac{\theta_x\theta_z}{2}$$

$$\frac{\partial v}{\partial x} = -\left(\theta_x \frac{\partial \theta_z}{\partial x} + \theta_z \frac{\partial \theta_x}{\partial x}\right)y + \frac{1}{2}\left(\theta_y \frac{\partial \theta_z}{\partial x} + \theta_z \frac{\partial \theta_y}{\partial x}\right)z$$

$$\frac{\partial w}{\partial x} = \frac{1}{2}\left(\theta_y \frac{\partial \theta_z}{\partial x} + \theta_z \frac{\partial \theta_y}{\partial x}\right)y - \left(\theta_x \frac{\partial \theta_x}{\partial x} + \theta_y \frac{\partial \theta_y}{\partial x}\right)z$$

Substituting these quantities in Equation A.45 and using abridged notation in which a prime represents differentiation with respect to x, its right-hand side becomes

$$\int_V \frac{\sigma_x}{2}(\delta(\theta_x\theta_y)'y + \delta(\theta_x\theta_z)'z)\,dV + \int_V \frac{\tau_{xy}}{2}(\delta(\theta_x\theta_y) - 2\delta(\theta_x\theta_x' + \theta_z\theta_z')y + \delta(\theta_y\theta_z)'z)\,dV$$
$$+ \int_V \frac{\tau_{xz}}{2}(\delta(\theta_x\theta_z) + \delta(\theta_y\theta_z)'y - 2\delta(\theta_x\theta_x' + \theta_y\theta_y')z)\,dV$$

Using Equation A.31 and noting that for bisymmetrical sections $\int_A \tau_{xy}y\,dA = \int_A \tau_{xz}z\,dA = 0$, we have

$$-\int_L \frac{M_z}{2}\delta(\theta_x\theta_y)'\,dx + \int_L \frac{M_y}{2}\delta(\theta_x\theta_z)'\,dx + \int_L \frac{F_y}{2}\delta(\theta_x\theta_y)\,dx$$
$$- \int_L \left(\frac{1-\alpha}{2}\right)M_x\delta(\theta_y\theta_z)'\,dx + \int_L \frac{F_z}{2}\delta(\theta_x\theta_z)\,dx + \int_L \frac{\alpha}{2}M_x\delta(\theta_y\theta_z)'\,dx \tag{A.46}$$

A.4.2.1 The Effect of α on the Internal Virtual Work Equation The net effect of the torsional shearing stress distribution on the internal virtual work expression for torsion-flexure interaction is represented in the sum of four integrals containing α—the two from Equation A.46 and two from Equation A.34. Thus

$$-\left(\frac{1-\alpha}{2}\right)\int_L M_x\delta(\theta_y\theta_z)'\,dx + \frac{\alpha}{2}\int_L M_x\delta(\theta_y\theta_z)'\,dx$$
$$- (1-\alpha)\int_L M_x\delta(v'w'')\,dx + \alpha\int_L M_x\delta(w'v'')\,dx$$

Combining the first two integrals, making the approximations $\theta_y = -w'$ and $\theta_z = v'$ (an approximation that is consistent with our replacement of rotation components by displacement derivatives in previous equations) and differentiating the result, we have

$$\left(\frac{1}{2} - \alpha\right)\int_L M_x\delta(v'w'' + w'v'')\,dx - (1-\alpha)\int_L M_x\delta(v'w'')\,dx + \alpha\int_L M_x\delta(w'v'')\,dx$$

Combining terms gives

$$\frac{1}{2}\int_L M_x\delta(-v'w'' + w'v'')\,dx$$

which is the same result as that obtained by assuming α to be 1/2, as in the case of pure St. Venant torsion (see Eq. A.35). Thus, to the order of approximation implicit in our analysis, the distribution of torsional shearing stress does not enter into the expression for the differential of internal virtual work. In numerical analysis using the assumed shape function approach it of course remains a factor, as we shall see when we consider the case of nonuniform torsion in Section A.5.

A.4.2.2 The Effect of Finite Rotations on Joint Equilibrium The apparent joint equilibrium imbalance illustrated in Figure A.5 can be explained and rectified by considering the remaining four finite rotation effect integrals of Equation A.46.

Integrating the bending moment terms by parts yields

$$-\frac{M_z}{2}\,\delta(\theta_x\theta_y)\bigg|_0^L + \int_L \frac{M_z'}{2}\,\delta(\theta_x\theta_y)\,dx + \frac{M_y}{2}\,\delta(\theta_x\theta_z)\bigg|_0^L - \int_L \frac{M_y'}{2}\,\delta(\theta_x\theta_z)\,dx + \int_L \frac{F_y}{2}\,\delta(\theta_x\theta_y)\,dx + \int_L \frac{F_z}{2}\,\delta(\theta_x\theta_z)\,dx$$

But, $M_z' = -F_y$ and $M_y' = F_z$. Therefore the integral terms cancel and the evaluation of the portion of Equation A.45 associated with the finite rotation effect reduces to

$$\left(-\frac{M_z}{2}\,\delta(\theta_x\theta_y) + \frac{M_y}{2}\,\delta(\theta_x\theta_z)\right)\bigg|_0^L \tag{A.47}$$

Letting

$$M_z = -M_{z1} + \frac{x}{L}(M_{z1} + M_{z2}) \qquad \text{and} \qquad M_y = -M_{y1} + \frac{x}{L}(M_{y1} + M_{y2})$$

this becomes

$$\tfrac{1}{2}(-M_{z1}\,\delta(\theta_{x1}\theta_{y1}) + M_{y1}\,\delta(\theta_{x1}\theta_{z1}) - M_{z2}\,\delta(\theta_{x2}\theta_{y2}) + M_{y2}\,\delta(\theta_{x2}\theta_{z2}))$$

This can be written in matrix form as $\lfloor \delta\boldsymbol{\theta} \rfloor [\mathbf{k_g}]\{\boldsymbol{\theta}\}$ in which

$$[\mathbf{k_g}] = \frac{1}{2}\begin{bmatrix} 0 & -M_{z1} & M_{y1} & 0 & 0 & 0 \\ & 0 & 0 & 0 & 0 & 0 \\ & & 0 & 0 & 0 & 0 \\ & \text{Sym.} & & 0 & -M_{z2} & M_{y2} \\ & & & & 0 & 0 \\ & & & & & 0 \end{bmatrix} \quad (\text{columns: } \theta_{x1}, \theta_{y1}, \theta_{z1}, \theta_{x2}, \theta_{y2}, \theta_{z2}) \tag{A.48}$$

These contributions have also been added to Equation 9.18 in the formation of Equation 9.19.

The significance of the terms of Equation A.48 is demonstrated in Example A.1, a frame studied in Reference A.6. Since the rectangular section is one in which warping resistance is essentially zero, the frame is also a test of the validity of Equation 9.19, an equation that only has provisions for handling St. Venant torsion. The elastic torsional flexural critical load of 1.093 N found in part 1 of the example (and shown to be attainable by the accompanying nonlinear analysis) is identical to that reported in Reference A.6 for the same four-element per member idealization. It is further supported by values ranging from 1.14 to 1.18 N determined in independent finite element analyses of multi-element models (Refs. A.6 and A.9). The somewhat higher values of the finite element analyses may be attributed to the fact that they account for the finite size of the junction at point b, whereas the frame model used in the example does not. When, as in part 2 of the example, the finite rotation effect terms are excluded from the geometric stiffness matrix used in the analysis the result is dramatically different. The critical load of 0.551 N, which is approximately half that found in the first part, must be considered incorrect in view of the independent corroboration of the higher value.

EXAMPLE A.1

Members *ab* and *bc* have the cross section shown. $E = 71{,}240\ \text{N/mm}^2$, $\nu = 0.3$.

1. (a) Determine the elastic critical load using Equation 8.7 and the $[\mathbf{k}_g]$ of Equation 9.19.
(b) Perform a nonlinear analysis with a lateral load of 0.01P added at point c.

2. Determine the elastic critical load omitting the contribution of Equation A.48 to 9.19.

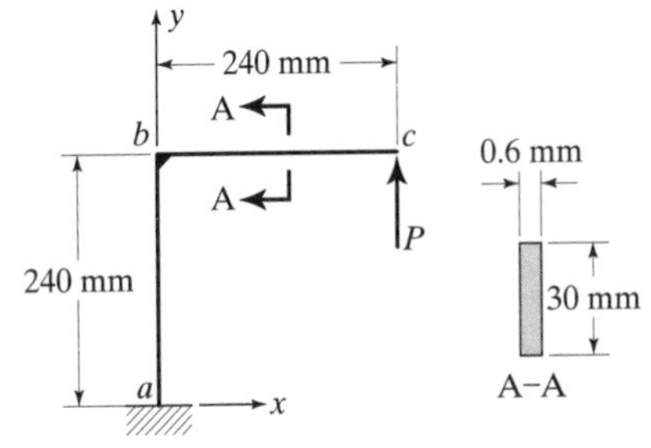

Model each member as four elements

1. (a) $P_{cr} = 1.093$ N. Buckling is in a torsional-flexural mode.
(b) In second-order analysis, a load-displacement plateau is reached at approximately the critical load.

2. Buckling in a torsional-flexural mode occurs at $P_{cr} = 0.551$ N.
Results:

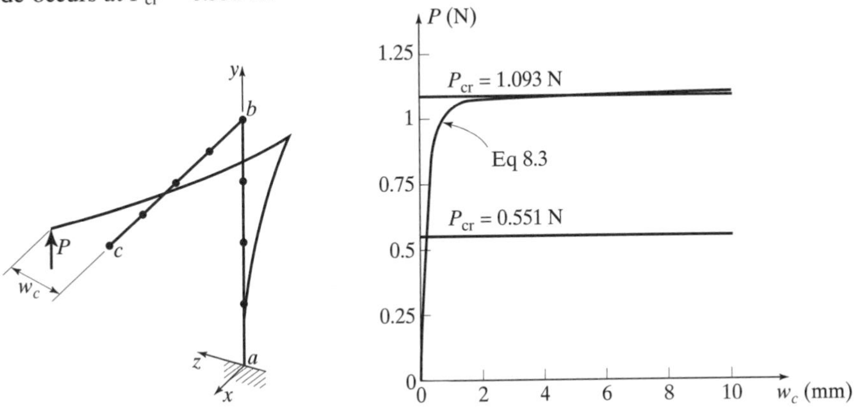

A.5 NONUNIFORM TORSION

The subject of nonuniform torsion was introduced in Section 7.4 and a stiffness matrix sufficient for the linear analysis of members subjected solely to torsion was developed there (Eq. 7.28). That matrix will be incorporated in the elastic stiffness matrix required for the application of Equations 8.3 and 8.7 to nonlinear and critical load analysis of problems in which nonuniform torsion is considered. First, however, the corresponding geometric stiffness matrix must be developed. For members of bisymmetrical section this may be done by adding two degrees of freedom to the framework element and using a cubic shape function for twist in virtual displacement integrals of Equations A.32 and A.34. Although approximate, it yields reliable results in typical test problems.

As shown in Figure A.8, the added degrees of freedom are the rates of twist at the element ends and their corresponding actions are the bimoments defined in Section 7.4. As in Section A.3, we will assume that the applied forces are deformation independent and express the external work as $\lfloor {}^{t+\Delta t}\mathbf{F} \rfloor \{\delta \mathbf{u}\}$, with

$$\lfloor {}^{t+\Delta t}\mathbf{F} \rfloor = \lfloor F_{x1}\ F_{y1}\ F_{z1}\ M_{x1}\ M_{y1}\ M_{z1}\ F_{x2}\ F_{y2}\ F_{z2}\ M_{x2}\ M_{y2}\ M_{z2}\ B_1\ B_2 \rfloor \tag{A.49}$$

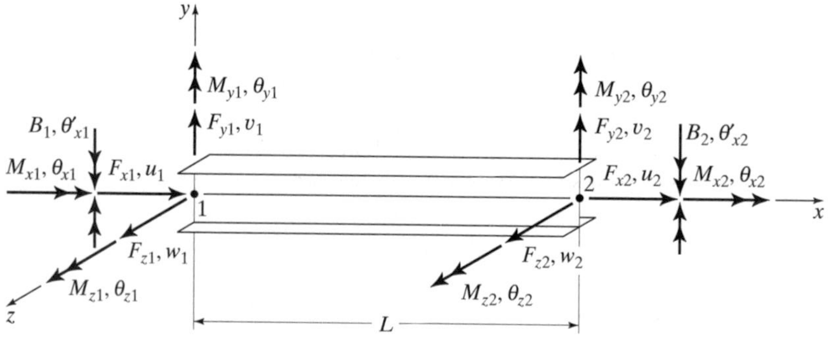

Figure A.8 Bisymmetrical 14 degree of freedom framework element.

and

$$\{\delta\mathbf{u}\} = \delta\lfloor u_1\ v_1\ w_1\ \theta_{x1}\ \theta_{y1}\ \theta_{z1}\ u_2\ v_2\ w_2\ \theta_{x2}\ \theta_{y2}\ \theta_{z2}\ \theta'_{x1}\ \theta'_{x2}\rfloor^{\mathrm{T}} \tag{A.50}$$

The relevant internal virtual work integrals are, from Equation A.32

$$\frac{1}{2}\frac{F_x I_\rho}{A}\int_L \delta\left(\frac{\partial\theta_x}{\partial x}\right)^2 dx - M_y\int_L \delta\left[\frac{\partial v}{\partial x}\frac{\partial\theta_x}{\partial x}\right]dx - M_z\int_L \delta\left[\frac{\partial w}{\partial x}\frac{\partial\theta_x}{\partial x}\right]dx \tag{A.51}$$

and from Equation A.34[8]

$$\begin{aligned}-\left(\frac{M_{z1}+M_{z2}}{L}\right)\int_L \delta\left[\theta_x\left(\frac{\partial w}{\partial x}\right)\right]dx - \left(\frac{M_{y1}+M_{y2}}{L}\right)\int_L \delta\left[\theta_x\frac{\partial v}{\partial x}dx\right]\\ -\frac{M_x}{2}\int_L \delta\left[\frac{\partial v}{\partial x}\frac{\partial^2 w}{\partial x^2}\right]dx + \frac{M_x}{2}\int_L \delta\left[\frac{\partial w}{\partial x}\frac{\partial^2 v}{\partial x^2}\right]dx\end{aligned} \tag{A.52}$$

Application of the shape function approach and analyzing terms in which w and θ_x are the real displacement components we have, proceeding as in the development of Equation A.37

$$\begin{aligned}&\lfloor\delta\boldsymbol{\theta}_x\rfloor\left[\frac{F_x I_\rho}{A}\int_L \{\overline{\mathbf{N}'_3}\}\lfloor\overline{\mathbf{N}'_3}\rfloor dx\right]\{\boldsymbol{\theta}\}\\ &+\lfloor\delta\boldsymbol{\theta}_x\rfloor\left[M_{z1}\int_L \{\overline{\mathbf{N}'_3}\}\lfloor\mathbf{N}'_3\rfloor dx - \left(\frac{M_{z1}+M_{z2}}{L}\right)\left(\int_L x\{\overline{\mathbf{N}'_3}\}\lfloor\mathbf{N}'_3\rfloor dx + \int_L \{\overline{\mathbf{N}_3}\}\lfloor\mathbf{N}_3\rfloor dx\right)\right]\{\mathbf{w}\}\\ &-\lfloor\delta\mathbf{v}\rfloor\left[\left(\frac{M_{x2}}{2}\right)\left(\int_L \{\overline{\mathbf{N}'_3}\}\lfloor\mathbf{N}''_3\rfloor dx + \int_L \{\overline{\mathbf{N}''_3}\}\lfloor\mathbf{N}'_3\rfloor dx\right)\right]\{\mathbf{w}\}\end{aligned} \tag{A.53}$$

in which

$$\begin{aligned}\theta_x &= \lfloor\overline{\mathbf{N}_3}\rfloor\{\boldsymbol{\theta}_x\}\\ &= \lfloor(1-3\xi^2+2\xi^3)\,(1-2\xi+\xi^2)x\,(3\xi^2-2\xi^3)\,-(\xi-\xi^2)x\rfloor\lfloor\theta_{x1}\theta'_{x1}\theta_{x2}\theta'_{x2}\rfloor^{\mathrm{T}}\end{aligned} \tag{A.54}$$

---

[8]Neither Equation A.32 nor Equation A.34 contains integrals representing internal virtual work attributable to warping resistance, the bimoment effect, and higher-order axial force-flexure interaction effects. Hence these effects are implicitly neglected in the development of Equation A.56. They are included in a geometric stiffness matrix developed in Reference A.8. Comparison of the results of typical problems indicates that the influence of the neglected terms on the response of bisymmetrical members of practical proportions is small.

and the other shape functions are as in Equation A.36. Integrating and combining terms, treating the integrals in which v and θ_z are the real displacement components in the same way, adding the unmodified terms of the basic 12-degree-of-freedom geometric stiffness matrix (Eq. 9.19), and including the finite rotation terms of Equation A.48 leads to the fourteen degree of freedom geometric stiffness matrix of Equation A.56. Preceding this is the corresponding elastic stiffness matrix in which the torsional terms are those of Equation 7.28 and the remainder those of Equation 4.34. (See pages 442 and 443 for Equations A.55 and A.56.)

Example A.2 illustrates the lateral buckling and nonlinear analysis of two simply supported beams subjected to pure bending and, in particular, the effect of pre-buckling deflection on the critical load. Warping at the ends is unrestrained in both cases. From Reference A.11, the critical bending moment is

$$M_{cr} = \frac{\dfrac{\pi}{L}\sqrt{EI_yGJ\left(1+\dfrac{\pi^2EC_w}{GJL^2}\right)}}{\sqrt{\left(1-\dfrac{I_y}{I_z}\right)\left[1-\dfrac{GJ}{EI_z}\left(1+\dfrac{\pi^2EC_w}{GJL^2}\right)\right]}} \tag{A.57}$$

in which the numerator is the value determined by the classical theory which neglects pre-critical displacements (Ref. A.12) and the denominator a term that accounts for them. In the example it is shown that this effect is significant in the shallow stocky member whereas in the deeper thinner one it is almost imperceptible. It is also demonstrated that, whereas critical loads determined by conventional linear stability analysis (Equation 8.7) do not account for precritical displacements, they are included in the incremental nonlinear analysis of a model subjected to a small minor axis perturbing moment.

EXAMPLE A.2

1. Use Equation A.57 to calculate the critical lateral buckling load for two sections: a W10×100 and a W27×102. $E = 29{,}000$ ksi, $\nu = 0.3$.
2. Compare with critical loads determined from Equation 8.7 and the stiffness matrices of Equations A.55 and A.56.
3. For each section perform a nonlinear analysis with a minor axis moment of 0.001 M applied at each end.

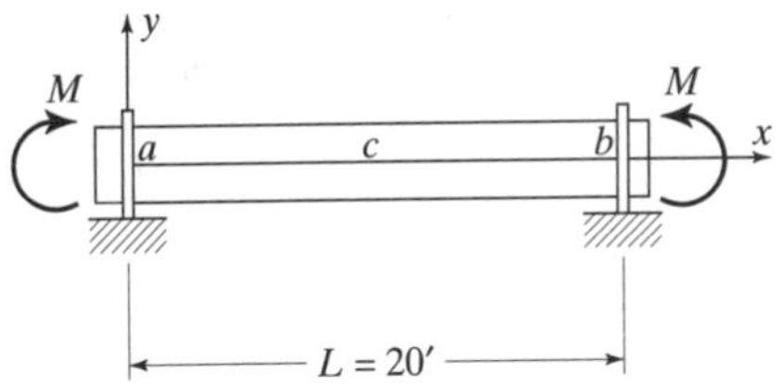

1. W10×100: $A = 29.4$ in.^{2}, $I_z = 623.0$ in.^{4}, $I_y = 207.0$ in.^{4}, $J = 10.9$ in.^{4}, $C_w = 5150$ in.^{6}

$$M_{cr} = \frac{\dfrac{\pi}{240}\sqrt{7.298\times10^{11}(1+0.2105)}}{\sqrt{(0.667)[0.992]}} = \frac{12{,}300}{0.813} = 15{,}100 \text{ in. kips}$$

W27×102: $A = 30.0$ in.^{2}, $I_z = 3620$ in.^{4}, $I_y = 139.0$ in.^{4}, $J = 5.29$ in.^{4}, $C_w = 24{,}000$ in.^{6}

$$M_{cr} = \frac{\dfrac{\pi}{240}\sqrt{2.378\times10^{11}(1+2.021)}}{\sqrt{(0.9616)[0.999]}} = \frac{11{,}100}{0.980} = 11{,}300 \text{ in. kips}$$

2. Use a four-element idealization. From Equation 8.7:

W10×100 $M_{cr} = 12{,}300$ in. kips

W27×102 $M_{cr} = 11{,}100$ in. kips

3. Comparative results, critical loads are from Equation 8.7 and incremental analysis by Equation 8.3.

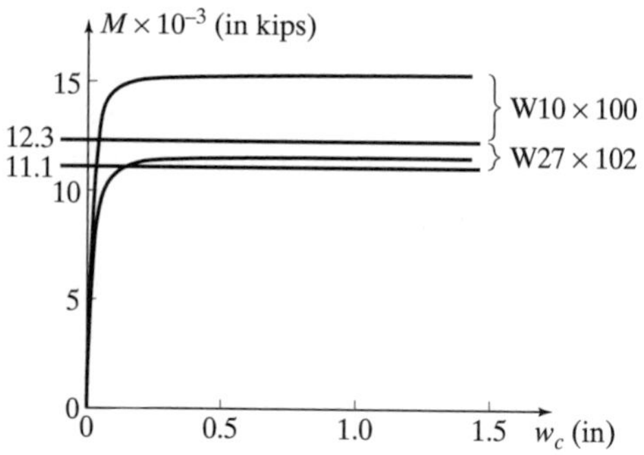

Example A.3 illustrates the phenomenon of elastic torsional-flexural instability of a beam-column. For the bisymmetrical section member subjected to equal major axis bending moments applied at the two ends (Case a of the example) it is shown in Reference A.12 that the critical axial force must be the smaller of the two forces that satisfy the equation

$$(P_{\mathrm{cf}} - P)(P_{\mathrm{ct}} - P) = \frac{A}{I_p} M^2 \tag{A.58}$$

in which $P_{\mathrm{cf}} = \pi^2 EI_y/L^2$ and $P_{\mathrm{ct}} = A(GJ + \pi^2 EC_w/L^2)/I_p$ are the minor axis and torsional critical loads, respectively. A comparable closed-form solution for unequal end moments does not exist, but it has been found that satisfactory approximate solutions can be obtained by replacing M in Equation A.58 by an *equivalent uniform moment*, M_{eq}. A simple commonly used expression is

$$M_{eq} = 0.6M_1 + 0.4M_2 \geq 0.4M_1 \tag{A.59}$$

in which M_1 is the larger of the two end moments in absolute value, and moments causing bending in the same direction have the same signs (Refs. A.13, A.14). In the example, it is seen that the results obtained by the application of Equations A.58[9] and A.59 are in good agreement with those from Eq. 8.7 for the two cases considered: equal end moments and a moment applied at one end only.

EXAMPLE A.3

The member shown is a W24×76 section:

$A = 22.4$ in.^{2}, $I_z = 2100$ in.^{4}, $I_y = 82.5$ in.^{4},
$J = 2.68$ in.^{4}, $C_w = 11{,}100$ in.^{6}, $E = 29{,}000$ ksi, $\nu = 0.3$

For each loading case:

1. Calculate the torsional-flexural critical load using Eqs. A.58 and A.59.
2. Compare with critical loads determined from Eq. 8.7 and the stiffness matrices of Equations A.55 and A.56.

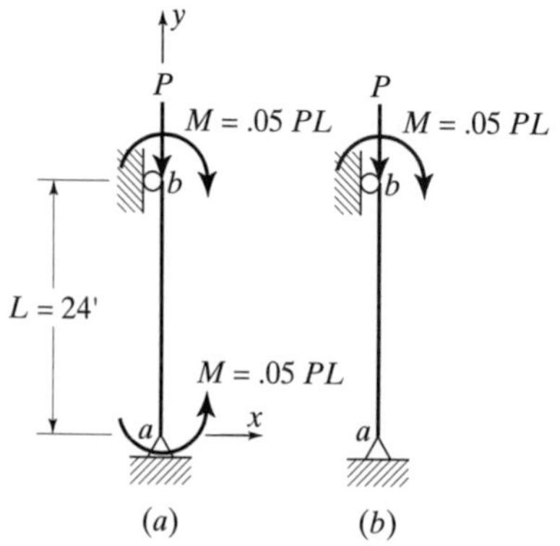

[9]Although not the purpose of the example, it may be noted that the negative sign of the second critical load in Case a demonstrates that buckling under a tensile force can occur when it is accompanied by bending of sufficient magnitude.

Case a

$$P_{cf} = \frac{\pi^2(29{,}000)(82.5)}{(288)^2} = 284.7 \text{ k} \qquad P_{ct} = 29{,}000\left(\frac{2.68}{2.6} + \frac{\pi^2(11{,}100)}{(288)^2}\right)\left(\frac{22.4}{2182}\right) = 699.7 \text{ k}$$

Equation A.59: $M_{eq} = M$

Equation A.58: $(284.7 - P)(699.7 - P) = \left(\frac{22.4}{2182}\right)(0.05 \times 288P)^2 = 2.129P^2$

$$\therefore P_{ctf} = \begin{cases} +169 \text{ k} \\ -1042 \text{ k} \end{cases}$$

Equations 8.7, A.55, A.56, using four-elemental idealization $P_{ctf} = 170$ k

Case b

Equation A.59: $M_{eq} = 0.6M$

Equation A.58: $(284.7 - P)(699.7 - P) = (0.36)(2.129\ P)^2$

$$P_{ctf} = \begin{cases} +213 \text{ k} \\ +4000 \text{ k} \end{cases}$$

Equation 8.7: $P_{ctf} = 225$ k

REFERENCES

A.1 L. E. Malvern, *Introduction to the Mechanics of a Continuous Medium*, Prentice-Hall, Englewood Cliffs, 1969.

A.2 K.-J. Bathe, *Finite Element Procedures*, Prentice-Hall, Englewood Cliffs, 1996.

A.3 Y.-B. Yang and S.-R. Kuo, *Theory and Analysis of Nonlinear Framed Structures*, Prentice-Hall, Englewood Cliffs, 1994.

A.4 A. E. H. Love, *A Treatise on the Mathematical Theory of Elasticity*, 4th edition, 1927 (1944 printing), Dover Press, New York.

A.5 Y. C. Fung, *A First Course in Continuum Mechanics*, Prentice-Hall, Englewood Cliffs, 1977.

A.6 J. H. Argyris, O. Hilpert, G. A. Malejannakis, and D. W. Scharf, "On the Geometric Stiffness of a Beam in Space—A consistent V. W. Approach," *Comp. Meth. in Appl. Mech. and Engr.*, Vol. 20, 1979.

A.7 Y.-B. Yang and W. McGuire, "Stiffness Matrix For Geometric Nonlinear Analysis," *Jl. Struct. Engr.*, ASCE, Vol. 112, Apr., 1986.

A.8 A. Conci and M. Gattass, "Natural Approach for Geometric Nonlinear Analysis of Thin-Walled Beams," *Intl. Jl. Num. Meth. in Engr.*, Vol. 30, 1990.

A.9 C.-S. Chen, "*Geometric Nonlinear Analysis of Three-Dimensional Structures*," Master of Science Thesis, Cornell University, Ithaca, NY, 1994.

A.10 L. Euler, "Nova Methodus Corporum Rigidorum Determinandi," *Novi Commentarii Acad. Scient. Imper. Petropolitanae* (Russia). Vol. 20, 1775.

A.11 P. Vacharajittiphan, S. T. Woolcock, and N. S. Trahair, "Effect of In-plane Deformation on Lateral Buckling," *Jl. Struct. Mech.*, Vol. 3, No. 1, 1974.

A.12 S. P. Timoshenko and J. M. Gere, *The Theory of Elastic Stability*, 2nd edition, McGraw-Hill, New York, 1951.

A.13 W. McGuire, *Steel Structures*, Prentice-Hall, Englewood Cliffs, N.J., 1968.

A.14 T. V. Galambos, editor, *Guide to Stability Design Criteria for Metal Structures*, 5th edition, John Wiley and Sons, New York, 1998.

$$
[\mathbf{k_e}] = \begin{bmatrix}
u_1 & v_1 & w_1 & \theta_{x1} & \theta_{y1} & \theta_{z1} & u_2 & v_2 & w_2 & \theta_{x2} & \theta_{y2} & \theta_{z2} & \theta'_{x1} & \theta'_{x2} \\
\frac{EA}{L} & 0 & 0 & 0 & 0 & 0 & -\frac{EA}{L} & 0 & 0 & 0 & 0 & 0 & 0 & 0 \\
 & \frac{12EI_z}{L^3} & 0 & 0 & 0 & \frac{6EI_z}{L^2} & 0 & -\frac{12EI_z}{L^3} & 0 & 0 & 0 & \frac{6EI_z}{L^2} & 0 & 0 \\
 & & \frac{12EI_y}{L^3} & 0 & -\frac{6EI_y}{L^2} & 0 & 0 & 0 & -\frac{12EI_y}{L^3} & 0 & -\frac{6EI_y}{L^2} & 0 & 0 & 0 \\
 & & & \left(\frac{6GJ}{5L} + \frac{12EC_w}{L^3}\right) & 0 & 0 & 0 & 0 & 0 & -\left(\frac{6GJ}{5L} + \frac{12EC_w}{L^3}\right) & 0 & 0 & \left(\frac{GJ}{10} + \frac{6EC_w}{L^2}\right) & \left(\frac{GJ}{10} + \frac{6EC_w}{L^2}\right) \\
 & & & & \frac{4EI_y}{L} & 0 & 0 & 0 & \frac{6EI_y}{L^2} & 0 & \frac{2EI_y}{L} & 0 & 0 & 0 \\
 & & & & & \frac{4EI_z}{L} & 0 & -\frac{6EI_z}{L^2} & 0 & 0 & 0 & \frac{2EI_z}{L} & 0 & 0 \\
 & & & & & & \frac{EA}{L} & 0 & 0 & 0 & 0 & 0 & 0 & 0 \\
 & & & & & & & \frac{12EI_z}{L^3} & 0 & 0 & 0 & -\frac{6EI_z}{L^2} & 0 & 0 \\
 & & & & & & & & \frac{12EI_y}{L^3} & 0 & \frac{6EI_y}{L^2} & 0 & 0 & 0 \\
 & & & \text{Sym} & & & & & & \left(\frac{6GJ}{5L} + \frac{12EC_w}{L^3}\right) & 0 & 0 & -\left(\frac{GJ}{10} + \frac{6EC_w}{L^2}\right) & -\left(\frac{GJ}{10} + \frac{6EC_w}{L^2}\right) \\
 & & & & & & & & & & \frac{4EI_y}{L} & 0 & 0 & 0 \\
 & & & & & & & & & & & \frac{4EI_z}{L} & 0 & 0 \\
 & & & & & & & & & & & & \left(\frac{2GJL}{15} + \frac{4EC_w}{L}\right) & -\left(\frac{GJL}{30} - \frac{2EC_w}{L}\right) \\
 & & & & & & & & & & & & & \left(\frac{2GJL}{15} + \frac{4EC_w}{L}\right)
\end{bmatrix} \qquad \text{(A.55)}
$$

$$
[\mathbf{k_g}] =
\begin{array}{cccccccccccccc}
u_1 & v_1 & w_1 & \theta_{x1} & \theta_{y1} & \theta_{z1} & u_2 & v_2 & w_2 & \theta_{x2} & \theta_{y2} & \theta_{z2} & \theta'_{x1} & \theta'_{x2} \\
\frac{F_{x2}}{L} & 0 & 0 & 0 & 0 & 0 & -\frac{F_{x2}}{L} & 0 & 0 & 0 & 0 & 0 & 0 & 0 \\
 & \frac{6F_{x2}}{5L} & 0 & \frac{11M_{y1} - M_{y2}}{10L} & \frac{M_{x2}}{L} & \frac{F_{x2}}{10} & 0 & -\frac{6F_{x2}}{5L} & 0 & -\frac{M_{y1} - 11M_{y2}}{10L} & -\frac{M_{x2}}{L} & \frac{F_{x2}}{10} & \frac{M_{y1}}{10} & -\frac{M_{y2}}{10} \\
 & & \frac{6F_{x2}}{5L} & \frac{11M_{z1} - M_{z2}}{10L} & -\frac{F_{x2}}{10} & \frac{M_{x2}}{L} & 0 & 0 & -\frac{6F_{x2}}{5L} & -\frac{M_{z1} - 11M_{z2}}{10L} & -\frac{F_{x2}}{10} & -\frac{M_{x2}}{L} & \frac{M_{z1}}{10} & -\frac{M_{z2}}{10} \\
 & & & \frac{6F_{x2}I_\rho}{5AL} & -\frac{2M_{z1} - M_{z2}}{5} & \frac{2M_{y1} - M_{y2}}{5} & 0 & -\frac{11M_{y1} - M_{y2}}{10L} & -\frac{11M_{z1} - M_{z2}}{10L} & -\frac{6F_{x2}I_\rho}{5A} & -\frac{2M_{z1} + M_{z2}}{10} & \frac{2M_{y1} + M_{y2}}{10} & \frac{F_{x2}I_\rho}{10A} & \frac{F_{x2}I_\rho}{10A} \\
 & & & & \frac{2F_{x2}L}{15} & 0 & 0 & -\frac{M_{x2}}{L} & \frac{F_{x2}}{10} & -\frac{M_{z1} + 2M_{z2}}{10} & -\frac{F_{x2}L}{30} & \frac{M_{x2}}{2} & -\frac{(3M_{z1} - M_{z2})L}{30} & \frac{M_{z1}L}{30} \\
 & & & & & \frac{2F_{x2}L}{15} & 0 & -\frac{F_{x2}}{10} & -\frac{M_{x2}}{L} & \frac{M_{y1} + 2M_{y2}}{10} & -\frac{M_{x2}}{2} & -\frac{F_{x2}L}{30} & \frac{(3M_{y1} - M_{y2})L}{30} & -\frac{M_{y1}L}{30} \\
 & & & & & & \frac{F_{x2}}{L} & 0 & 0 & 0 & 0 & 0 & 0 & 0 \\
 & & & & & & & \frac{6F_{x2}}{5L} & 0 & \frac{M_{y1} - 11M_{y2}}{10L} & \frac{M_{x2}}{L} & -\frac{F_{x2}}{10} & -\frac{M_{y1}}{10} & \frac{M_{y2}}{10} \\
 & & & & & & & & \frac{6F_{x2}}{5L} & \frac{M_{z1} - 11M_{z2}}{10L} & \frac{F_{x2}}{10} & \frac{M_{x2}}{L} & -\frac{M_{z1}}{10} & \frac{M_{z2}}{10} \\
 & & & & & & & & & \frac{6F_{x2}I_\rho}{5AL} & \frac{M_{z1} - 2M_{z2}}{5} & -\frac{M_{y1} - 2M_{y2}}{5} & -\frac{F_{x2}I_\rho}{10A} & -\frac{F_{x2}I_\rho}{10A} \\
 & & & & & & & & & & \frac{2F_{x2}L}{15} & 0 & -\frac{M_{z2}L}{30} & -\frac{(M_{z1} - 3M_{z2})L}{30} \\
 & & & & & & & & & & & \frac{2F_{x2}L}{15} & \frac{M_{y2}L}{30} & \frac{(M_{y1} - 3M_{y2})L}{30} \\
 & & & & & & & & & & & & \frac{2F_{x2}I_\rho}{15A} & -\frac{F_{x2}I_\rho}{30A} \\
 & & & & & & & & & & & & & \frac{2F_{x2}I_\rho}{15A}
\end{array}
\tag{A.56}
$$

Appendix B

On Rigid Body Motion and Natural Deformation

Throughout the text we have been concerned with both rigid body motion—displacement without change of size or shape of a body—and displacement resulting from its deformation. For example, in Chapter 3 it was shown that both are included in Equation 3.7, the basic global stiffness equation. In Example 8.1, the effect of nondeformative rotation of a rigid member on the stability of a system was demonstrated. And in Section 9.1 a method for incorporating both deformation and rigid body rotation in nonlinear analysis was developed.

The need to distinguish between the two effects is critical in general finite element analysis and, in particular, to the efficient solution of large displacement problems. In the moderate displacement framework analysis problems with which we are concerned, the rigid body effect is generally small, and the simple force recovery procedure of Section 12.5.2 is adequate. But to assist in the appreciation of its limitations as well as to provide further insight into the processes of nonlinear analysis this review of a method for separating rigid body and deformation effects is offered. It is based on further analysis of the element considered in Section 12.5.2. Several equations of that section will be used, either literally or in modified form. For the coherence of the present discussion they will be repeated here with numbers associated with this appendix and any modifications noted.

B.1 THE NATURE OF THE PROBLEM

In linear analysis, rigid body motion is of concern only to the extent of ensuring that the internal forces or support reactions are sufficient to provide *kinematic stability*, that is, to prevent incipient rigid body motion of all or a part of the structure. This may be addressed by testing the $[\mathbf{K}_{ff}]$ matrix of the global stiffness equation for singularity as in Problems 3.7 and 3.12, by the application of a Gauss-Jordan elimination procedure to the system equilibrium equations (Section 6.4 of the first edition of this text), or by a test such as that described in Section 11.2.2. Once kinematic stability has been assured, displacements are determined, and member forces calculated in the routine way by substitution of the relevant displacements in the element stiffness equation

$$\{\mathbf{F}\} = [\mathbf{k}]\{\boldsymbol{\Delta}\} \tag{B.1}$$

The vector $\{\mathbf{F}\}$ is a vector of forces in equilibrium on the undeformed configuration. It is therefore free of any rigid body motion effect. Its validity is only a function of the underlying theory used in the development of the element stiffness equation.

Kinematic stability is also a precondition for nonlinear analysis.[1] And in each linear step of the typical nonlinear analysis, displacement increments are calculated by the solution of global stiffness equations using methods that are in principle the same as those of linear analysis. In this case the equation for calculating internal forces in nonlinear elastic analysis may be written as

$$\{^2\mathbf{F}\} = \{^1\mathbf{F}\} + [\mathbf{k_e} + \mathbf{k_g}]\{\mathbf{d}\boldsymbol{\Delta}_n\} \tag{B.2}$$

in which $\{^1\mathbf{F}\}$ is a vector of element forces and $[\mathbf{k_e} + \mathbf{k_g}]$ the element tangent stiffness matrix at the start of the step. The vector $\{\mathbf{d}\boldsymbol{\Delta}_n\}$ consists of displacements determined by extracting the rigid body motion effects from the total incremental set and $\{^2\mathbf{F}\}$ is the resulting vector of member forces at the end of the step. Equation B.2 is similar to Equation 12.31c but it differs in two respects. The first is that whereas the equation of Section 12.5.2 anticipated a two-step process in which the end-of-step forces were first referred to the initial configuration and then transformed to the final local axes, here a direct determination of $\{^2\mathbf{F}\}$ is implied. The second is the difference in the definition of the displacement increments.

We will examine Equation B.2 and, in doing so, we'll assume that equilibrium on the correct deformed configuration at the start of the load step, as represented by the forces of $\{^1\mathbf{F}\}$, has been achieved. At the end of the step the forces of $\{^2\mathbf{F}\}$ are in equilibrium on the deformed configuration—to the order of the terms included in the geometric stiffness matrix (Equation 9.19). But the validity of that equilibrium state is affected not only by any finite element approximation made in the analysis but by the way in which the incremental displacements are used in the force recovery process. Those calculated in the global analysis include the effects of both deformation and rigid body motion. However, any change in the distribution of the components of force on the member ends is solely a function of its deformation. Thus if, in force recovery, the rigid body motion effect is not extracted from the displacements calculated in the global analysis—or some equivalent device employed—the deformed equilibrium configuration will be invalid in principle.

B.2 DISTINGUISHING BETWEEN RIGID BODY MOTION AND NATURAL DEFORMATION

In framework analysis, the main problem is to distinguish between the *natural deformation* of an element due to stretching, flexure, and twisting, and its *rigid body rotation*. There are various ways to do this. We will illustrate features that are common to most by considering the application of Equation B.2 to the case of a prismatic element of bisymmetrical cross section subjected to an axial force and bending about its z axis. Shearing deformation will be neglected. The approach followed will be that of Reference B.6. In the summary presented here there are some steps that should be intuitively correct but are not fully substantiated. Formal justification is contained in Reference B.6 as well as in parts of References B.1–B.5. Examples of the extension of the natural deformation approach to three-dimensional elements, twisting, and inelastic deformation may also be found in these references.

Forces on the element considered are shown in Figure B.1 and deformation and displacement information in Figure B.2. As in Section 12.5.2, to simplify the notation we have omitted all subscripts and superscripts except those needed to identify the

[1]There are cases, as in some suspension systems, in which there is a useful, kinematically stable configuration remote from an originally unstable one. If a method for locating and defining the stable state exists or can be devised, that state can be taken as the initial configuration for structural analysis.

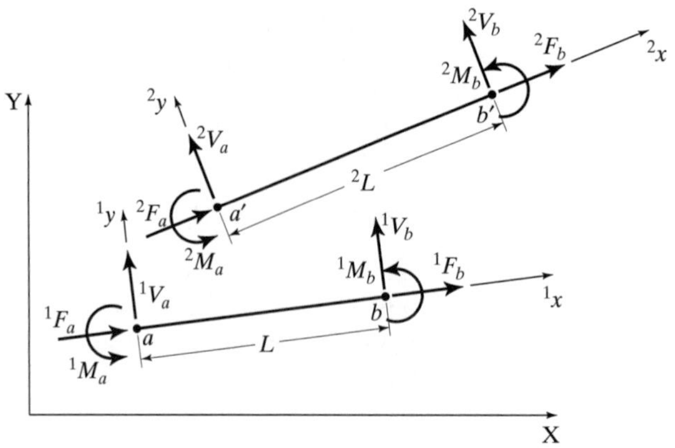

Figure B.1 Element forces.

ends of the member and its orientation at the beginning and end of the load step. In applying Equation B.2 we will use the elastic and geometric stiffness matrices of Equations 4.34 and 9.18 and the following vectors:

1. The forces at the start of the load step:

$$\{^1\mathbf{F}\} = \lfloor {}^1F_a \; {}^1V_a \; {}^1M_a \; {}^1F_b \; {}^1V_b \; {}^1M_b \rfloor^{\mathrm{T}} \tag{B.3a}$$

2. The increment of forces:

$$\{\mathbf{dF}\} = \lfloor dF_a \; dV_a \; dM_a \; dF_b \; dV_b \; dM_b \rfloor^{\mathrm{T}} \tag{B.3b}$$

3. The forces at the end of the load step referred to the final configuration:

$$\{^2\mathbf{F}\} = \lfloor {}^2F_a \; {}^2V_a \; {}^2M_a \; {}^2F_b \; {}^2V_b \; {}^2M_b \rfloor^{\mathrm{T}} \tag{B.3c}$$

4. The displacement increments transformed to the initial configuration:

$$\{\mathbf{d\Delta}\} = \lfloor u_a \; v_a \; \theta_a \; u_b \; v_b \; \theta_b \rfloor^{\mathrm{T}} \tag{B.3d}$$

5. Natural displacement increments referred to the final configuration:

$$\{\mathbf{d\Delta}_n\} = \lfloor 0 \; 0 \; \theta_{an} \; u_n \; 0 \; \theta_{bn} \rfloor^{\mathrm{T}} \tag{B.3e}$$

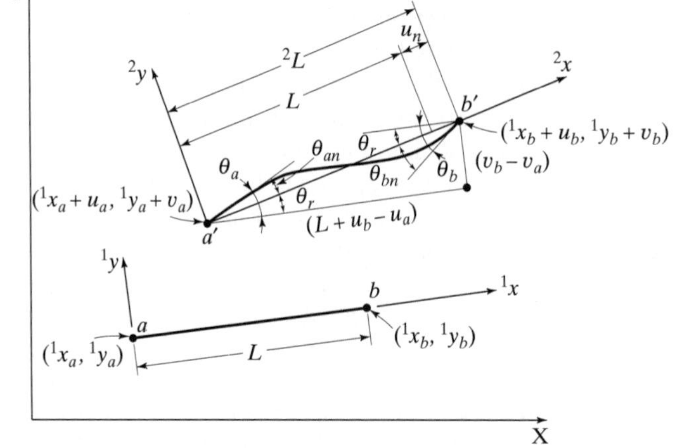

Figure B.2 Element displacements.

As may be seen from Figure B.2, the relationships between the components of the two displacement vectors are

$$\theta_{an} = \theta_a - \theta_r = \theta_a - \tan^{-1}\frac{(v_b - v_a)}{(L + u_b - u_a)} \tag{B.4a}$$

$$\theta_{bn} = \theta_b - \theta_r = \theta_b - \tan^{-1}\frac{(v_b - v_a)}{(L + u_b - u_a)} \tag{B.4b}$$

and

$$u_n = {}^2L - L = \sqrt{(L + u_b - u_a)^2 + (v_b - v_a)^2} - L \tag{B.4c}$$

in which, for incremental steps sufficiently small that ${}^2L \approx L$

$$u_n \approx (u_b - u_a) + \frac{(u_b - u_a)^2 + (v_b - v_a)^2}{2L} \tag{B.4d}$$

The rigid body motion consists of a translation and a rotation. The magnitude of the translation is irrelevant but the rotation, θ_r, is a function of the initial length of the element and the displacement increments determined in the global analysis (Eqs. B4a and B.4b). No work is required in this process. Therefore the components of $\{{}^1\mathbf{F}\}$, which are in equilibrium at the start, remain so. Their magnitude and their distribution on the ends of the rigidly displaced element are unchanged. But they rotate with the element and thus the directions of the axial and transverse components follow the rotation.

The deformation that takes place is a function of the stretch of the member and the rotation of the two ends with respect to the chord drawn between them. The changes in end force that result from this natural straining are the components of $\{\mathbf{dF}\}$. They are in the same direction as the rotated components of $\{{}^1\mathbf{F}\}$ and, if we use the stiffness method approach, they are equal to the tangent stiffness matrix at the start of the step multiplied by the vector of natural displacements, that is, $\{\mathbf{dF}\} = [\mathbf{k_e} + \mathbf{k_g}]\{\mathbf{d\Delta}_n\}$ in which the elements of $\{\mathbf{d\Delta}_n\}$ are determined from the relationships of Equation B.4. Thus, since $\{{}^2\mathbf{F}\} = \{{}^1\mathbf{F}\} + \{\mathbf{dF}\}$, we have Equation B.2.

With the updated coordinates, element length, and $\{{}^2\mathbf{F}\}$ in hand, the next step or iteration can be undertaken with these quantities used in determining the starting point.

B.3 CRITIQUE OF FORCE RECOVERY METHODS

B.3.1 The Natural Deformation Approach

To illustrate the application of Equation B.2 and test its validity, it is useful to expand the matrices and check the equilibrium of the elements of $\{{}^2\mathbf{F}\}$. Consider $\{\mathbf{dF}\}$ first. From Equations 4.34 and 9.18:

$$\begin{Bmatrix} dF_a \\ dV_a \\ dM_a \\ dF_b \\ dV_b \\ dM_b \end{Bmatrix} = \frac{1}{L}\begin{bmatrix} (EA + F_b) & 0 & 0 & -(EA + F_b) & 0 & 0 \\ & \left(\frac{12EI}{L^2} + \frac{6F_b}{5}\right) & \left(\frac{6EI}{L} + \frac{F_bL}{10}\right) & 0 & -\left(\frac{12EI}{L^2} + \frac{6F_b}{5}\right) & \left(\frac{6EI}{L} + \frac{F_bL}{10}\right) \\ & & \left(4EI + \frac{2F_bL^2}{15}\right) & 0 & -\left(\frac{6EI}{L} + \frac{F_bL}{10}\right) & \left(2EI - \frac{F_bL^2}{30}\right) \\ & & & (EA + F_b) & 0 & 0 \\ & \text{Sym.} & & & \left(\frac{12EI}{L^2} + \frac{6F_b}{5}\right) & -\left(\frac{6EI}{L} + \frac{F_bL}{10}\right) \\ & & & & & \left(4EI + \frac{2F_bL^2}{15}\right) \end{bmatrix}\begin{Bmatrix} 0 \\ 0 \\ \theta_{an} \\ u_n \\ 0 \\ \theta_{bn} \end{Bmatrix}$$

or

$$\{\mathbf{dF}\} = \begin{Bmatrix} dF_a \\ dV_a \\ dM_a \\ dF_b \\ dV_b \\ dM_b \end{Bmatrix} = \frac{1}{L}\begin{Bmatrix} -(EA + F_b)u_n \\ \left(\dfrac{6EI}{L} + \dfrac{F_bL}{10}\right)(\theta_{an} + \theta_{bn}) \\ \left(4EI + \dfrac{2F_bL^2}{15}\right)\theta_{an} + \left(2EI - \dfrac{F_bL^2}{30}\right)\theta_{bn} \\ (EA + F_b)u_n \\ -\left(\dfrac{6EI}{L} + \dfrac{F_bL}{10}\right)(\theta_{an} + \theta_{bn}) \\ \left(2EI - \dfrac{F_bL^2}{30}\right)\theta_{an} + \left(4EI + \dfrac{2F_bL^2}{15}\right)\theta_{bn} \end{Bmatrix} \quad \text{(B.5)}$$

It is obvious the $\{\mathbf{dF}\}$ is a vector of forces in equilibrium in the ${}^2x$ and ${}^2y$ directions and since $\{{}^1\mathbf{F}\}$ is also, it is clear that total equilibrium in those directions is satisfied. To check moment equilibrium, we take moments about point a of the deformed element (see Fig. B.1)

$$\Sigma M_a = (M_a + dM_a) + (M_b + dM_b) + (V_b + dV_b)\,{}^2L$$

Noting that $(M_a + M_b + V_bL) = 0$ and $(dM_a + dM_b + dV_bL) = 0$, this reduces to

$$\Sigma M_a = V_bu_n + dV_bu_n \neq 0$$

The first unbalanced moment term results from neglect of a degree of flexure-axial force interaction we believe to be insignificant in the analysis of small strain, moderate displacement problems (see the development of Equation A.34 in Appendix A). The effect is included in References B.3, B.6, and B.7 with the result that the equilibrant of this term appears in the geometric stiffness matrices of each. Since the second unbalanced moment term is normally smaller than the first, it follows that we believe that it, too, is insignificant and that all of the equilibrium conditions are substantially satisfied.

It may be noted that, except for a minor difference in the length used for the computation of shear force increments, Equation B.5 is identical to the force recovery procedure proposed in reference B.3.

B.3.2 The Elementary Approach

The elementary or conventional approach to force recovery described in Section 12.5.2 may be symbolized as

$$\{{}^2\mathbf{F}\} = [{}^2_1\mathbf{\Gamma}]\{\{{}^1\mathbf{F}\} + [\mathbf{k}_e + \mathbf{k}_g]\{\mathbf{d\Delta}\}\} \quad \text{(B.6)}$$

in which the terms are those of Equations 12.32. The difference between this and Equation B.2 is in the definition of the incremental displacement vector and the employment of a transformation matrix in Equation B.6. Since it is convenient to use the elementary approach in small strain, moderate displacement framework analysis, it is desirable to compare it with the more rigorous natural deformation method. We will do this by considering the beam element of Figure B.1 situated as in Figure B.3: It is located along the global X axis at the start of the load step, rigid body translation is neglected, and it is constrained to rotate about the left end during the step (for simplicity, forces are not shown). An informative qualitative comparison can be obtained by casting parts of Equations B.2 and B.6 in similar terms.

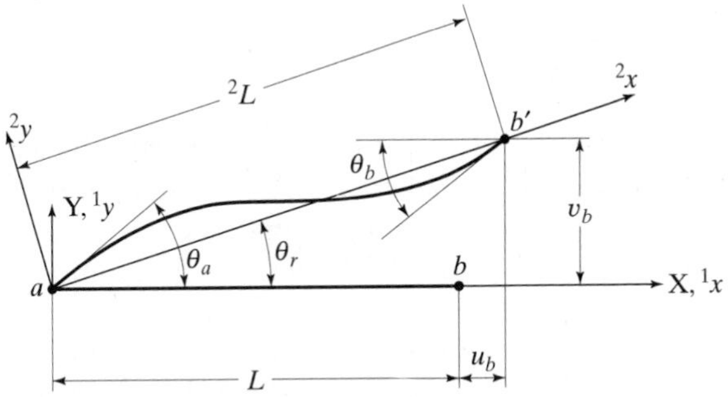

Figure B.3 Constrained element.

First, using Equations B.4a and B.4b to eliminate θ_{an} and θ_{bn} from Equation B.5:

$$
\begin{Bmatrix} dF_a \\ dV_a \\ dM_a \\ dF_b \\ dV_b \\ dM_b \end{Bmatrix} = \frac{1}{L}\begin{Bmatrix} -(EA + F_b)u_n \\ \left(\dfrac{6EI}{L} + \dfrac{F_bL}{10}\right)(\theta_a + \theta_b) - \left(\dfrac{12EI}{L} + \dfrac{F_bL}{5}\right)\theta_r \\ \left(4EI + \dfrac{2F_bL^2}{15}\right)\theta_a + \left(2EI - \dfrac{F_bL^2}{30}\right)\theta_b - \left(6EI + \dfrac{F_bL^2}{10}\right)\theta_r \\ (EA + F_b)u_n \\ -\left(\dfrac{6EI}{L} + \dfrac{F_bL}{10}\right)(\theta_a + \theta_b) + \left(\dfrac{12EI}{L} + \dfrac{F_bL}{5}\right)\theta_r \\ \left(2EI - \dfrac{F_bL^2}{30}\right)\theta_a + \left(4EI + \dfrac{F_bL^2}{15}\right)\theta_b - \left(6EI + \dfrac{F_bL^2}{10}\right)\theta_r \end{Bmatrix} \tag{B.7}
$$

Similarly, using the global displacement components to determine the force increments of Equation B.6:

$$
\begin{Bmatrix} dF_a \\ dV_a \\ dM_a \\ dF_b \\ dV_b \\ dM_b \end{Bmatrix} = \frac{1}{L}\begin{Bmatrix} -(EA + F_b)u_b \\ \left(\dfrac{6EI}{L} + \dfrac{F_bL}{10}\right)(\theta_a + \theta_b) - \left(\dfrac{12EI}{L} + \dfrac{6F_bL}{5}\right)\dfrac{v_b}{L} \\ \left(4EI + \dfrac{2F_bL^2}{15}\right)\theta_a + \left(2EI - \dfrac{F_bL^2}{30}\right)\theta_b - \left(6EI + \dfrac{F_bL^2}{10}\right)\dfrac{v_b}{L} \\ (EA + F_b)u_b \\ -\left(\dfrac{6EI}{L} + \dfrac{F_bL}{10}\right)(\theta_a + \theta_b) + \left(\dfrac{12EI}{L} + \dfrac{6F_bL}{5}\right)\dfrac{v_b}{L} \\ \left(2EI - \dfrac{F_bL^2}{30}\right)\theta_a + \left(4EI + \dfrac{2F_bL^2}{15}\right)\theta_b - \left(6EI + \dfrac{F_bL^2}{10}\right)\dfrac{v_b}{L} \end{Bmatrix} \tag{B.8}
$$

Also, for the element considered, $[{}^2_1\mathbf{\Gamma}] = \begin{bmatrix} [{}^2_1\boldsymbol{\gamma}] & \mathbf{0} \\ \mathbf{0} & [{}^2_1\boldsymbol{\gamma}] \end{bmatrix}$, in which

$$
[{}^2_1\boldsymbol{\gamma}] = \begin{bmatrix} \cos\theta_r & \sin\theta_r & 0 \\ -\sin\theta_r & \cos\theta_r & 0 \\ 0 & 0 & 1 \end{bmatrix} \tag{B.9}
$$

Thorough comparison of Equations B.2 and B.6 would require systematic analysis of the influence of the independent variables in these three equations, but the following practically useful conclusions can be drawn from their visual inspection:

1. The total rotation (and thus the rigid body rotation) or members of practical structures will invariably be small, even at the limit of resistance. Therefore, for most cases $[{}_1^2\Gamma]$ will approach a unit matrix.
2. The only difference in the incremental end moment terms of Equations B.7 and B.8 is the presence of θ_r in one and v_b/L in the other. Even for θ_r equal to 10 degrees—a very large value—the difference is insignificant.
3. For small stretching, $u_n \approx u_b$ (see Eq. B.4d) and the difference in the computed axial force increments of Equations B.7 and B.8 will be small.
4. Aside from the negligible difference in the representation of θ_r, the incremental end shear recovery terms of Equations B.7 and B.8 differ in an axial force contribution: $F_b L/5$ vs. $6F_b L/5$. This is the result of the rigid body rotational effect of $F_b \times v_b$ which has been eliminated from Equation B.7 but is present in Equation B.8 (see discussion of the *external stiffness matrix* in Reference B.6). It may be seen, however, that letting $\sin\theta_r = v_b/L$ in the transformation matrix $[{}_1^2\Gamma]$ of Equation B.9 when transforming initial forces to the final configuration by the elementary approach results in a shearing component $F_b V_b/L$ that essentially nullifies the difference.

REFERENCES

B.1 T. Belytschko and B. J. Hsieh, "Non-Linear Transient Finite Element Analyses with Convected Coordinates," *Intl. Jl. Num. Meth. in Engr.*, Vol. 7, No. 3, 1973.

B.2 J. H. Argyris et al., "Finite Element Method—The Natural Approach," *Comp. Meth. in Appl. Mech. and Engr.*, Vol. 17/18, 1979.

B.3 M. Gattass and J. F. Abel, "Equilibrium Considerations of the Updated Lagrangian Formulation of Beam-Columns with Natural Concepts, *Intl. Jl. Num. Meth. in Engr.*, Vol. 24, 1987, pp. 2119–2141.

B.4 A. Conci and M. Gattass, "Natural Approach for Thin-Walled Beam-Columns with Elastic-Plasticity," *Intl. Jl. Num. Meth. in Engr.*, Vol. 29, 1990, pp. 1653–1679.

B.5 A. Conci and M. Gattass, "Natural Approach for Geometric Non-Linear Analysis of Thin-Walled Frames," *Intl. Jl. Meth. in Engr*, Vol. 30, 1990, pp. 207–231.

B.6 Y.-B. Yang and S.-R. Kuo, *Theory and Analysis of Nonlinear Framed Structures*, Prentice-Hall, Englewood Cliffs, N.J., 1994.

B.7 Y.-B. Yang and W. McGuire, "Stiffness Matrix for Geometric Nonlinear Analysis," *Jl. Struct. Div., ASCE*, Vol. 112, No. 4, April, 1986.

Author Index

Pages in italics indicate an end-of-chapter reference to a published work.

Subject Index